Linux Kernel Prog

Second Edition

A comprehensive and practical guide to kernel internals, writing modules, and kernel synchronization

Kaiwan N. Billimoria

BIRMINGHAM—MUMBAI

Linux Kernel Programming

Second Edition

Senior Publishing Product Manager: Aaron Tanna
Acquisition Editor – Peer Reviews: Gaurav Gavas
Project Editor: Rianna Rodrigues
Content Development Editor: Shikha Parashar
Copy Editor: Safis Editing
Technical Editor: Aniket Shetty
Proofreader: Safis Editing
Indexer: Manju Arasan
Presentation Designer: Rajesh Shirsath
Developer Relations Marketing Executive: Meghal Patel

First published: March 2021
Second edition: February 2024

Production reference: 1270224

Published by Packt Publishing Ltd.
Grosvenor House
11 St Paul's Square
Birmingham
B3 1RB, UK.

ISBN 978-1-80323-222-5

`www.packt.com`

Contributors

About the author

Kaiwan N. Billimoria learned to code on his Dad's IBM PC (1983). He was programming in C and Assembly on DOS until he discovered Unix, and soon after, Linux! Kaiwan has worked extensively on the Linux system programming stack, including drivers and embedded Linux. He has actively worked on several commercial/FOSS projects. His contributions include drivers on Linux and many smaller projects hosted on GitHub. His passion for Linux helps when teaching these topics to engineers, which he has done for well over two decades now. He considers his major contribution to be his books: *Hands-On System Programming with Linux*, *Linux Kernel Programming* (and its *Part 2* book), and *Linux Kernel Debugging*. He is a recreational runner too.

First, to my wonderful family: my parents, Nads and Diana, my wife, Dilshad, my kids, Sheroy and Danesh, my bro, Darius, and the rest of the family. Thanks for being there! The Packt team has shepherded me through this work with patience and excellence, as usual. A special call out to Rianna Rodrigues, Aaron Tanna, and Aniket Shetty – thanks for all your timely support throughout!

About the reviewer

Chi-Thanh Hoang is currently working as a Senior Lead Radio SW Architect at Mavenir Systems, currently developing O-RAN 5G radios. He has over 30 years of software development experience, specializing mostly in embedded networking systems (switches, routers, Wi-Fi, and mobile networks) from chipsets up to communication protocols and of course kernel/RTOS. His first experience with the Linux kernel was in 1993. He still does hands-on debugging inside the kernel. He has a bachelor's degree in electrical engineering from Sherbrooke University, Canada. He is also an avid tennis player and invariably tinkers with electronics and software.

Learn more on Discord

To join the Discord community for this book – where you can share feedback, ask questions to the author, and learn about new releases – follow the QR code below:

`https://packt.link/SecNet`

Table of Contents

Chapter 5: Writing Your First Kernel Module – Part 2 187

Chapter 6: Kernel Internals Essentials – Processes and Threads 257

Chapter 7: Memory Management Internals – Essentials 301

Chapter 8: Kernel Memory Allocation for Module Authors – Part 1 369

Chapter 9: Kernel Memory Allocation for Module Authors – Part 2 433

Chapter 11: The CPU Scheduler – Part 2 549

Chapter 12: Kernel Synchronization – Part 1 601

Preface

This book, in its second edition now, has been explicitly written with a view to helping you learn Linux kernel development in a practical, hands-on fashion, along with the necessary theoretical background to give you a well-rounded view of this vast and interesting topic area. It deliberately focuses on kernel development via the powerful **Loadable Kernel Module** (**LKM**) framework; this is because the vast majority of real-world/industry kernel projects and products, which includes device driver development, are done in this manner.

The focus is kept on both working hands-on with, and understanding at a sufficiently deep level, the internals of the Linux OS. In this regard, we cover everything from building the Linux kernel from source to understanding and working with complex topics such as synchronization within the kernel.

To guide you on this exciting journey, we divide this book into three sections. The first section covers the basics – setting up an appropriate workspace for kernel development, building the modern kernel from source, and writing your first kernel module.

The next section, a key one, will help you understand essential kernel internals details; its coverage includes the Linux kernel architecture, the task structure, user - and kernel-mode stacks, and memory management. Memory management is a key and interesting topic – we devote three whole chapters to it (covering the internals to a sufficient extent, and importantly, how exactly to efficiently allocate and free kernel memory). The internal working and deeper details of CPU (task) scheduling on the Linux OS round off this section.

The last section of the book deals with the more advanced topic of kernel synchronization – a necessity for professional design and code on the Linux kernel. We devote two whole chapters to covering key topics here.

The book uses the kernel community's **6.1 Long Term Support** (**LTS**) Linux kernel. It's a kernel that will be maintained (both bug and security fixes) from December 2022 right through to December 2026. Moreover, the **CIP** (**Civil Infrastructure Project**) has adopted 6.1 as an **SLTS** (**Super LTS**) release and plans to maintain it for 10 years, until August 2033! This is a key point, ensuring that this book's content remains current and valid for years to come!

We very much believe in a hands-on approach: some 40 kernel modules (besides several user apps and shell scripts, double that of the first edition!) in this book's GitHub repository make the learning come alive, making it fun, practical, interesting, and useful.

We highly recommend you also make use of this book's companion guide, *Linux Kernel Programming Part 2 – Char Device Drivers and Kernel Synchronization: Create user-kernel interfaces, work with peripheral I/O, and handle hardware interrupts*. It's an excellent industry-aligned beginner's guide to writing `misc` character drivers, performing I/O on peripheral chip memory, and handling hardware interrupts. You can get this book for free along with your print copy; alternately, you can also find this eBook in the GitHub repository at `https://github.com/PacktPublishing/Linux-Kernel-Programming/tree/master/Linux-Kernel-Programming-(Part-2)`.

We really hope you learn from and enjoy this book. Happy reading!

Who this book is for

This book is primarily for those of you beginning your journey in the vast arena of learning and understanding modern Linux kernel architecture and internals, Linux kernel module development and, to some extent, Linux device driver development. It's also very much targeted at those of you who have already been working on Linux modules and/or drivers, who wish to gain a much deeper, well-structured understanding of Linux kernel architecture, memory management, task scheduling, cgroups, and synchronization. This level of knowledge about the underlying OS, covered in a properly structured manner, will help you no end when you face difficult-to-debug real-world situations.

What's been added in the second edition?

A pretty *huge* amount of new material has been added into this, the *Second Edition* of the *Linux Kernel Programming* book. As well, being based on the very recent (as of this writing) 6.1 LTS release, its information and even code will remain industry-relevant for many, many years to come.

Here's a quick chapter-wise summarization of what's new in this second edition:

- Materials updated for the 6.1 LTS kernel, maintained until December 2026, and until August 2033 via the CLP (6.1 SLTS)!
- Updated, new, and working code for the 6.1 LTS kernel
- Several new info-rich sections added to most chapters, many new diagrams, and new code examples to help explain concepts better
- *Chapter 1, Linux Kernel Programming – A Quick Introduction*
    - Introduction to the book
- *Chapter 2, Building the 6.x Linux Kernel from Source – Part 1*
    - The new LTS kernel lifetime mandate
    - More details on the kernel's Kconfig+Kbuild system
    - Updated approaches on configuring the kernel
- *Chapter 3, Building the 6.x Linux Kernel from Source – Part 2*
    - More details on the `initramfs` (`initrd`) image
    - Cross-compiling the kernel on an x86_64 host to an AArch64 target

- *Chapter 4, Writing Your First Kernel Module – Part 1*
    - new-ish `printk` indexing feature covered
    - Powerful kernel dynamic debug feature introduced
    - Rate-limiting macros updated (deprecated ones not used)
- *Chapter 5, Writing Your First Kernel Module – Part 2*
    - A better, 'better' Makefile (v0.2)
- *Chapter 6, Kernel Internals Essentials – Processes and Threads*
    - New linked list demo module
- *Chapter 7, Memory Management Internals – Essentials*
    - New coverage on how address translation works (including diagrams)
- *Chapter 8, Kernel Memory Allocation for Module Authors – Part 1*
    - Coverage on using the "exact" page allocator API pair
    - FAQs regarding (slab) memory usage and their answers
    - The graphing demo (via `gnuplot`) is now automated and even saved to an image file, via a helper script
    - Finding internal fragmentation (wastage) within the kernel
- *Chapter 9, Kernel Memory Allocation for Module Authors – Part 2*
    - Extracting useful information regarding slab caches
    - A word on slab shrinkers
    - Better coverage on the OOM killer (and `systemd-oomd`) and how it's triggered; includes a flowchart depicting demand-paging and possible OOM killer invocation
    - Better coverage on kernel page reclaim, as well as the new MGLRU and DAMON technologies
- *Chapter 10, The CPU Scheduler – Part 1*
    - New coverage on CFS scheduling period and timeslice. Coverage on the thread_info structure as well
    - New: the preempt dynamic feature
    - Enhanced coverage on exactly how and when `schedule()` is invoked
- *Chapter 11, The CPU Scheduler – Part 2*
    - Much more depth in the powerful cgroups (v2) coverage plus an interesting script to let you explore its content
    - Leveraging the cgroups v2 CPU controller via both `systemd` and manually to perform CPU bandwidth allocation
    - A note on Google's ghOSt OS

- *Chapter 12, Kernel Synchronization – Part 1*
    - A new intro to the LKMM (Linux Kernel Memory Model)
    - More on locking plus deadlock avoidance guidelines
- *Chapter 13, Kernel Synchronization – Part 2*
    - Expanded coverage on CPU caching and cache effects
    - New coverage on the powerful lock-free RCU synchronization technology
- *Online Chapter, Kernel Workspace Setup*
    - Fixed errors in package names and versions
    - Ubuntu-based helper script that auto-installs all required packages

Most (if not all) earlier code errors, typos, and URLs are now fixed, based on prompt feedback, raising Issues/PRs on the book's GitHub repo, from you, our wonderful readers!

What this book covers

Chapter 1, Linux Kernel Programming – A Quick Introduction, briefs you about the exciting journey in the sections of the book, which cover everything from building the Linux kernel from source to understanding and working with complex topics such as synchronization within the kernel.

Chapter 2, Building the 6.x Linux Kernel from Source – Part 1, is the first part of explaining how to build the modern Linux kernel from scratch with its source code. In this part, you will first be given necessary background information – the kernel version nomenclature, the different source trees, and the layout of the kernel source. Next, you will be shown in detail how exactly to download a stable vanilla Linux kernel source tree onto your Linux VM (Virtual Machine). We shall then learn a little regarding the layout of the kernel source code, getting, in effect, a "10,000-foot view" of the kernel code base. The actual work of extracting and configuring the Linux kernel then follows. Creating and using a custom menu entry for kernel configuration is also explained in detail.

Chapter 3, Building the 6.x Linux Kernel from Source – Part 2, is the second part on performing kernel builds from source code. In this part, you will continue from the previous chapter, now actually building the kernel, installing kernel modules, understanding what exactly the `initramfs` (`initrd`) image is and how to generate it, and setting up the bootloader (for the x86_64). Also, as a valuable add-on, this chapter then explains how to cross-compile the kernel for a typical embedded ARM target (using the popular Raspberry Pi 4 64-bit as a target device). Several tips and tricks on kernel builds, and even kernel security (hardening) tips, are detailed.

Chapter 4, Writing Your First Kernel Module – Part 1, is the first of two parts that cover a fundamental aspect of Linux kernel development – the LKM framework and how it is to be understood and used by you, the "module user," the kernel module or device driver programmer. It covers the basics of the Linux kernel architecture and then, in great detail, every step involved in writing a simple "Hello, world" kernel module, compiling, inserting, checking, and removing it from kernel space.

We also cover kernel logging via the ubiquitous `printk` API in detail. This edition also covers `printk` indexing and introduces the powerful dynamic debug feature.

Chapter 5, Writing Your First Kernel Module – Part 2, is the second part that covers the LKM framework. Here, we begin with something critical – learning how to use a "better" Makefile, which will help you generate more robust code (this so-called 'better' Makefile helps by having several targets for code-checking, code-style correction, static analysis, and so on. This edition has a superior version of it). We then show in detail the steps to successfully cross-compile a kernel module for an alternate architecture, how to emulate "library-like" code in the kernel (via both the linking and module-stacking approaches), and how to pass parameters to your kernel module. Additional topics include how to perform auto-loading of modules at boot, important security guidelines, and some information on the kernel coding style and upstream contribution. Several example kernel modules make the learning more interesting.

Chapter 6, Kernel Internals Essentials – Processes and Threads, delves into some essential kernel internals topics. We begin with what is meant by the execution of kernel code in process and interrupt contexts, and minimal but required coverage of the process user **virtual address space** (VAS) layout. This sets the stage for you; you'll then learn about Linux kernel architecture in more depth, focusing on the organization of process/thread task structures and their corresponding stacks – user- and kernel-mode. We then show you more on the kernel task structure (a "root" data structure), how to practically glean information from it (via the powerful 'current' macro), and even how to iterate over various (task) lists (there's sample code too!). Several kernel modules make the topic come alive.

Chapter 7, Memory Management Internals – Essentials, a key chapter, delves into essential internals of the Linux memory management subsystem, to the level of detail required for the typical module author or driver developer. This coverage is thus necessarily more theoretical in nature; nevertheless, the knowledge gained here is crucial to you, the kernel developer, both for deep understanding and usage of appropriate kernel memory APIs, as well as for performing meaningful debugging at the level of the kernel. We cover the VM split (and how it's defined on various actual architectures), gaining deep insight into the user VAS as well as the kernel VAS (our `procmap` utility will prove to be an eye-opener here!). We also cover more on how address translation works. We then briefly delve into the security technique of memory layout randomization ([K]ASLR), and end this chapter with a discussion on physical memory organization within Linux.

Chapter 8, Kernel Memory Allocation for Module Authors –Part 1, gets our hands dirty with the kernel memory allocation (and, obviously, deallocation) APIs. You will first learn about the two allocation "layers" within Linux – the slab allocator that's layered above the kernel memory allocation "engine," the page allocator (or BSA). We shall briefly learn about the underpinnings of the page allocator algorithm and its "freelist" data structure; this information is valuable when deciding which layer to use. Next, we dive straight into the hands-on work of learning about the usage of these key APIs. The ideas behind the slab allocator (or slab cache) and the primary kernel allocator APIs – the `kzalloc()`/`kfree()` pair (and friends) – are covered. Importantly, the size limitations, downsides, and caveats when using these common APIs are covered in a lot of detail as well.

Also, especially useful for driver authors, we cover the kernel's modern resource-managed memory allocation APIs (the `devm_*()` routines). Finding where internal fragmentation (wastage) occurs is another interesting area we delve into.

Chapter 9, Kernel Memory Allocation for Module Authors– Part 2, goes further, in a logical fashion, from the previous chapter. Here, you will learn how to create custom slab caches (useful for high-frequency (de)allocations for, say, your custom driver). Next, you'll learn how to extract useful information regarding slab caches as well as understanding slab shrinkers (new in this edition). We then move onto understanding and using the `vmalloc()` API (and friends). Very importantly, having covered many APIs for kernel memory (de)allocation, you will now learn how to pick and choose an appropriate API given the real-world situation you find yourself in. This chapter is rounded off with important coverage of the kernel's memory reclamation technologies and the dreaded **Out Of Memory (OOM)** "killer" framework. Understanding OOM and related areas will also lead to a much deeper understanding of how user space memory allocation really works, via the demand paging technique. This edition has more and better coverage of kernel page reclaim, as well as the new MGLRU and DAMON technologies.

Chapter 10, The CPU Scheduler – Part 1, the first of two chapters on this topic, covers a useful mix of theory and practice regarding CPU (task) scheduling on the Linux OS. The minimal necessary theoretical background on what the KSE (kernel schedulable entity) is – it's the thread! – and available kernel scheduling policies, are some of the initially covered topics. Next, we cover how to visualize the flow of a thread via tools like `perf`. Sufficient kernel internal details on CPU scheduling are then covered to have you understand how task scheduling on the modern Linux OS works. Along the way, you will learn about thread scheduling attributes (policy and real-time priority) as well. This edition includes new coverage on CFS scheduling periods/timeslices, enhanced coverage on exactly how and when the core scheduler code is invoked, and coverage of the new preempt dynamic feature.

Chapter 11, The CPU Scheduler – Part 2, the second part on CPU (task) scheduling, continues to cover the topic in more depth. Here, you learn about the CPU affinity mask and how to query/set it, controlling scheduling policy and priority on a per-thread basis – such powerful features! We then come to a key and very powerful Linux OS feature – control groups (cgroups). We understand this feature along with learning how to practically explore it (a custom script is also built). We further learn the role the modern systemd framework plays with cgroups. An interesting example on controlling CPU bandwidth allocation via cgroups v2 is then seen, from different angles. Can you run Linux as an RTOS? Indeed you can! We cover an introduction to this other interesting area...

Chapter 12, Kernel Synchronization – Part 1, first covers the really key concepts regarding critical sections, atomicity, data races (from the LKMM point of view), what a lock conceptually achieves, and, very importantly, the 'why' of all this. We then cover concurrency concerns when working within the Linux kernel; this moves us naturally on to important locking guidelines, what deadlock means, and key approaches to preventing deadlock. Two of the most popular kernel locking technologies – the mutex lock and the spinlock – are then discussed in depth along with several (simple device driver based) code examples. We also point out how to work with spinlocks in interrupt contexts and close the chapter with common locking mistakes (to avoid) and deadlock-avoidance guildelines.

Chapter 13, Kernel Synchronization – Part 2, continues the journey on kernel synchronization. Here, you'll learn about key locking optimizations – using lightweight atomic and (the more recent) refcount operators to safely operate on integers, RMW bit operators to safely perform bit ops, and the usage of the reader-writer spinlock over the regular one. What exactly the CPU caches are and the inherent risks involved when using them, such as cache "false sharing," are then discussed. We then get into another key topic – lock-free programming techniques with an emphasis on per-CPU data and **Read Copy Update** (**RCU**) lock-free technologies. Several module examples illustrate the concepts! A critical topic – lock debugging techniques, including the usage of the kernel's powerful "lockdep" lock validator – is then covered. The chapter is rounded off with a brief look at locking statistics and memory barriers.

Online Chapter – Kernel Workspace Setup

The online chapter on *Kernel Workspace Setup*, published online, guides you on setting up a full-fledged Linux kernel development workspace (typically, as a fully virtualized guest system). You will learn how to install all required software packages on it. (In this edition, we even provide an Ubuntu-based helper script that auto-installs all required packages.) You will also learn about several other open-source projects that will be useful on your journey to becoming a professional kernel/driver developer. Once this chapter is done, you will be ready to build a Linux kernel as well as to start writing and testing kernel code (via the loadable kernel module framework). In our view, it's very important for you to actually use this book in a hands-on fashion, trying out and experimenting with code. The best way to learn something is to do so empirically – not taking anyone's word on anything at all, but by trying it out and experiencing it for yourself. This chapter has been published online; do read it, here:

You can read more about the chapter online using the following link: `http://www.packtpub.com/sites/default/files/downloads/9781803232225_Online_Chapter.pdf`.

To get the most out of this book

To get the most out of this book, we expect the following:

- You need to know your way around a Linux system, on the command line (the shell).
- You need to know the C programming language.
- It's not mandatory, but experience with Linux system programming concepts and technologies will greatly help.

The details on hardware and software requirements, as well as their installation, are covered completely and in depth in *Online Chapter, Kernel Workspace Setup*. It's critical that you read it in detail and follow the instructions therein.

Also, we have tested all the code in this book (it has its own GitHub repository) on these platforms:

- x86_64 Ubuntu 22.04 LTS and guest OS (running on Oracle VirtualBox 7.0)
- x86_64 Ubuntu 23.04 LTS guest OS (running on Oracle VirtualBox 7.0)
- x86_64 Fedora 38 (and 39) on a native (laptop) system
- ARM Raspberry Pi 4 Model B (64-bit, running both its "distro" kernel as well as our custom 6.1 kernel); lightly tested

We assume that, when running Linux as a guest (VM), the host system is either Windows 10 or later (of course, even Windows 7 will work), a recent Linux distribution (for example, Ubuntu or Fedora), or even macOS.

If you are using the digital version of this book, we advise you to type the code yourself or, much better, access the code via the GitHub repository (link available in the next section). Doing so will help you avoid any potential errors related to the copying and pasting of code.

I strongly recommend that you follow the *empirical approach: not taking anyone's word on anything at all, but trying it out and experiencing it for yourself*. Hence, this book gives you many hands-on experiments and kernel code examples that you can and must try out yourself; this will greatly aid you in making real progress, deepening your understanding of the various aspects of Linux kernel development.

Download the example code files

You can download the example code files for this book from GitHub at `https://github.com/PacktPublishing/Linux-Kernel-Programming_2E`. If there's an update to the code, it will be updated on the existing GitHub repository. (So be sure to regularly do a "`git pull`" as well to stay up to date.)

We also have other code bundles from our rich catalog of books and videos available at `https://github.com/PacktPublishing/`. Check them out!

Download the color images

We also provide a PDF file that has color images of the screenshots/diagrams used in this book. You can download it here: `https://packt.link/gbp/9781803232225`.

Conventions used

There are a number of text conventions used throughout this book.

`CodeInText`: Indicates code words in text, database table names, folder names, filenames, file extensions, pathnames, dummy URLs, user input, and Twitter handles. Here is an example: "The `ioremap()` API returns a KVA of the `void *` type (since it's an address location)."

A block of code is set as follows:

```
static int __init miscdrv_init(void)
{
    int ret;
    struct device *dev;
```

When we wish to draw your attention to a particular part of a code block, the relevant lines or items are set in bold:

```
#if LINUX_VERSION_CODE < KERNEL_VERSION(5, 8, 0)
    vrx = __vmalloc(42 * PAGE_SIZE, GFP_KERNEL, PAGE_KERNEL_RO);
    if (!vrx) {
        pr_warn("__vmalloc failed\n");
```

```
        goto err_out5;
    }
[ … ]
```

Any command-line input or output is written as follows:

```
pi@raspberrypi:~ $ sudo cat /proc/iomem
```

Bold: Indicates a new term, an important word, or words that you see onscreen. For instance, words in menus or dialog boxes appear in the text like this. Here is an example: "Select **System info** from the **Administration** panel."

Warnings or important notes appear like this.

Tips and tricks appear like this.

Get in touch

Feedback from our readers is always welcome.

General feedback: Email feedback@packtpub.com and mention the book's title in the subject of your message. If you have questions about any aspect of this book, please email us at questions@packtpub.com.

Errata: Although we have taken every care to ensure the accuracy of our content, mistakes do happen. If you have found a mistake in this book, we would be grateful if you reported this to us. Please visit http://www.packtpub.com/submit-errata, click **Submit Errata**, and fill in the form. We shall strive to update all reported errors/omissions in a **Known Errata** section on this book's GitHub repo as well, allowing you to quickly spot them.

Piracy: If you come across any illegal copies of our works in any form on the internet, we would be grateful if you would provide us with the location address or website name. Please contact us at copyright@packtpub.com with a link to the material.

If you are interested in becoming an author: If there is a topic that you have expertise in and you are interested in either writing or contributing to a book, please visit http://authors.packtpub.com.

Share your thoughts

Once you've read *Linux Kernel Programming, Second Edition,* we'd love to hear your thoughts! Scan the QR code below to go straight to the Amazon review page for this book and share your feedback.

`https://packt.link/r/1803232226`

Your review is important to us and the tech community and will help us make sure we're delivering excellent quality content.

Download a free PDF copy of this book

Thanks for purchasing this book!

Do you like to read on the go but are unable to carry your print books everywhere?

Is your eBook purchase not compatible with the device of your choice?

Don't worry, now with every Packt book you get a DRM-free PDF version of that book at no cost.

Read anywhere, any place, on any device. Search, copy, and paste code from your favorite technical books directly into your application (using the book's GitHub repo is even better!).

The perks don't stop there, you can get exclusive access to discounts, newsletters, and great free content in your inbox daily

Follow these simple steps to get the benefits:

1. Scan the QR code or visit the link below

`https://packt.link/free-ebook/9781803232225`

2. Submit your proof of purchase
3. That's it! We'll send your free PDF and other benefits to your email directly

1

Linux Kernel Programming – A Quick Introduction

Hello, and welcome to this book on learning Linux kernel programming! This book, now in its second edition, has been explicitly written with a view to helping you learn Linux kernel development in a practical, hands-on fashion, along with the necessary theoretical background to give you a well-rounded view of this vast and interesting topic. It deliberately focuses on kernel development via the powerful **Loadable Kernel Module** (**LKM**) framework. This is because many, if not most, real-world/industry projects and products that involve kernel feature and/or device driver development, are done in this manner.

The focus is kept on both working hands-on with, and understanding at a sufficiently deep level, the internals of the Linux OS. In this regard, we cover everything from building the Linux kernel from source through to understanding and working with complex topics such as synchronization within the kernel.

To guide you on this exciting journey, we have divided this book into three sections. The first section covers the basics of setting up an appropriate workspace for kernel development, building the modern kernel from source, and writing your first kernel module.

The next section, a key one, will help you understand the essential details of kernel internals. Its coverage includes the Linux kernel architecture, the task structure, user and kernel-mode stacks, and memory management. Memory management is a key and interesting topic to which we have devoted three whole chapters (covering the internals to a sufficient extent, and importantly, how exactly to efficiently allocate and free kernel memory). The internal workings and deeper details of CPU (task) scheduling on the Linux OS round off this section.

The last section of the book deals with the more advanced topic of kernel synchronization – a necessity for professional design and code on the Linux kernel. We devote two whole chapters to covering key topics within this.

The book uses the kernel community's **6.1 Long Term Support** (**LTS**) Linux kernel. It's a kernel that will be maintained (with both bug and security fixes) from December 2022 right through December 2026. Moreover, the **Civil Infrastructure Project** (**CIP**) has adopted 6.1 as an **SLTS** (**Super LTS**) release and plans to maintain it for 10 years, until August 2033. This is a key point, ensuring that this book's content will remain current and valid for years to come.

We very much believe in a hands-on approach, with some 40 kernel modules (besides several user apps and shell scripts, double that of the first edition!) on this book's GitHub repository to bring the learning to life, making it fun, practical, interesting, and useful. (Link: `https://github.com/PacktPublishing/Linux-Kernel-Programming_2E`)

We highly recommend you also make use of this book's companion guide, *Linux Kernel Programming Part 2 – Char Device Drivers and Kernel Synchronization: Create user-kernel interfaces, work with peripheral I/O, and handle hardware interrupts*. It's an excellent industry-aligned beginner's guide to writing `misc` character drivers, performing I/O on peripheral chip memory, and handling hardware interrupts. You can get the eBook version for free along with your print copy, or alternately you can also find this eBook in the GitHub repository at: `https://github.com/PacktPublishing/Linux-Kernel-Programming/tree/master/Linux-Kernel-Programming-(Part-2)`.

We really hope you learn from and enjoy this book. Right, let's get on to the topic at hand now and study how to set up an appropriate kernel workspace, so that you can get the most out of this book with some practical hands-on sessions.

Kernel workspace setup

To get the most out of this book, it is very important that you first set up the workspace environment that we will be using throughout the book. This chapter will teach you exactly how to do this and get started.

Keeping size constraints in mind, the full and enhanced content of this chapter can be found within this book's GitHub repository, here: `http://www.packtpub.com/sites/default/files/downloads/9781803232225_Online_Chapter.pdf`. I request that you download and read it in full. What follows here is an introduction to this first chapter of ours. Thanks!

In this chapter, we will install a recent Linux distribution (or "distro") as a **Virtual Machine** (**VM**), and set it up to include all the required software packages. We will also clone this book's code via its GitHub repository. Furthermore, this chapter also introduces a few useful projects that will help you along this exciting journey into the Linux kernel.

Right at the outset, something I would like to emphasize is that the best way to learn something is to do so *empirically, hands-on,* not taking anyone's word on anything at all but trying it out and experiencing it for yourself. Hence, this book gives you many hands-on experiments and kernel code examples that you can and indeed must try out yourself. This will greatly aid you in making real progress, learning deeply, and understanding the various aspects of Linux kernel and driver development. So, remember:

Be empirical! Also, be bold, be daring, and try things out!

We will cover the following topics in this chapter in detail, which will help you to set up your working environment:

- Running Linux as a guest VM
- Installing an x86_64 Linux guest
- Additional useful projects

Technical requirements

You will need a modern and preferably powerful desktop PC or laptop. Ubuntu Desktop specifies some recommended minimum system requirements for the installation and usage of the distribution here: `https://help.ubuntu.com/community/Installation/SystemRequirements`. I would definitely suggest you go with a system well beyond the minimum recommendations – as powerful a system as you can afford to use. This is because performing tasks such as building a Linux kernel from source is a very memory- and CPU-intensive process. It should be pretty obvious that the more RAM, CPU power, and disk space the host system has, the better!

Like any seasoned kernel developer, I would say that working on a native Linux system is best. However, for this book, we cannot assume that you will always have a dedicated native Linux box available to you. So, we shall assume that you are working on a Linux guest. Working within a guest VM also adds an additional layer of isolation and thus safety. Of course, the downside is performance; working on a high-spec native Linux box can be up to twice as fast as compared to working on a VM!

Cloning this book's code repository

The complete source code for this book is freely available on GitHub at `https://github.com/PacktPublishing/Linux-Kernel-Programming_2E`. You can clone and work on it by cloning the `git` tree, like so:

```
git clone https://github.com/PacktPublishing/Linux-Kernel-Programming_2E
```

The source code is organized chapter-wise. Each chapter is represented as a directory – for example, `ch1/` has the source code for this chapter. The root of the source tree has some code that is common to all chapters, such as the `convenient.h` and `klib.c` source files, among others.

For efficient code browsing, I would strongly recommend that you always index the code base(s) with `ctags` and/or `cscope`. For example, to set up the `ctags` index on a source tree, just `cd` to the root of the source tree and type `ctags -R`. (If you haven't already, please invest the time in learning code browsing tools like `cscope` and `ctags`.)

Unless noted otherwise, the code output we show in the book is the output as seen on an x86_64 Ubuntu 22.04 LTS guest VM (running under Oracle VirtualBox 7.0). You should realize that due to (usually minor) distribution differences – and even within the same distributions but differing versions – the output shown in the book may not perfectly match what you see on your Linux system.

With that, I now leave it to you to download (from its GitHub repo) and read the full *Chapter 1* of this book, then move forward with the kernel workspace setup described therein.

Once you're done, I'll assume you have the work environment for the coming chapters ready. Awesome. So, next, let's move on and explore the brave world of Linux kernel development; your kernel journey is about to begin! The next two chapters will teach you how to download, extract, configure, and build a Linux kernel from source.

Learn more on Discord

To join the Discord community for this book – where you can share feedback, ask questions to the author, and learn about new releases – follow the QR code below:

`https://packt.link/SecNet`

2

Building the 6.x Linux Kernel from Source – Part 1

Building the Linux kernel from source code is an interesting way to begin your kernel development journey! Be assured, the journey's a long and arduous one, but that's the fun of it, right? The topic of kernel building itself is large enough to merit being divided into two chapters, this one and the next.

Recall what we have learned till now in the extended first chapter that has been published online (`http://www.packtpub.com/sites/default/files/downloads/9781803232225_Online_Chapter.pdf`): primarily, how to set the workspace for Linux kernel programming. You have also been introduced to user and kernel documentation sources and several useful projects that go hand-in-hand with kernel/driver development. By now, I assume you've completed the chapter that has been published online, and thus the setup of the workspace environment; if not, please do so before proceeding forward.

The primary purpose of this chapter and the next is to describe in detail how exactly you can build a modern Linux kernel from scratch using source code. In this chapter, you will first learn about the required basics: the kernel version nomenclature, development workflow, and the different types of source trees. Then, we'll get hands-on: you'll learn how to download a stable vanilla Linux kernel source tree onto a guest Linux **Virtual Machine** (**VM**). By "vanilla kernel," we mean the plain and regular default kernel source code released by the Linux kernel community on its repository, `https://www.kernel.org`. After that, you will learn a little bit about the layout of the kernel source code – getting, in effect, a 10,000-foot view of the kernel code base. The actual kernel build recipe then follows.

Before proceeding, a key piece of information: *any modern Linux system, be it a supercomputer or a tiny, embedded device, has three required components*:

- A bootloader
- An **Operating System** (**OS**) kernel
- A root filesystem

It additionally has two optional components:

- If the processor family is ARM or PPC (32- or 64-bit), a **Device Tree Blob (DTB)** image file
- An `initramfs` (or `initrd`) image file

In these two chapters, we concern ourselves only with the building of the OS (Linux) kernel from source code. We do not delve into the root filesystem details. In the next chapter, we will learn how to minimally configure the x86-specific GNU GRUB bootloader.

The complete kernel build process – for x86[_64] at least – requires a total of six or seven steps. Besides the required preliminaries, we cover the first three here and the remaining in the next chapter.

In this chapter, we will cover the following topics:

- Preliminaries for the kernel build
- Steps to build the kernel from source
- Step 1 – Obtaining a Linux kernel source tree
- Step 2 – Extracting the kernel source tree
- Step 3 – Configuring the Linux kernel
- Customizing the kernel menu, Kconfig, and adding our own menu item

You may wonder: what about building the Linux kernel for another CPU architecture (like ARM 32 or 64 bit)? We do precisely this as well in the following chapter!

Technical requirements

I assume that you have gone through *Online Chapter*, *Kernel Workspace Setup*, and have appropriately prepared an x86_64 guest VM running Ubuntu 22.04 LTS (or equivalent) and installed all the required packages. If not, I highly recommend you do this first.

To get the most out of this book, I also strongly recommend you clone this book's GitHub repository (`https://github.com/PacktPublishing/Linux-Kernel-Programming_2E`) for the code and work on it in a hands-on fashion.

Preliminaries for the kernel build

It's important to understand a few things right from the outset that will help you as we proceed on our journey of building and working with a Linux kernel. Firstly, the Linux kernel and its sister projects are completely decentralized – it's a virtual, online open-source community! The closest we come to an "office" for Linux is this: stewardship of the Linux kernel (as well as several dozen related projects) is in the capable hands of the **Linux Foundation** (`https://linuxfoundation.org/`); further, it manages the Linux Kernel Organization, a private foundation that distributes the Linux kernel to the public free of charge (`https://www.kernel.org/nonprofit.html`).

Did you know? The terms of the GNU GPLv2 license – under which the Linux kernel's released and will continue to be held for the foreseeable future – do not in any manner prevent original developers from charging for their work! It's just that Linus Torvalds, and now the Linux Foundation, makes the Linux kernel software available to everybody for free. This doesn't prevent commercial organizations from adding value to the kernel (and their products bundled with it) and charging for it, whether via an initial upfront charge or today's typical SaaS/IaaS subscription model.

Thus, open source is definitely viable for business, as has been, and continues to be, proved daily. Customers whose core business lies elsewhere simply want value for money; businesses built around Linux can provide that by providing the customer with expert-level support, detailed SLAs, and upgrades.

Some of the key points we'll discuss in this section include the following:

- The kernel release, or version number nomenclature
- The typical kernel development workflow
- The existence of different types of kernel source trees within the repository

With this information in place, you will be better armed to move through the kernel build procedure. All right let's go over each of the preceding points.

Understanding the Linux kernel release nomenclature

To see the kernel version number, simply run `uname -r` on your shell. How do you precisely interpret the output of `uname -r`? On our x86_64 Ubuntu distribution version 22.04 LTS guest VM, we run `uname`, passing the `-r` option switch to display just the current kernel release or version:

```
$ uname -r
5.19.0-40-generic
```

Of course, by the time you read this, the Ubuntu 22.04 LTS kernel will very likely have been upgraded to a later release; that's perfectly normal. The 5.19.0-40-generic kernel was the one I encountered with the Ubuntu 22.04.2 LTS at the time of writing this chapter.

The modern Linux kernel release number nomenclature is as follows:

```
major#.minor#[.patchlevel][-EXTRAVERSION]
```

This is also often written or described as `w.x[.y][-z]`.

The square brackets around the `patchlevel` and `EXTRAVERSION` (or the `y` and `-z`) components indicate that they are optional. The following table summarizes the meaning of the components of the release number:

Release # component	Meaning	Example numbers
Major # (or `w`)	Main or major number; currently, we are on the 6.x kernel series, thus the major number is 6.	`2, 3, 4, 5, 6`
Minor # (or `x`)	The minor number, hierarchically under the major number.	`0` onward
`[patchlevel]` (or `y`)	Hierarchically under the minor number – also called the ABI or revision – applied on occasion to the stable kernel when significant bug/security fixes are required.	`0` onward
`[-EXTRAVERSION]` (or `-z`)	Also called `localversion`; typically used by distribution kernels and vendors to track their internal changes.	Varies; Ubuntu uses `w.x.y-<z>-generic`

Table 2.1: Linux kernel release nomenclature

So, we can now interpret our Ubuntu 22.04 LTS distribution's kernel release number, `5.19.0-40-generic`:

- **Major # (or w)**: `5`
- **Minor # (or x)**: `19`
- **[patchlevel] (or y)**: `0`
- **[-EXTRAVERSION] (or -z)**: `-40-generic`

Note that distribution kernels may not precisely follow these conventions; it's up to them. The regular or vanilla kernels released on `https://www.kernel.org/` do follow these conventions (at least until Linus decides to change them).

Historically, in kernels before 2.6 (IOW, ancient stuff now), the minor number held a special meaning; if it was an even number, it indicated a stable kernel release, and if odd, an unstable or beta release. This is no longer the case.

As part of an interesting exercise configuring the kernel, we will later change the localversion (aka the `-EXTRAVERSION`) component of the kernel we build.

Fingers-and-toes releases

Next, it's important to understand a simple fact: with modern Linux kernels, when the kernel major and/or minor number changes, it does *not* imply that some tremendous or key new design, architecture, or feature has come about; no, it is simply, in the words of Linus, *organic evolution*.

The currently used kernel version nomenclature is a *loosely time-based* one, not feature-based. Thus, a new major number will pop up every so often. How often exactly? Linus likes to call it the **"fingers and toes" model**; when he runs out of fingers and toes to count the minor number (the *x* component of the *w.x.y* release), he updates the major number from *w* to *w+1*. Hence, after iterating over 20 minor numbers – from 0 to 19 – we end up with a new major number.

This has practically been the case since the 3.0 kernel; thus, we have the following:

- 3.0 to 3.19 (20 minor releases)
- 4.0 to 4.19 (20 minor releases)
- 5.0 to 5.19 (20 minor releases)
- 6.0 to ... (it's still moving along; you get the idea!)

Take a peek at *Figure 2.1* to see this. Each minor-to-next-minor release takes approximately between 6 to 10 weeks.

Kernel development workflow – understanding the basics

Here, we provide a brief overview of the typical kernel development workflow. Anyone like you who is interested in kernel development should at least minimally understand the process.

A detailed description can be found within the official kernel documentation here: `https://www.kernel.org/doc/html/latest/process/2.Process.html#how-the-development-process-works`.

A common misconception, especially in its baby years, was that the Linux kernel is developed in an ad hoc fashion. This is not true at all! The kernel development process has evolved to become a mostly well-oiled system with a thoroughly documented process and expectations of what a kernel contributor should know to use it well. I refer you to the preceding link for the complete details.

In order for us to take a peek into a typical development cycle, let's assume we've cloned the latest mainline Linux Git kernel tree on to our system.

The details regarding the use of the powerful **Git Source Code Management** (**SCM**) tool lie beyond the scope of this book. Please see the *Further reading* section for useful links on learning how to use Git. Obviously, I highly recommend gaining at least basic familiarity with using Git.

As mentioned earlier, at the time of writing, the **6.1 kernel** is a **Long-Term Stable** (**LTS**) version with the furthest projected EOL date from now (December 2026), so we shall use it in the materials that follow.

So, how did it come to be? Obviously, it has evolved from earlier **release candidate** (**rc**) kernels and the previous stable kernel release that preceded it, which in this case would be the *v6.1-rc'n'* kernels and the stable *v6.0* one before it. Let's view this evolution in two ways: via the command line and graphically via the kernel's GitHub page.

Viewing the kernel's Git log via the command line

We use the `git log` command as follows to get a human-readable log of the tags in the kernel Git tree ordered by date. Here, as we're primarily interested in the release of the 6.1 LTS kernel, we've deliberately truncated the following output to highlight that portion:

The `git log` command (that we use in the following code block, and in fact any other `git` sub-commands) *will only work on a Git tree*. We use the following one purely to demonstrate the evolution of the kernel. A bit later, we will show how you can clone a Git kernel source tree.

```
$ git log --date-order --tags --simplify-by-decoration \
--pretty=format:'%ai %h %d'
2023-04-23 12:02:52 -0700 457391b03803  (tag: v6.3)
2023-04-16 15:23:53 -0700 6a8f57ae2eb0  (tag: v6.3-rc7)
2023-04-09 11:15:57 -0700 09a9639e56c0  (tag: v6.3-rc6)
2023-04-02 14:29:29 -0700 7e364e56293b  (tag: v6.3-rc5)
[ … ]
2023-03-05 14:52:03 -0800 fe15c26ee26e  (tag: v6.3-rc1)
2023-02-19 14:24:22 -0800 c9c3395d5e3d  (tag: v6.2)
2023-02-12 14:10:17 -0800 ceaa837f96ad  (tag: v6.2-rc8)
[ … ]
2022-12-25 13:41:39 -0800 1b929c02afd3  (tag: v6.2-rc1)
2022-12-11 14:15:18 -0800 830b3c68c1fb  (tag: v6.1)
2022-12-04 14:48:12 -0800 76dcd734eca2  (tag: v6.1-rc8)
2022-11-27 13:31:48 -0800 b7b275e60bcd  (tag: v6.1-rc7)
2022-11-20 16:02:16 -0800 eb7081409f94  (tag: v6.1-rc6)
2022-11-13 13:12:55 -0800 094226ad94f4  (tag: v6.1-rc5)
2022-11-06 15:07:11 -0800 f0c4d9fc9cc9  (tag: v6.1-rc4)
2022-10-30 15:19:28 -0700 30a0b95b1335  (tag: v6.1-rc3)
2022-10-23 15:27:33 -0700 247f34f7b803  (tag: v6.1-rc2)
2022-10-16 15:36:24 -0700 9abf2313adc1  (tag: v6.1-rc1)
2022-10-02 14:09:07 -0700 4fe89d07dcc2  (tag: v6.0)
2022-09-25 14:01:02 -0700 f76349cf4145  (tag: v6.0-rc7)
[ … ]
2022-08-14 15:50:18 -0700 568035b01cfb  (tag: v6.0-rc1)
2022-07-31 14:03:01 -0700 3d7cb6b04c3f  (tag: v5.19)
2022-07-24 13:26:27 -0700 e0dccc3b76fb  (tag: v5.19-rc8)
[ … ]
```

In the preceding output block, you can first see that, at the time I ran this `git log` command (late April 2023), the 6.3 kernel was just released! You can also see that seven rc kernels led up to this release, numbered as 6.3-rc1, 6.3-rc2, ..., 6.3-rc7.

Delving further, we find what we're after – you can clearly see that the **stable 6.1 (LTS) kernel** initial release date was 11 December 2022, and its predecessor, the 6.0 tree, was released on 2 October 2022. You can also verify these dates by looking up other useful kernel resources, such as `https://kernelnewbies.org/LinuxVersions`.

For the development series that ultimately led to the 6.1 kernel, this latter date (2 October 2022) marks the start of what is called the **merge window** for the next stable kernel for a period of approximately two weeks. In this period, developers are allowed to submit new code to the kernel tree. In reality, the actual work would have been going on from a lot earlier; the fruit of this work is now merged into mainline at this time, typically by subsystem maintainers.

We attempt to diagram a timeline of this work in *Figure 2.1*; you can see how the earlier kernels (from 3.0 onward) had 20 minor releases. More detail is shown for our target kernel: 6.1 LTS.

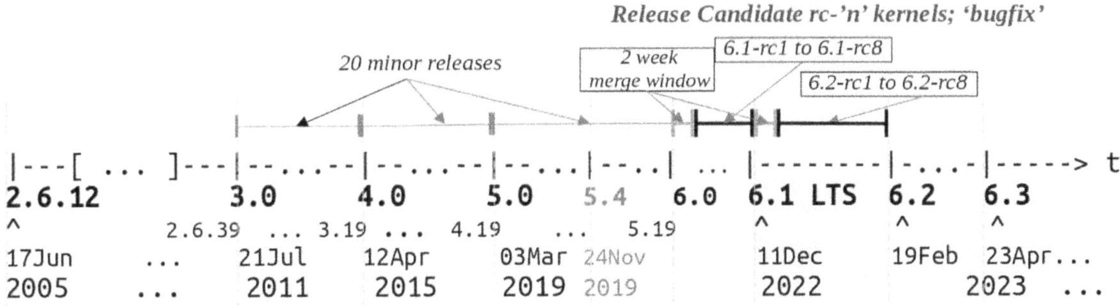

Figure 2.1: A rough timeline of modern Linux kernels, highlighting the one we'll primarily work with (6.1 LTS)

Two weeks from the start of the merge window (2 October 2022) for the 6.1 kernel, on 16 October 2022, the merge window was closed and the rc kernel work started, with `6.1-rc1` being the first of the rc versions, of course. The `-rc` (also known as *prepatch*) trees work primarily on merging patches and fixing (regression and other) bugs, ultimately leading to what is determined by the chief maintainers (Linus Torvalds and Andrew Morton) to be a "stable" kernel tree.

The number of rc kernels or prepatches varies; typically, though, this "bugfix" window takes anywhere between 6 to 10 weeks, after which the new stable kernel is released. In the preceding output block, we can see that eight release candidate kernels (from 6.1-rc1 to 6.1-rc8) finally resulted in the stable release of the v6.1 tree on 11 December 2022, taking a total of 70 days, or 10 weeks. See the 6.x section on `https://kernelnewbies.org/LinuxVersions` to confirm this.

Why does *Figure 2.1* begin at the 2.6.12 kernel? The answer is simple: this is the first version from which Git kernel history is maintained.

Another question may be, why show the 5.4 (LTS) kernel in the figure? Because it too is an LTS kernel (projected EOL is December 2025), and it was the kernel we used in the first edition of this book!

Viewing the kernel's Git log via its GitHub page

The kernel log can be seen more visually via the releases/tags page at Linus's GitHub tree here – `https://github.com/torvalds/linux/tags`:

github.com/torvalds/linux/tags?after=v6.2-rc4
Incognito
t
v6.2-rc3
on Jan 8 b7bfaa7 zip tar.gz
v6.2-rc2
on Jan 2 88603b6 zip tar.gz
v6.2-rc1
on Dec 26, 2022 1b929c0 zip tar.gz
v6.1
on Dec 12, 2022 830b3c6 zip tar.gz
v6.1-rc8
on Dec 5, 2022 76dcd73 zip tar.gz
v6.1-rc7
on Nov 28, 2022 b7b275e zip tar.gz

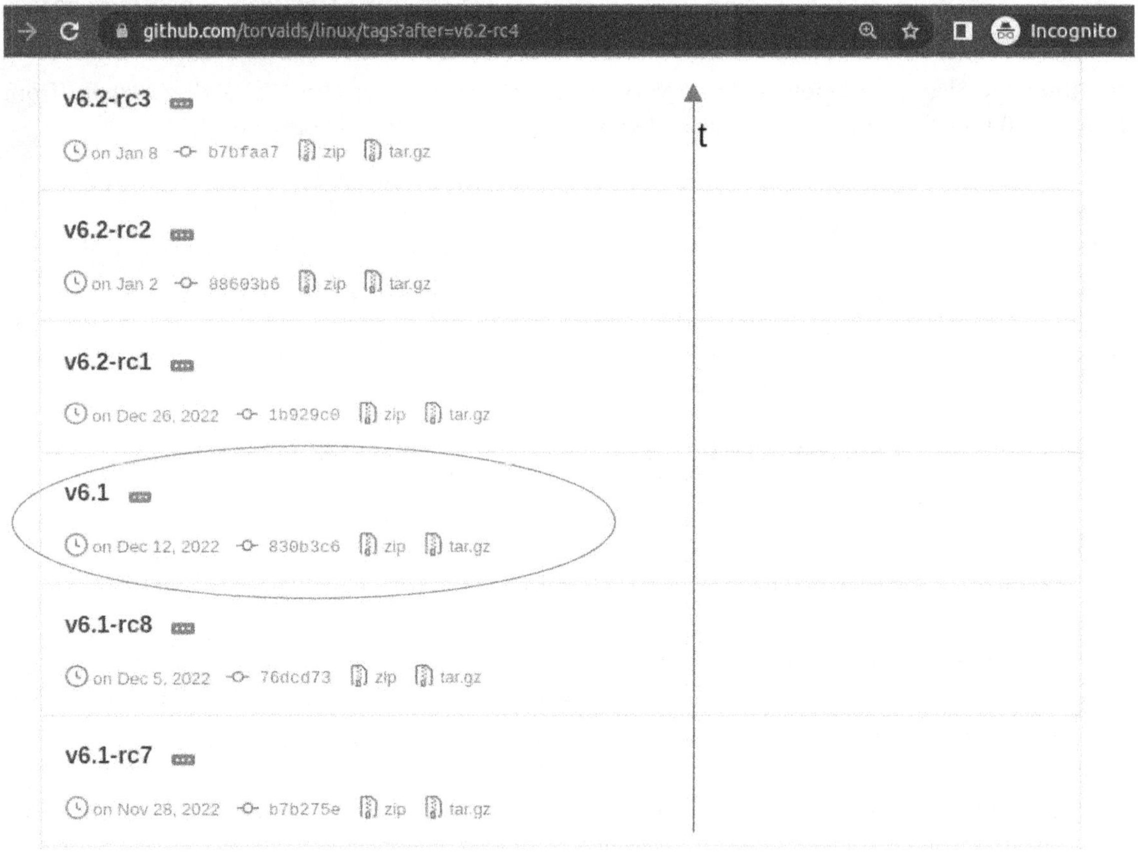

Figure 2.2: See the v6.1 tag with the -rc'n' release candidates that ultimately result in the v6.1 kernel seen above it (read it bottom-up)

Figure 2.2 shows us the v6.1 kernel tag in this truncated screenshot. How did we get there? Clicking on the **Next** button (not shown here) several times leads to the remaining pages where the v6.1-rc'n' release candidate kernels can be spotted. Alternatively, simply navigate to `https://github.com/torvalds/linux/tags?after=v6.2-rc4`.

The preceding screenshot is a partial one showing how two of the various *v6.1-rc'n'* release candidate kernels, 6.1-rc7 and 6.1-rc8, ultimately resulted in the release of the LTS 6.1 tree on 12 December 2022.

The work *never really stops*: as can be seen, by early January 2023, the *v6.2-rc3* release candidate went out, ultimately resulting in the v6.2 kernel on 19 February 2023. Then, the 6.3-rc1 kernel came out on 6 March 2023, and six more followed, ultimately resulting in the release of the stable 6.3 kernel on 23 April 2023. And so it continues...

Again, by the time you're reading this, the kernel will be well ahead. But that's okay – the 6.1 LTS kernel will be maintained for a relatively long while (recall, the projected EOL is December 2026) and it's thus a very significant release to products and projects!

The kernel dev workflow in a nutshell

Generically, taking the 6.x kernel series as an example, the kernel development workflow is as follows. You can simultaneously refer to *Figure 2.3*, where we diagram how the 6.0 kernel evolves into 6.1 LTS.

1. A 6.x stable release is made (for our purposes, consider x to be 0). Thus, the 2-week *merge window* for the *6.x+1* mainline kernel is opened.
2. The merge window remains open for about two weeks and new patches are merged into the mainline kernel by the various subsystem maintainers, who have been carefully accepting patches from contributors and updating their trees for a long while.
3. When around 2 weeks have elapsed, the merge window is closed.
4. Now, the "bugfix" period ensues; the rc (or mainline, prepatch) kernels start. They evolve as follows: *6.x+1-rc1, 6.x+1-rc2, ..., 6.x+1-rcn* are released. This process can take anywhere between 6 to 8 weeks.
5. A "finalization" period ensues, typically about a week long. The stable release arrives and the new *6.x+1* stable kernel is released.
6. The release is handed off to the "stable team":
    - Significant bug or security fixes result in the release of *6.x+1.y : 6.x+1.1, 6.x+1.2, ... , 6.x+1.n*. We're going to primarily work upon the 6.1.25 kernel, making the *y* value 25.
    - The release is maintained until the next stable release or **End Of Life** (**EOL**) date is reached, which for 6.1 LTS is projected as December 2026. Bug and security fixes will be applied right until then.

...and the whole process repeats.

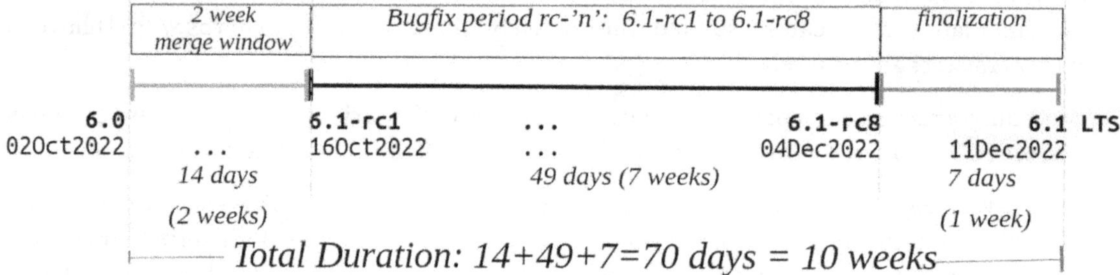

Figure 2.3 : How a 6.x becomes a 6.x+1 kernel (example here particularly for how 6.0 evolves into 6.1)

So, let's say you're trying to submit a patch but miss the merge window. Well, there's no help for it: you (or more likely the subsystem maintainer who's pushing your patch series as a part of several others) will just have wait until the next merge window arrives in about 2.5 to 3 months. Hey, that's the way it is; we aren't in that great a rush.

Exercise

Go through what we did – following the kernel's evolution – for the current latest stable kernel. (Quick tip: one more way to see the history of kernel releases: `https://en.wikipedia.org/wiki/Linux_kernel_version_history`).

So, when you now see Linux kernel releases, the names and the process involved will make sense. Let's now move on to looking at the different types of kernel source trees out there.

Exploring the types of kernel source trees

There are several types of Linux kernel source trees. A key one is the **Long Term Support** (**LTS**) kernel. It's simply a "special" release in the sense that the kernel maintainers will continue to backport important bugs and security fixes upon it until a given EOL date. By convention, the next kernel to be "marked" as an LTS release is the last one released each year, typically in December.

The "life" of an LTS kernel will usually be a minimum of 2 years, and it can be extended to go for several more. The **6.1.y LTS kernel** that we will use throughout this book is the 23rd LTS kernel and has a **projected lifespan of 4 years – from December 2022 to December 2026.**

This image snippet from the Wikipedia page on *Linux kernel version history* (`https://en.wikipedia.org/wiki/Linux_kernel_version_history`) says it all:

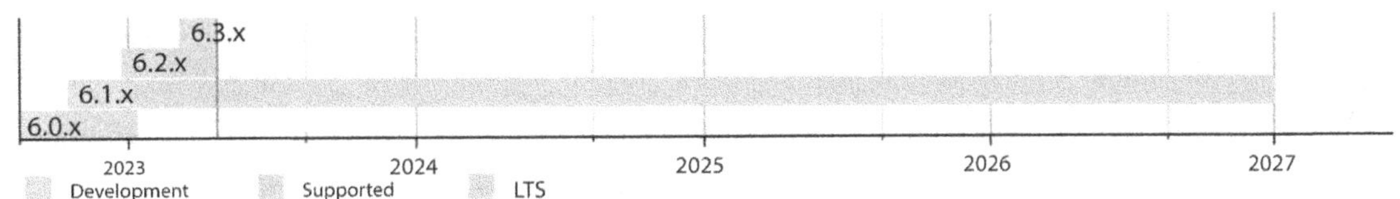

Figure 2.4: A look at the dev, supported (stable), and stable LTS kernels and their EOL dates; 6.1 LTS is what we work with (image credit: Wikipedia)

Interestingly, the 5.4 LTS kernel will be maintained until December 2025, and 5.10 LTS will be maintained for the same period as 6.1 LTS, up to December 2026.

Note, though, that there are several types of release kernels in the repository. Here, we mention an incomplete list, ordered from least to most stable (thus, their lifespan is from the shortest to longest time):

- **-next trees:** This is indeed *the bleeding edge* (the tip of the arrow!), subsystem trees with new patches collected here for testing and review. This is what an upstream kernel contributor will work on. If you intend to upstream your patches to the kernel (contribute), you must work with the latest *-next* tree.
- **Prepatches, also known as -rc or mainline:** These are release candidate kernels that get generated prior to a release.
- **Stable kernels:** As the name implies, this is the business end. These kernels are typically picked up by distributions and other projects (at least to begin with). They are also known as vanilla kernels.
- **Distribution and LTS kernels:** Distribution kernels are (obviously) the kernels provided by the distributions. They typically begin with a base vanilla/stable kernel. LTS kernels are the specially-maintained-for-a-longer-while kernels, making them especially useful for industry/production projects and products. Especially with regard to enterprise class distros, you'll often find that many of them seem to be using "old" kernels. This can even be the case with some Android vendors. Now, even if *uname -r* shows that the kernel version is, say, 4.x based, it does not necessarily imply it's old. No, the distro/vendor/OEM typically has a kernel engineering team (or outsources to one) that periodically updates the old kernel with new and relevant patches, especially critical security and bug fixes. So, though the version might appear outdated, it isn't necessarily the case!

Note, though, that this entails a huge workload on the kernel engineering team's part, and it still is very difficult to keep up with the latest stable kernel; thus, it's really best to work with vanilla kernels.

In this book, we will work throughout with one of the latest LTS kernels with the longest lifetime.

As of the time of writing, it's the **6.1.x LTS kernel**, with a projected EOL date of **December 2026**, thus keeping this book's content current and valid for years to come!

Two more types of kernel source trees require a mention here: **Super LTS (SLTS)** and chip (SoC) vendor kernels. SLTS kernels are maintained for even longer durations than the LTS kernels, by the *Civil Infrastructure Platform* (`https://www.cip-project.org/`), a Linux Foundation project. Quoting from their site:

> *"... The CIP project is focused on establishing an open source "base layer" of industrial grade software to enable the use and implementation of software building blocks in civil infrastructure projects. Currently, civil infrastructure systems are built from the ground up, with little re-use of existing software building blocks. The CIP project intends to create reusable building blocks that meet the safety, reliability, and other requirements of industrial and civil infrastructure."*

In fact, as of this writing, the latest CIP kernels – SLTS v6.1 and SLTS v6.1-rt – are based on the 6.1 LTS kernel and their projected EOL date is August 2033 (10 years)! As well, the SLTS 5.10 and SLTS 5.10-rt kernels have a projected EOL of January 2031. See the Wiki site here for the latest info: `https://wiki.linuxfoundation.org/civilinfrastructureplatform/start#kernel_maintainership`.

LTS kernels – the new mandate

A quick and important update! In September 2023 at the *Open Source Summit* in Bilbao, Spain, Jonathan Corbet, in his famous "kernel report" talk, made an important announcement: LTS kernels will from now on be maintained for a period of **only two years**.

Here's some resources, in case you'd like to look into it for yourself:

- *The Kernel Report*, Jon Corbet, September 2023, OSS EU, Spain; YouTube: `https://osseu2023.sched.com/event/4a09976cdb50e939601160ecdb45bc05#video_stream`; Slides (PDF): `https://lwn.net/talks/2023/kr-osseu.pdf`
- *Long-term support for Linux kernels is about to get a lot shorter*, Liam Proven, The Register, Sept 2023: `https://www.theregister.com/2023/09/26/linux_kernel_report_2023/`

This might come as something of a surprise to many. *Why only two years?* Briefly, two reasons were given:

- First, why maintain a kernel series for many years when people aren't really using them? Many enterprise kernel vendors, as well as SoC ones, maintain their own kernels.

- Next, an unfortunate and serious issue: *maintainer fatigue*. It's hard for the kernel community to keep all these LTS kernels – there are 7 major LTS versions as of now (4.14, 4.19, 5.4, 5.10, 5.15, 6.1, and 6.6) – continuously maintained! Furthermore, the burden only increases with time. For example, the 4.14 LTS series has had about 300 updates and close to 28,000 commits. Besides, old bugs eventually surface, and a lot of work ensues to fix them in modern LTS kernels. Not just that, but new bugs that surface in later kernels must be fixed and then back-ported to the older still-maintained LTS kernels. It all adds up.

As of this time, it appears that the kernel we work with here – 6.1 LTS – will be maintained for the usual time, until December 2026.

Moving along, let's now briefly discuss the aforementioned second type of kernel source tree: silicon/chipset SoC vendor kernels tend to maintain their own kernels for various boards/silicon that they support. They typically base their kernel on an existing vanilla LTS kernel (not necessarily the latest one!) and then build upon it, adding their vendor-specific patches, **Board Support Package** (**BSP**) stuff, drivers, and so on. Of course, as time goes by, the differences – between their kernel and the latest stable one – can become quite significant, leading to difficult maintenance issues, like the need to constantly backport critical security/bugfix patches to them.

When on a project using such silicon, the perhaps-best approach is to base your work on existing industry-strength solutions like the **Yocto Project** (`https://www.yoctoproject.org/`), which does a great job in keeping recent LTS kernels maintained with vendor layers applied in sync with key security/bugfix patches. For example, as of this writing, the latest stable Yocto release – Nanbield 4.3 – supports both the 6.1 LTS as well as the more recent 6.5 non-LTS kernel; of course, the particular version can vary with the architecture (processor family).

So, the types of kernel source trees out there are aplenty. Nevertheless, I refer you to kernel.org's *Releases* page to obtain details on the type of release kernels: `https://www.kernel.org/releases.html`. Again, for even more detail, visit *How the development process works* (`https://www.kernel.org/doc/html/latest/process/2.Process.html#how-the-development-process-works`).

Querying the repository, `https://www.kernel.org/`, in a non-interactive scriptable fashion can be done using `curl`. The following output is the state of Linux as of 06 December 2023:

```
$ curl -L https://www.kernel.org/finger_banner
The latest stable version of the Linux kernel is:              6.6.4
The latest mainline version of the Linux kernel is:            6.7-rc4
The latest stable 6.6 version of the Linux kernel is:          6.6.4
The latest stable 6.5 version of the Linux kernel is:          6.5.13 (EOL)
The latest longterm 6.1 version of the Linux kernel is:        6.1.65
The latest longterm 5.15 version of the Linux kernel is:       5.15.141
The latest longterm 5.10 version of the Linux kernel is:       5.10.202
The latest longterm 5.4 version of the Linux kernel is:        5.4.262
The latest longterm 4.19 version of the Linux kernel is:       4.19.300
```

```
The latest longterm 4.14 version of the Linux kernel is:        4.14.331
The latest linux-next version of the Linux kernel is:           next-20231206
$
```

By the time you read this, it's extremely likely – certain, in fact – that the kernel has evolved further, and later versions show up. For a book such as this, the best I can do is pick close to the latest stable LTS kernel with the longest projected EOL date at the time of writing: 6.1.x LTS.

Of course, it's happened already! The 6.6 kernel was released on 29 October 2023 and, as of the time of writing (just before going to print), it has in fact been marked as an LTS kernel, with a projected EOL date of December 2026 (the same as that of 6.1 LTS).

To help demonstrate this point, note that the first edition of this book used the 5.4.0 kernel, as 5.4 was the LTS kernel series with the longest lifetime at the time of its writing. Today, as I pen this, 5.4 LTS is still maintained with a projected EOL in December 2025 and the latest stable version is 5.4.268.

Which kernel should I run?

So, with all the types and types of kernels we've seen, it really does beg the question, *which kernel should I run?* The answer is nuanced, since it depends on the environment. Is it for embedded, desktop or server usage? Is it for a new project or a legacy one? What's the intended maintenance period? Is it an SoC kernel that is maintained by a vendor? What's the security requirement? Still, the "right answer," straight from the mouths of senior kernel maintainers, is this:

> ***Run the latest stable update.*** *That is the most stable, the most secure, the best kernel we know how to create at this time. That's the very best we can do. You should run that.*
>
> *– Jon Corbet, September 2023*

Tip: Point your browser to `https://kernel.org/`; do you see the big yellow button with the kernel release number inside it? That's the latest stable kernel as of today.

> *You have to take all of the stable/LTS releases in order to have a secure and stable system. If you attempt to cherry-pick random patches you will not fix all of the known, and unknown, problems, but rather you will end up with a potentially more insecure system, and one that contains known bugs.*
>
> *– Greg Kroah-Hartman*

Greg Kroah-Hartman's practical blog article *What Stable Kernel Should I Use*, August 2018 (`http://kroah.com/log/blog/2018/08/24/what-stable-kernel-should-i-use/`), echoes these thoughts.

Right, now that we're armed with the knowledge of kernel version nomenclature and types of kernel source trees, it's definitely time to begin our journey of building our kernel.

Steps to build the kernel from source

As a convenient and quick reference, the following are the main, key steps required to build a Linux kernel from source. As the explanation for each of them is pretty detailed, you can refer back to this summary to see the big picture. The steps are as follows:

1. *Obtain* a Linux kernel source tree through either of the following options:
   - Downloading a specific kernel source tree as a compressed file
   - Cloning a (kernel) Git tree
2. *Extract* the kernel source tree into some location in your home directory (skip this step if you obtained a kernel by cloning a Git tree).
3. *Configure*: Get a starting point for your kernel config (the approach varies). Then edit it, selecting the kernel support options as required for the new kernel. The recommended way of doing this is with `make menuconfig`.
4. *Build* the kernel image, the loadable modules, and any required **Device Tree Blobs (DTBs)** with `make [-j'n'] all`. This builds the compressed kernel image (`arch/<arch>/boot/[b|z|u]{Ii}mage`), the uncompressed kernel image - `vmlinux`, the `System.map` file, the kernel module objects, and any configured DTB files.
5. *Install* the just-built kernel modules (on x86) with `sudo make [INSTALL_MOD_PATH=<prefix-dir>] modules_install`. This step installs kernel modules by default under `/lib/modules/$(uname -r)/` (the `INSTALL_MOD_PATH` environment variable can be leveraged to change this).
6. *Bootloader (x86)*: Set up the GRUB bootloader and the `initramfs`, earlier called `initrd`, image: `sudo make [INSTALL_PATH=</new/boot/dir>] install`
   - This creates and installs the `initramfs` or `initrd` image under `/boot` (the `INSTALL_PATH` environment variable can be leveraged to change this).
   - It updates the bootloader configuration file to boot the new kernel (first entry).
7. *Customize* the GRUB bootloader menu (optional).

This chapter, being the first of two on this kernel build topic, will cover *steps 1 to 3*, with a lot of required background material thrown in as well. The next chapter will cover the remaining steps, *4 to 7*. So, let's begin with *step 1*.

Step 1 – Obtaining a Linux kernel source tree

In this section, we will see two broad ways in which you can obtain a Linux kernel source tree:

- By downloading and extracting a specific kernel source tree from the Linux kernel public repository: `https://www.kernel.org`.
- By cloning Linus Torvalds' source tree (or others') – for example, the `linux-next` Git tree.

How do you decide which approach to use? For most developers working on a project or product, the decision has already been made – the project uses a very specific Linux kernel version. You will thus download that particular kernel source tree, quite possibly apply project-specific patches to it as required, and use it.

For folks whose intention is to contribute or upstream code to the mainline kernel, the second approach – cloning the Git tree – is the way to go. Of course, there's more to it; we described some details in the *Exploring the types of kernel source trees* section.

In the following section, we demonstrate both approaches to obtaining a kernel source tree. First, we describe the approach where a particular kernel source tree (not a Git tree) is downloaded from the kernel repository. We choose the **6.1.25 LTS Linux kernel** for this purpose. So, for all practical purposes for this book, this is the approach to use. In the second approach, we clone a Git tree.

Downloading a specific kernel tree

Firstly, where is the kernel source code? The short answer is that it's on the public kernel repository server visible at `https://www.kernel.org`. The home page of this site displays the latest stable Linux kernel version, as well as the latest `longterm` and `linux-next` releases. The following screenshot shows the site as of 25 April 2023. It shows dates in the format `yyyy-mm-dd`:

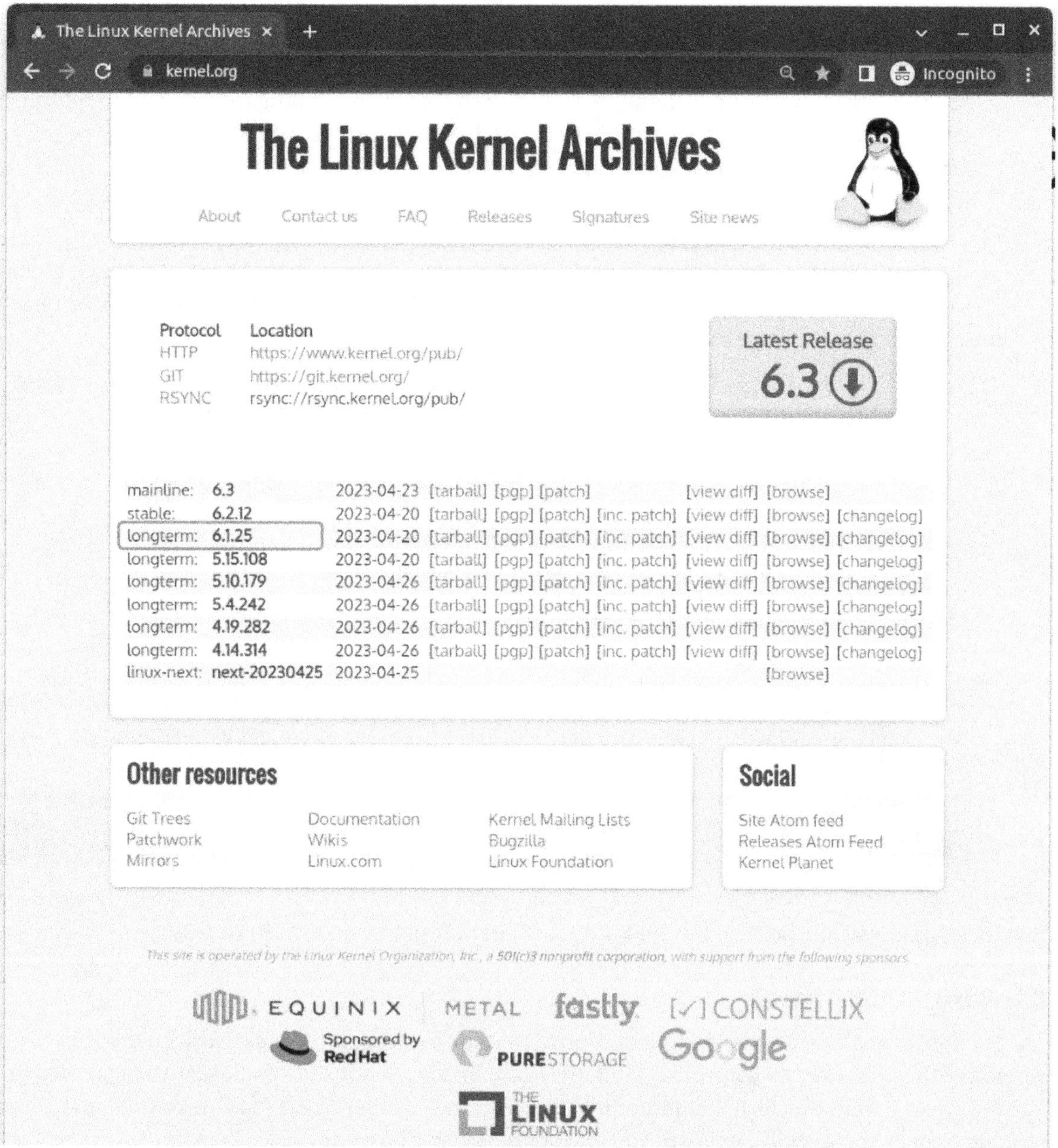

Figure 2.5: The kernel.org site (as of 25 April 2023) with the 6.1 LTS kernel highlighted

A quick reminder: we also provide a PDF file that has the full-color images of the screenshots/diagrams used in this book. You can download it here: `https://packt.link/gbp/9781803232225`.

There are many ways to download a compressed kernel source file from this server and/or its mirrors. Let's look at two of them:

- An interactive, and perhaps the simplest way, is to visit the preceding website and simply click on the appropriate `tarball` link within your web client. The browser will download the image file in `.tar.xz` format to your system.
- You can also download any kernel source tree in compressed form by navigating to `https://mirrors.edge.kernel.org/pub/linux/kernel/` and selecting the major version; practically speaking, for the major # 6 kernels, the URL is `https://mirrors.edge.kernel.org/pub/linux/kernel/v6.x/`; browse or search within this page for the kernel you want. For example, check out the following screenshot:

```
linux-6.1.24.tar.xz          13-Apr-2023 15:09    129M
linux-6.1.25.tar.gz          20-Apr-2023 10:43    207M
linux-6.1.25.tar.sign        20-Apr-2023 10:43     989
linux-6.1.25.tar.xz          20-Apr-2023 10:43    129M
linux-6.1.3.tar.gz           04-Jan-2023 10:37    206M
```

Figure 2.6: Partial screenshot from kernel.org highlighting the (6.1.25 LTS) kernel we'll download and work with

  The `tar.gz` and `tar.xz` files have identical content; it's just the compression type that differs. You can see that it's typically quicker to download the `.tar.xz` files as they're smaller.

- Alternatively, you can download the kernel source tree from the command line using the `wget` utility. We can also use the powerful `curl` utility to do so. For example, to download the stable 6.1.25 LTS kernel source compressed file, we type the following in one line:

```
wget –https-only -O ~/Downloads/linux-6.1.25.tar.xz  https://mirrors.edge.kernel.org/pub/linux/kernel/v6.x/linux-6.1.25.tar.xz
```

This will securely download the 6.1.25 compressed kernel source tree to your computer's `~/Downloads` folder. So, go ahead and do this, get the 6.1.25 (LTS) kernel source code onto your system!

Cloning a Git tree

For developers working on and looking to contribute code upstream, you *must* work on the very latest version of the Linux kernel code base. Well, there are fine gradations of what exactly constitutes the latest version within the kernel community. As mentioned earlier, the `linux-next` tree, and some specific branch or tag within it, is the one to work on for this purpose.

In this book, though, we do not intend to delve into the gory details of setting up a `linux-next` tree. This process is already very well documented, see the *Further reading* section of this chapter for detailed links. The detailed page on how exactly you should clone a `linux-next` tree is here: *Working with linux-next*, `https://www.kernel.org/doc/man-pages/linux-next.html`, and, as mentioned there, the *linux-next tree*, `http://git.kernel.org/cgit/linux/kernel/git/next/linux-next.git`, is the holding area for patches aimed at the next kernel merge window. If you're doing bleeding-edge kernel development, you likely want to work from that tree rather than Linus Torvalds' mainline tree or a source tree from the general kernel repository at `https://www.kernel.org`.

For our purposes, cloning the *mainline* Linux Git repository (in effect, Linus Torvalds' Git tree) is more than sufficient. Do so like this:

```
git clone https://git.kernel.org/pub/scm/linux/kernel/git/torvalds/linux.git
cd linux
```

Note that cloning a complete Linux kernel tree is a time-, network-, and disk-consuming operation! Ensure you have sufficient disk space free (at least a few gigabytes worth).

Performing `git clone --depth n <...>`, where n is an integer value, can be useful to limit the depth of history (commits) and thus keep the download/disk usage lower. As the man page on `git-clone` mentions for the `--depth` option: *"Create a shallow clone with a history truncated to a specified number of commits."* Also, FYI, to undo the "shallow fetch" and fetch everything, just do a `git pull --unshallow`.

The `git clone` command can take a while to finish. Further, you can specify that you want the latest *stable* version of the kernel Git tree by running `git clone` like shown below; for now, and only if you intend to work on this mainline Git tree, we'll just bite the bullet and clone the stable kernel Git tree with all its storied history (again, type this on one line):

```
git clone git://git.kernel.org/pub/scm/linux/kernel/git/stable/linux-stable.git
```

Now switch to the directory it got extracted to:

```
cd linux-stable
```

Again, if you intend to work on this Git tree, please skip the *Step 2 – extracting the kernel source tree* section as the `git clone` operation will, in any case, extract the source tree. Instead, continue with the *Step 3 – configuring the Linux kernel* section that follows it. This does imply, though, that the kernel source tree version you're using will be much different from the 6.1.25 one that we use in this book. Thus, I'd suggest you treat this portion as a demo of how to obtain the latest stable Git tree via `git`, and leave it at that.

Finally, yet another way to download a given kernel is provided by the kernel maintainers who offer a script to safely download a given Linux kernel source tree, verifying its PGP signature. The script is available here: `https://git.kernel.org/pub/scm/linux/kernel/git/mricon/korg-helpers.git/tree/get-verified-tarball`.

Step 2 – Extracting the kernel source tree

In the previous section, in step 1, you learned how exactly you can obtain a Linux kernel source tree. One way – and the one we follow in this book – is to simply download a compressed source file from the kernel.org website (or one of its mirror sites). Another way is to use Git to clone a recent kernel source tree.

So, I'll assume that by now you have obtained the 6.1.25 (LTS) kernel source tree in compressed form onto your Linux box. With it in place, let's proceed with *step 2*, a simple step, where we learn how to extract it.

As mentioned earlier, this section is meant for those of you who have downloaded a particular compressed Linux kernel source tree from the repository, `https://www.kernel.org`, and aim to build it. In this book, we work primarily on the **6.1 longterm kernel** series, particularly, on the 6.1.25 LTS kernel.

On the other hand, if you have performed `git clone` on the mainline Linux Git tree, as shown in the immediately preceding section, you can safely skip this section and move on to the next one – *Step 3 – Configuring the Linux kernel*.

Right; now that the download is done, let's proceed further. The next step is to extract the kernel source tree – remember, it's a tar-ed and compressed (typically `.tar.xz`) file. At the risk of repetition, we assume that by now you have downloaded the Linux kernel version 6.1.25 code base as a compressed file into the `~/Downloads` directory:

```
$ cd ~/Downloads ; ls -lh linux-6.1.25.tar.xz
-rw-rw-r-- 1 c2kp c2kp  129M  Apr 20 16:13  linux-6.1.25.tar.xz
```

The simple way uncompress and extract this file is by using the ubiquitous `tar` utility to do so:

```
tar xf ~/Downloads/linux-6.1.25.tar.xz
```

This will extract the kernel source tree into a directory named `linux-6.1.25` within the `~/Downloads` directory. But what if we would like to extract it into another folder, say, `~/kernels`? Then, do it like so:

```
mkdir -p ~/kernels
tar xf ~/Downloads/linux-6.1.25.tar.xz \
  --directory=~/kernels/
```

This will extract the kernel source into the `~/kernels/linux-6.1.25/` folder. As a convenience and good practice, let's set up an *environment variable* to point to the location of the root of our shiny new kernel source tree:

```
export LKP_KSRC=~/kernels/linux-6.1.25
```

Note that, going forward, we will assume that this variable `LKP_KSRC` holds the location of our 6.1.25 LTS kernel source tree.

While you could always use a GUI file manager application, such as **Nautilus**, to extract the compressed file, I strongly urge you to get familiar with using the Linux CLI to perform these operations.

Don't forget `tldr` when you need to quickly lookup the most frequently used options to common commands! Take `tar`, for example: simply do `tldr tar` to look at common tar commands, or look it up here: `https://tldr.inbrowser.app/pages/common/tar`.

Did you notice? We can extract the kernel source tree into *any* directory under our home directory, or elsewhere. This is unlike in the old days, when the tree was always extracted into a root-writeable location, often /usr/src/.

If all you wish to do now is proceed with the kernel build recipe, skip the following section and move along. If you're interested (I certainly hope so!), the next section is a brief but important digression into looking at the structure and layout of the kernel source tree.

A brief tour of the kernel source tree

Imagine! The entire Linux kernel source code is now available on your system! Awesome – let's take a quick look at it:

```
$ export LKP_KSRC=~/kernels/linux-6.1.25
$ cd $LKP_KSRC
$ pwd
/home/c2kp/kernels/linux-6.1.25
$
$ ls
arch/    CREDITS         fs/         ipc/     lib/         mm/     samples/  tools/
block/   crypto/         include/    Kbuild   LICENSES/    net/    scripts/  usr/
certs/   Documentation/  init/       Kconfig  MAINTAINERS  README  security/ virt/
COPYING  drivers/        io_uring/   kernel/  Makefile     rust/   sound/
$
```

Figure 2.7: The root of the pristine 6.1.25 Linux kernel source tree

Great! How big is it? A quick du -h . issued within the root of the uncompressed kernel source tree reveals that this kernel source tree (recall, its version is 6.1.25) is approximately 1.5 gigabytes in size!

FYI, the Linux kernel has grown to be big and is getting bigger in terms of **Source Lines of Code (SLOCs)**. Current estimates are close to 30 million SLOCs. Of course, do realize that not all this code will get compiled when building a kernel.

How do we know which version exactly of the Linux kernel this code is by just looking at the source? That's easy: one quick way is to just check out the first few lines of the project's Makefile. Incidentally, the kernel uses Makefiles all over the place; most directories have one. We will refer to this Makefile, the one at the root of the kernel source tree, as the *top-level Makefile*:

```
$ head Makefile
# SPDX-License-Identifier: GPL-2.0
VERSION = 6
PATCHLEVEL = 1
SUBLEVEL = 25
EXTRAVERSION =
NAME = Hurr durr I'ma ninja sloth
```

```
# *DOCUMENTATION*
# To see a list of typical targets execute "make help"
# More info can be located in ./README
```

Clearly, it's the source of the 6.1.25 kernel. We covered the meaning of the `VERSION`, `PATCHLEVEL`, `SUBLEVEL`, and `EXTRAVERSION` tags - corresponding directly to the `w.x.y.z` nomenclature - in the *Understanding the Linux kernel release nomenclature* section. The `NAME` tag is simply a nickname given to the release (looking at it here - well, what can I say: that's kernel humor for you. I personally preferred the `NAME` for the 5.x kernels - it's *"Dare mighty things"!*).

Right, let's now get for ourselves a zoomed-out 10,000-foot view of this kernel source tree. The following table summarizes the broad categorization and purpose of the more important files and directories within the root of the Linux kernel source tree. Cross-reference it with *Figure 2.7*:

File or directory name	Purpose
Top-level files	
`README`	The project's `README` file. It informs us as to where the official kernel documentation is kept - spoiler alert! it's in the directory called `Documentation` - and how to begin using it. The modern kernel documentation is online as well and very nicely done: `https://www.kernel.org/doc/html/latest/`. The documentation is really important; it's the authentic thing, written by the kernel developers themselves. Do read this short `README` file first! See more below, point [1].
`COPYING`	This file details the license terms under which the kernel source is released. The vast majority of kernel source files are released under the well-known GNU GPL v2 (written as GPL-2.0) license. The modern trend is to use easily grep-pable industry-aligned SPDX license identifiers. Here's the full list: `https://spdx.org/licenses/`. See more below, point [2].
`MAINTAINERS`	*FAQ: something's wrong in kernel component (or file) XYZ - who do I contact to get some support?* That is precisely what this file provides - the list of all kernel subsystems along with its maintainer(s). This goes all the way down to the level of individual components, such as a particular driver or file, as well as its status, who is currently maintaining it, the mailing list, website, and so on. Very helpful! There's even a helper script to find the person or team to talk to: `scripts/get_maintainer.pl`. See more, point [3].
`Makefile`	This is the kernel's top-level *Makefile*; the kernel's *Kbuild* build system as well as kernel modules use this `Makefile` for the build.

Major subsystem directories	
`kernel/`	Core kernel subsystem: the code here deals with a large number of core kernel features including stuff like process/thread life cycle management, CPU task scheduling, locking, cgroups, timers, interrupts, signaling, modules, tracing, RCU primitives, [e]BPF, and more.
`mm/`	The bulk of the **memory management** (**mm**) code lives here. We will cover a little of this in *Chapter 6, Kernel Internals Essentials – Processes and Threads*, and some related coverage in *Chapter 7, Memory Management Internals – Essentials*, and *Chapter 8, Kernel Memory Allocation for Module Authors – Part 1*, as well.
`fs/`	The code here implements two key filesystem features: the abstraction layer – the kernel **Virtual Filesystem Switch** (**VFS**) – and the individual filesystem drivers (for example, `ext[2|4]`, `btrfs`, `nfs`, `ntfs`, `overlayfs`, `squashfs`, `jffs2`, `fat`, `f2fs`, `isofs`, and so on).
`block/`	The underlying block I/O code path to the VFS/FS. It includes the code implementing the page cache, a generic block IO layer, IO schedulers, the new-ish blk-mq features, and so on.
`net/`	Complete implementation of the network protocol stack, to the letter of the **Request For Comments** (**RFCs**) – `https://whatis.techtarget.com/definition/Request-for-Comments-RFC`. Includes high-quality implementations of TCP, UDP, IP, and many, many more networking protocols. Want to see the code-level implementation of TCP/IP for IPv4? It's here: *net/ipv4/*, see the `tcp*.c` and `ip*.c` source, besides others.
`ipc/`	The **Inter-Process Communication** (**IPC**) subsystem code; the implementation of IPC mechanisms such as SysV and POSIX message queues, shared memory, semaphores, and so on.
`sound/`	The audio subsystem code, aka the **Advanced Linux Sound Architecture** (**ALSA**) layer.
`virt/`	The *virtualization* (hypervisor) code; the popular and powerful **Kernel Virtual Machine** (**KVM**) is implemented here.
Arch/Infrastructure/Drivers/Miscellaneous	
`Documentation/`	The official kernel documentation resides right here; it's important to get familiar with it. The `README` file refers to its online version.
`LICENSES/`	The text of all licenses, categorized under different heads. See point [2].
`arch/`	The arch-specific code lives here (by the word *arch*, we mean CPU). Linux started as a small hobby project for the i386. It is now very probably the most ported OS ever. See the *arch* ports in point [4] of the list that follows this table.
`certs/`	Support code for generating signed modules; this is a powerful security feature, which when correctly employed ensures that even malicious rootkits cannot simply load any kernel module they desire.

`crypto/`	This directory contains the kernel-level implementation of ciphers (as in encryption/decryption algorithms, or transformations) and kernel APIs to serve consumers that require cryptographic services.
`drivers/`	The kernel-level device drivers code lives here. This is considered a non-core region; it's classified into many types of drivers. This tends to be the region that's most often being contributed to; as well, this code accounts for the most disk space within the source tree.
`include/`	This directory contains the arch-independent kernel headers. There are also some arch-specific ones under `arch/<cpu>/include/....`
`init/`	The arch-independent kernel initialization code; perhaps the closest we get to the kernel's `main` function is here: `init/main.c:start_kernel()`, with the `start_kernel()` function within it being considered the early C entry point during kernel initialization. (Let's leverage Bootlin's superb web code browsing tool to see it, for 6.1.25, here: `https://elixir.bootlin.com/linux/v6.1.25/source/init/main.c#L936`.)
`io_uring/`	Kernel infrastructure for implementing the new-ish `io_uring` fast I/O framework; see point [5].
`lib/`	The closest equivalent to a library for the kernel. It's important to understand that the kernel does *not* support shared libraries as user space apps do. Some of the code here is auto-linked into the kernel image file and hence is available to the kernel at runtime. Various useful components exist within `lib/`: [un]compression, checksum, bitmap, math, string routines, tree algos, and so on.
`rust/`	Kernel infrastructure for supporting the Rust programming language; see point [6].
`samples/`	Sample code for various kernel features and mechanisms; useful to learn from!
`scripts/`	Various scripts are housed here, some of which are used during kernel build, many for other purposes like static/dynamic analysis, debugging, and so on. They're mostly Bash and Perl scripts. (FYI, and especially for debugging purposes, I have covered many of these scripts in *Linux Kernel Debugging, 2022*.)
`security/`	Houses the kernel's **Linux Security Module** (**LSM**), a **Mandatory Access Control** (**MAC**) framework that aims at imposing stricter access control of user apps to kernel space than the default kernel does. The default model is called **Discretionary Access Control** (**DAC**). Currently, Linux supports several LSMs; well-known ones are SELinux, AppArmor, Smack, Tomoyo, Integrity, and Yama. Note that LSMs are "off" by default.
`tools/`	The source code of various *user mode* tools is housed here, mostly applications or scripts that have a "tight coupling" with the kernel, and thus require to be within the particular kernel codebase. *Perf*, a modern CPU profiling tool, eBPF tooling, and some tracing tools, serve as excellent examples.

`usr/`	Support code to generate and load the *initramfs* image; this allows the kernel to execute user space code during kernel init. This is often required; we cover initramfs in *Chapter 3, Building the 6.x Linux Kernel from Source – Part 2,* section *Understanding the initramfs framework.*

Table 2.2: Layout of the Linux kernel source tree

The following are some important explanations from the table:

1. **README:** This file also mentions the document to refer to for info on the minimal acceptable versions of software to build and run the kernel: `Documentation/process/changes.rst`. Interestingly, the kernel provides an Awk script (`scripts/ver_linux`) that prints the versions of current software on the system it's run upon, helping you to check whether the versions you have installed are acceptable.
2. **Kernel licensing:** Without getting stuck in the legal details (needless to say, I am not a lawyer), here's the pragmatic essence of the thing. As the kernel is released under the GNU GPL-2.0 license (**GNU GPL** is the **GNU General Public License**), any project that directly uses the kernel code base automatically falls under this license. This is the "derivative work" property of the GPL-2.0. Legally, these projects or products must now release their kernel software under the same license terms. Practically speaking, the situation on the ground is a good deal hazier; many commercial products that run on the Linux kernel do have proprietary user- and/or kernel-space code within them. They typically do so by refactoring kernel (most often, device driver) work in **Loadable Kernel Module** (**LKM**) format. It is possible to release the kernel module (LKM) under a *dual-license* model. The LKM is the subject matter of *Chapter 4, Writing Your First Kernel Module – Part 1*, and *Chapter 5, Writing Your First Kernel Module – Part 2*, and we cover some information on the licensing of kernel modules there.

   Some folks, preferring proprietary licenses, manage to release their kernel code within a kernel module that is not licensed under GPL-2.0 terms; technically, this is perhaps possible, but is at the very least considered as being terribly anti-social and can even cross the line to being illegal. The interested among you can find more links on licensing in the *Further reading* document for this chapter.

3. **MAINTAINERS:** Just peek at this file in the root of your kernel source tree! Interesting stuff... To illustrate how it's useful, let's run a helper Perl script: `scripts/get_maintainer.pl`. Do note that, pedantically, it's meant to be run on a Git tree only. Here, we ask the script to show the maintainers of the kernel CPU task scheduling code base by specifying a file or directory via the `-f` switch:

```
$ scripts/get_maintainer.pl --nogit -f kernel/sched
Ingo Molnar <mingo@redhat.com> (maintainer:SCHEDULER)
Peter Zijlstra <peterz@infradead.org> (maintainer:SCHEDULER)
Juri Lelli <juri.lelli@redhat.com> (maintainer:SCHEDULER)
Vincent Guittot <vincent.guittot@linaro.org> (maintainer:SCHEDULER)
Dietmar Eggemann <dietmar.eggemann@arm.com> (reviewer:SCHEDULER)
```

```
Steven Rostedt <rostedt@goodmis.org> (reviewer:SCHEDULER)
Ben Segall <bsegall@google.com> (reviewer:SCHEDULER)
Mel Gorman <mgorman@suse.de> (reviewer:SCHEDULER)
Daniel Bristot de Oliveira <bristot@redhat.com> (reviewer:SCHEDULER)
Valentin Schneider <vschneid@redhat.com> (reviewer:SCHEDULER)
linux-kernel@vger.kernel.org (open list:SCHEDULER)
```

4. **Linux arch (CPU) ports:** As of 6.1, the Linux OS has been ported to all these processors. Most have MMUs, You can see the arch-specific code under the `arch/` folder, each directory representing a particular CPU architecture:

```
$ cd ${LKP_KSRC} ; ls arch/
alpha/      arm64/   ia64/     m68k/     nios2/    powerpc/
sh/         x86/     arc/      csky/     Kconfig   microblaze/
openrisc/   riscv/   sparc/    x86_64/   arm/      hexagon/
loongarch/  mips/    parisc/   s390/     um/       xtensa/
```

In fact, when cross-compiling, the `ARCH` environment variable is set to the name of one of these folders, in order to compile the kernel for that architecture. For example, when building target "foo" for the AArch64, we'd typically do something like `make ARCH=arm64 CROSS_COMPILE=<...> foo`.

As a kernel or driver developer, browsing the kernel source tree is something you will have to get quite used to (and even grow to enjoy!). Searching for a particular function or variable can be a daunting task when the code is in the ballpark of 30 million SLOCs though! Do learn to use efficient code browser tools. I suggest the `ctags` and `cscope` **Free and Open Source Software** (**FOSS**) tools. In fact, the kernel's top-level `Makefile` has targets for precisely these: `make [ARCH=<cpu>] tags ; make [ARCH=<cpu>] cscope` A must-do! (With `cscope`, with `ARCH` set to null (the default), it builds the index for the x86[_64]. To, for example, generate the tags relevant to the AArch64, run `make ARCH=arm64 cscope`.) Also, FYI, several other code-browsing tools exist of course; another good one is opengrok.

5. **io_uring:** It's not an exaggeration to say that `io_uring` and *eBPF* are considered to be two of the new(-ish) "magic features" that a modern Linux system provides (the `io_uring` folder here is the kernel support for this feature)! The reason database/network-like folks are going ga-ga over `io_uring` is simple: performance. This framework dramatically improves performance numbers in real-world high I/O situations, for both disk and network workloads. Its shared (between user and kernel-space) ring buffer architecture, zero-copy schema, and ability to use much fewer system calls compared to typical older AIO frameworks, including a polled mode operation, make it an enviable feature. So, for your user space apps to get on that really fast I/O path, check out `io_uring`. The *Further reading* section for this chapter carries useful links.

6. **Rust** in the kernel: Yes, indeed, there was a lot of hoopla about the fact that basic support for the Rust programming language has made it into the Linux kernel (6.0 first). Why? Rust does have a well-advertised advantage over even our venerable C language: ***memory-safety***. The fact is, even today, one of the biggest programing-related security headaches for code written in C/C++ – for both OS/drivers as well as user space apps – have at their root memory-safety issues (like the well-known **BoF** (**Buffer Overflow**) defect). These can occur when developers generate memory corruption defects (bugs!) in their C/C++ code. This leads to vulnerabilities in software that clever hackers are always on the lookout for and exploit! Having said all that, at least as of now, Rust has made a very minimal entry into the kernel – no core code uses it. The current Rust support within the kernel is to support writing modules in Rust in the future. (There is a bit of sample Rust code, of course, here: `samples/rust/`.) Rust usage in the kernel will certainly increase in time.... The *Further reading* section has some links on this topic – do check it out, if interested.

We have now completed *step 2*, the extraction of the kernel source tree! As a bonus, you also learned the basics regarding the layout of the kernel source. Let's now move on to *step 3* of the process and learn how to *configure* the Linux kernel prior to building it.

Step 3 – Configuring the Linux kernel

Configuring the kernel is perhaps the *most* critical step in the kernel build process. One of the many reasons Linux is a critically acclaimed OS is its versatility. It's a common misconception to think that there is a separate Linux kernel code base for an enterprise-class server, a data center, a workstation, and a tiny, embedded Linux device – no, they all use the very same unified Linux kernel source! Thus, carefully *configuring* the kernel for a particular use case (server, desktop, embedded, or hybrid/custom) is a powerful feature and a requirement. This is precisely what we are delving into here.

You'll probably find that this topic – arriving at a working kernel config – tends to be a long and winding discussion, but it's ultimately worth it; do take the time and trouble to read through it.

Also, to build a kernel, you *must* carry out the kernel configuration step regardless. Even if you feel you do not require any changes to the existing or default config, it's very important to run this step at least once as part of the build process. Otherwise, certain headers that are auto-generated here will be missing and cause issues later. At the very least, the `make old[def]config` step should be carried out. This will set up the kernel config to that of the existing system with answers to config options being requested from the user only for any new options.

Next, we cover a bit of required background on the kernel build system.

Note: you might find that if you're completely new to configuring the kernel, the following details might feel a bit overwhelming at first. In this case, I suggest you skip it on first reading, move ahead with the practicalities of configuring the kernel, and then come back to this section.

Minimally understanding the Kconfig/Kbuild build system

The infrastructure that the Linux kernel uses to configure the kernel is known as the **Kconfig** system, and to build it, there is the **Kbuild** infrastructure. Without delving into the gory details, the Kconfig + Kbuild system ties together the complex kernel configuration and build process by separating out the work into logical streams:

- *Kconfig* – the infrastructure to configure the kernel; it consists of two logical parts:
    - The *Kconfig language*: it's used to specify the syntax within the various `Kconfig[.*]` files, which in effect specify the "menus" where kernel config options are selected.
    - The *Kconfig parsers*: tools that intelligently parse the `Kconfig[.*]` files, figure out dependencies and auto-selections, and generate the menu system. Among them is the commonly used `make menuconfig`, which internally invokes the `mconf` tool (the code's under `scripts/kconfig`).
- *Kbuild* – the support infrastructure to build the source code into kernel binary components. It mainly uses a *recursive make* style build, originating at the kernel top-level *Makefile*, which, in turn recursively parses the content of hundreds of Makefiles embedded into sub-directories within the source (as required).

A diagram that attempts to convey information regarding this Kconfig/Kbuild system (in a simplified way) can be seen in *Figure 2.8*. Several details aren't covered yet; still, you can keep it in mind while reading the following materials.

To help you gain a better understanding, let's look at a few of the key components that go into the Kconfig/Kbuild system:

- The `CONFIG_FOO` symbols
- The menu specification file(s), named `Kconfig[.*]`
- The `Makefile(s)`
- The overall kernel config file – `.config` – itself.

The purposes of these components are summarized as follows:

Kconfig/Kbuild component	Purpose in brief
`Kconfig`: Config symbol: `CONFIG_FOO`	Every kernel configurable `FOO` is represented by a `CONFIG_FOO` macro. Depending on the user's choice, the macro will resolve to one of `y`, `m`, or `n`: • `y=yes`: this implies building the config or feature `FOO` into the kernel image itself • `m=module`: this implies building it as a separate object, a kernel module (a `.ko` file) • `n=no`: this implies not building the feature Note that `CONFIG_FOO` is an alphanumeric string. We will soon see a means to look up the precise config option name via the `make menuconfig` UI.
`Kconfig`: `Kconfig.*` files	This is where the `CONFIG_FOO` symbol is defined. The Kconfig syntax specifies its type (Boolean, tristate, [alpha]numeric, and so on) and dependency tree. Furthermore, for the menu-based config UI (invoked via one of `make [menu|g|x]config`), it specifies the menu entries themselves. We will, of course, make use of this feature later.
`Kbuild: Makefile(s)`	The Kbuild system uses a *recursive make* Makefile approach. The `Makefile` in the root of the kernel source tree is called the *top-level Makefile*, typically with a `Makefile` within each sub-folder to build the source there. The 6.1 kernel source has over 2,700 Makefiles in all!
The `.config` file	Ultimately, the kernel configuration distills down to this file; `.config` is the final kernel config file. It's generated and stored within the kernel source tree root folder as a simple ASCII text file. Keep it safe, as it's a key part of your product. Note that the config filename can be overridden via the environment variable `KCONFIG_CONFIG`.

Table 2.3: Major components of the Kconfig+Kbuild build system

How the Kconfig+Kbuild system works – a minimal take

Now that we know a few details, here's a simplified take on how it's tied together and works:

- First, the user (you) configures the kernel using some kind of menu system provided by *Kconfig*.
- The kernel config directives selected via this menu system UI are written into a few auto-generated headers and a final `.config` file, using a `CONFIG_FOO={y|m}` syntax, or, `CONFIG_FOO` is simply commented out (implying "don't build `FOO` at all").

- Next, the *Kbuild* per-component *Makefiles* (invoked via the kernel top-level Makefile) typically specify a directive FOO like this:

```
obj-$(CONFIG_FOO) += FOO.o
```

- A `FOO` component could be anything - a core kernel feature, a device driver, a filesystem, a debug directive, and so on. Recall, the value of `CONFIG_FOO` may be `y`, or `m`, or not exist; this accordingly has the build either build the component `FOO` into the kernel (when its value is y), or as a module (when its value is `m`)! If commented out, it isn't built at all, simple. In effect, the above-mentioned Makefile directive, at build time, expands into one of these three for a given kernel component `FOO`:

```
obj-y += FOO.o      # build the feature FOO into the kernel image
obj-m += FOO.o     # build the feature FOO as a discrete kernel module (a
#  foo.ko file)
<if CONFIG_FOO is null>        # do NOT build feature FOO
```

To see an instance of this in action, check out the *Kbuild* file (more details on *Kconfig* files are in the *Understanding the Kconfig* files* section) in the root of the kernel source tree:

```
$ cat Kconfig
…
# Kbuild for top-level directory of the kernel
…
# Ordinary directory descending
# -------------------------------------------------------------------

obj-y                   += init/
obj-y                   += usr/
obj-y                   += arch/$(SRCARCH)/
obj-y                   += $(ARCH_CORE)
obj-y                   += kernel/
[ … ]
obj-$(CONFIG_BLOCK)     += block/
obj-$(CONFIG_IO_URING)  += io_uring/
obj-$(CONFIG_RUST)      += rust/
obj-y                   += $(ARCH_LIB)
[ … ]
obj-y                   += virt/
obj-y                   += $(ARCH_DRIVERS)
```

Interesting! We can literally see how the top-level Makefile will descend into other directories, with the majority being set to `obj-y`; in effect, build it in (in a few cases it's parametrized, becoming `obj-y` or `obj-m` depending on how the user selected the option).

Great. Let's move along now; the key thing to do is to get ourselves a working `.config` file. How can we do so? We do this iteratively. We begin with a "default" configuration – the topic of the following section – and carefully work our way up to a custom config.

Arriving at a default configuration

So, how do you decide on the initial kernel configuration to begin with? Several techniques exist; a few common ones are as follows:

- Don't specify anything; Kconfig will pull in a default kernel configuration (as all kernel configs have a default value)
- Use the existing distribution's kernel configuration
- Build a custom configuration based on the kernel modules currently loaded in memory

The first approach has the benefit of simplicity. The kernel will handle the details, giving you a default configuration. The downside is that the default config can be very large (this is the case when building Linux for an x86_64-based desktop or server-type system); a huge number of options are turned on by default, just in case you need it, which can make the build time very long and the kernel image size very large. Typically, of course, you are then expected to manually configure the kernel to the desired settings.

This brings up the question, *where is the default kernel config stored?* The Kconfig system uses a priority list fallback scheme to retrieve a default configuration if none is specified. The priority list and its order (the first being the highest priority) is as follows:

- `.config`
- `/lib/modules/$(uname -r)/.config`
- `/etc/kernel-config`
- `/boot/config-$(uname -r)`
- `ARCH_DEFCONFIG          (if defined)`
- `arch/${ARCH}/defconfig`

From the list, you can see that the Kconfig system first checks for the presence of a `.config` file in the root of the kernel source tree; if found, it picks up all the config values from there. If it doesn't exist, it next looks at the path `/lib/modules/$(uname -r)/.config`. If found, the values found in that file will be used as the defaults. If not found, it checks the next one in the preceding priority list, and so on... You can see this shown in *Figure 2.8*.

For a more detailed look at the kernel's Kconfig and Kbuild infrastructure, we suggest you refer to the following excellent docs:

- Exploring the Linux kernel: The secrets of Kconfig/kbuild, Cao Jin, opensource.com, October 2018: `https://opensource.com/article/18/10/kbuild-and-kconfig`
- Slide presentation: A Dive Into Kbuild, Cao Jin, Fujitsu, August 2018: `https://events19.linuxfoundation.org/wp-content/uploads/2017/11/A-Dive-into-Kbuild-Cao-Jin-Fujitsu.pdf`

A diagram (inspired by Cao Jin's articles) that attempts to communicate the kernel's Kconfig/Kbuild system is shown here. The diagram conveys more information than has been covered by now; worry not, we'll get to it.

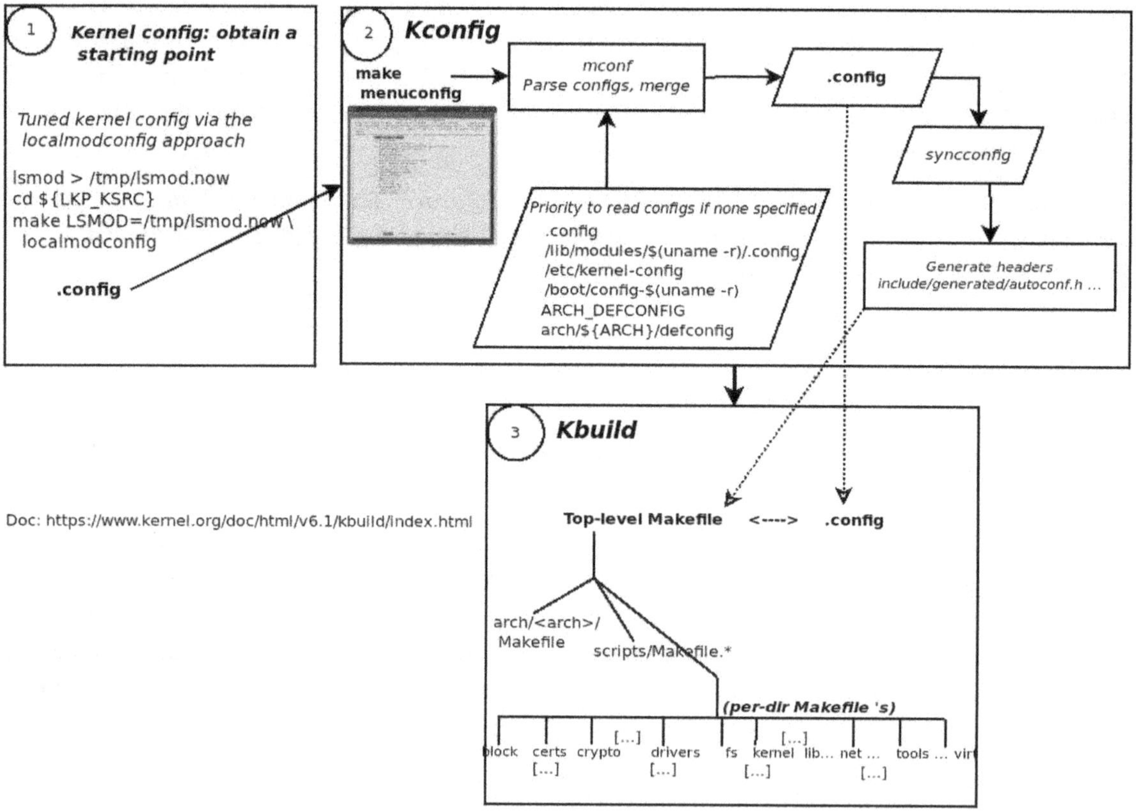

Figure 2.8: The kernel's Kconfig/Kbuild system in a simplified form

Right, let's now get to figuring out how exactly to get a working kernel config!

Obtaining a good starting point for kernel configuration

This brings us to a **really important point**: while playing around with the kernel configuration is okay to do as a learning exercise, for a production system it's critical that you base your custom config on a proven – known, tested, and working – kernel configuration.

Here, to help you understand the nuances of selecting a valid starting point for kernel configuration, we will see three approaches to obtaining a starting point for a typical kernel configuration:

- First, an easy (but sub-optimal) approach where you simply emulate the existing distribution's kernel configuration.
- Next, a more optimized approach where you base the kernel configuration on the existing system's in-memory kernel modules. This is the `localmodconfig` approach.
- Finally, a word on the approach to follow for a typical embedded Linux project.

Let's examine each of these approaches in a bit more detail. In terms of configuring the kernel you've downloaded and extracted in the previous two steps, don't do anything right now; read the sections that follow, and then, in the *Getting going with the localmodconfig approach* section, we'll have you actually get started.

Kernel config using distribution config as a starting point

The typical target system for using this approach is a x86_64 desktop or server Linux system. Let's configure the kernel to all defaults:

```
$ make mrproper
  CLEAN   scripts/basic
  CLEAN   scripts/kconfig
  CLEAN   include/config include/generated .config
```

To ensure we begin with a clean slate, we run `make mrproper` first; be careful, it cleans pretty much everything, including the `.config` if it exists.

Next, we perform the `make defconfig` step, which, as the `make help` command output shows (try it out! See *Figure 2.10*), gives us a new config:

```
$ make defconfig
  HOSTCC  scripts/basic/fixdep
  HOSTCC  scripts/kconfig/conf.o
 [ … ]
 HOSTLD  scripts/kconfig/conf
*** Default configuration is based on 'x86_64_defconfig'
#
# configuration written to .config
#
```

The building of the mconf utility itself is performed first (under `scripts/kconfig`), and then the config is generated. Here it is:

```
$ ls -l .config
-rw-rw-r-- 1 c2kp c2kp 136416 Apr 29 08:12 .config
```

There, done: we now have an "all-defaults" kernel config saved in `.config`!

What if there's no `defconfig` file for your arch under `arch/${ARCH}/configs`? Then, at least on x86_64, you can simply copy in the existing distro default kernel config:

```
cp /boot/config-$(uname -r) ${LKP_KSRC}/.config
```

Here, we simply copy the existing Linux distribution's (here, it's our Ubuntu 22.04 LTS guest VM) config file into the `.config` file in the root of the kernel source tree, thereby making the distribution config the starting point, which can then be further edited. As already mentioned, the downside of this quick approach is that the config tends to be large, thus resulting in a large-footprint kernel image.

Also, FYI, once the kernel config is generated in any manner (like via `make defconfig`), every kernel config `FOO` is shown as an empty file within `include/config`.

Tuned kernel config via the localmodconfig approach

The typical target system for using this approach is a (typically x86_64) desktop or server Linux system.

This second approach is more optimized than the previous one - a good one to use when the goal is to begin with a kernel config that is based on your existing running system and *is thus (usually) relatively compact compared to the typical default config* on a desktop or server Linux system.

Here, we provide the Kconfig system with a snapshot of the kernel modules currently running on the system by simply redirecting the output of `lsmod` into a temporary file and then providing that file to the build. This can be achieved as follows:

```
lsmod > /tmp/lsmod.now
cd ${LKP_KSRC}
make LSMOD=/tmp/lsmod.now localmodconfig
```

The `lsmod` utility simply lists all the kernel modules currently residing in system kernel memory. We will see more on this in *Chapter 4, Writing Your First Kernel Module - Part 1.*

We save its output in a temporary file, which we then pass via the `LSMOD` environment variable to the `Makefile's localmodconfig` target. The job of this target is to configure the kernel in a manner as to only include the base functionality plus the functionality provided by these kernel modules and leave out the rest, in effect giving us a reasonable facsimile of the current kernel (or of whichever kernel the `lsmod` output represents). We use precisely this technique to configure our 6.1 kernel in the upcoming *Getting going with the localmodconfig approach* section. We also show only this approach as *step 1 (1 in the circle)* in *Figure 2.8*.

Kernel config for typical embedded Linux systems

The typical target system for using this approach is usually a small embedded Linux system. The goal here is to begin with a proven – a known, tested, and working – kernel configuration for our embedded Linux project. Well, how exactly can we achieve this?

Before going further, let me mention this: the initial discussion here will be shown to be the older approach to configuring (the AArch32 or ARM-32 arch) embedded Linux; we shall then see the "correct" and modern approach for modern platforms.

Interestingly, for the AArch32 at least, the kernel code base itself contains known, tested, and working kernel configuration files for various well-known hardware platforms.

Assuming our target is ARM-32 based, we merely must select the one that matches (or is the nearest match to) our embedded target board. These kernel config files are present within the kernel source tree in the `arch/<arch>/configs/` directory. The config files are in the format `<platform-name>_defconfig`.

A quick peek is in order; see the following screenshot showing, for the ARM-32, the existing board-specific kernel code under `arch/arm/mach-<foo>` and platform config files under `arch/arm/configs` on the v6.1.25 Linux kernel code base:

```
$ ls arch/arm/
boot/              mach-artpec/      mach-gemini/     mach-mstar/      mach-rockchip/  mach-versatile/
common/            mach-asm9260/     mach-highbank/   mach-mv78xx0/    mach-rpc/       mach-vt8500/
configs/           mach-aspeed/      mach-hisi/       mach-mvebu/      mach-s3c/       mach-zynq/
crypto/            mach-at91/        mach-hpe/        mach-mxs/        mach-s5pv210/   Makefile
include/           mach-axxia/       mach-imx/        mach-nomadik/    mach-sa1100/    mm/
Kbuild             mach-bcm/         mach-iop32x/     mach-npcm/       mach-shmobile/  net/
Kconfig            mach-berlin/      mach-ixp4xx/     mach-nspire/     mach-socfpga/   nwfpe/
Kconfig.assembler  mach-clps711x/    mach-keystone/   mach-omap1/      mach-spear/     plat-orion/
Kconfig.debug      mach-cns3xxx/     mach-lpc18xx/    mach-omap2/      mach-sti/       probes/
Kconfig-nommu      mach-davinci/     mach-lpc32xx/    mach-orion5x/    mach-stm32/     tools/
kernel/            mach-digicolor/   mach-mediatek/   mach-oxnas/      mach-sunplus/   vdso/
lib/               mach-dove/        mach-meson/      mach-pxa/        mach-sunxi/     vfp/
mach-actions/      mach-ep93xx/      mach-milbeaut/   mach-qcom/       mach-tegra/     xen/
mach-airoha/       mach-exynos/      mach-mmp/        mach-rda/        mach-uniphier/
mach-alpine/       mach-footbridge/  mach-moxart/     mach-realtek/    mach-ux500/
$
$ ls arch/arm/configs/
am200epdkit_defconfig      gemini_defconfig           multi_v5_defconfig        s5pv210_defconfig
aspeed_g4_defconfig        h3600_defconfig            multi_v7_defconfig        sama5_defconfig
aspeed_g5_defconfig        h5000_defconfig            mv78xx0_defconfig         sama7_defconfig
assabet_defconfig          hackkit_defconfig          mvebu_v5_defconfig        shannon_defconfig
at91_dt_defconfig          hisi_defconfig             mvebu_v7_defconfig        shmobile_defconfig
axm55xx_defconfig          imxrt_defconfig            mxs_defconfig             simpad_defconfig
badge4_defconfig           imx_v4_v5_defconfig        neponset_defconfig        socfpga_defconfig
bcm2835_defconfig          imx_v6_v7_defconfig        netwinder_defconfig       sp7021_defconfig
cerfcube_defconfig         integrator_defconfig       nhk8815_defconfig         spear13xx_defconfig
clps711x_defconfig         iop32x_defconfig           omap1_defconfig           spear3xx_defconfig
cm_x300_defconfig          ixp4xx_defconfig           omap2plus_defconfig       spear6xx_defconfig
cns3420vb_defconfig        jornada720_defconfig       orion5x_defconfig         spitz_defconfig
colibri_pxa270_defconfig   keystone_defconfig         oxnas_v6_defconfig        stm32_defconfig
colibri_pxa300_defconfig   lart_defconfig             palmz72_defconfig         sunxi_defconfig
collie_defconfig           lpc18xx_defconfig          pcm027_defconfig          tct_hammer_defconfig
corgi_defconfig            lpc32xx_defconfig          pleb_defconfig            tegra_defconfig
davinci_all_defconfig      lpd270_defconfig           pxa168_defconfig          trizeps4_defconfig
dove_defconfig             lubbock_defconfig          pxa255-idp_defconfig      u8500_defconfig
dram_0x00000000.config     magician_defconfig         pxa3xx_defconfig          versatile_defconfig
dram_0xc0000000.config     mainstone_defconfig        pxa910_defconfig          vexpress_defconfig
dram_0xd0000000.config     milbeaut_m10v_defconfig    pxa_defconfig             vf610m4_defconfig
ep93xx_defconfig           mini2440_defconfig         qcom_defconfig            viper_defconfig
eseries_pxa_defconfig      mmp2_defconfig             realview_defconfig        vt8500_v6_v7_defconfig
exynos_defconfig           moxart_defconfig           rpc_defconfig             xcep_defconfig
ezx_defconfig              mps2_defconfig             s3c2410_defconfig         zeus_defconfig
footbridge_defconfig       multi_v4t_defconfig        s3c6400_defconfig
$
```

Figure 2.9: The contents of arch/arm and arch/arm/configs on the 6.1.25 Linux kernel

Whoah, quite a bit! The directories `arch/arm/mach-<foo>` represent hardware platforms (boards or *machines*) that Linux has been ported to (typically by the silicon vendor); the board-specific code is within these directories.

Similarly, working default kernel config files for these platforms are also contributed by them and are under the `arch/arm/configs` folder of the form `<foo>_defconfig`; as can clearly be seen in the lower portion of *Figure 2.9*.

Thus, for example, if you find yourself configuring the Linux kernel for a hardware platform having, say, an i.MX 7 SoC from NXP on it, please don't start with an x86_64 kernel config file as the default. It won't work. Even if you manage it, the kernel will not build/work cleanly. Pick the appropriate kernel config file: for our example here, perhaps the `imx_v6_v7_defconfig` file would be a good starting point. You can copy this file into `.config` in the root of your kernel source tree and then proceed to fine-tune it to your project-specific needs.

As another example, the Raspberry Pi (`https://www.raspberrypi.org/`) is a very popular hobbyist and production platform. The kernel config file – within its kernel source tree – used as a base for it is this one: `arch/arm/configs/bcm2835_defconfig`. The filename reflects the fact that Raspberry Pi boards use a Broadcom 2835-based SoC. You can find details regarding kernel compilation for the Raspberry Pi here: `https://www.raspberrypi.org/documentation/linux/kernel/building.md`. Hang on, though, we will be covering at least some of this in *Chapter 3, Building the 6.x Linux Kernel from Source – Part 2*, in the *Kernel build for the Raspberry Pi* section.

The modern approach – using the Device Tree

Okay, a word of caution! For AArch32, as we saw, you'll find the platform-specific config files under `arch/arm/configs` as well as the board-specific kernel code under `arch/arm/mach-<foo>`, where `foo` is the platform name. Well, the reality is that this approach – keeping board-specific config and kernel source files within the Linux OS code base – is considered to be exactly the wrong one for an OS! Linus has made it amply clear – the "mistakes" made in older versions of Linux, which happened when the ARM-32 was the popular arch, must never be repeated for other arches. Then how does one approach this? The answer, for the modern (32 and 64-bit) ARM and PPC architectures, is to use the modern **Device Tree** (**DT**) approach.

Very basically, the DT holds all platform hardware topology details. It is effectively the board or platform layout; it's not code, it's a description of the hardware platform analogous to VHDL. BSP-specific code and drivers still need to be written, but now have a neat way to be "discovered" or enumerated by the kernel at boot when it parses the **DTB** (**Device Tree Blob**) that's passed along by the bootloader. The DTB is generated as part of the build process, by invoking the **DTC** (**Device Tree Compiler**) on the DT source files for the platform.

So, nowadays, you'll find that the majority of embedded projects (for ARM and PPC at least) will use the DT to good effect. It also helps OEMs and ODMs/vendors by allowing usage of essentially the same kernel with platform/model-specific tweaks built into the DT. Think of the dozens of Android phone models from popular OEMs with mostly the same stuff but just a few hardware differences; one kernel will typically suffice! This dramatically eases the maintenance burden. For the curious – the DT sources – the `.dts` files can be found here: `arch/<arch>/boot/dts`.

The lesson on keeping board-specific stuff outside the kernel code base as far as is possible seems to have been learned well for the AArch64 (ARM-64). Compare its clean, well-organized, and uncluttered config and DTS folders (`arch/arm64/configs/`: it has only one file, a `defconfig`) to AArch32. Even the DTS files (look under `arch/arm64/boot/dts/`) are well organized compared to AArch32:

```
6.1.25 $ ls arch/arm64/configs/
defconfig
6.1.25 $ ls arch/arm64/boot/dts/
actions/     amazon/     apm/        bitmain/     exynos/     intel/
marvell/     nuvoton/    realtek/    socionext/   tesla/      xilinx/
allwinner/   amd/        apple/      broadcom/    freescale/  lg/
mediatek/    nvidia/     renesas/    sprd/        ti/         altera/
amlogic/     arm/        cavium/     hisilicon/   Makefile    microchip/
qcom/        rockchip/   synaptics/  toshiba/
```

So, with modern embedded projects and the DT, how is one to go about kernel/BSP tasks? Well, there are several approaches; the BSP work is:

- Carried out by an in-house BSP or platform team.
- Provided as a BSP "package" by a vendor (often the silicon vendor that you've partnered with) along with reference hardware.
- Outsourced to an external company or consultant that you've partnered with. Several companies exist in this space – among them are Siemens (ex Mentor Graphics), Timesys, and WindRiver.
- Often nowadays, with the project being built and integrated via sophisticated builder software like *Yocto* or *Buildroot*, the vendors contribute BSP layers, which are then integrated into the product by the build team.

Design: A bit off-topic, but I think it's important: when working on projects (especially embedded), teams have repeatedly shown the undesirable tendency to directly employ vendor SDK APIs to perform device-specific work in their apps and drivers. Now, at first glance, this might seem fine; it can become a huge burden when the realization dawns that, hey, requirements change, devices themselves change, and thus your tightly coupled software simply breaks! You had the apps (and drivers) tie into the device hardware with hardly any separation.

The solution, of course, is to use a **loosely coupled architecture**, with what's essentially a **HAL** (**Hardware Abstraction Layer**) to allow apps to interface with devices seamlessly. This also allows for the device-specific code to be changed without affecting the higher layers (apps). Designing this way might seem obvious in the abstract, essentially leveraging the *information-hiding* idea, but can be difficult to ensure in practice; do always keep this in mind. The fact is, Linux's device model encourages this loose coupling. The *Further reading* section has some good links on these design approaches, within the *Generic online and book resources...* section (here: `https://github.com/PacktPublishing/Linux-Kernel-Programming_2E/blob/main/Further_Reading.md#generic-online-and-book-resources--miscellaneous-very-useful`).

Right, that concludes the three approaches to setting up a starting point for kernel configuration.

Seeing all available config options

As a matter of fact, with regard to kernel config, we have just scratched the surface. Many more techniques to explicitly generate the kernel configuration in a given manner are encoded into the Kconfig system itself! How? Via configuration targets to `make`. See them by running `make help` in the root of your kernel source; they're under the `Configuration targets` heading:

```
$ pwd
/home/c2kp/kernels/linux-6.1.25
$ make help
Cleaning targets:
  clean           - Remove most generated files but keep the config and
                    enough build support to build external modules
  mrproper        - Remove all generated files + config + various backup files
  distclean       - mrproper + remove editor backup and patch files

Configuration targets:
  config          - Update current config utilising a line-oriented program
  nconfig         - Update current config utilising a ncurses menu based program
  menuconfig      - Update current config utilising a menu based program
  xconfig         - Update current config utilising a Qt based front-end
  gconfig         - Update current config utilising a GTK+ based front-end
  oldconfig       - Update current config utilising a provided .config as base
  localmodconfig  - Update current config disabling modules not loaded
                    except those preserved by LMC_KEEP environment variable
  localyesconfig  - Update current config converting local mods to core
                    except those preserved by LMC_KEEP environment variable
  defconfig       - New config with default from ARCH supplied defconfig
  savedefconfig   - Save current config as ./defconfig (minimal config)
  allnoconfig     - New config where all options are answered with no
  allyesconfig    - New config where all options are accepted with yes
  allmodconfig    - New config selecting modules when possible
  alldefconfig    - New config with all symbols set to default
  randconfig      - New config with random answer to all options
  yes2modconfig   - Change answers from yes to mod if possible
  mod2yesconfig   - Change answers from mod to yes if possible
  mod2noconfig    - Change answers from mod to no if possible
  listnewconfig   - List new options
  helpnewconfig   - List new options and help text
  olddefconfig    - Same as oldconfig but sets new symbols to their
                    default value without prompting
  tinyconfig      - Configure the tiniest possible kernel
  testconfig      - Run Kconfig unit tests (requires python3 and pytest)

Other generic targets:
  all             - Build all targets marked with [*]
```

Figure 2.10: Output from make help on an x86_64 (6.1.25 kernel) with interesting lines highlighted

Let's experiment with a couple of other approaches as well – the `oldconfig` one to begin with:

```
$ make mrproper
[ … ]
$ make oldconfig
  HOSTCC  scripts/basic/fixdep
  HOSTCC  scripts/kconfig/conf.o
  HOSTCC  scripts/kconfig/confdata.o
  [ … ]
 HOSTLD  scripts/kconfig/conf
 #
 # using defaults found in /boot/config-5.19.0-41-generic
 #
 [ … ]
 * Restart config...
 [ … ]
 *
 Control Group support (CGROUPS) [Y/?] y
  Favor dynamic modification latency reduction by default   (CGROUP_FAVOR_
DYNMODS) [N/y/?] (NEW)                         <waits for input here>
 [ … ]
```

It works of course, but you'll have to press the *Enter* key (a number of times, perhaps) to accept defaults for any and every newly detected kernel config... (or you can specify a value explicitly; we shall see more on these new kernel configs in the section that follows). This is quite normal, but a bit of an annoyance. There's an easier way : using the `olddefconfig` target; as its help line says – "Same as oldconfig but sets new symbols to their default value without prompting":

```
$ make olddefconfig
#
# using defaults found in /boot/config-5.19.0-41-generic
#
.config:10301:warning: symbol value 'm' invalid for ANDROID_BINDER_IPC
.config:10302:warning: symbol value 'm' invalid for ANDROID_BINDERFS
#
# configuration written to .config
#
Done.
```

Viewing and setting new kernel configs

Another quick experiment: we clean up and then copy in the distro kernel config. Keep a backup if you want your existing `.config`.

```
$ make mrproper
$ cp /boot/config-5.19.0-41-generic .config
cp: overwrite '.config'? y
$
```

We have the distro defaults in `.config`. Now, think on it: we're currently running the 5.19.0-41-generic distro kernel but are intending to build a new kernel, 6.1.25. So, the new kernel's bound to have at least a few new kernel configs. In cases like this, when you attempt to configure the kernel, the Kconfig system will question you: it will display every single new config option and the available values you can set it to, with the default one in square brackets, in the console window. You're expected to select the values for the new config options it encounters. You will see this as a series of questions and a prompt to answer them on the command line.

The kernel provides two interesting mechanisms to see all new kernel configs:

- `listnewconfig` – list new options
- `helpnewconfig` – list new options and help text

Running the first merely lists every new kernel config variable:

```
$ make listnewconfig
[ … ]
CONFIG_CGROUP_FAVOR_DYNMODS=n
CONFIG_XEN_PV_MSR_SAFE=y
CONFIG_PM_USERSPACE_AUTOSLEEP=n
[ … ]
CONFIG_TEST_DYNAMIC_DEBUG=n
```

There are lots of them – I got 108 new configs – so I've truncated the output here.

We can see all the new configs, though the output's not very helpful in understanding what exactly they mean. Running the `helpnewconfig` target solves this – you can now see the "Help" (from the config option's `Kconfig` file) for every new kernel config:

```
$ make helpnewconfig
[ … ]
CONFIG_CGROUP_FAVOR_DYNMODS:
This option enables the favordynmods mount option by default, which reduces
the latencies of dynamic cgroup modifications such as task migrations and
controller on/offs at the cost of making hot path operations such as forks and
exits more expensive.
Say N if unsure.
```

```
Symbol: CGROUP_FAVOR_DYNMODS [=n]
Type  : bool
Defined at init/Kconfig:959
  Prompt: Favor dynamic modification latency reduction by default
[ … ]
CONFIG_TEST_DYNAMIC_DEBUG:
This module registers a tracer callback to count enabled pr_debugs in a do_
debugging function, then alters their enablements, calls the function, and
compares counts.
If unsure, say N.
Symbol: TEST_DYNAMIC_DEBUG [=n]
[ … ]
$
```

Don't worry about understanding the `Kconfig` syntax for now; we shall cover it in the *Customizing the kernel menu, Kconfig, and adding our own menu item* section.

The LMC_KEEP environment variable

Also, did you notice in *Figure 2.10* that the `localmodconfig` and `localyesconfig` targets can optionally include an environment variable named `LMC_KEEP` (LMC is LocalModConfig)?

Its meaning is straightforward: setting `LMC_KEEP` to some colon-delimited values has the Kconfig system preserve the original configs for the specified paths. An example might look like this: `"drivers/usb:drivers/gpu:fs"`. In effect, it says, "Keep these modules enabled."

This is a feature introduced in the 5.8 kernel (the commit's here: `https://github.com/torvalds/linux/commit/c027b02d89fd42ecee911c39e9098b9609a5ca0b`). So, to make use of it, you could run the config command like this:

```
make LSMOD=/tmp/mylsmod \
  LMC_KEEP="drivers/usb:drivers/gpu:fs" \
        localmodconfig
```

Tuned config via the streamline_config.pl script

Interestingly, the kernel provides many helper scripts that can perform useful housekeeping, debugging, and other tasks within the `scripts/` directory. A good example with respect to what we're discussing here is the `scripts/kconfig/streamline_config.pl` Perl script. It's ideally suited to situations where your distro kernel has too many modules or built-in kernel features enabled, and you just want the ones that you're using right now – the ones that the currently loaded modules provide, like `localmodconfig`. Run this script with all the modules you want loaded up, saving its output to, ultimately, `.config`. Then run `make oldconfig` and the config's ready!

As a sidebar, here's how the original author of this script – Steven Rostedt – describes what it does (`https://github.com/torvalds/linux/blob/master/scripts/kconfig/streamline_config.pl`):

```
[…]
# Here's what I did with my Debian distribution.
#
#    cd /usr/src/linux-2.6.10
#    cp /boot/config-2.6.10-1-686-smp .config
#    ~/bin/streamline_config > config_strip
#    mv .config config_sav
#    mv config_strip .config
#    make oldconfig
[…]
```

You can try it if you wish.

Getting going with the localmodconfig approach

Now (finally!) let's get hands-on and create a reasonably sized base kernel configuration for our 6.1.25 LTS kernel by using the `localmodconfig` technique. As mentioned, this existing-kernel-modules-only approach is a good one when the goal is to obtain a starting point for kernel config on an x86-based system by keeping it tuned to the current host.

Don't forget: the kernel configuration being performed right now is appropriate for your typical x86_64 desktop/server systems, as a learning approach. This approach merely provides a starting point, and even that might not be okay. For actual projects, you'll have to carefully check and tune every aspect of the kernel config; having an audit of your precise hardware and software to support is key. Again, for embedded targets, the approach is different (as we discussed in the *Kernel config for typical embedded Linux systems* section).

Before going any further, it's a good idea to clean up the source tree, especially if you ran the experiments we worked on previously. Be careful: this command will wipe everything, including the `.config`:

```
make mrproper
```

As described previously, let's first obtain a snapshot of the currently loaded kernel modules, and then have the build system operate upon it by specifying the `localmodconfig` target, like so:

```
lsmod > /tmp/lsmod.now
cd ${LKP_KSRC}
make LSMOD=/tmp/lsmod.now localmodconfig
```

Now, when you run the `make [...] localmodconfig` command just shown, it's entirely possible, indeed probable, that there will be a difference in the configuration options between the kernel you are currently configuring (version 6.1.25) and the kernel you are currently running on the build machine (for myself, the host kernel is `$(uname -r) = 5.19.0-41-generic`). In such cases, as explained in the *Viewing and setting new kernel configs* section, the Kconfig system will question you for each new config; pressing Enter accepts the default.

Now let's make use of the `localmodconfig` command.

The prompt will be suffixed with `(NEW)`, in effect telling you that this is a new kernel config option and that it wants your answer as to how to configure it.

Enter the following commands (if not already done):

```
$ uname -r
5.19.0-41-generic
$ lsmod > /tmp/lsmod.now
$ make LSMOD=/tmp/lsmod.now localmodconfig
  HOSTCC  scripts/basic/fixdep
  HOSTCC  scripts/kconfig/conf.o
  [ ... ]
using config: '/boot/config-5.19.0-41-generic'
System keyring enabled but keys "debian/canonical-certs.pem" not found.
Resetting keys to default value.
*
* Restart config...
*
* Control Group support
*
Control Group support (CGROUPS) [Y/?] y
  Favor dynamic modification latency reduction by default (CGROUP_FAVOR_
DYNMODS) [N/y/?] (NEW) ↵
  Memory controller (MEMCG) [Y/n/?] y
[ … ]
Userspace opportunistic sleep (PM_USERSPACE_AUTOSLEEP) [N/y/?] (NEW) ↵
[ … ]
Multi-Gen LRU (LRU_GEN) [N/y/?] (NEW) ↵
[ … ]
```

```
 Rados block device (RBD) (BLK_DEV_RBD) [N/m/y/?] n
 Userspace block driver (Experimental) (BLK_DEV_UBLK) [N/m/y/?] (NEW)  ↵
[ … ]
  Pine64 PinePhone Keyboard (KEYBOARD_PINEPHONE) [N/m/y/?] (NEW)  ↵
[ … ]
  Intel Meteor Lake pinctrl and GPIO driver (PINCTRL_METEORLAKE) [N/m/y/?]
(NEW)  ↵
[ … ]
Test DYNAMIC_DEBUG (TEST_DYNAMIC_DEBUG) [N/m/y/?] (NEW)  ↵
[ … ]
#
# configuration written to .config
#
$ ls -l .config
-rw-rw-r-- 1 c2kp c2kp 170136 Apr 29 09:56 .config
```

After pressing the *Enter* key (↵) many times – we've highlighted this in the output block just above, showing just a few of the many new options encountered – the interrogation mercifully finishes and the `Kconfig` system writes the newly generated configuration to a file named `.config` in the current working directory. Note that we truncated the previous output as it's simply too voluminous, and unnecessary, to reproduce fully.

The preceding steps take care of generating the `.config` file via the `localmodconfig` approach. Before we conclude this section, here are a few additional points to note:

- To ensure a completely clean slate, run `make mrproper` or `make distclean` in the root of the kernel source tree, useful when you want to restart the kernel build procedure from scratch; rest assured, it will happen one day! Note that doing this deletes the kernel configuration file(s) too. Keep a backup before you begin, if required.
- Here, in this chapter, all the kernel configuration steps and the screenshots pertaining to it have been performed on an x86_64 Ubuntu 22.04 LTS guest VM, which we use as the host to ultimately build a brand-spanking-new 6.1 LTS Linux kernel. The precise names, presence, and content of the menu items seen, as well as the look and feel of the menu system (the UI), can and do vary based on (a) the architecture (CPU) and (b) the kernel version.
- As mentioned earlier, on a production system or project, the platform or **BSP** team, or indeed the embedded Linux BSP vendor company if you have partnered with one, will provide a good known, working, and tested kernel config file. Use this as a starting point by copying it into the `.config` file in the root of the kernel source tree. Alternatively, builder software like Yocto or Buildroot might be employed.

As you gain experience with building the kernel, you will realize that the effort in setting up the kernel configuration correctly the first time is higher; and, of course, the time required for the very first build is a lengthy one. Once done correctly, though, the process typically becomes much simpler – a recipe to run repeatedly.

Now, let's learn how to use a useful and intuitive UI to fine-tune our kernel configuration.

Tuning our kernel configuration via the make menuconfig UI

Okay, great, we now have an initial kernel config file (`.config`) generated for us via the `localmodconfig` Makefile target, as shown in detail in the previous section, which is a good starting point. Typically, we now further fine-tune our kernel configuration. One way to do this – in fact, the recommended way – is via the `menuconfig` Makefile target. This target has the Kbuild system generate a pretty sophisticated C-based program executable (`scripts/kconfig/mconf`), which presents to the end user a neat menu-based UI. This is step 2 in *Figure 2.8*. In the following output block, when (within the root of our kernel source tree), we invoke the command for the first time, the Kbuild system builds the `mconf` executable and invokes it:

```
$ make menuconfig
  UPD scripts/kconfig/.mconf-cfg
  HOSTCC scripts/kconfig/mconf.o
  HOSTCC scripts/kconfig/lxdialog/checklist.o
  […]
  HOSTLD  scripts/kconfig/mconf
```

Of course, a picture is no doubt worth a thousand words, so here's what the `menuconfig` UI looks like on my VM.

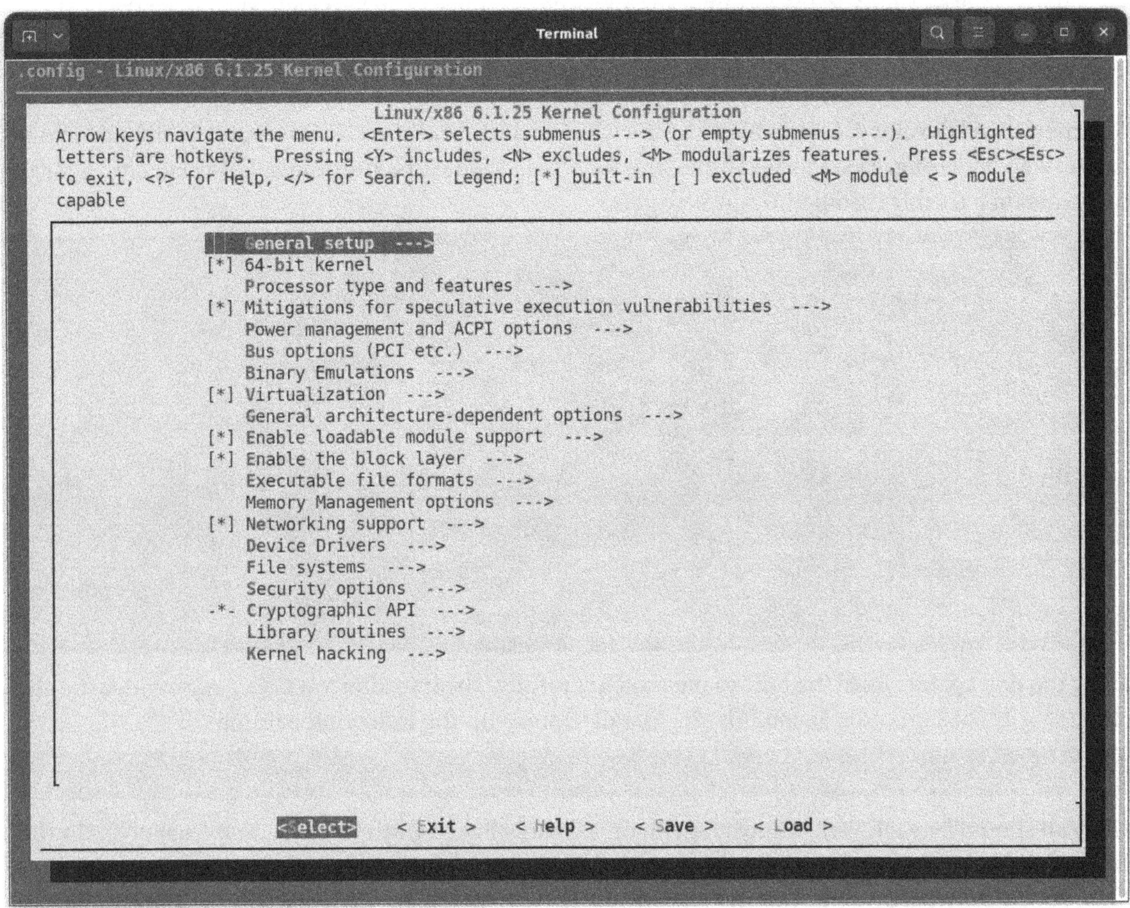

Figure 2.11: The main menu for kernel configuration via make menuconfig (on x86-64)

By the way, you don't need to be running your VM in GUI mode to use this approach; it even works on the terminal window when employing an SSH login shell from the host as well – another advantage of this UI approach to edit our kernel config!

As experienced developers, or indeed anyone who has sufficiently used a computer, well know, things can and do go wrong. Take, for example, the following scenario – running `make menuconfig` for the first time on a freshly installed Ubuntu system:

```
$ make menuconfig
  UPD     scripts/kconfig/.mconf-cfg
  HOSTCC  scripts/kconfig/mconf.o
  YACC    scripts/kconfig/zconf.tab.c
/bin/sh: 1: bison: not found
scripts/Makefile.lib:196: recipe for target 'scripts/kconfig/zconf.tab.c'
failed
make[1]: *** [scripts/kconfig/zconf.tab.c] Error 127
Makefile:539: recipe for target 'menuconfig' failed
make: *** [menuconfig] Error 2
$
```

Hang on, don't panic. Read the failure messages carefully. The line after `YACC [...]` provides the clue: `/bin/sh: 1: bison: not found`. Ah! So, install `bison` with the following command:

```
sudo apt install bison
```

Now, all should be well. Well, almost; again, on a freshly baked Ubuntu guest, `make menuconfig` then complains that `flex` wasn't installed. So, we installed it (you guessed it: via `sudo apt install flex`). Also, specifically on Ubuntu, you need the `libncurses5-dev` package installed. On Fedora, do `sudo dnf install ncurses-devel`.

If you read and followed *Online Chapter, Kernel Workspace Setup*, you would have all these prerequisite packages already installed. If not, please refer to it now and install all required packages. Remember, *as ye sow...*

Quick tip: running the `<book_src>/ch1/pkg_install4ubuntu_lkp.sh` Bash script will (on an Ubuntu system) install all required packages.

Moving along, the Kconfig+Kbuild open-source framework provides clues to the user via its UI. Look at *Figure 2.11*; you'll often see symbols prefixed to the menus (like `[*]`, `<>`, `-*-`, `()`, and so on); these symbols and their meaning are as follows:

- `[.]`: In-kernel feature, *Boolean* option. It's either On or Off; the '.' shown will be replaced by * or a space:
    - `[*]`: On, feature compiled and built in to the kernel image (y)
    - `[ ]`: Off, not built at all (n)
- `<.>`: A feature that could be in one of three states. This is known as *tristate*; the . shown will be replaced by *, `M`, or a space):
    - `<*>`: On, feature compiled and built in the kernel image (y)
    - `<M>`: Module, feature compiled and built as a kernel module (an LKM) (m)
    - `< >`: Off, not built at all (n)
- `{.}`: A *dependency* exists for this config option; hence, it's required to be built or compiled as either a module (m) or to the kernel image (y).
- `-*-`: A *dependency* requires this item to be compiled in (y).
- `(...)`: *Prompt*: an alphanumeric input is required. Press the *Enter* key while on this option and a prompt box appears.
- `<Menu name>  --->`: A *sub-menu* follows. Press *Enter* on this item to navigate to the sub-menu.

Again, the empirical approach is key. Let's perform a few experiments with the `make menuconfig` UI to see how it works. This is precisely what we'll learn in the next section.

Sample usage of the make menuconfig UI

To get a feel for using the Kbuild menu system via the convenient *menuconfig* target, let's turn on a quite interesting kernel config. It's named `Kernel .config support` and **allows one to see the content of the kernel config while running that kernel!** Useful, especially during development and testing. For security reasons, it's typically turned off in production, though.

A couple of nagging questions remain:

- Q. Where is it?

  A. It's located as an item under the **General Setup** main menu (we'll see it soon enough).
- Q. What is it set to by default?

  A. To the value `<M>`, meaning it will be built as a kernel module by default.

As a learning experiment, we'll set it to the value `[*]` (or `y`), building it into the very fabric of the kernel. In effect, it will be always on. Okay, let's get it done!

1. Fire up the kernel config UI:

```
make menuconfig
```

You should see a terminal UI as in *Figure 2.11*. The very first item is usually a submenu labeled **General Setup --->** ; press the *Enter* key while on it; this will lead you into the **General Setup** submenu, within which many items are displayed; navigate (by pressing the down arrow) to the item named `Kernel .config support`:

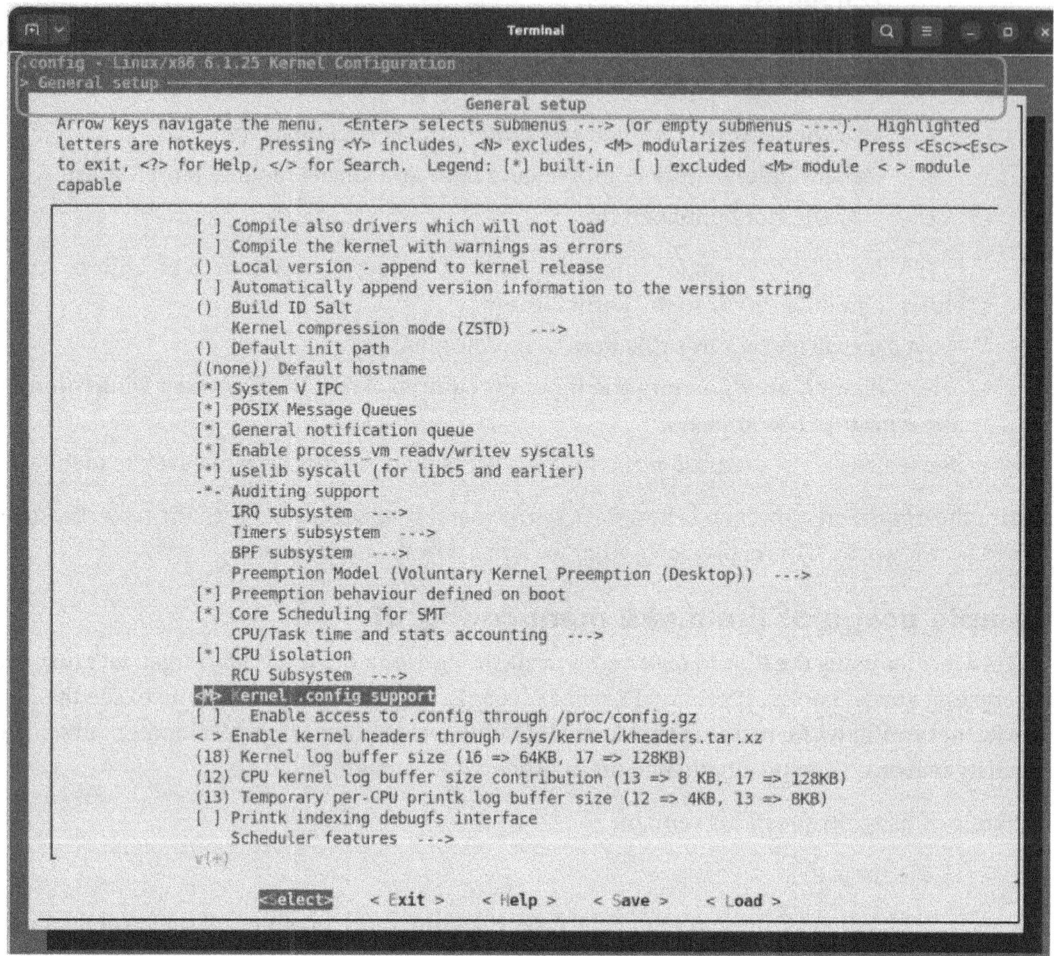

Figure 2.12: A screenshot of the General Setup menu items, with the relevant one highlighted (on x86-64)

2. We can see in the preceding screenshot that we're configuring the 6.1.25 kernel on an x86, the highlighted menu item is `Kernel .config` support, and, from its `<M>` prefix, that it's a tristate menu item that's set to the choice `<M>` for "module," to begin with (by default).

3. Keeping this item (`Kernel .config support`) highlighted, use the right arrow key to navigate to the < **Help** > button on the bottom toolbar and press the *Enter* key while on the < **Help** > button. Or, simply press *?* while on an option! The screen should now look something like this:

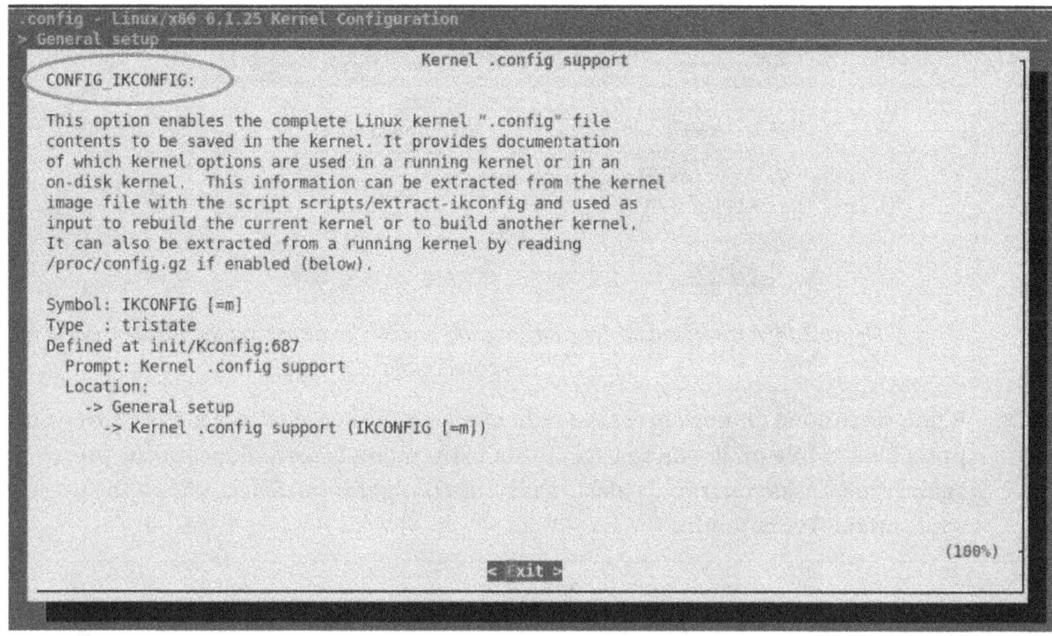

Figure 2.13: Kernel configuration via make menuconfig; an example Help screen (with the name of the kernel config macro highlighted)

The help screen is quite informative. Indeed, several of the kernel config help screens are very well populated and helpful. Unfortunately, some just aren't.

4. Okay, next, press *Enter* while on the < **Exit** > button so that we go back to the previous screen.
5. Change the value by pressing the spacebar; doing this has the current menu item's value toggle between `<*>` (always on), `< >` (off), and `<M>` (module). Keep it on `<*>`, meaning "always on."
6. Next, though it's now turned on, the ability to actually view the kernel config is provided via a pseudofile under `procfs`; the very next item below this one is the relevant one:

```
[ ]  Enable access to .config through /proc/config.gz
```

7. You can see it's turned off by default ([ ]); turn it on by navigating to it and pressing the spacebar. It now shows as [*]:

```
[*] CPU isolation
    RCU Subsystem  --->
<*> Kernel .config support
[*]   Enable access to .config through /proc/config.gz
< > Enable kernel headers through /sys/kernel/kheaders.tar.xz
(18) Kernel log buffer size (16 => 64KB, 17 => 128KB)
(12) CPU kernel log buffer size contribution (13 => 8 KB, 17 => 128KB)
(13) Temporary per-CPU printk log buffer size (12 => 4KB, 13 => 8KB)
[ ] Printk indexing debugfs interface
    Scheduler features  --->
v(+)

      <Select>    < Exit >    < Help >    < Save >    < Load >
```

Figure 2.14: A truncated screenshot showing how we've turned on the ability to view the kernel config

8. Right, we're done for now; press the right arrow or *Tab* key, navigate to the < **Exit** > button, and press *Enter* while on it; you're back at the main menu screen. Repeat this, pressing < **Exit** > again; the UI asks you if you'd like to save this configuration. Select < **Yes** > (by pressing **Enter** while on the **Yes** button):

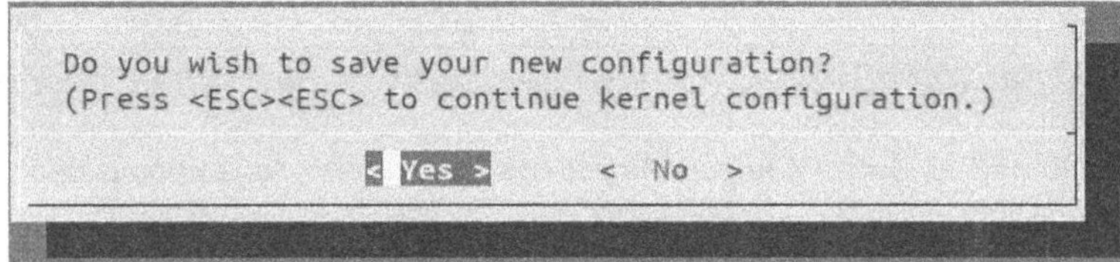

Figure 2.15: Save the modified kernel config prompt

9. The new kernel config is now saved within the `.config` file. Let's quickly verify this. I hope you noticed that the exact name of the kernel configs we modified – which is the macro as seen by the kernel source – is:

    a. `CONFIG_IKCONFIG` for the *Kernel .config support* option.

    b. `CONFIG_IKCONFIG_PROC` for the *Enable access to .config through /proc/config.gz* option.

    How do we know? It's in the top-left corner of the Help screen! Look again at *Figure 2.13*.

Done. Of course, the actual effect won't be seen until we build and boot from this kernel. Now, what exactly does turning this feature on achieve? When turned on, the currently running kernel's configuration settings can be looked up at any time in two ways:

- By running the `scripts/extract-ikconfig` script.
- By directly reading the content of the `/proc/config.gz` pseudofile. Of course, it's `gzip` compressed; first uncompress it, and then read it. `zcat /proc/config.gz` does the trick!

As a further learning exercise, why not further modify the default kernel config (of our 6.1.25 Linux kernel for the x86-64 architecture) of a couple more items? For now, don't stress out regarding the precise meaning of each of these kernel config options; it's just to get some practice with the Kconfig system. So, run `make menuconfig`, and within it make changes by following the format seen just below.

Format:

- Kernel config we're working with:
  - What it means
  - Where to navigate
  - Name of menu item (and kernel config macro `CONFIG_FOO` within parentheses)
  - Its default value
  - Value to change it to

Right, here's what to try out; let's begin with:

Local version:

- Meaning: the string to append to the kernel version. Take `uname -r` as an example; in effect, it's the "z" or `EXTRAVERSION` component in the w.x.y.z kernel version nomenclature).
- Navigate to: *General Setup*.
- Menu item: *Local version – append to kernel release (CONFIG_LOCALVERSION)*; press *Enter* once here and you'll get a prompt box.
- Default value: NULL.
- Change to: anything you like; prefixing a hyphen to the localversion is considered good practice; for example, `-lkp-kernel`.

Next:

- Timer frequency. You'll learn the details regarding this tunable in *Chapter 10, The CPU Scheduler – Part 1*:
  - Meaning: the frequency at which the timer (hardware) interrupt is triggered.
  - Navigate to: *Processor type and features | Timer frequency (250 HZ)* ---> . Keep scrolling until you find the second menu item.
  - Menu item: *Timer frequency (CONFIG_HZ)*.
  - Default value: 250 HZ.
  - Change to: 300 HZ.

Look up the *Help* screens for each of the kernel configs as you work with them. Great; once done, save and exit the UI.

Verifying the kernel config within the config file

But where's the new kernel configuration saved? This is repeated as it's important: the kernel configuration is written into a simple ASCII text file in the root of the kernel source tree, named `.config`. That is, it's saved in `${LKP_KSRC}/.config`.

As mentioned earlier, every single kernel config option is associated with a config variable of the form `CONFIG_<FOO>`, where `<FOO>`, of course, is replaced with an appropriate name. Internally, these become *macros* that the build system and indeed the kernel source code uses.

Thus, to verify whether the kernel configs we just modified will take effect, let's appropriately `grep` the kernel config file:

```
$ grep -E "CONFIG_IKCONFIG|CONFIG_LOCALVERSION|CONFIG_HZ_300" .config
CONFIG_LOCALVERSION="-lkp-kernel"
# CONFIG_LOCALVERSION_AUTO is not set
CONFIG_IKCONFIG=y
CONFIG_IKCONFIG_PROC=y
CONFIG_HZ_300=y
$
```

Aha! The configuration file now reflects the fact that we have indeed modified the relevant kernel configs; the values too show up.

Caution: it's best to NOT attempt to edit the `.config` file manually. There are several inter-dependencies you may not be aware of; always use the Kbuild menu system (we suggest using `make menuconfig`) to edit it.

Having said that, there is also a non-interactive way to do so, via a script. We'll learn about this later. Still, using the `make menuconfig` UI is really the best way.

So, by now, I expect you've modified the kernel config to suit the values just seen.

During our quick adventure with the Kconfig/Kbuild system so far, quite a lot has occurred under the hood. The next section examines some remaining points: a little bit more regarding Kconfig/Kbuild, searching within the menu system, cleanly visualizing the differences between the original and modified kernel configuration files, using a script to edit the config, security concerns and tips on addressing them; plenty still to learn!

Kernel config – exploring a bit more

The creation of, or edits to, the `.config` file within the root of the kernel source tree via the `make menuconfig` UI or other methods is not the final step in how the Kconfig system works with the configuration. No, it now proceeds to internally invoke a hidden target called `syncconfig`, which was earlier misnamed `silentoldconfig`. This target has Kconfig generate a few header files that are further used in the setup to build the kernel.

These files include some meta-headers under `include/config`, as well as the `include/generated/autoconf.h` header file, which stores the kernel config as C macros, thus enabling both the kernel Makefiles and kernel code to make decisions based on whether a kernel feature is available.

Now that we've covered sufficient ground, take another look at *Figure 2.8*, the high-level diagram (inspired by Cao Jin's articles) that attempts to communicate the kernel's Kconfig/Kbuild system. This diagram, in the Kconfig portion, only shows the common `make menuconfig` UI; note that several other UI approaches exist, which are `make config` and `make {x|g|n}config`. Those are not shown here.

Searching within the menuconfig UI

Moving along, what if – when running make menuconfig – you are looking for a particular kernel configuration option but are having difficulty spotting it? No problem: the `menuconfig` UI system has a `Search Configuration Parameter` feature. Just as with the famous `vi` editor (yes, [g]vi[m] is still our favorite text editor!), press the / (forward slash) key to have a search dialog pop up, then enter your search term with or without `CONFIG_` preceding it, and select the **< Ok >** button to have it go on its way.

The following couple of screenshots show the search dialog and the result dialog. As an example, we searched for the term `vbox`:

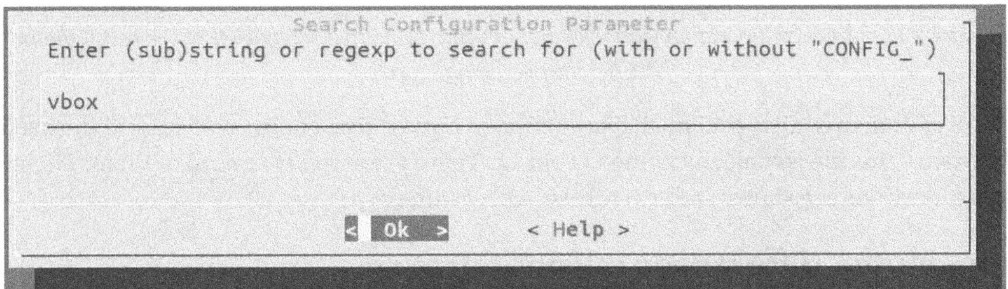

Figure 2.16: Kernel configuration via make menuconfig: searching for a config parameter

The result dialog in *Figure 2.17* for the preceding search is interesting. It reveals several pieces of information regarding the configuration options:

- The config directive. Just prefix `CONFIG_` onto whatever it shows in `Symbol:`.
- The `Type` of config (Boolean, tristate, alphanumeric, and so on).
- The `Prompt` string.
- Importantly, so you can find it, its `Location` in the menu system.
- Its internal dependencies (`Depends on:`) if any.
- The Kconfig file and line number *n* within it (`Defined at <path/to/foo.Kconfig*:n>`) where this particular kernel config is defined. We'll cover more on this in coming sections.
- Any config option it auto-selects (`Selects:`) if it itself is selected.

The following is a partial screenshot of the result dialog:

```
.config - Linux/x86 6.1.25 Kernel Configuration
> Search (vbox) ─────────────────────────────────
                              Search Results
 Symbol: DRM_VBOXVIDEO [=m]
 Type  : tristate
 Defined at drivers/gpu/drm/vboxvideo/Kconfig:2
   Prompt: Virtual Box Graphics Card
   Depends on: HAS_IOMEM [=y] && DRM [=m] && X86 [=y] && PCI [=y]
   Location:
     -> Device Drivers
       -> Graphics support
 (1)     -> Virtual Box Graphics Card (DRM_VBOXVIDEO [=m])
 Selects: DRM_KMS_HELPER [=m] && DRM_VRAM_HELPER [=m] && DRM_TTM [=m] && DRM_TTM_HELPER [=m] && GENERIC_

 Symbol: VBOXGUEST [=m]
 Type  : tristate
 Defined at drivers/virt/vboxguest/Kconfig:2
   Prompt: Virtual Box Guest integration support
   Depends on: VIRT_DRIVERS [=y] && X86 [=y] && PCI [=y] && INPUT [=y]
   Location:
     -> Device Drivers
       -> Virtualization drivers (VIRT_DRIVERS [=y])
 (2)     -> Virtual Box Guest integration support (VBOXGUEST [=m])
```

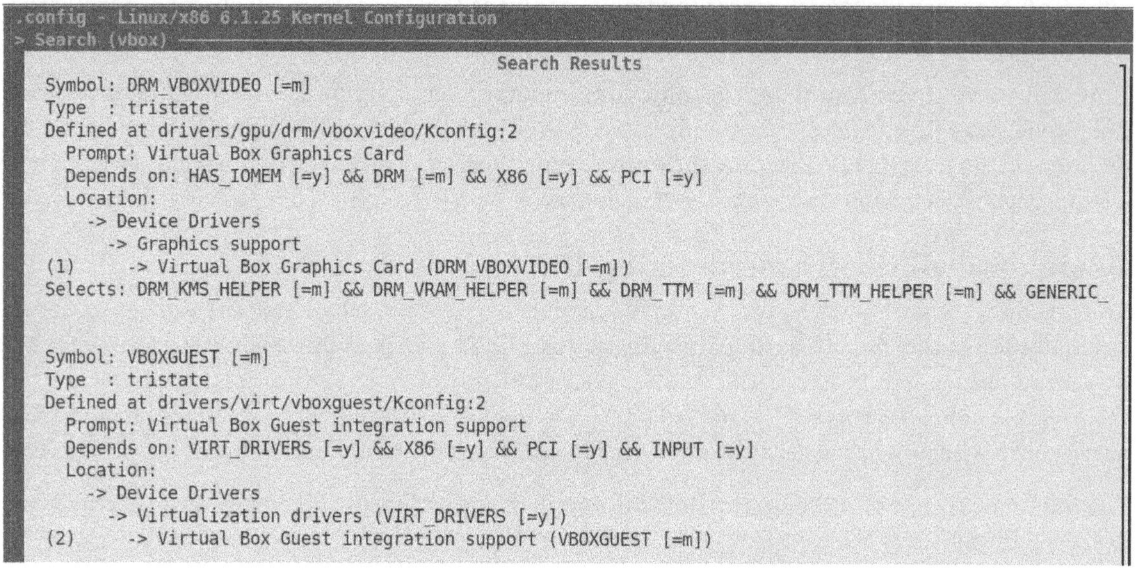

Figure 2.17: Kernel configuration via make menuconfig: truncated screenshot of the result dialog from the preceding search

All the information driving the menu display and selections is present in an ASCII text file used by the Kbuild system – this file is typically named `Kconfig`. There are actually several of them. Their precise names and locations are shown in the `Defined at ...` line.

Looking up the differences in configuration

The moment the `.config` kernel configuration file is to be written to, the Kconfig system checks whether it already exists, and if so, it backs it up with the name `.config.old`. Knowing this, we can always differentiate the two to see the changes we have just wrought. However, using your typical `diff` utility to do so makes the differences quite hard to interpret. The kernel helpfully provides a better way, a console-based script that specializes in doing precisely this. The `scripts/diffconfig` script within the kernel source tree is useful for this. Pass it the `--help` parameter to see a usage screen.

Let's try it out:

```
$ scripts/diffconfig .config.old .config
HZ 250 -> 300
HZ_250 y -> n
HZ_300 n -> y
LOCALVERSION "" -> "-lkp-kernel"
$
```

If you modified the kernel configuration changes as shown in the preceding section, you should see an output like that shown in the preceding code block via the kernel's `diffconfig` script. It clearly shows us exactly which kernel config options we changed and how. In fact, you don't even need to pass the `.config*` parameters; it uses these by default.

Using the kernel's config script to view/edit the kernel config

On occasion, there's a need to edit or query the kernel configuration directly, checking for or modifying a given kernel config. We've learned to do so via the super `make menuconfig` UI. Here we learn that there's perhaps an easier, and more importantly, non-interactive and thus *scriptable*, way to achieve the same – via a Bash script within the kernel source: `scripts/config`.

Running it without any parameters will result in a useful help screen being displayed; do check it out. An example will help regarding its usage.

The ability to ***look up the current kernel config's*** very useful, so let's ensure these kernel configs are turned on. Just for this example, let's first explicitly disable the relevant kernel configs and then enable them:

```
$ scripts/config --disable IKCONFIG --disable IKCONFIG_PROC
$ grep IKCONFIG .config
# CONFIG_IKCONFIG is not set
# CONFIG_IKCONFIG_PROC is not set

$ scripts/config --enable IKCONFIG --enable IKCONFIG_PROC
$ grep IKCONFIG .config
CONFIG_IKCONFIG=y
CONFIG_IKCONFIG_PROC=y
```

Voila, done.

Careful though: this script can modify the `.config` but there's no guarantee that what you ask it to do is actually correct. The validity of the kernel config will only be checked when you next build it. When in doubt, first check all dependencies via the `Kconfig*` files or by running `make menuconfig`, then use `scripts/config` accordingly, and then test the build to see if all's well.

Configuring the kernel for security

Before we finish, a quick note on something critical: *kernel security*. While user-space-security-hardening technologies have vastly grown, kernel-space-security-hardening technologies are playing catch-up. Careful configuration of the kernel's config options does indeed play a key role in determining the security posture of a given Linux kernel; the trouble is, there are so many options and opinions that it's often hard to check what's a good idea security-wise and what isn't.

Alexander Popov has written a very useful Python script named `kconfig-hardened-check`. It can be run to check and compare a given kernel configuration, via the usual config file, to a set of predetermined hardening preferences sourced from various Linux kernel security projects:

- The **Kernel Self Protection Project** (**KSPP**; link: `https://kernsec.org/wiki/index.php/Kernel_Self_Protection_Project`)
- The last public *grsecurity* patch
- The CLIP OS
- The security lockdown LSM
- Direct feedback from Linux kernel maintainers

You can clone the `kconfig-hardened-check` project from its GitHub repository at `https://github.com/a13xp0p0v/kconfig-hardened-check` and try it out! FYI, my *Linux Kernel Debugging* book does cover using this script in more detail. Below, a screenshot of its help screen will help you get started with it:

```
$ kconfig-hardened-check -h
usage: kconfig-hardened-check [-h] [--version] [-m {verbose,json,show_ok,show_fail}] [-c CONFIG]
                              [-l CMDLINE] [-p {X86_64,X86_32,ARM64,ARM}]
                              [-g {X86_64,X86_32,ARM64,ARM}]

A tool for checking the security hardening options of the Linux kernel

options:
  -h, --help            show this help message and exit
  --version             show program's version number and exit
  -m {verbose,json,show_ok,show_fail}, --mode {verbose,json,show_ok,show_fail}
                        choose the report mode
  -c CONFIG, --config CONFIG
                        check the security hardening options in the kernel Kconfig file (also
                        supports *.gz files)
  -l CMDLINE, --cmdline CMDLINE
                        check the security hardening options in the kernel cmdline file
  -p {X86_64,X86_32,ARM64,ARM}, --print {X86_64,X86_32,ARM64,ARM}
                        print the security hardening recommendations for the selected
                        microarchitecture
  -g {X86_64,X86_32,ARM64,ARM}, --generate {X86_64,X86_32,ARM64,ARM}
                        generate a Kconfig fragment with the security hardening options for the
                        selected microarchitecture
$
```

Figure 2.18: A screenshot showing the super kconfig-hardened-check script's help screen

A quick, useful tip: using the `kconfig-hardened-check` script, one can easily generate a security-conscious kernel config file like this (here, as an example, for the AArch64):

```
kconfig-hardened-check -g ARM64 > my_kconfig_hardened
```

(The output file is called the `config fragment`.) Now, practically speaking, what if you have an already existing kernel config file for your product? Can we merge both? Indeed we can! The kernel provides a script to do just this: `scripts/kconfig/merge_config.sh`. Run it, passing as parameters the pathname to the original (perhaps non-secure) kernel config file and then the path to the just-generated secure kernel config fragment; the result is the merger of both (additional parameters to `merge_config.sh` allow you to control it further; do check it out.

An example can be found here: https://github.com/a13xp0p0v/kconfig-hardened-check#generating-a-kconfig-fragment-with-the-security-hardening-options).

Also, you're sure to come across the fact that new-ish GCC plugins exist (CONFIG_GCC_PLUGINS) providing some cool arch-specific security features. For example, auto-initialization of local/heap variables, entropy generation at boot, and so on. However, they often don't even show up in the menu. Typically they're here: **General architecture-dependent options | GCC plugins**, as the support isn't installed by default. On x86 at least, try installing the gcc-<ver#>-plugin-dev package, where **ver#** is the GCC version number, and then retry configuring.

Miscellaneous tips – kernel config

A few remaining tips follow with regard to kernel configuration:

- When building the x86 kernel for a VM using VirtualBox (as we are here), when configuring the kernel, you might find it useful to set CONFIG_ISO9660_FS=y; it subsequently allows VirtualBox to have the guest mount the *Guest Additions* virtual CD and install the (pretty useful!) guest additions. Typically stuff that improves performance in the VM and allows better graphics, USB capabilities, clipboard and file sharing, and so on.
- When building a custom kernel, we at times want to write/build eBPF programs (an advanced topic not covered here) or stuff similar to it. In order to do so, some in-kernel headers are required. You can explicitly ensure this by setting the kernel config CONFIG_IKHEADERS=y (or to m; from 5.2 onward). This results in a /sys/kernel/kheaders.tar.xz file being made available, which can be extracted elsewhere to provide the headers.
    - Further, while talking about eBPF, modern kernels have the ability to generate some debug information, called **BPF Type Format** (**BTF**) metadata. This can be enabled by selecting the kernel config CONFIG_DEBUG_INFO_BTF=y. This also requires the *pahole* tool to be installed. More on the BTF metadata can be found within the official kernel documentation here: https://www.kernel.org/doc/html/next/bpf/btf.html.
    - Now, when this option is turned on, another kernel config – CONFIG_MODULE_ALLOW_BTF_MISMATCH – becomes relevant when building kernel modules. This is a topic we cover in depth in the following two chapters. If CONFIG_DEBUG_INFO_BTF is enabled, it's a good idea to set this latter config to **Yes**, as otherwise, your modules may not be allowed to load up if the BTF metadata doesn't match at load time.
- Next, the kernel build should, in theory at least, generate no errors or even warnings. To ensure this, to treat warnings as errors, set CONFIG_WERROR=y. Within the now familiar make menuconfig UI, it's under **General Setup | Compile the kernel with warnings as errors**, and is typically off by default.
- There's an interesting script here: scripts/get_feat.pl; its help screen shows how you can leverage it to list the kernel feature support matrix for the machine or for a given architecture. For example, to see the kernel feature support matrix for the AArch64, do this:

```
scripts/get_feat.pl --arch arm64 ls
```

- Next, an unofficial "database" of sorts, of all available kernel configs – and the kernel versions they're supported upon – is available at the **Linux Kernel Driver database (LKDDb)** project site here: `https://cateee.net/lkddb/web-lkddb/`.
- **Kernel boot config**: At boot, you can always override some kernel features via the powerful kernel command-line parameters. They're thoroughly documented here: `https://www.kernel.org/doc/html/latest/admin-guide/kernel-parameters.html`. While this is very helpful, sometimes we need to pass along more parameters as *key-value* pairs, essentially of the form `key=value`, extending the kernel command line. This can be done by populating a small kernel config file called the boot config. This boot config feature depends on the kernel config's `BOOT_CONFIG` being `y`. It's under the **General Setup** menu and is typically on by default.
  - It can be used in two ways: by attaching a boot config to the initrd or initramfs image (we cover initrd in the following chapter) or by embedding a boot config into the kernel itself. For the latter, you'll need to create the boot config file, pass the directive `CONFIG_BOOT_CONFIG_EMBED_FILE="x/y/z"` in the kernel config, and rebuild the kernel. Note that kernel command-line parameters will take precedence over the boot config parameters. On boot, if enabled and used, the boot config parameters are visible via `/proc/bootconfig`. Details regarding the boot config are in the official kernel documentation here: `https://elixir.bootlin.com/linux/v6.1.25/source/Documentation/admin-guide/bootconfig.rst`.

You're sure to come across many other useful kernel config settings and scripts, including those for hardening the kernel; keep a keen eye out.

Alright! You have now completed the first three steps of the Linux kernel build – quite a thing. Of course, we will complete the remaining four steps of the kernel build process in the following chapter. We will end this chapter with a final section on learning another useful skill – how to customize the kernel UI menu.

Customizing the kernel menu, Kconfig, and adding our own menu item

So, let's say you have developed a device driver, an experimental new modular scheduling class, a custom `debugfs` (debug filesystem) callback, or some other cool kernel feature. You will one day. How will you let others on the team – or, for that matter, your customer or the community – know that this fantastic new kernel feature exists? You will typically set up a new kernel config (macro) and allow folks to select it as either a built-in or as a kernel module, and thus build and make use of it. As part of this, you'll need to define the new kernel config and insert *a new menu item* at an appropriate place in the kernel configuration menu.

To do so, it's useful to first understand a little more about the various `Kconfig*` files and where they reside. Let's find out.

Understanding the Kconfig* files

The `Kconfig*` files contain metadata interpreted by the kernel's config and build system – *Kconfig/Kbuild* – allowing it to build and (conditionally) display the menus you see when you run the `menuconfig` UI, accept selections, and so on.

For example, the `Kconfig` file at the root of the kernel source tree is used to fill in the initial screen of the `menuconfig` UI. Take a peek at it; it works by sourcing various other `Kconfig` files in different folders of the kernel source tree. There are many `Kconfig*` files within the kernel source tree (over 1,700 for 6.1.25)! Each of them typically defines a single menu, helping us realize how intricate the build system really is.

As a real example, let's look up the `Kconfig` entry that defines the following items in the menu: **Google Devices** | **Google Virtual NIC (gVNIC) support**. Google Cloud employs a virtual **Network Interface Card (NIC)**; it's called the *Google Virtual NIC*. It's likely that their Linux-based cloud servers will make use of it. Its location within the `menuconfig` UI is here:

-> Device Drivers

-> Network device support

-> Ethernet driver support

-> Google Devices

Here's a screenshot showing these menu items:

```
.config - Linux/x86 6.1.25 Kernel Configuration
> Device Drivers > Network device support > Ethernet driver support
                              Ethernet driver support
  Arrow keys navigate the menu.  <Enter> selects submenus ---> (or empty submenus
  letters are hotkeys.  Pressing <Y> includes, <N> excludes, <M> modularizes featu
  to exit, <?> for Help, </> for Search.  Legend: [*] built-in  [ ] excluded  <M>
  capable
               < >   Dave ethernet support (DNET)
               [*]   Digital Equipment devices
               [*]     DEC - Tulip devices
               < >       Early DECchip Tulip (dc2104x) PCI support
               < >       DECchip Tulip (dc2114x) PCI support
               < >       Winbond W89c840 Ethernet support
               < >       Davicom DM910x/DM980x support
               < >       ULi M526x controller support
               [*]   D-Link devices
               < >     DL2000/TC902x/IP1000A-based Gigabit Ethernet support
               < >     Sundance Alta support
               [*]   Emulex devices
               < >     ServerEngines' 10Gbps NIC - BladeEngine
               [*]   Engleder devices
               < >     TSN endpoint support
               [*]   EZchip devices
               [*]   Fungible devices
               < >     Fungible Ethernet device driver
               [*]   Google Devices
               < >     Google Virtual NIC (gVNIC) support
```

Figure 2.19: Partial screenshot showing the Google Devices item in the make menuconfig UI (for x86/6.1.25)

How do we know which `Kconfig` file defines these menu items? The Help screen for a given config reveals it! So, while on the relevant menu item, select the < **Help** > button and press *Enter*; here, the *Help* screen says (among other things):

```
Defined at drivers/net/ethernet/google/Kconfig:5
```

That's the `Kconfig` file describing this menu! Let's look it up:

```
$ cat drivers/net/ethernet/google/Kconfig
#
# Google network device configuration
#
config NET_VENDOR_GOOGLE
    bool "Google Devices"
    default y
    help
      If you have a network (Ethernet) device belonging to this class, say Y.
[ … ]
```

This is a nice and simple `Kconfig` entry; notice the `Kconfig` language keywords: `config`, `bool`, `default`, and `help`. We've highlighted them in reverse colors. You can see that this device is enabled by default. We'll cover the syntax shortly.

The following table summarizes the more important `Kconfig*` files and which submenu they serve in the Kbuild UI:

Menu	Kconfig file location for it
The main menu, the initial screen of the menuconfig UI	`Kconfig`
General setup + Enable loadable module support	`init/Kconfig`
Processor types and features + Bus options + Binary Emulations (this menu title tends to be arch-specific; here,its wrt the x86[_64]; in general, the Kconfig file is here: `arch/<arch>/Kconfig`)	`arch/<arch>/Kconfig*`
Power management	`kernel/power/Kconfig*`
Firmware drivers	`drivers/firmware/Kconfig*`
Virtualization	`arch/<arch>/kvm/Kconfig*`
General architecture-dependent options	`arch/Kconfig*`
Enable the block layer + IO Schedulers	`block/Kconfig*`
Executable file formats	`fs/Kconfig.binfmt`
Memory management options	`mm/Kconfig*`
Networking support	`net/Kconfig, net/*/Kconfig*`

Device drivers	`drivers/Kconfig, drivers/*/Kconfig*`
Filesystems	`fs/Kconfig, fs/*/Kconfig*`
Security options	`security/Kconfig, security/*/Kconfig*`
Cryptographic API	`crypto/Kconfig, crypto/*/Kconfig*`
Library routines	`lib/Kconfig, lib/*/Kconfig*`
Kernel hacking (implies Kernel debugging)	`lib/Kconfig.debug, lib/Kconfig.*`

Table 2.4: Kernel config (sub) menus and the corresponding Kconfig file(s) defining them*

Typically, a single `Kconfig` file drives a single menu, though there could be multiple. Now, let's move on to actually adding a menu item.

Creating a new menu item within the General Setup menu

As a trivial example, let's add our own Boolean dummy config option within the **General Setup** menu. We want the config name to be, shall we say, `CONFIG_LKP_OPTION1`. As can be seen from the preceding table, the relevant `Kconfig` file to edit is the `init/Kconfig` one as it's the meta-file that defines the **General Setup** menu.

Let's get to it (we assume you're in the root of the kernel source tree):

1. Optional: to be safe, always make a backup copy of the `Kconfig` file you're editing:

```
cp init/Kconfig init/Kconfig.orig
```

2. Now, edit the `init/Kconfig` file:

```
vi init/Kconfig
```

Scroll down to an appropriate location within the file; here, we choose to insert our custom menu entry between the `LOCALVERSION_AUTO` and the `BUILD_SALT` ones. The following screenshot shows our new entry (the `init/Kconfig` file being edited with `vim`):

```
          which is done within the script "scripts/setlocalversion".)

config LKP_OPTION1
        bool "Test case for LKP 2e book/Ch 2: creating a new menu item in kernel config"
        default n
        help
          This option is merely a dummy or 'test' one; it's simply to have readers
          of this book - 'Linux Kernel Programming', 2nd Ed, Kaiwan NB, Packt -
          try out the creation of a few menu items within the kernel config.

          Within the 'make menuconfig', you can experiment: set this option to
          'y' (on), save and exit, and see the effect this has by doing:
          grep "CONFIG_LKP_OPTION1" .config

          If unsure, say N

config BUILD_SALT
        string "Build ID Salt"
```

Figure 2.20: Editing the 6.1.25:init/Kconfig and inserting our own menu entry (highlighted)

FYI, I've provided the preceding experiment as a patch to the original 6.1.25 `init/Kconfig` file in our book's GitHub source tree. Find the patch file here: `ch2/Kconfig.patch`.

The new item starts with the `config` keyword followed by the `FOO` part of your new `CONFIG_LKP_OPTION1` config variable. For now, just read the statements we have made in the `Kconfig` file regarding this entry. More details on the `Kconfig` language/syntax are in the *A few details on the Kconfig language* section that follows.

3. Save the file and exit the editor.
4. (Re)configure the kernel: run `make menuconfig`. Then navigate to our cool new menu item under **General Setup | Test case for LKP 2e book/Ch 2: creating ...**. Turn the feature *on*. Notice how, in *Figure 2.21*, it's highlighted and *off* by default, just as we specified via the `default n` line.

```
make menuconfig
[...]
```

Here's the relevant output:

```
.config - Linux/x86 6.1.25 Kernel Configuration
> General setup ─────────────────────────────────────
                              General setup
  Arrow keys navigate the menu.  <Enter> selects submenus ---> (or empty submenus ----).  Highlighted
  letters are hotkeys.  Pressing <Y> includes, <N> excludes, <M> modularizes features.  Press <Esc><Esc>
  to exit, <?> for Help, </> for Search.  Legend: [*] built-in  [ ] excluded  <M> module  < > module
  capable

              [ ] Compile also drivers which will not load
              [ ] Compile the kernel with warnings as errors
              (-lkp-kernel) Local version - append to kernel release
              [ ] Automatically append version information to the version string
              [ ] Test case for LKP 2e book/Ch 2: creating a new menu item in kernel config (NEW)
              ()  Build ID Salt
                  Kernel compression mode (ZSTD)  --->
```

Figure 2.21: Kernel configuration via make menuconfig showing our new menu entry (before turning it on)

5. Now turn it *on* by toggling it with the space bar, then save and exit the menu system.

While there, try pressing the < **Help** > button. You should see the "help" text we provided within the `init/Kconfig` file.

6. Check whether our feature has been selected:

```
$ grep "LKP_OPTION1" .config
CONFIG_LKP_OPTION1=y
$ grep "LKP_OPTION1" include/generated/autoconf.h
$
```

We find that indeed it has been set to *on* (y) within our `.config` file, but is not yet within the kernel's internal auto-generated header file. This will happen when we build the kernel.

Now let's check it via the useful non-interactive `config` script method. We covered this in the *Using the kernel's config script to view/edit the kernel config* section.

```
$ scripts/config -s LKP_OPTION1
y
```

Ah, it's on, as expected (the `-s` option is the same as `--state`). Below, we disable it via the `-d` option, query it (`-s`), and then re-enable it via the `-e` option, and again query it (just for learning's sake!):

```
$ scripts/config -d LKP_OPTION1 ; scripts/config -s LKP_OPTION1
n
$ scripts/config -e LKP_OPTION1 ; scripts/config -s LKP_OPTION1
y
```

7. Build the kernel. Worry not; the full details on building the kernel are found in the next chapter. You can skip this for now, or you could always cover *Chapter 3, Building the 6.x Linux Kernel from Source – Part 2,* and then come back to this point.

```
make -j4
```

Further, in recent kernels, after the build step, every kernel config option that's enabled (either y or m) appears as an empty file within `include/config`; this happens with our new config as well, of course:

```
$ ls -l include/config/LKP_*
-rw-r--r-- 1 c2kp c2kp 0 Apr 29 11:56 include/config/LKP_
OPTION1
```

8. Once done, recheck the `autoconf.h` header for the presence of our new config option:

```
$ grep LKP_OPTION1 include/generated/* 2>/dev/null
include/generated/autoconf.h:#define CONFIG_LKP_OPTION1 1
include/generated/rustc_cfg:--cfg=CONFIG_LKP_OPTION1
include/generated/rustc_cfg:--cfg=CONFIG_LKP_OPTION1="y"
```

It worked (from 6.0 even Rust knows about it!). Yes; however, when working on an actual project or product, in order to leverage this new kernel config of ours, we would typically require a further step, setting up our config entry within the `Makefile` relevant to the code that uses this config option.

Here's a quick example of how this might look. Let's imagine we wrote some kernel code in a C source file named `lkp_options.c`. Now, we need to ensure it gets compiled and built into the kernel image! How? Here's one way: in the kernel's top-level (or within its own) `Makefile`, the following line will ensure that it gets compiled into the kernel at build time; add it to the end of the relevant Makefile:

```
obj-${CONFIG_LKP_OPTION1}  +=  lkp_option1.o
```

Don't stress about the fairly weird kernel Makefile syntax for now. The next few chapters will certainly shed some light on this. Also, we did cover this particular syntax in the *How the Kconfig+Kbuild system works – a minimal take* section.

Further, you should realize that the very same config can be used as a normal C macro within a piece of kernel code; for example, we could do things like this within our in-tree kernel (or module) C code:

```
#ifdef CONFIG_LKP_OPTION1
    do_our_thing();
#endif
```

Then again, it's very much worth noting that the Linux kernel community has devised and strictly adheres to certain rigorous *coding style guidelines*. In this context, the guidelines state that conditional compilation should be avoided whenever possible; if it's required to use a `Kconfig` symbol as a conditional, then please do it this way:

```
if (IS_ENABLED(CONFIG_LKP_OPTION1))
    do_our_thing();
```

The Linux kernel coding style guidelines can be found here: `https://www.kernel.org/doc/html/latest/process/coding-style.html`. I urge you to refer to them often, and, of course, to follow them!

A few details on the Kconfig language

Our usage of the Kconfig language so far (*Figure 2.20*) is just the tip of the proverbial iceberg. The fact is, the Kconfig system uses the `Kconfig` language (or syntax) to express and create menus using simple ASCII text directives. The language includes menu entries, attributes, dependencies, visibility constraints, help text, and so on.

The kernel documents the `Kconfig` language constructs and syntax here: `https://www.kernel.org/doc/html/v6.1/kbuild/kconfig-language.html#kconfig-language`. Do refer to this document for complete details.

A brief mention of the more common `Kconfig` constructs is given in the following table:

Construct	Meaning
`config <FOO>`	Specifies the menu entry name of the form `CONFIG_FOO` here; just use the `FOO` part after the `config` keyword.
Menu attributes	
`bool ["<description>"]`	Specifies the config option as a *Boolean*; its value in `.config` will be either `y` (built into the kernel image) or will not exist (will show up as a commented-out entry).
`tristate ["description>"]`	Specifies the config option as *tristate*; its value in `.config` will be either `y` or `m` (built as a kernel module) or will not exist (will show up as a commented-out entry).
`int ["<description>"]`	Specifies the config option as taking an *integer* value.
`range x-y`	For an integer whose valid range is from `x` to `y`.
`default <value>`	Specifies the default value; use `y`, `m`, `n`, or other, as required.

`prompt "<description>" [if <expr>]`	An input prompt with a describing sentence (can be made conditional); a menu entry can have at most one prompt.
`depends on "expr"`	Defines a dependency for the menu item; can have several with the `depends on FOO1 && FOO2 && (FOO3 || FOO4)` type of syntax.
`select <config> [if "expr"]`	Defines a reverse dependency.
`help "my awesome help text … bleh bleh bleh "`	Text to display when the < **Help** > button is selected.

Table 2.5: Kconfig, a few constructs

To help understand the syntax, a few examples from `lib/Kconfig.debug` (the file that describes the menu items for the `Kernel Hacking` submenu – it means kernel *debugging*, really – of the UI) follow (don't forget, you can browse it online as well: `https://elixir.bootlin.com/linux/v6.1.25/source/lib/Kconfig.debug`):

1. We will start with a simple and self-explanatory one (the `CONFIG_DEBUG_INFO` option):

   ```
   config DEBUG_INFO
           bool
           help
             A kernel debug info option other than "None" has been selected
             in the "Debug information" choice below, indicating that debug
             information will be generated for build targets.
   ```

2. Next, let's look at the `CONFIG_FRAME_WARN` option. Notice the `range` and the conditional default value syntax, as follows:

   ```
   config FRAME_WARN
       int "Warn for stack frames larger than"
       range 0 8192
       default 0 if KMSAN
           default 2048 if GCC_PLUGIN_LATENT_ENTROPY
           default 2048 if PARISC
           default 1536 if (!64BIT && XTENSA)
           default 1280 if KASAN && !64BIT
           default 1024 if !64BIT
           default 2048 if 64BIT
           help
             Tell the compiler to warn at build time for stack frames larger
   than this.
   ```

```
        Setting this too low will cause a lot of warnings.
        Setting it to 0 disables the warning.
```

3. Next, the `CONFIG_HAVE_DEBUG_STACKOVERFLOW` option is a simple Boolean; it's either on or off (the kernel either has the capability to detect kernel-space stack overflows or doesn't). The `CONFIG_DEBUG_STACKOVERFLOW` option is also a Boolean. Notice how it depends on two other options, separated with a Boolean AND (&&) operator:

```
config HAVE_DEBUG_STACKOVERFLOW
        bool

config DEBUG_STACKOVERFLOW
        bool "Check for stack overflows"
        depends on DEBUG_KERNEL && HAVE_DEBUG_STACKOVERFLOW
        help
          Say Y here if you want to check for overflows of kernel, IRQ
          and exception stacks (if your architecture uses them). This
          option will show detailed messages if free stack space drops
          below a certain limit. [...]
```

Another useful thing: while configuring the kernel (via the usual `make menuconfig` UI), clicking on < **Help** > not only shows some (usually useful) help text, but it also displays the **current runtime values** of various config options. The same can be seen by simply searching for a config option (via the slash key, /, as mentioned earlier). So, for example, type / and search for the kernel config named `KASAN`; this is what I see when doing so.

```
.config - Linux/x86 6.1.25 Kernel Configuration
> Search (KASAN)
                              Search Results
Symbol: KASAN [=n]
Type  : bool
Defined at lib/Kconfig.kasan:34
  Prompt: KASAN: dynamic memory safety error detector
  Depends on: ((HAVE_ARCH_KASAN [=y] && CC_HAS_KASAN_GENERIC [=y] || HAVE_ARCH_KASAN_SW_TAGS [=n] && CC
  Location:
    -> Kernel hacking
      -> Memory Debugging
(1)     -> KASAN: dynamic memory safety error detector (KASAN [=n])
Selects: STACKDEPOT_ALWAYS_INIT [=n]
```

Figure 2.22: Partial screenshot showing the KASAN config option; you can see it's off by default

If you're unaware, **KASAN** is the **Kernel Address SANitizer** – it's a brilliant compiler-based technology to help catch memory corruption defects; I cover it in depth in the book *Linux Kernel Debugging*.

Look carefully at the **Depends on:** line; it shows the dependencies as well as their current value. The important thing to note is that *the menu item won't even show in the UI unless the dependencies are fulfilled.*

Alright! This completes our coverage of the `Kconfig` files, creating or editing a custom menu entry in the kernel config, a little Kconfig language syntax, and indeed this chapter.

Summary

In this chapter, you first learned about the Linux kernel's release (or version) nomenclature (remember, Linux kernel releases are time- and not feature-based!), the various types of Linux kernels (`-next` trees, `-rc/mainline` trees, stable, LTS, SLTS, distributions, custom embedded), and the basic kernel development workflow. You then learned how to obtain for yourself a Linux kernel source tree and how to extract the compressed kernel source tree to disk. Along the way, you even got a quick 10,000-foot view of the kernel source tree so that its layout is clearer.

After that, critically, you learned how to approach the kernel configuration step and perform it – a key step in the kernel build process! Furthermore, you learned how to customize the kernel menu, adding your own entries to it, and a bit about the Kconfig/Kbuild system and the associated `Kconfig` files it uses, among others.

Knowing how to fetch and configure the Linux kernel is a useful skill to possess. We have just begun this long and exciting journey. You will realize that with more experience and knowledge of kernel internals, drivers, and the target system hardware, your ability to fine-tune the kernel to your project's purpose will only get better.

We're halfway to building a custom kernel; I suggest you digest this material, try out the steps in this chapter in a hands-on fashion, work on the questions/exercises, and browse through the *Further reading* section. Then, in the next chapter, let's actually build the 6.1.25 kernel and verify it!

Exercise

Following pretty much exactly the steps you've learned in this chapter, I'd like you to now do the same for some other kernel, say, **the 6.0.y Linux kernel**, where **y** is the highest number (as of this writing, it's 19)! Of course, if you wish, feel free to work on any other kernel:

1. Navigate to `https://mirrors.edge.kernel.org/pub/linux/kernel/v6.x/` and look up the 6.0.y releases.
2. Download the latest v6.0.y Linux kernel source tree.
3. Extract it to disk.
4. Configure the kernel (begin by using the `localmodconfig` approach, then tweak the kernel config as required. As an additional exercise, you could run the `streamline_config.pl` script as well).
5. Show the "delta" – the differences between the original and the new kernel config file (tip: use the kernel's `diffconfig` script to do so).

Questions

As we conclude, here is a list of questions for you to test your knowledge regarding this chapter's material: https://github.com/PacktPublishing/Linux-Kernel-Programming_2E/blob/main/questions/ch2_qs_assignments.txt. You will find some of the questions answered in the book's GitHub repo: https://github.com/PacktPublishing/Linux-Kernel-Programming_2E/tree/master/solutions_to_assgn.

Further reading

To help you delve deeper into the subject, we provide a rather detailed list of online references and links (and, at times, even books) in a *Further reading* document in this book's GitHub repository. It's available here: https://github.com/PacktPublishing/Linux-Kernel-Programming_2E/blob/master/Further_Reading.md.

Learn more on Discord

To join the Discord community for this book – where you can share feedback, ask questions to the author, and learn about new releases – follow the QR code below:

https://packt.link/SecNet

3

Building the 6.x Linux Kernel from Source – Part 2

This chapter continues from where the previous chapter left off. In the previous chapter, in the *Steps to build the kernel from source* section, we covered the first three steps of building our kernel. There, you learned how to download and extract the kernel source tree or even use `git clone` to get one (*steps 1* and *2*). We then proceeded to understand the kernel source tree layout, and, very importantly, the various approaches to correctly arrive at a starting point to configure the kernel (*step 3*). We even added a custom menu item to the kernel configuration menu.

In this chapter, we will continue our quest to build the kernel by covering the remaining four steps to build it. First, of course, we will build it (*step 4*). You will then learn how to properly install the kernel modules that get generated as part of the build (*step 5*). Next, we will run a simple command that sets up the **GRUB** (**Grand Unified Bootloader**) bootloader and generates the `initramfs` (or `initrd`) image (*step 6*). The motivation for using an `initramfs` image and how it's generated are discussed as well. Some details on configuring the GRUB bootloader (for x86) are then covered (*step 7*).

By the end of the chapter, we'll boot the system with our new kernel image and verify that it's built as expected. We'll then finish off by learning how to *cross-compile* a Linux kernel for a foreign architecture (for the AArch 32/64, the board in question is the well-known Raspberry Pi).

Briefly, these are the areas that will be covered in this chapter:

- Step 4 – building the kernel image and modules
- Step 5 – installing the kernel modules
- Step 6 – generating the initramfs image and bootloader setup
- Understanding the initramfs framework
- Step 7 – customizing the GRUB bootloader
- Verifying our new kernel's configuration
- Kernel build for the Raspberry Pi
- Miscellaneous tips on the kernel build

Technical requirements

Before we begin, I assume that you have downloaded, extracted (if required), and configured the kernel, thus having a `.config` file ready. If you haven't already, please refer to *Chapter 2, Building the 6.x Linux Kernel from Source – Part 1*, for the details on how exactly this is done. We can now proceed to build it.

Step 4 – building the kernel image and modules

Performing the build from the end user's point of view is straightforward. In its simplest form, just ensure you're at the root of the configured kernel source tree and type `make`. That's it – the kernel image and any kernel modules (and, on an embedded system, possibly a **Device Tree Blob** (**DTB**) binary) will get built. Grab a coffee! The first time around, it could take a while.

Of course, there are various `Makefile` targets we can pass to `make`. A quick `make help` command issued on the command line reveals quite a bit. Remember, we used this earlier, in fact, to see all possible configuration targets (revisit *Chapter 2, Building the 6.x Kernel from Source – Part 1*, and particularly the *Seeing all available config options* section if you'd like). Here, we use it to see what gets built by default with the `all` target:

```
$ cd ${LKP_KSRC}      # recall that the env var LKP_KSRC holds the pathname to
                      # the 'root' directory of our 6.1 LTS kernel source tree
$ make help
 [...]
Other generic targets:
  all - Build all targets marked with [*]
* vmlinux  - Build the bare kernel
* modules - Build all modules
 [...]
Architecture specific targets (x86):
 * bzImage - Compressed kernel image (arch/x86/boot/bzImage)
 [...]
$
```

Okay, so here's what to notice: performing `make all` will get us the preceding three targets (the ones prefixed with the * symbol) built; what do they mean? Let's see:

- `vmlinux` matches the name of the uncompressed kernel image file.
- The `modules` target implies that all kernel config options marked as `m` (for module) will be built as kernel modules (`.ko` files) within the kernel source tree (details on what exactly a kernel module is and how to program one are the subject matter of the following two chapters).
- `bzImage` is an architecture-specific kernel image file. On an x86[_64] system, this is the name of the compressed kernel image – the one the bootloader will actually load into RAM, uncompress in memory, and boot into; in effect, it's the (compressed) kernel image file.

So, an FAQ: if `bzImage` is the actual kernel image file that we use to boot and initialize the system, then what's `vmlinux` for? Notice that `vmlinux` is the uncompressed kernel image file. It can be large (even very large, in the presence of kernel symbols generated during a debug build). While we never boot via `vmlinux`, it's nevertheless important – priceless, in fact. Do keep it around for kernel debugging purposes (my *Linux Kernel Debugging* book has you covered in this respect!).

Now, with the `kbuild` system (that the kernel uses), just running `make` equates to `make all`.

The modern Linux kernel code base is enormous. Current estimates are that recent kernels have in the region of 25 to 30 million **source lines of code** (**SLOC**)! Thus, building the kernel is indeed a *very memory- and CPU-intensive job*. Indeed, some folks use the kernel build as a stress test! (You should also realize that not all lines of code are going to be compiled during a particular build run). The modern `make` utility is powerful and multi-process capable. We can request it to spawn multiple processes to handle different (unrelated) parts of the build in parallel, leading to higher throughput and thus shorter build times. The relevant option is `-jn`, where `n` is the upper limit on the number of tasks to spawn and run in parallel. A heuristic (rule of thumb) used to determine this is as follows:

```
n = number-of-CPU-cores * factor;
```

Here, `factor` is 2 (or 1.5 on very high-end systems with hundreds or thousands of CPU cores). Also, technically, we require the cores to be internally "threaded" or using **Simultaneous Multi-Threading** (**SMT**) – what Intel calls *hyper-threading* – for this heuristic to be useful.

More details on parallelized `make` and how it works can be found on the man page of `make` (invoked with `man 1 make`) in the PARALLEL MAKE AND THE JOBSERVER section.

Another FAQ: how many CPU cores *are* there on my system? There are several ways to determine this, an easy way being to use the `nproc` utility:

```
$ nproc
4
```

A quick word regarding `nproc` and related utilities:

Performing `strace` on `nproc` reveals that it works by essentially using the `sched_getaffinity()` system call. We shall mention more on this and related system calls in *Chapter 10, The CPU Scheduler – Part 1*, and *Chapter 11, The CPU Scheduler – Part 2*, on CPU scheduling.

FYI, the *lscpu* utility yields the number of cores as well as additional useful CPU info. Try them out on your Linux system.

Clearly, my guest VM has been configured with four CPU cores, so let's keep `n=4*2=8`. So, off we go and build the kernel. The following output is from our trusty x86_64 Ubuntu 22.04 LTS guest system configured to have 2 GB of RAM and four CPU cores.

Remember, prior to building it, the kernel must first be correctly configured. For details, refer to *Chapter 2, Building the 6.x Linux Kernel from Source – Part 1.*

Again, when you begin, the kernel build may emit a warning, although non-fatal in this case:

```
$ make -j8
scripts/kconfig/conf --syncconfig Kconfig
 UPD include/config/kernel.release
warning: Cannot use CONFIG_STACK_VALIDATION=y, please install libelf-dev,
libelf-devel or elfutils-libelf-devel
 [...]
```

So, to address this, we break off the build with *Ctrl* + *C*, then follow the output's advice and install the `libelf-dev` package. On our Ubuntu box, `sudo apt install libelf-dev` is sufficient. Do realize that if you followed the detailed setup in *Online Chapter, Kernel Workspace Setup* (or ran our `ch1/pkg_install4ubuntu_lkp.sh` script), this will not happen.

Precisely because the kernel build is very CPU- and RAM-intensive, carrying this out on a guest VM is going to be a lot slower than on a native Linux system. It helps to conserve RAM by at least booting your guest at run level 3 (multiuser with networking, no GUI): `https://www.if-not-true-then-false.com/2012/howto-change-runlevel-on-grub2/`.

Right, let's run it with a few quick but very useful tips:

1. Due to the sheer high CPU and RAM usage during kernel build, I sometimes find that when running the build in a VM in graphical mode, errors can occur; the system can run short on memory, triggering weird failures, and even logging one out at times! To mitigate this, I recommend you boot into your VM at run level 3 – or what `systemd` typically calls `multi-user.target` – multi-user mode with networking but no graphics. To do so, you can edit the kernel command line from the GRUB menu and append 3 to it (we cover stuff like this in the *Step 7 – customizing the GRUB* section). Alternatively, if already in graphical mode (what `systemd` calls `graphical.target`), you can switch to `multi-user.target` with the command `sudo systemctl isolate multi-user.target`.
2. Then again, with the cost of RAM being low (and possibly dropping), just getting more RAM is a great and quick performance hack!
3. Especially when it's running in console mode, I personally definitely prefer to **ssh** into the VM and work from there.

4. When building, by leveraging the tee utility, we can easily save both the standard output and standard error into a file (tee lets us simultaneously see the output on the console as well):

```
$ sudo systemctl isolate multi-user.target
[...]
$ cd ${LKP_KSRC}
$ make -j8 2>&1 | tee out.txt
SYNC    include/config/auto.conf.cmd
  HOSTCC  scripts/basic/fixdep
  HOSTCC  scripts/kconfig/conf.o
  HOSTCC  scripts/kconfig/confdata.o
  HOSTCC  scripts/kconfig/expr.o
  LEX     scripts/kconfig/lexer.lex.c
  YACC    scripts/kconfig/parser.tab.[ch]
  HOSTCC  scripts/kconfig/preprocess.o
[...]
```

Good, the kbuild system should now ensure that our new kernel – and the components we've configured as modules – get built. Of course, at times, things can and do go wrong. We'll look at several of these circumstances as we progress through this chapter and, as far as possible, look at how to fix them, starting with the following section.

Getting over a cert config issue on Ubuntu

Particularly on recent Ubuntu systems, we run make and all seems well with the kernel build, until we (often) hit this issue:

```
$ make ...
[...]
  EXTRACT_CERTS   certs/signing_key.pem
  CC      certs/system_keyring.o
  CC      arch/x86/entry/vdso/vclock_gettime.o
  EXTRACT_CERTS
  CC      certs/common.o
  CC      arch/x86/entry/vdso/vgetcpu.o
make[1]: *** No rule to make target 'debian/canonical-revoked-certs.pem',
needed by 'certs/x509_revocation_list'.  Stop.
make[1]: *** Waiting for unfinished jobs....
  CC      certs/blacklist.o
[...]
```

However, you'll notice that make continues to execute the other parallel tasks – at least until it figures it's quite useless – and the build then terminates (it can take a while though), ultimately unsuccessfully; no kernel is actually built.

So, what's the problem? Here, it turns out to be a kernel config named `CONFIG_SYSTEM_REVOCATION_KEYS` that has been added to recent 5.x kernels; check it out:

```
$ grep CONFIG_SYSTEM_REVOCATION_KEYS .config
CONFIG_SYSTEM_REVOCATION_KEYS="debian/canonical-revoked-certs.pem"
```

At least on Ubuntu systems, this particular config setting seems to cause the build to fail; the quick and simple fix is to simply turn it off. You can do so with:

```
scripts/config --disable SYSTEM_REVOCATION_KEYS
```

Recheck with `grep`; you'll find that it's now unset.

FYI, this Q&A on the *Ask Ubuntu* forum covers it: `https://askubuntu.com/a/1329625/245524`. For the curious, the config seems to have made its way into the 5.13 kernel; here's the actual commit: `https://github.com/torvalds/linux/commit/d1f044103dad70c1cec0a8f3abdf00834fec8b98`.

Run the `make` command again (you might need to respond via *Enter* to a prompt or two); it should now succeed!

```
[...]
KSYMS   .tmp_vmlinux.kallsyms2.S
AS      .tmp_vmlinux.kallsyms2.S
LD      vmlinux
BTFIDS  vmlinux
SORTTAB vmlinux
SYSMAP  System.map
MODPOST modules-only.symvers
CC      arch/x86/boot/a20.o
AS      arch/x86/boot/bioscall.o
CC      arch/x86/boot/cmdline.o
[...]
GEN     Module.symvers
LDS     arch/x86/boot/compressed/vmlinux.lds
AS      arch/x86/boot/compressed/kernel_info.o
CC [M]  arch/x86/crypto/aesni-intel.mod.o
CC [M]  arch/x86/crypto/crc32-pclmul.mod.o
[...]
LD      arch/x86/boot/setup.elf
OBJCOPY arch/x86/boot/setup.bin
BUILD   arch/x86/boot/bzImage
Kernel: arch/x86/boot/bzImage is ready  (#3)
```

Ah, it's done!

An aside: in general, what to check if the build fails?

- Check, then recheck, that you've done everything correctly; do blame yourself and not the kernel (community/code)!
- Have all the required and up-to-date packages been installed? For example, if the kernel config has `CONFIG_DEBUG_INFO_BTF=y` (mine does), it requires `pahole` 1.16 or later to be installed.
- Is the kernel config sane?
- Is it a hardware issue? Errors like *internal compiler error: Segmentation fault* usually indicate this; is there sufficient RAM and swap space allocated? Try the build on another VM, or better, on a native Linux system.
- Start (or restart) from scratch; in the root of the kernel source tree, do a `make mrproper` (careful: it will clean *everything*, even deleting any `.config` file), and perform all the steps carefully.
- Hey, when all else fails, Google the error message!

The build *should* run cleanly, without any errors or warnings. Well, at times, compiler warnings are seen, but we shall blithely ignore them. What if you encounter compiler errors and thus a failed build during this step? How can we put this politely? Oh well, we cannot – it's very likely your fault, not the kernel community's. As just mentioned, please check and recheck every step, redoing it from scratch with a `make mrproper` command if all else fails! Very often, a failure to build the kernel implies either kernel configuration errors (randomly selected configs that might conflict), outdated versions of the toolchain, or incorrect patching, among other things. (FYI, we cover more, quite specific tips in the *Miscellaneous tips on the kernel build* section).

Okay, we shall assume that the kernel build step worked. The compressed kernel image (here, for the x86[_64], it's called `bzImage`) and the uncompressed one, the `vmlinux` file, have successfully been built by stitching together the various object files generated, as can be seen in the preceding output – the last line in the preceding block confirms this fact (the #3 implies that, for me, this is the third time the kernel was built). As part of the build process, the `kbuild` system also proceeds to finish building all kernel modules.

A quick tip: if you'd like to time how long a command takes to execute, prefix the `time` command to it (so, here: `time make -j8 2>&1 | tee out.txt`). It works but the `time(1)` utility provides only a (very) coarse-grained idea of the time taken by the command that follows it.

If you'd like accurate CPU profiling and time stats, learn how to use the powerful `perf` utility. Here, you can try it out with the `perf stat make -j8 ...` command. I suggest you try this out on a distro kernel as, otherwise, `perf` itself will have to be manually built for your custom kernel.

Also, in the previous output, as we're doing a parallelized build (via `make -j8`, implying up to eight processes performing the build in parallel), all the build processes write to the same `stdout` location – the console or terminal window. Hence, the output may be out of order or mixed up.

Assuming it goes off well, as indeed it should, by the time this step terminates, three key files (among many) have been generated by the `kbuild` system. In the root of the kernel source tree, we will now have the following files:

- The uncompressed kernel image file, `vmlinux` (used for debugging purposes)
- The symbol-address mapping file, `System.map`
- The compressed bootable kernel image file, `bzImage` (see the following output)

Let's check them out! We make the output (specifically the file size) more human-readable by passing the `-h` option to `ls`:

```
$ ls -lh vmlinux System.map
-rw-rw-r-- 1 c2kp c2kp 4.8M May 16 16:12 System.map
-rwxrwxr-x 1 c2kp c2kp 704M May 16 16:12 vmlinux
$ file vmlinux
vmlinux: ELF 64-bit LSB executable, x86-64, version 1 (SYSV), statically
linked, BuildID[sha1]=e4<...>, with debug_info, not stripped
```

As you can see, the `vmlinux` file is huge. This is because it contains all the kernel symbols as well as extra debug information encoded into it. (FYI, the `vmlinux` and `System.map` files are used in the kernel debug context; keep them around.) The useful `file` utility shows us more detail regarding this image file.

The actual kernel image file that the bootloader loads up and boots into will always be in the generic location `arch/<arch>/boot/`; hence, for the x86 architecture, we have the following:

```
$ ls -lh arch/x86/boot/bzImage
-rw-rw-r-- 1 c2kp c2kp 12M May 16 16:12 arch/x86/boot/bzImage
$ file arch/x86/boot/bzImage
arch/x86/boot/bzImage: Linux kernel x86 boot executable bzImage, version
6.1.25-lkp-kernel (c2kp@osboxes) #3 SMP PREEMPT_DYNAMIC Tue [...], RO-rootFS,
swap_dev 0XB, Normal VGA
```

So, here, our compressed kernel image version `6.1.25-lkp-kernel` for the x86_64 is approximately 12 MB in size. The `file` utility again clearly reveals that indeed it is a Linux kernel boot image for the x86 architecture.

By the way, the kernel's top-level `Makefile` has a few simple (yet useful) targets to verify stuff like the kernel version string, and so on; let's take a peek (it's toward the end of the large `Makefile`):

```
cd ${LKP_KSRC}
cat Makefile
[ … ]
kernelrelease:
        @echo "$(KERNELVERSION)$$($(CONFIG_SHELL) $(srctree)/scripts/
setlocalversion $(srctree))"
```

```
kernelversion:
        @echo $(KERNELVERSION)

image_name:
        @echo $(KBUILD_IMAGE)
[ … ]
```

Let's try out these targets:

```
$ make kernelrelease kernelversion image_name
6.1.25-lkp-kernel
6.1.25
arch/x86/boot/bzImage
$
```

Nice.

The kernel documents several tweaks and switches that can be performed during the kernel build by setting various environment variables. This documentation can be found within the kernel source tree at `Documentation/kbuild/kbuild.rst`. We shall in fact use the `INSTALL_MOD_PATH`, `ARCH`, and `CROSS_COMPILE` environment variables in the material that follows.

Great! Our kernel image and modules are ready! Read on as we install the kernel modules as part of our next step.

Step 5 – installing the kernel modules

In the previous step, all the kernel config options that were marked as `m` – in effect, all the kernel modules, the `*.ko` files – have by now been built within the source tree. As you shall learn, that's not quite enough: they must now be installed into a known location on the system. This section covers these details.

Locating the kernel modules within the kernel source

As you just learned, the previous step – building the kernel image and modules – resulted in the compressed and uncompressed kernel images being generated, as well as all the kernel modules (as specified by our kernel config). Kernel modules are identified as files that always have a `.ko` (for *kernel object*) suffix. These modules are very useful; they give us kernel functionality in a modular manner (we can decide to plug them in or out of kernel memory at will; the following two chapters will go into great detail on the topic).

For now, knowing that the previous step generated all the kernel module files as well, let's locate them within the kernel source tree. To that end, we use the `find` command to locate them within the kernel source folder:

```
$ cd ${LKP_KSRC}
$ find . -name "*.ko"
./crypto/crypto_simd.ko
./crypto/cryptd.ko
[...]
./fs/binfmt_misc.ko
./fs/vboxsf/vboxsf.ko
```

But merely building the kernel modules is not enough; why? They need to be *installed* into a well-known location within the root filesystem so that, at boot, the system *can actually find and load them* into kernel memory. This is why we need the following step, module installation (see the following *Getting the kernel modules installed* section). The "well-known location within the root filesystem" where they get installed is `/lib/modules/$(uname -r)/`, where `$(uname -r)` yields the kernel version number, of course.

Getting the kernel modules installed

Performing the kernel module installation is simple; after the build step, just invoke the `modules_install` Makefile target. Let's do so:

```
$ cd ${LKP_KSRC}
$ sudo make modules_install
[sudo] password for c2kp:
 INSTALL /lib/modules/6.1.25-lkp-kernel/kernel/arch/x86/crypto/aesni-intel.ko
  SIGN    /lib/modules/6.1.25-lkp-kernel/kernel/arch/x86/crypto/aesni-intel.ko
[ … ]
 INSTALL /lib/modules/6.1.25-lkp-kernel/kernel/sound/soundcore.ko
  SIGN    /lib/modules/6.1.25-lkp-kernel/kernel/sound/soundcore.ko
  DEPMOD  /lib/modules/6.1.25-lkp-kernel
$
```

A few points to notice:

- Notice that we use `sudo` to perform the module installation process as root (superuser). This is required as the default install location (under `/lib/modules/`) is only root-writable. The `modules_install` target causes the kernel modules to be copied across to the correct installation location under `/lib/modules/` (the work that shows up in the preceding output block as `INSTALL /lib/modules/6.1.25-lkp-kernel/<...>`).
- Next, the module is possibly "signed." On systems configured with cryptographic signing of kernel modules (`CONFIG_MODULE_SIG`: a useful security feature, as is the case here), the `SIGN` step has the kernel "sign" the modules.

Briefly, when the config option `CONFIG_MODULE_SIG_FORCE` is turned on (it's off by default), only correctly signed modules will be allowed to be loaded into kernel memory at runtime.

- Next, after all modules are copied across (and possibly signed), the `kbuild` system runs a utility called `depmod`. Its job essentially is to resolve dependencies between kernel modules and encode them (if they exist) into some metafiles.

Now let's see the result of the module installation step:

```
$ ls /lib/modules
5.19.0-40-generic/  5.19.0-41-generic/  5.19.0-42-generic/  6.1.25-lkp-kernel/
```

In the preceding output, we can see that for each (Linux) kernel installed on the system, there will be a folder under `/lib/modules/` whose name is the kernel release, as expected. Let's look within the folder of interest – that of our new kernel (`6.1.25-lkp-kernel`). There, under the `kernel/` sub-directory – within various directories – live the just-installed kernel modules:

```
$ ls /lib/modules/6.1.25-lkp-kernel/kernel/
arch/  crypto/  drivers/  fs/  lib/  net/  sound/
```

Incidentally, the `/lib/modules/<kernel-ver>/modules.builtin` file has the list of all installed kernel modules (that live under `/lib/modules/<kernel-ver>/kernel/`).

Overriding the default module installation location

A final key point: during the kernel build, we can install the kernel modules into a location that *we* specify, overriding the (default) `/lib/modules/<kernel-ver>` location. This is done by setting the environment variable `INSTALL_MOD_PATH` to the required location. As an example, here, we set up an environment variable, `STG_MYKMODS`, to hold the location where we want our kernel modules installed, and then run the `modules_install` command:

```
export STG_MYKMODS=../staging/rootfs/my_kernel_modules
make INSTALL_MOD_PATH=${STG_MYKMODS} modules_install
```

With this, all our kernel modules will get installed into the `${STG_MYKMODS}/` folder. Note how, perhaps, `sudo` is not required if `INSTALL_MOD_PATH` refers to a location that does not require *root* for writing.

This technique – overriding the kernel modules' install location – can be especially useful when building a Linux kernel and kernel modules for an embedded target. We must not overwrite the host system's kernel modules with that of the embedded target; that would be disastrous! The reality, of course, is that we can all make mistakes like this once in a while (I know I have!); the usefulness of VMs – especially one that's checkpointed so that we can quickly revert to a good state – becomes readily apparent!

The next step is to generate the so-called `initramfs` (or `initrd`) image and set up the bootloader. We also need to clearly understand what exactly this `initramfs` image is and the motivation behind using it. The section after the following one delves into these details.

Step 6 – generating the initramfs image and bootloader setup

Firstly, please note that this discussion is highly biased toward the x86[_64] architecture, perhaps the most common one in use. Nevertheless, the concepts learned here can be directly applied to other architectures (like ARM), though the precise commands may vary. Typically, unlike on the x86, and at least for ARM-based Linux, there's no direct command to generate the initramfs image; it has to be done manually, "by hand." Embedded builder projects like Yocto and Buildroot do provide ways to automate this.

For the typical x86 desktop or server kernel build procedure, this step is internally divided into two distinct parts:

- Generating the `initramfs` (formerly called `initrd`) image
- GRUB setup for the new kernel image

The reason it's encapsulated in a single step here is that, on the x86 architecture, convenience scripts perform both tasks, giving the appearance of a single step.

Are you wondering what exactly this initramfs (or initrd) image file is? Please see the *Understanding the initramfs framework* section for details. We'll get there soon.

For now, let's just go ahead and generate the **initramfs** (short for **initial RAM filesystem**) image file as well as update the bootloader. By the way, now might also be a good time to checkpoint your VM (or take a backup) so that, worst case, even if the root filesystem is corrupted (it shouldn't be), you have a means to revert to a good state and continue working. Performing this on x86[_64] Ubuntu is easily done in one simple step:

```
$ sudo make install
  INSTALL /boot
run-parts: executing /etc/kernel/postinst.d/dkms 6.1.25-lkp-kernel /boot/
vmlinuz-6.1.25-lkp-kernel
 * dkms: running auto installation service for kernel 6.1.25-lkp-kernel
[ OK ]
run-parts: executing /etc/kernel/postinst.d/initramfs-tools 6.1.25-lkp-kernel /
boot/vmlinuz-6.1.25-lkp-kernel
update-initramfs: Generating /boot/initrd.img-6.1.25-lkp-kernel
[ … ]
```

```
run-parts: executing /etc/kernel/postinst.d/xx-update-initrd-links 6.1.25-lkp-
kernel /boot/vmlinuz-6.1.25-lkp-kernel
I: /boot/initrd.img.old is now a symlink to initrd.img-5.19.0-42-generic
I: /boot/initrd.img is now a symlink to initrd.img-6.1.25-lkp-kernel
run-parts: executing /etc/kernel/postinst.d/zz-update-grub 6.1.25-lkp-kernel /
boot/vmlinuz-6.1.25-lkp-kernel
Sourcing file `/etc/default/grub'
Sourcing file `/etc/default/grub.d/init-select.cfg'
Generating grub configuration file ...
Found linux image: /boot/vmlinuz-6.1.25-lkp-kernel
Found initrd image: /boot/initrd.img-6.1.25-lkp-kernel
[ … ]
Found linux image: /boot/vmlinuz-5.19.0-42-generic
Found initrd image: /boot/initrd.img-5.19.0-42-generic
[ … ]
done
```

Notice that, again, we prefix the `make install` command with `sudo`. Quite obviously, this is because we require *root* permission to write the concerned files and folders; they're written into the `/boot` directory (which may be set up as part of the root filesystem or as a separate partition).

What if we don't want to save the output artifacts – the initamfs image and the bootloader files – in `/boot`? You can always override the destination directory via the `INSTALL_PATH` environment variable; this is often the case when building Linux for an embedded system. The kernel documentation mentions this here: `https://docs.kernel.org/kbuild/kbuild.html#install-path`.

So that's it, we are done: a brand new 6.1 kernel, along with all requested kernel modules and the `initramfs` image have been generated and installed, and GRUB has been updated to reflect the presence of the new kernel and initramfs images. All that remains is to reboot the system, select the new kernel image on boot (from the bootloader menu screen), boot up, log in, and verify that all is okay.

In this step, we had the `initramfs` image generated. The question is, what did the `kbuild` system perform under the hood when we did this? Read on to find out.

Generating the initramfs image – under the hood

When you ran the `sudo make install` command on the x86, under the hood, the kernel Makefile invoked this script: `scripts/install.sh`. It's a wrapper script that loops over all possible arch-specific install scripts that may or may not be present, running them (with appropriate parameters) if they exist. To be more precise, these are the locations of the possible arch-specific scripts that will be executed if they exist (in this order):

- `${HOME}/bin/${INSTALLKERNEL}`
- `/sbin/${INSTALLKERNEL}`

- `${srctree}/arch/${SRCARCH}/install.sh`
- `${srctree}/arch/${SRCARCH}/boot/install.sh`

Again, focusing on the x86[_64] here, the `arch/x86/boot/install.sh` script internally, as part of its work, copies the following files into the `/boot` folder, with the name format typically being `<filename>-$(uname -r)-kernel`:

```
/boot/config-6.1.25-lkp-kernel
/boot/System.map-6.1.25-lkp-kernel
/boot/initrd.img-6.1.25-lkp-kernel
/boot/vmlinuz-6.1.25-lkp-kernel
```

The `initramfs` image is built as well. On x86 Ubuntu Linux, a shell script named `update-initramfs` performs this task (which is itself a convenience wrapper over another script called `mkinitramfs` that performs the actual work).

But how exactly is the image built? Simplistically, the initramfs image is nothing but a `cpio` file built using the so-called `newc` format. The `cpio` utility (copy-in, copy-out) is an old one, used in creating an archive – a simple collection of files; `tar` is a well-known user of `cpio` internally. A simplistic way to build the initramfs image, from the content of a given directory (let's call it `my_initramfs`), is to do this:

```
find my_initramfs/ | sudo cpio -o --format=newc -R root:root | gzip -9 >
initramfs.img
```

Note how the image is typically compressed using `gzip`.

Once built, the `initramfs` image is also copied into the `/boot` directory, seen as the `/boot/initrd.img-6.1.25-lkp-kernel` file in the preceding output snippet.

If a file being copied into `/boot` already exists, it is backed up as `<filename>-$(uname -r).old`. The file named `vmlinuz-<kernel-ver>-kernel` is a copy of the `arch/x86/boot/bzImage` file. In other words, *it is the compressed kernel image* – the image file that the bootloader will be configured to load into RAM, uncompress, and jump to its entry point, thus handing over control to the kernel!

Why do they have the names `vmlinux` (recall, this is the uncompressed kernel image file stored in the root of the kernel source tree) and `vmlinuz`? It's an old Unix convention that the Linux OS is quite happy to follow: on many Unix flavors, the kernel was called `vmunix`, so Linux calls it `vmlinux` and the compressed one `vmlinuz`; the `z` in `vmlinuz` is to hint at the (by default) `gzip` compression it endures. By the way, using `gzip` for compressing modern kernels is quite outdated; the default on modern x86 is to use the superior (and faster) ZSTD compression, though the file naming conventions remain..

Also, the GRUB configuration file located at `/boot/grub/grub.cfg` is updated to reflect the fact that a new kernel is now available for boot.

Again, it's worth emphasizing the fact that all this is *very architecture-specific*. The preceding discussion concerns building the kernel on an Ubuntu Linux x86_64 system. While conceptually similar, the details of the kernel image filenames, their locations, and especially the bootloader, vary on different architectures and even on different distros.

You can skip ahead to the *Step 7 – customizing the GRUB* section if you wish. If you are curious (I'm hoping so), read on. In the following section, we describe in some more detail the *hows* and *whys* of the `initramfs` (earlier called `initrd`) framework.

Understanding the initramfs framework

A bit of a mystery remains! What exactly is this **initramfs (initial RAM filesystem)** or **initrd (initial RAM disk)** image for? Why is it there?

Firstly, using this feature is a choice – the kernel config directive is called `CONFIG_BLK_DEV_INITRD`. It's set to `y` and hence on by default. In brief, for systems that do not know in advance certain things such as the boot disk host adapter or controller type (SCSI, RAID, and so on), the exact filesystem type that the root filesystem is formatted as (is it `ext2`, `ext4`, `btrfs`, `f2fs`, or something else?), or for those systems where these functionalities are always built as kernel modules, we require the `initramfs` capability. Exactly why will become clear in a moment. Also, as mentioned earlier, `initrd` is now considered an older term. Nowadays, we more often use the term `initramfs` in its place.

But what exactly is the difference between the older `initrd` and newer `initramfs`? The key difference is in how they're generated. To build an (older) `initrd` image with the current directory content, we can do:

```
find . | sudo cpio -R root:root | gzip -9 > initrd.img
```

Whereas to build the (newer) initramfs image with the current directory content, we do it like this (using the `newc` format):

```
find . | sudo cpio -o --format=newc -R root:root | gzip -9 > initramfs.img
```

(Tip: these details will become clearer once you read the following section.)

Why the initramfs framework?

The `initramfs` framework is essentially a kind of middle-man between the early kernel boot and user mode. It allows us to run user space applications (or scripts) before the actual (real) root filesystem has been mounted, before the kernel has finished initializing the system. This is useful in many circumstances, a few of which are detailed in the following list. The key point here is that `initramfs` allows us to run user-mode apps that the kernel cannot normally run during boot time.

Practically speaking, among various uses, this framework allows us to do some interesting things, including the following:

- Set up a console font.
- Customize keyboard layout settings.
- Print a custom welcome message on the console device.
- Accept a password (required for encrypted disks).
- Load up kernel modules as required.
- Spawn a "rescue" shell if something fails.
- And many more!

Imagine for a moment that you are in the business of building and maintaining a new Linux distribution. Now, at installation time, the end user of your distribution might decide to format their SSD disk with, say, the *f2fs* (fast flash filesystem) filesystem. The thing is, you cannot know in advance exactly what choice the end user will make – it could be one of any number of filesystems. So, you decide to pre-build and supply a large variety of kernel modules that will fulfill almost every possibility. Fine, when the installation is complete and the user's system boots up, the kernel will, in this scenario, require the `f2fs.ko` kernel module to successfully mount the (f2fs) root filesystem and proceed further.

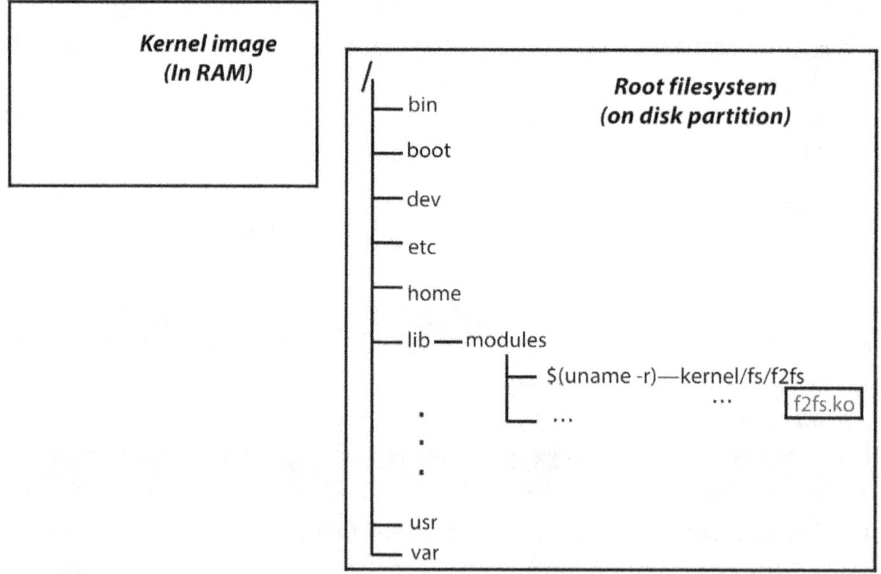

Figure 3.1: The kernel image is in RAM, and the root filesystem is on the disk and yet to be mounted

When the bootloader finishes its work, the Linux kernel takes over, running its code to initialize and prepare the system. So, think about this: the kernel is now running in RAM (conceptually seen at the upper left of *Figure 3.1*), but the modules it will soon require are still on secondary storage, the disk (or flash chip) conceptually seen at the lower right of *Figure 3.1*; more importantly, they're not accessible, as the root filesystem is yet to be mounted.

But wait, think about this, we now have a classic *chicken-and-egg* problem: in order for the kernel to mount the root filesystem (and thus gain access to required modules on it), it requires the `f2fs.ko` kernel module file to be loaded into RAM (as it contains the necessary code to be able to mount and then work with the filesystem). But that file is embedded inside the f2fs root filesystem itself – to be precise, here: `/lib/modules/<kernel-ver>/kernel/fs/f2fs/f2fs.ko` (see *Figure 3.1*).

One of the primary purposes of the `initramfs` framework is to solve this chicken-and-egg problem. The `initramfs` image file is a compressed `cpio` archive (`cpio` is a flat-file format used by `tar`). As we mentioned in the previous section, the `update-initramfs` script internally invokes the `mkinitramfs` script (on x86 Ubuntu at least, this is the case). These scripts *build a minimal temporary root filesystem* containing the kernel modules as well as supporting infrastructure such as the `/etc` and `/lib` folders in a simple `cpio` file format, which is then usually gzip-compressed. This now forms the so-called `initramfs` (or `initrd`) image file; it will be placed in a file named `/boot/initrd.img-<kernel-ver>`. Well, how does that help?

At boot, and, of course, assuming the `initramfs` feature is enabled, the bootloader will, as part of its work, perform an extra step: after uncompressing and loading the kernel image in RAM, it will also load the specified `initramfs` image file into RAM. Now, when the kernel runs and detects the presence of the `initramfs` image, it uncompresses it, and using its content (via scripts), it loads up the required kernel modules into RAM (see *Figure 3.2*):

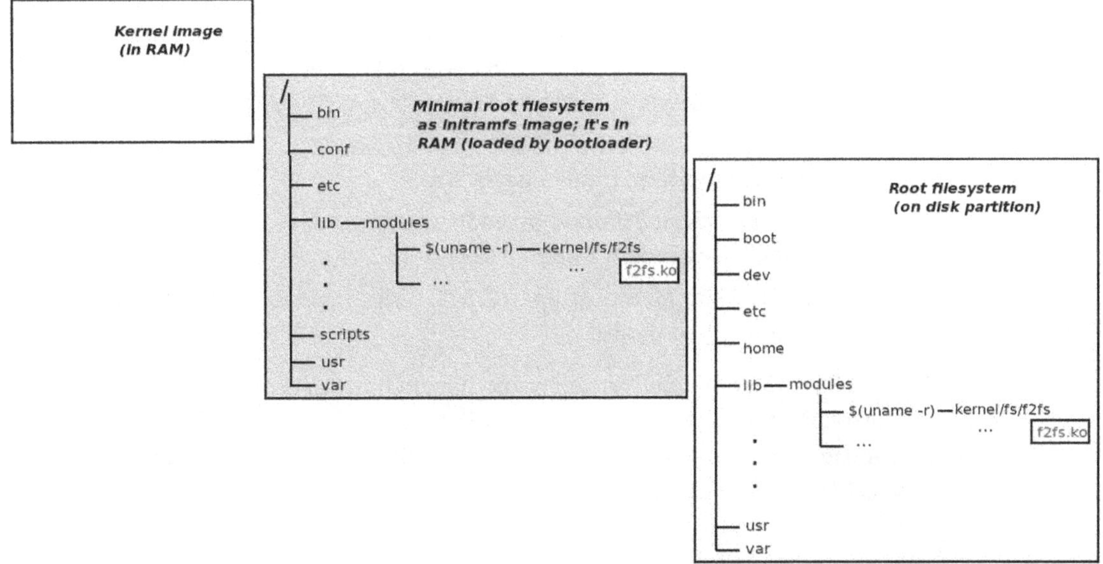

Figure 3.2: The initramfs image serves as a middleman between early kernel and actual root filesystem availability

Now that the required modules are available within the main memory (RAM, in a format called a "RAM disk"), the kernel can load the required modules (here, `f2fs.ko`), getting their functionality, and thus actually be able to mount the "real" root filesystem and proceed forward! Some more details on both the boot process (on x86) and the initramfs image can be found in the following sections.

Understanding the basics of the boot process on the x86

In the following list, we provide a brief overview of the typical boot process on an x86[_64] desktop (or laptop), workstation, or server:

1. Firstly, there are broadly two ways in which systems boot. First, the legacy way (specific to the x86) is via the **BIOS** (short for **Basic Input Output System** – essentially, the *firmware* on the x86). After carrying out basic system initialization and diagnostics (**POST** – **Power On Self Test**), it loads up the first sector of the first bootable disk into RAM and jumps to its entry point. This forms what is often referred to as the *stage one* bootloader, which is very small (typically just 1 sector, 512 bytes); its primary job is to load the *stage two* (larger) bootloader code – also present on the bootable disk – into memory and jump to it.

   The modern, more powerful way to boot is via the newer **UEFI** (**Unified Extensible Firmware Interface**) standard: on many modern systems, the UEFI framework is used as a superior and more secure way to boot the system.

   So what's the difference between the legacy BIOS and newer UEFI? Very briefly:

   - UEFI is a modern framework not limited to the x86 (the BIOS is x86-only); for example, ARM-based systems leverage UEFI as part of their famed Secure Boot capabilities.
   - UEFI is much more secure; it will only allow "signed" operating systems (what it calls "apps") to be booted via it (this can cause issues with dual-booting at times).
   - UEFI requires a separate special partition called the ESP (EFI System Partition); this holds a `.efi` file that contains the initialization code and data, unlike the BIOS, where it's written in firmware (typically in an EEPROM chip). Thus, updating UEFI is straightforward.
   - UEFI boot time is faster than that of legacy BIOS.
   - UEFI lets you run modern 32- or 64-bit code, thus attractive GUIs can be used for the interface; the BIOS works with 16-bit code only.
   - Drive size: the BIOS supports only up to 2.2 TB disks, whereas UEFI can support disks up to 9 ZB (zettabytes) in size!

2. Regardless of UEFI/BIOS, once the kernel image is loaded in RAM, the stage two bootloader code takes control. Its main job is to *load the actual (stage three) bootloader* from the filesystem into memory and jump to its entry point. On the x86, GRUB is typically the bootloader employed (an older one is **LILO** (**Linux Loader**)).

3. GRUB will be passed both the compressed kernel image file (`/boot/vmlinuz-<kernel-ver>`) as well as the compressed `initramfs` image file (`/boot/initrd.img-<kernel-ver>`) as parameters (via its config file, which we'll see in coming sections). The bootloader will (simplistically) do the following:
   - Perform low-level hardware initialization.
   - Load these images into RAM, uncompressing the kernel image to a certain extent.
   - It will then jump to the kernel entry point.
4. The Linux kernel, now having full control of the machine, will initialize the hardware and software environment. It typically makes no assumptions regarding the earlier work performed by the bootloader. Well, it does depend upon the BIOS or UEFI to set up things like PCI addressing and interrupt line assignment via ACPI tables (or, on ARM/PPC, via the Device Tree).
5. Upon completing most of the hardware and software initialization, if it notices that the `initramfs` feature is turned on (`CONFIG_BLK_DEV_INITRD=y`), it will locate (and if required, uncompress) the `initramfs` (`initrd`) image in RAM (see *Figure 3.2*).
6. It will then mount it as a temporary root filesystem in RAM itself, within a RAM disk.
7. We now have a base, minimal, and temporary root filesystem set up in memory. Thus, the `initramfs-based` startup scripts now run, performing, among other tasks, the loading of the required kernel modules into RAM (in effect, loading the root filesystem drivers, including, in our scenario, the `f2fs.ko` kernel module; again, see *Figure 3.2*).
8. When initramfs runs, it first invokes `/sbin/init` (this could be a binary executable or a script); besides other housekeeping tasks, it performs a key job: a *pivot-root*, *unmounting* the temporary initramfs root filesystem, freeing its memory, and mounting the real root filesystem. It's now possible because the kernel module providing that filesystem support is indeed available (in RAM).
9. Once the (actual disk or flash-based) root filesystem is successfully mounted, system initialization can proceed. The kernel continues, ultimately invoking the first user space process (PID 1), typically `/sbin/init` (when using the older SysV init framework), or, more likely these days, via the more powerful `systemd` init framework.
10. The `init` framework now proceeds to initialize the system, bringing up system services as configured.

A few things to note:

On modern Linux systems, the traditional (read: old/legacy) *SysV* (read as System Five) `init` framework has largely been replaced with a modern optimized framework called **systemd (system daemon)**. Thus, on many (if not most) modern Linux systems, including embedded ones, the traditional `/sbin/init` has been replaced with *systemd* (or is simply a symbolic link to its executable file). The systemd framework is considered superior, with the ability to fine-tune the boot process and optimize boot time, along with many, many more features. Find out more about systemd in the *Further reading* section.

The generation of the initramfs root filesystem itself is not covered in detail in this book; the official kernel documentation does cover some of it – *Using the initial RAM disk (initrd)*: `https://docs.kernel.org/admin-guide/initrd.html`.

Also, as a simple example of generating a root filesystem, you could look at the code of the SEALS project (at `https://github.com/kaiwan/seals`) that I mentioned in *Online Chapter, Kernel Workspace Setup*; it has a Bash script that generates a very minimal, or skeleton, root filesystem from scratch.

Now that you understand the motivation behind `initramfs`, we'll complete this section by providing a bit of a deeper look into `initramfs` in the following section. Do read on!

More on the initramfs framework

Another place where the `initramfs` framework helps is in bringing up computers whose disks are *encrypted*.

Are you working on a laptop with unencrypted disks? That's not a great idea; if your device is ever lost or stolen, hackers can easily gain access to your data by simply booting into it using a Linux-based USB pen drive and can then typically access all its disk partitions. Your login credentials don't help here. Modern distros can certainly and painlessly encrypt disk partitions; this is now a routine part of the installation. Besides volume-based encryption (provided by LUKS, dm-crypt, eCryptfs, and so on), several tools exist to individually encrypt and decrypt files. See these links for more:

Top 10 file and disk encryption tools for Linux, January 2022: `https://www.fosslinux.com/50005/top-10-file-and-disk-encryption-tools-for-linux.htm`, and *Best File & Disk Encryption Tools for Linux*, Day, May 2022: `https://linuxsecurity.com/features/top-8-file-and-disk-encryption-tools-for-linux`.

Assuming a system with encrypted filesystems, quite early in the boot process, the kernel will have to query the user for the password, and if correct, proceed with decrypting and mounting the disks, and so on.

But think about this: how can we run a C program executable that is, say, requesting a password, without having a C runtime environment in place – a root filesystem containing libraries, the loader program (*ld-linux...*), required kernel modules (for the crypto support perhaps), and so on?

Remember, the kernel *itself* hasn't yet completed initialization; how can user space apps run? Again, the `initramfs` framework solves this issue by setting up a pretty complete, though temporary, user-space runtime environment complete with the required root filesystem containing libraries, the loader, kernel modules, some scripts, and so on, in primary memory.

Peeking into the initramfs image

So, we just stated that the initramfs image is a temporary yet pretty complete one, containing system libraries, the loader program, minimally required kernel modules, some scripts, and so on. Can we verify this? Yes, indeed we can! Let's take a peek into the `initramfs` image file. The `lsinitramfs` script on Ubuntu serves exactly this purpose (on Fedora, the equivalent is called `lsinitrd` instead):

```
$ ls -lh /boot/initrd.img-6.1.25-lkp-kernel
-rw-r--r-- 1 root root 26M Jun 13 11:08 /boot/initrd.img-6.1.25-lkp-kernel
$ lsinitramfs /boot/initrd.img-6.1.25-lkp-kernel|wc -l
364
$ lsinitramfs /boot/initrd.img-6.1.25-lkd-kernel
.
kernel
[ ... ]
bin
[ ... ]
conf/initramfs.conf
etc
etc/console-setup
[ ... ]
etc/default/console-setup
etc/default/keyboard
etc/dhcp
[ ... ]
etc/modprobe.d
etc/modprobe.d/alsa-base.conf
[ ... ]
etc/udev/udev.conf
[ ... ]
lib64
libx32
run
sbin
```

```
scripts
scripts/functions
scripts/init-bottom
[ ... ]
usr
usr/bin
usr/bin/cpio
usr/bin/dd
usr/bin/dmesg
[ ... ]
usr/lib/initramfs-tools
usr/lib/initramfs-tools/bin
[ ... ]
usr/lib/modprobe.d/systemd.conf
usr/lib/modules
[ ... ]
usr/lib/modules/6.1.25-lkp-kernel/kernel/crypto/crc32_generic.ko
[ ... ]
usr/lib/modules/6.1.25-lkp-kernel/kernel/drivers/net/ethernet/intel/e1000/
e1000.ko
[ ... ]
usr/lib/modules/6.1.25-lkp-kernel/kernel/fs/f2fs/f2fs.ko
[ ... ]
usr/sbin/modprobe
[ ... ]
var/lib/dhcp
```

There's quite a bit in there: we truncate the output to show a few select snippets. We can see a minimal root filesystem with support for the required runtime libraries, kernel modules, the `/etc`, `/bin`, `/sbin`, `/usr`, and more directories, along with their utilities.

The details of constructing the `initramfs` (or `initrd`) image go beyond what we wish to cover here. We did mention the basics; you can generate an initramfs image like this: `find my_initramfs/ | sudo cpio -o --format=newc -R root:root | gzip -9 > initramfs.img`. I suggest you peek into these scripts to reveal their inner workings (on Ubuntu): `/usr/sbin/update-initramfs`, a wrapper script over the `/usr/sbin/mkinitramfs` shell script. See the *Further reading* section for more.

Also, modern systems feature what is sometimes referred to as a hybrid `initramfs`: an `initramfs` image that consists of an early `ramfs` image prepended to the regular or main `ramfs` image. The reality is that we require special tools to unpack/pack (uncompress/compress) these images. Ubuntu provides the `unmkinitramfs` and `mkinitramfs` scripts, respectively, to perform these operations.

As a quick experiment, let's unpack our brand-new `initramfs` image (the one generated in the previous section) into a temporary directory. Again, recall that this has been performed on our x86_64 Ubuntu 22.04 LTS guest VM. We view its output, truncated for readability, with `tree`:

```
$ TMPDIR=$(mktemp -d)
$ unmkinitramfs /boot/initrd.img-6.1.25-lkp-kernel ${TMPDIR}
$ tree ${TMPDIR}
/tmp/tmp.6JIg9JfKNQ
├── early
│   └── kernel
│       └── x86
│           └── microcode
│               └── AuthenticAMD.bin
├── early2
│   └── kernel
│       └── x86
│           └── microcode
│               └── GenuineIntel.bin
└── main
    ├── bin -> usr/bin
    ├── conf
    │   ├── arch.conf
    │   ├── conf.d
    │   │   └── zz-resume-auto
    │   └── initramfs.conf
    ├── etc
    │   ├── console-setup
    │   │   ├── cached_UTF-8_del.kmap.gz
    │   │   └── Uni2-Fixed16.psf.gz
    │   ├── default
    │   │   ├── console-setup
    │   │   └── keyboard
    │   ├── dhcp
    │   │   ├── dhclient.conf
    │   │   ├── dhclient-enter-hooks.d
    │   │   │   └── config
    │   │   └── dhclient-exit-hooks.d
    │   │       └── rfc3442-classless-routes
    │   ├── fstab
    │   ├── ld.so.cache
    │   ├── ld.so.conf
    │   ├── ld.so.conf.d
    │   │   ├── fakeroot-x86_64-linux-gnu.conf
    │   │   ├── libc.conf
    │   │   ├── x86_64-linux-gnu.conf
    │   │   └── zz_i386-biarch-compat.conf
    │   ├── modprobe.d
    │   │   ├── alsa-base.conf
    │   │   ├── amd64-microcode-blacklist.conf
    │   │   ├── blacklist-ath_pci.conf
    │   │   ├── blacklist.conf
```

Figure 3.3: Screenshot showing the partial directory tree of our 6.1 initramfs image

We've truncated the screenshot to show just a bit; plenty of output follows... I urge you to try this out and see for yourself.

This concludes our (rather lengthy!) discussion on the `initramfs` framework and the basics of the boot process on the x86[_64]. The good news is that now, armed with this knowledge, you can further customize your product by tweaking the `initramfs` image as required – an important skill! You'll find more on how exactly you can customize the initramfs image in the *Further reading* section.

Exercise

Take your existing initamfs image, extract it, and customize it to include a new app or script (which you can run via one of the startup scripts).

As an example (and as mentioned earlier), with *security* being a key factor on modern systems, being able to encrypt a disk at the block level is a powerful security feature; doing this very much involves tweaking the `initramfs` image. (Again, the *Further reading* section has links on how to go about disk encryption).

Now, let's finally complete the x86_64 kernel build procedure with some simple customization of the (x86) GRUB bootloader's boot script.

Step 7 – customizing the GRUB bootloader

We have now completed *steps 1* to *6* as outlined in *Chapter 2, Building the 6.x Linux Kernel from Source – Part 1*, in the *Steps to build the kernel from source* section. You can now reboot the system; of course, do first save and close all your apps and files. By default, though, the modern **GRUB** does not even show us any menu on reboot; it will by default boot into the newly built kernel (do remember that, here, we're describing this process *only* for x86[_64] systems running Ubuntu; the default kernel booted into can also vary with the distro).

On x86[_64] you can always get to the GRUB menu during early system boot. Just ensure you keep the *Shift* key pressed down during boot. Again, this behavior does depend on other factors – on systems with newer UEFI/BIOS firmware enabled, or when running within a nested VM, you may require other ways to force-see the GRUB menu at boot (try pressing *Esc* too).

What if we would like to see and customize the GRUB menu every time we boot the system, thus allowing us to possibly select an alternate kernel/OS to boot from (or even pass some kernel parameters along)? This is often very useful during development and debugging, so let's find out how we can do this.

Customizing GRUB – the basics

Customizing GRUB is quite easy to do; we can always, as root, edit its config file: `/etc/default/grub`. Do note the following:

- The following steps are to be carried out on the "target" system itself (not on the host) – in our case, within the x86_64 Ubuntu 22.04 guest VM. Of course, if working on Linux natively, you can go ahead on the same system.
- This procedure has been tested and verified on our x86_64 Ubuntu 22.04 LTS guest system only.

So what exactly are we setting out to do here? We want GRUB to show us its menu at boot, before running our favorite OS, allowing us to customize it further. Here's a quick series of steps to do this:

1. First, let's be safe and keep a backup copy of the GRUB bootloader config file:

   ```
   sudo cp /etc/default/grub /etc/default/grub.orig
   ```

2. Edit it. You can use `vi` or your editor of choice:

   ```
   sudo vi /etc/default/grub
   ```

3. To always show the GRUB prompt at boot, insert this line:

   ```
   GRUB_HIDDEN_TIMEOUT_QUIET=false
   ```

   On some Linux distros, you might instead have the `GRUB_TIMEOUT_STYLE=hidden` directive; simply change it to `GRUB_TIMEOUT_STYLE=menu` to achieve the same effect. Always showing the bootloader menu at boot is good during development and testing and is typically turned off in production for both speed and security.

   Talking about security, always ensure that access to the firmware (BIOS/UEFI) and bootloader is password-protected.

4. Set the timeout to boot the default OS (in seconds) as required. The default is `10` seconds; here, we set it to 3 seconds:

   ```
   GRUB_TIMEOUT=3
   ```

   Setting the preceding timeout value to the following values will produce the following outcomes:

   - `0`: Boot the system immediately without displaying the menu.
   - `-1`: Wait indefinitely.

   Furthermore, if a `GRUB_HIDDEN_TIMEOUT` directive is present in the config file, just comment it out:

   ```
   #GRUB_HIDDEN_TIMEOUT=1
   ```

5. Finally, run the `update-grub` program as *root* to have your changes take effect:

   ```
   sudo update-grub
   ```

The preceding command will typically cause the `initramfs` image to be refreshed (regenerated). Once done, you're ready to reboot the system. Hang on a second, though! The following section shows you how you can modify GRUB's configuration to boot by default into a kernel of your choice.

Selecting the default kernel to boot into

The GRUB default kernel is preset to be the number zero (via the `GRUB_DEFAULT=0` directive). This will ensure that the "first kernel" – the most recently added one – boots by default (upon timeout).

This may not be what we want; as a real example, on our x86 Ubuntu 22.04 LTS guest VM, we can set it to the default Ubuntu *distro kernel* by, as earlier, editing the `/etc/default/grub` file (as root, of course), like so:

```
GRUB_DEFAULT="Advanced options for Ubuntu>Ubuntu, with Linux 5.19.0-43-generic"
```

Of course, this implies that if your distro is updated or upgraded, you must again manually change the preceding line to reflect the new distro kernel that you wish to boot into by default, and then run **`sudo update-grub`**.

Right, our freshly edited GRUB configuration file is shown as follows:

```
$ cat /etc/default/grub
 [...]
#GRUB_DEFAULT=0
GRUB_DEFAULT="Advanced options for Ubuntu>Ubuntu, with Linux 5.19.0-43-generic"
 #GRUB_TIMEOUT_STYLE=hidden
GRUB_HIDDEN_TIMEOUT_QUIET=false
GRUB_TIMEOUT=3
GRUB_DISTRIBUTOR=`lsb_release -i -s 2> /dev/null || echo Debian`
GRUB_CMDLINE_LINUX_DEFAULT="quiet splash"
 GRUB_CMDLINE_LINUX="quiet splash"
 [...]
```

As in the previous section, don't forget: if you make any changes here, run the `sudo update-grub` command to have your changes take effect.

Additional points to note:

In addition, you can add "pretty" tweaks, such as changing the background image (or color) via the **`BACKGROUND_IMAGE="<img_file>"`** directive.

On Fedora, the GRUB bootloader config file is a bit different; run this command to show the GRUB menu at every boot:

```
sudo grub2-editenv - unset menu_auto_hide
```

The details can be found in the *Fedora wiki – Changes/HiddenGrubMenu*: `https://fedoraproject.org/wiki/Changes/HiddenGrubMenu`.

Unfortunately, GRUB2 (the latest version is now 2) seems to be implemented differently on pretty much every Linux distro, leading to incompatibilities when trying to tune it in one given manner.

All done! Let's (finally!) reboot the system, get into the GRUB menu, and boot our new kernel:

```
$ sudo reboot
 [sudo] password for c2kp:
```

Once the system completes its shutdown procedure and reboots, you should soon see the GRUB bootloader menu (the following section shows several screenshots too). Be sure to interrupt it by pressing any keyboard key!

Though always possible, **I recommend you don't delete the original distro kernel image(s) (and associated `initrd`, `System.map` files, and so on)**. What if your brand-new kernel fails to boot? (*If it can happen to the Titanic...!*) By keeping our original images, we thus have a fallback option: boot from the original distro kernel, fix our issue(s), and retry. Note that there's almost always automatically a *Recovery...* menu option as well, which will try and at least log you into a root shell. Close to worst case, you usually get logged in to an initramfs-based shell (working from RAM).

As a worst-case scenario, what if all other kernels/`initrd` images have been deleted (or are corrupt) and your single new kernel fails to boot successfully? Ahem. Well, you can always boot into a *recovery mode* Linux via a USB pen drive; a bit of googling regarding this will yield many links and video tutorials. (FYI, here's Ubuntu's link to do precisely this, via a neat GUI app that's typically preinstalled: `https://ubuntu.com/tutorials/create-a-usb-stick-on-ubuntu#1-overview`.)

Booting our VM via the GNU GRUB bootloader

Now our guest VM (here, using the *Oracle VirtualBox hypervisor*) is about to come up; once its (emulated) BIOS / UEFI routines are done, the GNU GRUB screen shows up first. (Note that the procedures explained here will work as well on a native Linux system.) This happens because we quite intentionally changed the `GRUB_HIDDEN_TIMEOUT_QUIET` GRUB configuration directive to the value of `false` (as explained in the previous section). See the following screenshot (*Figure 3.4*).

The particular styling seen in the screenshot is how it's customized to appear for the Ubuntu distro:

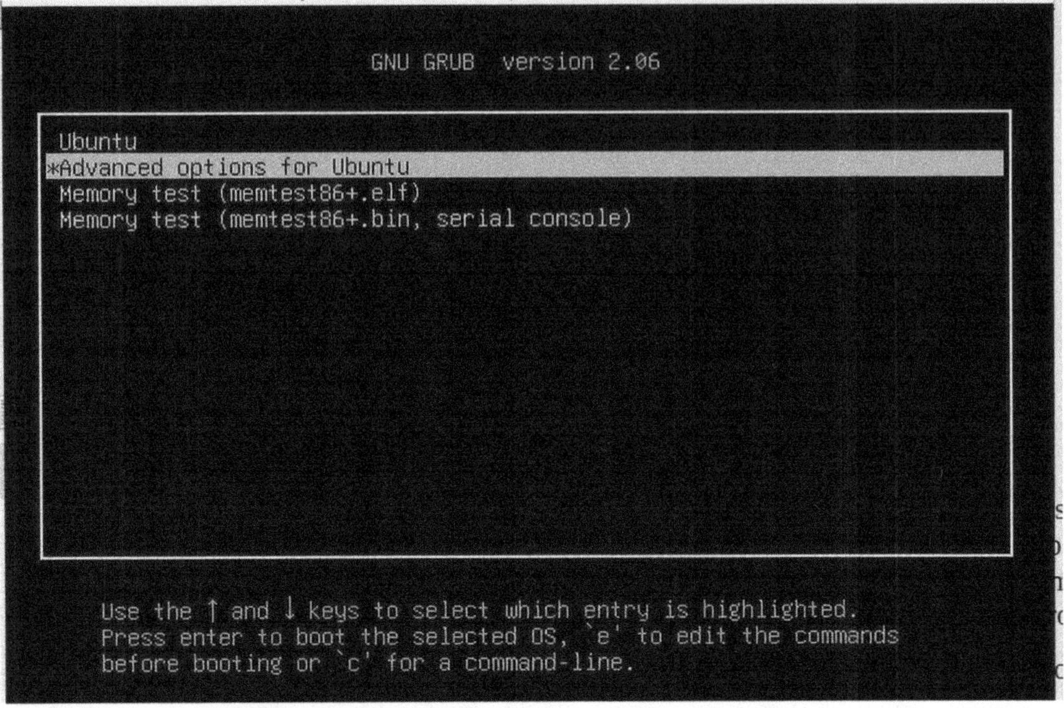

Figure 3.4: The GRUB2 menu – paused on system startup

Now let's go straight into booting our VM:

1. Press any keyboard key (except *Enter*) to ensure the default kernel is not booted once the timeout expires (recall, we set it to 3 seconds).
2. If not already there, scroll to the `Advanced options for Ubuntu` menu, highlighting it (as seen in *Figure 3.4*), and press *Enter*.
3. Now you'll see a menu similar, but likely not identical, to the following screenshot (*Figure 3.5*). For each kernel that GRUB has detected and can boot into, there are two lines shown – one for the kernel itself and one for the special *recovery mode* boot option for that kernel:

```
GNU GRUB  version 2.06

*Ubuntu, with Linux 6.1.25-lkp-kernel
 Ubuntu, with Linux 6.1.25-lkp-kernel (recovery mode)
 Ubuntu, with Linux 6.1.25-lkp-kernel.old
 Ubuntu, with Linux 6.1.25-lkp-kernel.old (recovery mode)
 Ubuntu, with Linux 5.19.0-43-generic
 Ubuntu, with Linux 5.19.0-43-generic (recovery mode)
 Ubuntu, with Linux 5.19.0-42-generic
 Ubuntu, with Linux 5.19.0-42-generic (recovery mode)

   Use the ↑ and ↓ keys to select which entry is highlighted.
   Press enter to boot the selected OS, `e' to edit the commands
   before booting or `c' for a command-line. ESC to return previous
   menu.
```

Figure 3.5: The GRUB2 Advanced options… menu showing available kernels to boot from

Notice how the kernel that will boot by default – in our case, `6.1.25-lkp-kernel` – is highlighted by default with an asterisk (*).

The preceding screenshot shows a few "extra" line items. This is because, at the time of taking this screenshot, I had built my 6.1.25 kernel a few times (resulting in the previous one being saved as the "`.old`" item; also, keeping Ubuntu updated results in newer distro kernels getting installed as well. We can spot the 5.19.0-43-generic kernel and its predecessor. No matter; we ignore them here.

An aside: for some variety, here's a screenshot of a UEFI-based system's boot selection menu screen (in this particular case, on an x86_64 (Dell) PC to show the boot menu, the UEFI hotkey happens to be *F12*); notice the rich GUI as well as mouse pointer support:

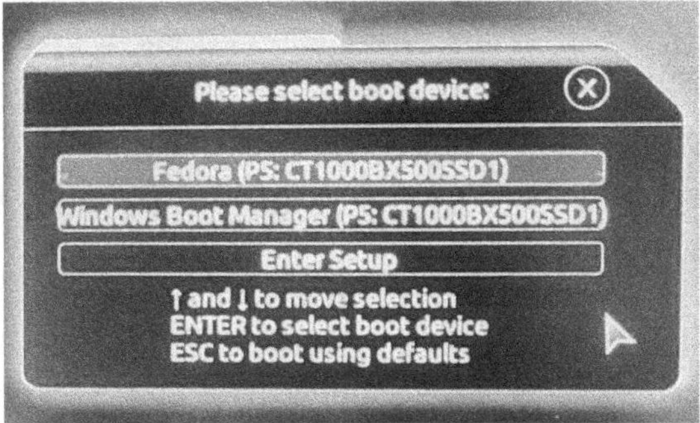

Figure 3.6: The boot selection menu on a UEFI-based x86_64 PC

4. Okay, back to our x86 Ubuntu VM with the GRUB: whatever the case, simply scroll to and highlight the entry of interest – here, our shiny new 6.1 LTS kernel, the menu entry labeled `*Ubuntu, with Linux 6.1.25-lkp-kernel` (as seen in *Figure 3.5*). Here, it's the very first line of the GRUB menu (as it's the most recent addition to the GRUB menu of bootable OSs).
5. Once you have highlighted the preceding menu item, press *Enter* and voilà! The bootloader will proceed to do its job, uncompressing and loading the kernel and initramfs (or `initrd`) images into RAM, and jumping to the Linux kernel's entry point, thus handing over control to Linux!

Right, if all goes well, as it should, you will have booted into your brand-new freshly baked 6.1.25 LTS Linux kernel! Congratulations on a task well done!

Then again, you could always do more – the following section shows you how you can further edit and customize GRUB's config at runtime (boot time). Again, this skill comes in handy now and then – for example, *forgot the root password?* Yes, indeed, you can actually *bypass it* using this technique! Read on to find out how.

Experimenting with the GRUB prompt

You could experiment further; instead of merely pressing *Enter* while on the `Ubuntu, with Linux 6.1.25-lkp-kernel` menu entry, ensure that this line is highlighted and press the *e* (for edit) key. We shall now enter GRUB's *edit screen*, wherein we are free to edit any value we like. Here's a screenshot after pressing the *e* key:

```
GNU GRUB  version 2.06

        insmod ext2
        set root='hd0,gpt2'
        if [ x$feature_platform_search_hint = xy ]; then
          search --no-floppy --fs-uuid --set=root --hint-bios=hd\
0,gpt2 --hint-efi=hd0,gpt2 --hint-baremetal=ahci0,gpt2  ae89c631-fbfd-44\
64-bd2a-f044d6f289fe
        else
          search --no-floppy --fs-uuid --set=root ae89c631-fbfd-\
4464-bd2a-f044d6f289fe
        fi
        echo        'Loading Linux 6.1.25-lkp-kernel ...'
_       linux       /vmlinuz-6.1.25-lkp-kernel root=UUID=b67edd\
                    ro  quiet splash $vt_handoff
        echo        'Loading initial ramdisk ...'
        initrd        /initrd.img-6.1.25-lkp-kernel

   Minimum Emacs-like screen editing is supported. TAB lists
   completions. Press Ctrl-x or F10 to boot, Ctrl-c or F2 for a
   command-line or ESC to discard edits and return to the GRUB
   menu.
```

Figure 3.7: The GRUB2 bootloader: editing our custom 6.1.25-lkp-kernel kernel menu entry

This screenshot was taken after scrolling down a few lines; look carefully – you can spot the cursor (an underscore-type one: "_") at the very beginning of the fourth line from the bottom of the edit box. (I put this key line into a box, just for emphasis. Also, the disk UUID value has been partially hidden.) This is the crucial line; it starts with the suitably indented keyword `linux`.

It specifies the path to the (compressed) kernel image, followed by the storage device ID where the image is present, and then the list of *kernel parameters* being passed via the GRUB bootloader to the Linux kernel. Here, the pathname to the kernel image is `/vmlinuz-6.1.25-lkp-kernel`; it starts at `/` as `/boot` is a separate partition.

Try experimenting a bit here:

- *Exercise 1*: As a simple exercise, after highlighting the key menu line beginning with `linux`, scroll to the right and delete the words `quiet` and `splash` from this entry, then press *Ctrl* + *X* or *F10* to boot. This time, the pretty Ubuntu splash screen does not appear; you are directly in the console seeing all kernel messages (emitted via the ubiquitous `printk()`-related APIs, as you'll soon learn) as they flash past, and later, systemd's console messages and so on.

  Next, a common question: what if we forget our password and thus cannot log in? Well, there are several approaches to tackle this. One is to attempt to reset it via the bootloader's help; here's how, let's do it as a small learning exercise.

- *Exercise 2*: Boot into the GRUB menu as we have done, go to the relevant kernel menu entry, press *e* to edit it, scroll down to the line beginning with `linux`, and append the word `single` (or just the number `1`) at the end of this entry, such that it looks like this:

```
            linux       /vmlinuz-6.1.25-lkp-kernel root=UUID=<...> ro
  quiet splash $vt_handoff single
```

  Now, when you boot, the kernel boots into single-user mode and gives you, the eternally grateful user, *a shell with root access*. It might prompt you to `Press Enter for maintenance`; please do so. Just run the `passwd <username>` command to change/reset your password. Exiting this root shell has the normal boot procedure continue.

The precise procedure to boot into single-user mode varies with the distro. Exactly what to edit in the GRUB2 menu is a bit different on Red Hat/Fedora/CentOS. See the *Further reading* section for a link on how to set it for these systems.

This teaches us something regarding *security*, doesn't it? A system is considered insecure when access to the bootloader menu (and even to the BIOS/UEFI) is possible without a password! In fact, in highly secured environments, even physical access to the console device must be restricted.

Great, by now, you have learned how to customize the GRUB and, I expect, have booted into your fresh 6.1 Linux kernel! Let's not just assume things; let's verify that the kernel is indeed the right one and configured as per our plan.

Verifying our new kernel's configuration

Okay, so back to our discussion. We have now booted into our newly built kernel. But hang on, please don't blindly assume that everything's just fine; let's actually verify that.

The *empirical approach* is always best; in this section, let's verify that we are indeed running the (6.1.25) kernel we just built and that it has indeed been configured as we intended. We start by examining the kernel version:

```
$ uname -r
6.1.25-lkp-kernel
```

Indeed, we are now running Ubuntu 22.04 LTS on our just-custom-built **6.1.25** LTS kernel! Further variations of the uname utility show us the machine hardware name and the OS is as planned: we're on the x86_64 running GNU/Linux:

```
$ uname -m ; uname -o
x86_64
GNU/Linux
```

Moving along, recall our discussion in *Chapter 2, Building the 6.x Linux Kernel from Source – Part 1*, in the *Sample usage of the make menuconfig UI* section, where we performed a couple of small changes to the kernel config as an exercise.

Now, we check that each configuration items we changed then has taken effect. Let's list them here, starting with the concerned `CONFIG_'FOO'` name, as follows:

- `CONFIG_LOCALVERSION`: The preceding output of `uname -r` clearly shows the `localversion` (or `-EXTRAVERSION`) part of the kernel version has been set to what we wanted: the `-lkp-kernel` string.
- `CONFIG_IKCONFIG`: This should have been enabled (see the previous chapter); it allows us to query the current kernel's config via the `/proc/config.gz` pseudo-file.
- `CONFIG_HZ`: We set this option to 300 as an experiment.

So, let's now check via the kernel `extract-ikconfig` script (also, recall that you are to set the `LKP_KSRC` environment variable to the root location of your 6.1 kernel source tree directory):

```
$ ${LKP_KSRC}/scripts/extract-ikconfig /boot/vmlinuz-6.1.25-lkp-kernel
#
# Automatically generated file; DO NOT EDIT.
 # Linux/x86 6.1.25 Kernel Configuration
[...]
CONFIG_HZ_300=y
 [...]
```

It works! We can see the entire kernel configuration via the `scripts/extract-ikconfig` script (we can do so as the `CONFIG_IKCONFIG[_PROC]` configs are enabled). We shall use this very script to grep the remainder of the config directives that we changed in the previous chapter (*Chapter 2, Building the 6.x Linux Kernel from Source – Part 1*, in the *Sample usage of the make menuconfig UI* section):

```
$ scripts/extract-ikconfig /boot/vmlinuz-6.1.25-lkp-kernel | grep -E
"LOCALVERSION|CONFIG_HZ"
```

```
CONFIG_LOCALVERSION="-lkp-kernel"
# CONFIG_LOCALVERSION_AUTO is not set
# CONFIG_HZ_PERIODIC is not set
# CONFIG_HZ_100 is not set
# CONFIG_HZ_250 is not set
CONFIG_HZ_300=y
# CONFIG_HZ_1000 is not set
CONFIG_HZ=300
```

Carefully looking through the preceding output, we can see that we got precisely what we wanted. Our new kernel's configuration settings match precisely the settings expected in *Chapter 2, Building the 6.x Linux Kernel from Source – Part 1*, in the *Sample usage of the make menuconfig UI* section. Perfect.

Alternatively, as we have enabled the `CONFIG_IKCONFIG_PROC` option, we could have achieved the same verification by looking up the kernel config via the (compressed) *proc* filesystem entry, `/proc/config.gz`, like this:

```
gunzip -c /proc/config.gz | grep -E "LOCALVERSION|CONFIG_HZ"
```

So, the kernel build is done! Fantastic. I urge you to refer back to *Chapter 2, Building the 6.x Linux Kernel from Source – Part 1*, in the *Steps to build the kernel from source* section, to again see the high-level overview of the steps for the entire process. Right, onto something quite interesting: in the next section, we will learn how to *cross-compile* the Raspberry Pi board kernel.

Kernel build for the Raspberry Pi

A popular and relatively inexpensive **Single-Board Computer** (**SBC**) to experiment and prototype with is the ARM-based Raspberry Pi. Hobbyists, tinkerers, and, to some extent, even pros find it very useful to try out and learn how to work with embedded Linux, especially as it has a strong community backing (with many Q&A forums) and excellent support. (You'll find a brief discussion and picture of a Raspberry Pi board in *Online Chapter, Kernel Workspace Setup*, in the *Experimenting with the Raspberry Pi* section. By the way, there are several well-known Raspberry Pi clones – like the Orange Pi – that work very well; the discussions should apply equally to them.)

There are two typical ways in which you can build a kernel for the target device or **Device Under Test** (**DUT**) that we'll use here, the Raspberry Pi 4 Model B (64-bit):

- Build the kernel on a powerful host system, typically an x86_64 (or Mac) desktop or laptop running a Linux distro.
- Perform the build on the target device itself.

We shall follow the first method – it's a lot faster and is considered the right way to perform embedded Linux development. (FYI, building the kernel on the target device is explained here: `https://www.raspberrypi.com/documentation/computers/linux_kernel.html#building-the-kernel-locally`.)

We shall assume (as usual) that we are running on our x86_64 Ubuntu 22.04 LTS guest VM. So, think about it; now, the host system is a guest Linux VM! Also, we're targeting building the kernel for the AArch64 architecture, to take full advantage of it (not for 32-bit).

Performing large downloads and kernel build operations on a guest VM isn't really ideal. Depending on the power and RAM of the host and guest, it will take a while. It could end up being twice as slow as building on a native Linux box. Nevertheless, assuming you have set aside sufficient disk space in the guest VM (and, of course, the host actually has this space available), this procedure works.

We will have to use an *x86_64-to-AArch64 (64-bit) cross-compiler* to build the kernel, or any component for that matter, for the DUT, the AArch64 Raspberry Pi 4 target. This implies installing an appropriate **cross-toolchain** as well to perform the build.

In the following few sections, we divide the work into three discrete steps:

1. Step 1 – cloning the Raspberry Pi kernel source tree
2. Step 2 – installing an x86_64-to-AArch64 cross-toolchain
3. Step 3 – configuring and building the Raspberry Pi AArch64 kernel.

So, let's begin!

Step 1 – cloning the Raspberry Pi kernel source tree

We arbitrarily select a *staging folder* (the place where the build happens) for the Raspberry Pi kernel source tree and the cross-toolchain, and assign it to an environment variable (to avoid hardcoding it):

1. Set up your workspace. We set an environment variable as `RPI_STG` (it's not required to use exactly this name for the environment variable; just pick a reasonable-sounding name and stick to it) to the staging folder's location – the place where we shall perform the work. Feel free to use a value appropriate to your system (I'm using `~/rpi_work`):

   ```
   export RPI_STG=~/rpi_work
   mkdir -p ${RPI_STG}/kernel_rpi
   ```

   Do ensure you have sufficient disk space available: the Raspberry Pi Git kernel source tree takes approximately 1.7 GB, and the (simpler) toolchain is just a little over 40 MB. You'll require at least 3 to 4 GB for working space. Further, if you build the images as Deb packages (we cover this shortly), it will require at least 1 GB of disk space (it's really best to keep around 7 to 8 GB free disk space to be safe).

2. Download the Raspberry Pi-specific kernel source tree (we clone it from the official source, the Raspberry Pi GitHub repository for the kernel tree, here: `https://github.com/raspberrypi/linux/`):

   ```
   cd ${RPI_STG}/kernel_rpi
   ```

```
git clone --depth=1 --branch=rpi-6.1.y \
  https://github.com/raspberrypi/linux.git
```

The kernel source tree gets cloned under a directory called `linux/` (that is, under `${RPI_STG}/kernel_rpi/linux`). It can take a while. Notice how, in the preceding command, we have the following:

- The particular Raspberry Pi kernel tree branch we have selected is not the very latest one (at the time of writing, the very latest is the rpi-6.8.y series), it's the 6.1 kernel; that's perfectly okay (6.1.y is an LTS kernel and matches our x86 one as well!).
- We pass the `--depth` parameter set to `1` to `git clone` to reduce the download and uncompress loads.

Now the Raspberry Pi kernel source tree is installed. Let's briefly verify this:

```
$ cd ${RPI_STG}/kernel_rpi/linux ; head -n5 Makefile
# SPDX-License-Identifier: GPL-2.0
VERSION = 6
PATCHLEVEL = 1
SUBLEVEL = 34
EXTRAVERSION =
```

Okay, it's the 6.1.34 Raspberry Pi kernel port. (The kernel version we use on the x86_64 is the 6.1.25 one; the slight variation is fine. Also, the release version (y=34) could change by the time you try it out, which again is fine.)

Step 2 – installing an x86_64-to-AArch64 cross-toolchain

Now it's time to install a *cross-toolchain* on your host system (recall, it's our x86 Ubuntu VM) that's appropriate for performing the actual build. The thing is, there are several working toolchains available. Here, I shall employ the easiest and best way of obtaining and installing an appropriate toolchain for the task at hand. (FYI, the first edition of this book specified a second more complex way, via the Raspberry Pi "tools" GitHub repo; that is now considered deprecated, thus we won't delve into it.)

Modern distros typically provide ready-to-use *cross-build* (or cross-compile) packages! To see a subset of them (just the GCC compiler packages for various arch's) on (x86) Debian/Ubuntu, type in the following and press the *Tab* key twice (to auto-complete):

```
sudo apt install gcc-<TAB><TAB>
Display all 398 possibilities? (y or n) y
gcc-10                                                     gcc-12-plugin-dev-alpha-
linux-gnu
gcc-10-aarch64-linux-gnu                      gcc-12-plugin-dev-arm-linux-gnueabi
gcc-10-aarch64-linux-gnu-base          gcc-12-plugin-dev-arm-linux-gnueabihf
gcc-10-alpha-linux-gnu                          gcc-12-plugin-dev-hppa-linux-gnu
gcc-10-alpha-linux-gnu-base                gcc-12-plugin-dev-i686-linux-gnu
gcc-10-arm-linux-gnueabi                      gcc-12-plugin-dev-m68k-linux-gnu
[ … ]
```

Nice! So many toys to play with. We'll also require the `binutils` package; so, let's install both the relevant ones:

```
$ sudo apt install gcc-aarch64-linux-gnu binutils-aarch64-linux-gnu
```

The cross tools are typically installed under `/usr/bin/` and are, therefore, already part of your `PATH`; you now can simply use them. For example, check out the AArch64 `GCC` compiler's location and version info, as follows:

```
$ aarch64-linux-gnu-
aarch64-linux-gnu-addr2line      aarch64-linux-gnu-gcc-nm-11        aarch64-linux-gnu-ld.bfd
aarch64-linux-gnu-ar             aarch64-linux-gnu-gcc-ranlib       aarch64-linux-gnu-ld.gold
aarch64-linux-gnu-as             aarch64-linux-gnu-gcc-ranlib-11    aarch64-linux-gnu-lto-dump-11
aarch64-linux-gnu-c++filt        aarch64-linux-gnu-gcov             aarch64-linux-gnu-nm
aarch64-linux-gnu-dwp            aarch64-linux-gnu-gcov-11          aarch64-linux-gnu-objcopy
aarch64-linux-gnu-elfedit        aarch64-linux-gnu-gcov-dump        aarch64-linux-gnu-objdump
aarch64-linux-gnu-gcc            aarch64-linux-gnu-gcov-dump-11     aarch64-linux-gnu-ranlib
aarch64-linux-gnu-gcc-11         aarch64-linux-gnu-gcov-tool        aarch64-linux-gnu-readelf
aarch64-linux-gnu-gcc-ar         aarch64-linux-gnu-gcov-tool-11     aarch64-linux-gnu-size
aarch64-linux-gnu-gcc-ar-11      aarch64-linux-gnu-gprof            aarch64-linux-gnu-strings
aarch64-linux-gnu-gcc-nm         aarch64-linux-gnu-ld               aarch64-linux-gnu-strip
$ aarch64-linux-gnu-^C
$
$ aarch64-linux-gnu-gcc -v
Using built-in specs.
COLLECT_GCC=aarch64-linux-gnu-gcc
COLLECT_LTO_WRAPPER=/usr/lib/gcc-cross/aarch64-linux-gnu/11/lto-wrapper
Target: aarch64-linux-gnu
Configured with: ../src/configure -v --with-pkgversion='Ubuntu 11.3.0-1ubuntu1~22.04.1' --with-bugurl=file:///usr/share
/doc/gcc-11/README.Bugs --enable-languages=c,ada,c++,go,d,fortran,objc,obj-c++,m2 --prefix=/usr --with-gcc-major-versio
n-only --program-suffix=-11 --enable-shared --enable-linker-build-id --libexecdir=/usr/lib --without-included-gettext -
-enable-threads=posix --libdir=/usr/lib --enable-nls --with-sysroot=/ --enable-clocale=gnu --enable-libstdcxx-debug --e
nable-libstdcxx-time=yes --with-default-libstdcxx-abi=new --enable-gnu-unique-object --disable-libquadmath --disable-li
bquadmath-support --enable-plugin --enable-default-pie --with-system-zlib --enable-libphobos-checking=release --without
-target-system-zlib --enable-multiarch --enable-fix-cortex-a53-843419 --disable-werror --enable-checking=release --buil
d=x86_64-linux-gnu --host=x86_64-linux-gnu --target=aarch64-linux-gnu --program-prefix=aarch64-linux-gnu- --includedir=
/usr/aarch64-linux-gnu/include --with-build-config=bootstrap-lto-lean --enable-link-serialization=2
Thread model: posix
Supported LTO compression algorithms: zlib zstd
gcc version 11.3.0 (Ubuntu 11.3.0-1ubuntu1~22.04.1)
$
$ which aarch64-linux-gnu-gcc
/usr/bin/aarch64-linux-gnu-gcc
$
```

Figure 3.8: Screenshot showing the AArch64 cross-compiler packages have been successfully installed (in /usr/bin)

Figure 3.8 shows us the tools in the toolchain! Notice how they're all installed under `/usr/bin/` and, more importantly, all of them are prefixed with `aarch64-linux-gnu-`. This is called the *cross-compiler or toolchain prefix* and is typically placed in the environment variable named `CROSS_COMPILE` (we'll get to this shortly).

Does the cross-compiler prefix mean anything? Yes, there is a naming convention followed, and it's of the following form:

`MACHINE-VENDOR-OS-`

or

`<CPU>-<MANUFACTURER>[-<KERNEL>]-<OS>-`

The first form is called the "target triplet" and it's the one our cross-toolchain prefix (`aarch64-linux-gnu-`) is using (as it has three components in the prefix, not four). So, here, we have `MACHINE as aarch64` (which, of course, implies that the target CPU is an ARM 64-bit), the `VENDOR` is simply set to `linux`, and the `OS` field is set to `gnu` (it's not perfect; you can think of the `VENDOR` field as being empty and the `OS` field as the value `linux-gnu`).

Also, do keep in mind that this toolchain is appropriate for building the kernel for the AArch64 (ARM 64-bit) architecture, not for 32-bit. If that's your intention, you will need to install an x86_64-to-AArch32 toolchain with `sudo apt install gcc-arm-linux-gnueabihf binutils-arm-linux-gnueabihf` (the `hf` suffix in the toolchain package name is for "hard float," as the Raspberry Pi target processor uses one).

As mentioned earlier, alternate toolchains can be used as well. For example, several toolchains for ARM development (for GNU/Linux, Windows, and macOS hosts) are available on the ARM developer site at `https://developer.arm.com/downloads/-/arm-gnu-toolchain-downloads`.

Step 3 – configuring and building the Raspberry Pi AArch64 kernel

Right, let's now configure the kernel (for the Raspberry Pi 4). Before we begin, it's very important to keep the following in mind:

- The `ARCH` environment variable must be set to the CPU – the *architecture* – for which the software is to be cross-compiled (that is, the compiled code will run on that CPU). The precise value to set `ARCH` to is the name of the directory under the `arch/` directory in the kernel source tree. So, for example, you'd set `ARCH` to `arm` for ARM-32, to `arm64` for the AArch64, to `powerpc` for the PowerPC, and to `openrisc` for the OpenRISC processor.
- The `CROSS_COMPILE` environment variable must be set to the cross-compiler (toolchain) prefix. Essentially, it's the first few common letters that precede every utility in the toolchain. In our example, all the toolchain utilities (the C compiler `gcc`, linker, C++, `objdump`, and so on) begin with `aarch64-linux-gnu-` (see *Figure 3.8*), so that's what we set `CROSS_COMPILE` to. Also note that the `Makefile` will always invoke the utilities as `${CROSS_COMPILE}<utility>`, hence invoking the correct toolchain executable. This does imply that the toolchain directory should be within the `PATH` variable.

Lets go through the necessary steps now:

1. Okay, let's proceed by setting up the default kernel configuration:

```
cd ${RPI_STG}/kernel_rpi/linux
make mrproper
KERNEL=kernel8
make ARCH=arm64 bcm2711_defconfig
```

The result of the above is a default kernel config, saved of course as the file `.config`. A quick explanation regarding the configuration target, `bcm2711_defconfig`: this key point was mentioned in *Chapter 2, Building the 6.x Linux Kernel from Source – Part 1*, in the *Kernel config for typical embedded Linux systems* section. We must ensure that we use an appropriate board-specific kernel config file as a starting point. Here, this – `bcm2711_defconfig` – is the correct kernel config file for the Broadcom SoC (System on Chip) on the Raspberry Pi 3, 3+, 4, 400, Zero 2 W, and Raspberry Pi compute modules 3, 3+, and 4, in order to generate the default 64-bit build configuration.

(Important: if you are building the kernel for another type of Raspberry Pi device, please see `https://www.raspberrypi.org/documentation/linux/kernel/building.md`.)

FYI, the `kernel8` value is as such because the processor is ARMv8-based (64-bit).

The variety of SoCs, their packaging, and their resulting naming creates a good deal of confusion; this link might help: `https://raspberrypi.stackexchange.com/questions/840/why-is-the-cpu-sometimes-referred-to-as-bcm2708-sometimes-bcm2835`.

2. If any further customization of the board's kernel config is required, you could always do so with the following:

```
make ARCH=arm64 menuconfig
```

If not, just skip this step and proceed. Also, remember, if you want to fine-tune the kernel config, first generate it correctly via *Step 1*, and only then perform this step.

3. Build (cross-compile) the kernel, the kernel modules, and the DTBs with the following:

```
make -j8 ARCH=arm64 CROSS_COMPILE=aarch64-linux-gnu- all
```

4. Adjust the `-jn` appropriately for your build host. What targets will the `all` target result in? The answer is that it's all targets prefixed with the * symbol in the kernel 's top-level Makefile's `help` target:

```
$ make ARCH=arm64 help |grep "^\*"
```

```
* vmlinux        - Build the bare kernel
* modules        - Build all modules
* dtbs               - Build device tree blobs for enabled boards
* Image.gz        - Compressed kernel image (arch/arm64/boot/Image.gz)
```

Once the build is successfully completed, among the many files generated, we can see the following relevant ones:

```
$ ls -lh vmlinux System.map arch/arm64/boot/Image* arch/arm64/boot/dts/
broadcom/bcm2711-rpi-4-b.dtb
-rw-rw-r-- 1 c2kp c2kp  54K Jun 21 13:02 arch/arm64/boot/dts/broadcom/
bcm2711-rpi-4-b.dtb
-rw-rw-r-- 1 c2kp c2kp  22M Jun 21 13:24 arch/arm64/boot/Image
-rw-rw-r-- 1 c2kp c2kp 7.9M Jun 21 13:24 arch/arm64/boot/Image.gz
-rw-rw-r-- 1 c2kp c2kp 3.6M Jun 21 13:24 System.map
-rwxrwxr-x 1 c2kp c2kp 237M Jun 21 13:24 vmlinux
```

5. Great! FYI, the file named `Image` is an uncompressed version of `Image.gz`, the compressed kernel image, the latter being the one you can boot with. The `vmlinux` file is the actual uncompressed kernel image; keep it around for kernel debugging! Also, the `arch/arm64/boot/dts/broadcom/bcm2711-rpi-4-b.dtb` file is the DTB for this target platform. The `file` command further verifies the `vmlinux` image:

```
$ file ./vmlinux
./vmlinux: ELF 64-bit LSB pie executable, ARM aarch64, version 1 (SYSV),
statically linked, BuildID[sha1]=03b7<...>, with debug_info, not stripped
```

Here, our purpose has been to show how a Linux kernel can be configured and built for an architecture other than the host system it's compiled upon, or, in other words, cross-compiled. The details of placing the kernel image, modules, and DTBs (including overlays) on the microSD card and so on aren't delved into here. I refer you to the complete documentation for the Raspberry Pi kernel build, which can be found here: `https://www.raspberrypi.com/documentation/computers/linux_kernel.html#cross-compiling-the-kernel`.

This completes our coverage on experimenting with a kernel cross-compilation for the Raspberry Pi. We'll end this chapter with a few miscellaneous but useful tips.

Miscellaneous tips on the kernel build

We complete this chapter on building the Linux kernel from source with a few tips. The reason, of course, is that everything may not always go as smoothly as we'd wish it to. Each of the following subsections encapsulates a tip for you to take note of.

Often a point of confusion for folks new to this: once we configure, build, and boot from a new Linux kernel, we notice that the **root filesystem** and any other mounted filesystems remain identical to what was on the original (distro or custom) system. Only the kernel (and the initramfs image) has changed.

This is entirely intentional, due to the Unix paradigm of having *loose coupling* between the kernel and the (real) root filesystem. Since it's the root filesystem that holds all the applications, system tools, and utilities, including libraries, in effect, we can have several kernels (to suit different product flavors, perhaps) for the same base system.

Minimum version requirements

To successfully build the kernel, you must ensure that your build system has the documented *bare minimum* versions of the various software pieces of the toolchain (and other miscellaneous tools and utilities). This information is clearly within the kernel documentation in the *Minimal requirements to compile the kernel* section, available at `https://github.com/torvalds/linux/blob/master/Documentation/process/changes.rst#minimal-requirements-to-compile-the-kernel`.

For example, as of the time of writing, the recommended minimum version of `gcc` is 5.1, Clang (an alternate and powerful compiler for Linux!) is 11.0.0, and that of `make` is 3.82 (there are many others specified, so take a look).

Building a kernel for another site

In our kernel build walk-through in this book, we built a Linux kernel on a certain system (here, it was an x86_64 guest) and booted the newly built kernel off the very same system. What if this isn't the case, as will often happen when you are building a kernel for another site or customer premises? While it's always possible to manually put the pieces in place on the remote system, there's a far easier and more correct way to do it – package the kernel and associated artifacts bundled along with it (the `initrd` image, the kernel modules collection, the kernel headers, and so on) into a well-known **package format** (Debian's `deb`, Red Hat's `rpm`, and so on)! A quick `help` command on the kernel's top-level `Makefile` reveals these package targets (we only show the relevant ones below):

```
$ make ARCH=arm64 help
 [ ... ]

Kernel packaging:
  rpm-pkg             - Build both source and binary RPM kernel packages
  binrpm-pkg          - Build only the binary kernel RPM package
  deb-pkg             - Build both source and binary deb kernel packages
  bindeb-pkg          - Build only the binary kernel deb package
  snap-pkg            - Build only the binary kernel snap package
                        (will connect to external hosts)
  dir-pkg             - Build the kernel as a plain directory structure
  tar-pkg             - Build the kernel as an uncompressed tarball
  targz-pkg           - Build the kernel as a gzip compressed tarball
[ ... ]
```

So, for example, with the same setup just covered (building for the AArch64 Raspberry Pi 4 target), to build and package the kernel and its associated files as Debian packages, simply do the following:

```
$ make -j8 ARCH=arm64 CROSS_COMPILE=aarch64-linux-gnu- bindeb-pkg
sh ./scripts/package/mkdebian
dpkg-buildpackage -r"fakeroot -u" -a$(cat debian/arch)  -b -nc -uc
dpkg-buildpackage: info: source package linux-upstream
dpkg-buildpackage: info: source version 6.1.34-v8+-2
dpkg-buildpackage: info: source distribution jammy
[ … ]
make KERNELRELEASE=6.1.34-v8+ ARCH=arm64        KBUILD_BUILD_VERSION=2 -f ./
Makefile
  CALL    scripts/checksyscalls.sh
  UPD     init/utsversion-tmp.h
  CC      init/version.o
[ … ]
  GZIP    arch/arm64/boot/Image.gz
make KERNELRELEASE=6.1.34-v8+ ARCH=arm64        KBUILD_BUILD_VERSION=2 -f ./
Makefile intdeb-pkg
sh ./scripts/package/builddeb
  INSTALL debian/linux-image/usr/lib/linux-image-6.1.34-v8+/broadcom/bcm2711-
rpi-400.dtb
  INSTALL debian/linux-image/usr/lib/linux-image-6.1.34-v8+/broadcom/bcm2711-
rpi-4-b.dtb
[ … ]
 INSTALL debian/linux-image/usr/lib/linux-image-6.1.34-v8+/overlays/wm8960-
soundcard.dtbo
  INSTALL debian/linux-image/lib/modules/6.1.34-v8+/kernel/arch/arm64/crypto/
aes-arm64.ko
  INSTALL debian/linux-image/lib/modules/6.1.34-v8+/kernel/arch/arm64/crypto/
aes-neon-blk.ko
[ … ]
 XZ      debian/linux-image/lib/modules/6.1.34-v8+/kernel/arch/arm64/crypto/
aes-arm64.ko.xz
  XZ      debian/linux-image/lib/modules/6.1.34-v8+/kernel/arch/arm64/crypto/
aes-neon-bs.ko.xz
[ … ]
 DEPMOD  debian/linux-image/lib/modules/6.1.34-v8+
dpkg-deb: building package 'linux-headers-6.1.34-v8+' in '../linux-headers-
6.1.34-v8+_6.1.34-v8+-2_arm64.deb'.
  HOSTCC  scripts/unifdef
  HDRINST usr/include/asm-generic/stat.h
[ … ]
  INSTALL debian/linux-libc-dev/usr/include
```

```
dpkg-deb: building package 'linux-libc-dev' in '../linux-libc-dev_6.1.34-v8+-2_
arm64.deb'.
dpkg-deb: building package 'linux-image-6.1.34-v8+' in '../linux-image-6.1.34-
v8+_6.1.34-v8+-2_arm64.deb'.
dpkg-deb: building package 'linux-image-6.1.34-v8+-dbg' in '../linux-image-
6.1.34-v8+-dbg_6.1.34-v8+-2_arm64.deb'.
 dpkg-genbuildinfo --build=binary -O../linux-upstream_6.1.34-v8+-2_arm64.
buildinfo
 dpkg-genchanges --build=binary -O../linux-upstream_6.1.34-v8+-2_arm64.changes
dpkg-genchanges: info: binary-only upload (no source code included)
 dpkg-source --after-build .
dpkg-buildpackage: info: binary-only upload (no source included)
$
```

A folder named debian is created and comprehensively populated within the kernel tree (use the *tree* command to see its content). The actual package files are written into the directory immediately above the kernel source directory. For example, from the command we just ran, here are the deb packages that were generated:

```
$ pwd
/home/c2kp/rpi_work/kernel_rpi/linux
$ ls -lh ../*.deb
-rw-r--r-- 1 c2kp c2kp 8.3M Jun 21 13:38 ../linux-headers-6.1.34-v8+_6.1.34-v8+-2_arm64.deb
-rw-r--r-- 1 c2kp c2kp 312M Jun 21 13:39 ../linux-image-6.1.34-v8+_6.1.34-v8+-2_arm64.deb
-rw-r--r-- 1 c2kp c2kp  74M Jun 21 13:41 ../linux-image-6.1.34-v8+-dbg_6.1.34-v8+-2_arm64.deb
-rw-r--r-- 1 c2kp c2kp 1.3M Jun 21 13:38 ../linux-libc-dev_6.1.34-v8+-2_arm64.deb
$
```

Figure 3.9: The Debian package files generated via our make … bindeb-pkg command

This is indeed very convenient! Now, you can install the packages – and thus the new kernel along with its headers and associated artifacts – on any other matching (in terms of CPU and Linux flavor) system with a simple sudo dpkg -i <package-name> command.

Watching the kernel build run

To see details – the commands and scripts that execute, the detailed GCC compiler flags, and so on – while the kernel build runs, pass the V=1 verbose option switch to make. The following is a fragment of the output when building the Raspberry Pi 4 kernel with the verbose switch set to *on (I've taken the liberty of adding newlines and continuation characters to make the output more readable)*:

```
$ make -j8 V=1 ARCH=arm64 CROSS_COMPILE=aarch64-linux-gnu- all
[ … ]
aarch64-linux-gnu-gcc -Wp,-MMD,drivers/net/ethernet/realtek/.r8169_main.o.d
-nostdinc \
-I./arch/arm64/include -I./arch/arm64/include/generated  -I./include -I./arch/
arm64/include/uapi -I./arch/arm64/include/generated/uapi -I./include/uapi
```

```
-I./include/generated/uapi -include ./include/linux/compiler-version.h -include
./include/linux/kconfig.h -include ./include/linux/compiler_types.h \

-D__KERNEL__ -mlittle-endian -DCC_USING_PATCHABLE_FUNCTION_ENTRY
-DKASAN_SHADOW_SCALE_SHIFT= -fmacro-prefix-map=./= \

-Wall -Wundef -Werror=strict-prototypes -Wno-trigraphs -fno-strict-aliasing
-fno-common -fshort-wchar -fno-PIE -Werror=implicit-function-declaration
-Werror=implicit-int -Werror=return-type -Wno-format-security \

-std=gnu11 -mgeneral-regs-only -DCONFIG_CC_HAS_K_CONSTRAINT=1 -Wno-psabi
-mabi=lp64 -fno-asynchronous-unwind-tables -fno-unwind-tables -mbranch-
protection=pac-ret+leaf \

-Wa,-march=armv8.5-a -DARM64_ASM_ARCH='"armv8.5-a"'
-DKASAN_SHADOW_SCALE_SHIFT= -fno-delete-null-pointer-checks -Wno-frame-address
-Wno-format-truncation -Wno-format-overflow -Wno-address-of-packed-member -O2
-fno-allow-store-data-races -Wframe-larger-than=2048 -fstack-protector-strong
-Wno-main -Wno-unused-but-set-variable -Wno-unused-const-variable
-fno-omit-frame-pointer -fno-optimize-sibling-calls -fno-stack-clash-protection
-fpatchable-function-entry=2 \

-Wdeclaration-after-statement -Wvla -Wno-pointer-sign
-Wcast-function-type -Wno-stringop-truncation -Wno-stringop-overflow -Wno-
restrict
-Wno-maybe-uninitialized -Wno-array-bounds -Wno-alloc-size-larger-than
-Wimplicit-fallthrough=5 -fno-strict-overflow -fno-stack-check -fconserve-stack
-Werror=date-time -Werror=incompatible-pointer-types -Werror=designated-init
-Wno-packed-not-aligned \

-g -mstack-protector-guard=sysreg -mstack-protector-guard-reg=sp_el0
-mstack-protector-guard-offset=1432  -DMODULE
-DKBUILD_BASENAME='"r8169_main"' -DKBUILD_MODNAME='"r8169"'
-D__KBUILD_MODNAME=kmod_r8169 \

-c -o drivers/net/ethernet/realtek/r8169_main.o drivers/net/ethernet/realtek/
r8169_main.c
[ … ]
```

This level of detail can help debug situations where the build fails, or to learn what compiler options are being leveraged! (The GCC man page comes in use now!)

Also, of course, the output can be simultaneously seen on the terminal window and redirected to a file by leveraging the shell and `tee`; for example, you can do this to save it into a file named `out.txt` while seeing the output as well:

```
make -j8 V=1 ARCH=arm64 CROSS_COMPILE=aarch64-linux-gnu- all 2>&1 | tee out.txt
```

Furthermore, and FYI, using `tee -a <filename>` ensures the output is appended to the file specified and doesn't overwrite it.

A shortcut shell syntax to the build procedure

A shortcut shell (Bash, typically) syntax to the build procedure – assuming the required packages are installed and the kernel configuration step is done (and assuming we're on the x86) – could be something like the following example, to be used in non-interactive build scripts, perhaps:

```
time make -j8 [ARCH=<...> CROSS_COMPILE=<...>] all && \
 sudo make modules_install && \
 sudo make install
```

In the preceding code, the && and || elements are the shell's (Bash's) convenience *conditional list* syntax:

- `cmd1 && cmd2` implies : run `cmd2` only if `cmd1` succeeds.
- `cmd1 || cmd2` implies : run `cmd2` only if `cmd1` fails.

In any case, a simple minimally-tested Bash script to perform the x86_64 kernel build is provided in the book repo here: `https://github.com/PacktPublishing/Linux-Kernel-Programming_2E/blob/main/ch3/kbuild.sh`. Check it out.

Dealing with missing OpenSSL development headers

In one instance, the kernel build on x86_64 on an Ubuntu box failed with the following error:

```
[...] fatal error: openssl/opensslv.h: No such file or directory
```

This is simply a case of missing OpenSSL development headers; this is mentioned in the *Minimal requirements to compile the kernel* document here: `https://github.com/torvalds/linux/blob/master/Documentation/process/changes.rst#openssl`. Specifically, it mentions that from v4.3 (and 3.7 if module signing is enabled) and higher, the `openssl` development packages are required. Thus, our `ch1/pkg_install4ubuntu_lkp.sh` installs the `libssl-dev` package; FYI, the equivalent package for the Fedora-type distros is `openssl-devel`.

FYI, this Q&A also shows how the installation of the `libssl-dev` package (or equivalent package that needs to be installed) solves the issue: *OpenSSL missing during ./configure. How to fix*, available at `https://superuser.com/questions/371901/openssl-missing-during-configure-how-to-fix`.

How can I check which distro kernels are installed?

Sometimes it's important to know which kernel packages (typically installed by the distro being used) are installed. This could be because, sometimes, you want to uninstall older kernels to recover disk space, in the /boot partition perhaps, as it's running short (as often happens on my OSBoxes install!). On Debian/Ubuntu, this is one way to look them up:

```
dpkg --list | grep linux-image
```

On Red Hat/Fedora/CentOS, you could use:

```
dnf list installed kernel
```

These are package-based kernels; they won't include manually built ones of course!

Several Linux "builder" projects exist, which are elaborate frameworks for building a Linux system or distribution in its entirety (typically used for embedded Linux projects). As of the time of writing, Yocto (https://www.yoctoproject.org/) is considered the industry standard Linux-builder project, with Buildroot (https://buildroot.org/) being an older but very much supported one; they are indeed well worth checking out.

Next, what if the kernel build, initramfs generation, and everything else goes well but the kernel fails to complete booting, ending with kernel panic and a (well-known to the initiated!) message like `[ end Kernel panic - not syncing: Attempted to kill init! …]` or `[Kernel panic - not syncing: No working init found. …]`. The root cause of such mishaps is often a misconfiguration of the kernel config, the kernel command-line parameters (especially that of the root device), or an inability (for whatever reason) to mount the actual root filesystem. The official kernel documentation does have some tips here: https://www.kernel.org/doc/Documentation/admin-guide/init.rst.

Finally, remember an almost guaranteed way to succeed: when you get those build and/or boot errors that you *just cannot fix*, copy the exact error message into the clipboard, go to Google (or another search engine), and type something akin to `linux kernel build <ver ...> fails with <paste-your-error-message-here>`. You might be surprised at how often this helps. If not, diligently do your research, and if you really cannot find any relevant/correct answers, post your well-thought-out-with-all-relevant-details question on an appropriate forum.

Well, with this, we conclude our kernel build chapters!

Summary

This chapter, along with the previous one, covered in a lot of detail the necessary preliminaries and how exactly to configure and build the Linux kernel from source.

In this chapter, we began with the actual kernel (and kernel modules') build process. Once built, we showed how the kernel modules are to be installed onto the system. We then moved on to the practicalities of generating the initramfs (or initrd) image and went on to explain the motivation behind it.

The final step in the kernel build was the (simple) customization of the bootloader (here, we focused only on x86 GRUB). We then showed how to boot the system via the newly baked kernel and verify that its configuration is as we expected. As a useful add-on, we then showed how we can even cross-compile the Linux kernel for another processor (AArch64, in this instance). Finally, we shared some additional tips to help you with the kernel build.

Again, if you haven't done so already, we urge you to carefully review and try out the procedures mentioned here and build your own custom Linux kernel.

So, congratulations on completing a Linux kernel build from scratch! You might well find that for an actual project (or product), you may *not* have to carry out each and every step of the kernel build procedure as we have tried hard to carefully show. Why? Well, one reason is that there might be a separate platform/BSP team that works on this aspect; another reason – increasingly likely, especially on embedded Linux projects – is that a Linux-builder framework such as *Yocto* (or *Buildroot*) is being used (or SoC vendor-supplied custom kernels are what you use). Yocto will typically take care of the mechanical aspects of the build (Yocto doesn't exactly have the gentlest learning curve, though). *However*, you need to be able to *configure* the kernel as required by the project; that still requires the knowledge and understanding gained here.

The next two chapters will take you squarely into the world of Linux kernel development, showing you how to write your first kernel module.

Questions

As we conclude, here is a list of questions for you to test your knowledge regarding this chapter's material: `https://github.com/PacktPublishing/Linux-Kernel-Programming_2E/blob/main/questions/ch3_qs_assignments.txt`. You will find some of the questions answered in the book's GitHub repo: `https://github.com/PacktPublishing/Linux-Kernel-Programming/tree/master/solutions_to_assgn`.

Further reading

To help you delve deeper into the subject with useful materials, we provide a rather detailed list of online references and links (and at times, even books) in a *Further reading* document in this book's GitHub repository. The *Further reading* document is available here: `https://github.com/PacktPublishing/Linux-Kernel-Programming/blob/master/Further_Reading.md`.

Learn more on Discord

To join the Discord community for this book – where you can share feedback, ask questions to the author, and learn about new releases – follow the QR code below:

`https://packt.link/SecNet`

4

Writing Your First Kernel Module – Part 1

In the previous two chapters, you have learned the ins and outs of obtaining the kernel source tree and configuring and building it (for the x86). Now, welcome to your journey of learning about a fundamental aspect of Linux kernel development – the **Loadable Kernel Module** (**LKM**) framework – and how it is to be used by the *module user or module author*, who is typically a kernel or device driver programmer. This topic is rather vast and hence is split into two chapters – this one and the next.

In this chapter, we'll begin by taking a quick look at the basics of the Linux kernel architecture, which will help us understand the LKM framework. Then, we'll look into why kernel modules are useful and write, build, and run our simple *Hello, world* LKM. We'll see how messages are written to the kernel log and understand and make use of the LKM Makefile. By the end of this chapter, you will have learned the basics of Linux kernel architecture and the LKM framework, applying it to write a simple yet complete piece of kernel code.

In this chapter, we will cover the following recipes:

- Understanding the kernel architecture – part 1
- Exploring LKMs
- Writing our very first kernel module
- Common operations on kernel modules
- Understanding kernel logging and printk
- Understanding the basics of a kernel module Makefile

Technical requirements

If you have already carefully followed *Online Chapter, Kernel Workspace Setup*, then a portion of the technical prerequisites that follow will already be taken care of. (The first chapter also mentions various useful open-source tools and projects; I definitely recommend that you browse through it at least once.) For your convenience, we summarize some key points here.

To build and use an external (or out-of-tree) kernel module on a Linux distribution (or custom system), you need:

- The kernel to be built with module support enabled (`CONFIG_MODULES=y`)
- At a minimum, the following two components to be installed:
    - **A toolchain:** This includes the compiler, assembler, linker/loader, C library, and various other bits and pieces. If building for the local system, as we assume for now, then any modern Linux distribution will have a native toolchain pre-installed. If not, simply installing the GCC package for your distribution should be sufficient; on an Ubuntu- or Debian-based Linux system, use this:

    ```
    sudo apt install gcc
    ```

    - **Kernel headers:** These headers will be used during the compilation of the LKM. In reality, you install a package geared to not only install the kernel headers but also other required bits and pieces (such as the kernel Makefile) onto the system. Again, any modern Linux distribution will/should have the kernel header pre-installed. If not (you can check using the `dpkg` utility, as shown here), simply install the package for your distribution; on an Ubuntu- or Debian-based Linux system, use this:

    ```
    $ sudo apt install linux-headers-generic
    $ dpkg -l | grep linux-headers | awk '{print $1, $2}'
    ii linux-headers-5.19.0-45-generic
    ii linux-headers-generic-hwe-22.04
    ```

Here, the second command using the `dpkg` utility is simply used to verify that the `linux-headers` packages are indeed installed.

This package may be named `kernel-headers-<ver#>` on some distributions. Also, for development directly on a Raspberry Pi, install the relevant kernel headers package named `raspberrypi-kernel-headers`.

Just in case you can't build modules, try this: prepare the kernel for external module builds by running `make modules_prepare` in the root of your kernel source tree. With typical modern distros, you should not need to do this; module building should just work.

The entire source tree for this book is available in its GitHub repository at `https://github.com/PacktPublishing/Linux-Kernel-Programming_2E`; we expect you to clone it:

```
git clone \
  https://github.com/PacktPublishing/Linux-Kernel-Programming_2E.git
```

The code for this chapter is under its directory namesake, `chn` (where `n` is the chapter number; so here, it's under `ch4/`).

Understanding the kernel architecture – part 1

In this section, we begin to deepen our understanding of the Linux kernel. More specifically, here, we delve into what user and kernel spaces are, and the major subsystems and various components that make up the kernel. This information is dealt with at a higher level of abstraction for now and is deliberately kept brief. We shall delve a lot deeper into understanding the fabric of the kernel in *Chapter 6, Kernel Internals Essentials – Processes and Threads.*

User space and kernel space

Modern microprocessors support code execution at a minimum of two privilege levels. As a real-world example, the Intel/AMD x86[-64] family supports four privilege levels (they call them *ring levels*), the AArch32 (ARM-32) microprocessor family supports up to seven modes (ARM calls them *execution modes*; six are privileged and one is non-privileged), and the AArch64 (ARM-64/ARMv8) microprocessor family supports four exception levels (EL0 to EL3, with EL0 being the least and EL3 being the most privileged).

The key point here is that for security and stability on the platform, all modern operating systems running on these processors will make use of (at least) two of the privilege levels (or modes) of execution, leading to the separation of the **Virtual Address Space** (**VAS**) into two clearly distinguished (virtual) **address spaces**:

User space: For *applications* to run in *unprivileged user mode*. All apps – processes and threads – will execute at this privilege in this space. So, right now, you're perhaps using a browser, editor, PDF reader, terminal, email client, and so on. They're all apps – ultimately, processes and threads; on Linux, they all execute in user space in user mode, which is unprivileged. We shall soon come to what exactly is meant by privileged versus unprivileged.

Kernel space: For the *kernel* (and all its components) to run in *privileged mode – kernel mode*. This is the domain of the OS and whatever's within it (like drivers, networking, I/O, and so on, including kernel modules). They all run with OS privilege; in effect, they can do anything they like! Note carefully that this privilege level is a hardware feature, and it's not the same as running as root or not (that's a pure software artifact); in many cases, running with kernel privilege can be thought of as effectively running as root.

The following figure shows this basic architecture:

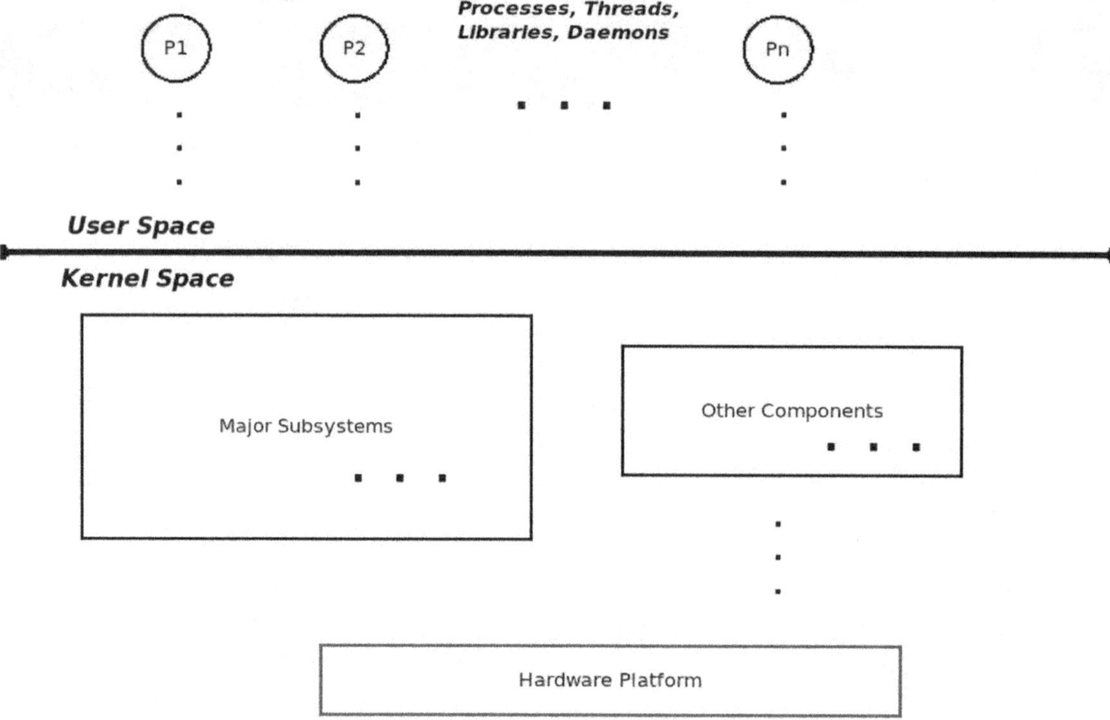

Figure 4.1: Basic architecture – two address spaces, two privilege modes

A few details on the Linux system architecture follow; do read on.

Library and system call APIs

User space applications often rely on **Application Programming Interfaces** (**APIs**) to perform their work. A *library* is essentially a collection or archive of APIs, allowing you to use a standardized, well-written, and well-tested interface (and leverage the usual benefits: not having to reinvent the wheel, portability, standardization, and so on). Linux systems have several libraries: even hundreds on enterprise-class systems are not uncommon. Of these, *all* user mode Linux applications (executables) are "auto-linked" into one important, always-used library: `glibc` – *the GNU standard C library*, as you shall learn. However, libraries are only ever available in user mode; the kernel does not use these user mode libraries (more on this in the following chapter).

Examples of library APIs are the well-known `printf(3)`, `scanf(3)`, `strcmp(3)`, `malloc(3)`, and `free(3)`. (Recall, from *Online Chapter*, *Kernel Workspace Setup*, the *Using the Linux man pages* section.)

Now, a key point: if the user and kernel are separate address spaces and at differing privilege levels, how can a user process – which, as we just learned, is confined to user space – *access* the kernel? The short answer is: *via system calls*. A **system call** is a special API, in the sense that it is the only legal (synchronous) way for user space processes (or threads) to access the kernel. In other words, system calls are the *only legal entry point into the kernel space*.

They (system calls) have the built-in ability to *switch* from non-privileged user mode to privileged kernel mode (more on this and the monolithic design in *Chapter 6, Kernel Internals Essentials – Processes and Threads*, under the *Process and interrupt contexts* section). Examples of system calls include `fork(2)`, `execve(2)`, `open(2)`, `read(2)`, `write(2)`, `socket(2)`, `accept(2)`, `chmod(2)`, and so on.

Look up all library and system call APIs in the man pages online:

Library APIs, man section 3: `https://linux.die.net/man/3/`

System call APIs, man section 2: `https://linux.die.net/man/2/`

The point being stressed here is that it's *only* via system calls that user applications (processes and threads) and the kernel communicate – that is the interface. In this book, we do not delve further into these details. If you are interested in knowing more, please refer to my earlier book *Hands-On System Programming with Linux*, Packt (specifically *Chapter 1, Linux System Architecture*).

This book is on kernel development, let's begin to check out the various components of the kernel.

Kernel space components

This book focuses entirely on the kernel space, of course. The Linux kernel today is a rather large and complex beast. Internally, it consists of a few major subsystems and several components. A broad enumeration of kernel subsystems and components yields the following list:

- **Core kernel:** This code handles the typical hardcore work of any modern operating system, including (user and kernel) process and thread creation/destruction, CPU scheduling, synchronization primitives, signaling, timers, interrupt handling, namespaces, cgroups, module support, crypto, and more.
- **Memory Management (MM):** This handles all memory-related work, including the setup and maintenance of kernel and process **Virtual Address Spaces (VASs)**.
- **VFS (for filesystem support):** The **Virtual Filesystem Switch (VFS)** is an abstraction layer over the actual filesystems implemented within the Linux kernel (for example, `ext[2|4]`, `vfat`, `ntfs`, `msdos`, `iso9660`, `f2fs`, `ufs`, and many more).
- **Block I/O:** The code paths implementing the actual file I/O, from the filesystems right down to the block device driver and everything in between (really, quite a lot!), are encompassed here.
- **Network protocol stack:** Linux is well known for its precise, to-the-letter-of-the-RFC, high-quality implementation of the well-known (and not-so-well-known) network protocols at all layers of the model, with TCP/IP being perhaps the most famous.
- **Inter-Process Communication (IPC) support:** The implementation of IPC mechanisms is done here; Linux supports message queues, shared memory, semaphores (both the older SysV and the newer POSIX ones), and other IPC mechanisms.
- **Sound support:** All the code that implements audio is here, from the firmware to drivers and codecs.

- **Virtualization support:** Linux has become extremely popular with large and small cloud providers alike, a big reason being its high-quality, low-footprint virtualization engine, **Kernel-based Virtual Machine (KVM)**.

All this forms the major kernel subsystems; in addition, we have these:

- Arch-specific (meaning CPU-specific) code
- Kernel initialization
- Security frameworks
- Many types of device drivers

Recall that in *Chapter 2, Building the 6.x Linux Kernel from Source – Part 1*, the *A brief tour of the kernel source tree* section gave the kernel source tree (code) layout corresponding to the major subsystems and other components.

It is a well-known fact that the Linux kernel follows the **monolithic kernel architecture** (as opposed to microkernel architecture). Essentially, a monolithic design is one in which all kernel components (that we mentioned in this section) live in and share the kernel VAS (or kernel segment). This can be clearly seen in the following diagram:

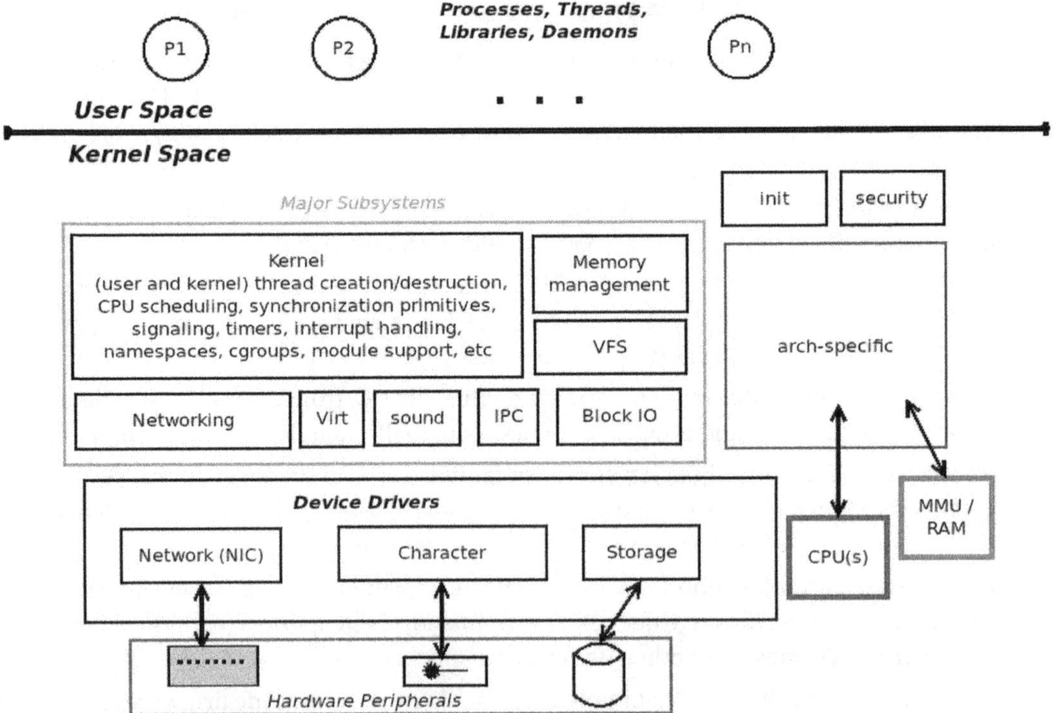

Figure 4.2: Linux kernel space – major subsystems and blocks

Another fact you should be aware of is that these address spaces are of course *virtual* address spaces and not physical. The kernel will *map*, at page granularity level, virtual pages to physical page frames leveraging hardware blocks such as the **Memory Management Unit** (**MMU**) and the processor plus **Translation Lookaside Buffer** (**TLB**) caches for gaining efficiencies. It does this by using a *master* kernel paging table to map kernel virtual pages to physical frames (RAM), and, for every single user space process that is alive, it maps the process's (user) virtual pages to physical page frames via individual paging tables for each process.

More in-depth coverage of the essentials of the kernel and memory management architecture and internals awaits you in *Chapter 6, Kernel Internals Essentials – Processes and Threads* (and more chapters that follow).

Now that we have a basic understanding of the user and kernel virtual address spaces, let's move on and begin our journey into the LKM framework.

Exploring LKMs

A kernel module is a means to provide kernel-level functionality without resorting to working within the kernel source tree and the static kernel image.

Visualize a scenario where you have to add a support feature to the Linux kernel – perhaps a new device driver in order to use a certain hardware peripheral chip, a new filesystem, or a new I/O scheduler. One way to do this is obvious: update the kernel source tree with the new code, and configure, build, test, and deploy it.

Though this may seem straightforward, it's a lot of work – every change in the code that we write, no matter how minor, will require us to rebuild the kernel image and then reboot the system in order to test it. There must be a cleaner, easier way; indeed, there is – *the LKM framework!*

The LKM framework

The LKM framework is a means to compile a piece of kernel code typically *outside of* the kernel source tree, often referred to as "out-of-tree" code, keeping it independent from the kernel in a limited sense, and then insert or *plug* in the generated "module object" *into* kernel memory, kernel VAS, have it run and perform its job, and then remove it (or *unplug* it) from kernel memory. (Note that the LKM framework can also be leveraged to generate *in-tree* kernel modules, as we did when we built the kernel. Here, we focus on the out-of-tree module).

The kernel module's source code, typically consisting of one or more C source files, header files, and a Makefile, is built (via `make` of course) into a *kernel module*. The kernel module itself is merely a binary object file and not a binary executable. In Linux 2.4 and earlier, the kernel module's filename had a `.o` suffix; on modern 2.6 Linux and later, it instead has a `.ko` (kernel object) suffix. Once built, you can insert this `.ko` file – the kernel module – into the live kernel at runtime, effectively making it a part of the kernel.

Note that not `all` kernel functionality can be provided via the LKM framework. Several core features, such as the core CPU (task) scheduler code, memory management, signaling, timer, interrupt management code paths, platform-specific drivers for driving pin controllers, clocks, and so on, can only be developed within the kernel itself. Similarly, a kernel module is only allowed access to a subset of the full kernel API and data variables; more on this later.

You might ask: how do I *insert* this module object into the kernel at runtime? Let's keep it simple – the answer is: via the `insmod` utility. For now, let's skip the details (these will be explained in the upcoming *Running the kernel module* section). The following figure provides an overview of first building and then inserting a kernel module into kernel memory:

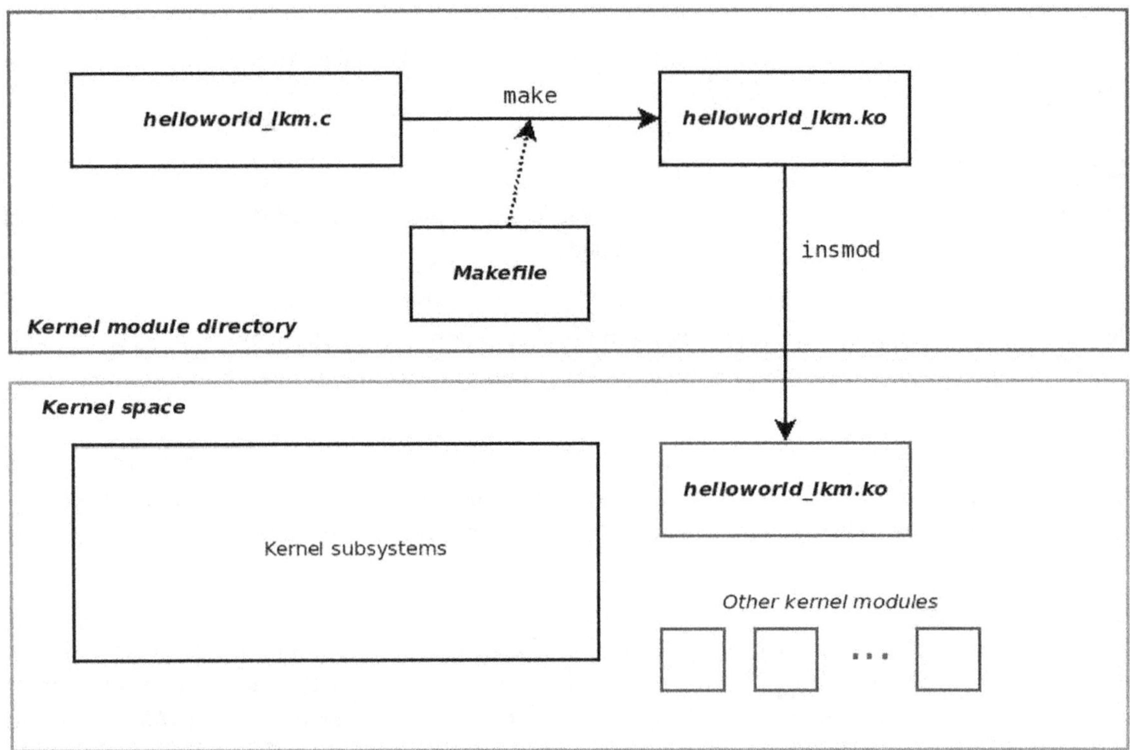

Figure 4.3: Building and then inserting a kernel module into kernel memory

Worry not: the actual code for both the kernel module C source as well as its Makefile is dealt with in deep detail in an upcoming section; for now, we want to gain a conceptual understanding only.

The kernel module is loaded into and lives in kernel memory – that is, the kernel VAS (the bottom half of *Figure 4.3*) in an area of space allocated for it by the kernel. Make no mistake, *it is kernel code and runs with kernel privilege*.

This way, you, the kernel (or driver) developer, do not have to reconfigure, rebuild, and reboot the system each time. All you have to do is edit the code of the kernel module, rebuild it, remove the old copy from memory (if it exists), and insert the new version. It saves time and increases productivity.

Having said that, in the real world, even kernel modules can (and do) cause system-level crashes, necessitating the need to reboot the system (another reason why running in an isolated VM is advantageous).

Another reason that kernel modules are advantageous is that they lend themselves to dynamic product configuration. For example, kernel modules can be designed to provide different features at differing price points; a script generating the final image for an embedded product could install a given set of kernel modules depending on the price the customer is willing to pay. Here's another example of how this technology is leveraged in a *debug* or troubleshooting scenario: a kernel module could be used to dynamically generate diagnostics and debug logs on an existing product. Technologies such as *Kprobes* and the like allow just this (my *Linux Kernel Debugging* book covers these in depth).

In effect, the LKM framework gives us a means of dynamically extending kernel functionality by allowing us to insert (and later remove) live code into (from) kernel memory. This ability to plug in and unplug kernel functionality at our whim makes us realize that the Linux kernel is not purely monolithic, it is also *modular*.

Kernel modules within the kernel source tree

In fact, the kernel module object isn't completely unfamiliar to us. In *Chapter 3, Building the 6.x Linux Kernel from Source – Part 2*, we built kernel modules as part of the kernel build process and had them installed.

Recall that these kernel modules are part of the kernel source and have been configured as modules by selecting `M` in the tristate kernel menu config prompt. They get installed in directories under `/lib/modules/$(uname -r)/`. So, to see a little bit regarding the kernel modules installed under our currently running x86_64 Ubuntu 22.04 LTS guest kernel, we can do this:

```
$ lsb_release -a 2>/dev/null |grep Description
Description:    Ubuntu 22.04.2 LTS
$ uname -r
5.19.0-45-generic
$ find /lib/modules/$(uname -r)/ -name "*.ko" | wc -l
6189
```

Okay, the folks at Canonical and elsewhere have been busy! Over six thousand kernel modules... Think about it, it makes sense: distributors cannot know in advance exactly what hardware peripherals a user will end up using (especially on generic computers like x86 PCs). Kernel modules serve as a convenient means to support huge amounts of hardware without insanely bloating the kernel image file (the `bzImage` or `Image` files, for example).

The installed kernel modules for our Ubuntu Linux system live within the `/lib/modules/$(uname -r)/kernel` directory, as seen here:

```
$ ls /lib/modules/5.19.0-45-generic/kernel/
arch/      block/    crypto/    drivers/       fs/  kernel/  lib/  mm/  net/
samples/  sound/    ubuntu/   v4l2loopback/  zfs/
$ ls /lib/modules/6.1.25-lkp-kernel/kernel/
arch/  crypto/  drivers/  fs/  lib/  net/  sound/
```

Here, looking at the top level of the `kernel/` directory under `/lib/modules/$(uname -r)` for the distro kernel (Ubuntu 22.04 LTS running the `5.19.0-45-generic` kernel), we see that there are many sub-folders and literally a few thousand kernel modules packed within. By contrast, for the kernel we built (refer to *Chapter 2, Building the 6.x Linux Kernel from Source – Part 1*, and *Chapter 3, Building the 6.x Linux Kernel from Source – Part 2*, for the details), there are much fewer. You will recall from our discussions in *Chapter 2, Building the 6.x Linux Kernel from Source – Part 1*, that we deliberately used the `localmodconfig` target to keep the build (relatively) small and fast. Thus, here, for example, our custom 6.1.25 kernel has just some 70-odd kernel modules built against it.

One area that sees heavy usage of kernel modules is that of *device drivers*. As an example, let's look at a network device driver that is architected as a kernel module. You can find several (with familiar brand names!) under the *distro* kernel's `kernel/drivers/net/ethernet` folder:

```
$ ls /lib/modules/5.19.0-45-generic/kernel/drivers/net/ethernet/
3com/        aquantia/  dec/        fungible/    microsoft/       pensando/  sis/         wiznet/
8390/        asix/      dlink/      google/      mscc/            qlogic/    smsc/        xilinx/
adaptec/     atheros/   dnet.ko     huawei/      myricom/         qualcomm/  stmicro/     xircom/
agere/       broadcom/  ec_bhf.ko   intel/       natsemi/         rdc/       sun/
alacritech/  brocade/   emulex/     jme.ko       neterion/        realtek/   synopsys/
alteon/      cadence/   engleder/   marvell/     netronome/       rocker/    tehuti/
altera/      cavium/    ethoc.ko    mellanox/    ni/              samsung/   ti/
amazon/      chelsio/   fealnx.ko   micrel/      nvidia/          sfc/       vertexcom/
amd/         cisco/     fujitsu/    microchip/   packetengines/   silan/     via/
$ _
```

Figure 4.4: Content of our distro kernel's Ethernet network drivers (kernel modules)

Popular on many Intel-based laptops is the Intel 1GbE **Network Interface Card** (**NIC**) Ethernet adapter. The network device driver that drives it is called the `e1000` driver (more recent systems perhaps use a later model, and hence the `e1000e` module driver).

A convenient way to look up all kernel modules currently in kernel memory is via the `lsmod` utility (read it as "list mod(ules)" – we'll learn more regarding its output format later). Our x86_64 Ubuntu 22.04 guest (running on an x86_64 host laptop) shows that it indeed uses this driver:

```
$ lsmod | grep e1000
e1000                          159744  0
```

Importantly for us right now, we can see that the `e1000` driver is a kernel module! How about obtaining some more information on this particular kernel module? That's quite easily done by leveraging the `modinfo` utility (for readability, we truncate its verbose output here):

```
$ ls -l /lib/modules/5.19.0-45-generic/kernel/drivers/net/ethernet/intel/e1000
total 364
-rw-r--r-- 1 root root 372185 Jun  7 19:53 e1000.ko
$ modinfo /lib/modules/5.19.0-45-generic/kernel/drivers/net/ethernet/intel/
e1000/e1000.ko
filename:       /lib/modules/5.19.0-45-generic/kernel/drivers/net/ethernet/
intel/e1000/e1000.ko
license:        GPL v2
description:    Intel(R) PRO/1000 Network Driver
author:         Intel Corporation, <linux.nics@intel.com>
srcversion:     F35F102C5522A6614A9D65C
alias:          pci:v00008086d00002E6Esv*sd*bc*sc*i*
alias:          pci:v00008086d000010B5sv*sd*bc*sc*i*
[ … ]
intree:         Y
name:           e1000
vermagic:       5.19.0-45-generic SMP preempt mod_unload modversions
sig_id:         PKCS#7
signer:         Build time autogenerated kernel key
[ … ]
parm:           SmartPowerDownEnable:Enable PHY smart power down (array of
int)
parm:           copybreak:Maximum size of packet that is copied to a new
buffer on receive (uint)
parm:           debug:Debug level (0=none,...,16=all) (int)
$
```

The `modinfo` utility allows us to peek into a kernel module's binary image and extract some details regarding it; more on using `modinfo` follows in the next section.

You might find the kernel module file is compressed (for example, `e1000.ko.xz`); its a feature, not a bug (more on this later). Another way to gain useful information on the system, including information on kernel modules that are currently loaded up, is via the `systool` utility. For an installed kernel module (details on *installing* a kernel module follow in the next chapter, in the *Auto-loading modules on system boot* section), doing `systool -m <module-name> -v` reveals information about it. Look up the `systool(1)` man page for usage details.

The bottom line is that kernel modules have come to be *the* pragmatic way to build and distribute some types of kernel components, with *device drivers* likely being the most frequent use case for them. Other uses include but aren't limited to filesystems, network firewalls, packet sniffers, and custom kernel code.

So, if you would like to learn how to write a Linux device driver, a filesystem, or a firewall, it's highly recommended you first learn how to write a kernel module, thus leveraging the kernel's powerful LKM framework. That's precisely what we will be doing next.

Writing our very first kernel module

When introducing a new programming language or topic, it has become a widely accepted computer programming tradition to mimic the original *Hello, world* program as the very first piece of code. I'm happy to follow this venerated tradition to introduce the Linux kernel's powerful LKM framework. In this section, you will learn the steps to code a simple LKM. We explain the code in detail.

Introducing our Hello, world LKM C code

Without further ado, here is some simple *Hello, world* C code, implemented to abide by the Linux kernel's LKM framework:

For reasons of readability and space constraints, only the key parts of most source code are displayed here. To view the complete source code (with all comments), build it, and run it, the entire source tree for this book is available in its GitHub repository here: `https://github.com/PacktPublishing/Linux-Kernel-Programming_2E`. We definitely expect you to clone it and use it:

```
git clone https://github.com/PacktPublishing/Linux-Kernel-Programming_2E.git
```

```
$ cat <LKP2E_src>/ch4/helloworld_lkm/helloworld_lkm.c
// ch4/helloworld_lkm/helloworld_lkm.c
#include <linux/init.h>
#include <linux/module.h>

/* Module stuff */
MODULE_AUTHOR("<insert your name here>");
MODULE_DESCRIPTION("LKP2E book:ch4/helloworld_lkm: hello, world, our first LKM");
MODULE_LICENSE("Dual MIT/GPL");
MODULE_VERSION("0.2");

static int __init helloworld_lkm_init(void)
{
    printk(KERN_INFO "Hello, world\n");
    return 0;    /* success */
}
static void __exit helloworld_lkm_exit(void)
```

```
{
    printk(KERN_INFO "Goodbye, world! Climate change has done us in...\n");
}
module_init(helloworld_lkm_init);
module_exit(helloworld_lkm_exit);
```

You can try out this simple *Hello, world* kernel module right away! Just `cd` to the correct source directory and use our helper `lkm` script to build and run it:

```
$ cd ~/Linux-Kernel-Programming_2E/ch4/helloworld_lkm
$ ../../lkm helloworld_lkm.c
Usage: lkm name-of-kernel-module-file ONLY (do NOT put any extension).

$ ../../lkm helloworld_lkm
Version info:
Distro:     Ubuntu 22.04.2 LTS
Kernel: 5.19.0-45-generic
[ … ]
make || exit 1
-----------------------------
make -C /lib/modules/5.19.0-45-generic/build/ M=/home/c2kp/Linux-Kernel-
Programming_2E/ch4/helloworld_lkm modules
[ … ]
sudo dmesg
-----------------------------
[ 4123.028252] Hello, world
$
```

The *hows and whys* will be explained in a lot of detail shortly. Though tiny, the code of this, our very first kernel module, requires careful perusal and understanding. Do read on!

Breaking it down

The following subsections explain pretty much each line of the preceding *Hello, world* C LKM code. Remember that although the program appears very small and trivial, there is a lot to be understood regarding it and the surrounding LKM framework. The rest of this chapter focuses on this and goes into detail. I highly recommend that you take the time to read through and understand these fundamentals first. This will help you immensely in later, possibly difficult-to-debug situations.

Kernel headers

First thing in the code, we use `#include` to (obviously) include a few header files. Unlike in user space C application development, these are *kernel headers* (as mentioned in the *Technical requirements* section). Recall from *Chapter 3, Building the 6.x Linux Kernel from Source – Part 2*, that kernel modules were installed under a specific root-writable branch.

Let's check it out again (here, we're running on our guest x86_64 Ubuntu VM with the 5.19.0-45-generic distro kernel):

```
$ ls -l /lib/modules/$(uname -r)
total 6712
lrwxrwxrwx  1 root root     40 Jun  7 19:53 build -> /usr/src/linux-headers-
5.19.0-45-generic
drwxr-xr-x  2 root root   4096 Jun 23 09:47 initrd
[ … ]
```

Notice the symbolic or soft link here named `build`. It points to the location of the kernel headers on the system. In the preceding code block, you can see that it's in the directory `/usr/src/linux-headers-5.19.0-45-generic/`! As you will see, we will supply this information to the Makefile used to build our kernel module. (Also, some systems have a similar soft link called `source`.)

The `kernel-headers` or `linux-headers` package unpacks a limited kernel source tree onto the system, typically under `/usr/src/`…. This kernel code base, however, isn't complete, hence our use of the phrase *limited* source tree. This is because the complete kernel source tree isn't required for the purpose of building modules – just the required components (the headers, the Makefiles, and so on) are what's packaged and extracted.

The first line of code in our *Hello, world* kernel module is `#include <linux/init.h>`.

The compiler resolves this line by searching for the previously mentioned kernel header file under `/lib/modules/$(uname -r)/build/include/`. Thus, by following the `build` soft link, we can see that it ultimately picks up this header file:

```
$ ls -l /usr/src/linux-headers-5.19.0-45-generic/include/linux/init.h
-rw-r--r-- 1 root root 11963 Aug  1  2022 /usr/src/linux-headers-5.19.0-45-
generic/include/linux/init.h
```

The same follows for the other kernel headers included in the kernel module's source code.

Module macros

Next, we have a few module macros of the form `MODULE_FOO();` (colloquially called "module stuff"). Most are quite intuitive:

- `MODULE_AUTHOR()`: Specifies the author(s) of the kernel module
- `MODULE_DESCRIPTION()`: Briefly describes the function or purpose of this LKM
- `MODULE_LICENSE()`: Specifies the license(s) under which this kernel module is released
- `MODULE_VERSION()`: Specifies the (local) version string of the kernel module

In the absence of the source code, how will this information be conveyed to the end user (or customer)? Ah, the `modinfo` utility does precisely that! These macros and their information might seem trivial, but they are important in projects and products.

This information is relied upon, for example, by a vendor establishing the (open source) licenses that code is running under by using `grep` on the `modinfo` output on all installed kernel modules. (These are the basic module macros; there are more that we shall cover as we go along.)

Entry and exit points

Never forget, kernel modules are, after all, *kernel code running with kernel privilege*. It's *not* an application and thus does not have its entry point as the familiar `main()` function (that we know well and love). This, of course, begs the question: what are the entry and exit points of the kernel module? Notice, at the bottom of our simple kernel module, the following lines:

```
module_init(helloworld_lkm_init);
module_exit(helloworld_lkm_exit);
```

The `module_{init|exit}()` code are macros specifying the entry and exit points, respectively. The parameter to each is a function pointer. With modern C compilers, we can just specify the name of the function. Thus, in our code, the following applies:

- The `helloworld_lkm_init()` function is the entry point.
- The `helloworld_lkm_exit()` function is the exit point.

You can almost think of these entry and exit points as a *constructor/destructor* pair for a kernel module. Technically, it's not the case of course, as this isn't object-oriented C++ code, it's plain C. Nevertheless, it's a useful analogy, perhaps.

Return values

Notice the signature of the `init` and `exit` functions is as follows:

```
static int  __init <modulename>_init(void);
static void __exit <modulename>_exit(void);
```

As good coding practice, we have used the naming format for the functions as `<modulename>_{init|exit}()`, where `<modulename>` is replaced with the name of the kernel module. You will realize that this naming convention is just that – it's merely a convention that is, technically speaking, unnecessary, but it is intuitive and thus helpful (remember, we humans must write code for *humans* to read and understand, not machines). Clearly, neither routine receives any parameter.

Marking both functions with the `static` qualifier implies that they are private to this kernel module. That is what we want.

Now let's move along to the important convention that is followed for a kernel module's `init` function's return value.

The 0/-E return convention

The kernel module's `init` function is to return an integer, a value of type `int`; this is a key aspect. The Linux kernel has evolved a *style* or convention, if you will, with regard to returning values from it (meaning from kernel space, where the module resides and runs, to the user space process).

To return a value, the LKM framework follows what is colloquially referred to as the `0/-E` convention:

- Upon success, return integer value `0`.
- Upon failure, return the negative of the value you would like the user space global uninitialized integer `errno` to be set to.

Be aware that `errno` is a global integer residing in a user process VAS within its uninitialized data segment. With very few exceptions, whenever a Linux system call fails, `-1` is returned and `errno` is set to a positive value, representing the failure code or diagnostic; this work is carried out by `glibc` "glue" code on the `syscall` return path.

Furthermore, the `errno` value is an index into a global table of English error messages (`const char * sys_errlist[]`); this is how routines such as `perror(3)`, `strerror[_r](3)`, and the like can print out failure diagnostics.

By the way, you can look up the **complete list of error (errno) codes** available to you from within these (kernel source tree) header files: `include/uapi/asm-generic/errno-base.h` and `include/uapi/asm-generic/errno.h`.

A quick example of how to return from a kernel module's `init` function will help make this key point clear: say our kernel module's `init` function is attempting to dynamically allocate some kernel memory (details on the `kmalloc()` API and so on will be covered in later chapters of course; please ignore it for now). Then, we could code it like so:

```
[...]
ptr = kmalloc(87, GFP_KERNEL);
 if (!ptr) {
    pr_warning("%s():%s():%d: kmalloc failed! Out of memory\n", __FILE__, __
func__, __LINE__);
    return -ENOMEM;
 }
 [...]
return 0;    /* success */
```

If the memory allocation does fail (very unlikely, but hey, it can happen on a bad day!), we do the following:

1. First, we emit a warning `printk` (don't worry, we'll cover these syntax details and much more on the `printk`). In this particular case – being "out of memory" – it's considered pedantic and unnecessary to emit a message. The kernel will certainly emit sufficient diagnostic information if a kernel-space memory allocation ever fails! See this link for more details: `https://lkml.org/lkml/2014/6/10/382`; we do so here merely as it's early in the discussion and for reader continuity.

2. Return the integer value `-ENOMEM`:
   - The layer to which this value will be returned in user space is actually `glibc`; it has some "glue" code that multiplies this value by `-1` and sets the global integer `errno` to it.
   - Now, the `[f]init_module()` system call will return `-1`, indicating failure (this is because `insmod` actually invokes the `finit_module()` (or, earlier, the `init_module()`) system call, as you will soon see).
   - `errno` will be set to `ENOMEM`, reflecting the fact that the kernel module insertion failed due to a failure to allocate memory.

Conversely, the framework expects the `init` function to return the value `0` upon success. In fact, in older kernel versions, failure to return `0` upon success would cause the kernel module to be abruptly and immediately unloaded from kernel memory. Nowadays, this removal of the kernel module does not happen; instead, the kernel emits a warning message regarding the fact that a suspicious non-zero value has been returned. Moreover, modern compilers typically catch the fact that you aren't returning a value when expected to, triggering an error message similar to this: `error: no return statement in function returning non-void [-Werror=return-type]`.

There's not much to be said for the cleanup routine. It receives no parameters and returns nothing (`void`). Its job is to perform any and all required cleanup (freeing memory objects, setting certain registers, perhaps, and so on, depending on what the module's designed to do) before the kernel module is unloaded from kernel memory.

Not including the `module_exit()` macro in your kernel module makes it impossible to ever unload it (notwithstanding a system shutdown or reboot, of course). Interesting... I suggest you try this out as a small exercise! Of course, it's never that simple: this behavior preventing the module from unloading is guaranteed only if the kernel is built with the `CONFIG_MODULE_FORCE_UNLOAD` flag set to `Disabled` (the default).

The ERR_PTR and PTR_ERR macros

On the discussion of return values, you now understand that the kernel module's `init` routine must return an integer. What if you wish to return a pointer instead? The `ERR_PTR()` inline function comes to our rescue, allowing us to return an integer *disguised* as a pointer simply by typecasting it as `void *`.

It gets better: you can check for an error using the `IS_ERR()` inline function (which really just figures out whether the value is in the range [-1 to -4095]), *encodes* a negative error value into a pointer via the `ERR_PTR()` inline function, and retrieves this value from the pointer using the converse routine `PTR_ERR()`.

As a simple example, see the callee code given here. This time, as an example, we have the (sample) function `myfunc()` return a pointer (to a structure named `mystruct`) and not an integer:

```
struct mystruct * myfunc(void)
```

```
{
    struct mystruct *mys = NULL;
    mys = kzalloc(sizeof(struct mystruct), GFP_KERNEL);
    if (!mys)
        return ERR_PTR(-ENOMEM);
    [...]
    return mys;
}
```

The caller code is as follows:

```
[...]
retp = myfunc();
if (IS_ERR(retp)) {
    pr_warn("myfunc() mystruct alloc failed, aborting...\n");
    stat = PTR_ERR(retp);     /* sets 'stat' to the value -ENOMEM */
    goto out_fail_1;
 }
 [...]
 out_fail_1:
    return stat;
 }
```

FYI, the inline `ERR_PTR()`, `PTR_ERR()`, and `IS_ERR()` functions all live within the (kernel header) `include/linux/err.h` file. One example of usage for these functions is here: `https://elixir.bootlin.com/linux/v6.1.25/source/arch/x86/kernel/cpu/sgx/ioctl.c#L269` (and the `ERR_PTR()` on line 31).

The __init and __exit keywords

Recall our simple module's `init` and `cleanup` functions:

```
static int __init helloworld_lkm_init(void)
{
    [ … ]
static void __exit helloworld_lkm_exit(void)
{
    [ … ]
```

A niggling leftover: what exactly are the `__init` and `__exit` macros we see within the preceding function signatures? These are merely specifying memory optimization linker attributes.

The `__init` macro defines an `init.text` section for code. Similarly, any data declared with the `__initdata` attribute goes into an `init.data` section. The whole point here is the code and data in the `init` function are used exactly once during initialization.

Once it's invoked, it will never be called again; so, once called, all the code and data in these `init` sections are freed up (via `free_initmem()`).

The deal is similar with the `__exit` macro, though, of course, this only makes sense with kernel modules. Once the `cleanup` function is called, all the memory is freed. If the code were instead part of the static kernel image (or if module support were disabled), this macro would have no effect.

Fine, but so far, we have still not explained some practicalities: how exactly can you build your new kernel module, get it into kernel memory and have it execute, and then unload it, plus several other operations you might wish to perform? Let's discuss these in the following section.

Common operations on kernel modules

Now let's delve into how exactly you can build, load, and unload a kernel module. Besides this, we'll also walk through the basics regarding the tremendously useful `printk()` kernel API, details on listing the currently loaded kernel modules with `lsmod`, and a convenient script for automating some common tasks during kernel module development. So, let's begin!

Building the kernel module

At the risk of repetition, we urge you to try out our simple *Hello, world* kernel module as an exercise (if you haven't already done so)! To do so, we assume you have cloned this book's GitHub repository (`https://github.com/PacktPublishing/Linux-Kernel-Programming_2E`) already. If not, please do so now (refer to the *Technical requirements* section for details).

Here, we show, step by step, how exactly you can build and then insert our very first kernel module into kernel memory. Again, a quick reminder: we have performed these steps on an x86_64 Linux guest VM (under Oracle VirtualBox 7.x) running the Ubuntu 22.04 LTS distribution (not that it should matter too much which distro you use; the general steps remain the same):

1. Change to this book's source code chapter directory and sub-directory. Our first kernel module lives in its own folder (as it should!) called `helloworld_lkm`:

```
cd <book-code-dir>/ch4/helloworld_lkm
```

`<book-code-dir>` is, of course, the folder into which you cloned this book's GitHub repository; on my system (see the screenshot in *Figure 4.5*), you can see that it's `/home/c2kp/kaiwanTECH/Linux-Kernel-Programming_2E`.

2. Now verify the code base:

```
$ pwd
<book-code-dir>/ch4/helloworld_lkm
$ ls -l
```

```
total 8
-rw-rw-r-- 1 c2kp c2kp 1238  Dec 18  12:38 helloworld_lkm.c
-rw-rw-r-- 1 c2kp c2kp   290  Oct 27  07:26 Makefile
```

(It's all right if the sizes don't perfectly match the one you have; software evolves.)

3. Build the kernel module with `make`:

```
$ uname -r
6.1.25-lkp-kernel
$ pwd
/home/c2kp/kaiwanTECH/Linux-Kernel-Programming_2E/ch4/helloworld_lkm
$ ls -l
total 8
-rw-rw-r-- 1 c2kp c2kp 1238 Dec 18 12:38 helloworld_lkm.c
-rw-rw-r-- 1 c2kp c2kp  290 Oct 27 07:26 Makefile
$
$ make
make -C /lib/modules/6.1.25-lkp-kernel/build/ M=/home/c2kp/kaiwanTECH/Linux-Kernel-Programming_2E/ch4/hello
world_lkm modules
make[1]: Entering directory '/home/c2kp/kernels/linux-6.1.25'
  CC [M]  /home/c2kp/kaiwanTECH/Linux-Kernel-Programming_2E/ch4/helloworld_lkm/helloworld_lkm.o
  MODPOST /home/c2kp/kaiwanTECH/Linux-Kernel-Programming_2E/ch4/helloworld_lkm/Module.symvers
  CC [M]  /home/c2kp/kaiwanTECH/Linux-Kernel-Programming_2E/ch4/helloworld_lkm/helloworld_lkm.mod.o
  LD [M]  /home/c2kp/kaiwanTECH/Linux-Kernel-Programming_2E/ch4/helloworld_lkm/helloworld_lkm.ko
make[1]: Leaving directory '/home/c2kp/kernels/linux-6.1.25'
$
$ ls -l ./helloworld_lkm.ko
-rw-rw-r-- 1 c2kp c2kp 114632 Dec 18 12:39 ./helloworld_lkm.ko
$
```

Figure 4.5: Listing and building our very first Hello, world kernel module

If you do come across it when building the module, you can for now quite safely ignore any warning messages related to "Skipping BTF generation for `<...>/helloworld_lkm.ko` due to the unavailability of vmlinux...."

The preceding screenshot shows that our kernel module has been successfully built; it's the file, the "kernel object," named `./helloworld_lkm.ko`.

Always use `make`, and thus the Makefile provided, to build the kernel module; don't try to manually build it by invoking `gcc` (or Clang) directly.

Also, note that we booted from, and thus have built the kernel module against, our custom 6.1.25 LTS kernel, built in earlier chapters. Now that we've successfully built our module, let's run it!

Running the kernel module

In order to have a kernel module run, you need to first load it into kernel memory space, of course. This is known as *inserting* the module into kernel memory.

Getting the kernel module into the Linux kernel segment or VAS (which is in RAM, of course) can be done in a few ways, which all ultimately boil down to invoking one of the `[f]init_module()` system calls.

For convenience, several wrapper utilities exist that will do so (or you can always write one). We will use the popular `insmod` (**"insert module"**) utility below; the parameter to `insmod` is the pathname to the kernel module to insert (FYI, `insmod`, under the hood, invokes the `finit_module()` system call to load the module into kernel memory):

```
$ insmod ./helloworld_lkm.ko
insmod: ERROR: could not insert module ./helloworld_lkm.ko: Operation not permitted
$
```

It fails! In fact, it should be obvious why. Think about it: inserting code into the kernel is, in a very real sense, even superior to being *root* (superuser) on the system – again, I remind you: *it's kernel code and will run with kernel privilege.* If any and every user is allowed to insert or remove kernel modules, hackers would have a field day! Deploying malicious code, that too at the level of the OS, would become a trivial affair. So, for security reasons, **only with root access can you insert or remove kernel modules.**

Technically, being `root` implies that the process's (or thread's) **Real** and/or **Effective UID** (**RUID/EUID**) value is the special value `zero`. Not only that, but the modern kernel "sees" a thread as having certain **capabilities** (via the modern and superior **POSIX Capabilities** model). Besides being root, an even better approach is to leverage this capabilities model; with it, only a process/thread with the `CAP_SYS_MODULE` capability can (un)load kernel modules. We refer you to the man page on `capabilities(7)` for more details.

So, let's again attempt to insert our kernel module into memory, this time with root privileges via `sudo`:

```
$ sudo insmod ./helloworld_lkm.ko
 [sudo] password for c2kp:
$ echo $?
0
```

Now it works (the `$?` result variable being `0` implies that the previous shell command was successful)! As alluded to earlier, the `insmod` utility works by invoking the `finit_module()` system call. When might the `insmod` utility fail? There are a few cases:

- The kernel config `CONFIG_MODULES` is not set to `y` – i.e., the kernel's built without support for loadable kernel modules.
- When not run as root, or a lack of the `CAP_SYS_MODULE` capability (`errno` is then set to the value `EPERM`).
- When the kernel tunable within the `proc` filesystem, `/proc/sys/kernel/modules_disabled`, is set to `1` (it defaults to `0`).
- When a kernel module with the same name is already in kernel memory (`errno` is then set to the value `EEXISTS`).

Okay, here, all looks good. That's great, but where on earth is our precious *Hello, world* message? Read on!

A quick first look at the kernel printk()

To emit a message, the user space C developer will often use the trusty `printf()` glibc API (or perhaps the `cout` when writing C++ code). However, it's important to understand that in kernel space, *there are no (user mode or other) libraries*. Hence, we simply do not have access to the good old `printf()` API. Instead, it has essentially been re-implemented *within* the kernel as the `printk()` kernel API. Curious as to where its code is? Well, it's defined here as a macro within the kernel source tree: `include/linux/printk.h:printk(fmt, ...)`. The actual function is here: `kernel/printk/printk.c:_printk()`. The reality is that the internal `printk` implementation is pretty complex, involving a wrapper layer for `printk` indexing, and much more; see the *Further reading* section for more.

Luckily, for the kernel or a kernel module to emit a message via the `printk()` API is simple and very much similar to doing so with `printf()`. In our simple kernel module, here's where the action occurs:

```
printk(KERN_INFO "Hello, world\n");
```

Though very similar to `printf` at first glance, the `printk` API in the kernel is really quite different. In terms of similarities, the API receives a format string as its parameter. The format string is pretty much identical to that of `printf`.

But the similarities end there. A key difference between `printf` and `printk` is this: the user space `printf()` library API works by formatting a text string as requested and invoking the `write()` system call, which in turn actually performs a write to the `stdout` *device*, which, by default, is the Terminal window (or console device).

The kernel `printk` API also formats its text string as requested, but its *output destination(s)* differs. It writes to at least one place – the first one in the following list – and possibly to a few more:

- A (volatile) kernel log buffer in RAM
- A (non-volatile) log file, the kernel log file
- The console device

For now, we shall skip the inner details regarding the workings of `printk`. Also, right now, please ignore the `KERN_INFO` token within the `printk` API; we shall cover all this soon enough.

Next, don't get overly attached to the `printk()` API; you'll soon learn that the modern and recommended way to use it is via macros of the form `pr_foo()` (we cover them in the upcoming *The pr_<foo> convenience macros* section).

When you emit a message via `printk`, it's guaranteed that the output goes into a log buffer in kernel memory (RAM). This, in effect, constitutes the (volatile) **kernel log**. It's important to note that you will never see the `printk` output directly when working in graphical mode with an X server (Xorg, Xwayland, or whatever) process running (the default environment when working on a typical Linux distro).

So, the obvious question here is: *how do you see the kernel log buffer content?* There are a few ways. For now, let's just make use of the quick and easy way.

Use the `dmesg` utility! By default, `dmesg` will dump the entire kernel log buffer content to stdout. Let's try it out!

```
$ dmesg
dmesg: read kernel buffer failed: Operation not permitted
```

We'll quite often come across what's termed a **sysctl**. Here, *sysctl* is short for *system control*, a tuning knob within the kernel, typically located under `/proc/sys/<dir>/*`; there are several directories, including `fs`, `kernel`, `net`, `user`, and `vm`. Typically, root access is required to write to a sysctl, while several can be read without special privileges. The official kernel documentation on the kernel *sysctls* is here, do have a look: `https://www.kernel.org/doc/Documentation/sysctl/kernel.txt`. In a similar manner, some of the other branches also have documentation; look here for them: `https://www.kernel.org/doc/Documentation/sysctl/`. Also look up this useful "sysctl explorer" website: `https://sysctl-explorer.net/`.

On several recent distros, a sysctl named `dmesg_restrict` (more on this later) prevents ordinary users from viewing the kernel log content; this is good for security (preventing info leaks). So, now let's redo this, but this time with root privilege; let's look up the last two lines of the kernel log buffer:

```
$ sudo dmesg | tail -n2
[39884.691954] Hello, world
```

There it is, finally: our *Hello, world* message!

You may see a message in the kernel log, something like `loading out-of-tree module taints kernel` and possibly `module verification failed: signature and/or required key missing - tainting kernel`. You can simply ignore them for now. For security reasons, most modern Linux distros will mark the kernel as *tainted* (literally, "contaminated" or "polluted") if a third party/out-of-tree/external (or non-signed) kernel module is inserted. (Well, it's really more of a pseudo-legal cover-up along the lines of "if something goes wrong from this point in time onward, we are not responsible, and so on..." - you get the idea!)

Within the kernel log, as displayed by the `dmesg` utility, the numbers within square brackets in the leftmost column are a simple timestamp, in `[seconds.microseconds]` format – the time elapsed since system boot (it's not recommended to treat it as being perfectly accurate, though). By the way, this timestamp is a `Kconfig` variable – a kernel config option – named `CONFIG_PRINTK_TIME`; it can be overridden by the `printk.time` kernel parameter (link: `https://elixir.bootlin.com/linux/v6.1.25/source/Documentation/admin-guide/kernel-parameters.txt#L4504`).

Listing the live kernel modules

Back to our kernel module. So far, we have built it, loaded it into the kernel, and verified that its entry point, the `helloworld_lkm_init()` function, got invoked, thus executing the `printk` API. So after that, what does it do? Well, nothing; the kernel module merely (happily?) sits in kernel memory doing absolutely nothing. We can easily look it up with the `lsmod` utility:

```
$ lsmod | head
helloworld_lkm         16384  0
isofs                  49152  1
vboxvideo              45056  0
vboxsf                 90112  0
binfmt_misc            24576  1
snd_intel8x0           49152  2
snd_ac97_codec        176128  1 snd_intel8x0
ac97_bus               16384  1 snd_ac97_codec
snd_pcm               151552  2 snd_intel8x0,snd_ac97_codec
```

`lsmod` shows all kernel modules currently residing (or *live*) in kernel memory, sorted in reverse chronological order. Its output is column formatted, with three columns and an optional fourth one. Let's look at each column separately:

- The first column displays the *name* of the kernel module.
- The second column is the (static) *size* in bytes that it's taking in the kernel.
- The third column is the module *usage count*.
- The optional fourth column indicates a dependent module(s) and is explained in the next chapter (in the *Module stacking* section). Also, on recent x86_64 Ubuntu distros, a minimum of 16 KB of kernel memory seems to be taken up by a kernel module, and around 12 KB on Fedora.

So, great! By now you've successfully built, loaded, and run your first kernel module in kernel memory and it basically works: what next? Well, nothing much really with this one! We simply learn how to unload it in the following section. There's a lot more to come of course... keep going!

Unloading the module from kernel memory

To unload the kernel module, we use the convenience utility `rmmod` (*remove module*):

```
$ rmmod
rmmod: ERROR: missing module name.

$ rmmod helloworld_lkm
rmmod: ERROR: ../libkmod/libkmod-module.c:799 kmod_module_remove_module() could
not remove 'helloworld_lkm': Operation not permitted
rmmod: ERROR: could not remove module helloworld_lkm: Operation not permitted
```

```
$ sudo rmmod helloworld_lkm
 [sudo] password for c2kp:

$ dmesg | tail -n2
[39884.691954] Hello, world
[40280.138269] Goodbye, world! Climate change has done us in...
 $
```

The parameter to `rmmod` is the *name* of the kernel module (as shown in the first column of `lsmod`), not the pathname. Clearly, just as with `insmod`, we need to run the `rmmod` utility as the *root* user for it to succeed (or with the `CAP_SYS_MODULE` capability).

Here, we can also see that, because of our `rmmod`, the LKM framework first invokes the exit routine (or "destructor") `helloworld_lkm_exit()` function of the kernel module being removed. It in turn invokes `printk`, which emits the *Goodbye, world...* message (which we looked up with `dmesg`).

When could `rmmod` (note that internally, it invokes the `delete_module()` system call) fail? Here are some cases:

- **Permissions:** If it is not run as root or there is a lack of the `CAP_SYS_MODULE` capability (`errno` is then set to the value `EPERM`).
- If the kernel module's code and/or data is being used by another module (if a dependency exists; this is covered in detail in the next chapter's *Module stacking* section) or the module is currently in use by a process (or thread), then the module usage count will be positive and `rmmod` will fail (`errno` is then set to the value `EBUSY`); makes sense.
- The kernel module did not specify an exit routine (or destructor) with the `module_exit()` macro *and* the `CONFIG_MODULE_FORCE_UNLOAD` kernel config option is disabled.

Several convenience utilities concerned with module management are nothing but symbolic (soft) links to the single `kmod` utility (analogous to what the popular *busybox* utility does). The wrappers are `lsmod`, `rmmod`, `insmod`, `modinfo`, `modprobe`, and `depmod`. Take a look at a few of them:

```
$ ls -l $(which insmod) ; ls -l $(which lsmod) ; ls -l $(which rmmod)
lrwxrwxrwx  1 root root 9 Apr 23  2022 /usr/sbin/insmod -> /bin/kmod
lrwxrwxrwx  1 root root 9 Apr 23  2022 /usr/sbin/lsmod  -> /bin/kmod
lrwxrwxrwx  1 root root 9 Apr 23  2022 /usr/sbin/rmmod -> /bin/kmod
```

Note that the precise location of these utilities (`/bin`, `/sbin`, or `/usr/sbin`) can vary with the distribution.

Our lkm convenience script

Let's round off this *first kernel module* discussion with a simple yet useful custom Bash script called `lkm` that helps you out by automating the kernel module build, load, `dmesg`, and unload workflow.

Here it is (the complete code is in the root of this book's source tree):

```
# !/bin/bash
# lkm : a silly kernel module dev - build, load, unload - helper wrapper script
 [...]
# Turn on unofficial Bash 'strict mode'! V useful
# "Convert many kinds of hidden, intermittent, or subtle bugs into immediate,
glaringly obvious errors" [...]
set -euo pipefail

unset ARCH
unset CROSS_COMPILE
name=$(basename "${0}")

# Display and run the provided command.
 # Parameter(s) : the command to run
runcmd()
{
    local SEP="------------------------------"
    [[ $# -eq 0 ]] && return
    echo "${SEP}
$*
${SEP}"
    eval "$@"
    [[ $? -ne 0 ]] && echo " ^--[FAILED]"
 }
```

```
### "main" here
[[ $# -ne 1 ]] && {
    echo "Usage: ${name} name-of-kernel-module-file (without the .c)"
    exit 1
}
[[ "${1}" = *"."* ]] && {
     echo "Usage: ${name} name-of-kernel-module-file ONLY (do NOT put any
extension)."
     exit 1
}
echo "Version info:"
 which lsb_release >/dev/null 2>&1 && {
     echo -n "Distro: "
     lsb_release -a 2>/dev/null |grep "Description" |awk -F':' '{print $2}'
 }
echo -n "Kernel: " ; uname -r
runcmd "sudo rmmod $1 2> /dev/null" || true
# runcmd "make clean" || true
runcmd "sudo dmesg -C > /dev/null" || true
runcmd "make || exit 1" || true

[[ ! -f "$1".ko ]] && {
     echo "[!] ${name}: $1.ko has not been built, aborting..."
     exit 1
}
```

```
runcmd "sudo insmod ./$1.ko && lsmod|grep $1" || true
# Ubuntu 20.10 onward has enabled CONFIG_SECURITY_DMESG_RESTRICT !
# That's good for security; so we need to 'sudo' dmesg
runcmd "sudo dmesg" || true
exit 0
```

Given the name of the kernel module as a parameter – without any extension part (such as `.c`) – the `lkm` script performs some validity checks, displays some version information (though, as of now, this works for Ubuntu only), and then uses a wrapper `runcmd()` Bash function to display the name of and run a given command, in effect getting the `build/load/lsmod/dmesg` workflow done painlessly. Let's be empirical and try it out on our first kernel module (note that the `/home/c2kp/lkp2e` is just a soft link to the full pathname where the books code resides):

```
$ pwd
/home/c2kp/lkp2e/ch4/helloworld_lkm
$ ../../lkm
Usage: lkm name-of-kernel-module-file (without the .c)
$ ls
helloworld_lkm.c  Makefile
$ ../../lkm helloworld_lkm.c
Usage: lkm name-of-kernel-module-file ONLY (do NOT put any extension).
$
$ ../../lkm helloworld_lkm
Version info:
Distro:         Ubuntu 22.04.2 LTS
Kernel: 6.1.25-lkp-kernel
------------------------------
sudo rmmod helloworld_lkm 2> /dev/null
------------------------------
 ^--[FAILED]
------------------------------
sudo dmesg -C
------------------------------
------------------------------
make || exit 1
------------------------------
make -C /lib/modules/6.1.25-lkp-kernel/build/ M=/home/c2kp/Linux-Kernel-Programming_2E/ch4/hello
world_lkm modules
make[1]: Entering directory '/home/c2kp/kernels/linux-6.1.25'
  CC [M]  /home/c2kp/Linux-Kernel-Programming_2E/ch4/helloworld_lkm/helloworld_lkm.o
  MODPOST /home/c2kp/Linux-Kernel-Programming_2E/ch4/helloworld_lkm/Module.symvers
  CC [M]  /home/c2kp/Linux-Kernel-Programming_2E/ch4/helloworld_lkm/helloworld_lkm.mod.o
  LD [M]  /home/c2kp/Linux-Kernel-Programming_2E/ch4/helloworld_lkm/helloworld_lkm.ko
make[1]: Leaving directory '/home/c2kp/kernels/linux-6.1.25'
------------------------------
sudo insmod ./helloworld_lkm.ko && lsmod|grep helloworld_lkm
------------------------------
helloworld_lkm         16384  0
------------------------------
sudo dmesg
------------------------------
[41052.797932] Hello, world
$
```

Figure 4.6: Screenshot showing our lkm convenience script performing the build, load, and dmesg for our kernel module

All done! Remember to unload the kernel module with `rmmod`.

Congratulations! You have now learned how to write and try out a simple *Hello, world* kernel module. Much work remains, though, before you rest on your laurels; the next section delves into more key details regarding kernel logging and the versatile `printk` API.

Understanding kernel logging and printk

There is still a lot to cover regarding the logging of kernel messages via the `printk` kernel API. This section delves into some of the details. It's important for a budding kernel/driver developer like you to clearly understand these topics.

We saw earlier, in the *A quick first look at the kernel printk()* section, the essentials of using the kernel `printk` API's functionality (have another look at it if you wish to). Here, we will explore a lot more with respect to the `printk()` API's usage. Let's get going!

Using the kernel memory ring buffer

The **kernel log buffer** is simply a memory buffer within the kernel virtual address space where the `printk` output is saved (logged). More technically, it's the global `__log_buf[]` variable. Its definition in the kernel source is as follows:

```
kernel/printk/printk.c
#define __LOG_BUF_LEN (1 << CONFIG_LOG_BUF_SHIFT)
#define LOG_BUF_LEN_MAX (u32)(1 << 31)
 static char __log_buf[__LOG_BUF_LEN] __aligned(LOG_ALIGN);
static char *log_buf = __log_buf;
static u32 log_buf_len = __LOG_BUF_LEN;
```

It's architected as a *ring buffer*; it has a finite size (`__LOG_BUF_LEN` bytes), and once it's full, it gets overwritten from byte zero. Hence, it's called a "ring" (or circular) buffer. Here, we can see that the size is based on the `Kconfig` variable `CONFIG_LOG_BUF_SHIFT` (`1 << n` in C implies `2^n`). This value is shown and can be overridden as part of the kernel `(menu)config` here: `General Setup | Kernel log buffer size`.

It's an integer value with a range of `12-25` (we can always search `init/Kconfig` and see its spec), with a default value of `18`. So, the size of the kernel log buffer by default is (1 << 18), which is 2^{18} = 256 KB. However, the actual runtime size is affected by other config directives as well – notably, `LOG_CPU_MAX_BUF_SHIFT`, which makes the size a function of the number of CPUs on the system. Furthermore, the relevant `Kconfig` file says, *"Also this option is ignored when the log_buf_len kernel parameter is used as it forces an exact (power of two) size of the ring buffer."* So, that's interesting; we can often override defaults by passing a kernel parameter (via the bootloader)!

As an aside, kernel parameters are useful, many, and varied, and are well worth checking out. See the official documentation here: `https://www.kernel.org/doc/html/latest/admin-guide/kernel-parameters.html`.

A snippet from the Linux kernel documentation on the `log_buf_len` kernel parameter reveals the details:

```
log_buf_len=n[KMG]      Sets the size of the printk ring buffer,
                    in bytes. n must be a power of two and greater
                    than the minimal size. The minimal size is defined
                    by LOG_BUF_SHIFT kernel config parameter. There is
                    also CONFIG_LOG_CPU_MAX_BUF_SHIFT config parameter
                    that allows to increase the default size depending
                    on the number of CPUs. See init/Kconfig for more details.
```

Whatever the size of the kernel log buffer, two issues when dealing with the `printk` API become obvious:

- Its messages are being logged in *volatile* memory (RAM); if the system crashes or power cycles in any manner, we will lose the precious kernel log (often severely limiting or even eliminating our ability to debug kernel issues).
- The log buffer isn't very large by default, typically just 256 KB (and/or perhaps 4 to 8 KB per CPU on the system); thus, it's clear that voluminous prints will overwhelm the ring buffer, making it wrap around, thus losing information.

How can we fix this? Read on...

Kernel logging and systemd's journalctl

Just as user space apps use logs, so does the kernel; the feature is called *kernel logging* and is required on all but the most resource-constrained systems. An obvious solution to the previously mentioned issues with kernel logging to a small and volatile memory buffer is to write (append, really) every single kernel `printk` *to a file in (non-volatile) secondary storage*. This is precisely how most modern Linux distributions are set up. The location of the kernel log file varies with the distro: conventionally, the Red Hat-based ones write into the `/var/log/messages` file and the Debian-based ones into `/var/log/syslog`. Traditionally, the kernel `printk` would hook into the user space *system logger daemon* (`syslogd`) to perform file logging, thus automatically getting the benefit of more sophisticated features, such as log rotation, compression, and archival.

Over the past several years, though, system logging has been completely taken over by the useful and powerful new framework for system initialization called **systemd** (it replaces, or often works in addition to, the old SysV init framework). Indeed, systemd is now routinely used on even embedded Linux devices. Within the systemd framework, logging is performed by a daemon process called `systemd-journal`, and the `journalctl` utility is the user interface to it.

The detailed coverage of systemd and its associated utilities is user mode stuff and thus not within the scope of this book. Please refer to the *Further reading* section of this chapter for links to (a lot) more on it.

One key advantage of using the journal to retrieve and interpret logs is that **all logs** - those from applications (processes and threads), libraries, system daemons, the kernel, drivers, and so on - are written (merged) here.

This way, we can see a (reverse) chronological timeline of events without having to manually piece together different logs into a timeline. The man page on the `journalctl(1)` utility covers its various options in detail. Here, we present some (hopefully) convenient aliases based on this utility (we assume it's within the `PATH` of course):

```
#--- a few journalctl(1) aliases
# jlog: current (from most recent) boot only, everything
alias jlog='journalctl -b --all --catalog -no-pager'

# jlogr: current (from most recent) boot only, everything,
 #  in *reverse* chronological order
alias jlogr='journalctl -b --all --catalog --no-pager --reverse'

# jlogall: *everything*, all time; --merge => _all_ logs merged
alias jlogall='journalctl --all --catalog --merge --no-pager'

# jlogf: *watch* the live log, akin to 'tail -f' mode; very useful to 'watch
live' logs;
# use 'journalctl -f -k' to only watch for kernel printk's
alias jlogf='journalctl -f'

# jlogk: only kernel messages, this (from most recent) boot
alias jlogk='journalctl -b -k --no-pager'
```

Note that the `-b` option implies "show logs for the current boot only" - in other words, the journal is displayed from the most recent system boot date at the present moment. A numbered listing of stored system (re)boots can be seen with `journalctl --list-boots`.

We deliberately use the `--no-pager` option as it allows us to further filter the output with standard Linux utilities like `grep`, `awk`, `sort`, and so on, as required. A simple example of using `journalctl` follows:

```
$ journalctl -b -k --no-pager | tail -n2
Jul 07 12:25:05 osboxes kernel: Goodbye, world! Climate change has done us
in...
Jul 07 12:37:57 osboxes kernel: Hello, world
```

Notice the default log format of the journal:

```
[timestamp] [hostname] [source-of-msg]: [... log message ...]
```

Here, [source-of-msg] is kernel for kernel messages or the name of the particular application or service that writes the message.

It's useful to see a couple of usage examples from the man page on journalctl(1):

```
Show all kernel logs from previous boot:
    journalctl -k -b -1

Show a live log display from a system service apache.service:
    journalctl -f -u apache
```

Furthermore, journalctl makes it easy to search the logs in a human-friendly time-based fashion – for example, to see all kernel logs that have been written in the last half hour:

```
journalctl -k --since="30 min ago"
```

The non-volatile logging of kernel messages into files is very useful, of course. Note, though, that there exist circumstances, often dictated by hardware constraints, that might render it impossible. For example, a tiny, highly resource-constrained embedded Linux device might use a small internal flash chip as its storage medium. Now, not only is it small, and all the space is pretty much used up by apps, libraries, the kernel, drivers, and the bootloader, but it is also a fact that flash-based chips have an effective limit on the number of erase-write cycles they can sustain before wearing out. Thus, writing to it a few million times might finish it off! So, sometimes, system designers deliberately and/or additionally use cheaper external flash memory such as (micro)SD/MMC (MultiMediaCard) cards for non-critical data to mitigate this impact, as they're easily (and cheaply) replaceable.

Let's move on to understanding printk log levels.

Using printk log levels

When you emit a message to the kernel log via the printk API (and friends), you typically must also specify the *logging level* at which the message is logged. To understand and use these printk log levels, let's begin by reproducing that single line of code – the first printk from our world-famous helloworld_lkm kernel module:

```
printk(KERN_INFO "Hello, world\n");
```

Let's now address the elephant in the room: what exactly does KERN_INFO mean? Firstly, be careful now; it's *not* what your knee-jerk reaction says it is – a parameter. No! Notice that there is no comma character between it and the *"Hello, world\n"* format string, just white space. KERN_INFO is one of **eight log levels** that a kernel printk can get logged at. A key thing to understand right away is that this log level is *not* a priority specifier of any sort; its presence allows us *to filter messages* based on the log level. The kernel defines eight possible log levels for printk; here they are:

```
// include/linux/kern_levels.h
```

```
#ifndef __KERN_LEVELS_H__
#define __KERN_LEVELS_H__

#define KERN_SOH      "\001"       /* ASCII Start Of Header */
#define KERN_SOH_ASCII  '\001'

#define KERN_EMERG    KERN_SOH "0"    /* system is unusable */
#define KERN_ALERT    KERN_SOH "1"    /* action must be taken immediately */
#define KERN_CRIT     KERN_SOH "2"    /* critical conditions */
#define KERN_ERR      KERN_SOH "3"    /* error conditions */
#define KERN_WARNING  KERN_SOH "4"    /* warning conditions */
#define KERN_NOTICE   KERN_SOH "5"    /* normal but significant condition */
#define KERN_INFO     KERN_SOH "6"    /* informational */
#define KERN_DEBUG    KERN_SOH "7"    /* debug-level messages */

#define KERN_DEFAULT           ""     /* the default kernel loglevel */
```

So, now we see that the `KERN_<FOO>` log levels are merely strings (`"0"`, `"1"`, `...`, `"7"`) that get prefixed to the kernel message being emitted by `printk`, nothing more. *This gives us the useful ability to filter messages based on log level.* The comment on the right of each of them clearly shows the developer when to use which log level.

What's `KERN_SOH`? That's the ASCII Start Of Header (SOH) value `\001`. See the man page on `ascii(7)`; the `ascii` utility dumps the ASCII table in various numerical bases. From here, we can clearly see that numeric `1` (or `\001`) is the `SOH` character, a convention that is followed here (Er, don't worry about it).

Now let's quickly look at a couple of actual `printk` usage examples from within the Linux kernel source tree. When the kernel's `hangcheck-timer` device driver (somewhat akin to a software watchdog) determines that a certain timer expiry (60 seconds by default) was delayed for over a certain threshold (by default, 180 seconds), it restarts the system! Here we show the relevant kernel code – the place where the `hangcheck-timer` driver emits a `printk` in this regard `https://elixir.bootlin.com/linux/v6.1.25/source/drivers/char/hangcheck-timer.c#L134`:

```
// drivers/char/hangcheck-timer.c
 [...]
if (hangcheck_reboot) {
     printk(KERN_CRIT "Hangcheck: hangcheck is restarting the machine.\n");
     emergency_restart();
} else {
   [...]
```

Check out how the `printk` API was called with the log level set to `KERN_CRIT`.

On the other hand, squeaking out an informational message might be just what the doctor ordered. Here, we see the generic parallel printer driver politely informing all concerned that the printer is on fire (rather understated, yes?). The relevant code is here: `https://elixir.bootlin.com/linux/v6.1.25/source/drivers/char/lp.c#L260`:

```
// drivers/char/lp.c
 [...]
if (last != LP_PERRORP) {
        last = LP_PERRORP;
        printk(KERN_INFO "lp%d on fire\n", minor);
 }
```

You'd think a device being on fire qualifies the `printk` to be emitted at the emergency logging level, yes? Well, at least `arch/x86/kernel/cpu/mce/p5.c:pentium_machine_check()` thinks so, `https://elixir.bootlin.com/linux/v6.1.25/source/arch/x86/kernel/cpu/mce/p5.c#L24`:

```
// arch/x86/kernel/cpu/mce/p5.c
 [...]
    pr_emerg("CPU#%d: Machine Check Exception: 0x%8X (type 0x%8X).\n",
            smp_processor_id(), loaddr, lotype);

    if (lotype & (1<<5)) {
        pr_emerg("CPU#%d: Possible thermal failure (CPU on fire ?).\n",
                smp_processor_id());
    }
[...]
```

The `pr_<foo>()` convenience macros are covered next.

An FAQ: If, within the `printk()`, the log level is not specified, what log level is the print emitted at? It's 4 by default – that is, `KERN_WARNING` (the *Writing to the console* section reveals why exactly this is). Note, though, that you are expected to always specify a suitable log level when using `printk`.

There's an easier way to specify the kernel message log level; it's what we will delve into next.

The pr_<foo> convenience macros

The convenience `pr_<foo>()` (often referred to as `pr_*()`) macros explained here ease coding pain. The clunky

```
printk(KERN_FOO "<format-str>", vars...);
```

is replaced with the elegant

```
        pr_foo("<format-str>", vars...);
```

where <foo> is the log level, one of *emerg, alert, crit, err, warn, notice, info,* or *debug*. Their use is encouraged:

```
// include/linux/printk.h:
[ ... ]
/**
 * pr_emerg - Print an emergency-level message
 * @fmt: format string
 * @...: arguments for the format string
 *
 * This macro expands to a printk with KERN_EMERG loglevel. It uses pr_fmt() to
 * generate the format string.
 */
#define pr_emerg(fmt, ...) \
    printk(KERN_EMERG pr_fmt(fmt), ##__VA_ARGS__)
/**
 * pr_alert - Print an alert-level message
 * @fmt: format string
 * @...: arguments for the format string
 *
 * This macro expands to a printk with KERN_ALERT loglevel. It uses pr_fmt() to
 * generate the format string.
 */
#define pr_alert(fmt, ...) \
    printk(KERN_ALERT pr_fmt(fmt), ##__VA_ARGS__)
 […]
#define pr_err(fmt, ...) \
    printk(KERN_ERR pr_fmt(fmt), ##__VA_ARGS__)
[ … ]
#define pr_warn(fmt, ...) \
    printk(KERN_WARNING pr_fmt(fmt), ##__VA_ARGS__)
[ … ]
#define pr_notice(fmt, ...) \
    printk(KERN_NOTICE pr_fmt(fmt), ##__VA_ARGS__)
[ … ]
#define pr_info(fmt, ...) \
    printk(KERN_INFO pr_fmt(fmt), ##__VA_ARGS__)
[ … ]
/**
 * pr_devel - Print a debug-level message conditionally
 [ … ]
```

```
 * This macro expands to a printk with KERN_DEBUG loglevel if DEBUG is
 * defined. Otherwise it does nothing.
 [ … ]
#ifdef DEBUG
#define pr_devel(fmt, ...) \
    printk(KERN_DEBUG pr_fmt(fmt), ##__VA_ARGS__)
[ … ]
```

While on the subject of using the `pr_*()` macros, there's one called `pr_cont()`. Its job is to act as a continuation string, continuing the previous `printk`! This can be useful… here's an example of its usage:

```
// kernel/module.c
    if (last_unloaded_module[0])
        pr_cont(" [last unloaded: %s]",
                last_unloaded_module);
    pr_cont("\n");
```

We typically ensure that only the final `pr_cont()` contains the newline character.

To avoid unpleasant surprises, like the `printk()` seeming to not take effect, make it a habit to end your `printk`'s with a newline character `"\n"` (this is true for the `printf()` as well).

From now on, we'll typically only use these `pr_foo() (or pr_*())` style macros when emitting `printk` messages to the kernel log.

Furthermore, and very importantly: **driver authors** must use the `dev_*()` macros. This implies passing along an additional first parameter, a pointer to the device in question (`struct device *`). You'll find them all defined here: `https://elixir.bootlin.com/linux/v6.1.25/source/include/linux/dev_printk.h#L137`. The reason to use them is that the driver core auto-prefixes useful details to the driver `printk`, thus enabling better logs (and one to quickly spot details). For example, when writing, say, an I2C driver, the `dev_*()` message will include details like the driver's name, I2C bus number, and the address the chip is on! Here's an example, from a (reworked) I2C RTC driver:

```
ks3231 1-0068: IRQ! #5
```

Here, `ks3231` is the driver's name; for the token `1-0068`, `1` is the I2C bus number and `0x68` is the address where the client (RTC) chip resides on that bus. The actual message is `IRQ! #5`. Useful.

The kernel allows us to pass `loglevel=n` as a kernel command-line parameter, where `n` is an integer between `0` and `7`, corresponding to the eight log levels mentioned previously. As expected (and as you shall soon learn), all `printk` instances with a log level less than that which was passed will be directed to the console device as well.

Writing a kernel message directly to the console device is at times very useful; the next section deals with the details of how we can achieve this.

Writing to the console

Recall that the `printk` output might go to up to three locations:

- The first is the kernel memory log buffer (always)
- The second is non-volatile log files (typical)
- The last one (that we'll address here) is the *console device*

Traditionally, the console device is a pure kernel feature, the initial Terminal window that the (super) user logs in to (`/dev/console`) in a non-graphical environment. Interestingly, on Linux, we can define several consoles – a **teletype terminal** (**tty**) window (such as `/dev/console`), a text-mode VGA console, a framebuffer, or even a serial port served over USB (this being common on embedded systems during development; see more on Linux consoles in the *Further reading* section of this chapter).

For example, when we connect a Raspberry Pi board to an x86_64 laptop via a USB-to-RS232 TTL UART (USB-to-serial) cable (see the *Further reading* section of this chapter for a blog article on this very useful accessory and how to set it up on the Raspberry Pi) and then use a Terminal emulator app like `minicom` (or `screen`), this is what shows up as the `tty` device on the Raspberry Pi – it's the serial port:

```
rpi # tty
/dev/ttyS0
```

The point here is that the console is often the target of *important enough* log messages, including those originating from deep within the kernel. Linux's printk uses a *sysctl*, a `proc`-based mechanism, for conditionally delivering its logs to the console device. To understand this better, let's first check out the relevant `proc` pseudofile (here, on our x86_64 Ubuntu guest):

```
$ cat /proc/sys/kernel/printk
4    4    1    7
```

We interpret the preceding four numbers as `printk` log levels (with `0` being the highest and `7` the lowest in terms of "urgency"). The preceding four-integer sequence's meaning is this:

- The current (console) log level. *The key implication is that all messages less than this value will be sent to the console device as well!*
- The default level for messages that lack an explicit log level.
- The minimum allowed log level.
- The boot-time default log level.

We can see that log level `4` corresponds to `KERN_WARNING`. Thus, with the first number being `4` (indeed, the typical default on a Linux distro), *all printk instances lower than log level 4 will be sent to the console device*, in addition to being logged to the kernel log buffer and a file, of course. In effect, with these settings, this applies to all kernel messages at the following log levels: `KERN_EMERG`, `KERN_ALERT`, `KERN_CRIT`, and `KERN_ERR` will, by default, be automatically sent to the console device as well.

Kernel messages at log level `0 [KERN_EMERG]` are always printed to the console, and indeed to all (non-GUI) Terminal windows and the kernel log buffer and log file, regardless of any settings.

It's worth noting that very often when working on embedded Linux or any kernel development, you will work on the console device, as is the case with the Raspberry Pi example just given. On my Raspberry Pi 4 model B running the stock distro (and kernel), here's the default `printk` log level setting:

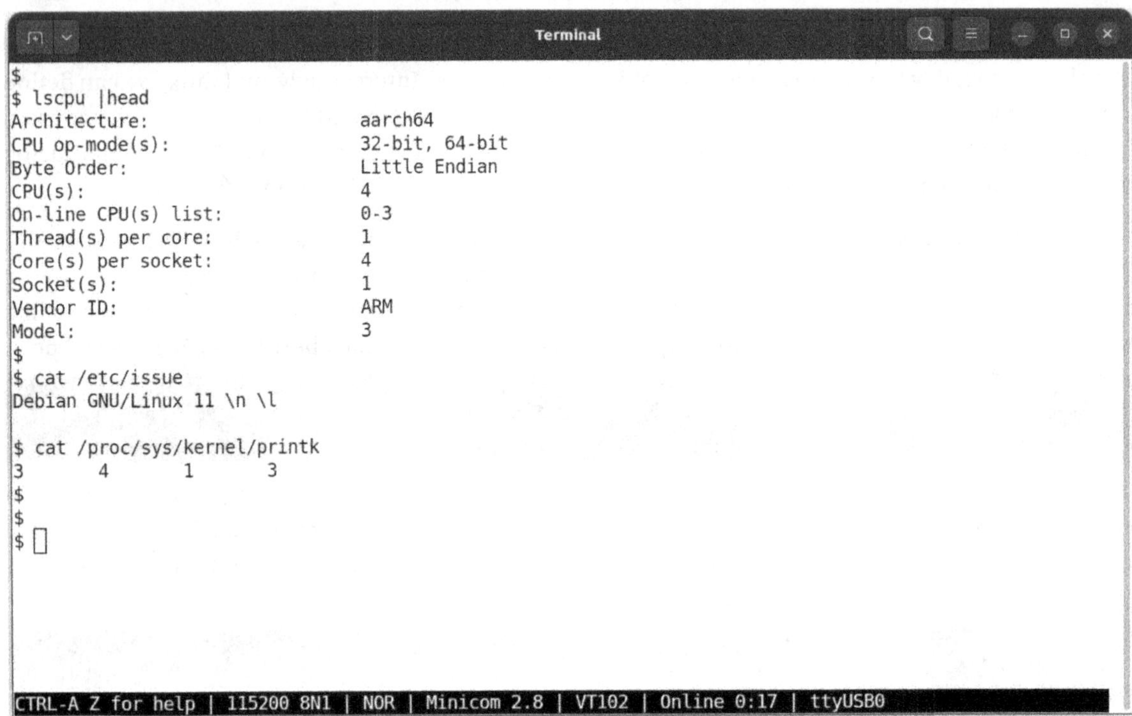

Figure 4.7: Screenshot of running minicom to connect to the Raspberry Pi board over a USB-to-serial cable, looking up a few things

Notice that the default `printk` logging levels (i.e., the `printk sysctl`) are different from those of the x86_64; it happens.

Now, the interesting bit: setting the `/proc/sys/kernel/printk` pseudofile's first integer value to 8 will *guarantee that all printk instances are sent directly to the console as well*, **thus making printk behave like a regular printf would!** Here, we show how the root user can easily set this up:

```
$ sudo sh -c "echo '8 4 1 7' > /proc/sys/kernel/printk"
$ cat /proc/sys/kernel/printk
8       4       1       7
```

Of course, we do this as root; do notice the shell syntax used. Setting the console output like this can be very convenient during development and testing!

However, realize that the changes made to this sysctl (via `proc`) are temporary; they apply for this session only. Once the system reboots or power cycles, the changes are lost. One way to make changes to a sysctl permanent is via the `sysctl(8)` utility – for example, running, as root, the following:

```
sysctl -w kernel.printk='8 4 1 7'
```

This makes the `printk sysctl` get set to these values on every boot.

On my Raspberry Pi (and other boards), I keep a startup script that contains the following line:

```
[[ $(id -u) -eq 0 ]] && echo "8 4 1 7" > /proc/sys/kernel/printk
```

Thus, when running it as root, this takes effect and all `printk` instances now directly appear on the `minicom` console, just as `printf` would; great for debugging! Using `sysctl`, you can make it permanent.

Talking about the versatile Raspberry Pi, the next section demonstrates running a kernel module on one.

Writing output to the Raspberry Pi console

On to our second kernel module! Here, we shall (from now onward) employ the `pr_*()` style macros to emit nine `printk` instances, one at each of the eight log levels, plus one via the `pr_devel()` macro (which is really nothing but the `KERN_DEBUG` log level). Let's check out the relevant code:

```
// ch4/printk_loglvl/printk_loglvl.c
[ … ]
static int __init printk_loglvl_init(void)
{
    pr_emerg("Hello, world @ log-level KERN_EMERG   [0]\n");
    pr_alert(  "Hello, world @ log-level KERN_ALERT   [1]\n");
    pr_crit(    "Hello, world @ log-level KERN_CRIT    [2]\n");
    pr_err(     "Hello, world @ log-level KERN_ERR     [3]\n");
    pr_warn(  "Hello, world @ log-level KERN_WARNING [4]\n");
    pr_notice("Hello, world @ log-level KERN_NOTICE  [5]\n");
    pr_info(    "Hello, world @ log-level KERN_INFO    [6]\n");
    pr_debug("Hello, world @ log-level KERN_DEBUG   [7]\n");
    pr_devel( "Hello, world via the pr_devel() macro"
        " (eff @KERN_DEBUG) [7]\n");

    return 0; /* success */
}
```

```
static void __exit printk_loglvl_exit(void)
{
    pr_info("Goodbye, world @ log-level KERN_INFO    [6]\n");
}
module_init(printk_loglvl_init);
module_exit(printk_loglvl_exit);
```

Now, we will discuss the output when running the preceding `printk_loglvl` kernel module on a Raspberry Pi device. If you don't possess one or it's not handy, that's not a problem; please go ahead and try it out on an x86_64 (guest VM or native system).

On the Raspberry Pi device (here, I used the Raspberry Pi 4 model B running the default Raspberry Pi OS), we log in and get ourselves a root shell via a simple `sudo -s`. We then build the kernel module.

If you have installed the default Raspberry Pi image on the device, all required development tools, kernel headers, and more will be pre-installed! So, log in to the board, build the module, and try it out. *Figure 4.8* is a screenshot of running our `printk_loglvl` kernel module on a Raspberry Pi board. Also, it's important to realize that we're running **on the console device** as we are using the aforementioned USB-to-serial cable over the `minicom` Terminal emulator app, and not simply over an SSH connection:

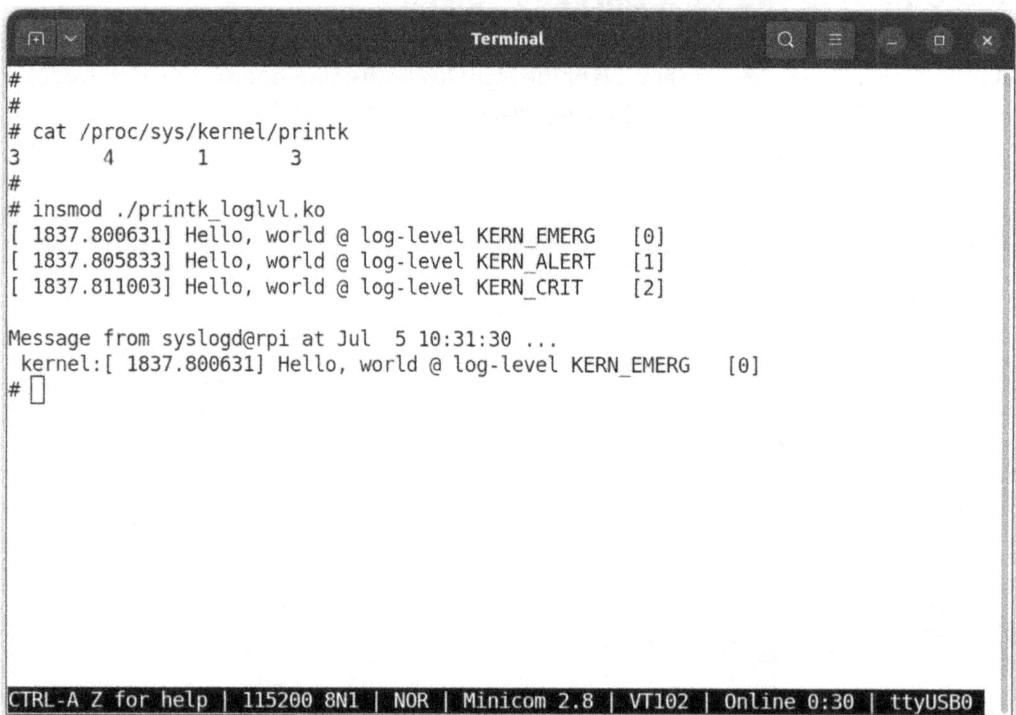

Figure 4.8: The minicom Terminal emulator app window – the console – with the printk_loglvl kernel module output

Notice something a bit different from the x86_64 environment: here, by default, the first integer in the output of `/proc/sys/kernel/printk` – the current console log level – is 3 (not 4). Okay, so this implies that all kernel `printk` instances at log level *less than log level 3* will appear directly on the console device. Look at the screenshot: this is indeed the case! Furthermore, and as expected, the `printk` instance at the "emergency" log level (`0, KERN_EMERG`) always appears on the console – indeed, on every open non-GUI Terminal window.

Now for the interesting part; let's set (as root, of course) **the current console log level** (remember, it's the first integer in the output of `/proc/sys/kernel/printk`) to the value `8`. This way, all `printk` instances should appear directly on the console. We test precisely this here:

```
Terminal
# cat /proc/sys/kernel/printk
3       4       1       3
# echo "8 4 1 3" > /proc/sys/kernel/printk
# cat /proc/sys/kernel/printk
8       4       1       3
#
# insmod ./printk_loglvl.ko
insmod: ERROR: could not insert module ./printk_loglvl.ko: File exists
# rmmod printk_loglvl
[ 2083.540591] Goodbye, world @ log-level KERN_INFO    [6]
#
# insmod ./printk_loglvl.ko
[ 2086.684939] Hello, world @ log-level KERN_EMERG   [0]
[ 2086.690143] Hello, world @ log-level KERN_ALERT   [1]
[ 2086.695526] Hello, world @ log-level KERN_CRIT    [2]

[ 2086.700826] Hello, world @ log-level KERN_ERR     [3]
Message[ 2086.706233] Hello, world @ log-level KERN_WARNING [4]
 from sy[ 2086.711999] Hello, world @ log-level KERN_NOTICE  [5]
slogd@rp[ 2086.717931] Hello, world @ log-level KERN_INFO    [6]
i at Jul  5 10:35:39 ...
 kernel:[ 2086.684939] Hello, world @ log-level KERN_EMERG   [0]
# 

CTRL-A Z for help | 115200 8N1 | NOR | Minicom 2.8 | VT102 | Online 0:35 | ttyUSB0
```

Figure 4.9: The minicom Terminal – in effect, the console – with the console log level set to 8

Of course, we first remove the earlier instance of the module from memory via `rmmod`. Indeed, as expected, we see all the `printk` instances on the console device itself, thus obviating the need to use `dmesg`.

Hang on a moment, though: whatever happened to the `pr_debug()` and `pr_devel()` macros emitting a kernel message at log level `KERN_DEBUG` (that is, the integer value 7)? It has not appeared on the console nor in the following `dmesg` output. We explain this shortly; please read on.

With `dmesg`, of course, all kernel messages – well, at least those still in the kernel log (ring) buffer in RAM – will be revealed. We see this to be the case here:

```
rpi # rmmod printk_loglvl
rpi # dmesg
```

```
 [...]
[ 2086.684939] Hello, world @ log-level KERN_EMERG   [0]
[ 2086.690143] Hello, world @ log-level KERN_ALERT   [1]
[ 2086.695526] Hello, world @ log-level KERN_CRIT    [2]
[ 2086.700826] Hello, world @ log-level KERN_ERR     [3]
[ 2086.706233] Hello, world @ log-level KERN_WARNING [4]
[ 2086.711999] Hello, world @ log-level KERN_NOTICE  [5]
[ 2086.717931] Hello, world @ log-level KERN_INFO    [6]
[ 2372.690250] Goodbye, world @ log-level KERN_INFO    [6]
rpi #
```

All `printk` instances – except the `KERN_DEBUG` ones – are seen as we are looking at the kernel log via the `dmesg` utility. So, how do we get a debug message displayed? That's covered in the next section.

Turning on debug-level kernel messages

We just saw that we can emit `printk` instances at various log levels; all showed up except for the debug-level messages. Ah yes, `pr_debug()` (as well as `dev_dbg()`) turns out to be a bit of a special case; unless the `DEBUG` symbol is *defined* for the kernel module, a `printk` instance at log level `KERN_DEBUG` does not show up. We can edit the kernel module's Makefile to enable this. There are (at least) two ways to set this up:

- Insert this line into the Makefile:

  ```
  CFLAGS_printk_loglvl.o := -DDEBUG
  ```

Generically, it's `CFLAGS_<filename>.o  :=  <value>`

- We could also – more commonly in fact – just insert this statement into the Makefile:

  ```
  ccflags_y += -DDEBUG
  ```

Traditionally, the Makefile variable to use for passing on more C compiler flags was named `EXTRA_CFLAGS`; it's now considered deprecated, hence we use the newer `ccflags_y` variable.

In our Makefile, we have deliberately kept the `-DDEBUG` commented out to begin with. Now, to try it out, uncomment one of the following commented-out lines (I'd suggest you uncomment the first one, the `ccflags_y` variable):

```
# Enable the pr_debug() as well (rm the comment from one of the lines below)
# (Note: EXTRA_CFLAGS deprecated; use ccflags-y)
#ccflags-y += -DDEBUG
#CFLAGS_printk_loglvl.o := -DDEBUG
```

Once edited, rebuild it, and insert it using our `lkm` script. A partial screenshot (*Figure 4.10*) of the `lkm` script's output clearly reveals the `dmesg` color-coding, with `KERN_ALERT / KERN_CRIT / KERN_ERR` background highlighted in red/in bold red typeface/in red foreground color, respectively, and `KERN_WARNING` in bold black typeface, helping us humans quickly spot important kernel messages:

```
------------------------------
sudo insmod ./printk_loglvl.ko && lsmod|grep printk_loglvl
------------------------------

Message from syslogd@rpi at Jul  5 10:53:15 ...
 kernel:[ 3142.614320] Hello, world @ log-level KERN_EMERG   [0]
printk_loglvl          16384  0
------------------------------
sudo dmesg
------------------------------
[ 3142.614320] Hello, world @ log-level KERN_EMERG   [0]
[ 3142.619525] Hello, world @ log-level KERN_ALERT   [1]
[ 3142.624670] Hello, world @ log-level KERN_CRIT    [2]
[ 3142.629825] Hello, world @ log-level KERN_ERR     [3]
[ 3142.635041] Hello, world @ log-level KERN_WARNING [4]
[ 3142.640176] Hello, world @ log-level KERN_NOTICE  [5]
[ 3142.645381] Hello, world @ log-level KERN_INFO    [6]
[ 3142.650525] Hello, world @ log-level KERN_DEBUG   [7]
[ 3142.655818] Hello, world via the pr_devel() macro (eff @KERN_DEBUG) [7]
$
$
Message from syslogd@rpi at Jul  5 10:57:46 ...
 kernel:[ 3414.117994] Hello, world @ log-level KERN_EMERG   [0]
```

Figure 4.10: Partial screenshot of the lkm script's output when running the module with DEBUG defined

We've highlighted the two `printk` instances emitted at `KERN_DEBUG`. Note that the behavior of `pr_debug()` is not identical when the "dynamic debug" feature (`CONFIG_DYNAMIC_DEBUG=y;` – covered next) is enabled.

As mentioned before, device driver authors should carefully note that you must use the special `dev_*()` macros for emitting `printk` instances, not the usual `pr_*()` ones used by the kernel and pure modules. This implies passing along an additional first parameter, a pointer to the device in question (`struct device *`). You'll find them all defined here: `https://elixir.bootlin.com/linux/v6.1.25/source/include/linux/dev_printk.h#L137`.

Also, `pr_devel()` is meant to be used for kernel-internal debug `printk` instances whose output should never be visible in production systems.

Now, back to the section on console output. So, for perhaps the purpose of kernel/driver debugging (if nothing else), is there any other guaranteed way to ensure that *all* `printk` instances are directed to the console? Yes, indeed – just pass the kernel (boot-time) parameter called `ignore_level`. For more details on this, do look up the description in the official kernel documentation: `https://www.kernel.org/doc/html/latest/admin-guide/kernel-parameters.html`.

Additionally, you can turn on the ignoring of printk log levels at runtime by echoing Y into the following pseudofile, thus allowing all printk instances to appear on the console device:

```
sudo bash -c "echo Y > /sys/module/printk/parameters/ignore_loglevel"
```

Conversely, you can turn it off at runtime by echoing N into the same pseudo-file. Useful!

The dmesg utility can also be used to control the enabling/disabling of kernel messages to the console device, as well as the console logging level (that is, the level numerically below which messages will appear on the console) via various option switches (in particular, the --console-level option). I'll leave it to you to browse through the man page on dmesg(1) for the details.

Important note: Using the printk (and friends) APIs/macros for the express purpose of debugging has been covered in depth in my *Linux Kernel Debugging* book (specifically in *Chapter 3, Debug via Instrumentation – printk and Friends*). It covers the usual printk stuff (which we have covered here), printk for debug tips and tricks, rate-limiting, and, importantly, using the powerful *dynamic debug* kernel framework. Please do check it out.

An introduction to the kernel's powerful dynamic debug feature

We, as programmers, need debug techniques. Perhaps the best known is the *instrumentation* one, where debug prints are sprinkled at strategic locations in the code base. Seeing the log, we then understand the control flow and can perhaps spot the bug.

With the Linux kernel, debug and other prints are emitted into the kernel log via various printk variants – the regular printk() macro, the preferred pr_*() macros, and the dev_*() macros for drivers; those that are emitted with log level KERN_DEBUG are the so-called debug prints. So, in effect, the following are all possible "debug-level" print call sites:

- `pr_debug("<fmt str>"[, args…]);`
- `dev_dbg(dev, "<fmt str>"[, args…]);`
- `printk(KERN_DEBUG "<fmt str>"[, args…]);`

It's very important to recall that *debug printk instances are always off by default*; only when the symbol DEBUG is defined do they actually take effect.

Now, this seems convenient; however, in actual production, what if you need to emit, say, a couple of debug prints from a given module? To do so, you'll have to define the symbol DEBUG (of course, you can do that within the Makefile), rebuild, unload, and reload the module. This approach, however, just isn't practical (or even allowed) on most production systems. Thus, we need a more dynamic approach, which is precisely what's offered by the kernel's **dynamic debug** feature.

With the kernel dynamic debug feature, every single printk call site that's emitted at log level KERN_DEBUG from within the kernel, as well as those from kernel modules, is compiled into the kernel. Within the make menuconfig UI, the dynamic debug option is here: *Kernel hacking > printk and dmesg options > Enable dynamic printk() support*. The config is a Boolean named CONFIG_DYNAMIC_DEBUG.

We'll assume you have it on (set to `Y`; even when so, the overhead when not tracing dynamic prints is considered minimal, which is why the feature was merged!).

As this book isn't focused on kernel debugging, we cover the basics here; please see the *Linux Kernel Debugging* book for details. Still, enough is certainly covered here to get you started and to appreciate this powerful feature.

This feature's "Help" (recall it will be within one of the `Kconfig*` files) is superb; let's just dump it, as reading it will make you understand how you're to use it (on the 6.1.25 kernel, it's here: `https://elixir.bootlin.com/linux/v6.1.25/source/lib/Kconfig.debug#L108`:

```
config DYNAMIC_DEBUG
        bool "Enable dynamic printk() support"
        default n
        depends on PRINTK
        depends on (DEBUG_FS || PROC_FS)
        select DYNAMIC_DEBUG_CORE
        help

          Compiles debug level messages into the kernel, which would not
          otherwise be available at runtime. These messages can then be
          enabled/disabled based on various levels of scope - per source file,
          function, module, format string, and line number. This mechanism
          implicitly compiles in all pr_debug() and dev_dbg() calls, which
          enlarges the kernel text size by about 2%.

          If a source file is compiled with DEBUG flag set, any
          pr_debug() calls in it are enabled by default, but can be
          disabled at runtime as below.  Note that DEBUG flag is
          turned on by many CONFIG_*DEBUG* options.

          Usage:

          Dynamic debugging is controlled via the 'dynamic_debug/control' file,
          which is contained in the 'debugfs' filesystem or procfs.
          Thus, the debugfs or procfs filesystem must first be mounted before
          making use of this feature.
          We refer the control file as: <debugfs>/dynamic_debug/control. This
          file contains a list of the debug statements that can be enabled. The
          format for each line of the file is:

                filename:lineno [module]function flags format

          filename : source file of the debug statement
          lineno : line number of the debug statement
          module : module that contains the debug statement
          function : function that contains the debug statement
          flags : '=p' means the line is turned 'on' for printing
          format : the format used for the debug statement

          From a live system:

                nullarbor:~ # cat <debugfs>/dynamic_debug/control
"~/kernels/linux-6.1.25/lib/Kconfig.debug" 2821 lines --5%--
```

Figure 4.11: Screenshot showing the "Help" for the kernel's powerful dynamic debug feature, part 1

As the full details don't fit, here's the next screenshot to cover it:

```
From a live system:

       nullarbor:~ # cat <debugfs>/dynamic_debug/control
       # filename:lineno [module]function flags format
       fs/aio.c:222 [aio]__put_ioctx =_ "__put_ioctx:\040freeing\040%p\012"
       fs/aio.c:248 [aio]ioctx_alloc =_ "ENOMEM:\040nr_events\040too\040high\012"
       fs/aio.c:1770 [aio]sys_io_cancel =_ "calling\040cancel\012"

Example usage:

       // enable the message at line 1603 of file svcsock.c
       nullarbor:~ # echo -n 'file svcsock.c line 1603 +p' >
                                       <debugfs>/dynamic_debug/control

       // enable all the messages in file svcsock.c
       nullarbor:~ # echo -n 'file svcsock.c +p' >
                                       <debugfs>/dynamic_debug/control

       // enable all the messages in the NFS server module
       nullarbor:~ # echo -n 'module nfsd +p' >
                                       <debugfs>/dynamic_debug/control

       // enable all 12 messages in the function svc_process()
       nullarbor:~ # echo -n 'func svc_process +p' >
                                       <debugfs>/dynamic_debug/control

       // disable all 12 messages in the function svc_process()
       nullarbor:~ # echo -n 'func svc_process -p' >
                                       <debugfs>/dynamic_debug/control

See Documentation/admin-guide/dynamic-debug-howto.rst for additional
information.
```

Figure 4.12: Screenshot showing the "Help" for the kernel's powerful dynamic debug feature, part 2

Do read both *Figure 4.11* and *Figure 4.12* carefully. They provide enough detail to get you started on using dynamic debug; the official kernel documentation on this feature is superb as well, so please do browse through it: `https://www.kernel.org/doc/html/v6.1/admin-guide/dynamic-debug-howto.html#dynamic-debug`.

To check whether the kernel's *dynamic debug* facility is available (on an x86), if it shows as *y* (as is the case here), it's available, else, it's not:

```
$ grep "DYNAMIC_DEBUG=y" /boot/config-$(uname -r)
CONFIG_DYNAMIC_DEBUG=y
```

So, our interface to work with dynamic debug is a pseudofile under `debugfs`: `<debugfs_mount_point>/dynamic_debug/control`. On many production systems, though, `debugfs` may not be configured (or is kept as "invisible"); in such cases, the control file is visible via `proc`: `/proc/dynamic_debug/control`. Both are identical.

Viewing the control file content shows all debug prints or call sites that are compiled into the kernel. A quick look helps:

```
$ head -n1 /proc/dynamic_debug/control
# filename:lineno [module]function flags format
```

The first line reveals the format. All fields are self-explanatory except the (key) one named `flags`; this is where the magic lies!

Here's the scoop on `flags`, in brief: if the value is `=_`, the debug print is off (typically, this is the case by default); for example, let's look up a few debug print call sites for the e1000 (module) driver:

```
# grep "e1000" /proc/dynamic_debug/control | head -n3
drivers/net/ethernet/intel/e1000/e1000_hw.c:385 [e1000]e1000_reset_hw =_
"Disabling MWI on 82542 rev 2.0\n"
drivers/net/ethernet/intel/e1000/e1000_hw.c:390 [e1000]e1000_reset_hw =_
"Masking off all interrupts\n"
drivers/net/ethernet/intel/e1000/e1000_hw.c:423 [e1000]e1000_reset_hw =_
"Issuing a global reset to MAC\n"
```

You can see that the debug prints are off (the `=_` confirms this). To turn it on, write the value `+p` into the `flags` field. There are several optional values you can write as well: `m`, `f`, `l`, and `t`; please see the documentation for their meanings.

A few examples follow, running the dynamic debug facility at different levels of scope. Do first ensure the basic prerequisite (`CONFIG_DYNAMIC_DEBUG=y`) is met. You'll need to run these as root, of course; also, you can always substitute the control file path of `/sys/kernel/debug/dynamic_debug/control` used below with the path `/proc/dynamic_debug/control`:

- File scope: enable all debug messages in all files where the pathname includes the string "`usb`":

  ```
  echo -n 'file *usb* +p' > /sys/kernel/debug/dynamic_debug/control
  ```

  You can try plugging in/unplugging a USB device while this is on while watching the kernel log with `journalctl -k -f` in another terminal. Turn it off with the `-p`:

  ```
  echo -n 'file *usb* -p' > /sys/kernel/debug/dynamic_debug/control
  ```

- File scope: enable all debug messages in all files where the pathname includes the string "`ip`":

  ```
  echo -n 'file *ip* +p' > /sys/kernel/debug/dynamic_debug/control
  ```

  Try pinging a website while watching the kernel log with `journalctl -k -f` in another terminal. Turn it off with the `-p`:

  ```
  echo -n 'file *ip* -p' > /sys/kernel/debug/dynamic_debug/control
  ```

- Module scope: enable all debug `printk` instances (all `pr_debug()`|`dev_dbg()` call sites) for all kernel modules currently in memory:

  ```
  echo -n 'module * +pflmt' > \
  ```

```
    /sys/kernel/debug/dynamic_debug/control
```

WARNING: This can often lead to very voluminous output!

Turn it off with:

```
echo -n 'module * -pflmt' > \
 /sys/kernel/debug/dynamic_debug/control
```

While these run, keep a keen eye on the kernel log output in another Terminal window with `journalctl -k -f`. Of course, the settings are volatile and will reset to defaults on a power cycle or reboot.

The kernel documentation has more examples; do check them out: `https://www.kernel.org/doc/html/v6.1/admin-guide/dynamic-debug-howto.html#examples`.

The next section deals with another very useful logging feature: rate limiting.

Rate limiting the printk instances

When we emit `printk` instances from a code path that is executed very often, the sheer volume of `printk` instances might quickly overflow the kernel log buffer (in RAM, remember that it's a circular buffer!), thus overwriting what might well be key information. Besides that, ever-growing non-volatile log files that repeat pretty much the same kernel log messages (almost) ad infinitum are not a great idea either and waste disk space, or worse, flash space. For example, think of a large-ish `printk` in an interrupt handler code path. What if the hardware interrupt is invoked at a frequency of, say, 100 Hz – that is, 100 times every single second?

To mitigate these issues, the kernel provides an interesting and useful alternative: the *rate-limited* `printk`. The `pr_<foo>_ratelimited()` macro (where `<foo>` is one of *emerg, alert, crit, err, warn, notice, info, or debug*) has identical syntax to the regular `printk`; the key point is that it effectively *suppresses* regular prints when certain conditions are fulfilled.

The kernel provides two `sysctl` files named `printk_ratelimit` and `printk_ratelimit_burst` via the `proc` filesystem for this purpose. Here, we directly reproduce the `sysctl` documentation (from `https://www.kernel.org/doc/Documentation/sysctl/kernel.txt`) that explains the precise meaning of these two (pseudo) files:

```
printk_ratelimit:
Some warning messages are rate limited. printk_ratelimit specifies
the minimum length of time between these messages (in jiffies), by
default we allow one every 5 seconds.
A value of 0 will disable rate limiting.
==============================================
printk_ratelimit_burst:
While long term we enforce one message per printk_ratelimit
seconds, we do allow a burst of messages to pass through.
printk_ratelimit_burst specifies the number of messages we can
send before ratelimiting kicks in.
```

On our x86_64 Ubuntu 22.04 LTS guest system, we find that their (default) values are as follows:

```
$ cat /proc/sys/kernel/printk_ratelimit /proc/sys/kernel/printk_ratelimit_burst
5
10
```

This implies that by default, up to 10 instances of the same message occurring within a 5-second time interval can make it through before rate limiting kicks in.

FYI, this a web tool to help you search out what a certain `sysctl` means; here is an example: `https://sysctl-explorer.net/kernel/printk_ratelimit/`. Useful.

The `printk` rate limiter, when it does suppress kernel `printk` instances, emits a helpful message mentioning exactly how many earlier `printk` callbacks were suppressed. Some sample output helps make this clear:

```
ratelimit_test_init: 41 callbacks suppressed
ratelimit_test:ratelimit_test_init():45: [51] ratelimited printk @ KERN_INFO
[6]
```

Rate-limiting macros to use

The kernel provides the following macros to help you rate-limit your prints/logging (you should use `#include <linux/kernel.h>`):

- `printk_ratelimited()`: Warning! Do **not** use it – the kernel warns against this.
- `pr_*_ratelimited()`: Where the wildcard * is replaced by the usual – one of *emerg, alert, crit, err, warn, notice, info,* or *debug*.
- `dev_*_ratelimited()`: Where the wildcard * is replaced by the usual – one of *emerg, alert, crit, err, warn, notice, info,* or *debug*.

Ensure you use the `pr_*_ratelimited()` macros in preference to `printk_ratelimited()`; driver authors should use the `dev_*_ratelimited()` macros.

An example kernel module to demo the rate-limited `printk` is available in the GitHub repo of my *Linux Kernel Debugging* book, here: `https://github.com/PacktPublishing/Linux-Kernel-Debugging/tree/main/ch3/ratelimit_test`. Do try it out.

A code-level example (among several within the kernel tree) of using the rate-limited (driver debug) `printk` is seen in an AMD GPU driver, `https://elixir.bootlin.com/linux/v6.1.25/source/drivers/gpu/drm/amd/amdgpu/sdma_v4_0.c#L2146`:

```
dev_dbg_ratelimited(adev->dev,
```

```
"[sdma%d] address:0x%016llx src_id:%u ring:%u vmid:%u "
"pasid:%u, for process %s pid %d thread %s pid %d\n",
instance, addr, entry->src_id, entry->ring_id, entry->vmid,
entry->pasid, task_info.process_name, task_info.tgid,
task_info.task_name, task_info.pid);
```

With the `dev_*_*()` macros, notice how the first parameter is always a pointer to the device.

Don't use the older (and now deprecated) `printk_ratelimited()` and `printk_ratelimit()` macros. Also, FYI, the actual rate-limiting code is in `lib/ratelimit.c:___ratelimit()`. Here is a kernel convention: a function/macro beginning with two or more underscores (`__foo()`) is considered an internal function/macro. Avoid using it directly; use it's wrapper `foo()`.

Can we generate kernel-level messages from user space? Sounds interesting; that's our next sub-topic.

Generating kernel messages from user space

A popular debug technique that we programmers use is to sprinkle prints at various points in the code, often allowing us to narrow down the source of an issue. This is indeed a useful debugging technique and is more formally called **instrumenting** the code. Kernel developers often use the venerable `printk` API (and friends) for just this purpose.

So, imagine you have written a kernel module and are in the process of debugging it (by adding several `printk` instances at appropriate points in the code). Your kernel code now emits these several `printk` instances, which, of course, you can see at runtime via `dmesg` (or `journalctl`, or whatever).

That's fine, but what if, especially because you're running some automated user space test script, you'd like to see the point at which the script initiated some action within our kernel module, by printing out a certain message? As a concrete example, say we want the log to look something like this:

```
test_script: @user msg 1 ; kernel_module: msg n, msg n+1, ..., msg n+m ; test_
script: @user msg 2 ; ...
```

We can have our user space test script write a message into the kernel log buffer, just like a kernel `printk` would, by writing said message into the special `/dev/kmsg` device file:

```
echo "test_script: @user msg 1" > /dev/kmsg
```

Well, hang on – doing so requires running with root access, of course. However, notice here that a simple sudo before echo just doesn't work:

```
$ sudo echo "test_script: @user msg 1" > /dev/kmsg
bash: /dev/kmsg: Permission denied
$ sudo bash -c "echo \"test_script: @user msg 1\" > /dev/kmsg"
 [sudo] password for c2kp:
```

```
$ dmesg |tail -n1
[55527.523756] test_script: @user msg 1
```

The syntax used in the second attempt works, but it's just simpler to get yourself a root shell (`sudo -s`) and carry out tasks such as this.

One more thing. The `dmesg` utility has several options designed to make the output more human-readable; we show some of them via our sample alias to `dmesg` here, after which we use it:

```
$ alias dmesg='sudo dmesg --decode --nopager --color --ctime'
$ dmesg | tail -n1
kern  :debug : [Mon Jul 10 08:14:32 2023] wlo1: Limiting TX power to 30 (30 -
0) dBm as advertised by b8:<...>
```

The message written to the kernel log via the special `/dev/kmsg` device file will be printed at the current default log level, typically, `4 : KERN_WARNING`. We can override this by prefixing the message with the required log level (as a number in string format). For example, to write from the user space into the kernel log at log level `6 : KERN_INFO`, use this:

```
$ sudo bash -c "echo \"<6>test_script: test msg at KERN_INFO\"   \
   > /dev/kmsg"
 $ dmesg | tail -n1
user  :info  : [Mon Jul 10 08:25:27 2023] test_script: test msg at KERN_INFO
```

We can see that our latter message is emitted at log level `6`, as specified within `echo`.

There really is no way to distinguish between a user-generated kernel message and a kernel `printk()`-generated one; they look identical. So, of course, it could be as simple as inserting or prefixing some special signature byte or string within the message, such as `@user ...`, to help you distinguish these user-generated prints from the kernel ones.

As a bit of an aside, something to note follows.

Interfacing between user and kernel space is a key and important topic, often something that driver authors find themselves needing to do. There are many ways to do so, always, at some level involving issuing a system call from user space as the means to switch into the kernel. Among the available technologies on Linux to interface between user and kernel space are to do so via `procfs, sysfs, debugfs, netlink sockets`, and the `ioctl()` system call. These are covered in depth in the *Linux Kernel Programming - Part 2* book in *Chapter 2, User-Kernel Communication Pathways.*

Let's now move on to seeing how we can easily standardize our kernel `printk` output format.

Standardizing printk output via the pr_fmt macro

Another important point regarding the kernel `printk`: pretty often, to give context to your `printk()` output (*where exactly did it occur?*), you might write the code like this, taking advantage of various GCC macros like `__FILE__`, `__func__`, and `__LINE__`, which of course emit the filename, function name and line number, respectively, from where the print was emitted:

```
pr_info("%s:%s:%s():%d: mywork XY failed!\n", OURMODNAME, __FILE__, __func__,
__LINE__);
```

This is fine as such; the problem is that if there are a lot of `printk` instances in your project, it can be painful to guarantee that a standard `printk` format (for example, first displaying the module name followed by the filename, then the function name and possibly the line number, as seen here) is always consistently and correctly followed by everyone working on the project.

Enter the `pr_fmt()` macro; defining this macro right at the beginning of your code (it must be even before the first `#include`), guarantees that every single subsequent `printk` in your code *will be prefixed with the format specified by this macro.*

Let's take an example (we show a snippet of code from the next chapter; worry not, it's really very simple and serves as a template for your future kernel modules):

```
// ch5/lkm_template/lkm_template.c
 [ ... ]
 */
#define pr_fmt(fmt) "%s:%s(): " fmt, KBUILD_MODNAME, __func__

#include <linux/init.h>
#include <linux/module.h>
#include <linux/kernel.h>
 [ ... ]
 static int __init lkm_template_init(void)
 {
     pr_info("inserted\n");
     [ ... ]
```

The `pr_fmt()` macro is highlighted; it uses the predefined `KBUILD_MODNAME` macro to substitute the name of your kernel module, and the GCC `__func__` specifier to display the name of the function we're currently running! You can add `%s` for `__FILE__`, and `%d` matched by the corresponding `__LINE__` macro to display the line number; it's really up to you.

So, the bottom line is that the `pr_info()` we emit in the `init` function of this LKM will display like this in the kernel log:

```
[381534.391966] lkm_template:lkm_template_init(): inserted
```

Notice how the module name and the function name are automatically prefixed to the `printk` message content! This is very useful and indeed very common; in the kernel, literally hundreds of source files begin with the `pr_fmt()`. A quick search on the 6.1.25 kernel code base revealed over 2,200 instances of this macro in the code base! We too shall follow this convention, though not in all our demo kernel modules.

The `pr_fmt()` also takes effect on the recommended `printk` usage for driver authors via the `dev_*()` macros.

Next, let's delve into the issue of writing portable code with reference to `printk`.

Portability and the printk format specifiers

There's a question to ponder regarding the versatile `printk` kernel API: how will you ensure that your `printk` output looks correct (is correctly formatted) and works equally well on any CPU regardless of bit width? The portability issue raises its head here; the good news is that getting familiar with the various format specifiers provided will help you a great deal in this regard, in effect allowing you to write arch-independent `printk` instances.

It's important to realize that the `size_t` - pronounced `size type` - is a `typedef` for an unsigned integer; similarly, `ssize_t` (*signed size type*) is a `typedef` for a signed integer.

Here are a few top-of-mind common `printk` format specifiers to keep in mind when writing portable code:

- For `size_t` and `ssize_t` (unsigned and signed) integers, use `%zu` and `%zd`, respectively.
- For kernel pointers, use `%pK` for *security* (hashed values), and `%px` for actual pointers (don't use this in production!). Additionally, use `%pa` for physical addresses (must pass it by reference). The `kptr_restrict` sysctl can be leveraged in this regard; we further delve into this aspect in *Chapter 5, Writing Your First Kernel Module – Part 2*, in the *Proc filesystem tunables affecting the system log* section.
- For a raw buffer as a string of hex characters, use `%*ph` (where * is replaced by the number of characters). Use it for buffers within 64 characters, and use the `print_hex_dump_bytes()` routine for more; variations are available (see the kernel documentation – the link follows).
- For IPv4 addresses, use `%pI4`, and for IPv6 addresses, use `%pI6` (variations here too).

An exhaustive list of `printk` format specifiers and an explanation of when to use each (with examples!) is part of the official kernel documentation here: `https://www.kernel.org/doc/Documentation/printk-formats.txt`.

The kernel also explicitly documents the fact that using the unadorned `%p` in a `printk()` statement **can lead to security issues** (`https://www.kernel.org/doc/html/latest/process/deprecated.html#p-format-specifier`). I urge you to browse through it!

Let's round off this section by checking out a recent feature: `printk` indexing.

Understanding the new printk indexing feature

So, you've written a kernel module (a driver, perhaps) that, as part of the job, emits some `printk` instances (of course, being a good person, you know to keep the number of prints as low as possible). So, let's say one of your log messages is *mydriver: detected crazy situation X*; okay, fine. Your project is deployed in production; lets say a user space monitoring daemon of some sort is (also) monitoring kernel log messages for abnormal conditions so that it can alert the human user. This makes us realize that `printk` instances aren't just for human consumption; programs too might be continuously monitoring them, which is often the case in large installations.

Now, a few months later, another developer on the project feels that your kernel log message isn't great and changes it to *mydriver: detected abnormal condition X* (come on, admit it, it's better). The trouble is that the log monitoring daemon process, still looking for your old message (perhaps by grepping for the *mydriver: detected crazy situation* string), will now completely miss this (new) key message, causing all kinds of problems.

To tackle situations like this (yes, they do arise), Chris Down proposed a kernel `printk indexing` feature (which has been merged into mainline in 5.15). Essentially what happens with this enabled is this: every `printk` instance's metadata – the actual format string (the message in effect), the location in the source from where it's emitted, the log level, and more – is saved into a structure (`struct pi_entry`), and all these structures are collated into a special section within the kernel `vmlinux` image (named `.printk_index`; the same goes for individual kernel modules!).

These messages are made visible via a `debugfs` entry (this does indicate a dependency on `debugfs` being enabled), `<debugfs_mount_point>/printk/index/<file>`, where `<file>` can be `vmlinux` as well as all kernel modules.

The kernel config in question (a Boolean) is named `CONFIG_PRINTK_INDEX`; it's off by default (you can see it via the usual *make menuconfig* UI here: **General setup | Printk indexing debugfs interface**).

With the kernel built with it turned on, here's a couple of quick examples (notice we run as root, as `debugfs` is accessible only as root):

```
# head -n3 /sys/kernel/debug/printk/index/vmlinux
# <level/flags> filename:line function "format"
<3> init/main.c:1591 console_on_rootfs "Warning: unable to open an initial
console.\n"
<3> init/main.c:1569 kernel_init "Default init %s failed (error %d)\n"
```

The format of each line is shown by the very first line (it's quite intuitive). Here's another example (let's look for "`fire`"):

```
# grep -Hn -i -w "fire" /sys/kernel/debug/printk/index/*
/sys/kernel/debug/printk/index/lp:13:<6> drivers/char/lp.c:262 lp_check_status
"lp%d on fire\n"
```

(Ha! We saw this very same example `printk` in the *Using printk log levels* section!) So, think about it: the user space monitoring daemon can now look up the kernel vmlinux, as well as any kernel module's log messages, via the printk index debugfs entries, checking that they're what's expected, and if not, can take remedial action (reporting an issue or whatever).

Further, with `PRINTK_INDEX` disabled, there's no overhead at all. However, with `printk` indexing on, at least two possible issues arise: first, the non-inclusion of `debugfs` in many production systems, and second, the extra memory used up by keeping thousands of format strings within the kernel image (on my 6.1.25 `vmlinux` image, there are close to 12,000 `printk` instances detected by this feature!). Still, it does solve, or at least mitigate, a problem; thus, it's in the mainline kernel.

Exercise: Build the kernel with this feature enabled and try it out.

FYI, here's the commit that merged in `PRINTK_INDEX`: `https://github.com/torvalds/linux/commit/337015573718b161891a3473d25f59273f2e626b`.

A quick thought: is this `printk` indexing feature pretty much identical to the dynamic kernel debug feature (that we covered in the section *An introduction to the kernel's powerful dynamic debug feature)?* No, realize that the dynamic debug covers *only all debug* `printk` call sites; `printk` indexing covers all `printk` call sites!

Having covered a good amount on the venerable kernel `printk` (and friends), you might be wondering about internal implementation details. The **LWN Kernel Index** is a great place to find detailed technical articles on this (and indeed all aspects of the kernel!): `https://lwn.net/Kernel/Index/#Kernel_messages`. Browsing these (as of this writing, July 2023), you'll notice a lot of churn in the works, the idea being to finally iron out old latency (and other) issues with the `printk`, clearing the path to `PREEMPT_RT` finally being merged (*Chapter 11, The CPU Scheduler – Part 2*, will provide an introduction to the Linux real-time effort).

Okay! Let's move toward completing this chapter by learning the basics of how the Makefile for your kernel module looks and works.

Understanding the basics of a kernel module Makefile

You may have noticed by now that we tend to follow a *one-kernel-module-per-directory* rule of sorts. Yes, that definitely helps keep things organized. So, let's take our second kernel module, the `ch4/printk_loglvl` one. To build it, we just `cd` to its folder, type `make`, and (fingers crossed!) voilà, it's done. We have the `printk_loglevel.ko` kernel module object freshly generated (which we can then apply `insmod`/`rmmod` to). But how exactly did it get built when we typed `make`? Explaining this is the purpose of this section.

First off, we do expect you to understand the basics regarding `make` and the Makefile. If not, don't fret, we've provided links to check this out, within the *Further reading* section (in the paragraph labeled *Makefiles: introductory stuff*) of this chapter. . Check it out! (Link: `https://github.com/PacktPublishing/Linux-Kernel-Programming_2E/blob/main/Further_Reading.md#chapter-4-writing-your-first-kernel-module-lkms-part-1---further-reading`).

Next, as this is our very first chapter that deals with the LKM framework and its corresponding Makefile, we will keep things nice and simple, especially regarding the Makefile. However, early on in *Chapter 5*, *Writing Your First Kernel Module – Part 2*, in the *A better Makefile template for your kernel modules* section, we shall introduce a more sophisticated, and (what we hope is) a better Makefile, that is still quite simple to understand. We shall then use this better Makefile in all subsequent code (but not right now); do look out for it and use it!

As you will know, the `make` command will by default look for a file named `Makefile` in the current directory; if it exists, it will parse it and execute command sequences as specified within it. Here's our simple Makefile for the `printk_loglevel` kernel module project (`I` use the `nl` – "number lines" – utility to show it with line numbers prefixed):

```
$ nl Makefile
     1	# ch4/printk_loglvl/Makefile
     2	PWD          := $(shell pwd)
     3	KDIR          := /lib/modules/$(shell uname -r)/build/
     4	obj-m       += printk_loglvl.o

     5	# Enable the pr_debug() and pr_devel() as well by removing the
comment from
     6	# one of the lines below
     7	# (Note: EXTRA_CFLAGS deprecated; use ccflags-y)
     8	#ccflags-y += -DDEBUG
     9	#CFLAGS_printk_loglvl.o := -DDEBUG

    10	all:
    11	    make -C $(KDIR) M=$(PWD) modules
```

```
12    install:
13        make -C $(KDIR) M=$(PWD) modules_install
14    clean:
15        make -C $(KDIR) M=$(PWD) clean
```

It should go without saying that the Unix Makefile syntax demands this basic format:

```
target: [dependent-source-file(s)]
    rule(s)
```

The `rule(s)` instances are always prefixed with a `[Tab]` character, *not* white space.

Let's gather the basics regarding how this (module) Makefile works. First off, a key point is this: the kernel's `Kbuild` system (which we've been mentioning and using since *Chapter 2, Building the 6.x Linux Kernel from Source – Part 1*), primarily uses two variable strings of software to build, chained up within the two variables `obj-y` and `obj-m`.

The `obj-y` string has the concatenated list of all objects to build and merge into the final kernel image files – the uncompressed `vmlinux` and the compressed (boot-able) `[b][z]Image` images. Think about it – it makes sense: the `y` in `obj-y` stands for *Yes*. All kernel built-in and `Kconfig` options that were set to `Y` during the kernel configuration process (or are `Y` by default) are chained together via this item, built, and ultimately woven into the final kernel image files by the `Kbuild` build system.

On the other hand, it's now easy to see that the `obj-m` string is a concatenated list of all kernel objects to build *separately, as kernel modules*! This is precisely why our Makefile has this all-important line (line 4):

```
obj-m += printk_loglvl.o
```

In effect, it tells the `Kbuild` system to include our code; more correctly, it tells it to implicitly compile the `printk_loglvl.c` source code into the `printk_loglvl.o` binary object, and then add this object to the `obj-m` list. Next, with the default rule for `make` being the `all` rule (lines 10 and 11), it is processed:

```
all:
    make -C $(KDIR) M=$(PWD) modules
```

The processing of this single statement (line 11) is quite involved; here's what transpires:

1. The `-C` option switch to `make` has the `make` process *change directory* (via the `chdir()` system call) to the directory name that follows `-C`. Thus, it changes the directory to the `$(KDIR)` directory, which is set (in line 3) to the kernel `build` symbolic link under `/lib/modules/$(uname -r)` (which, as we covered earlier, points to the location of the limited kernel source tree that got installed via the `kernel-headers` package). To remind you, here it is (on my Ubuntu guest):

   ```
   $ ls -l /lib/modules/$(uname -r)/build
   lrwxrwxrwx 1 root root    31 May  5 10:51 build -> /home/c2kp/kernels/
   linux-6.1.25/
   ```

2. So, clearly, the `make` process changes directory to the folder `~/kernels/linux-6.1.25/`, which, in *this* case, points back to our original 6.1.25 kernel source tree (as we're running off the custom kernel that we built earlier). Once there, it automatically *parses in* the content of the *kernel's top-level* Makefile – that is, the Makefile that resides there, in the root of this kernel source tree. This is a key point. The kernel top-level Makefile is a rather large and sophisticated one (on 6.1.25, it's well over 2,000 lines) and contains key build details and variables. This way, being parsed in every time even an out-of-tree module is being built, it's guaranteed that all kernel modules are tightly coupled to the kernel that they are being built against (more on this a bit later). This also guarantees that kernel modules are built with the exact same set of rules – that is, the compiler/linker configurations (the `*CFLAGS*` options, the compiler option switches, and so on) – as the kernel image itself is. *This is required for binary compatibility.*
3. Next, in the Makefile, still on line 11, you can see the initialization of the variable named `M` (to the present working directory), and that the target specified is `modules`; hence, the `make` process now changes the directory to that specified by the `M` variable, to `$(PWD)` – the very folder we started from (line 2: `PWD := $(shell pwd)` in the Makefile initializes it to the correct value)!

So, interestingly, it's a recursive build: the build process, having – very importantly – parsed the kernel top-level Makefile, now switches back to the kernel module's directory and builds the module(s) therein.

Lines 12 and 13 make up the `install` target (which we cover in the next chapter), and lines 14 and 15 make up the `clean` target.

Did you notice that when a kernel module is built, a fair number of intermediate working files are generated as well? Among them are `modules.order`, `<file>.mod.c`, `<file>.o`, `Module.symvers`, `<file>.mod.o`, `.<file>.o.cmd`, `.<file>.ko.cmd`, a folder called `.tmp_versions/`, and, of course, the kernel module binary object itself, `<file>.ko` – the whole point of the build exercise. Further, there are several hidden files generated as well. Getting rid of all these temporary build artifacts, including the target (the kernel module object itself) is easy: just perform `make clean`. The `clean` rule cleans it all up. We shall delve into the `install` target in the following chapter.

A screenshot helps convey the output seen on building and then cleaning up:

```
$ ls
Makefile  printk_loglvl.c
$ uname -r
6.1.25-lkp-kernel
$
$ ls -l /lib/modules/6.1.25-lkp-kernel/build
lrwxrwxrwx 1 root root 31 May  5 10:51 /lib/modules/6.1.25-lkp-kernel/build -> /home/c2kp/kernels
/linux-6.1.25/
$
$ make
make -C /lib/modules/6.1.25-lkp-kernel/build/ M=/home/c2kp/Linux-Kernel-Programming_2E/ch4/printk
_loglvl modules
make[1]: Entering directory '/home/c2kp/kernels/linux-6.1.25'
  CC [M]  /home/c2kp/Linux-Kernel-Programming_2E/ch4/printk_loglvl/printk_loglvl.o
  MODPOST /home/c2kp/Linux-Kernel-Programming_2E/ch4/printk_loglvl/Module.symvers
  CC [M]  /home/c2kp/Linux-Kernel-Programming_2E/ch4/printk_loglvl/printk_loglvl.mod.o
  LD [M]  /home/c2kp/Linux-Kernel-Programming_2E/ch4/printk_loglvl/printk_loglvl.ko
make[1]: Leaving directory '/home/c2kp/kernels/linux-6.1.25'
$
$ ls -a
./                  Module.symvers         printk_loglvl.mod         printk_loglvl.o
../                 .Module.symvers.cmd    printk_loglvl.mod.c       .printk_loglvl.o.cmd
Makefile            printk_loglvl.c        .printk_loglvl.mod.cmd
modules.order       printk_loglvl.ko       printk_loglvl.mod.o
.modules.order.cmd  .printk_loglvl.ko.cmd  .printk_loglvl.mod.o.cmd
$
$
$ make clean ; ls -a
make -C /lib/modules/6.1.25-lkp-kernel/build/ M=/home/c2kp/Linux-Kernel-Programming_2E/ch4/printk
_loglvl clean
make[1]: Entering directory '/home/c2kp/kernels/linux-6.1.25'
  CLEAN   /home/c2kp/Linux-Kernel-Programming_2E/ch4/printk_loglvl/Module.symvers
make[1]: Leaving directory '/home/c2kp/kernels/linux-6.1.25'
./  ../  Makefile  printk_loglvl.c
$
```

Figure 4.13: Screenshot showing the build (make) and "make clean" of our printk_loglvl kernel module (on our x86_64 Ubuntu 22.04 guest running our custom-built 6.1.25 LTS kernel)

You can look up what the `modules.order` and `modules.builtin` files (and other files) are meant for within the kernel documentation here: `https://elixir.bootlin.com/linux/v6.1.25/source/Documentation/kbuild/kbuild.rst`. Also, as mentioned previously, we shall, in the following chapter, introduce and use a more sophisticated Makefile variant – a "better" Makefile. It is designed to help you, the kernel module/driver developer, improve code quality by running targets related to kernel coding style checks, static analysis, simple packaging, and (a dummy target) for dynamic analysis.

With that, we conclude this chapter. Well done – you are now well on your way to learning Linux kernel development!

Summary

In this chapter, we covered the basics of Linux kernel architecture and the LKM framework. You learned what a kernel module is and why it's useful. We then wrote a simple yet complete kernel module, a very basic *Hello, world*. We then delved further into how it works and the coding conventions to follow, along with the practicalities of how exactly to build, load, see the module listing, and unload it. Kernel logging with `printk` (and friends) was covered in some detail, along with explanations regarding the `printk` logging levels, controlling output to the console(s), and more. Details on how to emit pure debug-level kernel messages, and more importantly, an introduction to using the kernel's *dynamic debug* feature were then dealt with. We then moved on to the rate-limiting `printk`, generating kernel messages from user space, standardizing its output format, and understanding the new `printk` indexing feature. We closed this chapter with an understanding of the basics of the kernel module Makefile and how it operates within the LKM framework.

I urge you to work on the sample code (via the book's GitHub repository), work on the questions/assignments, and then proceed on to the next chapter, where we continue our coverage on writing a Linux kernel module!

Questions

As we conclude, here is a list of questions for you to test your knowledge regarding this chapter's material: `https://github.com/PacktPublishing/Linux-Kernel-Programming_2E/blob/main/questions/ch4_qs_assignments.txt`. You will find some of the questions answered in the book's GitHub repo: `https://github.com/PacktPublishing/Linux-Kernel-Programming_2E/tree/master/solutions_to_assgn`.

Further reading

To help you delve deeper into the subject with useful materials, we provide a rather detailed list of online references and links (and at times, even books) in a *Further reading* document in this book's GitHub repository. The *Further reading* document is available here: `https://github.com/PacktPublishing/Linux-Kernel-Programming_2E/blob/master/Further_Reading.md`.

Leave a review!

Enjoying this book? Help readers like you by leaving an Amazon review. Scan the QR code below for a 20% discount code.

**Limited Offer*

5

Writing Your First Kernel Module – Part 2

This chapter is the second half of our coverage regarding the **Loadable Kernel Module (LKM)** framework and how to write kernel modules using it. To get the most out of it, I expect you to complete the previous chapter and try out the code and questions/exercises there before tackling this one.

In this chapter, we will continue from the point where we left off in the previous one. Here, we cover making use of a "better" Makefile for LKMs, cross-compiling a kernel module for the ARM platform (as a typical example), what module stacking is and how to do it, and how to set up and use module parameters. Along the way, among several other things, you will learn about the kernel API/ABI stability (or rather, the lack thereof!), the key differences between writing user-space and kernel code, auto-loading a kernel module at system boot, and security concerns and how they can be addressed. We end with information on the kernel documentation (including coding style) and how one gets started on contributing to the mainline kernel.

In brief, we will cover the following topics in this chapter:

- A "better" Makefile template for your kernel modules
- Cross-compiling a kernel module
- Gathering minimal system information
- Licensing kernel modules
- Emulating "library-like" features for kernel modules
- Passing parameters to a kernel module
- Floating point not allowed in the kernel
- Auto-loading modules on system boot
- Kernel modules and security - an overview
- Coding style guidelines for kernel developers
- Contributing to the mainline kernel

Technical requirements

The technical requirements – the software packages required – for this chapter are identical to what was shown in the *Technical requirements* section in *Chapter 4, Writing Your First Kernel Module – Part 1*; please refer to it. As always, you can find the source code for this chapter in this book's GitHub repository. Clone it with the following:

```
git clone https://github.com/PacktPublishing/Linux-Kernel-Programming_2E
```

The code displayed in the book is often just a relevant snippet. Follow along with the full source code from the repository.

A "better" Makefile template for your kernel modules

The preceding chapter introduced you to the Makefile used to generate the kernel module from the source code, to install and clean it up. However, as we briefly mentioned there, I will now introduce what is, in my opinion, a superior, so-called "better" Makefile, and explain how it's better.

Ultimately, we all must write better and more secure code – both user- and kernel-space. The good news is that there are several tools to help improve your code's robustness and security posture, static and dynamic analyzers being among them (as several have already been mentioned in *Online Chapter, Kernel Workspace Setup,* I won't repeat them here).

I have devised a simple yet useful Makefile "template" of sorts for kernel modules that includes several targets that help you run these tools. These targets allow you to perform valuable checks and analysis very easily; *stuff you might otherwise forget, ignore, or put off forever!* These targets include the following:

- The "usual" ones – the `build (all)`, `install`, and `clean` targets (taking into account a "debug" setting, stripping the module if debug is off).
- Kernel coding style generation and checking (via `indent` and the kernel's `checkpatch.pl` script).
- Kernel static analysis targets (`sparse`, `gcc`, and `flawfinder`), with a mention of **Coccinelle**.
- A couple of dummy kernel dynamic analysis targets pointing out how you should invest time in configuring and building a "debug" kernel and using it to catch bugs (via `KASAN` and `LOCKDEP` / `CONFIG_PROVE_LOCKING`; more on this follows shortly)
- A simple `tarxz-pkg` target that tars and compresses the source files into the parent directory. This enables you to transfer the compressed `tar-xz` file to any other Linux system, then extract and build the LKM there.

You can find the code (along with a `README` file as well) in the `ch5/lkm_template` directory. To help you understand its use and power and get started, the following figure simply shows the output and a screenshot of the output the code produces when you do `make <tab><tab>` and run `make help`:

```
lkm_template $ make <tab><tab>
all                    clean              help           install    sa
sa_flawfinder  sa_sparse
```

```
checkpatch      code-style      indent      nsdeps    sa_cppcheck  sa_gcc
tarxz-pkg
```

```
lkm_template $ make help
=== Makefile Help : additional targets available ===

TIP: Type make <tab><tab> to show all valid targets
FYI: KDIR=/lib/modules/6.5.6-200.fc38.x86_64/build ARCH= CROSS_COMPILE= ccflags-y="-UDEBUG -DDYNAMIC_DEBUG_MODULE" MY
DEBUG=n DBG_STRIP=n

--- usual kernel LKM targets ---
typing "make" or "all" target : builds the kernel module object (the .ko)
install     : installs the kernel module(s) to INSTALL_MOD_PATH (default here: /lib/modules/6.5.6-200.fc38.x86_64/).
            : Takes care of performing debug-only symbols stripping iff MYDEBUG=n and not using module signature
nsdeps      : namespace dependencies resolution; for possibly importing namespaces
clean       : cleanup - remove all kernel objects, temp files/dirs, etc

--- kernel code style targets ---
code-style : "wrapper" target over the following kernel code style targets
 indent     : run the indent utility on source file(s) to indent them as per the kernel code style
 checkpatch : run the kernel code style checker tool on source file(s)

--- kernel static analyzer targets ---
sa          : "wrapper" target over the following kernel static analyzer targets
 sa_sparse     : run the static analysis sparse tool on the source file(s)
 sa_gcc        : run gcc with option -W1 ("Generally useful warnings") on the source file(s)
 sa_flawfinder : run the static analysis flawfinder tool on the source file(s)
 sa_cppcheck   : run the static analysis cppcheck tool on the source file(s)
TIP: use Coccinelle as well: https://www.kernel.org/doc/html/v6.1/dev-tools/coccinelle.html

--- kernel dynamic analysis targets ---
da_kasan   : DUMMY target: this is to remind you to run your code with the dynamic analysis KASAN tool enabled; requi
res configuring the kernel with CONFIG_KASAN On, rebuild and boot it
da_lockdep : DUMMY target: this is to remind you to run your code with the dynamic analysis LOCKDEP tool (for deep lo
cking issues analysis) enabled; requires configuring the kernel with CONFIG_PROVE_LOCKING On, rebuild and boot it
TIP: Best to build a debug kernel with several kernel debug config options turned On, boot via it and run all your te
st cases

--- misc targets ---
tarxz-pkg  : tar and compress the LKM source files as a tar.xz into the dir above; allows one to transfer and build t
he module on another system
        TIP: When extracting, to extract into a directory with the same name as the tar file, do this:
             tar -xvf lkm_template.tar.xz --one-top-level
help       : this help target
```

Figure 5.1: The output of the help target from our "better" Makefile

As we can see in *Figure 5.1*, we type `make`, immediately followed by pressing the *Tab* key twice, thus having it display all available targets (you might need the Bash completion package installed for stuff like this to work; it's usually installed on most distros). Do study this carefully and use it! For example, running `make  sa` (see the *sa* target in *Figure 5.1* along with the other ones) will cause it to run all the **static analysis** (sa) targets on your code.

Note how the line beginning with `FYI: KDIR=...` (highlighted) shows the Makefile's current understanding of various variables and "settings." We've reproduced it here:

```
FYI: KDIR=/lib/modules/6.1.25-lkp-kernel/build ARCH= CROSS_COMPILE= ccflags-
y="-UDEBUG -DDYNAMIC_DEBUG_MODULE" MYDEBUG=n DBG_STRIP=n
```

In the Makefile, we use a variable named MYDEBUG to determine whether we perform a 'debug' build. Clearly, with the `MYDEBUG` variable being set to `n` by default, we're not performing a so-called debug build (the next section delves into what exactly this means).

Also, our variable DBG_STRIP=n essentially says to make : don't strip the module object of debug 'symbols.' Actually, it's set to y by default; our Makefile has the intelligence to strip debug symbols (don't ever strip anything more from a module!) only if MYDEBUG is n (off) and if the kernel config is not using the module-signing feature (which needs symbols; we cover this aspect later in this chapter). The KDIR variable is the path to the (limited) kernel source tree, the kernel headers really, and is the usual value; the ARCH and CROSS_COMPILE variables are null by default as we aren't cross-compiling the module.

Okay, let's use this "better" Makefile to simply build our "template" module, then insert it, remove it, and see that the kernel printk's do show up (the screenshot in *Figure 5.2* reveals the output):

```
$ ls
lkm_template.c  Makefile  README
$ make

--- Building : KDIR=/lib/modules/6.1.25-lkp-kernel/build ARCH= CROSS_COMPILE= ccflags-y="-UDEBUG -DDYNAMIC_DEBUG_MODULE" MYDEBUG=n DBG_STRIP=n ---
gcc (Ubuntu 11.3.0-1ubuntu1~22.04.1) 11.3.0

make -C /lib/modules/6.1.25-lkp-kernel/build M=/home/c2kp/Linux-Kernel-Programming_2E/ch5/lkm_template modules
make[1]: Entering directory '/home/c2kp/kernels/linux-6.1.25'
  CC [M]  /home/c2kp/Linux-Kernel-Programming_2E/ch5/lkm_template/lkm_template.o
  MODPOST /home/c2kp/Linux-Kernel-Programming_2E/ch5/lkm_template/Module.symvers
  CC [M]  /home/c2kp/Linux-Kernel-Programming_2E/ch5/lkm_template/lkm_template.mod.o
  LD [M]  /home/c2kp/Linux-Kernel-Programming_2E/ch5/lkm_template/lkm_template.ko
make[1]: Leaving directory '/home/c2kp/kernels/linux-6.1.25'
if [ "n" = "y" ]; then \
   strip --strip-debug lkm_template.ko ; \
fi
$ ls -lh ./lkm_template.ko
-rw-rw-r-- 1 c2kp c2kp 108K Oct 14 10:36 ./lkm_template.ko
$
$ sudo dmesg -C
$ sudo insmod ./lkm_template.ko
$ lsmod |head -n2
Module                  Size  Used by
lkm_template           16384  0
$
$ sudo rmmod lkm_template
$ sudo dmesg
[ 2012.653246] lkm_template:lkm_template_init(): inserted
[ 2029.253820] lkm_template:lkm_template_exit(): removed
$
```

Figure 5.2: Building our lkm_template module with the "better" Makefile and trying it out (on our x86_64 Ubuntu guest)

It's important to note that to fully exploit this "better" Makefile, you will need to have installed a few packages/apps on the system; these include (for a base Ubuntu system) `indent(1)`, `linux-headers-$(uname -r)`, `sparse(1)`, `flawfinder(1)`, `cppcheck(1)`, and `tar(1)`. (*Online Chapter, Kernel Workspace Setup*, already specified that these should be installed. Also, running our `ch1/pkg_install4ubuntu_lkp.sh` script (on Ubuntu) will ensure they're installed.)

Also, note that the so-called **dynamic analysis (da)** targets mentioned in the Makefile are merely dummy targets that don't do anything other than print a message. They are there *to remind you* to thoroughly test your code by running it on an appropriately configured "debug" kernel!

Speaking of a "debug" kernel, the next section shows you the basics of how to configure one.

Configuring a "debug" kernel

For details on configuring and building the kernel, look back to *Chapter 2, Building the 6.x Linux Kernel from Source – Part 1*, and *Chapter 3, Building the 6.x Linux Kernel from Source – Part 2.*

Running your code on a *debug kernel* can help you uncover hard-to-spot bugs and issues. I highly recommend doing so, typically during development and testing!

In effect, you should have two kernels to run and test your work on: the carefully-configured-for-optimization regular production kernel, as well as a debug kernel, deliberately configured such that many kernel debug options are enabled (perhaps unoptimized but the idea is to use it to catch bugs).

Here, I minimally expect you to configure your custom 6.1 kernel to have the following kernel debug config options turned on (that is, set to `y`; you'll find them within the `make menuconfig` UI, and you will find most of them under the `Kernel Hacking` sub-menu; the following list is with respect to Linux 6.1.25):

- `CONFIG_DEBUG_KERNEL` and `CONFIG_DEBUG_INFO`
- `CONFIG_DEBUG_MISC`
- Generic kernel debugging instruments:
  - `CONFIG_MAGIC_SYSRQ` (the magic SysRq hotkeys feature)
  - `CONFIG_DEBUG_FS` (the `debugfs` pseudo filesystem)
  - `CONFIG_KGDB` (kernel GDB; optional, recommended)
  - `CONFIG_UBSAN` (the undefined behaviour sanity checker)
  - `CONFIG_KCSAN` (the dynamic data race detector)

- Memory debugging:
  - `CONFIG_SLUB_DEBUG`
  - `CONFIG_DEBUG_MEMORY_INIT`
  - `CONFIG_KASAN`: The powerful **Kernel Address Sanitizer** (**KASAN**) memory checker
- `CONFIG_DEBUG_SHIRQ`
- `CONFIG_SCHED_STACK_END_CHECK`
- `CONFIG_DEBUG_PREEMPT`
- Lock debugging:
  - `CONFIG_PROVE_LOCKING`: The very powerful `lockdep` feature to catch locking bugs! This turns on several other lock debug configs as well, explained in *Chapter 13*, *Kernel Synchronization – Part 2*.
  - `CONFIG_LOCK_STAT`.
  - `CONFIG_DEBUG_ATOMIC_SLEEP`.
- `CONFIG_BUG_ON_DATA_CORRUPTION`
- `CONFIG_STACKTRACE`
- `CONFIG_DEBUG_BUGVERBOSE`
- `CONFIG_FTRACE` (`ftrace`: Within its sub-menu, turn on at least a couple of "tracers," including the 'Kernel Function [Graph] Tracer')
- `CONFIG_BUG_ON_DATA_CORRUPTION`
- `Arch-specific` (shows up under "x86 Debugging" on the x86):
  - `CONFIG_EARLY_PRINTK` (arch-specific)
  - `CONFIG_DEBUG_BOOT_PARAMS`
  - `CONFIG_UNWINDER_FRAME_POINTER` (selects `FRAME_POINTER` and `CONFIG_STACK_VALIDATION`)

A few things to note:

a) Don't worry too much right now if you don't get what all the previously mentioned kernel debug config options do; by the time you're done with this book, many of them will be clear.

b) Turning on some `Ftrace` tracers (or plugins), such as `CONFIG_IRQSOFF_TRACER`, would be useful as we actually make use of them in our *Linux Kernel Programming – Part 2* companion volume book in the *Handling Hardware Interrupts* chapter (note that while Ftrace itself may be enabled by default, all its tracers aren't).

c) Detailed coverage on tuning and building a debug kernel can be found in the *Linux Kernel Debugging* book.

Note that turning on these config options *does* entail a performance hit, but that's okay. We're running a "debug" kernel of this sort for the express purpose of *catching errors and bugs* (especially the hard-to-uncover kind!). It can indeed be a life-saver! On your project, *your workflow should involve your code being tested and run on both of the following*:

- The *debug* kernel system, where all required kernel debug config options are turned on (as previously shown minimally)
- The *production* kernel system (where perhaps most of the preceding kernel debug options will be turned off; note that having a few so-called "debug" kernel configs enabled even in production can be extremely helpful).

Needless to say, we will be using the preceding "better" Makefile in all the subsequent LKM code in this book.

Alright, now that you're all set, let's dive into an interesting and practical scenario – compiling your kernel module(s) for another target (typically ARM).

Cross-compiling a kernel module

In *Chapter 3, Building the 6.x Linux Kernel from Source – Part 2*, in the *Kernel build for the Raspberry Pi* section, we showed you how we can cross-compile the Linux kernel for a "foreign" target architecture (such as ARM[64], PowerPC, MIPS, and so on). Essentially, the same can be done for a kernel module as well; you can easily cross-compile a kernel module by setting up a cross-toolchain as well as the "special" `ARCH` and `CROSS_COMPILE` environment variables appropriately.

For example, let's imagine we are working on an embedded Linux product; the target device on which our code will run has an AArch64 (ARM-64) CPU. Why not take an actual example: let's cross-compile our `ch5/lkm_template` kernel module for the Raspberry Pi 4 **Single-Board Computer** (**SBC**)!

This is interesting. You will find that although it appears simple and straightforward, we will end up taking four iterations before we succeed. Why? Read on to find out.

FYI, I have summarized what you'll learn in the following sections during four attempts to cross-build and load the kernel module in the upcoming section *Summarizing what went wrong with the module cross-build/load and how it was fixed*. (I do suggest that, at least on first reading, you first read the full notes to understand the details before looking at the summary).

First, though, we definitely require an appropriate cross-toolchain setup, so let's cover that now.

Setting up the system for cross-compilation

The prerequisites to cross-compile a kernel module are quite clear:

- We need the *kernel source tree for the target system* installed as part of the workspace on our host system, typically an x86_64 desktop (for our example, using the Raspberry Pi as a target; please refer to the official Raspberry Pi documentation here: `https://www.raspberrypi.org/documentation/linux/kernel/building.md`).

- We now need a cross toolchain (host-to-target). Typically, the host system is an x86_64 and here, as the target is an ARM-64, we will need an *x86_64-to-ARM64 cross toolchain*. Again, as is clearly mentioned in *Chapter 3, Building the 6.x Linux Kernel from Source – Part 2*, in the *Kernel build for the Raspberry Pi* section, you must install the Raspberry Pi-specific x86_64-to-ARM toolchain as part of the host system workspace (if you haven't already set up the toolchain, refer to the just-mentioned location to learn how to install it).

Quick tip: If you haven't already, install the `aarch64-linux-gnu` cross-compiler with `sudo apt install gcc-aarch64-linux-gnu binutils-aarch64-linux-gnu`.

Okay, from this point on, I will assume that you have these prerequisites among the things we covered in *Chapter 3, Building the 6.x Linux Kernel from Source – Part 2,* in the section *Kernel build for the Raspberry Pi*: you have the Raspberry Pi 6.1.34 kernel source tree and an x86_64-to-ARM64 cross toolchain installed. Thus, I shall also assume the *toolchain prefix* is `aarch64-linux-gnu`; we can quickly check that the toolchain is installed and its binaries added to the path by trying to invoke the `gcc` cross-compiler:

```
$ aarch64-linux-gnu-gcc
aarch64-linux-gnu-gcc: fatal error: no input files
compilation terminated.
```

It works – it's just that we have not passed any C program as a parameter to compile, hence it complains.

You can certainly look up the compiler version as well, with the `aarch64-linux-gnu-gcc --version` command (and more with the `-v` option switch). As an aside, a modern alternative to GCC is the Clang toolchain; it's quite heavily used these days (including by Android AOSP) and is even superior to GCC in some respects (see the *Further reading* section for more on cross-compilation with Clang). Here, we continue to use GCC, native and cross, compilers.

Okay, let's now dive into cross-compiling our module for the Raspberry Pi device!

Attempt 1 – setting the ARCH and CROSS_COMPILE environment variables

Cross-compiling the kernel module is very easy (or so we think!). First, ensure that you set the "special" `ARCH` and `CROSS_COMPILE` environment variables appropriately. Follow along with these steps:

1. Let's re-build our just discussed `lkm_template` kernel module for the Raspberry Pi target; this also has the advantage of employing the so-called "better" Makefile. Here's how to build it:

To do so without corrupting the original code, make a new folder called `cross` with a copy of the code to begin with. (BTW, the codebase already has it set up, here: `ch5/cross`.)

```
cd <book-dir>/ch5; mkdir cross ; cd cross
cp ../lkm_template/lkm_template.c ../lkm_template/Makefile .
```

Here, `<book-dir>` is the root of this book's GitHub source tree.

2. Now, run the following command (we assume the cross-compiler tools are in the path):

```
make ARCH=arm64 CROSS_COMPILE=aarch64-linux-gnu-
```

(By the way, the book codebase has a small wrapper script, `ch5/cross/buildit`, that performs a couple of validity checks and runs this command for you.) But it doesn't work (or it may work; please see the following materials) straight off the bat. We get compile failures, as seen here:

```
$ make ARCH=arm64 CROSS_COMPILE=aarch64-linux-gnu-

--- Building : KDIR=~/arm64_prj/kernel/linux ARCH=arm64 CROSS_
COMPILE=aarch64-linux-gnu- ccflags-y="-UDEBUG -DDYNAMIC_DEBUG_MODULE"
MYDEBUG=n ---
aarch64-linux-gnu-gcc (Ubuntu 11.3.0-1ubuntu1~22.04.1) 11.3.0

make -C ~/arm64_prj/kernel/linux M=/home/c2kp/Linux-Kernel-
Programming_2E/ch5/cross modules
make[1]: *** /home/c2kp/arm64_prj/kernel/linux: No such file or
directory.  Stop.
make: *** [Makefile:93: all] Error 2
$
```

(Notice how, this time, the `ARCH` and `CROSS_COMPILE` environment variables are correctly defined.) Why did it fail?

The clue as to why the preceding cross-compilation attempt failed lies in the fact that it is attempting to use – *build against* – the kernel source tree of the *current host system* and not the target's kernel source tree. So, *we need to modify the Makefile to point it to the correct kernel source tree, the one for the target*. It's quite easy to do so. In the following code, we see the typical way that the (corrected) Makefile code is written:

```
# ch5/cross/Makefile:
# To support cross-compiling for kernel modules:
# For architecture (cpu) 'arch', invoke make as:
```

```
# make ARCH=<arch> CROSS_COMPILE=<cross-compiler-prefix>
[ ... ]
else ifeq ($(ARCH),arm64)
  # *UPDATE* 'KDIR' below to point to the ARM64 (Aarch64) Linux kernel source
  # tree on your box
  #KDIR ?= ~/arm64_prj/kernel/linux
  KDIR ?= ~/rpi_work/kernel_rpi/linux
else ifeq ($(ARCH),powerpc)
  [ ... ]
else
  [ … ]
endif
[ ... ]
# IMPORTANT :: Set FNAME_C to the kernel module name source filename (without
.c)
FNAME_C := lkm_template

PWD           := $(shell pwd)
obj-m         += ${FNAME_C}.o
[ ... ]
all:
    @echo
    @echo '--- Building : KDIR=${KDIR} ARCH=${ARCH} CROSS_COMPILE=${CROSS_
COMPILE} ccflags-y="${ccflags-y}" MYDEBUG=${MYDEBUG} DBG_STRIP=${DBG_STRIP}
---'
    @echo
    make -C $(KDIR) M=$(PWD) modules
[...]
```

Look carefully at the (new and "better," as explained in the preceding section) Makefile and you will see how it works:

- Most importantly, we conditionally set the `KDIR` variable to point to the correct kernel source tree, depending on the value of the `ARCH` environment variable (of course, I've used a sample pathname to kernel source trees for the ARM-32 and PowerPC as examples; do substitute the pathname with the actual path to your kernel source trees).
- As usual, we set `obj-m` to the object filename.
- You can set the variable `ccflags-y` (`CFLAGS_EXTRA` is considered deprecated) to add the `DEBUG` symbol (so that the `DEBUG` symbol is defined in our LKM and the `pr_debug()/dev_dbg()` macros work). The 'debug' build's off by default.
- The `@echo '<...>'` lines are equivalent to the shell's `echo` command; it just emits some useful information while building (the `@` prefix hides the echo statement itself from displaying).

- Finally, we have the "usual" Makefile targets: `all`, `install`, and `clean`. Pay attention to how, within them, when invoking `make`, we ensure we **change directory** via the `-C` switch to the value of `KDIR`!
- Though not shown in the preceding code, this "better" Makefile – at the risk of repetition – has several additional useful targets. You should definitely take the time to explore and use them (as explained in the preceding section; to start, simply type `make help`, study the output, and try things out).

Having done this, let's retry the module cross-compile with this version and see how it goes.

Attempt 2 – pointing the Makefile to the correct kernel source tree for the target

So now, with the *enhanced and fixed* Makefile described in the previous section, it *should* work. In our new directory where we will try this out – `cross` (as we're cross-compiling, not that we're angry!) – follow along with these steps:

1. Attempt the build (for a second time) with the `make` command appropriate for cross-compilation:

```
$ make ARCH=arm CROSS_COMPILE=arm-linux-gnueabihf-
[ ... ]

  CC [M]  /home/c2kp/Linux-Kernel-Programming_2E/ch5/cross/lkm_template.o
  MODPOST /home/c2kp/Linux-Kernel-Programming_2E/ch5/cross/Module.symvers
ERROR: modpost: "_printk" [/home/c2kp/Linux-Kernel-Programming_2E/ch5/
cross/lkm_template.ko] undefined!
make[2]: *** [scripts/Makefile.modpost:126: /home/c2kp/Linux-Kernel-
Programming_2E/ch5/cross/Module.symvers] Error 1
[ ... ]
```

   Oops. This time we have a `modpost` error. In this module build stage, `MODPOST`, the build system must be able to access and check all exported kernel symbols. This very information is stored in the file `Module.symvers` (in the root of the kernel source tree). It's typically generated just after the modules are built (during the kernel build procedure). We can get this `modpost` failure – as we just did – when this file isn't present. So, to fix this in one shot, I simply cleaned up my (Raspberry Pi) kernel source tree (with `make mrproper`) and then rebuilt it; the `Module.symvers` file now did indeed turn up and the build went through:

```
[ … ]
rpi $  make ARCH=arm64 CROSS_COMPILE=aarch64-linux-gnu-

--- Building : KDIR=~/rpi_work/kernel_rpi/linux ARCH=arm64 CROSS_
COMPILE=aarch64-linux-gnu- ccflags-y="-UDEBUG -DDYNAMIC_DEBUG_MODULE"
MYDEBUG=n ---
aarch64-linux-gnu-gcc (Ubuntu 11.3.0-1ubuntu1~22.04.1) 11.3.0
```

```
make -C ~/rpi_work/kernel_rpi/linux M=/home/c2kp/Linux-Kernel-
Programming_2E/ch5/cross modules
make[1]: Entering directory '/home/c2kp/rpi_work/kernel_rpi/linux'
  CC [M]  /home/c2kp/Linux-Kernel-Programming_2E/ch5/cross/lkm_template.o
  MODPOST /home/c2kp/Linux-Kernel-Programming_2E/ch5/cross/Module.symvers
  CC [M]  /home/c2kp/Linux-Kernel-Programming_2E/ch5/cross/lkm_template.
mod.o
  LD [M]  /home/c2kp/Linux-Kernel-Programming_2E/ch5/cross/lkm_template.
ko
make[1]: Leaving directory '/home/c2kp/rpi_work/kernel_rpi/linux'
if [ "n" != "y" ]; then \
    sudo aarch64-linux-gnu-strip --strip-debug lkm_template.ko ; \
fi
```

2. Ah, the cross-compile build did work! Let's see the freshly built module:

```
$ ls -l ./lkm_template.ko
-rw-rw-r-- 1 c2kp c2kp          [ … ]               ./lkm_template.ko
$ file ./lkm_template.ko
./lkm_template.ko: ELF 64-bit LSB relocatable, ARM aarch64, version 1
(SYSV), BuildID[sha1]=03<...>4, not stripped
```

The reality of course is that the build might fail for several reasons, not just the ones that happen to be seen here... In another instance, it failed as the target kernel source tree that we're compiling our kernel module against was still in a "virgin" state. It perhaps does not even have the `.config` file present (as well as other required headers) in its root directory, which it requires to (at least) be configured. To fix this, you'll have to configure and cross-compile the kernel in the usual manner, then retry the module build.

Attempt 3 – cross-compiling our kernel module

Now that we have generated the cross-compiled kernel module using a properly configured Raspberry Pi kernel source tree (on the host system with the `Module.symvers` file present; see the *Attempt 2 – pointing the Makefile to the correct kernel source tree for the target* section), it *should* work on the board (hey, we're optimists).

Of course, the proof of the pudding is in the eating. So, we fire up our Raspberry Pi, `scp` across our cross-compiled kernel module object file to it, and as follows (within an `ssh` session on the Raspberry Pi), try it out (the following output is directly from the device):

```
$ sudo insmod ./lkm_template.ko
insmod: ERROR: could not insert module ./lkm_template.ko: Invalid module format
```

Clearly, the `insmod` has failed! *It's important to understand why.* It's because there is a *mismatch between the kernel version* that we're attempting to load the module on and the kernel version the module has been compiled against.

While logged in to the Raspberry Pi, let's print out the current Raspberry Pi kernel version we're running on and use the `modinfo` utility to print out details regarding the kernel module itself:

```
rpi $ cat /proc/version
Linux version 6.1.21-v8+ (dom@buildbot) (aarch64-linux-gnu-gcc-8 (Ubuntu/Linaro
8.4.0-3ubuntu1) 8.4.0, GNU ld (GNU Binutils for Ubuntu) 2.34) #1642 SMP PREEMPT
Mon Apr  3 17:24:16 BST 2023
rpi $ modinfo ./lkm_template.ko
filename:       /home/pi/lkp2e/ch5/cross/./lkm_template.ko
version:        0.2
license:        Dual MIT/GPL
description:    a simple LKM template; do refer to the (better) Makefile as
well
author:         Kaiwan N Billimoria
srcversion:     606276CA0788B10170FC6D5
depends:
name:           lkm_template
vermagic:       6.1.34-v8+ SMP preempt mod_unload modversions aarch64
rpi $
```

From the preceding output, clearly, on the ARM64 Raspberry Pi board here, we're running the `6.1.21-v8+` kernel. This, in fact, is the kernel I inherited when I installed the *default* Raspberry Pi OS on the device's microSD card (it's a deliberate scenario introduced here, at first *not* using the 6.1 kernel we built earlier for the Raspberry Pi). The kernel module, on the other hand, reveals that it's been compiled against the `6.1.34-v8+` Linux kernel (the `vermagic` string from `modinfo(8)` reveals this). *Clearly, there's a mismatch.* Well, so what? Read on, the next section reveals this.

Examining Linux kernel ABI compatibility issues

The Linux kernel has a rule, part of the kernel **Application Binary Interface** (**ABI**): it will only ever insert a kernel module into kernel memory if that kernel module has been built against it – the precise kernel version, build flags, and even the kernel configuration options matter!

The *built-against* kernel is the kernel whose source location you specified in the Makefile (we did so via the `KDIR` variable previously).

In other words, **kernel modules are not binary-compatible with kernels other than the one they have been built against.** For example, if we build a kernel module on, say, an x86_64 Ubuntu 22.04 LTS box, then it will *only work on a system running this precise environment* (hardware, libraries, and kernel)! It will *not* work (load) on a Fedora 38 or an RHEL 8.x, a Raspberry Pi, and so on, or even an x86_64 Ubuntu 22.04 box running a different kernel.

Now, importantly – and again, think about this – this does not mean that kernel modules are completely incompatible and non-portable. No, they are *source-compatible across different architectures* (at least they can or *should* be written that way). So, assuming you have the source code, you can always *rebuild* a kernel module on a given system and then it will work on that system. It's just that the *binary image* (the `.ko` file) is incompatible with kernels other than the precise one it's built against. (On some versions of the kernel, the kernel log will spell this out.) So, now it's clear; here, there is a mismatch – between the binary module and the kernel it's built against, and the kernel we're attempting to run (load) it on – and hence the failure to insert the kernel module.

Though we don't use this approach here, there is a way to ensure the successful build and deployment of third-party out-of-tree kernel modules (if their source code is available), via a framework called **DKMS (Dynamic Kernel Module Support)**. The following is a quote directly from it:

Dynamic Kernel Module Support (DKMS) is a program/framework that enables generating Linux kernel modules whose sources generally reside outside the kernel source tree. The concept is to have DKMS modules automatically rebuilt when a new kernel is installed.

Note well: the key phrase is "*... to have DKMS modules automatically rebuilt when a new kernel is installed,*" which proves the point: kernel modules need to be built against precisely the kernel on which they will be deployed. As an example of DKMS usage, the Oracle VirtualBox hypervisor (when running on a Linux host) uses DKMS to auto-build and keep up to date its kernel modules.

By the way, what if you don't have a Raspberry Pi handy to try out these experiments? No worries, you can use other ARM-based boards or even virtualization! (FYI, the SEALS project mentioned back in *Chapter 1* allows very simple virtualization of an ARM-32, ARM-64, and x86_64 PC via QEMU.)

Attempt 4 – cross-compiling our kernel module

So, now that we have understood the binary compatibility issue, there are a few possible solutions:

- We must configure, (cross) build, and boot the device with the required custom kernel for the product and build all our kernel modules against that kernel source tree.
- Use the DKMS approach.
- Alternatively, we could rebuild the kernel module to match the current kernel the Raspberry Pi device happens to be running.

Now, in typical embedded Linux projects, you will almost certainly have a custom-configured kernel for the target device, one that you must work with. All kernel modules for the product will/must be built against it. Thus, we follow the first approach – we must boot the device with our custom-configured and built (6.1.34) kernel, and since our kernel module is built against it, it should certainly work now.

We covered the kernel build for the Raspberry Pi in *Chapter 3, Building the 6.x Linux Kernel from Source – Part 2,* in the *Kernel build for the Raspberry Pi* section. Refer back there for the details if required.

Okay, I will have to assume that you've followed the steps (covered in *Chapter 3, Building the 6.x Linux Kernel from Source – Part 2*, in the *Kernel build for the Raspberry Pi* section) and have by now configured and built a 6.x kernel for the Raspberry Pi. The nitty-gritty details regarding how to copy our custom kernel image, DTB, and module files onto the microSD card of the device and so on isn't covered; I refer you to the official Raspberry Pi documentation here: `https://www.raspberrypi.org/documentation/linux/kernel/building.md`.

Nevertheless, we will point out a convenient way to switch between kernels on the device (here, I assume the device is a Raspberry Pi 4B running a 64-bit kernel):

1. Copy your custom-built `Image` kernel binary into the device's microSD card's `/boot` partition as `kernel8.img`. (For safety, first ensure you save the original (default) Raspberry Pi kernel image as `kernel8.img.orig`).
2. Copy (`scp`) the just-cross-compiled kernel module (`lkm_template.ko` for ARM64, built in the previous section) from your host system onto the microSD card (into `/home/pi` is fine).
3. [Optional] On the Raspberry Pi boards, you can specify the kernel to boot with as follows; this is not required in our case: the bootloader will automatically pick the file `kernel8.img` as the kernel to boot by default. If you'd like to specify a different kernel to boot with, on the device's microSD card, edit the `/boot/config.txt` file, setting the kernel image to boot via the `kernel=xxx` line.
4. Once saved and rebooted, we log in to the device and retry our kernel module. *Figure 5.3* is a screenshot showing the just-cross-compiled `lkm_template.ko` LKM being used on the Raspberry Pi 4 B device:

```
rpi $ cat /proc/version
Linux version 6.1.34-v8+ (c2kp@osboxes) (aarch64-linux-gnu-gcc (Ubuntu 11.3.0-1ubuntu1~22.04.1
) 11.3.0, GNU ld (GNU Binutils for Ubuntu) 2.38) #1 SMP PREEMPT Mon Oct  9 17:03:41 IST 2023
rpi $
rpi $ modinfo ./lkm_template.ko
filename:       /home/pi/lkp2e/ch5/cross/./lkm_template.ko
version:        0.2
license:        Dual MIT/GPL
description:    a simple LKM template; do refer to the (better) Makefile as well
author:         Kaiwan N Billimoria
srcversion:     606276CA0788B10170FC6D5
depends:
name:           lkm_template
vermagic:       6.1.34-v8+ SMP preempt mod_unload modversions aarch64
rpi $
rpi $ sudo dmesg -C
rpi $ sudo rmmod lkm_template 2>/dev/null
rpi $ sudo insmod ./lkm_template.ko
rpi $ dmesg
[  850.778496] lkm_template:lkm_template_init(): inserted
rpi $ lsmod |grep lkm_template
lkm_template           16384  0
rpi $
rpi $ sudo rmmod lkm_template 2>/dev/null
rpi $ dmesg
[  850.778496] lkm_template:lkm_template_init(): inserted
[  875.330843] lkm_template:lkm_template_exit(): removed
rpi $ _
```

Figure 5.3: The cross-compiled LKM being used on a Raspberry Pi 4B; see how the kernel versions, the hardware, and kernel config perfectly match

Ah, it worked! Notice how, this time, the current kernel version running on the board (`6.1.34-v8+`) precisely matches that of the kernel the module was built against – in the `modinfo` output's `vermagic` line, we can see it's `6.1.34-v8+`. Whew, finally, it's done.

If you do see an issue with `rmmod` throwing a non-fatal error (though the cleanup hook is still called), the reason is that you haven't yet fully set up the newly built kernel on the device. You will have to copy in all the kernel modules (under `/lib/modules/<kernel-ver>/`) and run the `depmod(8)` utility there (as root). Here, we will not delve further into these details – as mentioned before, the official documentation for the Raspberry Pi covers all these steps.

Of course, the Raspberry Pi board is a pretty powerful system; you can install the (default) Raspberry Pi OS (a Debian derivative) along with development tools and kernel headers and thus compile kernel modules on the board itself! (No cross-compile required.) Here, though, we have followed the cross-compile approach as this is typical and required when working on embedded Linux projects.

A summary of what we've just learned with regard to cross-compiling kernel modules follows.

Summarizing what went wrong with the module cross-buildd/load and how it was fixed

Let's summarize what we learned in this interesting section on cross-compiling a kernel module:

Attempt #	Cause of failure to cross-compile or run	Fix
1	Module build failed because we were attempting to build the target kernel module against the kernel source tree of the current host (x86_64) system (or an invalid path specified) and not against the target's kernel source tree. `KDIR ?= /lib/modules/$(shell uname -r)/build`	Modify the module Makefile to point the KDIR variable to the correct kernel source tree, that of the target system's: `KDIR ?= ~/rpi_work/kernel_rpi/linux`
2	Module build failed in the `MODPOST` stage (due to the absence of the `Module.symvers` file in the target's kernel source tree).	Implies the target kernel tree hasn't been configured and/or built; so do so... then this file appears and it (cross) builds the module.

3	Cross-compiled module fails to load on the device (the Raspberry Pi) with the kernel citing an "Invalid module format" error.	This is due to the kernel ABI rule: the kernel will only ever insert a kernel module into kernel memory if that kernel module has been built against it. So modules have no binary compatibility across systems but can be written to be source portable. (Attempt #4 fixes it.)
4	<Same as above, Attempt #3>	The fix: we boot the device with our custom 6.1 kernel AND cross-compile our module against the same kernel (referencing it via the KDIR Makefile variable), copy the binary module onto the target's root filesystem, and then `insmod` it. It works!

Table 5.1: Summary of our attempts to cross-compile and run a kernel module from an x86_64 host to an AArch64 (Raspberry Pi 4) target

Done.

The LKM framework is a rather large piece of work. Plenty more remains to be explored. Let's get to it. In the next section, we will examine how you can obtain some minimal system information from within a kernel module.

Gathering minimal system information

At times, especially when writing a module that's meant to be portable across various architectures (CPUs), we need to conditionally perform work based on the actual processor family we're running upon. The kernel provides a few macros and ways to figure this out; here, we build a simple demo module (`ch5/min_sysinfo/min_sysinfo.c`) that, though still quite simplistic, shows a few ways to "detect" some system details (such as the CPU family, bit-width, and endian-ness). In the following code snippet, we show only the relevant function:

```
// ch5/min_sysinfo/min_sysinfo.c
[ ... ]
void llkd_sysinfo(void)
{
    char msg[128];

    memset(msg, 0, 128);
    my_snprintf_lkp(msg, 47, "%s(): minimal Platform Info:\nCPU: ", __func__);

    /* Strictly speaking, all this #if...#endif is considered ugly and should
be
     * isolated as far as is possible */
```

```
#ifdef CONFIG_X86
#if(BITS_PER_LONG == 32)
    strncat(msg, "x86-32, ", 9);
#else
    strncat(msg, "x86_64, ", 9);
#endif
#endif
#ifdef CONFIG_ARM
    strncat(msg, "AArch32 (ARM-32), ", 19);
#endif
#ifdef CONFIG_ARM64
    strncat(msg, "AArch64 (ARM-64), ", 19);
#endif
#ifdef CONFIG_MIPS
    strncat(msg, "MIPS, ", 7);
#endif
#ifdef CONFIG_PPC
    strncat(msg, "PowerPC, ", 10);
#endif
#ifdef CONFIG_S390
    strncat(msg, "IBM S390, ", 11);
#endif

#ifdef __BIG_ENDIAN
    strncat(msg, "big-endian; ", 13);
#else
    strncat(msg, "little-endian; ", 16);
#endif

#if(BITS_PER_LONG == 32)
    strncat(msg, "32-bit OS.\n", 12);
#elif(BITS_PER_LONG == 64)
    strncat(msg, "64-bit OS.\n", 12);
#endif
    pr_info("%s", msg);

  show_sizeof();
 /* Word ranges: min & max: defines are in include/linux/limits.h */
 [ ... ]
}
EXPORT_SYMBOL(llkd_sysinfo);
```

Additional details that this LKM shows you – like the size of various primitive data types plus word ranges – are not shown here; please do refer to the source code from our GitHub repository and try it out for yourself.

The preceding kernel module code is instructive as it helps demonstrate how you can write portable code. Remember, the kernel module itself is a binary non-portable object file, but its source code could (perhaps, should, depending on your project) be written in such a manner so that it's portable across various architectures. A simple build on (or for) the target architecture would then have it ready for deployment.

For now, please ignore the `EXPORT_SYMBOL()` macro used here. We will cover its usage shortly. Also, `my_snprintf_lkp()` is a simple wrapper over `snprintf()`, for better security, and is, for now, simply defined in the current source file (`min_sysinfo.c`; do check it out. FYI, once we cover more on emulating 'library-like' functionality, we shall define the `snprintf_lkp()` wrapper elsewhere; don't worry about it now). The next section delves more into the security aspect, as do several places in the book.

Building and running it on our now familiar x86_64 Ubuntu 22.04 LTS guest, we get this output (from the `sudo dmesg` command onward):

```
[13892.202097] min_sysinfo:min_sysinfo_init(): inserted
[13892.202105] min_sysinfo:llkd_sysinfo(): llkd_sysinfo(): minimal Platform Info:
               CPU: x86_64, little-endian; 64-bit OS.
[13892.202108] min_sysinfo:llkd_sysinfo2(): llkd_sysinfo2(): minimal Platform Info:
               CPU: x86_64, little-endian; 64-bit OS.
[13892.202111] min_sysinfo:show_sizeof(): sizeof: (bytes)
                  char =  1   short int =  2            int =  4
                  long =  8   long long =  8         void * =  8
                 float =  4      double =  8    long double = 16
[13892.202117] min_sysinfo:llkd_sysinfo2(): Word [U|S][8|16|32|64] ranges: unsigned max, signed max,
signed min:
               U8_MAX =                    255 = 0x               ff,  S8_MAX =                 127 =
0x                 7f,  S8_MIN =                   -128 = 0x         ffffff80
               U16_MAX =                 65535 = 0x             ffff, S16_MAX =               32767 =
0x               7fff, S16_MIN =                 -32768 = 0x         ffff8000
               U32_MAX =            4294967295 = 0x         ffffffff, S32_MAX =          2147483647 =
0x           7fffffff, S32_MIN =            -2147483648 = 0x         80000000
               U64_MAX = 18446744073709551615 = 0xffffffffffffffff, S64_MAX = 9223372036854775807 =
0x7fffffffffffffff, S64_MIN = -9223372036854775808 = 0x8000000000000000
               PHYS_ADDR_MAX = 18446744073709551615 = 0xffffffffffffffff
```

Figure 5.4: Kernel output from our simple yet interesting min_sysinfo module (on our x86_64 guest)

The output from the `llkd_sysinfo()` function that we just saw is highlighted in a rectangle in *Figure 5.4*. It's followed by the output of our `llkd_sysinfo2()` function (which is the same of course; its just more security-conscious, the next section covers it). It goes on to use the `sizeof()` operator to show the byte size of various data types; finally, our module shows the word ranges for that architecture (for all unsigned and signed 8, 16, 32, and 64 bit words).

Great! Similarly (as demonstrated earlier), we can *cross-compile* this kernel module for an AArch64 target, then transfer (`scp`) the cross-compiled kernel module and run it there (the following output is from a Raspberry Pi 4B running our custom 64-bit 6.1.34-v8+ kernel):

```
$ sudo insmod ./min_sysinfo.ko
min_sysinfo:min_sysinfo_init(): inserted
min_sysinfo:llkd_sysinfo(): llkd_sysinfo(): minimal Platform Info:
               CPU: AArch64 (ARM-64), little-endian; 64-bit OS.
min_sysinfo:llkd_sysinfo2(): llkd_sysinfo2(): minimal Platform Info:
               CPU: AArch64 (ARM-64), little-endian; 64-bit OS.
min_sysinfo:show_sizeof(): sizeof: (bytes)
                 char =  1   short int =  2          int =  4
                 long =  8   long long =  8       void * =  8
                float =  4      double =  8   long double = 16
min_sysinfo:llkd_sysinfo2(): Word [U|S][8|16|32|64] ranges: unsigned max, 
signed max, signed min:
                U8_MAX =                    255 = 0x             ff,  S8_MAX = 
127 = 0x            7f,  S8_MIN =                    -128 = 0x        ffffff80
               U16_MAX =                  65535 = 0x           ffff, S16_MAX = 
32767 = 0x          7fff, S16_MIN =                  -32768 = 0x        ffff8000
               U32_MAX =             4294967295 = 0x       ffffffff, S32_MAX
=           2147483647 = 0x         7fffffff, S32_MIN =          -2147483648 =
0x          80000000
               U64_MAX = 18446744073709551615 = 0xffffffffffffffff, S64_MAX
=  9223372036854775807 = 0x7fffffffffffffff, S64_MIN = -9223372036854775808 =
0x8000000000000000
               PHYS_ADDR_MAX = 18446744073709551615 = 0xffffffffffffffff
```

Again, we can see the details this time, relevant to the AArch64 platform!

FYI, the powerful *Yocto Project* (`https://www.yoctoproject.org/`) is one (industry-standard) way to generate a complete embedded Linux BSP for the Raspberry Pi. Alternatively (and much easier to quickly try), Ubuntu provides a custom Ubuntu 64-bit kernel and root filesystem for the device (`https://wiki.ubuntu.com/ARM/RaspberryPi`).

I urge you to go through the code and try out this module for yourself.

Being a bit more security-aware

Security, of course, is a key concern these days. Professional developers are expected to write secure code. In recent years, there have been many known exploits against the Linux kernel (see the *Further reading* section for more on this). In parallel, many efforts toward improving Linux kernel security are in place.

In our preceding kernel module (`ch5/min_sysinfo/min_sysinfo.c`), be wary of using older-style routines (like the `sprintf`, `strlen`, and so on; yes, they're present within the kernel)! *Static analyzers* can greatly aid in catching potential security-related and other bugs; we highly recommend you use them. *Online Chapter, Kernel Workspace Setup,* mentions several useful static analysis tools for the kernel. In the following output block, we use one of the `sa` targets within our "better" Makefile to run a relatively simple static analyzer: `flawfinder` (written by David Wheeler) against the source code in the `ch5/min_sysinfo/` folder:

```
$ make [tab][tab]
all         clean       help    install  sa          sa_flawfinder  sa_sparse
checkpatch  code-style  indent  nsdeps   sa_cppcheck  sa_gcc         tarxz-pkg
$ make sa_flawfinder
make clean
make[1]: Entering directory '<...>/Linux-Kernel-Programming_2E/ch5/min_sysinfo'

--- cleaning ---

[...]

--- static analysis with flawfinder ---

flawfinder *.[ch]
Flawfinder version 2.0.19, (C) 2001-2019 David A. Wheeler.
Number of rules (primarily dangerous function names) in C/C++ ruleset: 222
Examining min_sysinfo.c

FINAL RESULTS:

min_sysinfo.c:54:  [2] (buffer) char:
  Statically-sized arrays can be improperly restricted, leading to potential
  overflows or other issues (CWE-119!/CWE-120). Perform bounds checking, use
  functions that limit length, or ensure that the size is larger than the
  maximum possible length.
[ … ]
min_sysinfo.c:136:  [1] (buffer) strncat:
  Easily used incorrectly (e.g., incorrectly computing the correct maximum
  size to add) [MS-banned] (CWE-120). Consider strcat_s, strlcat, snprintf,
  or automatically resizing strings. Risk is low because the source is a
  constant string.
min_sysinfo.c:138:  [1] (buffer) strncat:
  Easily used incorrectly (e.g., incorrectly computing the correct maximum
[ ... ]
```

Look carefully at the above warning emitted by `flawfinder` regarding the `strncat()` function (among the many it generates!). Following it's advice, we use the `strlcat()` in its place, rewriting the code of the `llkd_sysinfo()` function as `llkd_sysinfo2()` (but why exactly? It's written to be safer, more secure; look up the details on its man page: `https://linux.die.net/man/3/strlcat`). Similarly, actually leveraging the power of the `snprintf()` API requires checking it's return value to see whether a (possible) buffer overflow occurred (the root cause of many vulnerabilities!); hence (as already mentioned), I wrote a simple wrapper over it, `my_snprintf_lkp()`.

The output of `flawfinder` often mentions the **CWE** number (here, CWE-119/120) of the *generalized class* of security issue that is being seen here (do Google it and you will see the details). In this instance, CWE-120 represents the 'Buffer Copy without Checking Size of Input ('Classic Buffer Overflow'): `https://cwe.mitre.org/data/definitions/120.html`. As you'll realize, Buffer Overflow defects form one of the largest classes of vulnerabilities that are frequently the target of hacking attacks (yes, even on the kernel!).

We also add a few lines of code to show the *range* (min, max) of both unsigned and signed variables on the platform (in both base 10 and 16). We leave it to you to read through. As a simple assignment, run this kernel module on your Linux box(es) and verify the output.

There's much more, of course, to Linux kernel security and hardening techniques. I have spoken about these concerns and possible kernel hardening mitigations here: *Mitigating Hackers with Hardening on Linux – An Overview for Developers, Focus on BoF, Kaiwan Billimoria* (for the Embedded IOT Summit, Prague, June 2023): `https://www.youtube.com/watch?v=KQa_XEiLGMc`.

The upcoming *Kernel modules and security – an overview* section has some coverage on this topic. Besides, be sure to check out the numerous resources in the *Further reading* section of this chapter (under the *Linux Kernel Security* heading).

Now, let's move on to discuss a little bit regarding the licensing of the Linux kernel and kernel module code.

Licensing kernel modules

As is well known, the Linux kernel code base itself is licensed under the GNU GPL v2 (aka GPL-2.0; **GPL** stands for **General Public License**), and as far as most people are concerned will remain that way. As briefly mentioned before, in *Chapter 4, Writing Your First Kernel Module – Part 1*, licensing your kernel code is required and important. Let's split up this short discussion on licensing into two portions: one, as it applies to inline kernel code (or the mainline kernel), and two, as it applies to writing third-party out-of-tree kernel modules (as many of us do).

Licensing of inline kernel code

We begin with the first, that is, inline kernel code. A key point with regard to licensing here: if your intention is to directly use Linux kernel code and/or contribute your code upstream into the mainline kernel (we cover more on this in an upcoming section), you *must* release the code under the same license that the Linux kernel is released under: the GNU GPL-2.0. The details are well documented; see the official take here: `https://docs.kernel.org/process/license-rules.html#linux-kernel-licensing-rules`.

When new to licensing, the plethora of open-source licenses, along with their do's and dont's, can be perplexing, no doubt. Do check out this site to help clear up the fog: `https://choosealicense.com/`. Also Bootlin's coverage on open-source licensing in their guide *Embedded Linux system development training* under the section *Open source licenses and compliance* is excellent (`https://bootlin.com/doc/training/embedded-linux/embedded-linux-slides.pdf`).

In order to be consistent, recent kernels have a rule: every single source file's first line must be an **SPDX (Software Package Data Exchange)** license identifier (see `https://spdx.org/` for details), a shorthand and concise format for expressing the license the code is released under. (Of course, scripts will require the first line to specify the interpreter.) Thus, the very first line of most C source files in the kernel is a comment regarding the SPDX license:

```
// SPDX-License-Identifier: GPL-2.0
```

Licensing of out-of-tree kernel modules

For an out-of-tree kernel module, the situation is still a bit "fluid," shall we say. No matter, to engage the kernel community and have them help (a huge plus), you should, or are expected to, release the code under the GNU GPL-2.0 license (though dual licensing is certainly possible and acceptable).

The license(s) that a kernel module is released under is specified in two ways:

1. Via the SPDX-License-Identifier tag as (a comment in) the very first source line. This strictly applies to modules within the source tree, not always to out-of-tree modules.
2. Via the `MODULE_LICENSE()` macro. Do note what the official kernel documentation clearly states (`https://docs.kernel.org/process/license-rules.html#id1`): "*Loadable kernel modules also require a MODULE_LICENSE() tag. This tag is neither a replacement for proper source code license information (SPDX-License-Identifier) nor in any way relevant for expressing or determining the exact license under which the source code of the module is provided.*"

   The sole purpose of this tag is to provide sufficient information whether the module is free software or proprietary for the kernel module loader and for user space tools.

The following comment, reproduced from the `include/linux/module.h` kernel header, clearly shows what license "idents" are acceptable (notice the dual licensing). Obviously, the kernel community would highly recommend releasing your kernel module under the GPL-2.0 (GPL v2) and/or another similar one, such as BSD/MIT/MPL. If you are intending to contribute code upstream to the kernel mainline, it goes without saying that the GPL-2.0 alone *is* the license to release under:

```
// include/linux/module.h
[...]
/*
 * The following license idents are currently accepted as indicating free
 * software modules
 *
 * "GPL"                          [GNU Public License v2 or later]
 * "GPL v2"                       [GNU Public License v2]
 * "GPL and additional rights"    [GNU Public License v2 rights and more]
 * "Dual BSD/GPL"                 [GNU Public License v2
 *                                 or BSD license choice]
 * "Dual MIT/GPL"                 [GNU Public License v2
 *                                 or MIT license choice]
 * "Dual MPL/GPL"                 [GNU Public License v2
 *                                 or Mozilla license choice]
 *
 * The following other idents are available
 *
 * "Proprietary" [Non free products]
 [ ... ]
```

It goes on to say that the usage of `MODULE_LICENSE()` is largely a historic and failed attempt to convey more information in the `MODULE_LICENSE` string. It's main purpose now is to be able to check for and flag `Proprietary` modules, thus restricting their usage of symbols exported under the GPL (we'll get to this). It also helps the community, end users, and vendors of modules to quickly vet whether the codebase is open-source or not.

FYI, the kernel source tree has a `LICENSES/` directory under which you will find detailed information on relevant licenses; a quick `ls` on this folder reveals the sub-folders therein:

```
$ ls <...>/linux-6.1.25/LICENSES/
deprecated/ dual/ exceptions/ preferred/
```

We'll leave it to you to take a look, and with this, we will leave the discussion on licensing at that; the reality is that it's a complex topic requiring legal knowledge. You would be well advised to consult specialist legal staff (lawyers) within your company (or hire them) regarding getting the legal angle right for your product or service.

FYI, some answers to FAQs on the GPL license are addressed here: `https://www.gnu.org/licenses/gpl-faq.html`.

More on licensing models, not abusing the `MODULE_LICENSE` macro, and particularly the multi-licensing/dual-licensing one, can be found in references provided in the *Further reading* section of this chapter.

Now, let's get back to the technical stuff. The next section explains how you can effectively emulate a library-like feature in kernel space.

Emulating "library-like" features for kernel modules

One of the major differences between user-mode and kernel-mode programming is the complete absence of the familiar "library" concept in the latter. Libraries are essentially a collection or archive of APIs, conveniently allowing developers to meet these important goals: *do not reinvent the wheel, software reuse, modularity, portability* and the like. But within the Linux kernel, libraries – in the traditional sense of the word – just do not exist. Having said that, the `lib/` folder within the kernel source tree contains library-like routines, several of which get built into the kernel image and are thus available to kernel/module developers at runtime.

The good news is that, broadly speaking, there are two techniques by which you can achieve "library-like" functionality in kernel space for your kernel modules:

- The first technique is by explicitly "linking in" multiple source files, including the so-called "library" code, to your kernel module object.
- The second is a technique called *module stacking*.

Do read on as we discuss these techniques in more detail. A spoiler, perhaps, but useful to know, is that the first of the preceding techniques is often superior to the second. Then again, it does depend on the project. Do read the details in the coming sections; we list out some pros and cons as we go along.

Performing library emulation via linking multiple source files

So far, we have dealt with very simple kernel modules that have had exactly one C source file. What about the (quite typical) real-world situation where there is *more than one C source file for a single kernel module*? All source files will have to be compiled and then linked together as a single `.ko` binary object.

For example, say we're building a kernel module project called `projx`. It consists of three C source files: `prj1.c`, `prj2.c`, and `prj3.c`. We want the final kernel module to be called `projx.ko`. The *Makefile* is where you specify these relationships, as shown:

```
obj-m       := projx.o
projx-objs := prj1.o prj2.o prj3.o
```

In the preceding code, note how the `projx` label has been used after the `obj-m` directive *and* as the prefix for the `-objs` directive on the next line. Of course, you can use any label in place of `projx`. Our preceding example Makefile will have the kernel build system first compile the three individual C source files into individual object (`.o`) files, and then *link them all together to form the final binary kernel module object file*, `projx.ko`, just as we desire.

Indeed, we leverage precisely this mechanism to build a small "library" of routines within our book's source tree (the source files for this 'kernel library' are in the root of this book's source tree here: `klib.h` and `klib.c`). The idea, of course, is that other kernel modules can use the functions within here by linking into them! For example, in the upcoming *Chapter 8, Kernel Memory Allocation for Module Authors – Part 1,* we have our `ch8/lowlevel_mem/lowlevel_mem.c` kernel module code invoke a function that resides in our library code, at this relative path from it: `../../klib.c`. This "linking into" our so-called library code is achieved by putting the following into the `lowlevel_mem` kernel module's Makefile (with the prefix highlighted in bold):

```
FNAME_C  := lowlevel_mem
[ … ]
PWD                  := $(shell pwd)
obj-m                += ${FNAME_C}_lkm.o
lowlevel_mem_lkm-objs := ${FNAME_C}.o ../../klib.o
```

The last line above specifies the source files to build (into object files); they are the code of the `lowlevel_mem.c` kernel module and the `../../klib.c` library code. Then, the build system *links* both into a single binary kernel module, `lowlevel_mem_lkm.ko`, achieving our objective. (Why not work on *Assignment 5.1* specified in the *Questions* section at the end of this chapter to better understand this approach?)

Next, let's understand something fundamental – the scope of functions and variables within a kernel module.

Understanding function and variable scope in a kernel module

Before delving further, a quick re-look at some basics is a good idea. When programming with C, you will understand the following:

- Variables declared locally within a function are obviously local to it and only have scope within that function (local variables 'disappear' when the function returns, hence they're referred to as "automatic." A common bug is to try and refer to a local or automatic variable outside its scope; this is often called the **Use After Scope (UAS)** or **Use After Return (UAR)** defect).
- Variables and functions prefixed with the `static` qualifier have scope only within the current "unit"; effectively, the file they have been declared within. This is good as it helps reduce namespace pollution. Static data variables declared within a function retain their value across function calls..

Prior to 2.6 Linux (that is, <= 2.4.x, ancient history now), kernel module static and global variables, as well as all functions, were automatically visible throughout the kernel. This was, in retrospect, obviously not a great idea. The decision was reversed from 2.5 (and thus 2.6 onward, modern Linux): **all kernel module variables (static and global data) and functions are by default scoped to be private to that kernel module only and are thus invisible outside it.** So, if two kernel modules, `lkmA` and `lkmB`, have a global (or static) variable named `maya`, it's unique to each of them; there is no clash.

To change the scope, the LKM framework provides the `EXPORT_SYMBOL()` macro. Using it, you can declare a data item or function to be *global* in scope – in effect, visible to any and all other kernel modules.

Let's take a simple example. We have a kernel module called `prj_core` that contains a global and a function:

```
static int my_glob = 5;
static long my_foo(int key)
{ [...]
}
```

Though both the variable `my_glob` and the function `my_foo()` are usable within this kernel module itself, neither can be seen outside it. This is intentional. To make them visible outside this kernel module, we can *export* them:

```
int my_glob = 5;
EXPORT_SYMBOL(my_glob);

long my_foo(int key)
{ [...]
}
EXPORT_SYMBOL(my_foo);
```

Now, both have scope outside this kernel module (notice how, in the preceding code block, the `static` keyword has been deliberately removed). *Other kernel modules can now "see" and use them.* Precisely, this idea is leveraged in two broad ways:

- First, the kernel *exports* (via the `EXPORT_SYMBOL*()` macros) a well-thought-out subset of global variables and functions that form a part of its core functionality, as well as that of other subsystems. Now, these globals and functions are visible and thus usable from kernel modules! We will see some sample uses shortly.
    - This idea also spawns a very important corollary: out-of-tree kernel modules can *only access* functions and variables within the kernel that have been explicitly exported.
- Second, kernel module authors (often device drivers) use this very notion to export certain data and/or functionality so that other kernel modules, at a higher abstraction level, perhaps, can leverage this design and use this data and/or functionality – this concept is called *module stacking* and we will delve into it shortly with an example.

With the first use case, for example, a device driver author might want to handle (trap into) a hardware interrupt from a peripheral device. A common way to do so is via the `request_threaded_irq()` API:

```
// kernel/irq/manage.c
int request_threaded_irq(unsigned int irq, irq_handler_t handler,
                         irq_handler_t thread_fn, unsigned long irqflags,
                         const char *devname, void *dev_id)
{
    struct irqaction *action;
```

```
[...]
        return retval;
}
EXPORT_SYMBOL(request_threaded_irq);
```

Precisely because the `request_threaded_irq()` function is *exported*, can it be called from within a device driver, which is very often written as a (often out-of-tree, as we do) kernel module.

Similarly, module developers often require some "convenience" routines - for example, string processing ones. The Linux kernel, in `lib/string.c`, provides an implementation of several common string processing functions (that you'd expect to be present): `str[n]casecmp()`, `str[n|l|s]cpy()`, `str[n|l]cat()`, `str[n]cmp()`, `strchr[nul]()`, `str[n|r]chr()`, `str[n]len()`, and so on. Of course, these are all *exported* via the `EXPORT_SYMBOL()` macro so as to make them visible and thus available to module authors.

Here, we used the `str[n|l|s]cpy()` notation to imply that the kernel provides the four functions: `strcpy()`, `strncpy()`, `strlcpy()`, and `strscpy()`. Note that some interfaces may be deprecated (`strcpy()`, `strncpy()`, and `strlcpy()`). In general, always avoid using deprecated stuff documented here: *Deprecated Interfaces, Language Features, Attributes, and Conventions* (`https://www.kernel.org/doc/html/latest/process/deprecated.html#deprecated-interfaces-language-features-attributes-and-conventions`).

On the other hand, let's glance at a (tiny) bit of the core **CFS** (**Completely Fair Scheduler**) CPU scheduling code deep within the kernel core. Here, the `pick_next_task_fair()` function is the one invoked by the scheduling code when we need to find another task to context-switch to:

```
// kernel/sched/fair.c
static struct task_struct *
pick_next_task_fair(struct rq *rq, struct task_struct *prev, struct rq_flags
*rf)
{
        struct cfs_rq *cfs_rq = &rq->cfs;
[...]
    update_idle_rq_clock_pelt(rq);
        return NULL;
}
```

We don't really want to study scheduling here (*Chapter 10, The CPU Scheduler – Part 1*, and *Chapter 11, The CPU Scheduler – Part 2*, will take care of it!); the point here is this: as the preceding function is *not* marked with any of the `EXPORT_SYMBOL*()` macros, it cannot ever be invoked by a kernel module. It remains *private* to the core kernel; this is, of course, a deliberate design decision.

You can also mark data structures as exported with the same macro(s). Also, it should be obvious that only globally scoped data – not local variables – can be marked as exported. (FYI, if you want to see how the `EXPORT_SYMBOL()` macro works, please refer to the *Further reading* section of this chapter.)

Recall our brief discussion on the licensing of kernel modules. The Linux kernel has a, shall we say, interesting, proposition: there is also a macro called `EXPORT_SYMBOL_GPL()`. It's just like its cousin, the `EXPORT_SYMBOL()` macro, except that the function or data item exported will only be visible to those kernel modules that include the word `GPL` within their `MODULE_LICENSE()` macro! Ah, the sweet revenge of the kernel community. It is indeed used in several places in the kernel code base. (I'll leave it as an exercise for you to find occurrences of this macro in the code; on the 6.1.25 kernel, a quick search with `cscope` revealed "just" 17,000-odd usage instances!)

To view all exported symbols, navigate to the root of your kernel source tree and issue the `make export_report` command. Note though that this works only upon a kernel tree that has been configured and built.

Let's now look at another key approach to realizing a library-like kernel feature: module stacking.

Understanding module stacking

The second important idea here – *module stacking* – is what we will now delve further into.

Module stacking is a concept that, to a degree, provides a "library-like" feature to kernel module authors. Here, we typically architect our project or product design in such a manner that we have one or more "core" kernel modules, whose job is to act as a library of sorts. It will include the data structures and functionality (functions/APIs) that will be *exported* to other kernel modules in the project (and to anyone else who wants to use them; the preceding section discussed this).

To better understand this concept, let's look at a couple of real examples. To begin with, on my host system, an Ubuntu 22.04.3 LTS native Linux system, I ran a guest VM(s) over the *Oracle VirtualBox 7.0* hypervisor application. Okay, performing a quick `lsmod` on the host system while filtering for the string `vbox` reveals the following:

```
$ lsmod | grep vbox
vboxnetadp             28672  0
vboxnetflt             28672  1
vboxdrv               614400  3 vboxnetadp,vboxnetflt
```

Recall from our earlier discussion that, of the four columns seen, the third column is the *usage count*. It's `0` in the first row but has a value of `3` in the third row. Not only that, the `vboxdrv` kernel module has two kernel modules listed to its right (after the usage count column). If any kernel modules show up after the third column, they represent **module dependencies**; read it this way: the kernel modules displayed on the right *depend on* the kernel module on the left.

So, in the preceding example, the `vboxnetadp` and `vboxnetflt` kernel modules depend on the `vboxdrv` kernel module. *Depend on it* in what way? They call functions (APIs) and/or use data structures that reside within the `vboxdrv` core kernel module, of course! In general, kernel modules showing up on the right of the third column imply they are calling one or more functions and/or using data structures of the kernel module on the left (leading to an increment in the usage count; this usage count is a good example of a *reference counter* (here, it's actually a 32-bit atomic variable), something we delve into in the last chapter). In effect, the `vboxdrv` kernel module is akin to a "library" (in a limited sense, with none of the usual user - space connotations associated with user-mode libraries except that it provides modular functionality). You can see that, in this snapshot, its usage count is `3` and the kernel modules that depend on it are stacked on top of it – literally! (You can see them in the preceding two lines of `lsmod` output.) Also, notice that the `vboxnetflt` kernel module has a positive usage count (`1`) but no kernel modules show up on its right; this still implies that something is using it at the moment, typically a process or thread.

FYI, the **Oracle VirtualBox** kernel modules we see in this example are actually the implementation of the **VirtualBox Guest Additions**. They are essentially a para-virtualization construct, helping to accelerate the working of the guest VM. Oracle VirtualBox provides similar functionality for Windows and macOS hosts as well (as do all the major virtualization vendors).

Another example of module stacking, as promised: the kernel has a generic core **HID** (**Human Interface Device**) driver, often run as a module; it has several other drivers that 'depend' on it that are stacked above it, leveraging precisely the "library-like" feature we have been discussing here; the screenshot makes it clear:

```
$ lsmod |grep hid
hid_multitouch          36864  0
mac_hid                 16384  0
intel_hid               24576  0
sparse_keymap           16384  2 intel_hid,dell_wmi
usbhid                  65536  0
hid_sensor_custom       28672  0
hid_sensor_hub          28672  1 hid_sensor_custom
intel_ishtp_hid         28672  0
hid_generic             16384  0
i2c_hid_acpi            16384  0
intel_ishtp             57344  2 intel_ishtp_hid,intel_ish_ipc
i2c_hid                 36864  1 i2c_hid_acpi
hid                    159744  6 i2c_hid,usbhid,hid_multitouch,hid_sensor_hub,intel_ishtp_hid,hid_generic
$ _
```

Figure 5.5: A screenshot showing how the kernel "hid" module has several other HID-related drivers depending upon it, demonstrating the "module-stacking" approach

Here's some quick scripting magic to see all kernel modules whose usage count is non-zero (they often – but not always – have some dependent kernel modules show up on their right; we further numerically sort the output in ascending order by usage count):

```
lsmod | awk '$3 > 0 {print $0}' | sort -k3n
```

An implication of module stacking: you can only successfully `rmmod` a kernel module if its usage count is `0`; that is, it is not in use. Thus, for the preceding first example, we can only remove the `vboxdrv` kernel module after removing the two dependent kernel modules that are stacked on it (thus getting the usage count down to `0`).

Trying out module stacking

Let's architect some very simple proof-of-concept code for module stacking. To do so, we will build two kernel modules:

- The first we will call `core_lkm`; its job is to act as a 'library' of sorts, making available to the kernel and other modules a couple of functions (we think of them as APIs provided by this 'core' module).
- Our second kernel module, `user_lkm`, is the 'user' (or consumer/client) of the 'library.' It will simply invoke the functions (and use some data) residing within the first.

To do so, our pair of kernel modules will need to do the following:

- The 'core' kernel module must use the `EXPORT_SYMBOL()` macro to mark some data and functions as being exported (you could use the `EXPORT_SYMBOL_GPL()` macro as well if you choose to).
- The 'user' kernel module must declare the data and/or functions that it expects to use as being external to it, via the C `extern` keyword (remember, exporting data or functionality merely sets up the appropriate linkage; the compiler still needs to know about the data and/or functions being invoked).
- With recent toolchains, marking the exported function(s) and data items as `static` is allowed. As this results in a compiler warning, we don't use the `static` keyword for exported symbols.
- Edit the custom Makefile to build both kernel modules.

The sample 'core' module – the 'library'

The code follows; first, the core or library kernel module. To (hopefully) make this more interesting, we will copy the code of one of our previous module's functions – `ch5/min_sysinfo/min_sysinfo.c:llkd_sysinfo2()` – into this kernel module and *export* it, thus making it visible to our second "user" LKM, which will invoke that function.

Here, we do not show the full code; you can refer to the book's GitHub repo for it:

```
// ch5/modstacking/core_lkm.c
#define pr_fmt(fmt) "%s:%s(): " fmt, KBUILD_MODNAME, __func__
#include <linux/init.h>
#include <linux/module.h>

#define THE_ONE     0xfedface
MODULE_LICENSE("Dual MIT/GPL");

int exp_int = 200;
```

```
EXPORT_SYMBOL_GPL(exp_int);

/* Functions to be called from other LKMs */
void llkd_sysinfo2(void)
{
[...]
}
EXPORT_SYMBOL(llkd_sysinfo2);

#if(BITS_PER_LONG == 32)
u32 get_skey(int p)
#else // 64-bit
u64 get_skey(int p)
#endif
{
#if(BITS_PER_LONG == 32)
    u32 secret = 0x567def;
#else // 64-bit
    u64 secret = 0x123abc567def;
#endif
    if (p == THE_ONE)
        return secret;
    return 0;
}
EXPORT_SYMBOL(get_skey);
[...]
```

The sample 'user' module – the 'library' client

Next is the `user_lkm` kernel module, the one "stacked" on top of the `core_lkm` kernel module, the 'user' or client of the core or 'library' module:

```
// ch5/modstacking/user_lkm.c
#define pr_fmt(fmt) "%s:%s(): " fmt, KBUILD_MODNAME, __func__

#if 1
MODULE_LICENSE("Dual MIT/GPL");
#else
MODULE_LICENSE("MIT");
#endif

extern void llkd_sysinfo2(void);
```

```
extern long get_skey(int);
extern int exp_int;

/* Call some functions within the 'core' module */
static int __init user_lkm_init(void)
{
#define THE_ONE 0xfedface
     pr_info("inserted\n");
     u64 sk = get_skey(THE_ONE);
     pr_debug("Called get_skey(), ret = 0x%llx = %llu\n", sk, sk);
     pr_debug("exp_int = %d\n", exp_int);
     llkd_sysinfo2();
     return 0;
}

static void __exit user_lkm_exit(void)
{
    pr_info("bids you adieu\n");
}
module_init(user_lkm_init);
module_exit(user_lkm_exit);
```

Trying them out

The Makefile remains largely identical to our earlier kernel modules, except that this time we need two kernel module objects (that are within the same directory) to be built, as follows:

```
obj-m      := core_lkm.o
obj-m      += user_lkm.o
```

Okay, let's try it out:

1. First, build the kernel modules:

```
$ make
--- Building : KDIR=/lib/modules/6.1.25-lkp-kernel/build ARCH= CROSS_
COMPILE= ccflags-y="-DDEBUG -g -ggdb -gdwarf-4 -Wall -fno-omit-frame-
pointer -fvar-tracking-assignments -DDYNAMIC_DEBUG_MODULE" MYDEBUG=y ---
gcc (Ubuntu 11.3.0-1ubuntu1~22.04.1) 11.3.0

make -C /lib/modules/6.1.25-lkp-kernel/build M=/home/c2kp/Linux-Kernel-
Programming_2E/ch5/modstacking modules
[ … ]
  LD [M]  /home/c2kp/Linux-Kernel-Programming_2E/ch5/modstacking/core_
lkm.ko
```

```
  CC [M]  /home/c2kp/Linux-Kernel-Programming_2E/ch5/modstacking/user_
lkm.mod.o
  LD [M]  /home/c2kp/Linux-Kernel-Programming_2E/ch5/modstacking/user_
lkm.ko
[ … ]
$ ls *.ko
core_lkm.ko  user_lkm.ko
```

2. Now, let's perform a quick series of tests to demonstrate the *module-stacking* proof of concept. Let's first do it *wrongly*: we will first attempt to insert the `user_lkm` kernel module before inserting the `core_lkm` module.

   This will fail – why? You will realize that the exported functionality (and data) that the `user_lkm` kernel module depends on is not (yet) available within the kernel. More technically, the symbols will not be located within the kernel's symbol table as the `core_lkm` kernel module that has them hasn't been inserted yet:

```
$ sudo dmesg -C
$ sudo insmod ./user_lkm.ko
insmod: ERROR: could not insert module ./user_lkm.ko: Unknown symbol in
module
$ sudo dmesg
[13204.476455] user_lkm: Unknown symbol exp_int (err -2)
[13204.476493] user_lkm: Unknown symbol get_skey (err -2)
[13204.476531] user_lkm: Unknown symbol llkd_sysinfo2 (err -2)
```

   As expected, as the required symbols are unavailable, `insmod` fails (the precise error message you see in the kernel log may vary slightly depending on the kernel version and debug config options set).

3. Now, let's do it right:

```
$
$
$ sudo insmod ./core_lkm.ko
$ sudo insmod ./user_lkm.ko
$ lsmod |egrep "core_lkm|user_lkm"
user_lkm               20480  0
core_lkm               16384  1 user_lkm
$
$
Oct 11 15:50:33 osboxes kernel: core_lkm:core_lkm_init(): inserted
                                Exported: get_skey(), llkd_sysinfo2() and exp_int
Oct 11 15:51:00 osboxes kernel: user_lkm:user_lkm_init(): inserted
Oct 11 15:51:00 osboxes kernel: core_lkm:get_skey(): /home/c2kp/Linux-Kernel-Programming_2E/ch5/modstackin
g/core_lkm.c:102: I've been called
Oct 11 15:51:00 osboxes kernel: user_lkm:user_lkm_init(): Called get_skey(), ret = 0x123abc567def = 200434
77188079
Oct 11 15:51:00 osboxes kernel: user_lkm:user_lkm_init(): exp_int = 200
Oct 11 15:51:00 osboxes kernel: core_lkm:llkd_sysinfo2(): llkd_sysinfo2(): minimal Platform Info:
                                CPU: x86_64, little-endian; 64-bit OS.
```

Figure 5.6: Screenshot showing running the module-stacking proof-of-concept demo: the top window is the shell where we're doing the insmod, and the bottom one is running journalctl -k -f to continually show the kernel log "live"

Study *Figure 5.6*; yes, now it works as expected!

Notice how, for the `core_lkm` kernel module, the usage count column has incremented to `1` *and* we can now see that the `user_lkm` kernel module appears on its right indicating that it *depends upon* the `core_lkm` one. (Recall that the kernel module(s) displayed in the extreme-right columns of `lsmod`'s output depend on the one in the extreme-left column.)

4. Now, let's remove the kernel modules. Removing the kernel modules has an *ordering dependency* as well (just as with insertion). Attempting to remove the `core_lkm` one first fails, as, obviously, there is another module still in the kernel memory relying upon its code/data; in other words, it's still in use:

```
$ sudo rmmod core_lkm
rmmod: ERROR: Module core_lkm is in use by: user_lkm
```

Note that if the modules are *installed* on the system, then you could use the `modprobe -r <modules...>` command to remove all related modules; we cover module installation in the *Auto-loading modules on system boot* section.

5. The preceding `rmmod` failure message is self-explanatory. So, let's do it right:

```
$ sudo rmmod user_lkm core_lkm
$ dmesg
[...]
 CPU: x86_64, little-endian; 64-bit OS.
[13889.717265] user_lkm:user_lkm_exit(): bids you adieu
[13889.732018] core_lkm:core_lkm_exit(): bids you adieu
```

There, done!

Next, you will notice that in the code of the `user_lkm` kernel module, the license we release it under is in a conditional `#if` statement:

```
#if 1
MODULE_LICENSE("Dual MIT/GPL");
#else
MODULE_LICENSE("MIT");
#endif
```

We can see that it's released (by default) under the *Dual MIT/GPL* license; well, so what? Think about it; in the code of the `core_lkm` kernel module, we have the following:

```
int exp_int = 200;
EXPORT_SYMBOL_GPL(exp_int);
```

The `exp_int` integer is *only visible to those kernel modules that run under a GPL license.* So, an exercise for you: change the `#if 1` statement in `core_lkm.c` to `#if 0`, thus now releasing it under an MIT-only license. Now, rebuild and retry. It *fails* at the build stage itself:

```
$ make
[...]
  MODPOST /home/c2kp/Linux-Kernel-Programming_2E/ch5/modstacking/Module.symvers
```

```
ERROR: modpost: GPL-incompatible module user_lkm.ko uses GPL-only symbol 'exp_
int'
[...]
```

Read the error message; the license does matter!

Did you notice that, in our module-stacking demo (under the `ch5/modstacking` folder), both the "core" and "user" modules' code were in the same directory, thus easing the Makefile rules and the build. What if this isn't the case? No problem; consider they're in different folders, something like this:

```
modstacking2
├── core_module
└── user_module
```

Everything remains the same except for the Makefile of course:

- For the "core" module's Makefile, we now only require one module to be built, the core one:

  ```
  obj-m          := core_module.o
  ```

- For the "user" module's Makefile, we of course require it to refer to both it and the core module, as before; it's just that, this time, we have to specify the correct (relative) path to the core module. Thus it becomes:

  ```
  obj-m          := user_module.o
  obj-m          += ../core_module/core_module.o
  ```

Build both, insert them into the kernel in the correct order, and it all works. (We leave actually doing this as an assignment for you.)

Before we wind up this section, here's a quick list of things that can go wrong with module stacking; that is, things to check:

- Trying to insert/remove kernel modules in the wrong order.
- Attempting to insert an exported routine that is already in kernel memory – a *namespace collision* issue. Here's an example:

  ```
  $ cd <book_src>/ch5/min_sysinfo ; make && sudo insmod ./min_sysinfo.ko
  [...]
  $ cd ../modstacking ; sudo insmod ./core_lkm.ko
  insmod: ERROR: could not insert module ./core_lkm.ko: Invalid module
  format
  $ sudo dmesg
  [...]
  [335.823472] core_lkm: exports duplicate symbol llkd_sysinfo2 (owned by
  min_sysinfo)
  $ sudo rmmod min_sysinfo
  $ sudo insmod ./core_lkm.ko        # now it's ok
  ```

- License issues caused by the usage of the `EXPORT_SYMBOL_GPL()` macro.

Tip: Always look up the kernel log (with `dmesg` or `journalctl`). It often helps to show what actually went awry.

Emulating 'library-like' features – summary and conclusions

So, let's briefly summarize the key points we learned for emulating library-like features within the kernel module space. We explored two techniques:

- The first technique we used works by *linking multiple source files together into a single kernel module*.
- This is as opposed to the *module-stacking* technique, where we actually build multiple kernel modules and "stack" them on top of each other.

Not only does the first technique work well, it also has these advantages:

- We do *not* have to explicitly mark (via `EXPORT_SYMBOL()`) every function and/or data item that we use as exported.
- The functions and data are only available to the kernel module to which they are actually linked (and not *to any and all other modules*). This is a good thing! All this at the cost of slightly tweaking the Makefile – well worth it.

A downside to the "linking" (first) approach: when linking multiple files, the size of the kernel module can grow to be quite large.

This concludes your learning about a powerful feature of kernel programming – the ability to link multiple source files together to form one kernel module, and/or leveraging the module-stacking design, both allowing you to develop more sophisticated kernel projects. (Do work on the sample code and the suggested assignments before moving on to the next topic area.)

Now, can one pass *parameters* to a kernel module? The following section shows you how to!

Passing parameters to a kernel module

A common debugging technique is to *instrument* your code; that is, insert prints at appropriate points such that you can follow the path the code takes. Within a kernel module, of course, we would use the versatile `printk` *(and friends)* functions for this purpose. So, let's say we do something like the following (pseudo-code):

```
#define pr_fmt(fmt) "%s:%s():%d: " fmt, KBUILD_MODNAME, __func__, __LINE__
[ ... ]
func_x() {
    pr_debug("At 1\n");
    [...]
    while (<cond>) {
```

```
            pr_debug("At 2: j=0x%x\n", j);
            [...]
    }
    [...]
}
```

Okay, great. But, hey, we don't want the debug prints to appear in a production (or release) version. That's precisely why we're using `pr_debug()`: it emits a printk only when the `DEBUG` symbol is defined! Indeed, but what if, interestingly, our customer is an engineering customer and wants to *dynamically turn on or turn off these debug prints*? There are several approaches you might take to make this happen; one is as in the following pseudo-code:

```
static int debug_level;     /* will be init to zero */
func_x() {
    if (debug_level >= 1)
        pr_debug("At 1\n");
    [...]
    while (<cond>) {
        if (debug_level >= 2)
            pr_debug("At 2: j=0x%x\n", j);
        [...]
    }
  [...]
}
```

Ah, that's nice. So, what we're getting at really is this: what if we can make the `debug_level` variable a parameter to our kernel module? Then – a powerful thing – the user of your kernel module has control over which debug messages appear and which do not.

This, being a contrived example, is fine... But, realistically, you can always – even in production! – use the easier and more powerful *dynamic debug* framework of the kernel to see debug prints whenever you'd like to. Recall we covered this key topic in brief in *Chapter 4, Writing Your First Kernel Module – Part 1*, in the *An introduction to the kernel's powerful dynamic debug feature* section.

Besides debug instrumentation, there are many more cases where module parameters prove invaluable (think: setting the initial volume level for an audio driver, the brightness level on a camera driver, and so on). So let's now jump into learning about the usage of module parameters.

Declaring and using module parameters

Module parameters are passed to a kernel module as *name-value* pairs at module insertion (`insmod`/`modprobe`) time.

For example, assume we have a *module parameter* named `mp_debug_level`; then, we could pass its value at `insmod` time, like this:

```
sudo insmod ./modparams1.ko mp_debug_level=2
```

Here, the `mp` prefix stands for *module parameter*. It's not required to name it that way, of course; it is pedantic, but consistently following such conventions helps by making the reading of large codebases a bit more intuitive.

That would be quite powerful. Now, the end user can decide at exactly what *verbosity* they want *debug-level* messages to be emitted at. We can easily arrange for the default value to be `0` and thus the module emits no debug messages by default.

You might wonder how it works: kernel modules have no `main()` function and hence no conventional `(argc, argv)` parameter list, so how exactly, then, do you pass parameters along? It's simply a bit of linker trickery... Just do this: declare your intended module parameter as a global (`static`) variable, then specify to the build system that it's to be treated as a module parameter by employing the `module_param()` macro.

This is easy to see with our first module parameter's demo kernel module (as usual, the full source code and Makefile can be found in the book's GitHub repo):

```
// ch5/modparams/modparams1/modparams1.c
[ ... ]
/* Module parameters */
static int mp_debug_level;
module_param(mp_debug_level, int, 0660);
MODULE_PARM_DESC(mp_debug_level,
"Debug level [0-2]; 0 => no debug messages, 2 => high verbosity");

static char *mp_strparam = "My string param";
module_param(mp_strparam, charp, 0660);
MODULE_PARM_DESC(mp_strparam, "A demo string parameter");
```

An aside: in the `static int mp_debug_level;` statement above, there is no harm in changing it to `static int mp_debug_level = 0;`, thus explicitly initializing the variable to 0, right? Well, no: the kernel's `scripts/checkpatch.pl` script output reveals that this is not considered a good coding style by the kernel community; if you do so, it triggers this "error":

```
ERROR: do not initialise statics to 0

#28: FILE: modparams1.c:28:

+static int mp_debug_level = 0;
```

In the preceding code block, we have declared two variables to be module parameters via the `module_param()` macro. The `module_param()` macro itself takes three parameters:

- The first parameter: The variable in our code that we would like to be treated as a module parameter. This should be declared using the `static` qualifier.
- The second parameter: Its data type.
- The third parameter: Permissions (really, its visibility via `sysfs`; this is explained shortly).

Next, the `MODULE_PARM_DESC()` macro allows us to "describe" what the parameter represents. Think about it: this is how you inform the end user of the kernel module (or driver) regarding what parameters are actually available. The lookup is performed via the `modinfo(8)` utility. Furthermore, you can specifically print only the information on parameters to a module by using the `-p` option switch, as shown:

```
$ cd <booksrc>/ch5/modparams/modparams1 ; make
[ … ]
$ modinfo -p ./modparams1.ko
parm:           mp_debug_level:Debug level [0-2]; 0 => no debug messages, 2 =>
high verbosity (int)
parm:           mp_strparam:A demo string parameter (charp)
```

The `modinfo` output displays available module parameters, if any. Here, we can see that our `modparams1.ko` kernel module has two parameters: their name, description, and data type (the last component, within parentheses; BTW, `charp` is "character pointer," a string) is then displayed.

Right, let's now give our demo kernel module a quick spin:

```
$ sudo dmesg -C
$ sudo insmod ./modparams1.ko
$ sudo dmesg
[630238.316261] modparams1:modparams1_init(): inserted
[630238.316685] modparams1:modparams1_init(): module parameters passed: mp_
debug_level=0 mp_strparam=My string param
```

Here, we see from the `dmesg` output that, as we did not explicitly pass any kernel module parameters, the module variables obviously retain their default (original) values. Let's redo this, this time passing explicit values to the module parameters:

```
sudo rmmod modparams1
sudo insmod ./modparams1.ko mp_debug_level=2 mp_strparam=\"Hello modparams1\"
sudo dmesg
[...]
[630359.270765] modparams1:modparams1_exit(): removed
[630373.572641] modparams1:modparams1_init(): inserted
[630373.573096] modparams1:modparams1_init(): module parameters passed: mp_
debug_level=2 mp_strparam=Hello modparams1
```

It works as expected. Isn't it interesting: the (parameter) variables have now been modified in kernel memory with the new values we passed from user space!

Now that we've seen how to declare and pass along some parameters to a kernel module, let's look at retrieving and even modifying them at runtime.

Getting/setting module parameters after insertion

Let's look carefully at the `module_param()` macro usage in our preceding `modparams1.c` source file again:

```
module_param(mp_debug_level, int, 0660);
```

Notice the third parameter, the *permissions* (or *mode*): it's `0660` (which, of course, is an *octal* number, implying read-write access for the owner and group and no access for others). It's a bit confusing until you realize that if this permissions *(or mode)* parameter is specified as non-zero, a pseudofile(s) gets created under the `sysfs` filesystem, representing the kernel module parameter(s), here: `/sys/module/<module-name>/parameters/`:

`sysfs` is usually mounted under `/sys`. Also, by default, all pseudofiles under it will have the owner and group as root.

1. So, for our `modparams1` kernel module (first ensure you `insmod` it into kernel memory!), let's look them up:

   ```
   $ ls /sys/module/modparams1/
   coresize  holders/  initsize  initstate  notes/  parameters/  refcnt
   sections/  srcversion  taint  uevent  version
   $ ls -l /sys/module/modparams1/parameters/
   total 0
   -rw-rw---- 1 root root 4096 Oct 11 17:09 mp_debug_level
   -rw-rw---- 1 root root 4096 Oct 11 17:09 mp_strparam
   $
   ```

   Indeed, there they are! Not only that, but the real beauty of it is also that these "parameters" can now be read and written at will, at any time (though, here, only with root permission, of course)!

2. Check it out; let's look up the current value of the `mp_debug_level` module parameter:

   ```
   $ cat /sys/module/modparams1/parameters/mp_debug_level
   cat: /sys/module/modparams1/parameters/mp_debug_level: Permission denied
   $ sudo cat /sys/module/modparams1/parameters/mp_debug_level
   ```

```
[sudo] password for c2kp:
2
```

The current value of our mp_debug_level kernel module parameter is indeed 2 (as this is precisely the value we assigned to it when we did the insmod).

3. Now let's dynamically change it to 0 by writing into it (as root), implying that no "debug" messages will be emitted by the modparams1 kernel module:

```
$ sudo sh -c "echo 0 > /sys/module/modparams1/parameters/mp_debug_level"
$ sudo cat /sys/module/modparams1/parameters/mp_debug_level
0
```

Voilà – done. You can similarly get and/or set the mp_strparam parameter; we will leave it to you to try this as a simple exercise. Do dwell on this concept. It's quite powerful stuff: you could write simple scripts to control a device's (or whatever's) behavior via kernel module parameters, get (or cut off) debug info, and anything else; the possibilities are quite endless.

Actually, coding the third parameter to module_param() as a literal octal number (such as 0660) is not considered the best programming practice in some circles. Specify the permissions of the sysfs pseudofile via appropriate macros (specified in include/uapi/linux/stat.h), for example:

```
module_param(mp_debug_level, int, S_IRUSR|S_IWUSR|S_IRGRP|S_IWGRP);
```

However, having said this, our "better" Makefile's *checkpatch* target (which, of course, invokes the kernel's scripts/checkpatch.pl "coding-style" Perl script checker) politely informs us that simply using octal permissions is better:

```
$ make checkpatch
[ ... ]
checkpatch.pl: /lib/modules/<ver>/build//scripts/checkpatch.pl --no-tree -f
*.[ch]
[ ... ]
WARNING: Symbolic permissions 'S_IRUSR|S_IWUSR|S_IRGRP|S_IWGRP' are not
preferred. Consider using octal permissions '0660'.
 #32: FILE: modparams1.c:32:
 +module_param(mp_debug_level, int, S_IRUSR|S_IWUSR|S_IRGRP|S_IWGRP);
```

So, the kernel community disagrees. Hence, we will just use the "usual" octal number notation of 0660. (Interestingly and FYI, here's a recent (6.3) commit that fixes precisely this for some drivers: *dm: avoid using symbolic permissions:* https://github.com/torvalds/linux/commit/6a808034724b5a36f8e0b712427bfbe9e667d296.)

Again, repeated for emphasis, a useful tip: from what we learned, for the purpose of providing "dynamic" debug instrumentation, using a `debug_level` type of module parameter seems to be a good approach. I'd argue, though, that the Linux kernel's built-in ***dynamic debug*** feature is much superior (`CONFIG_DYNAMIC_DEBUG`)! We did briefly cover this key topic in *Chapter 4, Writing Your First Kernel Module – Part 1*, in the *An introduction to the kernel's powerful dynamic debug feature* section. You can learn more about it from the official kernel documentation here (`https://www.kernel.org/doc/html/v6.1/admin-guide/dynamic-debug-howto.html`) and from my *Linux Kernel Debugging* book's third chapter.

Learning module parameter data types and validation

In our preceding simple kernel module, we set up two parameters of the integer (`int`) and string data types (`charp`). What other data types can be used for module parameters? Several, as it turns out: the `moduleparam.h` include file reveals all (within a comment, duplicated as follows):

```
// include/linux/moduleparam.h
[...]
 * Standard types are:
 *  byte, hexint, short, ushort, int, uint, long, ulong
 *  charp: a character pointer
 *  bool: a bool, values 0/1, y/n, Y/N.
 *  invbool: the above, only sense-reversed (N = true).
```

All right, that's pretty comprehensive. You can even define your own data types if required. Usually, though, the standard types more than suffice.

Validating kernel module parameters

All kernel module parameters are *optional* by default; the user may or may not explicitly pass them. But what if our project requires that the user *must explicitly pass* a value for a given kernel module parameter? We address this here: let's enhance our previous kernel module, creating another (`ch5/modparams/modparams2`), the key difference being that we set up an additional parameter called `control_freak`. Here, let's say we *require* that the user *must* mandatorily pass this parameter along at module insertion time:

1. Let's set up the new module parameter in code:

   ```
   static int control_freak;
   module_param(control_freak, int, 0660);
   MODULE_PARM_DESC(control_freak, "Set to the project's control level [1-
   5]. MANDATORY");
   ```

2. How can we achieve this "mandatory parameter passing"? Well, it's a bit of a hack really: at insertion time, just check whether the (parameter) variable value is the default (`0`, here). If so, it of course implies that the end user hasn't overridden it by explicitly passing it; so abort with an appropriate message (we also do a simple validity check to ensure that the integer passed is within a given range). Here's the `init` code of `ch5/modparams/modparams2/modparams2.c`:

```
static int __init modparams2_init(void)
{
    pr_info("inserted\n");
    if (mp_debug_level > 0)
        pr_info("module parameters passed: "
                "mp_debug_level=%d mp_strparam=%s\n control_freak=%d\n",
                mp_debug_level, mp_strparam, control_freak);

    /* param 'control_freak': if it hasn't been passed (implicit guess),
     * or is the same old value, or isn't within the right range,
     * it's Unacceptable!  :-)
     */
    if ((control_freak < 1) || (control_freak > 5)) {
      pr_warn("I'm a control freak; thus, you *Must* pass along module
parameter 'control_freak', value in the range [1-5]; aborting...\n");
        return -EINVAL;
    }
    return 0;    /* success */
}
```

So let's do the unthinkable and load up this module *without* passing the `control_freak` parameter (shudder)!

```
$ sudo dmesg -C
[sudo] password for c2kp:
$ sudo insmod ./modparams2.ko
insmod: ERROR: could not insert module ./modparams2.ko: Invalid
parameters
$ sudo dmesg
[ … ]
kern  :info  : [632275.085408] modparams2:modparams2_init(): inserted
kern  :warn  : [632275.085866] modparams2:modparams2_init(): I'm a
control freak; thus, you *Must* pass along module parameter 'control_
freak', value in the range [1-5]; aborting...
$
```

It works as designed, aborting the module (by returning a negative errno, -EINVAL here) as the parameter's not explicitly passed.

3. Also, as a quick demo, notice how we emit a printk, showing the module parameter values only if `mp_debug_level` is positive.
4. Finally, on this topic, the kernel framework provides a more rigorous way to "get/set" kernel (module) parameters and perform validity checking on them via the `module_param_cb()` macro (`cb` for callbacks). We will not delve into this here; I refer you to a blog article mentioned in the *Further reading* section for details on using it.

Now, let's move on to how (and why) we can override a module parameter's name.

Overriding the module parameter's name

To explain this feature, let's take an example from the (6.1.25) kernel source tree: the direct mapping buffered I/O library driver, `drivers/md/dm-bufio.c`, has a need to use (as one example) the `dm_bufio_current_allocated` variable as a module parameter. However, this name is really that of an *internal variable* and is not highly intuitive to a user of this driver. The authors of this driver would therefore much prefer to use another more intuitive and meaningful name for the parameter – `current_allocated_bytes` – as an *alias* or *name override*. Precisely this can be achieved via the `module_param_named()` macro, overriding and completely equivalent to the internal variable name, as follows:

```
module_param_named(current_allocated_bytes, dm_bufio_current_allocated, ulong,
S_IRUGO);
MODULE_PARM_DESC(current_allocated_bytes, "Memory currently used by the
cache");
```

You can look this up, and several similar usages of the `module_param_named()` macro at the end of this file, here: `https://elixir.bootlin.com/linux/v6.1.25/source/drivers/md/dm-bufio.c#L2169`.

So, when the user performs `insmod` on this driver, they can do stuff like the following:

```
sudo insmod <path/to/>dm-bufio.ko current_allocated_bytes=4096 ...
```

Internally, the actual variable, `dm_bufio_current_allocated`, will now be assigned the value `4096`.

Hardware-related kernel parameters

For security reasons, module or kernel parameters that specify hardware-specific values have a separate macro: `module_param_hw[_named|array]()`. David Howells submitted a patch series for these new hardware parameters' kernel support on 1 December 2016. The patch email (`https://lwn.net/Articles/708274/`) mentions the following:

```
Provided an annotation for module parameters that specify hardware
parameters (such as io ports, iomem addresses, irqs, dma channels, fixed
dma buffers and other types).
```

```
This will enable such parameters to be locked down in the core parameter
parser for secure boot support.  [...]
```

This concludes our discussion on kernel module parameters. Let's now move on to a slightly peculiar aspect of writing kernel code – that of floating-point usage within the kernel.

Floating point not allowed in the kernel

Years ago, a young and relatively inexperienced lad working on a temperature sensor device driver (me) had an amusing experience (though it wasn't quite so amusing at the time). Attempting to express a temperature value in millidegrees Celsius as a "regular" temperature value in degrees Celsius correct to three decimal places, he did something like the following:

```
int temperature;
double temperature_fp;
 [... processing …]
temperature_fp = temperature / 1000.0;
printk(KERN_INFO "temperature is %.3f degrees C\n", temperature_fp);
```

It all went downhill from there!

The venerable LDD (*Linux Device Drivers*, by *Corbet, Rubini, and Kroah-Hartman*) book pointed out my error – **floating-point** (**FP**) arithmetic is not allowed in kernel space! It's a conscious design decision; saving processor (FP) state, turning on the FP unit, working on it, then turning off and restoring the FP state is just not considered a worthwhile thing to do while in the kernel. The kernel (or driver) developer is well advised to just not attempt to perform FP work while in kernel space.

Well, then, you ask, how can you do the (as in my example) temperature conversion? Simple: pass the *integer* millidegrees Celsius value *to user space* and perform the FP work there!

In quite a few cases, it's just plain wrong to perform certain kinds of work in the kernel; **user space** is the right place for such work (there are a few exceptions, as with every rule). This includes performing floating-point arithmetic, file I/O (yes, do *not* try doing file reading and writing within the kernel via system call-like code paths), and running applications, though the kernel does provide the (fairly surprising but very useful at times) ability to do so via the **UMH** (**user-mode helper**) APIs like the `call_usermodehelper*()` ones (use with caution!), and so on.

Forcing FP in the kernel

While we said that FP isn't allowed in kernel space, there does exist a way to force the kernel to perform floating point arithmetic: put your floating-point code between the `kernel_fpu_begin()` and `kernel_fpu_end()` macros. There are a few places where precisely this technique is used within the kernel code base (typically, some code paths covering crypto/AES, CRC, and so on).

The recommendation is clear: the typical module (or driver) developer should *perform only integer arithmetic within the kernel.*

Nevertheless, to test this whole scenario (always remember, *the empirical approach – actually trying things out – is the only realistic way forward!*), we write a simple kernel module that attempts to perform some FP work. The key part of the code is shown here:

```
// ch5/fp_in_kernel/fp_in_kernel.c
static double num = 22.0, den = 7.0, mypi;
static int __init fp_in_lkm_init(void)
{
    [...]
    kernel_fpu_begin();
    mypi = num/den;
    kernel_fpu_end();
#if 1
    pr_info("%s: PI = %.4f = %.4f\n", OURMODNAME, mypi, num/den);
#endif
    return 0;     /* success */
}
```

It actually works, until we attempt to display the FP value via `printk()`! At that point, it goes quite berserk. See the following screenshot:

```
------------------------------
sudo insmod ./fp_in_lkm.ko && lsmod|grep fp_in_lkm
------------------------------
fp_in_lkm              16384  0
------------------------------
sudo dmesg
------------------------------
[633848.557056] fp_in_lkm:fp_in_lkm_init(): inserted
[633848.557529] ------------[ cut here ]------------
[633848.557992] Please remove unsupported %f in format string
[633848.558534] WARNING: CPU: 2 PID: 583793 at lib/vsprintf.c:2638 format_decode+0x3a6/0x430
[633848.559337] Modules linked in: fp_in_lkm(OE+) modparams1(OE) pl2303 usbserial mmc_block cpuid cdc_acm tls cdc_
mbim cdc_wdm cdc_ncm cdc_ether usbnet mii snd_usb_audio uas snd_usbmidi_lib usb_storage netlink_diag procmap(OE) c
cm rfcomm xt_conntrack nft_chain_nat xt_MASQUERADE nf_nat nf_conntrack_netlink nf_conntrack nf_defrag_ipv6 nf_defr
ag_ipv4 xfrm_user xfrm_algo xt_addrtype nft_compat nf_tables libcrc32c nfnetlink br_netfilter bridge stp llc snd_c
tl_led snd_hda_codec_realtek snd_hda_codec_generic vboxnetadp(OE) vboxnetflt(OE) cmac vboxdrv(OE) algif_hash algif
_skcipher af_alg bnep overlay nvidia_uvm(POE) nvidia_drm(POE) snd_sof_pci_intel_cnl nvidia_modeset(POE) snd_sof_in
tel_hda_common soundwire_intel intel_tcc_cooling soundwire_generic_allocation soundwire_cadence snd_sof_intel_hda
x86_pkg_temp_thermal intel_powerclamp snd_sof_pci snd_sof_xtensa_dsp snd_sof snd_sof_utils snd_soc_hdac_hda snd_hd
a_ext_core snd_soc_acpi_intel_match snd_soc_acpi soundwire_bus snd_hda_codec_hdmi
[633848.559369]  snd_soc_core snd_compress ac97_bus snd_pcm_dmaengine snd_hda_intel snd_intel_dspcfg coretemp crct
10dif_pclmul snd_intel_sdw_acpi ghash_clmulni_intel snd_hda_codec aesni_intel snd_hda_core crypto_simd dell_laptop
 cryptd nvidia(POE) mei_pxp mei_hdcp intel_rapl_msr snd_hwdep btusb i915 snd_pcm kvm_intel btrtl dell_smm_hwmon uv
cvideo btbcm videobuf2_vmalloc snd_seq_midi iwlmvm btintel videobuf2_memops binfmt_misc snd_seq_midi_event btmtk k
vm mac80211 videobuf2_v4l2 snd_rawmidi dell_wmi drm_buddy ledtrig_audio libarc4 videobuf2_common ttm snd_seq bluet
ooth dell_smbios iwlwifi drm_display_helper videodev spi_nor processor_thermal_device_pci_legacy input_leds cec sn
d_seq_device rapl dell_wmi_sysman dcdbas intel_cstate nls_iso8859_1 rc_core ecdh_generic serio_raw processor_therm
al_device snd_timer firmware_attributes_class dell_wmi_descriptor joydev mei_me intel_wmi_thunderbolt mc wmi_bmof
mtd mxm_wmi ecc ee1004 cfg80211 drm_kms_helper processor_thermal_rfim hid_multitouch
[633848.567739]  i2c_algo_bit snd_mei fb_sys_fops processor_thermal_mbox syscopyarea sysfillrect processor_thermal
_rapl soundcore sysimgblt intel_rapl_common intel_pch_thermal intel_soc_dts_iosf mac_hid int3403_thermal int340x_t
hermal_zone int3400_thermal intel_hid sparse_keymap acpi_thermal_rel acpi_pad sch_fq_codel msr parport_pc ppdev lp
 parport drm efi_pstore ip_tables x_tables autofs4 usbhid hid_sensor_custom hid_sensor_hub intel_ishtp_hid hid_gen
eric rtsx_pci_sdmmc crc32_pclmul psmouse nvme i2c_i801 spi_intel_pci ucsi_acpi e1000e i2c_smbus spi_intel thunderb
olt rtsx_pci intel_lpss_pci intel_ish_ipc typec_ucsi intel_lpss xhci_pci nvme_core i2c_hid_acpi xhci_pci_renesas i
dma64 intel_ishtp i2c_hid typec hid_wmi video pinctrl_cannonlake [last unloaded: min_sysinfo]
[633848.582947] CPU: 2 PID: 583793 Comm: insmod Tainted: P      U     OE      5.19.0-50-generic #50-Ubuntu
[633848.583867] Hardware name: Dell Inc. Precision 7550/01PXFR, BIOS 1.25.0 08/22/2023
[633848.584682] RIP: 0010:format_decode+0x3a6/0x430
[633848.585140] Code: c6 03 03 44 29 e0 e9 2e fd ff ff c6 43 05 08 e9 e7 fd ff ff 0f be 30 48 c7 c7 78 cf a6 a9 c6
 05 27 2c c2 01 01 e8 f3 15 6e 00 <0f> 0b 48 8b 45 e0 eb bf 80 f9 6c 74 61 80 f9 68 0f 85 84 fd ff ff
[633848.587036] RSP: 0018:ffffb3329269b990 EFLAGS: 00010046
[633848.587559] RAX: 0000000000000000 RBX: ffffb3329269b9d8 RCX: 0000000000000000
```

Figure 5.7: The output of WARN_ONCE() when we try and print an FP number in kernel space

The key line here is `Please remove unsupported %f in format string`.

This tells us the story. The system does not actually crash or panic as this is a mere `WARNING`, spat out to the kernel log via the `WARN_ONCE()` macro. Do realize, though, that on a production system, the `/proc/sys/kernel/panic_on_warn` sysctl (pseudofile) will, perhaps, be set to the value `1`, causing the kernel to (quite rightly) panic.

Though not visible in the preceding screenshot (*Figure 5.7*), further down, the section beginning with `Call Trace:` is, of course, a peek into the current state of the *kernel-mode stack* of the process or thread that was "caught" in the preceding `WARN_ONCE()` code path (hang on, you will learn key details regarding the user- and kernel-mode stacks and so on in *Chapter 6, Kernel Internals Essentials – Processes and Threads*).

One can interpret the kernel stack by reading it in a bottom-up fashion, greatly aiding the debug effort. FYI, here's a truncated portion of the `Call Trace:` output (continuing the output from *Figure 5.7*):

```
[ … ]
[633848.592829] CR2: 000056086ea0dbd8 CR3: 0000000775ed8003 CR4:
00000000007706e0
[633848.593607] PKRU: 55555554
[633848.593889] Call Trace:
[633848.594148]  <TASK>
[633848.594374]  vsnprintf+0x71/0x560
[633848.594717]  vprintk_store+0x19a/0x5a0
[633848.595101]  vprintk_emit+0x8e/0x1f0
[633848.595468]  ? 0xffffffffc0a33000
[633848.595811]  vprintk_default+0x1d/0x30
[633848.596194]  vprintk+0x67/0xb0
[633848.596587]  ? 0xffffffffc0a33000
[633848.596927]  _printk+0x58/0x81
[633848.597246]  fp_in_lkm_init+0x62/0x1000 [fp_in_lkm]
[633848.597735]  do_one_initcall+0x46/0x230
[633848.598163]  ? kmem_cache_alloc_trace+0x1a6/0x330
[633848.598638]  do_init_module+0x52/0x220
[633848.599022]  load_module+0xb56/0xd40
[633848.599388]  ? security_kernel_post_read_file+0x5c/0x80
[ … ]
```

Here, following along, reading it from the bottom up, we can see the `do_one_initcall()` function called `fp_in_lkm_init()` (which belongs to the kernel module in square brackets, `[fp_in_lkm_init]`, our kernel module of course), which then calls `[_]printk()`, which then ends up causing all kinds of trouble as it attempts to print an FP quantity! (By the way, simply ignore any call frames beginning with the ? symbol.)

The moral here is clear: as far as possible, *avoid using floating-point math within kernel space.*

Exercise: What if the offending `printk()` is removed? Will it then work? Try it.

Let's now move on to the practical topic of how you can install and auto-load kernel modules on system startup.

Auto-loading modules on system boot

Until now, we have written simple "out-of-tree" kernel modules that reside in their own private directories and have to be manually loaded up, typically via the `insmod` or `modprobe(8)` utilities. In most real-world projects and products, you will require your out-of-tree kernel module(s) *to be auto-loaded at boot*. This section covers how you can achieve this.

Consider we have a kernel module named `foo.ko`. We assume we have access to its source code and Makefile. In order to have it *auto-load* on system boot, you need to first *install* the kernel module to a known location on the system. To do so, we expect that the Makefile for the module contains an `install` target, typically:

```
install:
        make -C $(KDIR) M=$(PWD) modules_install
```

This is not something new; we have been placing the `install` target within the `Makefiles` of our demo kernel modules. Further, as it writes into a root-owned directory, we can always modify the rule to run via `sudo` to ensure it succeeds (or, the user must invoke it via `sudo`, either way):

```
install:
        sudo make -C $(KDIR) M=$(PWD) modules_install
```

To demonstrate this "auto-load" procedure, we have shown the set of steps to follow in order to *install and auto-load on boot* our `ch5/min_sysinfo` kernel module:

1. First, change directory to the module's source directory:

    ```
    cd <book_src>/ch5/min_sysinfo
    ```

2. Next, it's important to first build the kernel module (with `make`), and, on success, install it (as you'll soon see, our "better" Makefile makes the process simpler by guaranteeing that the build is done first, followed by the `install` and the `depmod`):

    ```
    make && sudo make install
    ```

    Assuming it builds successfully, the `sudo make install` command then *installs* the kernel module here: `/lib/modules/<kernel-ver>/extra/` as is expected (do see the following info box and tips as well). Let's give it a shot:

    ```
    $ cd <book_src>/ch5/min_sysinfo
    ```

```
$ make              # <-- ensure it's first built 'locally'
                    # generating the min_sysinfo.ko kernel module object
[...]
$ sudo make install
[sudo] password for c2kp:

--- installing ---
 [First, invoking the 'make' ]
make
[ … ]
 [Now for the 'sudo make install' ]
sudo make -C /lib/modules/6.1.25-lkp-kernel/build M=/home/c2kp/Linux-
Kernel-Programming_2E/ch5/min_sysinfo modules_install
make: Entering directory '/home/c2kp/kernels/linux-6.1.25'
  INSTALL /lib/modules/6.1.25-lkp-kernel/extra/min_sysinfo.ko
  SIGN    /lib/modules/6.1.25-lkp-kernel/extra/min_sysinfo.ko
  DEPMOD  /lib/modules/6.1.25-lkp-kernel
make: Leaving directory '/home/c2kp/kernels/linux-6.1.25'
sudo depmod
 [If !debug, stripping debug info from  /lib/modules/6.1.25-lkp-kernel/
extra/min_sysinfo.ko]
if [ "n" != "y" ]; then \
    sudo strip --strip-debug /lib/modules/6.1.25-lkp-kernel/extra/min_
sysinfo.ko ; \
fi

$ ls -l /lib/modules/6.1.25-lkp-kernel/extra/
total 16
-rw-r--r-- 1 root root            <...>    min_sysinfo.ko
$
```

For now, simply ignore the `SIGN` line; it's saying that the module has been "signed" by the kernel, a security feature that we'll cover shortly.

During sudo make install, it's possible you might see (non-fatal) errors regarding SSL; they can be safely ignored. They indicate that the system failed to "sign" the kernel module (here, it worked). More on this in the upcoming section on security.

Also, just in case you find that sudo make install fails, try the following approaches:

a) Switch to a root shell (sudo -s) and, within it, run the make ; make install commands.

b) A useful reference: *Makefile: installing external Linux kernel module, Stack Overflow, June 2016* (https://unix.stackexchange.com/questions/288540/makefile-installing-external-linux-kernel-module).

3. Another module utility, called depmod(8), is then typically invoked by default within the sudo make install (as can be seen from the preceding output). Just in case (for whatever reason) this has not occurred, you can always manually invoke depmod (which our Makefile does); its job is essentially to resolve module dependencies (see its man page for details): sudo depmod. Once you install the kernel module, you can see the effect of depmod with its --dry-run option switch:

```
$ sudo depmod --dry-run | grep min_sysinfo
extra/min_sysinfo.ko:
alias symbol:llkd_sysinfo min_sysinfo
alias symbol:llkd_sysinfo2 min_sysinfo
```

4. Ensure the just-installed module is auto-loaded on boot. One way to do so is to create the /etc/modules-load.d/<foo>.conf config file (where <foo> is the name of the module. Of course, you will need root access to create this file). The simple case is really simple: just put the kernel module's name inside the file and that's it. Any line starting with a # character is treated as a comment and ignored. For our min_sysinfo example, we have the following:

```
$ sudo vim /etc/modules-load.d/min_sysinfo.conf
# Auto load kernel module demo for the LKP2E book: ch5/min_sysinfo
min_sysinfo
<save & exit>

$ cat /etc/modules-load.d/min_sysinfo.conf
# Auto load kernel module demo for the LKP2E book: ch5/min_sysinfo
min_sysinfo
$
```

On modern Linux, the de facto initialization framework is **systemd.** An (even simpler) way to autoload your module at boot is to inform systemd to load it up: simply enter the *name* of the (already installed) module into the (pre-existing) `/etc/modules-load.d/modules.conf` file.

5. Reboot the system with `sync; sudo reboot`.

   Once the system is up, use `lsmod` to spot the module and look up the kernel log (with `dmesg` or `journalctl`). You should see relevant info pertaining to the kernel module loading up (in our example, `min_sysinfo`):

```
[... system boots up, we log in and check …]
```

```
$ w
 12:21:24 up 2 min,  1 user,  load average: 0.02, 0.02, 0.00
USER     TTY      FROM             LOGIN@   IDLE   JCPU   PCPU WHAT
c2kp     pts/0    192.168.1.25     12:19    3.00s  0.06s  0.01s w
$
$ lsmod |grep min_sysinfo
min_sysinfo            16384  0
$ sudo dmesg |grep -A1 min_sysinfo
[    4.141769] min_sysinfo: loading out-of-tree module taints kernel.
[    4.142348] min_sysinfo:min_sysinfo_init(): inserted
[    4.142567] min_sysinfo:llkd_sysinfo(): llkd_sysinfo(): minimal Platform Info:
               CPU: x86_64, little-endian; 64-bit OS.
--
[    4.142984] min_sysinfo:llkd_sysinfo2(): llkd_sysinfo2(): minimal Platform Info:
               CPU: x86_64, little-endian; 64-bit OS.
--
[    4.143866] min_sysinfo:show_sizeof(): sizeof: (bytes)
                 char =  1   short int =  2              int =  4
--
[    4.145253] min_sysinfo:llkd_sysinfo2(): Word [U|S][8|16|32|64] ranges: unsigned max, signed max, signed min:
               U8_MAX =                  255 = 0x               ff,  S8_MAX =                 127 = 0x
     7f,  S8_MIN =                -128 = 0x          ffffff80
$
```

Figure 5.8: Screenshot showing our min_sysinfo module has been auto-loaded into kernel memory at boot

There, it's done: our `min_sysinfo` kernel module has indeed been auto-loaded into kernel space on boot! (The "platform details" – the output of our module – do indeed follow in the kernel log; it's just that all output lines aren't matched here via the `grep` we did.)

Remember, you must *first* build your kernel module and then perform the *install*; to help automate this, our "better" Makefile has the following in its module installation `install` target:

```
// ch5/min_sysinfo/Makefile
[ ... ]
install:
	@echo
	@echo "--- installing ---"
	@echo " [First, invoking the 'make' ]"
```

```
        make
        @echo
        @echo " [Now for the 'sudo make install' ]"
        sudo make -C $(KDIR) M=$(PWD) modules_install
        sudo depmod
    @echo " [If !debug and !(module signing), stripping debug info from
${KMODDIR}/extra/${FNAME_C}.ko]"
    if [ "${DBG_STRIP}" = "y" ]; then \
            sudo ${STRIP} --strip-debug ${KMODDIR}/extra/${FNAME_C}.ko ; \
    fi
```

It ensures that, first, the build is done, followed by the install and (explicitly) the depmod. (Our Makefile also has the intelligence to *strip* the module of debug symbols if it's built as a release binary and if module signing (coming up) isn't enabled.)

An alternate way to auto-load your module at boot is to "manually" load it via a startup script that invokes the modprobe(8) utility, a more intelligent version of insmod (more on this in the following section).

What if your auto-loaded kernel module requires some (module) parameters passed at load time? There are two ways to ensure that this happens: via a so-called modprobe config file (under /etc/modprobe.d/) or, if the module's built into the kernel, via the kernel command line.

Here we'll show the first way: simply set up your modprobe configuration file (as an example here, we use the name mykmod as the name of our LKM; again, you require root access to create this file): /etc/modprobe.d/mykmod.conf; in it, you can pass parameters like this:

```
options <module-name> <parameter-name>=<value>
```

As an example, the /etc/modprobe.d/alsa-base.conf modprobe config file on my x86_64 Ubuntu 22.04 LTS system contains the lines (among several others):

```
# Ubuntu #62691, enable MPU for snd-cmipci
options snd-cmipci mpu_port=0x330 fm_port-0x388
```

Here, we understand it as: if the snd-cmipci module is required, it's loaded into kernel memory along with the module parameters mpu_port=0x330 fm_port=0x388. A few more points on kernel module auto-loading-related items follow.

Module auto-loading – additional details

Once a kernel module has been installed on a system (via sudo make install, as shown previously), you can also insert it into the kernel interactively (or via a script) simply by using a "smarter" version of the insmod utility, called modprobe(8). For our example, we could first rmmod the module and then do the following:

```
sudo modprobe min_sysinfo
```

As an interesting aside, consider the following. In cases where there are several kernel module objects to load (for example, with the *module-stacking* design), the modprobe utility has the intelligence to load them up in the right order (we covered the need for this in the section *Understanding module stacking*) so that module dependencies are served and it all just works!

How does `modprobe` know the right *order* in which to load up kernel modules? When performing a build locally, the build process generates a file called `modules.order`. It tells utilities such as `modprobe` the order in which to load up kernel modules such that all dependencies are resolved. When kernel modules are *installed* into the kernel (that is, into the `/lib/modules/$(uname -r)/extra/`, or a similar, location), the `depmod` utility generates a `/lib/modules/$(uname -r)/modules.dep` file. This contains the dependency information – it specifies whether a kernel module depends on another. Using this information, modprobe then loads them up in the required order. To flesh this out, let's *install* our module-stacking example modules:

```
$ cd <...>/ch5/modstacking
$ sudo make install
[ ... ]
  INSTALL /lib/modules/6.1.25-lkp-kernel/extra/core_lkm.ko
  SIGN    /lib/modules/6.1.25-lkp-kernel/extra/core_lkm.ko
  INSTALL /lib/modules/6.1.25-lkp-kernel/extra/user_lkm.ko
  SIGN    /lib/modules/6.1.25-lkp-kernel/extra/user_lkm.ko
  DEPMOD  /lib/modules/6.1.25-lkp-kernel
make: Leaving directory '/home/c2kp/kernels/linux-6.1.25'
sudo depmod
[ ... ]
$ ls -l /lib/modules/6.1.25-lkp-kernel/extra/
total 248
-rw-r--r-- 1 root root 118377 Oct 12 16:37 core_lkm.ko
-rw-r--r-- 1 root root 12672 Oct 12 15:51 min_sysinfo.ko
-rw-r--r-- 1 root root 117097 Oct 12 16:37 user_lkm.ko
```

Clearly, the two kernel modules from our module-stacking example (`core_lkm.ko` and `user_lkm.ko`) are now installed under the expected location, `/lib/modules/$(uname -r)/extra/` (and we can see that depmod was run).

Now, check this out:

```
$ grep user_lkm /lib/modules/6.1.25-lkp-kernel/* 2>/dev/null
/lib/modules/6.1.25-lkp-kernel/modules.dep:extra/user_lkm.ko: extra/core_lkm.ko
$
```

Aha, `depmod` has arranged for the `modules.dep` file to include the dependency! It shows that the `extra/user_lkm.ko` kernel module (this time, the one on the left) *depends on* the `extra/core_lkm.ko` kernel module (via the `<k1.ko>: <k2.ko>...` notation, implying that the `k1.ko` module depends on the `k2.ko` module).

Thus, modprobe, parsing the /lib/modules/$(uname -r)/modules.order file, notices this and thus loads them in the required order, avoiding any issues.

FYI, while on this topic, the generated /lib/modules/$(uname -r)/modules.symbols file has information on all exported module symbols. Similarly, within the module directory, the Module.symvers file contains the same. (As well, all symbols in the kernel symbol table – exported and private – along with their (virtual) addresses, can be seen via the /proc/kallsyms pseudofile (as root; see the man page on nm(1) to understand the notation)).

Next, on modern Linux systems, it's actually *systemd* that takes care of auto-loading kernel modules at system boot, by parsing the content of files such as /etc/modules-load.d/* (the systemd service responsible for this is systemd-modules-load.service(8) – for details, refer to the man page on modules-load.d(5)).

Conversely, you might sometimes find that a certain auto-loaded kernel module is misbehaving – causing lockups or delays, or it simply doesn't work – and so you want to disable it being auto-loaded at boot. This can be done by *blacklisting* the module. You can specify this either on the kernel command line (convenient when all else fails!) or within the (previously mentioned) /etc/modules-load.d/<foo>.conf config file. On the kernel command line, via module_blacklist=mod1,mod2,..., the kernel docs shows us the syntax/explanation:

```
module_blacklist=  [KNL] Do not load a comma-separated list of
                   modules. Useful for debugging problem modules.
```

You can look up the current kernel command line by doing cat /proc/cmdline.

While on the topic of the kernel command line, several other useful options exist, enabling us to use the kernel's help for debugging issues concerned with kernel initialization. As a few examples, among several others, the kernel provides the following parameters in this regard (sourced from the official kernel docs here: https://www.kernel.org/doc/html/latest/admin-guide/kernel-parameters.html):

```
debug            [KNL] Enable kernel debugging (events log level).
[...]
initcall_debug   [KNL] Trace initcalls as they are executed. Useful
                 for working out where the kernel is dying during
                 startup.
[...]
ignore_loglevel [KNL] Ignore loglevel setting - this will print /all/
                 kernel messages to the console. Useful for
                 debugging. We also add it as printk module
```

```
        parameter, so users could change it dynamically,
        usually by /sys/module/printk/parameters/ignore_loglevel.
```

Fastboot! Also, the `initcall_debug` kernel parameter can help you pinpoint kernel functionality that's taking a long time to run during boot; this is a part of getting the system to "fast boot," a requirement for some kinds of projects. (On this note, be sure to check out `systemd-analyze`, the Perl script in the kernel tree `scripts/bootgraph.pl`, and more. See the *Further reading* section for more.)

FYI, and as mentioned earlier in this chapter, there is an alternate framework for third-party kernel module auto-rebuilding, called **Dynamic Kernel Module Support** (**DKMS**). The *Further reading* document for this chapter also provides some helpful links. In conclusion, auto-loading kernel modules into memory on system startup is a useful and often required functionality in a product.

Building high-quality products requires a keen understanding of – and the knowledge to build in – *security*; that's the focus of the next section.

Kernel modules and security – an overview

An ironic reality is that enormous efforts spent on improving *user-space* security considerations have resulted in a large payoff over recent years. A malicious user performing a **Buffer Overflow** (**BoF**) attack was relatively straightforward a couple of decades back, but today it is really hard to pull off well. Why? Because there are many layers of beefed-up security mechanisms to prevent many of these attack classes.

To quickly name a few countermeasures: compiler protections (`-fstack-protector[...]`, `-Wformat-security`, `-D_FORTIFY_SOURCE=3`), partial/full RELRO, better sanity and security checker tools (`checksec.sh`, address sanitizers, PaxTest, static analysis tools, and so on), secure libraries, hardware-level protection mechanisms (NX, SMEP, SMAP, and so on), [K]ASLR, better testing (fuzzing), and so on. Whew.

The irony is that *kernel-space* attacks have become increasingly common over the last several years! It has been demonstrated that revealing even a single valid kernel (virtual) address (and its corresponding symbol) to a clever attacker can allow them to figure out the location of some key internal kernel structures, paving the way to carry out all kinds of **privilege escalation** (**privesc**) attacks. Coming up, we will enumerate and briefly describe a few security/hardening features that the Linux kernel provides. However, ultimately, as a kernel developer – you have a large role to play: writing secure code, to begin with! Using our "better" Makefile is a great way to get started – several targets within it are concerned with security (all the static analysis ones, for example). Within this section on security and kernel modules, we'll cover a few pertinent topics; let's begin with looking at some key proc-based `sysctl`.

Proc filesystem tunables affecting the system log

We directly refer you to the man page on `proc(5)` – very valuable! – to glean information on these two security-related tunables or `sysctl` (Link: `https://man7.org/linux/man-pages/man5/proc.5.html`):

- dmesg_restrict
- kptr_restrict

A quick word on the dmesg_restrict sysctl

The dmesg_restrict sysctl is a means to determine what minimal privileges are required to look up the kernel log. Recall that even something as seemingly innocent as this is actually valuable information to a determined hacker! The relevant entry for it within the man page on proc(5) is as follows:

```
dmesg_restrict
/proc/sys/kernel/dmesg_restrict (since Linux 2.6.37)
The value in this file determines who can see kernel syslog contents. A  value
of 0 in this file imposes no restrictions. If the value is 1, only privileged
users can read the kernel syslog. (See syslog(2) for more details.) Since Linux
3.4, only users with the CAP_SYS_ADMIN capability may change the value in this
file.
```

(Recall that a "CAP_SYS_ADMIN capability" user is equivalent to root.) The default values of the dmesg_restrict tunable on (x86_64):

- Ubuntu 22.04 (running a 5.19-based kernel) is 1. That's good; until the 20.04 release, Ubuntu had the default value as 0 (weaker).
- Fedora 38 (running a 6.2-based kernel) is 0.

How can I see the current value on my Linux box? Easy; here are two ways to do so:

```
$ cat /proc/sys/kernel/dmesg_restrict
0
$ sysctl -a 2>/dev/null |grep -w dmesg_restrict
kernel.dmesg_restrict = 0
```

Clearly, it's 0, implying there aren't any restrictions (this is on a Fedora 39 workstation; recent Ubuntu distros set it to 1).

FYI, Linux kernels use the powerful fine-granularity POSIX *capabilities* model. The CAP_SYS_ADMIN capability essentially is a catch-all for what is traditionally *root (superuser/sysadmin)* access. The CAP_SYSLOG capability gives the process (or thread) the capability to perform privileged syslog(2) operations.

A quick word on the kptr_restrict sysctl

As already mentioned, "leaking" a kernel address and the symbol it's associated with might result in an info-leak-based attack. To help prevent these, kernel and module authors are advised to, firstly, simply avoid printing kernel addresses. If you must print them, then always use a new-ish (and kernel-specific) printf-style format: instead of the familiar %p or %px to print a kernel (virtual) address, you should use the %pK format specifier for printing an address. Using the %px format specifier ensures the actual address is printed; while perhaps useful for debugging, you'll definitely want to avoid this in production. How do these this printk format specifiers help? Read on...

The `kptr_restrict` sysctl (2.6.38 onward) affects the `printk()` output when printing kernel addresses; doing `printk("&var = %pK\n", &var);`and not the good old `printk("&var = %p\n", &var);` is considered a security best practice. Understanding how exactly the `kptr_restrict` tunable works is key to this:

```
kptr_restrict
/proc/sys/kernel/kptr_restrict (since Linux 2.6.38)
The value in this file determines whether kernel addresses are exposed via
/proc files and other interfaces. A value of 0 in this file imposes no
restrictions. If the value is 1, kernel pointers printed using the %pK format
specifier will be replaced with zeros unless the user has the CAP_SYSLOG
capability. If the value is 2, kernel pointers printed using the %pK format
specifier will be replaced with zeros regardless of the user's capabilities.
The initial default value for this file was 1, but the default was changed to 0
in Linux 2.6.39. Since Linux 3.4, only users with the CAP_SYS_ADMIN capability
can change the value in this file.
```

Again, the default values of this `sysctl` on our recent-enough Ubuntu and Fedora platforms are 1 and 0 respectively. You can leverage the `sysctl` utility to see (and, as root, change) kernel sysctls; on our Ubuntu 22.04 VM:

```
$ sysctl kernel.dmesg_restrict kernel.kptr_restrict
kernel.dmesg_restrict = 1
kernel.kptr_restrict = 1
```

Here's a table summarizing how the `kptr_restrict` sysctl behaves:

kptr_restrict value	**Unprivileged user and printk uses %pK**	**Privileged user (root) and printk uses %pK**
0	No restrictions (kernel addresses are exposed)	
1	Kernel addresses replaced with zeroes (unless the caller has `CAP_SYSLOG` capability)	No restrictions (kernel addresses are exposed)
2	Kernel addresses replaced with zeroes	

Table 5.2: A summary of how the kptr_restrict sysctl's value affects the display of kernel addresses in different circumstances

You can – rather, *must* – change these sysctls on production systems to a secure value (1 or 2), a best practice for security.

a) As `procfs` is, of course, a volatile filesystem, you can always make the changes permanent by using the `sysctl(8)` utility with the `-w` option switch (or by directly updating the `/etc/sysctl.conf` file).

b) For the purpose of debugging, if you must print an actual kernel (unmodified) address, you're advised to use the `%px` format specifier; do remove these prints on production systems!

c) Detailed kernel documentation on `printk` format specifiers can be found at `https://www.kernel.org/doc/html/latest/core-api/printk-formats.html#how-to-get-printk-format-specifiers-right`; do browse through it.

Of course, security measures only work when developers make use of them; as of the 6.1.25 kernel source tree, there is an approximate total of 87,000-odd uses of the `printk()` (~ 33k) and friends (`pr_*()` : ~ 54k) routines. Of these, a measly 24 employ the security-aware `%pK` format specifier! (To find out, I grepped the kernel source tree's `*.[ch]` files with the regular expressions `printk\(.*\%pK` and `pr_.*\(.*\%pK`. Though unlikely to be perfect, this gives a decent approximation. The first edition of this book found just 14 users of the `%pK` format specifier in the 5.4.0 kernel source tree; so, it's improving, slowly.)

With the advent of hardware-level defects in early 2018 (the now well-known *Meltdown, Spectre,* and other processor speculation security issues), there was a sense of renewed urgency in *detecting information leakage,* thus enabling developers and administrators to block it off.

A useful Perl script, `scripts/leaking_addresses.pl`, was released in mainline in 4.14 (in November 2017; I am happy to have lent a hand in this important work: `https://github.com/torvalds/linux/commit/1410fe4eea22959bd31c05e4c1846f1718300bde`), with more checks being made for detecting leaking kernel addresses.

Let's now look at an exciting and powerful security feature, especially for module authors!

Understanding the cryptographic signing of kernel modules

Once a malicious attacker gets a foothold on a system, they will typically attempt some kind of privesc vector in order to gain root access. Once this is achieved, the typical next step is to install a *rootkit*: essentially, a collection of scripts and kernel modules that will pretty much take over the system in a stealthy manner (by "hijacking" system calls, setting up backdoors, keyloggers, and so on) so that the attacker can reenter the system at will.

Of course, it's not easy – the security posture of a modern production-quality Linux system, replete with **Linux Security Modules (LSMs)** and various user/kernel hardening features, means it's not at all a trivial thing to do, but for a skilled and motivated attacker, anything's possible (for sensitive installations, that's the right mindset to have). Assuming they have a sufficiently sophisticated rootkit installed, the system is now considered compromised.

An interesting idea is this: *even* with root access, do not allow anything – the insmod (or modprobe, or even the underlying [f]init_module() system calls) – to insert kernel modules into kernel address space **unless they are cryptographically signed with a security key** that is in the kernel's keyring. This powerful security feature was introduced with the 3.7 kernel, over a decade back! (The relevant initia commit is here: https://git.kernel.org/pub/scm/linux/kernel/git/torvalds/linux.git/commit/?id=106a4ee258d14818467829bf0e12aeae14c16cd7.)

A few relevant kernel configuration options concerned with this feature are CONFIG_MODULE_SIG, CONFIG_MODULE_SIG_FORCE, CONFIG_MODULE_SIG_ALL, and so on. You can see and set them up from the usual make menuconfig UI: navigate to Enable loadable module support > Module signature verification. Most are off by default; I turned Module signature verification on first. A partial screenshot showing these options follows:

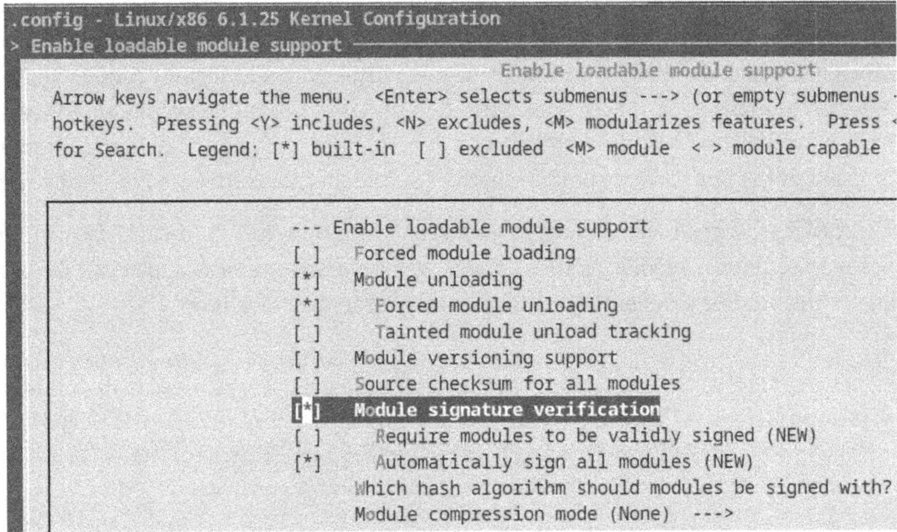

Figure 5.9: Partial screenshot showing the various module signature options (among others)

So, now we understand better: when installing modules, we can arrange appropriate kernel configs so that:

- Installed modules can be signed (CONFIG_MODULE_SIG=y).
- All installed modules will always be signed (CONFIG_MODULE_SIG_ALL=y).
- Installed modules will not be loaded into kernel memory unless they are signed (aka "sig force" (CONFIG_MODULE_SIG_FORCE=y)!
- Modules are (also) compressed when being installed (CONFIG_MODULE_COMPRESS_XZ=y, among others).

To help understand what exactly the first one, CONFIG_MODULE_SIG=y, means, let's see its Kconfig file's "help" section, as follows (from init/Kconfig):

```
config MODULE_SIG
 bool "Module signature verification"
```

```
  depends on MODULES
  select SYSTEM_DATA_VERIFICATION
  help
   Check modules for valid signatures upon load: the signature is simply
   appended to the module. For more information see
   <file:Documentation/admin-guide/module-signing.rst>. Note that this
   option adds the OpenSSL development packages as a kernel build
   dependency so that the signing tool can use its crypto library.

   You should enable this option if you wish to use either
   CONFIG_SECURITY_LOCKDOWN_LSM or lockdown functionality imposed via
   another LSM - otherwise unsigned modules will be loadable regardless of the
   lockdown policy.

      !!!WARNING!!! If you enable this option, you MUST make sure that the
      module DOES NOT get stripped after being signed. This includes the
      debuginfo strip done by some packagers (such as rpmbuild) and
      inclusion into an initramfs that wants the module size reduced.
```

Do read the above *Help* screen carefully. Our "better" Makefile takes this warning seriously and does *not* strip (debug) symbols if any of the `CONFIG_MODULE_SIG*` kernel configs are on.

The two module-signing modes

Module signing works in two possible modes:

- **Permissive** (the default): When `CONFIG_MODULE_SIG_FORCE` is off. Allows modules that aren't signed (or where the key isn't available) to be loaded. As a side effect, the kernel is marked as being "tainted" (with the E bit being set for it and the concerned module). This setting – *permissive* – tends to be the default on a typical modern Linux distribution.
- **Restrictive**: When `CONFIG_MODULE_SIG_FORCE` is on (set to `y`, or the kernel command line includes `module.sig_enforce=1`). Now only valid signed modules (which are verified by a key in the kernel) will be allowed to load. Further, any modules with a signature block that can't be parsed by the kernel or is mismatched will refuse to be loaded. This is good for added security. (Recall, though, that the module(s) must *not* be stripped in any manner.)

Tainted kernel: The word "tainted" means polluted or dirtied... the kernel is marked as tainted in some situations (like an Oops, a proprietary or unsigned module being loaded, and several more). The kernel maintains a tainted bitmask; it has no material effect, but it can help when investigating what happened. See details here: `https://docs.kernel.org/admin-guide/tainted-kernels.html`.

In the following output block, I look up these kernel configs on my x86_64 Fedora 38 system (running a recent 6.5-based distro kernel):

```
$ grep MODULE_SIG /boot/config-6.5.6-200.fc38.x86_64
CONFIG_MODULE_SIG_FORMAT=y
CONFIG_MODULE_SIG=y
# CONFIG_MODULE_SIG_FORCE is not set
CONFIG_MODULE_SIG_ALL=y
[ ... ]
```

The `CONFIG_MODULE_SIG_ALL` being set to `y` ensures that *all* kernel modules, when installed, are signed! (Using module signing does imply that the OpenSSL developer packages must be installed on the system, as it has the tool that performs the signing.) This being the default setting seems to be good for security. However, precisely for security, it's then also definitely recommended that the keypair (public/private) being used should be one other than the kernel default (these are configurable via the `certs/x509.genkey`, which will be generated as required). Thus, it's recommended you explicitly set up this file and don't use the kernel defaults.

It's also possible to manually sign your kernel modules by leveraging the `scripts/sign-file` program (which gets built from its C source file). All of this, and more, is very well documented in the official kernel documentation: `https://www.kernel.org/doc/html/v6.1/admin-guide/module-signing.html`; do refer to it. (FYI, Ubuntu has its own kernel module-signing tool, *kmodsign*.)

Thus, the cryptographic signing and load-only-if-signed behavior of kernel modules is encouraged on production systems (in recent years, with (I)IoT edge devices becoming more prevalent, security is a key concern).

Disabling kernel modules altogether

Paranoid folks might want to completely disable the loading of kernel modules. Rather drastic, but hey, this way you can completely lock down the kernel space of a system (as well as render any rootkits pretty much harmless). This can be achieved in two broad ways:

- First, while configuring your kernel, set the `CONFIG_MODULES` kernel config to off (it's on, of course, by default). Doing this *is* pretty drastic – it makes the decision a permanent one!
- Second, assuming `CONFIG_MODULES` is turned on, module loading can be dynamically turned off at runtime via the `modules_disabled sysctl`; take a look at this:

  ```
  $ cat /proc/sys/kernel/modules_disabled
  0
  ```

  It's *off* (`0`) by default, of course. As usual, the man page on `proc(5)` tells us the story:

  ```
  /proc/sys/kernel/modules_disabled (since Linux 2.6.31)
  A toggle value indicates if modules are allowed to be loaded in an
  otherwise modular kernel. This toggle defaults to off (0), but can be set
  true (1).
  ```

```
Once true, modules can be neither loaded nor unloaded, and the toggle
cannot be set back to false. The file is present only if the kernel is
built with the CONFIG_MODULES option enabled.
```

The kernel lockdown LSM – an introduction

An even more powerful kernel hardening feature is a relatively recent (from 5.4) one, an LSM (Linux Security Module) named ***lockdown***; it can be quite strict and invasive and is thus disabled by default. In short, it disables user-mode apps/scripts from making modifications to the running kernel (via its so-called *Integrity* mode). An even more strict lockdown mode, *Confidential*, in addition, prevents the extraction of kernel information considered to be confidential (thus helping prevent info leaks and more).

Note that the *lockdown* feature can be auto-enabled when using **Secure Boot** mode on (U)EFI x86 or AArch64! This, in turn, can lead to your modules failing to load (it happened to me once). Security can – and does – tend to be a double-edged sword.

More on the *lockdown* LSM can be found in the *Further reading* section.

In conclusion, of course, kernel security hardening and malicious attacks are a cat-and-mouse game. For example, a security hardening measure called (K)ASLR (we talk about what (K)ASLR means in the chapters to come on Linux memory management) is quite regularly defeated. Also, see this article – *Effectively bypassing kptr_restrict on Android*: `http://bits-please.blogspot.com/2015/08/effectively-bypassing-kptrrestrict-on.html`. Security is not easy; it's always a work in progress. It (almost) goes without saying: developers – in both user and kernel space – need to be clued into the required and actual security posture of the system and *must* write code that is security-aware, using tools and testing to verify it on a continuous basis. *(More work, but hey, we have to do it.)*

Let's complete this chapter with a bit on the kernel community – coding style guidelines, and quick how-to-style pointers on upstream contribution.

Coding style guidelines for kernel developers

Many large projects specify their own set of coding guidelines; so does the Linux kernel community. Adhering to the Linux kernel *coding style* guidelines is a really good idea when writing kernel and/or kernel module code. You can see the style guidelines officially documented here: `https://www.kernel.org/doc/html/latest/process/coding-style.html` (please do read them!).

Furthermore, as part of the (quite exhaustive) code-submission checklist(s) for developers like you wanting to upstream your code, you are expected to run your patch through a Perl script that checks your code for congruence with the Linux kernel coding style: `scripts/checkpatch.pl`.

By default, this script only runs on a well-formatted `git` patch. It's possible to run it against standalone C code (as in your out-of-tree kernel module code), as follows (as our "better" Makefile indeed does):

```
<kernel-src>/scripts/checkpatch.pl --no-tree -f <filename>.c
```

Doing this as a habit with your kernel code is helpful, enabling you to catch those annoying little issues – plus more serious ones! – that might otherwise hold your patch up. Again, we'll remind you: our "better" Makefile's `indent` and `checkpatch` targets are geared toward this.

Exercise: `cd` to your module directory (the one that has the "better" Makefile) and do this: `make checkpatch`. This will, via the Makefile, have the `checkpatch` target execute on your code, which in turn invokes the kernel's `checkpatch.pl` script on it. Carefully read any warnings and errors generated, fix them, and repeat. (Note: Typically, the first warning emitted can be ignored as it pertains to specifying the "SPDX-License" tag as the first line.)

Besides coding style guidelines, you will find that every now and then, you need to dig into the elaborate and useful kernel documentation. A gentle reminder: we covered locating and using the kernel documentation in *Online Chapter, Kernel Workspace Setup*, under the *Locating and using the Linux kernel documentation* section. It pays to get yourself familiar with the kernel documentation.

We will now complete this chapter by making a brief mention of how you can get started on a noble objective: contributing code to the mainline Linux kernel project.

Contributing to the mainline kernel

In this book, we typically perform kernel development *outside* the kernel source tree, via the LKM framework. What if you are writing code *within* the kernel tree, with the explicit goal of *upstreaming* your code to the mainline kernel? This is a laudable goal indeed – the whole basis of open source stems from the community's willingness to put in work and contribute upstream to the project.

Getting started with contributing to the kernel

The most frequently asked question, of course, is *how do I get started*? To help you precisely with this, a long and detailed answer lies within the official kernel documentation here: HOWTO do Linux kernel development: `https://www.kernel.org/doc/html/latest/process/howto.html#howto-do-linux-kernel-development`.

As a matter of fact, you can generate the full Linux kernel documentation (via the `make pdfdocs` command, in the root of the kernel source tree); once successful, you will find this PDF document – the same *HOWTO do Linux kernel development* one – here: `<root-of-kernel-source-tree>/Documentation/output/latex/development-process.pdf`.

It is a very detailed guide to the Linux kernel development process, including guidelines for code submission. A cropped screenshot of this document is shown here:

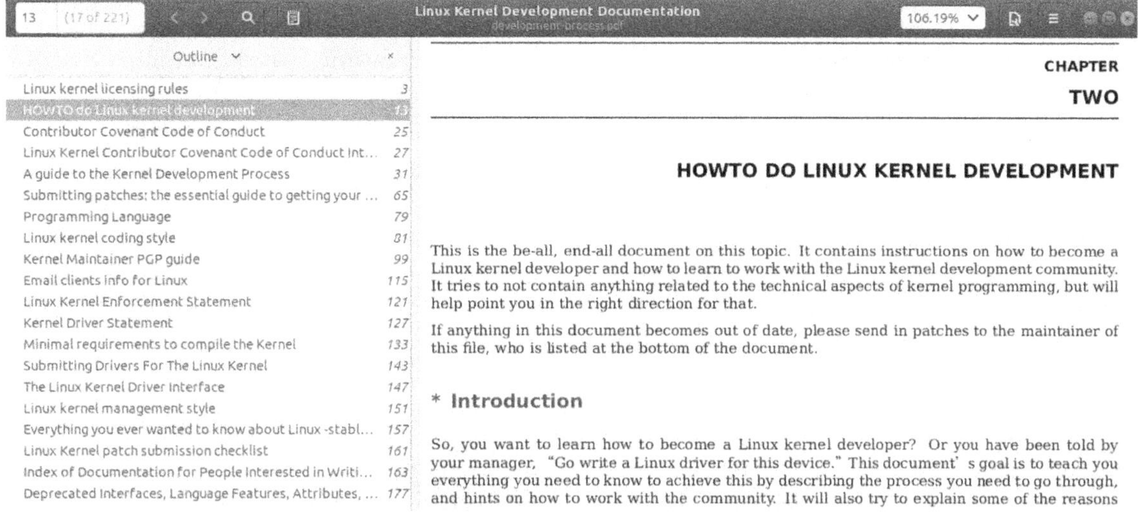

Figure 5.10: (Partial) screenshot of the kernel development docs just generated

As a part of the kernel development process, and to maintain quality standards, a rigorous, *must-be-followed* checklist – a long recipe of sorts! – is very much a part of the kernel patch submission process (the current latest version of this checklist has 24 points, no less). The official checklist resides here: ***Linux Kernel patch submission checklist***: `https://www.kernel.org/doc/html/latest/process/submit-checklist.html#linux-kernel-patch-submission-checklist`.

Though it may seem an onerous task for a kernel newbie, carefully following this checklist lends both rigor and credibility to your work and ultimately results in superior code. I strongly encourage you to read through the kernel patch submission checklist and try out the procedures mentioned therein.

Is there a practical hands-on tip, an almost guaranteed way, to become a kernel hacker? Of course, keep reading this book! Ha ha, yes, besides that, do partake in the simply awesome **Eudyptula Challenge** (`http://www.eudyptula-challenge.org/`) Oh, hang on, it's – very unfortunately, and as of the time of writing – closed down.

Fear not; here's a site with all the challenges (and solutions, but don't cheat!) posted. Do check it out and try the challenges. This will greatly accelerate your kernel hacking skills: `https://github.com/agelastic/eudyptula`.

Summary

In this chapter, the second one on writing kernel modules via the LKM framework, we covered several (remaining) areas pertaining to this important topic: among them, using a "better" Makefile for your kernel module, tips on configuring a debug kernel (it's very important!), cross-compiling a kernel module, gathering some minimal platform information from within a kernel module, and even a bit on the licensing of kernel modules. We also looked at emulating library-like features with two different approaches (first – the typically preferred one, which is the linking approach, and second, the module-stacking approach), using module parameters, avoiding floating-point arithmetic, the auto-loading of your kernel modules at boot, and so on. Security concerns, especially regarding modules and how they can be addressed, are important and have been covered. Finally, we wrapped up this chapter by covering kernel coding style guidelines, and how you can get started with contributing to the mainline kernel. So, congratulations! You now know how to develop a kernel module and can even get started on the journey to kernel upstream contribution.

In the next chapter (the first of our *Part 2 – Understanding and Working with the Kernel*, topics), we will delve into an interesting and necessary topic. We will begin our exploration in some depth into the *internals* of the Linux kernel, with respect to its architecture, how it encapsulates processes and threads, and more.

Questions

As we conclude, here is a list of questions for you to test your knowledge regarding this chapter's material: `https://github.com/PacktPublishing/Linux-Kernel-Programming_2E/blob/main/questions/ch5_qs_assignments.txt`. You will find some of the questions answered in the book's GitHub repo: `https://github.com/PacktPublishing/Linux-Kernel-Programming_2E/tree/main/solutions_to_assgn`.

Further reading

To aid you in delving deeper into the subject with useful materials, we provide a rather detailed list of online references and links (and at times even books) in a *Further reading* markdown document – organized by chapter – in this book's GitHub repository. The *Further reading* document is available here: `https://github.com/PacktPublishing/Linux-Kernel-Programming_2E/blob/main/Further_Reading.md`.

Learn more on Discord

To join the Discord community for this book – where you can share feedback, ask questions to the author, and learn about new releases – follow the QR code below:

`https://packt.link/SecNet`

6

Kernel Internals Essentials – Processes and Threads

With the previous chapter, you're now in a good position to understand and write simple kernel modules. In this chapter, we begin our exploration of Linux kernel internals, a vast and complex topic. In this book, we do not intend to delve very deep into the details of kernel and memory internals. At the same time, I would like to provide sufficient and requisite background knowledge for a budding kernel module author or device driver developer like you to successfully tackle the key topics necessary to understand kernel architecture in terms of how processes, threads, and their stacks are managed. Armed with this knowledge, you'll be able to better understand the coming chapters on correctly and efficiently managing dynamic kernel memory. As a side benefit, you will find yourself becoming more proficient at *debugging* both user and kernel space code.

I have divided the discussion on essential kernel internals into two chapters, this one and the next. This chapter covers key aspects of the architecture of the Linux kernel, especially concerning how processes and threads are managed within it. The following chapter will focus on memory management internals, another critical aspect of understanding and working with the Linux kernel. Of course, the reality is that all these things don't really get covered in a chapter or two but are spread out across this book (for example, details on the CPU scheduling of processes/threads will be found in later chapters – similarly for memory internals, synchronization topics, and so on).

Briefly, these are the topics covered in this chapter:

- Understanding process and interrupt contexts
- Understanding the basics of the process **Virtual Address Space** (**VAS**)
- Organizing processes, threads, and their stacks – user and kernel space
- Understanding and accessing the kernel task structure
- Working with the task structure via 'current'
- Iterating over the kernel's task lists

Technical requirements

I assume that you have gone through *Online Chapter, Kernel Workspace Setup*, and have appropriately prepared a guest **Virtual Machine** (**VM**) running Ubuntu 22.04 LTS (or a later stable release) and installed all the required packages. If not, I recommend you to do this first.

Also, if you haven't yet, do clone this book's GitHub repository for the code (found here: `https://github.com/PacktPublishing/Linux-Kernel-Programming_2E`); let's work on it in a hands-on fashion.

I assume that you are familiar with basic virtual memory concepts, the user-mode process **Virtual Address Space** (**VAS**) layout of segments (mappings), the stack, and so on. Nevertheless, we devote a few pages to explaining these basics (in the *Understanding the basics of the process VAS* section that soon follows).

Understanding process and interrupt contexts

In *Chapter 4, Writing Your First Kernel Module – Part 1*, we presented a brief section entitled *Understanding kernel architecture – part 1* (if you haven't read it yet, I suggest you do so before continuing). We will now expand on this discussion.

First off, modern processors execute code at different levels of privilege. For example, the x86-based ones offer four levels (or rings) of privilege, with Ring 0 being the most privileged and Ring 3 being the least. Similarly, the ARM-32 (AArch32) has seven execution modes, six of which are privileged. ARM64 (AArch64) uses the notion of exception levels (EL0 to EL3, with EL0 being the least and EL3 being the most privileged). Realistically though, and a key point: all modern OSs employ just two of the available CPU privilege levels – a privileged level and an unprivileged one at which code executes; we refer to them as kernel and user mode, respectively.

It's also critical to understand that most modern OS are **monolithic** in design. The word *monolithic* literally means a *single large piece of stone*. We shall defer a little later to how exactly this applies to our favorite OS! For now, just understand *monolithic* as meaning this: when a process or thread issues a system call, it switches to (privileged) kernel mode, executes kernel code, and possibly works on kernel data. Yes, there is no kernel or kernel thread executing code on its behalf; the process (or thread) issuing the system call is switched to kernel mode and *itself* executes kernel code. Thus, we say that kernel code executes within the context of a user-space process or thread – we call this execution context as **process context**. Think about it, significant portions of the kernel execute precisely this way, including a large portion of the code of device drivers. (FYI, even the handling of processor *exceptions* – like that of a page fault or system call – as well as CPU scheduling, is carried out in process context).

Well, you may ask, now that I understand this, how else – besides process context – can kernel code execute? There is another way: when a hardware interrupt (from a peripheral device – the keyboard, a network card, a disk, and so on) fires, the CPU's control unit saves the current context and immediately re-vectors the CPU to run the code of the interrupt handler (the **interrupt service routine** (**ISR**)). Now, this code runs in kernel (privileged) mode too – in effect, this is another, asynchronous, way to switch to kernel mode (unless you were already there)! The interrupt code path of many device drivers is executed like this; we now say that the kernel/driver code being executed in this manner is executing in **interrupt context**.

(Again, FYI, several modern drivers use a threaded interrupt model where the majority of interrupt processing takes place in the context of a kernel thread, in effect, in process context).

So, any and every piece of kernel (or module/driver) code is entered by and executed in one of two contexts:

- **Process context:** Kernel-space is entered because a process or thread issued a system call or a processor *exception* (such as a page fault) occurred, kernel code is executed, and kernel data is worked upon; it's typically synchronous.
- **Interrupt context:** Kernel-space is entered because a peripheral chip asserted a hardware interrupt, kernel (and/or driver) code is executed, and kernel data is worked upon; it's asynchronous.

Figure 6.1 shows the conceptual view: user-mode processes and threads execute in an unprivileged user context; the user-mode thread can switch to privileged kernel mode by issuing a *system call*. The diagram also shows us that pure *kernel threads* exist as well within Linux; they're very similar to user-mode threads, with the key difference that they only execute in kernel space; they can't even 'see' the user VAS. A synchronous switch to kernel mode via a system call (or processor exception) has the task now running kernel code in *process context*. (Kernel threads also run kernel code in process context).

Hardware interrupts, though, are a different ball game – they preempt anything, including kernel code, causing execution to asynchronously switch to kernel privilege (if not already in it). The code they execute (typically the kernel's or a device driver's interrupt handler) runs in the so-called *interrupt context*.

Figure 6.1 shows more details – interrupt context top and bottom halves, kernel threads, and work queues; these details (and much more!) are covered in this book's companion volume, *Linux Kernel Programming – Part 2*.

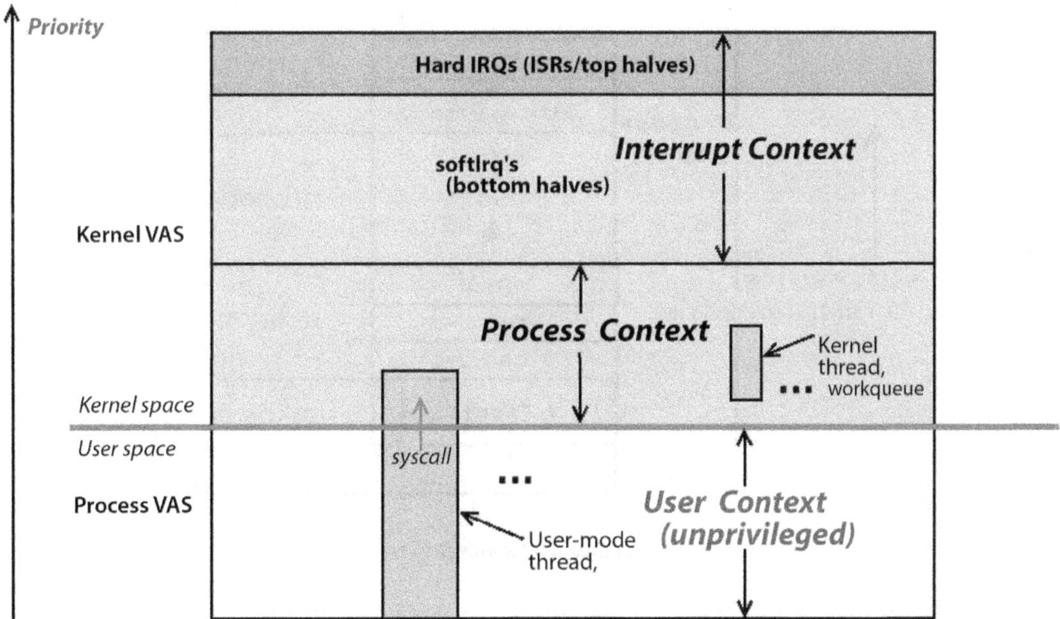

Figure 6.1: Conceptual diagram showing unprivileged user-mode execution and privileged kernel-mode execution with both process and interrupt contexts

Further on (in the *Determining the context* section), we shall show you how exactly you can check *in which context (process or interrupt)* your kernel (or driver) code is currently running. Read on!

Understanding the basics of the process Virtual Address Space (VAS)

A fundamental 'rule' of virtual memory is this: all potentially addressable memory is in a box; that is, it's *sandboxed*. We think of this 'box' as the *process image* or the process VAS. Looking 'outside' the box is impossible.

Here, we provide only a quick overview of the process user VAS (which should be sufficient). For more details, please refer to my earlier book *Hands-On System Programming with Linux*.

The user VAS is divided into homogeneous memory regions called ***segments*** or, more technically, ***mappings*** *(as they're internally constructed via the mmap() system call). Figure 6.2* shows the minimal mappings (segments) that every single Linux (user space) process will have:

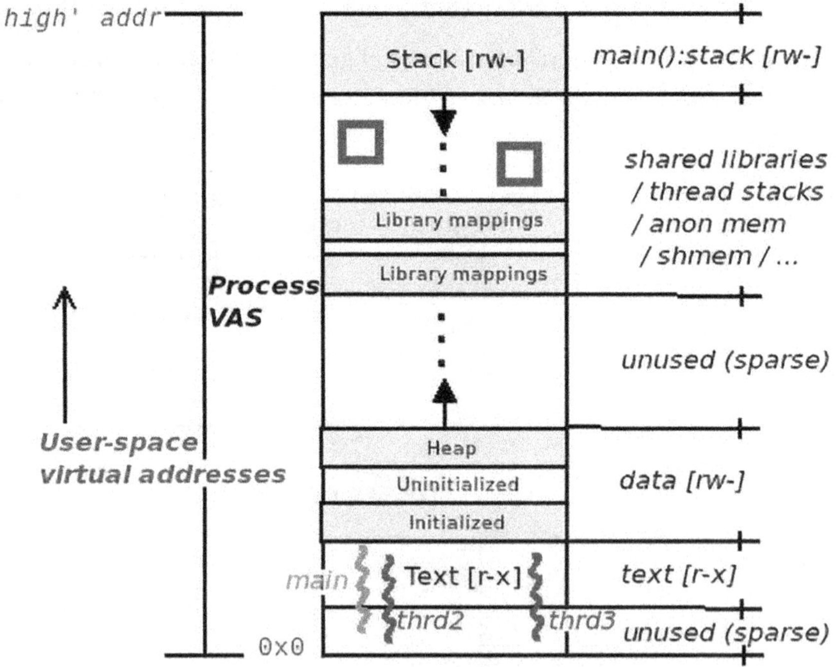

Figure 6.2: Linux user space process VAS

Let's go over a quick breakdown of these segments or mappings (from the bottom up):

- **Text segment***:* This is where the machine code is stored; it's where the processor core's instruction pointer (or equivalent) register points while a thread of the process executes code – static/fixed size (mode: `r-x`). (Note that the text segment does not start at virtual address `0x0`; it's some distance above that. In fact, the very first virtual page – the one encapsulating the NULL address (`0x0`) – is called the 'null trap' page. More on the null trap later.)
- **Data segment(s)***:* Immediately above the text mapping. This is where the global and static data variables are stored (mode: `rw-`). In reality, there are three distinct data segments:
  - **Initialized data segment***:* Pre-initialized global/static variables are stored here – static/fixed size.
  - **Uninitialized data segment**: Uninitialized global/static variables are stored here (they are auto-initialized to `0` at runtime; this region is sometimes called the *bss*) – static / fixed size.
  - **Heap segment***:* The *standard C library APIs* for memory allocation and freeing (the familiar `malloc()` family of routines) get memory from here. That's also not completely true. On modern glibc, only `malloc()` calls for memory below `MMAP_THRESHOLD` bytes (128 KB by default) get their memory from the heap. Any higher size requested is allocated as a separate mapping in the process VAS (via the powerful `mmap()` system call), called an anonymous (or anon) mapping. The heap is a dynamic segment (it can grow/shrink in size). We colloquially say that the heap 'grows up' towards higher virtual addresses. The last legally reference-able location on the heap is referred to as the *program break* (and can be retrieved by calling sbrk(0)).
- **Shared libraries (text, data)**: All shared libraries that a process dynamically links into are mapped (at runtime, via the loader calling mmap()) into the process VAS – somewhere between the top of the heap and the stack of the `main()` thread (mode: `r-x`/`rw-`). This general region – between the heap and stack also holds any other thread's stack memory (besides that of main()), anonymous memory, and shared memory regions.
- **Stack***:* A region of memory that uses the **Last In, First Out** (**LIFO**) semantics; the stack is used to implement *a high-level language's function-calling* mechanism and, in effect, holds the thread's execution context. The stack (frame) includes the job of parameter passing, local variable instantiation (and destruction), and return value propagation. It is a dynamic segment. On all modern processors (including the x86 and ARM families), *the stack 'grows down' toward lower virtual addresses* (called a fully descending stack). Every time a function is called, a *stack frame* (or call frame) is allocated and initialized as required; the precise layout of a stack frame is very CPU-dependent (you must refer to the respective CPU **Application Binary Interface** (**ABI**) document for this – see the *Further reading* section for references). The processor core's **Stack pointer** (**SP**) register (or equivalent) always points to the current frame, the top of the stack; as stacks grow towards lower (virtual) addresses, the top of the stack is actually the lowest (virtual) address! It's non-intuitive but true (mode: `rw-`). (FYI, there's more to it of course; stack frames aren't really allocated and initialized when a function's called, that would be too slow. Similarly, they aren't deallocated when a function returns; you'll learn more as you progress.)

Of course, you will understand that processes must contain at least one *thread* of execution (a thread is an execution path within a process); the one guaranteed thread is of course the `main()` function. In *Figure 6.2*, as an example, we show three threads of execution – `main`, `thrd2`, and `thrd3`. Also, as expected, every thread shares everything in the process VAS *except* for the stack; as you'll know, *every thread has its own private stack*. The stack of `main` is shown at the very top of the process (user) VAS; the stacks of the `thrd2` and `thrd3` threads can be allocated anywhere between the library mappings and the stack of `main` is illustrated as such via the two (blue) squares in this region.

I have designed and implemented what I feel is a pretty useful learning/teaching and debugging utility called **procmap** (`https://github.com/kaiwan/procmap`); it's a console-based *process VAS visualization utility*. It can actually show you the complete process VAS (in quite a bit of detail); we shall commence using it in the next chapter. Don't let that stop you from trying it out right away though; do clone it and give it a spin on your Linux system (it does require root access).

Now that you have understood the basics of the process VAS, it's time to delve quite a bit deeper into the kernel internals regarding the process VAS, the user and kernel address spaces, and their threads and stacks.

Organizing processes, threads, and their stacks – user and kernel space

The traditional **UNIX process model** – *Everything is a process; if it's not a process, it's a file* – has a lot going for it. The very fact that it is still *the* model followed by operating systems after a span of over five decades amply validates this. Of course, nowadays, the **thread** is considered the atomic execution context; *a thread is an execution path within a process*. Threads share all process resources, including the user VAS, open files, signal dispositions, IPC objects, credentials, paging tables, and so on, *except for the stack*. Every thread has its own private stack region (this makes perfect sense; if not, how could threads truly run in parallel, as it's the stack that holds execution context).

The other reason we focus on the *thread* and not the process is made clearer in *Chapter 10, The CPU Scheduler – Part 1*. For now, we shall just say this: *the thread, not the process, is the kernel schedulable entity* (aka the KSE) – it's what gets scheduled to run on a CPU core. This is a fallout of a key aspect of the Linux OS architecture. On Linux, every thread – including kernel threads – maps to a kernel metadata structure called the **task structure**. The task structure (also known as the *process descriptor*) is essentially a large kernel data structure that the kernel uses as a per-thread attribute structure. For every *thread* alive, the kernel maintains a corresponding *task structure* (see *Figure 6.3*, and worry not, we shall cover more on the task structure in the coming sections).

The next key point to grasp is that we *require one stack per thread per privilege level supported by the CPU.* On modern OSs such as Linux, we support two CPU privilege levels – the unprivileged user mode (or user space) and the privileged kernel mode (or kernel space). Thus, on Linux, every user space thread alive has two stacks:

- **A user space stack:** This stack is in play when the thread executes user-mode code paths.
- **A kernel space stack:** This stack is in play when the thread switches to kernel mode (via a system call or processor exception) and executes kernel code paths (in process context).

Of course, every good rule has an exception: kernel threads (kthreads, for short) are threads that live purely within the kernel and thus have a "view" of only kernel (virtual) address space; they cannot "see" userland. Hence, these kthreads, as they will only ever execute kernel space code paths, they have just one stack per kthread – a kernel space stack.

Also, this might have you wonder what stack's used when a hardware interrupt's handler is being processed; though arch-dependant, the kernel typically maintains one IRQ stack per core for just this purpose.

A quick example to help make this key point clear: when you execute the classic K&R C 'Hello, world' process, the kernel creates the process; this has the kernel set up and initialize several objects – among them, the process task structure, its process VAS, including the user mode stack, and a unique kernel mode stack for it as well. Once it runs, the process, of course, first executes the `printf()` API in user mode, thus making use of the user space stack (of `main()`). The `printf()`, after setting things up, issues the `write()` system call! This has our process switch to kernel mode and actually write the `Hello, world\n` string (via the `tty` layer code paths within the kernel) to the stdout device. As it now executes kernel code (in kernel mode), it makes use of it's kernel space stack.

Figure 6.3 divides the address space into two – user space and kernel space. In the upper part of the diagram – user space – you can see the conceptual view of several processes and their *user VASes*. In the bottom part – kernel space (a large monolithic space shared by all user mode processes) – you can see, corresponding to every user mode thread, a kernel metadata structure `struct task_struct`, which we shall cover a bit later in detail) and the kernel mode stack of that thread.

In addition, we see (at the very bottom), as an example, three kernel threads (labeled `kthrd1`, `kthrd2`, and `kthrdn`); as expected, they too have a `task_struct` metadata structure representing their innards (attributes) and a kernel mode stack:

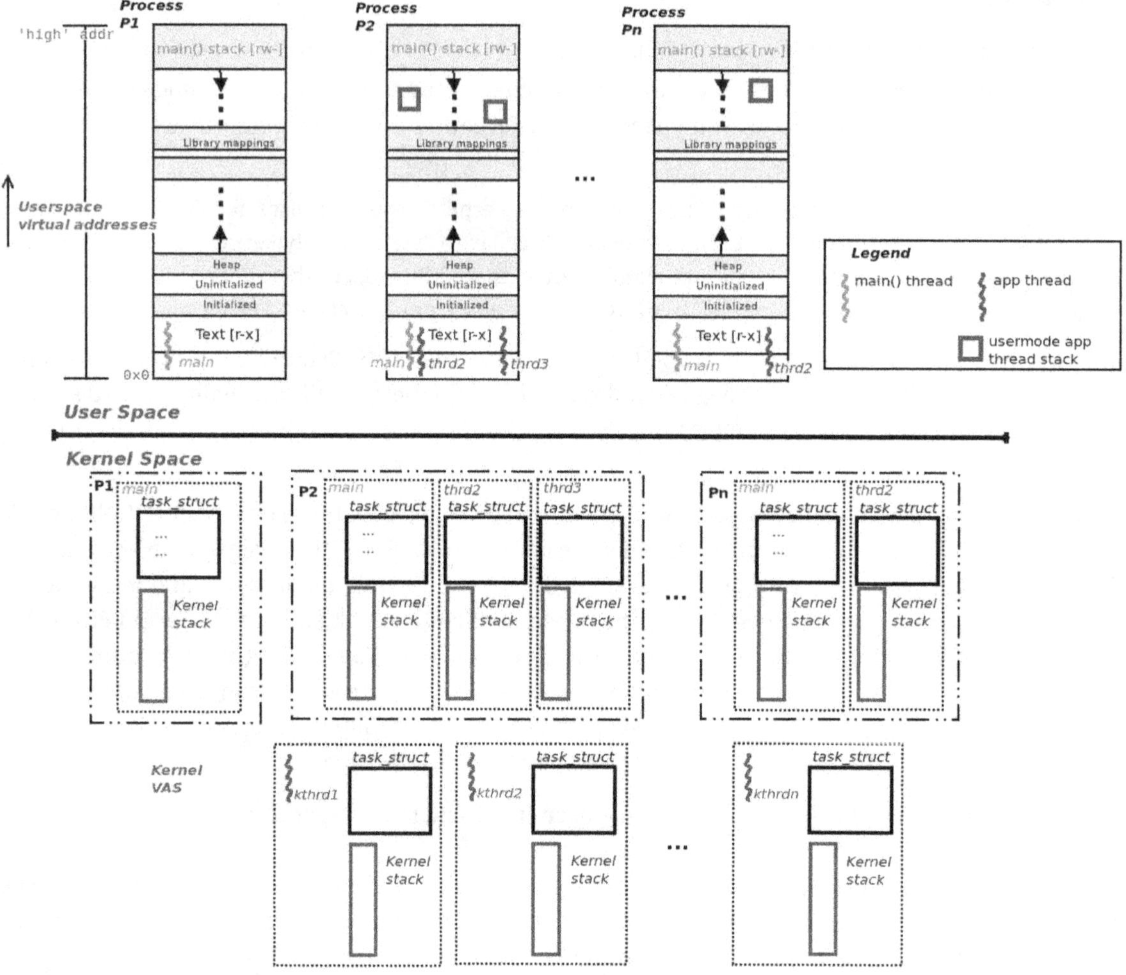

Figure 6.3: User and kernel VASes – processes, threads, stacks, and task structures (task_struct)

Running a small script to see the number of processes and threads alive

To help make this discussion on the user and kernel virtual address spaces concrete, let's execute a simple Bash script (`ch6/countem.sh`) that counts the number of processes and threads currently alive. I did this on my native x86_64 Ubuntu 22.04 LTS box; see the following resulting output (the initial portion of the output may differ slightly on other distros; no matter):

```
$ cd <booksrc>/ch6
$ ./countem.sh
```

```
System release info:
Distributor ID: Ubuntu
Description:    Ubuntu 22.04.1 LTS
Release:        22.04
Codename:       jammy

Total # of processes alive                =      234
Total # of threads alive                  =      514
Total # of kernel threads alive           =      116
Thus, total # of user mode threads alive =       398
$
```

How does the script get the number of processes and threads? Simple, its just based on ps, using appropriate option switches, filtering the output and a bit of calculation. (The following sections flesh out more details.) I'll leave it to you to look up the code of this simple script here: `ch6/countem.sh`. Study the (bottom portion of the) preceding output and understand it. You will realize, of course, that this is a snapshot of the situation at a certain point of time. It can and does change.

In the following sections, we divide up the discussion into two parts (corresponding to the two address spaces) – that of what we see in *Figure 6.3* in user space and what is seen in the kernel space. Let's begin with the user space components.

User space organization

With reference to the `countem.sh` Bash script that we ran in the preceding section, we will now break it down and discuss some key points, confining ourselves to the *user space portion* of the process VAS for now. Please take care to read and understand this (the numbers we refer to in the following discussion are regarding our sample run of our `countem.sh` script in the *Running a small script to see the number of processes and threads alive* section). For the sake of better understanding, I've placed the user space portion of the diagram here:

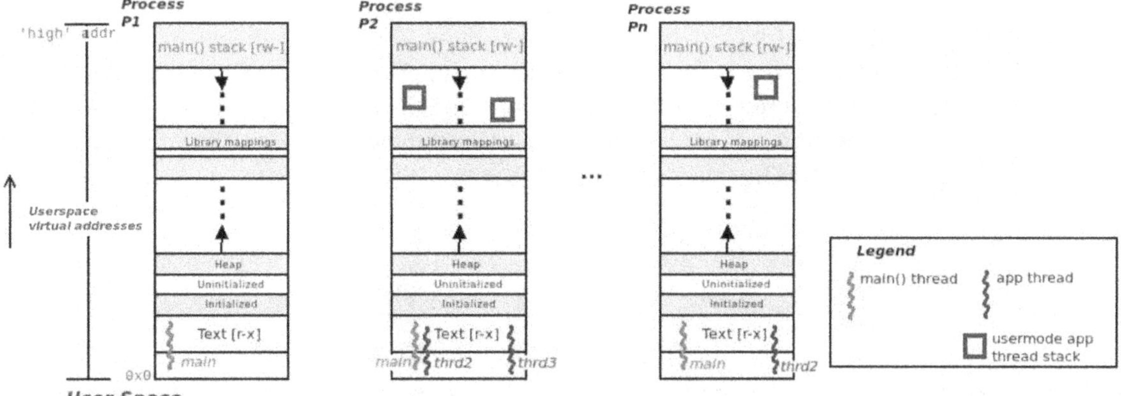

Figure 6.4: The user space portion of the overall picture seen in Figure 6.3

Here (*Figure 6.4*) you can see three individual user mode processes. Every process has at least one thread of execution (the `main()` thread). Here, we show three processes, `P1`, `P2`, and `Pn`, with one, three, and two threads in them, respectively, including `main()`. In our preceding example run (the one shown in the *Running a small script to see the number of processes and threads alive* section) of the `countem.sh` script, `Pn` would have *n*=234.

Do note that these diagrams are purely conceptual. For example, in reality, the process with PID 2 is typically a single-threaded kernel thread called *kthreadd*.

Each process consists of several segments (technically, mappings). Broadly, the user mode segments (or mappings) are as follows:

- **Text:** Code; `r-x`
- **Data segments:** `rw-`; consists of three distinct mappings – the initialized data segment, the uninitialized data segment (or `bss`), and an 'upward-growing' `heap`
- **Library mappings:** For the text and data of each shared library, the process dynamically links to:
- **Downward-growing stack(s).**

Regarding these stacks, we saw from our preceding sample run that there are 398 user-mode threads currently alive on the system. This implies that there are 398 user space stacks as well, as there will exist one user mode stack for every user-mode thread alive. Of these user-space thread stacks, we can say the following:

- One user-space stack is always present for the `main()` thread and it will be located close to the very top – the high end – of the user VAS. If the process is single-threaded (only a `main()` thread), then it will have just one user-mode stack; the P1 process in *Figure 6.4* shows this case.

An aside, but an important one: on Linux, any `foo()` system call will typically become the `sys_foo()` function within the kernel. Also, often but not always, this `sys_foo()` function is a wrapper that invokes the 'real' code `do_[*]_foo()`. A further detail: in the kernel code, you might see macros of the type `SYSCALL_DEFINEn(foo, ...)`; the macro becomes the `sys_foo()` routine. The number appended, `n`, is in the range [0,6]; it's the number of parameters being passed to the kernel from user space via the system call.

- If the process is multithreaded, it will have one user-mode thread stack per thread alive (including `main()`); processes `P2` and `Pn` in *Figure 6.4* illustrate this case. The stacks are allocated either at the time of calling `fork(2)` (for `main()`) or `pthread_create()` (for the remaining threads within the process), which results in this code path being executed in process context within the kernel (in `kernel/fork.c`): `sys_fork() --> kernel_clone()`

  (FYI, earlier, the worker routine for the `fork()` system call implementation within the kernel was called `_do_fork()`; from the 5.10 kernel, this function has been renamed to `kernel_clone()`; commit # `cad6967ac108`).

- Also, FYI, the Pthreads creation library API, `pthread_create()`, on Linux invokes the (very Linux-specific) `clone()` system call (the code within the kernel is here: `kernel/fork.c:sys_clone()`). This system call ends up calling `kernel_clone()`; the parameter – particularly the `flags` value – passed along informs the kernel as to how exactly to create the custom process – in other words, a thread!
- The user space stacks are of course dynamic; they can grow (and shrink as well) up to the stack size resource limit `RLIMIT_STACK` (typically 8 MB – you can use the `prlimit` utility to look it up).

FYI: With 6.6 Linux, a new feature for the x86 family called "user space **shadow stacks**" has finally been merged. It's essentially a security hardening feature, helping protect apps against dangerous **Return Oriented Programming** (**ROP**) style attacks. These attacks work by modifying the call stack/function return address. Using a shadow stack allows (only) the kernel implementation to save the return address in the (hardware) shadow stack, which can't be modified, and then, on return, pop its value and compare it with that of the regular stack. If they differ, it signals a possible exploit under way and has the processor raise a general protection fault to stop the attack in its tracks. The current implementation (on 6.6) is for x86_64 and only for user-mode stacks. See more here: *User-space shadow stacks (maybe) for 6.4, Jon Corbet, LWN, March 2023*: `https://lwn.net/Articles/926649/`.

Having seen and understood the user space portion, now let's delve into the kernel space side of things.

Kernel space organization

Continuing our discussion with reference to the `countem.sh` Bash script (that we ran in the *Running a small script to see the number of processes and threads alive* section), we will now break it down and discuss some key points, confining ourselves to the *kernel space portion* of the VAS. Please take care to carefully read and understand this (while reading the numbers that were output in our preceding sample run of the `countem.sh` script).

For the sake of better understanding, I have placed the kernel space portion of the diagram here (*Figure 6.5*):

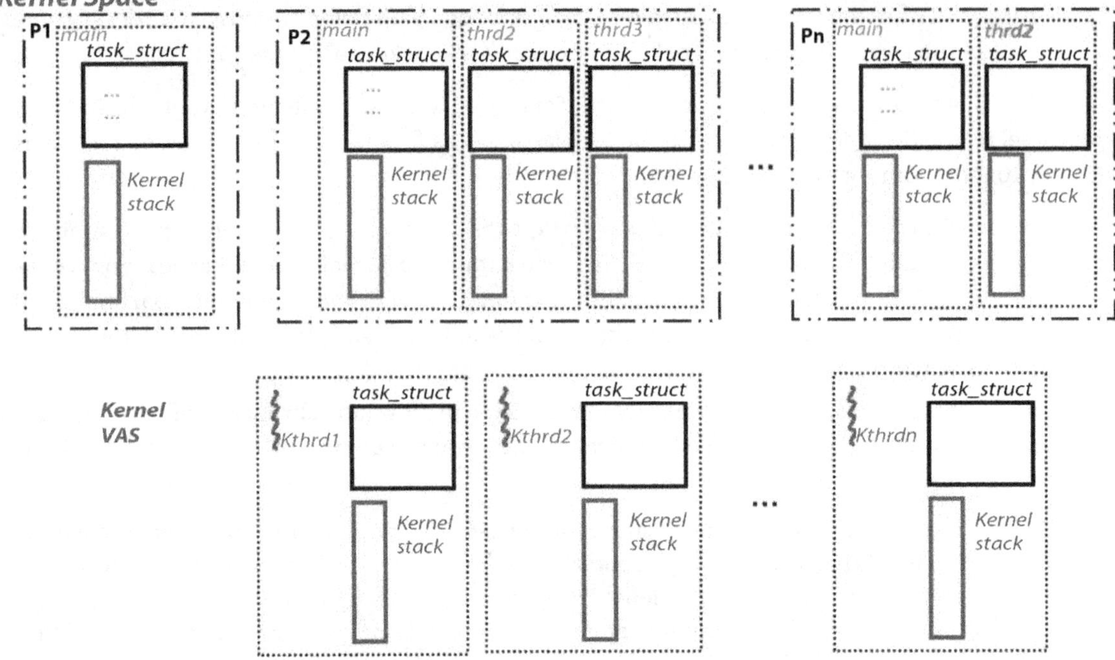

Figure 6.5: Kernel space portion of the overall picture seen in Figure 6.3

Again, from our preceding sample run, you can see that there are 398 user-mode threads and 116 kernel threads currently alive on the system. This yields a total of 514 kernel space stacks. How? As mentioned earlier, every user-mode thread has two stacks – one user-mode stack and one kernel-mode stack. Thus, we'll have 398 kernel-mode stacks for each of the user-mode threads, plus 116 kernel-mode stacks for the (pure) kernel threads (recall, kernel threads have *only* a kernel-mode stack – they cannot "see" user space at all), resulting in a total of (398+116=) 514 kernel space stacks. Let's list a few characteristics of kernel-mode stacks:

- There will be one kernel-mode stack for each user-mode thread alive, including `main()`.
- Kernel-mode stacks are fixed in size (static) and are quite small. Practically speaking, their size is 2 pages on 32-bit and 4 pages on 64-bit OSs (with a page typically being 4 KB in size).

Don't simply assume the page size is always 4 KB – in user space, use the `getpagesize()` system call to query its value; in kernel space, the `PAGE_SIZE` macro yields the same.

- They are allocated at thread creation time (it usually boils down to the kernel code here: `kernel_clone() --> copy_process() --> dup_task_struct()`).

Again, let's be crystal clear on this: each user-mode thread has two stacks – a user-mode stack and a kernel-mode stack. The exception to this rule is kernel threads; they only have a kernel-mode stack (as they possess no user mapping and thus no user space segments). In the lower part of *Figure 6.5*, we show three *kernel threads* – `kthrd1`, `kthrd2`, and `kthrdn` (in our preceding sample run, `kthrdn` would have *n*=116). Further, each kernel thread has a task structure and a kernel-mode stack allocated to it at creation time.

A kernel-mode stack is similar in most respects to its user-mode counterpart – every time a function within kernel space is called, a *stack frame* is set up (the frame layout is particular to the architecture and forms a part of the CPU ABI document; see the *Further reading* section for more on these details). The CPU has a register to track the current location of the stack (usually called a **Stack Pointer** (**SP**)), and the stack "grows" toward *lower* virtual addresses. But, unlike the dynamic user-mode stack, *the kernel-mode stack is fixed in size and small.*

An important implication of the pretty small (two-page or four-page) kernel-mode stack size for the kernel/driver developer – **be very careful to not overflow your kernel stack** by performing stack-intensive work (such as using large local variables or recursion).

There exists a kernel configurable (`CONFIG_FRAME_WARN`) to warn you about high kernel stack usage at compile time; here's the relevant text from the `lib/Kconfig.debug` file:

```
config FRAME_WARN:
  int "Warn for stack frames larger than"
  range 0 8192
  [...]
  default 2048 if 64BIT
  help
     Tell gcc to warn at build time for stack frames larger than this.
     Setting this too low will cause a lot of warnings.
     Setting it to 0 disables the warning.
```

Summarizing the kernel with respect to threads, task structures, and stacks

Okay, great, let's now summarize our learning and findings from our preceding discussions and sample run of the `countem.sh` script (in the *Running a small script to see the number of processes and threads alive* section):

- **Task structures:**
    - Every thread alive (user or kernel) has a corresponding task structure (`struct task_struct`) in the kernel; this is how the kernel tracks and manages it. Also, all the thread's attributes are stored here (you'll learn more on this in the *Understanding and accessing the kernel task structure* section)

- With respect to our sample run of our `ch6/countem.sh` script (in the *Running a small script to see the number of processes and threads alive* section):
    - As there are a total of 514 threads (both user and kernel) alive on the system, this implies a total of 514 *task (metadata) structures* in kernel memory (in the code, it's `struct task_struct`), of which we can say the following:
    - 398 of these task structures represent user threads.
    - The remaining (514 – 398 =) 116 task structures represent kernel threads.

- Stacks:
    - Every user space thread has two stacks:
        - A user mode stack (which is in play when the thread executes user-mode code paths)
        - A kernel mode stack (which is in play when the thread executes kernel-mode code paths)
        - Also, a separate per-core IRQ stack is present for use when hardware interrupt handlers execute their code paths
    - *Exception case*: a kernel thread has only one stack, it's kernel mode stack
    - Thus, with respect to our sample run of our `ch6/countem.sh` script, we have:
        - 398 user space stacks (in user land).
        - Above, plus 398 kernel space stacks (in kernel memory).
        - Above, plus 116 kernel space stacks (for the 116 kernel threads that are alive).
        - This comes together for a grand total of 398 + 398 + 116 = 912 stacks! (At 4 pages per kernel-mode stack on a 64-bit Linux, and assuming a page size of 4 KB, that's 4*4096*912 = 14.25 MB of RAM taken up for stack memory).
        - FYI, the command `grep "KernelStack" /proc/meminfo` will show how much memory's currently being used for kernel stacks. (Look up the man page on proc(5) for details).

As briefly mentioned already, many architectures (including x86 and ARM64) support a separate per-CPU stack for *interrupt handling (called the IRQ stack)*. When an external hardware interrupt occurs, the CPU's control unit immediately re-vectors control to, ultimately, the interrupt handling code (perhaps within a device driver). A separate per-CPU interrupt stack is used to hold the stack frame(s) for the interrupt code path(s); this helps avoid putting too much pressure on the existing (small) kernel-mode stack of the process/thread that got interrupted. The IRQ stack size will be the same as the kernel mode stack size for that architecture. (As well, architectures like the x86_64 support even more types of stacks, but we won't delve further).

Okay, now that you understand the overall organization of the user and kernel spaces in terms of processes/threads and their stacks, let's move on to seeing how you can actually view the content of both the kernel and user space stacks. Besides being useful for learning purposes, this knowledge can greatly aid you in debugging situations.

Viewing the user and kernel stacks

The *stack* is often the key to a debug session. It is the stack that holds the *current execution context* of the thread – in which function is it executing code right now, and, key, how it got here – which allows us to infer the *history* (what it's doing and what happened). Being able to see and interpret the thread's *call stack (aka call chain / call trace / backtrace)* crucially allows us to understand how exactly we got here. All this precious information resides in the stack. But wait, there are two stacks for every thread – the user space and the kernel space stack. How do we view their content?

Here, we shall show two broad ways of viewing the kernel and user-mode stacks of a given process or thread, firstly via the traditional approach, and then a more recent modern approach (via eBPF). Do read on.

Traditional approach to viewing the stacks

Let's first learn to view both the kernel and user-mode stacks of a given process or thread using what we shall call the traditional approach. Let's begin with the kernel-mode stack.

Viewing the kernel space stack of a given thread or process

Good news; this is easy. The Linux kernel makes a given thread's kernel stack visible via the usual mechanism to expose kernel internals to user space – the powerful and versatile `proc` filesystem interfaces. Just read the content of the pseudofile `/proc/PID/stack`.

An example always helps; so, let's look up the kernel-mode stack of our *Bash* process. Let's say that on our x86_64 Ubuntu guest, our Bash process' PID is `2459`:

On modern kernels, to avoid information leakage, viewing the kernel-mode stack of a process or thread requires root access as a security requirement.

```
$ sudo cat /proc/2549/stack
[<0>] do_wait+0x184/0x340
[<0>] kernel_wait4+0xaf/0x150
[<0>] __do_sys_wait4+0x89/0xa0
[<0>] __x64_sys_wait4+0x1e/0x30
[<0>] do_syscall_64+0x5c/0x90
[<0>] entry_SYSCALL_64_after_hwframe+0x63/0xcd
$
```

In the preceding output, each line represents a *call frame (or stack frame)* on the stack. To help decipher a kernel stack backtrace, it's worth knowing the following points:

- The name shown is that of the function that was called – for example, the second one seen here is named `do_syscall_64()` (because we always read stack traces from the bottom-up).

- The function call graph ordering is from bottom to top; hence, this output should be read in a bottom-up fashion. So, here, it implies that (ignoring the first one at the very bottom) the call graph is like this: `do_syscall_64() --> __x64_sys_wait4() --> __do_sys_wait4() --> kernel_wait4() --> do_wait()`.
- Each line of output represents a *call frame* – in effect, a function in the call chain.
- If a call frame is prefixed with one or more ? symbols, it implies that the kernel cannot reliably interpret this stack frame. Ignore it, it's the kernel saying that it's very likely an invalid stack frame (a blip left behind); the kernel backtrace code is usually dead right! (Realize that, very often, the same portions of stack memory are continually reused; this can leave behind blips from previous call stacks that have no relevance to the current one.)
- As already mentioned, on Linux, any `foo()` system call will typically become a `sys_foo()` function within the kernel. Also, often (but not always), `sys_foo()` is a wrapper that invokes the "real" code `do_[*]_foo()`.

Now, look again at the preceding output. It should be quite clear: our *Bash* process is currently executing the `do_wait()` function within the kernel; the call graph clearly shows us that it got there via a system call, the `wait4()` system call! This is quite right; the shell works by forking off a child process and then waiting for its demise via the `wait4(2)` system call.

Curious readers (you!) should note that the `[<0>]` in the leftmost column of each stack frame displayed in the preceding command line output snippet are the placeholders for the text (code) address of that function. Again, for security reasons – to prevent information leakage – it is zeroed out on modern kernels. (Another security measure related to the kernel and process layout is discussed in *Chapter 7, Memory Management Internals – Essentials,* in the *Randomizing the memory layout – KASLR* and *User-mode ASLR* sections.)

Next, what's with the `<func>+x/y` syntax – for example, the line `do_wait+0x184/0x340` – I hear you ask? This information can be *really* useful when debugging:

- The first number (x, and always in hex) is the byte offset from the beginning of the function where the execution is currently at.
- The second number (y, again in hex) is what the kernel has deemed to be the length of this function; it's usually correct.

So, here, `do_wait+0x184/0x340` implies that the `do_wait()` function was executing its machine code at an offset of `0x184` (`388` decimal) bytes from the start of the function, and the length of the function is `0x340` (`832` decimal) bytes! (FYI, my *Linux Kernel Debugging* book covers these aspects in a lot of detail.)

Viewing the user space stack of a given thread or process

Ironically, viewing the user space stack of a process or thread seems harder to do on a typical Linux distro (as opposed to viewing the kernel-mode stack, as we just saw in the previous section). There is a utility to do so: *gstack*. In reality, it's just a simple wrapper over a script that invokes the venerable GDB debugger in batch mode, getting GDB to invoke its `backtrace` command.

Unfortunately, on Ubuntu (23.10 and below at least), there seems to be an issue; the *gstack* program was not found in any native package. (Ubuntu does have an older pstack utility, but it works only on x86 32-bit Linux systems.)

A simple workaround is to simply use GDB in non-interactive or batch mode (you can always run GDB's `attach <PID>` command and then issue the `[thread apply all] bt` command to view the user mode stack(s)). This, in fact, is pretty much precisely what a web page aptly named the "poor man's profiler" (`https://poormansprofiler.org/`) tells us to do (as indeed, the `pstack` utility on Fedora does)! Thus, the following simple (and simplistic) wrapper script can display the user-mode stack frames of any given process (where `$1` is the PID of the process/thread, and provided you run it as root of course):

```
sudo gdb \
  -ex "set pagination 0" \
  -ex "thread apply all bt" \
  --batch -p $1
```

I put this simple code into a script here: *ch6/ustack*. Right, time to (minimally) test it! We unleash our script on our unsuspecting (and newest) *Bash* process:

```
$ ./ustack $(pgrep --newest bash)
[sudo] password for c2kp:
0x00007fadd3109c3a in __GI___wait4 (pid=-1, stat_loc=0x7fffe3a3a1e0,
options=10, usage=0x0) at ../sysdeps/unix/sysv/linux/wait4.c:27
27     ../sysdeps/unix/sysv/linux/wait4.c: No such file or directory.

Thread 1 (process 2549):
#0  0x00007fadd3109c3a in __GI___wait4 (pid=-1, stat_loc=0x7fffe3a3a1e0,
options=10, usage=0x0) at ../sysdeps/unix/sysv/linux/wait4.c:27
#1  0x0000555b98cd4f03 in ?? ()
#2  0x0000555b98cd6373 in wait_for ()
#3  0x0000555b98cc3216 in execute_command_internal ()
#4  0x0000555b98cc39da in execute_command ()
#5  0x0000555b98cab665 in reader_loop ()
#6  0x0000555b98ca9ef9 in main ()
[Inferior 1 (process 2549) detached]
```

And there we are! The user mode stack of Bash, along with all its call frames (read it bottom-up) shows up! Again, each line under `Thread 1 (process …)` (*Thread 1* is the `main()` thread) represents a call frame; read it bottom-up. Clearly, *Bash* executes a command and ends up invoking the `__GI___wait4()` function (on modern Linux systems, it's simply a glibc wrapper over the actual `wait4()` system call)! The leftmost column (for example, `#3`) represents the call frame number; call frame `#0` is considered the top of the stack – that is, the latest function to execute (it's the last one in the call chain).

Again, simply ignore any call frames labeled or prefixed with ??. (Incidentally, the *gstack* utility worked just fine on my x86_64 Fedora 38 / 39 guest VMs).

Being able to peek into the kernel and user space stacks (as shown in the preceding snippets) and using utilities including *strace* and *ltrace* for tracing system and library calls of a process (or thread), respectively, can be a tremendous aid when debugging! Don't ignore them.

Incidentally, you might wonder what's the meaning of the `... at ../sysdeps/unix/sysv/linux/wait4.c:27` string that shows up in frame 0 of the call stack is. Well, debug information present in the `libthread_db` library allows GDB to figure out the precise line of source code where the `__GI___wait4()` function is being invoked! Of course, you'll require the source code for that particular version of glibc to look it up.

Now for a modern approach to viewing stacks.

eBPF – the modern approach to viewing both stacks

Now – more exciting ! – let's learn the very basics of using a powerful modern approach, leveraging (as of the time of writing) recent technology called the **extended Berkeley Packet Filter** (**eBPF**). We did mention the eBPF project in *Online Chapter, Kernel Workspace Setup*, under the *Additional useful projects* section). The older BPF has been around a long time and has been used for network packet tracing; eBPF is a relatively recent innovation, available only as of 4.x Linux kernels (which, of course, implies that you will need to be on a 4.x or more recent Linux system to use this approach).

Directly using the underlying kernel-level BPF bytecode technology is (extremely) difficult to do; thus, the good news is that there are several easy-to-use frontends (tools and scripts) to this technology. Among the frontends, the **BPF Compiler Collection** (BCC) and **bpftrace** are considered very useful. (A diagram showing the current BCC performance analysis tools can be found at `https://github.com/iovisor/bcc/blob/master/images/bcc_tracing_tools_2019.png`; a list of the eBPF frontends can be found at `http://www.brendangregg.com/ebpf.html#frontends`.)

Here, we shall simply provide a quick demonstration using a BCC tool called `stackcount` (well, on Ubuntu at least, these eBPF tools are suffixed with the string `-bpfcc`, so this one's named `stackcount-bpfcc`). Another advantage is that using this tool allows you to see both the kernel and user-mode stacks at once; there's no need for separate tools.

You can install the BCC tools for your host Linux distro by reading the installation instructions here: `https://github.com/iovisor/bcc/blob/master/INSTALL.md`. What about installing them on our guest Linux VM while running our custom 6.1 kernel? You can (though, in earlier kernel versions, this could present a problem and require running a distro kernel such as an Ubuntu- or Fedora-supplied kernel).

Figure 6.6 is a screenshot of the short and sweet help screen thrown up by this utility when run with the -h parameter:

```
ch6 $ stackcount-bpfcc -v
usage: stackcount-bpfcc [-h] [-p PID] [-c CPU] [-i INTERVAL] [-D DURATION] [-T] [-r] [-s] [-P]
 [-K] [-U] [-v] [-d] [-f] [--debug] pattern
stackcount-bpfcc: error: the following arguments are required: pattern
ch6 $ stackcount-bpfcc -h
usage: stackcount-bpfcc [-h] [-p PID] [-c CPU] [-i INTERVAL] [-D DURATION] [-T] [-r] [-s] [-P]
 [-K] [-U] [-v] [-d] [-f] [--debug] pattern

Count events and their stack traces

positional arguments:
  pattern               search expression for events

options:
  -h, --help            show this help message and exit
  -p PID, --pid PID     trace this PID only
  -c CPU, --cpu CPU     trace this CPU only
  -i INTERVAL, --interval INTERVAL
                        summary interval, seconds
  -D DURATION, --duration DURATION
                        total duration of trace, seconds
  -T, --timestamp       include timestamp on output
  -r, --regexp          use regular expressions. Default is "*" wildcards only.
  -s, --offset          show address offsets
  -P, --perpid          display stacks separately for each process
  -K, --kernel-stacks-only
                        kernel stack only
  -U, --user-stacks-only
                        user stack only
  -v, --verbose         show raw addresses
  -d, --delimited       insert delimiter between kernel/user stacks
  -f, --folded          output folded format
  --debug               print BPF program before starting (for debugging purposes)

examples:
    ./stackcount submit_bio         # count kernel stack traces for submit_bio
    ./stackcount -d ip_output       # include a user/kernel stack delimiter
    ./stackcount -s ip_output       # show symbol offsets
    ./stackcount -sv ip_output      # show offsets and raw addresses (verbose)
    ./stackcount 'tcp_send*'        # count stacks for funcs matching tcp_send*
    ./stackcount -r '^tcp_send.*'   # same as above, using regular expressions
    ./stackcount -Ti 5 ip_output    # output every 5 seconds, with timestamps
    ./stackcount -p 185 ip_output   # count ip_output stacks for PID 185 only
    ./stackcount -c 1 put_prev_entity   # count put_prev_entity stacks for CPU 1 only
    ./stackcount -p 185 c:malloc    # count stacks for malloc in PID 185
    ./stackcount t:sched:sched_fork # count stacks for sched_fork tracepoint
    ./stackcount -p 185 u:node:*    # count stacks for all USDT probes in node
    ./stackcount -K t:sched:sched_switch   # kernel stacks only
    ./stackcount -U t:sched:sched_switch   # user stacks only

ch6 $ ▮
```

Figure 6.6: Screenshot of the useful output from stackcount-bpfcc -h

Do check it out. I'd definitely recommend you look at the *stackcount[-bpfcc]* examples by Brendan Gregg and other contributors here: `https://github.com/iovisor/bcc/blob/master/tools/stackcount_example.txt`. They're worth looking through.

In the following example, we use the `stackcount-bpfcc` BCC tool (on my x86_64 Ubuntu 22.04 LTS host system running our custom 6.1 kernel) to look up various aspects of the call stacks while the `ping` process is running. To set it up, I ran `ping yahoo.com` in the background and a few instances of `stackcount_bpfcc` with different and appropriate parameters via a simple wrapper script: `ch6/stackcount_eg.sh`; do check it out.

The kernel symbols, even for many production systems, are typically sufficiently enabled to allow one to see (at least) the function names; in fact, the `kptr_restrict` sysctl has a definite effect (look it up here: `https://sysctl-explorer.net/kernel/kptr_restrict/`). On the other hand, kind of ironically, it's usually applications that aren't compiled with symbols (as they often use the `-fomit-frame-pointer` compiler option), thus their function names may not even show up in the user mode call stack. Having them compiled with `-fno-omit-frame-pointer` or using debug symbol packages can solve the issue (when debugging).

With `stackcount[-bpfcc]`, you must specify a function (or functions) of interest (interestingly, you can specify either a user space or kernel space function and use "wildcards" or even a regular expression when doing so!); only when those functions are invoked will the stacks be traced and reported. *Figure 6.7* is a partial screenshot of our demo stackcount script in action; while `ping` pings away, you can clearly see the kernel call stack and the not-so-clear user mode call stack as well (as user-space symbols aren't available):

```
------------------------------------------------------------------
Show dev_hard_start_xmit() call stacks:
------------------------------------------------------------------
Tracing 1 functions for "dev_hard_start_xmit"... Hit Ctrl-C to end.
  b'dev_hard_start_xmit'
  b'__dev_xmit_skb'
  b'__dev_queue_xmit'
  b'dev_queue_xmit'
  b'neigh_hh_output'
  b'ip_finish_output2'
  b'__ip_finish_output'
  b'ip_finish_output'
  b'ip_output'                                  kernel-mode stack
  b'ip_push_pending_frames'
  b'ping_v4_sendmsg'
  b'inet_sendmsg'
  b'sock_sendmsg'
  b'__sys_sendto'
  b'__x64_sys_sendto'
  b'do_syscall_64'
  b'entry_SYSCALL_64_after_hwframe'
    --
  b'__libc_sendto'
  b'[unknown]'
    b'ping' [3760920]                           user-mode stack
    1
```

Figure 6.7: Partial screenshot of our ch6/stackcount_eg.sh script in execution, with the kernel and user stacks artificially highlighted between the --delimiter

The `--delimited` option switch passed prints the delimiter `--`; it denotes the boundary between the kernel-mode and the user-mode stack of the process. (Unfortunately, as most production user-mode apps will have their symbolic information stripped, most user-mode stack frames simply show up as `[unknown]`.) On this system at least, the kernel stack frames are very clear; with the `--verbose` option passed, even the virtual address of the text (code) function in question is printed on the left. Again, note that the `stackcount-bpfcc` tool works only with Linux 4.6+, and requires root access. Do see its man page for details.

Note that eBPF programs might fail due to the relatively new kernel *lockdown* feature being enabled (it's disabled by default though). It's a **Linux Security Module** (**LSM**) that enables an extra hard level of security on Linux systems. Of course, security is a double-edged sword; having a very secure system implicitly means that certain things will not work as expected, and this includes some eBPF programs. Do refer to the *Further reading* section for more on kernel lockdown.

As an interesting and useful aside: Brendan Gregg has built (a while back) some powerful visualization tooling, aptly named ***Flame Graphs***; it's stack trace visualization on profiled code paths, allowing one to quickly spot frequently-on-the-cpu code (with features like the ability to zoom in or out of call paths!). We refer you to the *Further reading* section for links on eBPF, BCC, bpftrace, and Flame Graphs.

We have merely scratched the surface here; eBPF tools such as BCC and *bpftrace* really are a modern, powerful approach to system app tracing and performance analysis on the Linux OS (and to observability in general). Other useful profiling tooling includes perf, Flame Graphs, and several more. Do take the time to learn how to use these powerful tools!

Right, good going, let's now conclude this section by zooming out and looking at an overview of what you have learned so far!

The 10,000-foot view of the process VAS

Before we conclude this section, it's important to take a step back and see the complete VASes of each process and how it looks for the system as a whole – in other words, to zoom out and see the "10,000-foot view" of the complete system address space. This is what we attempt to do with the following rather large and detailed diagram (*Figure 6.8*), an extension or superset of our earlier *Figure 6.3*.

For those of you reading a hard copy of the book (good for you!), I'd definitely recommend you view this book's figures in full color within this PDF document at `https://packt.link/gbp/9781803232225`.

Besides what you have learned about and seen – the process user space segments, the user and kernel threads, and the kernel-mode stacks – don't forget that there is a lot of other metadata within the kernel: the per-thread task structures, kernel threads, per-process memory descriptor metadata structures, open file metadata structures, IPC metadata structures, and so on. They all are very much a part of the *kernel VAS*, which is often called the *kernel segment*.

There's more to the kernel segment than tasks and stacks. It also contains (obviously!) the static kernel (core) code and data – in effect, all the major (and minor) *subsystems* of the kernel, the arch-specific code, and so on (that we spoke about in *Chapter 4, Writing Your First Kernel Module – Part 1*, under the *Kernel space components* section).

As just mentioned, the following diagram presents an attempt to sum up and present all (well, much) of this information in one place:

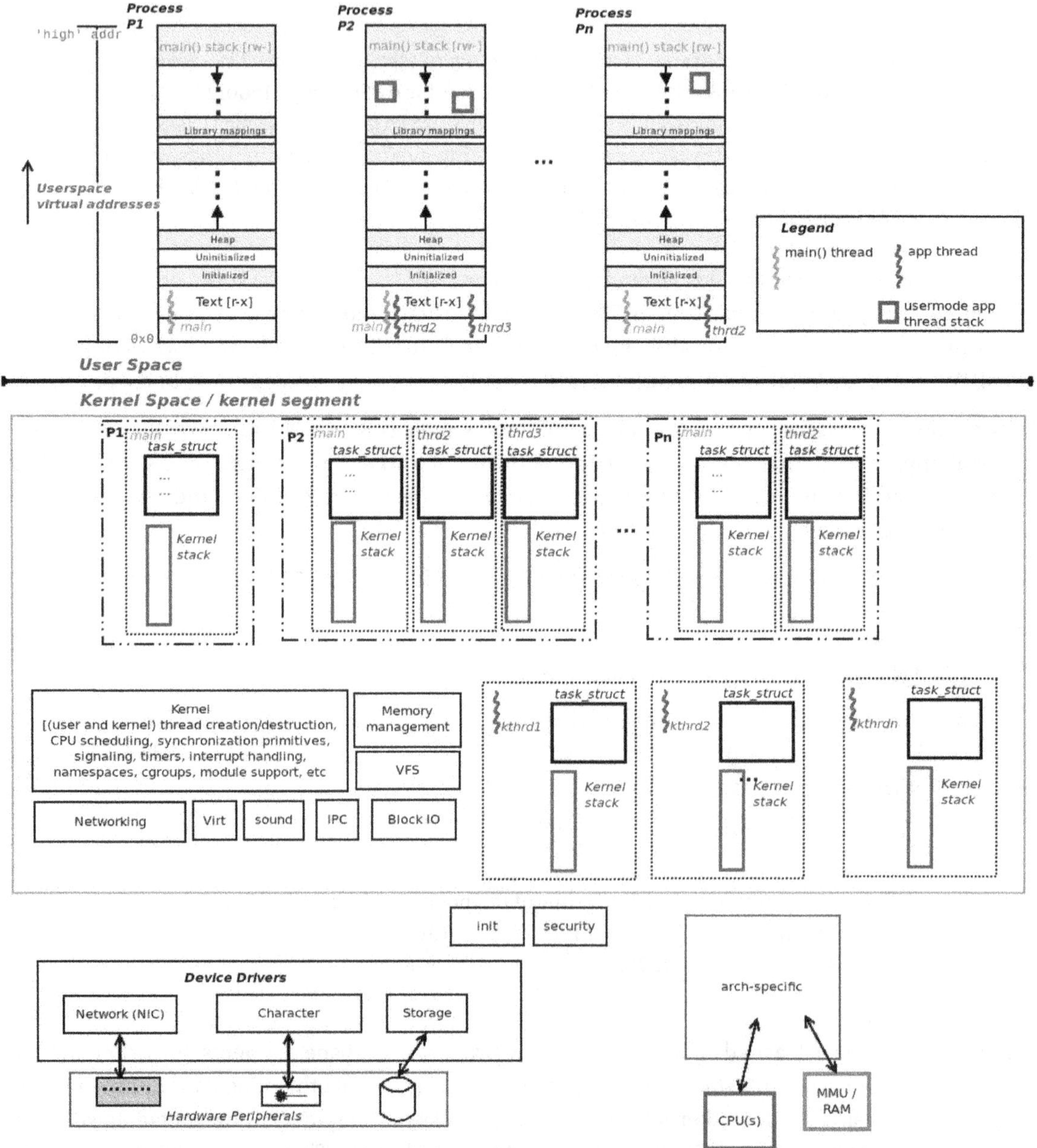

Figure 6.8: The 10,000-foot view of the processes, threads, stacks, task structures, and more, of the user and kernel VASes

Whew, quite a thing, isn't it? The large (red) box in the kernel segment of the preceding diagram (conceptually) encompasses the *core kernel code and data* – the major kernel subsystems; it shows the per-thread task structures and kernel-mode stacks as well. The rest of it is considered non-core stuff; this includes device drivers. (The arch-specific code can arguably be viewed as core code; we just show it separately here). Also, don't let the preceding information overwhelm you; just focus on what we're here for right now – the processes, threads, task structures, and stacks. If you're still unclear about it, be sure to re-read the preceding material.

Now, let's move on to actually understanding and learning how to reference the key or "root" metadata structure for every single thread alive – the *task structure*.

Understanding and accessing the kernel task structure

As you have learned by now, every single user and kernel space thread is internally represented within the Linux kernel by a metadata structure containing all its attributes – the **task structure**. The task structure is represented within the kernel code here: `include/linux/sched.h:struct task_struct`.

To view any version of the kernel code online, a superb (searchable) system is in place here: `https://elixir.bootlin.com/linux/latest/source`. For the 6.1.25 LTS kernel release, for example, here's the task structure definition: `https://elixir.bootlin.com/linux/v6.1.25/source/include/linux/sched.h#L737`.

It's often, unfortunately, referred to as the "process descriptor," causing no end of confusion! Thankfully, the phrase *task structure* is so much better; it represents a runnable task – in effect, a *thread*.

So, there we have it: in the Linux design, every process consists of one or more threads and *each thread maps to a kernel metadata structure called a task structure* (`struct task_struct`).

The task structure is the "root" metadata structure for the thread – it encapsulates all the information required by the OS for that thread. This includes information on its memory (segments/mappings setup, paging tables, usage info, and more), CPU scheduling details, all files it currently has open, its credentials, capability bitmasks, timers, locks, **Asynchronous I/O** (**AIO**) contexts, hardware context info, signal dispositions, IPC objects, resource limits, (optional) audit, security and profiling info, and many more such details.

Figure 6.9 is a conceptual representation of the Linux kernel *task structure* and most of the information (metadata) it contains:

The task structure

Inherited across fork()

OFDT (open files)
VFS data
Hardware Context
Credentials / Caps
Paging Tables (PTEs)
Signal Handling
Sched info
Resource Limits
Namespaces
Thread TLS
Profiling info
Security - LSMs - seccomp
IPC Structures

Not inherited across fork()

Pending/blocked signals
PID, PPID
AIO
Locks
Timers, Alarms
Audit Info
semop adj

Figure 6.9: Linux kernel task structure: struct task_struct

As can be seen from *Figure 6.9*, the task structure holds a huge quantity of information regarding every single task (thread) alive on the system (again, I reiterate: this includes kernel threads as well). We show – in a compartmentalized conceptual format in *Figure 6.9* – the different kinds of attributes encapsulated within this data structure. Also, as you can see, certain attributes will be *inherited* by a child process or thread upon `fork()` (or `pthread_create()`) while some attributes won't be inherited and will be merely reset to default values.

For now, at least, suffice it to say that the kernel understands whether a task is a process or a thread. We'll later demonstrate a kernel module (`ch6/foreach/thrd_showall`) that reveals exactly how we can determine this (hang on, we'll get there!).

Now, let's start to understand in more detail some of the more important members of the huge task structure; read on!

Here, I only intend to give you a feel for the kernel task structure; we do not delve deep into the details, as it's not required for now. You will find that, in later parts of this book, we delve into specific areas as required.

Looking into the task structure

Firstly, recall that the task structure is essentially the root data structure of the process or thread – it holds all attributes of the task (as we saw earlier). Thus, it's rather large; the powerful `crash` utility (used to analyze Linux crash dump data or investigate a live system) reports its size on x86_64 for the 6.1.25 kernel to be 13,120 bytes, as does the C `sizeof` operator.

The task structure is defined in the `include/linux/sched.h` kernel header (it's a rather key header). In the following code, we show its definition with the caveat that we display only a few of its many members. (Also, the annotations in **<< double angle brackets like this >>** are used to very briefly explain the member(s)):

```
// include/linux/sched.h
struct task_struct {
#ifdef CONFIG_THREAD_INFO_IN_TASK
    /*
     * For reasons of header soup (see current_thread_info()), this
     * must be the first element of task_struct.
     */
    struct thread_info  thread_info;    << important flags and status bits >>
#endif
    unsigned int            __state;
    [...]
    void      *stack;          << the location of the kernel-mode stack >>
    [...]
    << the members that follow are to do with CPU scheduling; some of them are
discussed in Ch 10 & Ch 11 on CPU Scheduling >>
    int on_rq;
    int prio;
    int static_prio;
    int normal_prio;
    unsigned int rt_priority;
    struct sched_entity se;
    struct sched_rt_entity rt;
    struct sched_dl_entity dl;
    const struct sched_class *sched_class;
    [...]
    unsigned int policy;
    int nr_cpus_allowed;
    const cpumask_t *cpus_ptr;
    [...]
```

Continuing with the task structure in the following code block, see the members relating to memory management (`mm`), the **Process ID (PID)** and **Thread Group ID (TGID)** values, the credentials' structure, signal handling, open files, and many more. Again, it's not the intention to delve into (all of) them in detail; where appropriate, in later sections of this chapter, and possibly in other chapters of this book, we shall (re)visit them:

```
    [...]
    struct mm_struct *mm;      << memory management info >>
    struct mm_struct *active_mm;
    [...]
    pid_t pid;    << task PID and TGID values; explained below >>
    pid_t tgid;
    [...]
    /* Context switch counts: */
    unsigned long nvcsw;
    unsigned long nivcsw;
    [...]
    /* Effective (overridable) subjective task credentials (COW): */
    const struct cred __rcu *cred;
    [...]
    /* Signal handlers: */
    struct signal_struct        *signal;
    struct sighand_struct __rcu        *sighand;
    sigset_t            blocked;
    sigset_t            real_blocked;
    /* Restored if set_restore_sigmask() was used: */
    sigset_t            saved_sigmask;
    struct sigpending        pending;
    [...]
    char comm[TASK_COMM_LEN];            << task name >>
    [...]
     /* Open file information: */
    struct files_struct *files;  << ptr to the 'open files' ds >>
    [...]
#ifdef CONFIG_VMAP_STACK
    struct vm_struct *stack_vm_area;
#endif
    [...]
#ifdef CONFIG_SECURITY
    /* Used by LSM modules for access restriction: */
    void *security;
```

```
#endif
    [...]
    /* CPU-specific state of this task: */
    struct thread_struct thread;  << task hardware context detail >>
    [...]
};
```

It's important to realize that the `struct task_struct` members in the preceding code blocks are shown with respect to the 6.1.25 kernel source *only*; on other kernel versions, the members can and do change! Of course, it should go without saying that this is true of the entire book – all code/data is presented with regard to the 6.1(.25) LTS Linux kernel (which will be maintained up to December 2026, and the CIP SLTS 6.1 kernel right until August 2033).

Okay, now that you have a better idea of the members within the task structure, how exactly do you access it and its various members? Read on.

Accessing the task structure with current

You will recall, in our sample run of the preceding `countem.sh` script (in the *Running a small script to see the number of processes and threads alive* section), we found that there are a total of 514 threads (both user and kernel) alive on the system. This implies that there will be a total of 514 task structure objects in kernel memory.

They need to be organized in a way that the kernel can easily access them as and when required. Thus, all the task structure objects in kernel memory are chained up on a *circular doubly linked list* called the **task list**. This kind of organization is required in order for various kernel code paths to iterate over them (commonly, the `procfs` code, among others). Even so, think about this: when a process or thread is running kernel code (in process context), how exactly can it find out which `task_struct` belongs to it among the perhaps hundreds or thousands that exist in kernel memory? This turns out to be a non-trivial task. The kernel developers have evolved a way to guarantee you can find the particular task structure representing the thread currently running the kernel code in an efficient manner. It's achieved via a macro called `current`. Think of it this way:

- Looking up `current` yields the pointer to `struct task_struct` of the thread that is running the kernel code right now – in other words, *the process context running right now on a processor core (the one performing the lookup!)*.
- `current` is analogous (but of course, not exactly) to what object-oriented languages refer to as the `this` pointer.

The implementation of the `current` macro is very architecture/CPU-specific. Here, we do not delve into all the gory implementation-level details. Suffice it to say that the implementation is carefully engineered to be fast (typically via an *O(1)* algorithm). For example, on some **Reduced Instruction Set Computer** (**RISC**) architectures with many general-purpose registers (such as the PowerPC and AArch64 processors), a register is dedicated to holding the value of `current`!

I urge you to browse the kernel source tree and see the implementation details of `current` (under `arch/<arch>/asm/current.h`). On ARM32, an *O(1)* calculation yields the result; on AArch64 and PowerPC, it's stored in a register (and thus the lookup is blazing fast). On x86_64 architectures, the implementation uses a lock-free technology – a `per-CPU` variable – to hold `current` (avoiding the use of costly locking; we cover this in *Chapter 13, Kernel Synchronization – Part 2* in the *Per-CPU variables* section). All that's required to use the `current` macro in your (module) code is the inclusion of the `<linux/sched.h>` header.

We can use `current` to dereference the task structure and cull information from within it; for example, the process (or thread) PID and name can be looked up as follows:

```
#include <linux/sched.h>
current->pid, current->comm
```

In the next section, you will see a full-fledged kernel module that iterates over the task list, printing out some details from each task structure it encounters along the way.

Determining the context

As you know by now, kernel code runs in one of two contexts:

- Process (or task) context
- Interrupt (an atomic) context

They are mutually exclusive – kernel code runs in either process (which is at times atomic) or **interrupt context (which is always atomic**) at any given point in time (we explain the term *atomic* very shortly).

Why is it important to be able to determine the context kernel or driver code's running in? A golden rule within the kernel is that ***you cannot sleep (or block) in any kind of atomic context; doing so causes a kernel bug***. It can lock up the system, typically causing a kernel panic.

Why? Well, realize firstly that sleeping implies context-switching – switching the CPU to run another task while the previous one goes to sleep. Thus sleeping implies invoking the scheduler code and a subsequent context switch (we cover this in detail in *Chapter 10, The CPU Scheduler – Part 1*, and *Chapter 11, The CPU Scheduler – Part 2*). This is indeed how any blocking API works. When running in an ***atomic context*** though – like a hardware interrupt (as well as a software interrupt, a softirq), or when holding a spinlock – one must complete the work without blocking, without sleeping, and without yielding the CPU. It's why the word "atomic" is used; it implies *running to completion without interruption*.

Now that we know this rule – don't sleep in atomic context – the question comes up: *How do I know whether my code path is (or isn't) running in an atomic context?* Here's how you can easily determine the context in which your kernel/driver code is currently executing in:

```
#include <linux/preempt.h>
if (in_task())
    foo(); /* runs in process context; usually safe to sleep or block */
else
```

```
    bar(); /* runs in an atomic context; unsafe to sleep or block! */
```

The `in_task()` macro returns a Boolean; `True` if your code is running in process (or task) context, where it's – usually – safe to sleep; it returning `False` implies you are in some kind of atomic context – perhaps in interrupt context – where it is never safe to sleep.

You might have come across the usage of the `in_interrupt()` macro – if it returns `True`, your code is within an interrupt context; if `False`, it isn't. However, the recommendation for modern code is to *not* rely on this macro (due to the fact that **Bottom Half** (**BH**) disabling can interfere with its working). Hence, we recommend using `in_task()` instead.

Hang on though! It can get a bit tricky: while `in_task()` returning `True` does imply that your code is in process context, this fact by itself does *not* guarantee whether it's currently *atomic*, or whether it's *safe to sleep*. For example, you could be running kernel or driver code in process context *but holding a spinlock* (a very common lock used within the kernel); here, the code between the lock and unlock – the so-called *critical section* – must run atomically! This implies that though your code may be in process (or task) context, it still will cause a kernel-level bug if it attempts to issue any possibly blocking (sleeping) APIs! Worry not, locking is covered in detail in the last two chapters of this book.)

Also, be careful: (by definition) usage of the `current` macro is only considered valid when running in *process context*.

Right, now, you have learned useful background information on the task structure, how it can be accessed via the `current` macro, and figuring out the context that your kernel or driver code is currently running in. So now, let's write some kernel module code to examine a bit of the kernel task structure!

Working with the task structure via 'current'

Here, we will write a simple kernel module to show a few members of the task structure. Further, I want you to think about this: who exactly is running the code of this (or any) kernel module's (or the kernel's) *init* and *cleanup* code paths? From what we've learned, it's not the kernel; as stated before, there's no grand overall "kernel" process as such... Then, who's running it?

The answer should be clear in a monolithic kernel such as the Linux OS: when a user space process (or thread) issues a system call, it switches to kernel mode and runs kernel (or module) code in *process context*. So, yes, it will be a process (or thread). Which one? The code we write will reveal the *process context* that our module's *init* and *cleanup* code paths run in. (We discuss this very point in more detail in the upcoming *Seeing that the Linux OS is monolithic* section.) To do so, we cook up a `show_ctx()` function that uses `current` to access a few members of the task structure and display their values. It's invoked from both the *init* as well as the *cleanup* methods, as follows:

For reasons of readability and space constraints, only key parts of the source code are displayed here. The entire source tree for this book is available in its GitHub repository; we expect you to clone and use it: `git clone https://github.com/PacktPublishing/Linux-Kernel-Programming_2E`.

```
/* code: ch6/current_affairs/current_affairs.c */
#define pr_fmt(fmt) "%s:%s(): " fmt, KBUILD_MODNAME, __func__
[ ... ]
#include <linux/sched.h>      /* current */
#include <linux/preempt.h>    /* in_task() */
#include <linux/cred.h>        /* current_{e}{u,g}id() */
#include <linux/uidgid.h>     /* {from,make}_kuid() */
[ ... ]
static void show_ctx(char *nm)
{
        /* Extract the task UID and EUID using helper methods provided */
        unsigned int uid = from_kuid(&init_user_ns, current_uid());
        unsigned int euid = from_kuid(&init_user_ns, current_euid());

        pr_info("\n"); /* shows mod & func names(due to pr_fmt()!) */
        if (likely(in_task())) {
        pr_info("we're running in process context ::\n"
            " name        : %s\n"
            " PID            : %6d\n"
            " TGID          : %6d\n"
            " UID            : %6u\n"
            " EUID          : %6u (%s root)\n"
            " state          : %c\n"
            " current (ptr to our process context's task_struct) :\n"
            "               0x%pK (0x%px)\n"
            " stack start : 0x%pK (0x%px)\n",
            current->comm,
            /* always better to use the helper methods provided */
        task_pid_nr(current), task_tgid_nr(current),
        /* ... rather than using direct lookups:
         * current->pid, current->tgid,
         */
        uid, euid,
        (euid == 0 ? "have" : "don't have"),
        task_state_to_char(current),
        /* Printing addresses twice- via %pK and %px
         * Here, by default, the values will typically be the same as
         * kptr_restrict == 1 and we've got root.
         */
```

```
        current, current, current->stack, current->stack);
} else
    pr_alert("Whoa! running in interrupt context [Should NOT Happen here!]\n");
```

As is highlighted in bold in the preceding code snippet, you can see that (for some members) we can simply dereference the `current` pointer to gain access to various `task_struct` structure members and display them (via the kernel log buffer).

Great! The preceding code snippet does indeed show you how to gain access to a few `task_struct` members directly via `current`; not all members, though, can or should be accessed directly. Rather, the kernel provides some *helper methods* to access them; let's briefly get into this next.

Built-in kernel helper methods and optimizations

In the preceding code example, we made use of a few kernel's *built-in helper or accessor methods* to extract various members of the task structure. This is the recommended approach; for example, we use `task_pid_nr()` to peek at the PID member instead of directly via `current->pid`. Similarly, the process credentials within the task structure (such as the effective UID (`EUID`) and similar members we showed in the preceding code) are abstracted within `struct cred`, and access to them is provided via helper routines, like the `from_kuid()` helper that we used in the preceding code. In a similar fashion, there are several other helper methods; look them up in `include/linux/sched.h` just below the `struct task_struct` definition (here onward: `https://elixir.bootlin.com/linux/v6.1.25/source/include/linux/sched.h#L1547`).

Why is this the case? Why not simply access task structure members directly via `current-><member-name>`? Well, there are various real reasons: one, perhaps the access requires a *lock* to be taken (we cover details on the key topic of locking and synchronization in the last two chapters of this book). Two, perhaps there's a more optimal way to access them; three, perhaps they're not to be directly accessed as they make sense only within the *namespace* the task belongs to, and so on... In fact, the key notions of kernel namespace and control groups (or cgroups) form the basis of *container* technology; we discuss cgroups (in detail) and namespaces (a little) in *Chapter 11, The CPU Scheduler – Part 2*, in the *An introduction to cgroups* section.

Also, as shown in the preceding code, we can easily figure out whether the kernel code (of our kernel module) is running in process or interrupt context by employing the `in_task()` macro – it returns `True` if in the process (or task) context, and `False` if otherwise.

Did you notice? We used an interesting macro – `likely()` – to span the conditional; what does it mean? It's, as usual, to do with performance; this macro becomes a compiler `__builtin_expect` attribute. It provides a hint to the compiler's branch prediction setup and optimizes the instruction sequence being fed into the CPU pipeline, thus keeping our code on the "fast path" (more on this micro-optimization with the `likely()`/`unlikely()` macros can be found in the *Further reading* section for this chapter). You will quite often see kernel code employing the `likely()`/`unlikely()` macros in situations where the developer "knows" whether the code path is likely or unlikely, respectively.

The preceding `[un]likely()` macros are a good example of micro-optimization, of how the Linux kernel quite deeply leverages the GCC or clang compilers. In fact, until recently, the Linux kernel could *only* be compiled with GCC; more recently, compiling the kernel with clang is a reality (FYI, the modern **Android Open Source Project** (**AOSP**) is compiled with `clang`).

Okay, now that you've understood the workings of our kernel module's `show_ctx()` function, why not try it out?

Trying out the kernel module to print process context info

We build our `current_affair.ko` kernel module (we don't show the build output here) and then insert it into kernel space (via `insmod` as usual). Now, let's view the kernel log with `dmesg`, then `rmmod` it and use `dmesg` again. The following screenshot shows this:

```
$ sudo dmesg -C
$ sudo insmod ./current_affairs.ko ; lsmod|grep current_affairs
current_affairs        16384  0
$ sleep 1
$ sudo rmmod current_affairs
$ sudo dmesg
[  295.072202] current_affairs:current_affairs_init(): inserted
[  295.072208] current_affairs:current_affairs_init(): sizeof(struct task_struct)=13120
[  295.072212] current_affairs:show_ctx():
[  295.072215] current_affairs:show_ctx(): we're running in process context ::
               name        : insmod
               PID         :   3303
               TGID        :   3303
               UID         :      0
               EUID        :      0 (have root)
               state       : R
               current (ptr to our process context's task_struct) :
                             0xffff88804d7c0000 (0xffff88804d7c0000)
               stack start : 0xffffc90003048000 (0xffffc90003048000)
[  300.789069] current_affairs:show_ctx():
[  300.789076] current_affairs:show_ctx(): we're running in process context ::
               name        : rmmod
               PID         :   3312
               TGID        :   3312
               UID         :      0
               EUID        :      0 (have root)
               state       : R
               current (ptr to our process context's task_struct) :
                             0xffff88801ce6a780 (0xffff88801ce6a780)
               stack start : 0xffffc90002768000 (0xffffc90002768000)
[  300.789085] current_affairs:current_affairs_exit(): removed
$
```

Figure 6.10: The output of the current_affairs.ko kernel module on our x86_64 Ubuntu VM running our custom 6.1.25 kernel

Clearly, as can be seen from *Figure 6.10*, the initial *process context* – the process (or thread) running the init kernel code of our kernel module – i.e., `current_affairs.c:current_affairs_init()` – is the `insmod` process (see the output: `name : insmod`). Similarly, the `current_affairs.c:current_affairs_exit()` process context executing the cleanup code is the `rmmod` process!

Notice how the timestamps in the left column (`[sec.usec]`) in the preceding figure help us understand that `rmmod` was called close to 6 seconds after `insmod`.

There's more to this small demo kernel module than perhaps first meets the eye. It's actually very helpful in understanding Linux kernel architecture. The following section explains how this is so.

Seeing that the Linux OS is monolithic

Besides the exercise of using the `current` macro, a key point behind this kernel module (`ch6/current_affairs`) is to clearly show you one aspect of the *monolithic nature of the Linux OS*. In the preceding code, we saw that when we performed the `insmod` operation on our kernel module file (`current_affairs.ko`), it got inserted into the kernel and its `init` code path ran; a really important question is: *Who ran it?* Ah, that question is answered by checking the output: the `insmod` process itself ran the module's `init` code in process context, thus proving the monolithic nature of the Linux kernel! (Ditto with the `rmmod` process and the `cleanup` code path; it was run by the `rmmod` process in process context.)

Note carefully and clearly: there is no "kernel" (or kernel thread) that executes the code of the kernel module; it's the user space process (or thread) itself that, by issuing system calls (recall that both the insmod and rmmod utilities issue system calls), switches into kernel space and executes the code of the kernel module. This is how it is with a monolithic kernel.

Just FYI, the *microkernel architecture* is perhaps the diametrically opposite approach to monolithic ones. It's approach is a message-passing one (no system calls), where messages are passed from the user app/process to server processes, which perform the work. They, in turn, and as required, talk to the small microkernel, again via messages. Done well, it also offers excellent performance; apparently, though, it's hard to design and implement a micro-kernel-based OS well (think of GNU Hurd). Of course, they very much exist: QNX, VxWorks, and ENEA are excellent real-world (and hard real-time) examples of a microkernel OS done well, while Tannenbaum's Minix is a (mostly) classroom microkernel.

Back to the monolithic approach and Linux: this type of execution of kernel code (as *Figure 6.10* reveals) is what we refer to as *running in process context*, as opposed to running in *interrupt context*. The Linux kernel, though, isn't strictly purely monolithic; if so, it would be a single hard-coded piece of memory. Instead, like all modern operating systems, Linux supports *modularization* (via the LKM framework).

As an aside, do note that you can create and run *kernel threads* within kernel space; they, just like user mode threads, are scheduled to run on the CPU and execute kernel code in process context.

Coding for security with printk

In our previous kernel module demo (`ch6/current_affairs/current_affairs.c`), you noticed, I hope, the usage of `printk` with the special `%pK` format specifier. We repeat the relevant code snippet here:

```
pr_info(
[...]
      " current (ptr to our process context's task_struct) :\n"
      "                   0x%pK (0x%px)\n"
      " stack start : 0x%pK (0x%px)\n",
      [...]
      current, current,
      current->stack, current->stack); [...]
```

Recall from our discussion in *Chapter 5, Writing Your First Kernel Module – Part 2*, in the *A quick word on the kptr_restrict sysctl* section, that when printing an address (firstly, you really shouldn't be printing addresses in production), I urged you to not use the usual `%p` (or `%px`) but the `%pK` format specifier instead. That's what we've done in the preceding code; *this is for security, to prevent a kernel information leak*. With a well-tuned (for security) system, `%pK` will result in a mere zeroed-out or a hashed value and not the actual address being displayed. To show this, we also display the actual kernel address via the `0x%px` format specifier, just for contrast.

Interestingly enough, here, `%pK` seems to have no effect – both addresses print identical values (as can be seen in *Figure 6.10*); this is expected as the `kptr_restrict` sysctl value defaults to 1 and we're running as root.

On production systems (embedded or otherwise), be safe: set `kernel.kptr_restrict` to `1` or, even better, to `2`, thus sanitizing pointers, and set `kernel.dmesg_restrict` to `1` (thus allowing only privileged users to read the kernel log).

Now, let's move on to something quite interesting: in the following section, you will learn how to iterate over the Linux kernel's *task lists*, thus in effect learning how to obtain kernel-level information on every single process and/or thread alive on the system.

Iterating over the kernel's task lists

As mentioned earlier, all the task structures are organized in kernel memory in a linked list called the *task list* (allowing them to be iterated over). The list data structure has evolved to become the very commonly used *circular doubly linked list.* In fact, the core kernel code to work with these lists has been factored out into a header called `list.h`; it's well-known and expected to be used for any list-based work.

The `include/linux/types.h:list_head` data structure forms the essential doubly linked circular list; as expected, it consists of two pointers, one to the `prev` member on the list and one to the `next` member:

```
struct list_head {
        struct list_head *next, *prev;
};
```

You can easily iterate over various lists concerned with tasks via conveniently provided macros in the `include/linux/sched/signal.h` header file for versions >= 4.11; note that for kernels 4.10 and older, the macros are in `include/linux/sched.h`.

Now, as usual, let's make this discussion empirical and hands-on! In the following sections, we will write kernel modules to iterate over the kernel task list in two ways:

- One: Iterate over the kernel task list and display all *processes* alive (thus showing only the `main()` or `T0` thread of each process).
- Two: Iterate over the kernel task list and display all *threads* alive.

We show the detailed code view for the latter case. Read on and be sure to try both out yourself!

Iterating over the task list I – displaying all processes

The kernel provides a convenient routine, the `for_each_process()` macro, which lets you easily iterate over every *process* in the task list:

```
// include/linux/sched/signal.h:
#define for_each_process(p) \
    for (p = &init_task ; (p = next_task(p)) != &init_task ; )
```

Clearly, the macro expands to a `for` loop, allowing us to loop over the circular list. `init_task` is a convenient "head" or starting point – it's always the pointer to the task structure of the CPU core's "idle" thread – the thread that gets scheduled to run when no one else wants to run (on that core).

Note that the `for_each_process()` macro is expressly designed to only iterate over the `main()` or T0 thread of every *process* and not any other (child or peer) threads.

A brief snippet of the beginning portion of our `ch6/foreach/prcs_showall` kernel module's output is shown here (*Figure 6.11*, when run on our x86_64 Ubuntu 22.04 LTS guest system running our custom 6.1.25 kernel):

```
[ 1685.208236] prcs_showall: inserted
[ 1685.208239] prcs_showall:         Name      |  TGID  |   PID  | RUID |  EUID
[ 1685.208241] prcs_showall: systemd           |       1|       1|     0|     0
[ 1685.208242] prcs_showall: kthreadd          |       2|       2|     0|     0
[ 1685.208243] prcs_showall: rcu_gp            |       3|       3|     0|     0
[ 1685.208244] prcs_showall: rcu_par_gp        |       4|       4|     0|     0
[ 1685.208245] prcs_showall: slub_flushwq      |       5|       5|     0|     0
[ 1685.208246] prcs_showall: netns             |       6|       6|     0|     0
[ 1685.208247] prcs_showall: kworker/0:0H      |       8|       8|     0|     0
[ 1685.208248] prcs_showall: mm_percpu_wq      |      10|      10|     0|     0
[ 1685.208249] prcs_showall: rcu_tasks_kthre   |      11|      11|     0|     0
[ 1685.208250] prcs_showall: rcu_tasks_rude_   |      12|      12|     0|     0
[ 1685.208251] prcs_showall: rcu_tasks_trace   |      13|      13|     0|     0
[ 1685.208252] prcs_showall: ksoftirqd/0       |      14|      14|     0|     0
[ 1685.208253] prcs_showall: rcu_preempt       |      15|      15|     0|     0
[ 1685.208254] prcs_showall: migration/0       |      16|      16|     0|     0
```

Figure 6.11: Partial screenshot – output from our prcs_showall kernel module

Notice how, in the preceding snippet (*Figure 6.11*) at least, the TGID and PID values of each process are always equal, proving that the `for_each_process()` macro only iterates over the `main` thread of every process (and not every thread). We explain the details regarding the TGID/PID members in the following section.

We'll leave the studying and trying out of the sample kernel module at `ch6/foreach/prcs_showall` as an exercise for you.

Iterating over the task list II – displaying all threads

To iterate over each *thread* that's alive and well on the system, we use the `do_each_thread() { ... } while_each_thread()` *pair* of helper macros; we write a sample kernel module to do just this (here: `ch6/foreach/thrd_showall/`).

Before diving into the code, let's see a portion of the output it emits (via `dmesg`); so, we build and `insmod` it (on our x86_64 Ubuntu 22.04 LTS guest). As displaying the complete output of this module isn't possible here – it's far too large – I've shown only the lower part of the output in the following screenshot (*Figure 6.12*). Also, we've reproduced the header so that you can make sense of what each column represents:

```
 Timestamp        TGID    PID   current             stack pointer     name          #threads

[ 1009.149105]    2225    2225  0xffff91503742b280  0xffffb6c9c1590000      kerneloops
[ 1009.149107]    2237    2237  0xffff915058f9b280  0xffffb6c9c31bc000  update-notifier     4
[ 1009.149108]    2237    2240  0xffff915058d1b280  0xffffb6c9c3164000            gmain
[ 1009.149109]    2237    2241  0xffff915058d1cbc0  0xffffb6c9c318c000            gdbus
[ 1009.149110]    2237    2245  0xffff91500e55cbc0  0xffffb6c9c31dc000      dconf worker
[ 1009.149112]    2301    2301  0xffff91500e603280  0xffffb6c9c326c000         dhclient     4
[ 1009.149113]    2301    2302  0xffff91505b1be500  0xffffb6c9c32d4000    isc-worker0000
[ 1009.149114]    2301    2303  0xffff91505b1b9940  0xffffb6c9c1c98000       isc-socket
[ 1009.149116]    2301    2304  0xffff915037429940  0xffffb6c9c32f4000        isc-timer
[ 1009.149117]    2422    2422  0xffff915037446500  0xffffb6c9c3304000             sshd
[ 1009.149119]    2537    2537  0xffff915003059940  0xffffb6c9c3664000             sshd
[ 1009.149121]    2538    2538  0xffff91500305b280  0xffffb6c9c368c000             bash
[ 1009.149122]    2573    2573  0xffff91505d65b280  0xffffb6c9c362c000               vi
[ 1009.149123]    2576    2576  0xffff9150063b9940  0xffffb6c9c3414000             sshd
[ 1009.149125]    2612    2612  0xffff91500dddcbc0  0xffffb6c9c369c000             sshd
[ 1009.149126]    2613    2613  0xffff91500e730000  0xffffb6c9c35bc000             bash
[ 1009.149127]    2664    2664  0xffff915020741940  0xffffb6c9c3904000 [     kworker/4:0]
[ 1009.149129]    3613    3613  0xffff915039dd1940  0xffffb6c9c427c000 [   kworker/u12:2]
[ 1009.149130]    3621    3621  0xffff915036960000  0xffffb6c9c4364000 [     kworker/1:0]
[ 1009.149131]    4120    4120  0xffff915037449940  0xffffb6c9c19d8000              lkm
[ 1009.149133]    4437    4437  0xffff9150035f4bc0  0xffffb6c9c1dd0000             sudo
[ 1009.149134]    4438    4438  0xffff915024df1940  0xffffb6c9c1d28000             sudo
[ 1009.149135]    4439    4439  0xffff9150369e4bc0  0xffffb6c9c1d18000           insmod
[ 1009.149135] thrd_showall: total # of threads on the system: 526
$
```

Figure 6.12: Output from our thrd_showall.ko kernel module; a multithreaded process (dhclient) is highlighted

As an interesting aside, in *Figure 6.12*, notice how all the (kernel-mode) stack start addresses (the fifth column) end in zeros, `0xffff .... .... .000`, implying that the stack region is *always aligned on a page boundary* (as `0x1000` is `4096` in decimal). This will be the case as kernel-mode stacks are always fixed in size and a multiple of the system page size (typically 4 KB).

Following convention (as per how the `ps aux` output looks) in our kernel module, we arrange things such that if the thread is a *kernel thread*, its name shows up within square brackets.

How does one interpret the (quite weird) `kthread` names (like the "[kworker/u12:2]" seen in *Figure 6.12*)? This link provides the answer: `https://unix.stackexchange.com/a/152865`.

Before continuing to the code, we first need to examine in a bit of detail the TGID and PID members of the task structure.

Differentiating between the process and thread – the TGID and the PID

Think about this: as the Linux kernel uses a unique task structure (`struct task_struct`) to represent every thread, and as the unique member within it is a **Process IDentifier** (**PID**), this implies that, within the Linux kernel, *every thread has a unique PID*. This gives rise to a question, a puzzle: in a multithreaded process, how then can the multiple threads of this same process share a common PID (as that's what the POSIX Threads POSIX.1c standard demands)? The reality is that they don't: every thread of a multithreaded process (indeed every single thread alive) on modern Linux has a unique PID! But hang on: doesn't this violate the Pthreads standard? Indeed, for a while (a long time ago), Linux *was* non-compliant with the standard, creating porting issues, among other things.

To fix this annoying user space standards issue, Ingo Molnar (of Red Hat) proposed and mainlined a patch way back in the 2.5 kernel series. A new member called the **Thread Group IDentifier** or **TGID** was slipped into the task structure.

This is how it works: if the process is single-threaded, the `tgid` and `pid` values are the same. If it's a multithreaded process, then the `tgid` value of the `main` thread is equal to the `main` thread's `pid` value; other threads of the process will inherit the `main` thread's `tgid` value but will retain their own unique `pid` values. To summarize it, this can be viewed as follows:

- Single-threaded process: `main()` (or T0) thread: one unique PID and TGID == PID
- Multithreaded process:
    - `main()` (or T0) thread: PID and TGID are the same: TGID == PID
    - not `main()` thread(s): several unique PIDs: TGID == PID of `main()`

To understand this better, let's take an actual example from the previous screenshot. In *Figure 6.12*, notice how, if a positive integer appears in the last column on the right, it represents the number of threads in the multithreaded process to its immediate left.

So, check out the `dhclient` process highlighted (by a rounded rectangle) in *Figure 6.12*; it has a TGID and PID value of 2301 representing its `main()` (T0) thread and a total of *four threads*. What are the four threads? First, of course, the `main()` thread is `dhclient`, and the three displayed below it are, by name, `isc-worker0000`, `isc-socket`, and `isc-timer`. How do we know this for sure? It's easy: look carefully at the second and third columns in *Figure 6.12*; they represent the TGID and PID respectively. If they are the same, it's the main thread of the process; *if the TGID repeats, the process is multithreaded* and the PID value represents the unique PIDs of its worker or peer threads.

As a matter of fact, it's entirely possible to see the kernel's TGID/PID representation in user space via the ubiquitous GNU `ps` command, by using its `-LA` option switches (among other ways to do so):

```
$ ps -LA
    PID    LWP TTY          TIME CMD
      1      1 ?        00:00:01 systemd
      2      2 ?        00:00:00 kthreadd
      3      3 ?        00:00:00 rcu_gp
[...]
   2301   2301 ?        00:00:00 dhclient
   2301   2302 ?        00:00:00 isc-worker0000
   2301   2303 ?        00:00:00 isc-socket
   2301   2304 ?        00:00:00 isc-timer
 [...]
```

The `ps` labels are interpreted as follows:

- The first column is `PID` – however, realize that this value actually represents the `tgid` member of the task structure within the kernel for this task!
- The second column is `LWP` (literally, a LightWeight Process or a thread) – this represents the `pid` member of the task structure within the kernel for this task.

Note that only with the GNU ps utility can you pass parameters (like `-LA`) and see the threads; this isn't possible with a lightweight implementation of `ps` like that of *busybox*. It isn't a problem though: you can always look up the same details by looking under procfs. In this example, under `/proc/2301/task`, you'll see sub-folders representing the worker threads. Guess what: this is actually how GNU `ps` works under the hood as well!

Right, on to the code now...

Iterating over the task list III – the code

Now let's see the (relevant) code of our `thrd_showall` kernel module:

```
// ch6/foreach/thrd_showall/thrd_showall.c */
[...]
#include <linux/sched.h>     /* current */
#include <linux/version.h>
#if LINUX_VERSION_CODE > KERNEL_VERSION(4, 10, 0)
#include <linux/sched/signal.h>
#endif
[...]

static int showthrds(void)
{
    struct task_struct *p, *t;       /* 'p' : process ptr; 't': thread ptr */
    [...]
    disp_idle_thread();
```

A few points to note regarding the preceding code:

- We use the `LINUX_VERSION_CODE` macro to conditionally include a header, as required.
- Please ignore the kernel-level *locking* work for now (our usage of the `task_{un}lock()` and/or RCU routines); we defer this discussion to the last two chapters.
- A key point: when working upon – *even reading* – the members of the task structure, we need to inform the rest of the kernel about this (so that it doesn't abruptly free it from under us); this is done by taking a reference to the structure via the `get_task_struct()` helper and later releasing the same via the `put_task_struct()`.
- Don't forget the CPU idle thread! Every CPU core has a dedicated idle thread (named `swapper/n`) that runs when no other thread wants to (`n` being the core number, starting with `0`). Unfortunately, the `do ... while` loop we run does not start at this thread (nor does `ps` ever show it). Thus, we include a small routine to display it, making use of the fact that the hardcoded task structure for the idle thread is available and exported at `init_task` (a detail: `init_task` always refers to the first CPU's – core # `0` – idle thread).

Let's continue to iterate over every thread alive, we need to use a *pair* of macros, forming a loop; the `do_each_thread(p,t) { ... } while_each_thread(p,t)` pair of macros do precisely this, allowing us to iterate over every *thread* alive on the system.

From the 6.6 (very recent as of this writing) kernel, the `do_each_thread()/while_each_thread()` style macros have been removed in favor of the simpler and more readable `for_each_process_thread()` macro. Our code takes this into account via the `LINUX_VERSION_CODE` and `KERNEL_VERSION` macros, thus becoming more portable as well.

The relevant portions of the code follows:

```
#if LINUX_VERSION_CODE < KERNEL_VERSION(6, 6, 0)
    do_each_thread(p, t) {      /* 'p' : process ptr; 't': thread ptr */
#else
    for_each_process_thread(p, t) {   /* 'p' : process ptr; 't': thread ptr */
#endif
         get_task_struct(t);     /* take a reference to the task struct */
        task_lock(t);
        snprintf_lkp(tmp1, TMPMAX-1, "%8d %8d  0x%px 0x%px",
                  p->tgid, t->pid, t, t->stack);

        [...]             << we cover this portion in the notes below >>

    total++;
    memset(tmp1, 0, sizeof(tmp1));        << cleanup >>
    memset(tmp2, 0, sizeof(tmp2));
    memset(tmp3, 0, sizeof(tmp3));

    task_unlock(t);

        memset(tmp, 0, sizeof(tmp));
        put_task_struct(t);     /* release reference to the task struct */

#if LINUX_VERSION_CODE < KERNEL_VERSION(6, 6, 0)
    } while_each_thread(p, t);
#else
    }
#endif

    return total;
}
```

Referring to the preceding code block, the `do_each_thread() { ... } while_each_thread()` pair of macros form a do-while loop, allowing us to iterate over every *thread* alive on the system. Some details on the code snippet follow:

- We follow a strategy of setting a few temporary variables (named `tmp1, tmp2, tmp3`) to fetch and set up the required data items, which we then print once on every loop iteration.
- To setup the data we want to report, C programmers often resort to using the `sprintf()` routine; note though, that this could certainly become a security issue (due to the well-known buffer overflow vulnerability and exploits related to it). Thus, we've been taught to instead use the `snprintf()`.

True, but we can improve it further: we write a small wrapper routine named `snprintf_lkp()` and define it in our `klib.c` 'library' source file. It explicitly it checks for and warns us about any buffer overflow (thus further bolstering our security posture).

- Obtaining the `TGID`, `PID`, `task_struct`, and `stack` start addresses is trivial – here, keeping it simple, we just use `current` to dereference them (of course, you could use the more sophisticated kernel helper methods we saw earlier in this chapter to do so as well; here, we wish to keep it simple).

  Also, notice that here we deliberately do *not* use the (safer) `%pK` printk format specifier but rather the generic `%px` specifier in order to display the *actual* kernel virtual addresses of the task structure and the kernel-mode stack (don't do this in production).
- Clean up as required before looping over (increment a counter of total threads, `memset()` the temporary buffers to `NULL`, and so on).
- On completion, we return the total number of threads we have iterated across.

In the following code block, we cover the portion of code that was deliberately left out in the preceding block. We retrieve the thread's name and print it within square brackets if it's a kernel thread. We also query the number of threads within the process. The explanation follows the code:

```
        if (!p->mm)          // kernel thread?
            snprintf_lkp(tmp2, TMPMAX-1, " [%16s]", t->comm);
        else
            snprintf_lkp(tmp2, TMPMAX-1, "  %16s ", t->comm);

        /* Is this the "main" thread of a multithreaded process?
         * We check by seeing if (a) it's a user space thread,
         * (b) its TGID == its PID, and (c), there are >1 threads in
         * the process.
         * If so, display the number of threads in the overall process
         * to the right..
         */
        nr_thrds = get_nr_threads(g);
        if (p->mm && (p->tgid == t->pid) && (nr_thrds > 1))
            snprintf_lkp(tmp3, TMPMAX-1, " %3d", nr_thrds);
    pr_info("%s%s%s\n", tmp1, tmp2, tmp3);
        [...]
```

On the preceding code block, we can say the following:

- A *kernel thread (kthread)* has no user space mapping. The `main()` thread's `current->mm` member is a pointer to a structure of type `mm_struct` and represents the entire process' *user space* mapping; if `NULL`, it stands to reason that this is a kernel thread (as kernel threads have no user space mappings). We check and print the name accordingly.

- We print the name of the thread as well (by looking up the `comm` member of the task structure). You might question why we don't use the `get_task_comm()` helper routine to obtain the task's name here; the short answer is that it causes a *(self) deadlock*! We shall explore this (and how to detect and avoid it) in detail in the later chapters on kernel synchronization. For now, again, we just do it the simple way.
- We fetch the number of threads in a given process conveniently via the `get_nr_threads()` macro; the rest is explained clearly in the code comment above the macro in the preceding block.

Finally, the code simply emits the temporary strings to the kernel log via one `printk` invocation.

Great! With this, we complete our discussion (for now) on Linux kernel internals and architecture with a primary focus on processes, threads, and their stacks.

Summary

In this chapter, we covered the key aspects of kernel internals that will help you as a kernel module or device driver author to better and more deeply understand the internal workings of the OS. We examined, in some detail, the organization of and relationships between the process and its threads and stacks (in both user and kernel space). We examined the kernel `task_struct` metadata structure and learned how to iterate over the *task list* in different ways via our custom written kernel modules.

Though it may not be obvious, the fact is that understanding these kernel internal details is a necessary and required step in your journey to becoming a seasoned kernel (and/or device driver) developer. The content of this chapter will help you debug many system programming scenarios and it lays the foundation for our deeper exploration into the Linux kernel, particularly that of memory management.

The next chapter and the couple that follow it are critical indeed: we'll cover what you need to understand regarding the deep and complex topic of memory management internals. I suggest you digest the content of this chapter first, browse through the *Further reading* links of interest, work on the exercises (and *Questions* section), and then, get to the next chapter!

Questions

As we conclude, here is a list of questions for you to test your knowledge regarding this chapter's material: `https://github.com/PacktPublishing/Linux-Kernel-Programming_2E/blob/main/questions/ch6_qs_assignments.txt`. You will find some of the questions answered in the book's GitHub repo: `https://github.com/PacktPublishing/Linux-Kernel-Programming_2E/tree/main/solutions_to_assgn`.

Further reading

To help you delve deeper into the subject with useful materials, we provide a rather detailed list of online references and links (and at times, even books) in a *Further reading* document in this book's GitHub repository. The *Further reading* document is available here: `https://github.com/PacktPublishing/Linux-Kernel-Programming_2E/blob/main/Further_Reading.md`.

Leave a review!

Enjoying this book? Help readers like you by leaving an Amazon review. Scan the QR code below for a 20% discount code.

**Limited Offer*

7

Memory Management Internals – Essentials

Kernel internals, especially regarding memory management, is a vast and complex topic. In this book, I do not intend to delve into the deep, gory details of kernel memory internals. At the same time, I would like to provide sufficient – and required – background knowledge for a budding kernel or device driver developer like you to successfully tackle this key topic.

Accordingly, this chapter will help you understand, to sufficient depth, the internals of how memory management is performed on the Linux OS; this includes delving into the **Virtual Memory** (**VM**) split, examining both the user-mode and kernel VAS of the process to a good level of depth, and covering the basics of how the kernel manages physical memory. In effect, you will come to understand the memory maps – both virtual and (to some extent) physical – of the process and the system.

This background knowledge will go a long way in helping you correctly and efficiently manage dynamic kernel memory (with a focus on writing kernel or driver code using the **Loadable Kernel Module** (**LKM**) framework; this aspect – dynamic memory management – in a practical fashion is the focal point of the next two chapters in the book). As an important side benefit, armed with this knowledge, you will find yourself becoming more proficient at the debugging of both user and kernel-space code. (The importance of this cannot be overstated! Debugging code is both an art and a science, as well as a reality.)

In this chapter, the areas we will cover include the following:

- Understanding the VM split
- Examining the process VAS
- Examining the kernel VAS
- Randomizing the memory layout – KASLR
- Understanding physical memory organization

Technical requirements

I assume that you have gone through *Online Chapter, Kernel Workspace Setup*, and have appropriately prepared a guest VM (or native system) running Ubuntu 22.04 LTS (or a later stable release) and installed all the required packages. If not, I recommend you do this first. To get the most out of this book, I strongly recommend you first set up the workspace environment, including cloning this book's GitHub repository for the code (`https://github.com/PacktPublishing/Linux-Kernel-Programming_2E`), and work on it in a hands-on fashion.

I assume that you are familiar with the basic virtual memory concepts, the user-mode process **Virtual Address Space** (**VAS**) layout of segments, user and kernel-mode stacks, the task structure, and so on. If you're unsure on this footing, I strongly suggest you read the preceding chapter first.

Understanding the VM split

In this chapter, we will broadly look at how the Linux kernel manages memory in two ways:

- The virtual memory-based approach, where memory is virtualized (the usual case)
- A view of how the kernel organizes physical memory (RAM pages)

First, let's begin with the virtual memory view and then discuss physical memory organization later in the chapter.

As we saw in *Chapter 6, Kernel Internals Essentials – Processes and Threads*, in the *Understanding the basics of the process Virtual Address Space (VAS)* section, a key property of the process VAS is that it is completely self-contained, a sandbox. You cannot look outside the box. In that chapter, in *Figure 6.2*, we saw that the process VAS ranges from virtual address `0x0` to what we simply termed as the "high address." What is the actual value of this "high" address? It's the highest extent of the VAS and thus depends on the number of bits used for addressing; so:

- On a Linux OS running on a 32-bit processor (or compiled for 32-bit), the highest virtual address will be 2^{32} = *4 GB*.
- On a Linux OS running on (and compiled for) a 64-bit processor, the highest virtual address will be 2^{64} = *16 EB*. (EB is short for exabyte, which is 1,152,921,504,606,846,976 bytes or 1,024 petabytes. Clearly, it's an enormous quantity; 16 EB is the number 16×10^{18}.)

For simplicity, to keep the numbers manageable, let's focus for now on the 32-bit address space (we will certainly cover 64-bit addressing as well). So, according to our discussions, on a 32-bit system, the process VAS ranges from 0 to 4 GB – this region consists of empty space (unused regions, called **sparse regions** or **holes**) and valid regions of memory, commonly termed **segments** (or more correctly, **mappings**) – text, data, library, and stack (all of this having been covered in some detail in *Chapter 6, Kernel Internals Essentials – Processes and Threads*).

On our journey to understanding virtual memory, it's useful to take the well-known `Hello, world` K&R C program and understand (to a decent extent) its inner workings on a Linux system; this is what the next section covers!

Looking under the hood – the Hello, world C program

Right, is there anyone here who knows how to code the canonical K&R C `Hello, world` program? Okay, very amusing, let's check out the one meaningful line therein:

```
printf("Hello, world.\n");
```

The process is calling the `printf()` function. Have you written the code of `printf()`? "No, of course not," you say, "it's within the standard C library, typically `glibc` (GNU `libc`) on Linux." Yes, you're right; but hang on, unless the code and data of `printf()` (and similarly all other library APIs) is actually *within* the process VAS, how can we ever access it? (Recall that you can't look *outside the box!*) For that, the code (and data) of `printf()` (in fact, of the entire `glibc` library) must be mapped within the process *box* – the process VAS. It *is* indeed mapped within the process VAS, in the library segments or mappings (as we saw in *Chapter 6, Kernel Internals Essentials – Processes and Threads, Figure 6.2*). How did this mapping happen?

The reality is that on application startup, as part of the C runtime environment setup, there is a small **Executable and Linkable Format** (**ELF**) binary (embedded into your `a.out` binary executable file) called the **loader** (`ld.so` or `ld-linux.so`). It is given control early in execution. It detects all required shared libraries and memory maps all of them – the library text (code), data, and any other required segments from the (library) file(s) into the process VAS by opening them and issuing the `mmap()` system call. So, now, once the code and data of the library are mapped within the process VAS, the process can indeed access it, and thus – wait for it – the `printf()` API can be successfully invoked! (We've skipped the gory details of memory mapping and linkage here.)

Further verifying this, the `ldd` script (the following output is from an x86_64 system) reveals that this is indeed the case:

```
$ gcc helloworld.c -o helloworld
$ ./helloworld
Hello, world
$ ldd ./helloworld
        linux-vdso.so.1 (0x00007fffcfce3000)
        libc.so.6 => /lib/x86_64-linux-gnu/libc.so.6 (0x00007feb7b85b000)
        /lib64/ld-linux-x86-64.so.2 (0x00007feb7be4e000)
$
```

A few quick points to note:

- Every single Linux process – automatically and by default – links to a minimum of two objects: the `glibc` shared library and the program loader (no explicit linker switch is required during compilation/link).
- The name of the loader program varies with the architecture. Here, on our x86_64 system, it's `ld-linux-x86-64.so.2`.

- In the preceding `ldd` output, the address within parentheses on the right is the (user space) virtual address of the location of the mapping. For example, in the preceding output, `glibc` is mapped into our process VAS at the **User Virtual Address** (**UVA**), which equals `0x00007feb7b85b000`. Note that it's runtime dependent (it also varies on every run when **Address Space Layout Randomization** (**ASLR**) semantics are enabled (ASLR is typically enabled by default; details seen later)).
- For security reasons (and on architectures besides x86), it's considered better to use the `objdump` utility to look up details like these.

Try performing `strace` on the `Hello, world` binary executable and you will see numerous `mmap()` system calls, mapping in `glibc` (and other) segments!

Now let's examine our simple `Hello, world` application more deeply.

Going beyond the printf() API

As you will know, the `printf()` API performs it's pretty formatting and invokes the `write()` system call, which, of course, writes the `"Hello, world"` string to `stdout` – here, by default, `stdout` will be the (pseudo) terminal window or the console device.

We also understand that as `write()` is a system call, this implies that the current process (or thread) running this code – the process context – must now switch to kernel mode and run the kernel code of `write()` (monolithic kernel architecture)! Indeed, it does. But hang on a second: the kernel code of `write()` is in the kernel VAS (refer to *Chapter 6, Kernel Internals Essentials – Processes and Threads, Figure 6.1*). The point here is a key one: if the kernel VAS is outside the box, then how in the world are we going to call it?

Well, it could be done by placing user and kernel VASs into two separate 4 GB spaces, but this approach results in very slow context switching (and expensive **Translation Lookaside Buffer** (TLB) flushing), so it's simply not done.

The way it is engineered is like this: **both user and kernel VASs "live" in the same "box" – the available VAS.** How exactly? By *splitting* the available address space between the user and kernel in some `User:Kernel :: U:K` ratio. This is called the **VM split** (the ratio `U:K` being typically expressed in megabytes, gigabytes, terabytes, or even petabytes).

On many ARM-32 (AArch32), and x86-32 systems, the default VM split is often 3:1 GB. The following diagram (*Figure 7.1*) is representative of a 32-bit Linux process having a *3:1* VM split (the unit is GB); that is, the total 4 GB process VAS is split into 3 GB of user space and 1 GB of kernel space. In other words, the split can be described as U:K :: 3:1.

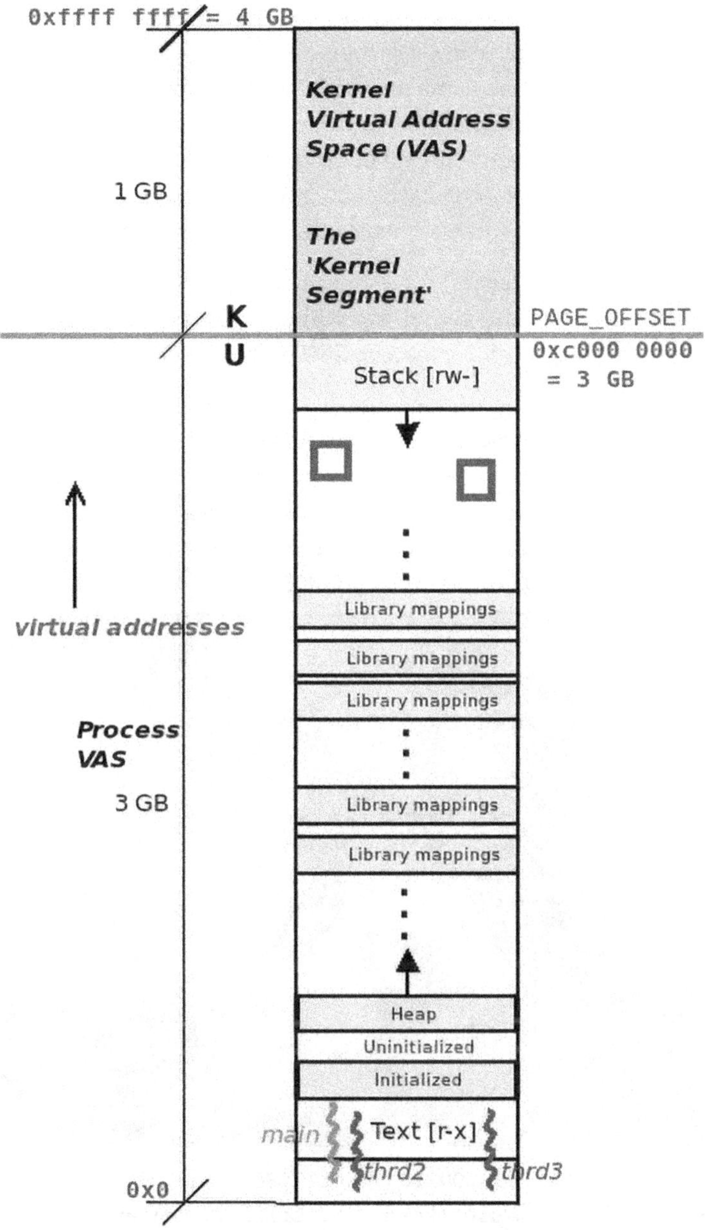

Figure 7.1: A process VAS with a User:Kernel :: 3:1 GB VM split on an AArch32 system running the Linux OS

The VM split can be configured at kernel config time on 32-bit (more on this follows. Other valid values for the VM split on an AArch32 (or x86-32) system are U:K :: 2:2 or even 1:3, though the last one would be very unusual in practice.)

So, now that the kernel VAS is "within the box," it's suddenly clear and critical to understand this: when a user-mode process or thread issues a system call, there is a context switch to the kernel's (in this example) 1 GB VAS within the very same process's VAS (this switch involves the update of various arch-specific CPU registers, including the stack pointer). The thread issuing the system call now runs the relevant kernel code paths in process context in privileged kernel mode (and works on kernel-space data). When done, it returns from the system call, context switching back into unprivileged user mode, and now runs user-mode code (and works on user-mode data) within the first 3 GB of VAS.

The exact virtual address where the kernel VAS begins (aka the **kernel segment**) is typically represented via the `PAGE_OFFSET` macro within the kernel. We will examine this, and some other key macros as well, in the *Macros and variables describing the kernel VAS layout* section.

Where is this decision regarding the precise location and size of the VM split taken? Ah, on 32-bit Linux, it's a kernel build-time configurable. It's done within the kernel build as part of the `make [ARCH=xxx] menuconfig` procedure – for example, when configuring the kernel for a Broadcom BCM2835 (or the BCM2837) **System on Chip (SoC)** (the Raspberry Pi family being a popular board series with this very SoC). Here's a snippet from the official kernel configuration file (the following output is from a 32-bit Raspberry Pi running the default 32-bit Raspberry Pi OS):

```
$ uname -r
5.15.76+
# gain access to /proc/config.gz by loading the 'configs' module
$ sudo modprobe configs
$ zcat /proc/config.gz | grep -C3 VMSPLIT
#
# Kernel Features
#
CONFIG_VMSPLIT_3G=y
# CONFIG_VMSPLIT_3G_OPT is not set
# CONFIG_VMSPLIT_2G is not set
# CONFIG_VMSPLIT_1G is not set
CONFIG_PAGE_OFFSET=0xC0000000
[ … ]
```

As seen in the preceding snippet, the `CONFIG_VMSPLIT_3G` kernel config option is set to `y` (for yes) **implying that the default VM split is** `user:kernel :: 3:1`. For 32-bit architectures, the VM split location is **tunable** (as can be seen in the preceding snippet, via the `CONFIG_VMSPLIT_[1|2|3]G macros; CONFIG_PAGE_OFFSET` gets set accordingly). With a 3:1 VM split, `PAGE_OFFSET` is set to the virtual address `0xC000 0000` (3 GB).

The default VM split for the IA-32 processor (the Intel x86-32) is 3:1 (GB). Interestingly, the (ancient) Windows 3.x OS running on the IA-32 had the same VM split, showing that these concepts are essentially OS-agnostic. Later in this chapter, we will cover several more architectures and their VM splits (on Linux), in addition to other details.

Configuring the VM split for 64-bit architectures is sometimes possible, depending on the specific CPU and kernel **Board Support Package** (**BSP**) (we mention examples of how this can be done for the AArch64 in the coming materials). So, now that we understand the VM split on 32-bit systems, let's move on to examining how it's done on 64-bit systems. But first, it's useful and important to understand at least the basics of how exactly virtual addresses get translated to physical addresses; let's get to it!

Virtual addressing and address translation

Before diving further into these details, it's very important to clearly understand a few key points.

Consider a small and typical code snippet from a C program:

```
int i = 5;
printf("address of i is 0x%x\n", &i);
```

When running this program on a rich modern OS (like Linux, Unix, Windows, or Mac), the address you see the `printf()` emit is (almost) always *a virtual address* and not a physical one. Further, we distinguish between two kinds of virtual addresses:

- If you run this code in a user space process, the address of variable `i` that you will see is a **User Virtual Address**, a **UVA** for short.
- If you run this code within the kernel, or a kernel module (of course, you'd then use the `printk()` (or similar) API in place of the `printf()`), the address of variable `i` that you will see is a **Kernel Virtual Address** (**KVA**).

Next, as is often thought, a virtual address is *not* an absolute value (an offset from `0`); it's a *bitmask* that's designed for and interpreted by the **MMU** (the **Memory Management Unit** that's within the silicon of modern microprocessors):

- On a 32-bit Linux OS, the 32 available bits are divided into what's called the **Page Global Directory** (**PGD**) value, the **Page Table** (**PT**) value, and the offset.
- These become indices into physical memory via which the MMU, with access to the kernel page tables for the current process context, performs address translation.

We do not intend to cover the deep details of MMU-level address translation here. It's also very arch-specific. Do refer to the *Further reading* section for useful links on this topic.

- As might be expected, on a 64-bit system, even with 48-bit addressing, there will be more fields within the virtual address bitmask.

Okay, if this 48-bit addressing is the typical case on the x86_64 processor, then how are the bits in a 64-bit virtual address laid out?

What happens to the unused 16 MSB bits? The following figure answers the question; it's a representation of the breakup of a 64-bit virtual address (bitmask) on an x86_64 Linux system:

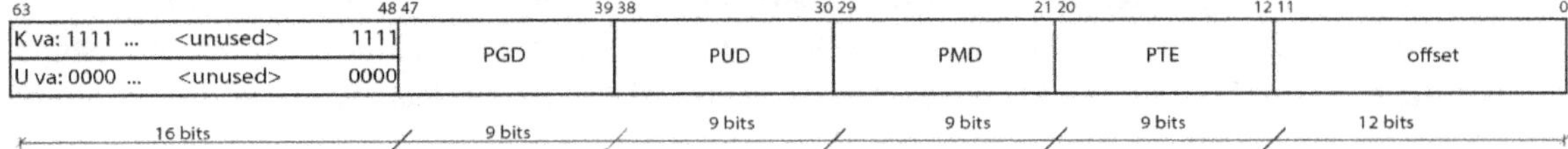

Figure 7.2: Breakup of a 64-bit virtual address on the Intel x86_64 processor with 4 KB pages

Essentially, with 48-bit addressing, we use bits 0 to 47 (the LSB 48 bits) and ignore the **MSB (Most Significant Bit)** 16 bits, treating it much as a sign extension. Not so fast though, the value of the unused MSB 16 bits – as shown below – varies with the address space being referenced:

- **In kernel VAS**, the MSB 16 bits are always set to `1`.
- **In user VAS**, the MSB 16 bits are always set to `0`.

This is useful information! Knowing this, by merely looking at a (full 64-bit) virtual address, you can immediately tell whether it's a KVA or a UVA; thus, on an x86_64 Linux system:

- KVAs always have the format `0xffff XXXX XXXX XXXX`
- UVAs always have the format `0x0000 XXXX XXXX XXXX`

A word of caution: the preceding format holds true only for processors (MMUs, really) that define virtual addresses as being KVAs or UVAs; the x86 and ARM family of processors (both 32- and 64-bit) do fall in this bracket. Here, the MSB (bit 63) acts as a selector for the paging tables; if set, the kernel paging table (at `swapper_pg_dir`) is used as a kernel virtual address is being referenced; if cleared, the process paging table (the physical address of the base of the process paging table is in the CR3 register on x86[_64] and the TTBR0 (Translation Table Base Register 0) register on ARM[64]) is used as a user virtual address is being referenced (while TTBR1 holds the base address of the kernel master PGD - `swapper_pg_dir`).

Another point: what does *N-level paging* mean? Relook at *Figure 7.2*; there are four levels of "indirection" – the **Page Global Directory** (PGD), **Page Upper Directory** (PUD), Page Middle Directory (PMD) and **Page Table Entry** (PTE) – before getting to the offset. This is a property of the paging schema – the number of levels of indirection; here it's 4, hence, we call this a 4-level paging (later, *Figure 7.6* will show you various N-level paging values).

Getting from virtual to physical address – a very brief overview

As can now be seen (and I reiterate here), the reality is that virtual addresses are not absolute addresses (absolute offsets from zero, as you might have mistakenly imagined) but are bitmasks. The fact is that memory management is a a complex area where the work is shared; let's reiterate the really key points:

- **Every process alive has a set of paging tables** that map the virtual pages to the physical ones (called *page frames*); the paging tables come into play whenever a virtual address – user or kernel – is accessed. The kernel too has its own paging tables.

- **The OS creates and manipulates the paging tables of each process as well as of the kernel;** the **toolchain** (compiler/linker) generates virtual addresses.
- The **processor MMU performs runtime address translation**, translating a given (user or kernel) virtual address to a physical (RAM) address.

So, again, without going into the gory details, here's the overall procedure for what's generally called **hardware paging** (here, for the x86) in brief (*Figure 7.3* gives a high level overview of the overall flow):

1. A process (or a thread within it) looks up *a virtual address* (a UVA or a KVA) – that is, it performs a read/write/execute operation upon a virtual address. (This is completely normal, expected behavior; for example, reading or writing a variable, or executing a machine instruction).
2. For the CPU to perform the work at that location in memory, we must now somehow **translate** this virtual address to its corresponding physical counterpart. Hang on, though: due to hardware optimization, this step can either be bypassed or done quicker (steps 3.1, 3.2); if it cannot, it must be translated "manually" via the MMU (slower; step 4).
3. Before going to the MMU, a couple of **hardware optimizations** can help us be quicker, short-circuiting the 'usual' slower path:

    i. First, the code or data being worked upon might already be residing within the **on-CPU caches** (L1/L2/L3/...); this is checked for first. If this is indeed the case, we have a *cache-hit*: the code/data item is worked upon within the CPU caches itself, and the job is done. If not, we have a CPU cache miss (technically, an **LLC (Last Level Cache) miss** (expensive)); so, we move to the next step. (We shall, in *Chapter 13, Kernel Synchronization – Part 2,* cover more on CPU caching and the question of cache coherency; ignore it for now.)

    ii. Is the virtual address already translated? Look up the CPU's **Translation Lookaside Buffer (TLB)**. If the translation is present, we have a **TLB hit**; if so, we have the physical address cached within the TLB: skip ahead to step 5; if not, we have a TLB miss (expensive).

    (An aside: it's not necessarily done in the order shown here; some micro-architectures use a *physical cache* model where the CPU caches lie between the MMU and RAM access, in effect, reversing the order of the above two steps).

4. The virtual address is sent to the MMU; it now "walks" the paging tables of the process (it knows where the base page table of the process (or kernel) is in physical RAM via a system register holding it's physical address) and ends up with the corresponding page frame and physical address (*Figure 7.4* depicts this step).
5. The physical address (obtained from one of the earlier steps) is now piggy-backed onto the CPU address lines, and the work is carried out.

Do ensure you carefully read and understand this. All steps shown, except for step 4, are depicted by *Figure 7.3*; step 4 – the translation-by-MMU one – is depicted via *Figure 7.4*. Do study them.

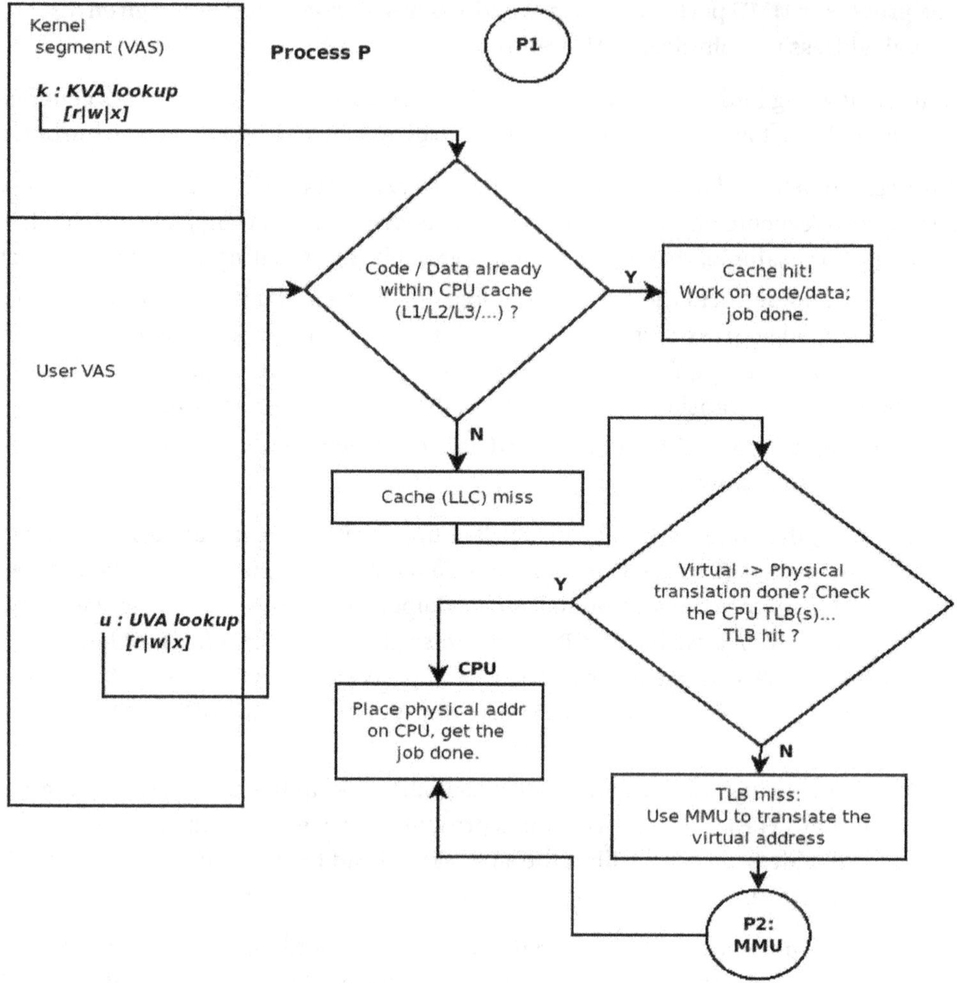

Figure 7.3: Translating VA to PA, part 1 (steps 1,2,3,5)

And here's the "P2: MMU" portion, where the MMU performs the runtime address translation, translating the user or kernel virtual address and passing into a physical address:

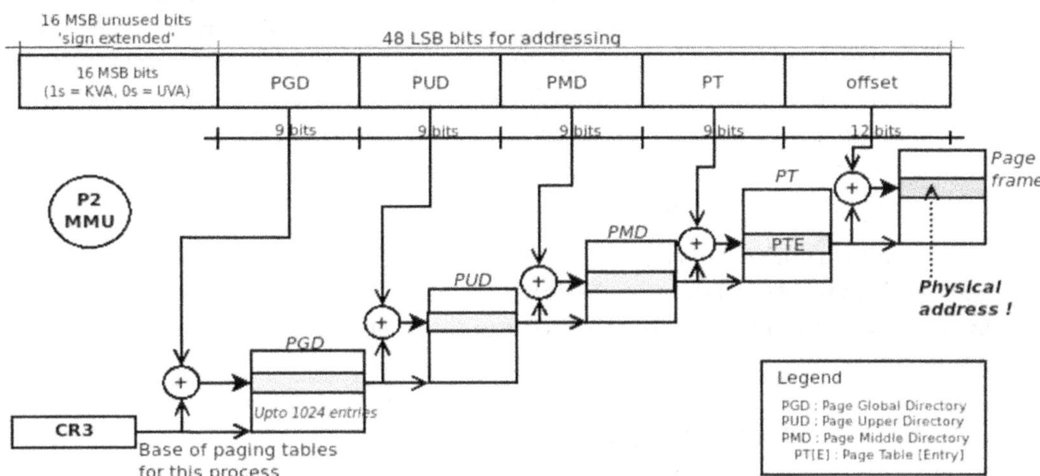

Figure 7.4: Translating VA to PA, part 2, MMU (steps 4, 5) on the x86_64 with 48-bit addresses and 4 KB pages

For the MMU address translation diagram (*Figure 7.4*) do note that here we use a very arch-specific example of the x86_64 with 4-level paging, 4 KB page size and 48-bit addressing. When passed a virtual address, the MMU has the ability to see it as a bitmask; it looks up the base physical address – it's in Control Register 3 (CR3) on the x86 (on the ARM families, it's the TTBR0 register for user processes and TTBR1 for the kernel paging tables) – and proceeds with translation. (Think on this: of course, the base address in CR3 is a physical address, else this step would become infinitely recursive!)

In a nutshell, this is what the MMU does to perform runtime address translation: it looks up the base CR3 address, adds the 9-bit value in the PGD portion of the virtual address to it, and looks up to that quantity. This is a pointer to the next table, where it repeats the previous step, but this time with the value from the 9-bit PUD field. This continues (over the PMD) until the Page Table, where it references the actual **Page Table Entry (PTE)**. This contains (among other micro-arch-specific bits) a pointer to the physical *page frame*; adding the 12-bit offset to the base address of the page frame yields the physical address.

The reality is more nuanced; a couple of key points deserve a mention here.

One, theoretically at least, when a virtual address is passed to the MMU, it "walks" the paging tables and the end result should be the physical address; in practice, the MMU address translation attempt can fail! One case is the obvious one: the virtual address supplied is incorrect (an unmapped address): in effect, we have a bug, a defect, causing the MMU to raise a fault (the OS fault handler will handle it appropriately). The other case is *demand paging* – the case where the virtual address is legal but physical memory hasn't been allocated (yet), causing the translation to fail (causing the MMU to raise a 'good' fault which has the system allocate the page frame). We precisely cover this in some detail in *Chapter 9, Kernel Memory Allocation for Module Authors – Part 2* in the section *Demand paging and OOM*; if you'd like, you can glance at *Figure 9.9* which shows what else can happen... of course, we'll cover the details there.

Two, the kernel, being the boss, can actually bypass using the MMU and perform address translation itself, in software. In practice, this is rarely done as, of course, it will be slower. One place where it is done is when performing IO (reads or writes) by leveraging an `mmap()` into the process VAS via the */proc/PID/maps* pseudo file. By employing this approach, one can **write to memory that's marked as read-only**! (*offlinemark* does a great job detailing this in this blog article: *Linux Internals: How /proc/self/mem writes to unwritable memory*, May 2021: `https://offlinemark.com/2021/05/12/an-obscure-quirk-of-proc/`; be sure to check it out.)

We will not delve into further details regarding hardware paging (and various hardware acceleration technologies, such as the **Translation Lookaside Buffer** (**TLB**)) in this book. These topics are well covered by various other excellent books and reference sites that are mentioned in the *Further reading* section of this chapter.

VM split on 64-bit Linux systems

First of all, it is worth noting that on 64-bit systems, all 64 bits are not used for addressing. A standard Linux OS configuration for the x86_64 with a (typical) 4 KB page size uses the **LSB** (**Least Significant Bit**) 48 bits for addressing. Why not the full 64 bits? It's simply too much! No existing computer comes close to having even half of the full $2^{64} = 18,446,744,073,709,551,616$ bytes, which is equivalent to 16 EB (16 exabytes, that's 16,384 petabytes) of RAM!

"Why," you might well wonder, "do we equate this virtual addressing with RAM?". Please read on – more material needs to be covered before this becomes clear. The *Examining the kernel VAS* section is where you will understand this fully.

As just stated, the available VAS on a 64-bit system is a mind-blowing gigantic $2^{64} = 16$ *EB* (16×10^{18} bytes!). It's perhaps likely that when Linus was working on the first 64-bit Linux port (to the DEC Alpha, the first commercial 64-bit processor), he would have had to decide how to lay out the process and kernel segments within this enormous VAS. The decision reached has more or less remained (conceptually), even on today's x86_64 Linux OS. This enormous 64-bit VAS is split into a VAS for user mode and a separate VAS for the kernel as follows. Here, we assume 48-bit addressing with a 4 KB page size:

- The so-called "canonical lower half" of the process VAS, for 128 TB: User VAS – the virtual addresses range from `0x0` to `0x0000 7fff ffff ffff`
- The so-called "canonical upper half" of the process VAS, for 128 TB: Kernel VAS – the virtual addresses range from `0xffff 8000 0000 0000` to `0xffff ffff ffff ffff`

The word *canonical* effectively means *as per the law or as per common convention*.

This 64-bit VM split on an x86_64 Linux platform can be seen in the following figure:

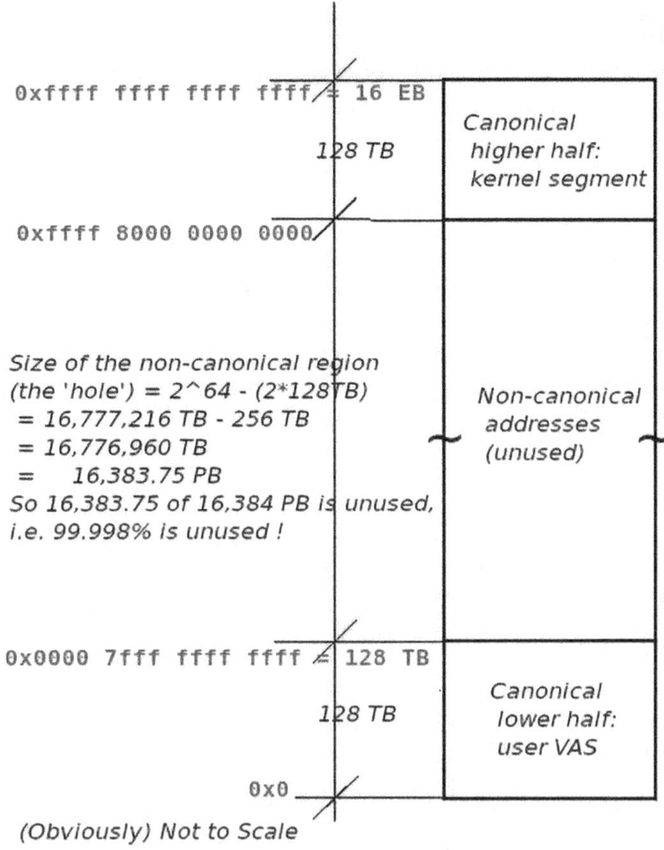

Figure 7.5: The Intel x86_64 (or AMD64) 16 EB VAS layout (with 48-bit addressing and a 4 KB page size); the VM split User:Kernel :: 128 TB:128 TB

In the preceding figure, you can clearly see the user VAS "anchored" at the bottom 128 terabytes and the kernel VAS "anchored" at the top 128 terabytes of the total existing 16 EB VAS.

What about the in-between unused region (here, from `0x0000 8000 0000 0000` to `0xffff 7fff ffff ffff`)? This is simply a hole or *sparse* region; it's also called the **non-canonical addresses** region. Interestingly, as the diagram informs us, with the typical 48-bit addressing scheme, most of the VAS (99.998%) is left unused! This is why we term the VAS as being very sparse.

The preceding figure is certainly not drawn to scale! Also, always keep in mind that this is all virtual memory space, not physical.

Common VM splits

To round off our discussion on the VM split, some common user:kernel VM split ratios for different CPU architectures are shown in the following figure (again, we assume an MMU page size of 4 KB, except for the last row, which uses 64 KB size pages):

Row #	Arch	N-Level	# Addr Bits	Total in-use VAS (2^ addr-bits)	VM Split U : K	User-space		Kernel-space
						Start UVA	End UVA	Start KVA (with End KVA always = 0xffff ffff ffff ffff)
1	IA-32	2	32	4 GB	3 GB : 1 GB	0x0	0xbfff ffff	0xc000 0000
2	ARM (AArch32)	2	32	4 GB	2 GB : 2 GB (or 3GB : 1GB)	0x0	0x7fff ffff	0x8000 0000
3	x86_64	4	48	256 TB	128 TB : 128TB	0x0	0x0000 7fff ffff ffff	0xffff 8000 0000 0000
4		5	57	128 PB	64 PB : 64 PB	0x0	0x00ff ffff ffff ffff	0xff00 0000 0000 0000
5	AArch64	3	40	1 TB	512 GB : 512GB	0x0	0x0000 7fff ffff ffff	0xffff ff80 0000 0000
6		4	49	512 TB	256 TB : 256TB	0x0	0x0000 ffff ffff ffff	0xffff 0000 0000 0000
7	ARMv8.2 LPA	3	53	8 PB	4 PB : 4 PB	0x0	0x0010 0000 0000 0000	0xfff0 0000 0000 0000

Figure 7.6: Common user:kernel VM split ratios for different CPU architectures on modern Linux

A few notes pertaining to *Figure 7.6* (with regard to the row number shown) follow:

- **Row #1** and **Row #2:** The 32-bit IA-32 (or x86-32) and AArch32 with a 3:1 and 2:2 (GB) VM split, respectively.
- **Row #3,** the common case: The x86_64 with 4-level paging and 48 (LSB) bits used for addressing.
- **Row #4:** 5-level paging is possible but requires 4.14 Linux or later.
- **Row #5,** AArch64 examples: With a standard Yocto 4.0.4 (Kirkstone) build for Poky distro, machine raspberrypi4-64 (with default config value `CONFIG_ARM64_VA_BITS=39`; thus 40 address bits configured, total VAS is 2^40 = 1 TB; thus, here, both the kernel and user VAS are 512 GB each).
- **Row #6,** AArch64: Above + config `CONFIG_ARM64_VA_BITS=48`; thus 49 (0 to 48) address bits configured, total VAS is 2^49 = 512 TB, thus, here, both the kernel and user VAS are 256 TB each.

- **Row #7**: This config requires (AArch64) ARMv8.2 **LPA (Large Physical Address)** extension + >= 5.4 Linux + 64K page size (for 3-level paging) enabled. It results in 53 address bits being configured, total VAS is 2^53 = 8 PB; thus, here, both the kernel and user VAS are 4 PB each.

A word on x86_64 Linux addressing

Using the x86_64 as an example, 2^47 is 128 TB; so why is the number of address bits 48 (in *Figure 7.6* for the x86_64) and not 47? This, of course, is as we totally require 2*128 TB = 256 TB of usable address space, 128 TB for user space, and another 128 TB for kernel space (and 2^48 = 256 TB).

In effect, the number of address bits (the fourth column in *Figure 7.6*) used for addressing determines the overall size of the total in-use virtual memory: user + kernel – with each VAS typically getting one-half of the total (the exceptions tend to be the 32-bit x86 and ARM-32, where the VM split can be deliberately unequal).

We highlight the third row in bold (red) as it's considered the common case: running Linux on the x86_64 (or AMD64) architecture, with a `user:kernel :: 128 TB:128 TB` VM split. The fourth column, *#Addr Bits,* shows us that, on 64-bit processors, no real-world processor actually uses all 64 bits for addressing.

Under the x86_64, there are two VM splits, as seen in *Figure 7.6* (rows 3 and 4):

- The first one (row #3 in *Figure 7.6*), *128 TB : 128 TB* (4-level paging) is the VM split being used by default on Linux x86_64 systems as of today (embedded systems, laptops, PCs, workstations, and servers). It limits the physical address space to 64 TB (of RAM).
- The second one (row #4 in *Figure 7.6*), *64 PB : 64 PB,* is, as of the time of writing at least, still purely theoretical; it comes with support for what is called 5-level paging from 4.14 Linux. The assigned VASs (with 57-bit addressing, we get an incredible total of 128 PB of VAS and 4 PB of physical address space!) are so enormous that, as of our knowledge at the time of writing this, no actual computer is (yet) using it.

A word on AArch64 Linux addressing

Note that the two rows for the AArch64 (ARM-64) architecture running on Linux are merely representative. The BSP vendor or platform team working on the product could well use differing splits (the notes following *Figure 7.6* mention the precise meaning here).

Furthermore, the last row in the table mentions the fact that later-generation (AArch64) ARMv8.2 processors provide the LPA extension. When enabled, and with a 64 KB page size, the address space can be expanded to 52 bits in each user and kernel address space (and thus 53 bits in total), providing for (2^53) 4 PB of VAS space each (along with an upper limit of 4 PB physical memory as well). This is supported from Linux 5.4 onward.

More details on AArch64 52-bit addressing can be obtained here:

- Official kernel doc: *Memory Layout on AArch64 Linux*: `https://www.kernel.org/doc/html/v6.1/arm64/memory.html`
- *Understanding 52-bit virtual address support in the Arm64 kernel*, B Sharma, Dec 2020: `https://opensource.com/article/20/12/52-bit-arm64-kernel`

Right, moving on; what's actually residing within the kernel VAS (aka the kernel segment)? All kernel code, kernel data structures (including the task structures, the lists, the kernel-mode stacks, paging tables, and so on), device drivers, kernel modules, and so on are within here (as the lower half of *Figure 6.3* in *Chapter 6, Kernel Internals Essentials – Processes and Threads*, showed; we cover precisely this in some detail in the upcoming *Understanding the kernel segment* section).

It's important to realize that, as a performance optimization on Linux, **kernel memory is always non-swappable**; that is, kernel memory can never be paged out to a swap partition. User space memory pages are always candidates for paging, unless locked (see the `mlock[all](2)` system calls).

With this background, you're now in a great position to understand the full process VAS layout on the Linux OS. Read on!

Understanding the process VAS – the full view

Once again, refer to *Figure 7.1*; it shows the actual and full process VAS layout for a single 32-bit process. The reality, of course, is that **all processes alive on the system have their own unique user-mode VAS but share the same kernel VAS**. The following figure attempts to conceptually convey just this; it shows the situation for a typical IA-32 (or it could be an AArch32) system, with a 3:1 (GB) VM split. Here, every process's user space is unique to it, with all processes *sharing* the very same kernel VAS:

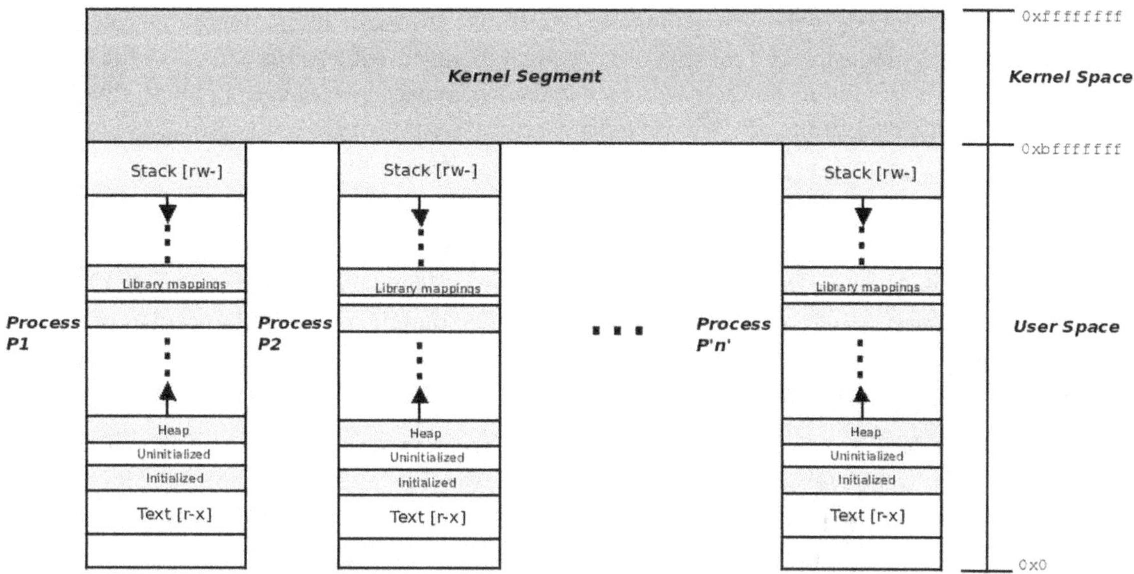

Figure 7.7: Processes have a unique user VAS but share the kernel VAS (32-bit OS); IA-32 (or AArch32) with a 3:1 (GB) VM split

Notice in the preceding figure how the virtual address space reflects a 3:1 (GB) VM split. The user VAS extends from `0` to `0xbfff ffff` (`0xc000 0000` is the 3 GB mark; this is what the `PAGE_OFFSET` macro is set to here), and the kernel VAS extends from `0xc000 0000` (3 GB) to `0xffff ffff` (4 GB).

Later in this chapter, we will cover the usage of a useful utility called `procmap`. It will help you literally visualize the VASs, both kernel and user, in detail, similar to how our preceding diagrams have been showing.

A few things to note:

- For the example shown in *Figure 7.7*, the value of the `PAGE_OFFSET` kernel macro is `0xc000 0000`.
- The figures and numbers we have shown here are not absolute and binding across all architectures; they tend to be very arch-specific and many highly vendor-customized Linux systems may change them (as *Figure 7.6* hints at).
- *Figure 7.7* details the VM layout on a 32-bit Linux OS. On 64-bit Linux, the *concepts* remain identical, it's just the numbers that (significantly) change. As shown in some detail in the preceding sections, the VM split on an x86_64 (with 48-bit addressing and 4K pages) Linux system becomes `User : Kernel :: 128 TB : 128 TB`.

Now that the fundamentals of the virtual memory layout of a process are understood, you will find that it greatly helps in deciphering and making progress in difficult-to-debug situations. As usual, there's still more to it; sections follow on the user space and kernel-space virtual memory map (the kernel VAS), and some coverage on the physical memory map as well. On, on!

Examining the process VAS

We have already covered the layout – the segments or mappings – that every process's VAS is made up of (see the *Understanding the basics of the process Virtual Address Space (VAS)* section in *Chapter 6, Kernel Internals Essentials – Processes and Threads*). We learned that the process VAS consists of various mappings or segments; among them are text (code), data segments, library mappings, and at least one stack. Here, we will expand greatly on that discussion.

Being able to dive deep into the kernel and see various runtime values is an important skill for a developer like you (as well as for the app user, QA, sysadmin, DevOps folks, and so on). The Linux kernel provides us with an amazing interface to do precisely this – it's, you guessed it, the `proc` filesystem (`procfs`).

This pseudo filesystem is always present on Linux (at least it should be) and is mounted under `/proc` by default. The `procfs` system has two primary jobs:

- To provide a unified set of (pseudo or virtual) files and directories, enabling you to look deep into the kernel and at hardware internal details.
- To provide a unified set of root-writeable files, allowing the sysad (or root user) to modify key kernel parameters. These are present under `/proc/sys/` and are termed `sysctl` – they are the *tuning knobs* of the Linux kernel.

Familiarity with the `proc` filesystem is indeed a must. I urge you to check it out and read the excellent man page on `proc(5)` (by typing `man 5 proc` in the terminal) as well. For example, simply doing `cat /proc/PID/status` (where `PID` is, of course, the unique process identifier of a given process or thread) yields a whole bunch of useful details from the process or thread's task structure!

Conceptually similar to `procfs` is the `sysfs` filesystem, mounted under `/sys` (and under it `debugfs`, typically mounted at `/sys/kernel/debug`). `sysfs` is a representation of >= 2.6 Linux's new device and driver model; it exposes a tree of all devices (and their drivers) on the system, as well as several kernel-tuning knobs. All these are pseudo filesystems; that is, they're mounted in RAM (and their content is thus volatile).

Examining the user VAS in detail

Let's begin by checking out the user VAS of any given process. A pretty detailed map of the user VAS is made available via `procfs`, particularly via the `/proc/PID/maps` pseudofile. Let's learn how to use this interface to peek into a process's user space (virtual) memory map. We will see two ways of doing so:

- Directly via the `procfs` interface's `/proc/PID/maps` pseudo file
- Using a few useful frontends (making the output more human-digestible)

Let's start with the first one.

Directly viewing the process memory map using procfs

Looking up the internal process details of any arbitrary process does require root access, whereas looking up details of a process under your ownership (including the caller process itself) does not. So, as a simple example, we will look up the calling process's VAS by using the `self` keyword in place of the PID. The following screenshot shows this (on an x86_64 Ubuntu 22.04 LTS guest):

```
$ cat /proc/self/maps
558822d64000-558822d66000 r--p 00000000 08:01 7340181                    /usr/bin/cat
558822d66000-558822d6a000 r-xp 00002000 08:01 7340181                    /usr/bin/cat
558822d6a000-558822d6c000 r--p 00006000 08:01 7340181                    /usr/bin/cat
558822d6c000-558822d6d000 r--p 00007000 08:01 7340181                    /usr/bin/cat
558822d6d000-558822d6e000 rw-p 00008000 08:01 7340181                    /usr/bin/cat
558823b90000-558823bb1000 rw-p 00000000 00:00 0                          [heap]
7f44c48f8000-7f44c491a000 rw-p 00000000 00:00 0
7f44c491a000-7f44c4e8d000 r--p 00000000 08:01 7340143                    /usr/lib/locale/locale-archive
7f44c4e8d000-7f44c4e90000 rw-p 00000000 00:00 0
7f44c4e90000-7f44c4eb8000 r--p 00000000 08:01 7342177                    /usr/lib/x86_64-linux-gnu/libc.so.6
7f44c4eb8000-7f44c504d000 r-xp 00028000 08:01 7342177                    /usr/lib/x86_64-linux-gnu/libc.so.6
7f44c504d000-7f44c50a5000 r--p 001bd000 08:01 7342177                    /usr/lib/x86_64-linux-gnu/libc.so.6
7f44c50a5000-7f44c50a9000 r--p 00214000 08:01 7342177                    /usr/lib/x86_64-linux-gnu/libc.so.6
7f44c50a9000-7f44c50ab000 rw-p 00218000 08:01 7342177                    /usr/lib/x86_64-linux-gnu/libc.so.6
7f44c50ab000-7f44c50b8000 rw-p 00000000 00:00 0
7f44c50c9000-7f44c50cb000 rw-p 00000000 00:00 0
7f44c50cb000-7f44c50cd000 r--p 00000000 08:01 7342169                    /usr/lib/x86_64-linux-gnu/ld-linux-x86-64.so.2
7f44c50cd000-7f44c50f7000 r-xp 00002000 08:01 7342169                    /usr/lib/x86_64-linux-gnu/ld-linux-x86-64.so.2
7f44c50f7000-7f44c5102000 r--p 0002c000 08:01 7342169                    /usr/lib/x86_64-linux-gnu/ld-linux-x86-64.so.2
7f44c5103000-7f44c5105000 r--p 00037000 08:01 7342169                    /usr/lib/x86_64-linux-gnu/ld-linux-x86-64.so.2
7f44c5105000-7f44c5107000 rw-p 00039000 08:01 7342169                    /usr/lib/x86_64-linux-gnu/ld-linux-x86-64.so.2
7fff477af000-7fff477d0000 rw-p 00000000 00:00 0                          [stack]
7fff477dd000-7fff477e1000 r--p 00000000 00:00 0                          [vvar]
7fff477e1000-7fff477e3000 r-xp 00000000 00:00 0                          [vdso]
ffffffffff600000-ffffffffff601000 --xp 00000000 00:00 0                  [vsyscall]
$
```

Figure 7.8: Output of the cat /proc/self/maps command on an x86_64 guest

In the preceding screenshot, you can actually see the layout of the user VAS of the `cat` process – a veritable memory map of the user VAS of that process! Also, notice that the preceding `procfs` output is sorted in ascending order by UVA.

Basic familiarity with using the powerful `mmap(2)` system call will greatly help in understanding further discussions. Do (at least) browse through its man page.

Interpreting the /proc/PID/maps output

To interpret the output of *Figure 7.8*, read it a line at a time. Each line represents a segment or mapping of the user-mode VAS of the process in question (in the preceding example, it's of the `cat` process). Each line consists of the following fields; to make it easier, I will show just one sample line of output (from *Figure 7.8*) whose fields we will label, refer to, and understand in the following notes:

```
start_uva   -   end_uva         mode,mapping  start-off  mj:mn inode#    image-name
558822d66000-558822d6a000  r-xp          00002000   08:01 7340181   /usr/bin/cat
```

Here, the entire line represents a segment, or, more correctly, a *mapping* within the process (user) VAS. `uva` is the user virtual address. The `start_uva` and `end_uva` for each mapping are displayed as the first two fields (or columns) and are separated by a hyphen character. So, the length of the mapping (segment) is easily calculated (`end_uva-start_uva` bytes). Thus, in the preceding line, `start_uva is 0x558822d66000` and `end_uva` is `0x558822d6a000` and the length can be calculated to be 16 KB; but what exactly does this segment represent within the process? Do read on...

The third field, `r-xp`, is a combination of two pieces of information:

- The first three letters represent the mode (permissions) of the segment (in the usual `rwx` notation).
- The next letter represents whether the mapping is a private one (`p`) or a shared one (`s`). Internally, this is set up by the fourth parameter to the `mmap()` system call, `flags`; it's really the `mmap()` system call that is internally responsible for creating every segment or mapping within a process!
- So, for the preceding sample segment shown, the third field being the value `r-xp`, we can now tell it's a text (code) segment and is a private mapping (as expected).

The fourth field `start-off` (here, it's the value `0x2000`) is the start offset from the beginning of the file whose contents have been mapped (for the size seen, 16 KB) into the process VAS. Obviously, this value is only valid for *file mappings*.

You can tell whether the current segment is a file mapping by glancing at the penultimate (sixth) field – the file inode number. For mappings that are not file-mapped – called **anonymous mappings** – it's always `0` (examples would be the mappings representing the heap or stack segments). In our preceding example line, it's a file mapping (that of `/usr/bin/cat`) and the offset from the beginning of that file is `0x2000` bytes (the length of the mapping, as we calculated in the preceding paragraph, is 16 KB).

The fifth field (`08:01`) is of the form `mj:mn`, where `mj` is the major number and `mn` is the minor number of the (block) device file where the image resides. Like the fourth field, it's only valid for file mappings, else it's simply shown as `00:00`; in our preceding example line, as it's a file mapping, the major and minor numbers (of the *block device* that represents the storage media that the file resides on) are `8` and `1`, respectively.

The sixth field (`7340181`) represents the inode number of the image file – the file whose contents are being mapped into the process VAS. The **inode** is the key data structure of the **VFS** (**Virtual File System**); it holds all metadata of the file object, everything except for its name (which is in the directory (or dot) file). Again, this value is only valid for file mappings and simply shows as `0` otherwise. This is, in fact, a quick way to tell whether the mapping is file-mapped or an anonymous mapping! In our preceding example mapping, clearly, it's a file mapping (that of `/usr/bin/cat`), and the inode number is `7340181`. Indeed, we can confirm this:

```
$ ls -i /usr/bin/cat
7340181 /usr/bin/cat
```

The seventh and last field represents the pathname of the file whose contents are being mapped into the user VAS. Here, as we're viewing the memory map of the `cat` process, the pathname (for the file-mapped segments) happens to be `/usr/bin/cat`. If the mapping represents a file, the file's inode number (the sixth field) shows up as a positive quantity; if not – meaning it's a pure memory or anonymous mapping with no backing store – the inode number shows up as `0` and the last field will be empty. (Of course, other files can show up as well: those of shared libraries, shared memory segments, and so on.)

It should by now be obvious, but we will point this out nevertheless as it is a key point: all the preceding addresses seen are *virtual*, not physical. Furthermore, they only belong to user space, hence they are termed UVAs and are always accessed (and translated) via the unique paging table metadata for that process. Also, the preceding screenshot was taken on a 64-bit (x86_64) Linux guest. Hence, here, we see 64-bit virtual addresses.

The way the virtual addresses are displayed here isn't as a full 64-bit number – for example, as 558822d66000 and not as 0000558822d66000. I want you to notice this because, as it's a UVA, the MSB 16 bits are zero! (Also, of course, the numbers are all in hexadecimal base.)

Right, while that covers how to interpret a particular segment or mapping (and the `[heap]` and `[stack]` lines are self-explanatory), there seem to be a few strange ones (glance at *Figure 7.8* again) – the `vvar`, `vdso`, and `vsyscall` mappings. Let's see what they mean.

The vsyscall page

Did you notice something a tad unusual in the output of *Figure 7.8*? The very last line there – the so-called `vsyscall` entry – maps a kernel page (by now, you know how we can tell: the MSB 16 bits of its start and end virtual addresses are set). Here, we just mention the fact that this is an (old) optimization for performing system calls. It works by alleviating the need to switch to kernel mode for a small subset of syscalls that don't really need to.

Currently, on the x86, these include the `gettimeofday()`, `time()`, and `getcpu()` system calls. Indeed, the `vvar` and `vdso` (aka Virtual Dynamic Shared Object) mappings above it are (slightly) modern variations on the same theme. If you are interested in finding out more about this, check out the *Further reading* section for this chapter.

A quick aside: a directory named `/proc/PID/map_files/` is another view; it displays only the *file* mappings within a process or thread. Here, each file-backed memory mapping (or segment) is displayed as a symbolic link to its corresponding file.

So, you've now learned how to examine and interpret the "raw" user space memory map of any given process by directly reading and interpreting the output of the `/proc/PID/maps` (pseudo) file for the process with the specified PID. There are other convenient frontends to do so; we'll now check out a few.

Frontends to view the process memory map

Besides the raw or direct format via `/proc/PID/maps` (which we saw how to interpret in the previous section), there are some wrapper utilities that help us more easily interpret the user-mode VAS. Among them are the additional (raw) `/proc/PID/smaps` pseudo file, the `pmap` and `smem` utilities, and my own utility (christened `procmap`).

The kernel provides detailed information on each segment or mapping via the `/proc/PID/smaps` pseudo file under `proc`. Do try running the command `cat /proc/self/smaps` to see this for yourself. You will notice that for each segment (mapping), a good amount of detailed information is provided on it. The man page on `proc(5)` helps explain the many fields seen.

For both the `pmap` and `smem` utilities, I refer you to the man pages on them for details. For example, with `pmap`, the man page informs us of the more verbose `-X` and `-XX` options:

```
$ pmap -h
[ … ]
-x, --extended show details
-X     show even more details
       WARNING: format changes according to /proc/PID/smaps
-XX    Show everything the kernel provides
[ … ]
```

Regarding the `smem` utility, the fact is that it does *not* show you the process VAS; rather, it's more about answering an FAQ: namely, ascertaining which process is taking up the most physical memory. It uses metrics such as **Resident Set Size (RSS)**, **Proportional Set Size (PSS)**, and **Unique Set Size (USS)** to provide a clearer picture. I will leave the further exploration of these utilities as an exercise to you, dear reader!

Incidentally, and FYI, we do cover a tiny bit on `smem` usage in *Chapter 8, Kernel Memory Allocation for Module Authors – Part 1*, in the *How can I see overall memory usage system-wide?* section.

Now, let's move on to exploring how we can leverage a useful utility – `procmap` – to view in quite a bit of detail both the kernel and user memory map of any given process.

The procmap process VAS visualization utility

As a small learning and teaching (and helpful during debug!) project, I have authored and hosted a project on GitHub going by the name of `procmap`, available here: `https://github.com/kaiwan/procmap` (do `git clone` it!). A snippet from its `README.md` file helps explain its purpose:

```
procmap is designed to be a console/CLI utility to visualize the complete
memory map of a Linux process, in effect, to visualize the memory mappings of
both the kernel and usermode Virtual Address Space (VAS).
```

```
It outputs a simple visualization, in a vertically-tiled format ordered
by descending virtual address (see screenshots below). The script has the
intelligence to show kernel and user space mappings as well as calculate and
show the sparse memory regions that will be present. Also, each segment or
mapping is scaled by relative size (and color-coded for readability). On 64-
bit systems, it also shows the so-called non-canonical sparse region or 'hole'
(typically close to 16,384 PB on the x86_64).
```

The section *Looking under the hood – the Hello, world C program*, covered a little on what actually goes on when the famous "Hello, world" K&R C program runs on Linux. Here, let's explore the *Hello, world* process's user VAS by leveraging the `procmap` utility!

My "Hello, world" C code is the usual one, except for the fact that after the `printf()`, we'd like the process to remain alive; so, I add the `pause()` system call (it keeps the caller blocking (asleep) until a signal is received):

```
#include <stdio.h>
#include <unistd.h>
int main()
{
    printf("Hello, world\n");
    pause();
}
```

(I also build with optimization, `-O2`, and *strip* the binary executable.) To add to the excitement, I perform this activity from within an SSH session on a Raspberry Pi 4 (Model B), running a standard Raspberry Pi OS (with the 5.15.84-v8+ kernel) configured as AArch64. Build it and then run it in the background:

```
$ ./helloworld &
[1] 835
$ Hello, world
```

Ah, it works. Right, with the process being alive, let's get down to business and run the *procmap* utility. How do we install and invoke it? Simple, see what follows (due to a lack of space, I won't show all the information, caveats, and more here; do try it out yourself):

```
$ git clone https://github.com/kaiwan/procmap
$ cd procmap
$ ./procmap
Usage: procmap [options] --pid=PID-of-process-to-show-memory-map-of
Options:
--only-user      : show ONLY the usermode mappings or segments (not kernel VAS)
--only-kernel    : show ONLY the kernel-space mappings or segments (not user
VAS)
  [default is to show BOTH]
--export-maps=filename
```

```
    write all map information gleaned to the file you specify in CSV
--export-kernel=filename
    write kernel information gleaned to the file you specify in CSV
--verbose         : verbose mode (try it! see below for details)
--debug           : run in debug mode
--version|--ver : display version info.
See the config file as well.
[…]
```

Do note that this procmap utility is not the same as the procmap utility provided by BSD Unix. Also, FYI, it requires the bc, smem, and yad utilities (among others) to be installed.

When I run the procmap utility with only --pid=<PID> (the one mandatory parameter), it will display both the kernel and user space VASs of the specified process. Now, as we have not yet covered the details regarding the kernel VAS (or segment), I won't show the kernel-space detailed output here; let's defer that to the upcoming *Examining the kernel VAS* section. As we proceed, you will find partial screenshots of only the user VAS output from the procmap utility. The complete output can be quite lengthy, depending, of course, on the process in question; do try it out for yourself.

As you'll see, procmap attempts to provide a basic visualization of the complete process memory map – both kernel and user space VASs in a vertically tiled format (as mentioned, here we just display truncated screenshots):

```
$ procmap --pid=$(pgrep helloworld)
[i] will display memory map for process PID=835

Detected machine type: ARM-64, 64-bit system & OS

[===================---      P R O C M A P      ---==================]
Process Virtual Address Space (VAS) Visualization utility
https://github.com/kaiwan/procmap

Fri Jan 13 09:54:12 IST 2023
[=====---  Start memory map for 835:helloworld  ---=====]
[Pathname: /home/kai1/kaiwanTECH/L1_sysprg_trg/helloworld/helloworld ]
+------------------ K E R N E L   V A S    end kva  ------------------+ ffffffffffffffff
|<... K sparse region ...> [   8.03 GB,--- ]                          |
|                                                                     |
```

Figure 7.9: Partial screenshot: the first line of the kernel VAS output from the procmap utility (running on an AArch64)

Notice, from the preceding (partial) screenshot, a few things:

- The procmap (Bash) script auto-detects that we're running on an ARM-64 (AArch64) system.
- Though we're not focused on it right now, the output of the kernel VAS appears first; this is natural as we show the output ordered by descending virtual address (*Figure 7.1*, *Figure 7.5*, and *Figure 7.7* reiterate this)

- You can see that the very first line (after the `KERNEL VAS ...` header) corresponds to a KVA at the very top of the VAS – the value `0xffff ffff ffff ffff` (as we're executing it on a 64-bit Linux).

Moving on to the next part of the `procmap` output, let's look at a truncated view of the upper end of the *user* VAS of the `helloworld` process:

```
+------------------          U S E R   V A S     end uva  ------------------+ 0000007fffffffff
|<... Sparse Region ...> [ 664.56 MB,---,-,0x0]                              |
|                                                                            |
|                                                                            |
|                                                                            |
|                                                                            |
|                                                                            |
+----------------------------------------------------------------------------+ 0000007fd6770000
|              [stack]   [ 132 KB,rw-,p,0x0]                                 |
|                                                                            |
+----------------------------------------------------------------------------+ 0000007fd674f000
|<... Sparse Region ...> [ 976.04 MB,---,-,0x0]                              |
|                                                                            |
|                                                                            |
|                                                                            |
|                                                                            |
|                                                                            |
+----------------------------------------------------------------------------+ 0000007f99743000
|/usr/lib/aarch64-linux-gnu/ld-2.31.so  [   8 KB,rw-,p,0x22000]              |
+----------------------------------------------------------------------------+ 0000007f99741000
|/usr/lib/aarch64-linux-gnu/ld-2.31.so  [   4 KB,r--,p,0x21000]              |
+----------------------------------------------------------------------------+ 0000007f99740000
|               [vdso]  [   4 KB,r-x,p,0x0]                                  |
+----------------------------------------------------------------------------+ 0000007f9973f000
|               [vvar]  [   8 KB,r--,p,0x0]                                  |
+----------------------------------------------------------------------------+ 0000007f9973d000
|<... Sparse Region ...>  [  48 KB,---,-,0x0]                                |
+----------------------------------------------------------------------------+ 0000007f99731000
|/usr/lib/aarch64-linux-gnu/ld-2.31.so  [ 136 KB,r-x,p,0x0]                  |
|                                                                            |
+----------------------------------------------------------------------------+ 0000007f9970f000
|          [-unnamed-]  [  20 KB,rw-,p,0x0]                                  |
+----------------------------------------------------------------------------+ 0000007f9970a000
|/usr/lib/aarch64-linux-gnu/libc-2.31.so  [   8 KB,rw-,p,0x160000]          |
+----------------------------------------------------------------------------+ 0000007f99708000
|/usr/lib/aarch64-linux-gnu/libc-2.31.so  [  16 KB,r--,p,0x15c000]          |
+----------------------------------------------------------------------------+ 0000007f99704000
|/usr/lib/aarch64-linux-gnu/libc-2.31.so  [  60 KB,---,p,0x15d000]          |
+----------------------------------------------------------------------------+ 0000007f996f5000
|/usr/lib/aarch64-linux-gnu/libc-2.31.so [   1.36 MB,r-x,p,0x0]              |
|                                                                            |
|                                                                            |
```

Figure 7.10: Partial screenshot: first few lines (high end) of the process user VAS output from the procmap utility (running on an AArch64)

Figure 7.10 is a partial screenshot of the `procmap` output and shows the user space VAS; at the very top of it, you can see the (high) end UVA.

On our AArch64 system (recall, this is arch-dependent), the (high) `end_uva` value is

`0x0000 007f ffff ffff` (exactly as seen; the `start_uva` is, of course, `0x0`; see the next screenshot). Notice, in *Figure 7.10*, some interesting mappings: the stack (of `main()`), sparse regions in the VAS (holes, illegal-to-access and unused regions, in effect), and mappings for the loader (`ld*.so`) and, of course, for the *glibc* shared library – the reason *Hello, world* works!

How does procmap figure out the precise address values? Ah, it's fairly sophisticated: for the kernel-space memory information, it uses a kernel module (an LKM!) to query the kernel and sets up an output file depending on the system architecture; user space details, of course, come from the /proc/PID/maps direct (raw) procfs pseudo file interface.

As an aside, the kernel component of procmap, a kernel module, sets up a way to interface with user space – the procmap scripts – by creating and setting up a debugfs pseudo file. Hey, it's open source; please go ahead and git clone it and learn how it works!

The following figure shows a partial screenshot of the low end of the user mode VAS for our "Hello, world" process, right down to the lowest UVA, 0x0:

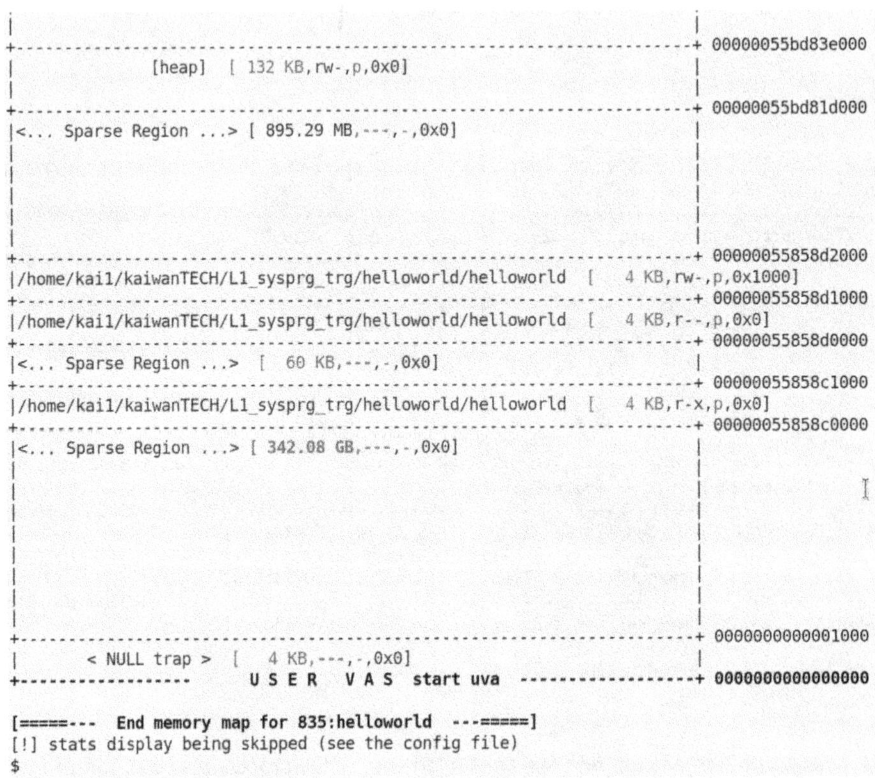

Figure 7.11: Partial screenshot: last few lines (low end) of the process user VAS output from the procmap utility (running on an AArch64)

The last mapping, a single page, is, as expected, the *null trap* page (from UVA 0x1000 to 0x0; we will explain its purpose in the upcoming *The null trap page* section). In this figure, notice the mappings for the heap and the helloworld program code/data.

The procmap utility, after displaying the memory map(s), if enabled in its config file, then calculates and displays a few statistics: this includes the sizes of both the kernel and user-mode VASs, the amount of memory taken up by sparse regions (on 64-bit, as in the preceding example, it's usually the vast majority of the space!) as an absolute number and as a percentage, the amount of physical RAM reported, and finally, the memory usage details for this particular process as reported by the ps and smem utilities.

You will find, in general, on a 64-bit system (see *Figure 7.5*), that the *sparse* (empty) memory regions of the process VAS can take up close to 100% of the available address space! This implies that, typically, well over 99% of the virtual memory space is sparse (empty)! This is the reality of the simply enormous VAS on a 64-bit system. Only a tiny fraction of its gigantic 16 EB of VAS is actually in use (the case on both the AArch64 and x86_64). Of course, the actual amounts of empty and used user-mode VAS depend on the size of the application process.

Being able to clearly visualize the process VAS can aid greatly when debugging or analyzing issues at a deeper level.

If you're reading this book in its hard-copy format, be sure to download the full-color PDF of diagrams/figures from the publisher's website: https://packt.link/gbp/9781803232225.

You will also see that the statistics printed out at the end of the output (they're enabled by default) show the number of **Virtual Memory Areas (VMAs)** set up for the target process. VMAs? The following section briefly explains what a VMA is. Let's get to it!

Understanding VMA basics

In the output of /proc/PID/maps, each line of the output is extrapolated from a kernel metadata structure called a *virtual memory area*, or a VMA. It's quite straightforward, really: the kernel uses the VMA data structure to abstract in code what we have been calling a *segment* or *mapping*. Thus, for every single mapping in the user VAS, there is a VMA object maintained by the OS. Please realize that *only user space mappings* are governed by the kernel metadata structure called the VMA; the kernel VAS itself has no VMAs.

So, how many VMAs will a given process have? Well, it's equal to the number of mappings (segments) in its user VAS. In my sample run of the helloworld process, it reported 15 segments or mappings, implying that there were 15 VMA metadata objects – representing the 15 user space segments or mappings – for this process in kernel memory.

Programmatically speaking, the kernel maintains a VMA "chain" (technically a red-black tree data structure for efficiency reasons) via the task structure rooted at current->mm->mmap. Why is the pointer called mmap? It's very deliberate: every time an mmap() system call – that is, a memory mapping operation – is performed, the kernel generates a mapping (or "segment") within the calling process's VAS, and thus, a VMA metadata object representing it.

The VMA metadata structure is akin to an umbrella encompassing the mapping and includes all required information for the kernel to perform various kinds of memory management: servicing page faults (very common), caching the contents of a file during I/O into (or out of) the kernel page cache, and so on.

Page fault handling is a very important OS activity, whose algorithm makes up quite a bit of usage of the kernel VMA objects; in this book, though, we don't delve into these details as it's largely transparent to kernel module/driver authors.

Just to give you a feel for it, we will show a few members of the kernel VMA data structure in the following code snippet: the comments alongside help explain their purpose:

```
// include/linux/mm_types.h
struct vm_area_struct {
 /* The first cache line has the info for VMA tree walking. */
    unsigned long vm_start;  /* Our start address within vm_mm. */
    unsigned long vm_end;    /* The first byte after our end
                                address within vm_mm. */
    struct mm_struct *vm_mm; /* The address space we belong to.*/
    [...]
    pgprot_t vm_page_prot;
    unsigned long vm_flags;      /* Flags, see mm.h. */
    [...]
    /* Function pointers to deal with this struct. */
    const struct vm_operations_struct *vm_ops;

    /* Information about our backing store: */
    unsigned long vm_pgoff; /* Offset (within vm_file) in
      PAGE_SIZE units */
    struct file * vm_file; /* File we map to (can be NULL). */
    [...]
} __randomize_layout;
```

It should now be clearer as to how `cat /proc/PID/maps` really works under the hood: when user space does, say, `cat /proc/self/maps`, a `read()` system call is (eventually) issued by the `cat` process; this results in it switching to kernel mode and running the `read()` system call code within the kernel with kernel privilege. Here, the kernel **Virtual Filesystem (VFS)** switch redirects control to the appropriate `procfs` callback handler (a function that is registered within a `proc_ops` structure in recent kernels, from 5.6.0 onward). This code iterates (loops) over every VMA metadata structure (for the process context running: in other words, for `current`, which is our `cat` process, of course), sending the relevant detail information from each VMA object back to user space. The `cat` process then faithfully dumps the data received via the read to `stdout`, and thus we see it: all the segments or mappings of the process – in effect, the memory map of the user-mode VAS!

Great, we now conclude this section; we've covered details on examining the process user VAS. This knowledge helps not only with understanding the precise layout of user-mode VAS but also with debugging user space issues!

Now, and about time, let's move on to understanding another critical aspect of memory management – the detailed layout of the kernel VAS!

Examining the kernel VAS

As we have talked about in the preceding chapter, and as seen in *Figure 7.7*, it's critical to understand that all processes have their own unique user VAS but *share* the kernel space – what we call the kernel segment or kernel VAS. Let's begin this section by starting to examine some common (arch-independent) regions of the kernel VAS.

The kernel VAS's memory layout is very arch (CPU)-dependent. Nevertheless, all architectures share some commonalities. The following basic diagram represents both the user VAS and the kernel VAS (in a horizontally tiled format), as seen on a typical x86_32 (or IA-32) with a 3:1 (GB) VM split:

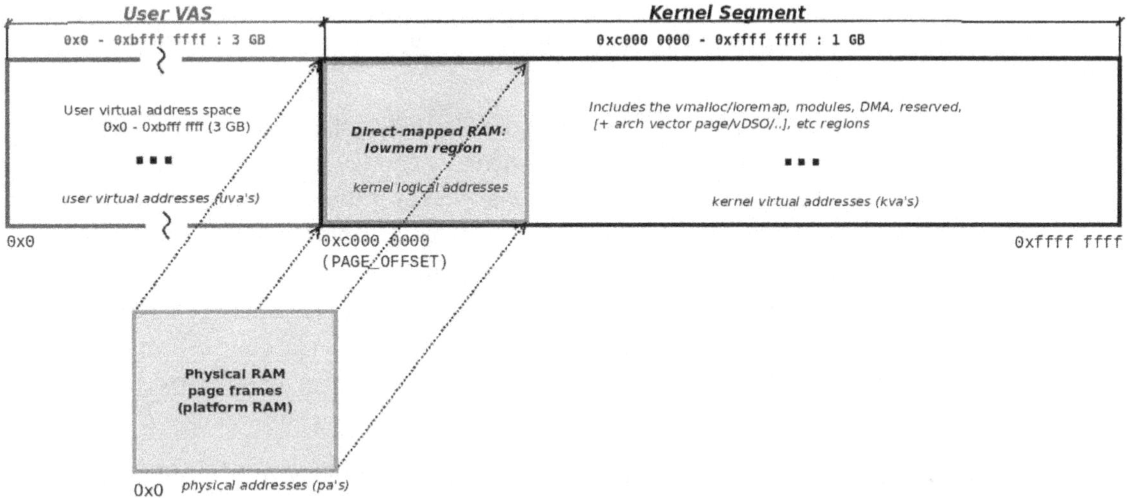

Figure 7.12: User and kernel VASs on an x86_32 with a 3:1 (GB) VM split with a focus on the lowmem region; this figure is deliberately simplistic

Let's go over each region of the process VAS (from left to right, as seen in *Figure 7.12*):

- **The user mode VAS**: This is the user VAS. We have covered it in detail in the preceding chapter as well as within earlier sections in this chapter; in this particular example, it takes 3 GB of VAS (its UVAs range from `0x0` to `0xbfff ffff`).
- **The kernel VAS (or kernel segment)**: In this particular example, we have 1 GB of kernel VAS (with its KVAs ranging from `0xc000 0000` to `0xffff ffff`); let's examine individual portions of it now.

- **The "lowmem" region:** This is the region into which the OS directly-maps platform (system) RAM into the kernel VAS. (*Figure 7.12* attempts to clearly convey this. We will cover this key topic in more detail in the *Direct-mapped RAM and address translation* section. If you feel it helps, you can read that section first and then return here). Skipping a bit ahead for now, let's just understand that the base location in the kernel VAS where platform RAM is mapped is specified by a kernel macro called `PAGE_OFFSET`. The precise value of this macro is very arch-dependent; we will leave this discussion to a later section. For now, we ask you to just take it on faith that on the x86_32 with a 3:1 (GB) VM split, the value of `PAGE_OFFSET` is `0xc000 0000`.

Obviously, the size of the kernel's so-called lowmem region is equal to the amount of RAM on the system. (Well, at least the amount of RAM as seen by the kernel; enabling the *kdump* facility, for example, has the OS reserve some specified amount of RAM very early on.) The virtual addresses that make up this region are termed **kernel logical addresses** as they are at a fixed offset from their physical counterparts. The core kernel and device drivers can allocate (physically contiguous!) memory from this region via various APIs (they include the page allocator APIs and the popular slab APIs – the `k{m|z}alloc()`. Patience please, we will cover precisely these APIs in detail in the following two chapters). The kernel static text (code), data, and BSS (uninitialized data) memory regions also reside within this lowmem region.

Right, following the lowmem region, other regions do exist; though not explicitly shown in *Figure 7.12* (for now), a couple of key ones are:

- **The kernel vmalloc region:** This is a region of the kernel VAS that is completely virtual. Core kernel and/or device driver code can allocate virtually contiguous memory from this region using the `vmalloc()` (and friends) API. Again, we will cover this in detail in *Chapter 8, Kernel Memory Allocation for Module Authors – Part 1*, and *Chapter 9, Kernel Memory Allocation for Module Authors – Part 2*. This is also the so-called `ioremap` space.
- **The kernel modules space:** A region of kernel VAS is set aside for memory taken up by the static text and data of LKMs. When you perform an `insmod` (or `modprobe`), the underlying kernel code of the resulting `[f]init_module()` system call allocates memory from this region (typically via the `vmalloc()` API) and loads the kernel module's (static) code and data there.

The preceding figure (*Figure 7.12*) is deliberately left simplistic and even a bit vague as the exact kernel virtual memory layout is very arch-dependent. We'll put off the temptation to draw a detailed diagram for a bit.

Instead, to make this discussion less pedantic, more practical, and useful, we'll present, in a soon-to-come section, a kernel module that queries and prints relevant information regarding the kernel VAS layout. Only then, once we have actual values for various regions of the kernel VAS for a particular architecture, will we present a detailed diagram depicting this.

Pedantically (and as can be seen in *Figure 7.10*), the addresses belonging to the kernel *lowmem* region are termed *kernel logical addresses* (they're at a fixed offset from their physical counterparts), whereas the addresses for the remainder of the kernel VAS are termed *KVAs*. Though this distinction is made here, please realize that, for all practical purposes, it's a rather pedantic one: we don't really differentiate one from the other and often simply refer to all addresses within the kernel VAS as KVAs.

Before we get to writing a module that interrogates the kernel VAS though, there are several other pieces of information to cover. Let's begin with a peculiarity, mostly brought about by the limitations of the 32-bit architecture: the so-called *high memory region* of the kernel VAS on 32-bit systems.

High memory on 32-bit systems

Keeping in mind the kernel *lowmem* region that we briefly discussed previously, an interesting observation ensues. On a 32-bit system with, say, a 3:1 (GB) VM split (just as *Figure 7.12* depicts), a system with (say) 512 MB of RAM will have this RAM direct-mapped into the kernel starting at `PAGE_OFFSET` (3 GB or KVA `0xc000 0000`) for 512 MB. This is quite clear.

But think about it: what would happen if the system had a lot more RAM, say, 2 GB? Now, it's obvious that we cannot direct-map the whole of RAM into the kernel lowmem region. It just cannot fit (as, in this example, the entire available kernel VAS is just a gigabyte and RAM is 2 GB)! So, on a 32-bit Linux OS, a certain amount of memory (typically 896 MB on the IA-32) is allowed to be direct-mapped and thus falls into the lowmem region. The remaining RAM is *indirectly mapped* into another memory "zone" called `ZONE_HIGHMEM` (we think of this as a high-memory region or *zone* as opposed to lowmem; more about memory zones will follow in a later section, *Zones*). More correctly, as the kernel now finds it impossible to direct-map all physical memory at once, it sets up a (virtual) region where it can set up and use temporary virtual mappings of that RAM (typically via calls to the `kmap()` and `kunmap()` APIs). This is the so-called high-memory region.

Don't get confused by the phrase "high memory." It's not necessarily placed "high" in the kernel VAS, nor is it the "high-memory" term used to describe memory over the hole at 640 KB on the PC. Instead, the `high_memory` global variable, valid only on 32-bit, represents the upper bound of the kernel's lowmem region. More on this follows in a later section, *Macros and variables describing the kernel VAS layout*.

Nowadays, though (and especially with 32-bit systems being used more and more infrequently), these concerns completely disappear on 64-bit Linux. Think about it: for example, on the x86_64 running a 64-bit Linux, the kernel VAS size is a whopping 128 TB (that's 131,072 GB!). As far as I know, no *single* system (or node) in existence has anywhere close to this much RAM.

As of this writing, NASA's Pleiades Supercomputer specifies a maximum of 128 GB of RAM per node (ref: `https://www.nas.nasa.gov/hecc/resources/pleiades.html`).

Next, when the sparse memory model is being employed (the usual case; the upcoming *An introduction to physical memory models* section covers this), the maximum supported bits in the physical address are given by the macro `MAX_PHYSMEM_BITS`. On x86_64, it's typically the value 46, implying the maximum amount of RAM supported on the machine is 2^{46} bytes – that is, 64 TB. Hence, all platform RAM can indeed (easily) be direct-mapped into the 128 TB kernel VAS and the need for `ZONE_HIGHMEM` (or equivalent silly workarounds) simply disappears. The fact is, deprecating this high memory region on 32-bit systems is something the majority of kernel folk would like to ultimately do, but as of now, it's still very much there, allowing older (typically ARM-32) systems with over a gigabyte of RAM to continue running on Linux (see this article for more: *An end to high memory?, LWN, Feb 2020*: `https://lwn.net/Articles/813201/`).

Again, the official kernel documentation provides details on this arcane "high-memory" region (relevant to 32-bit); take a look if interested: `https://www.kernel.org/doc/Documentation/vm/highmem.txt`.

Okay, let's now tackle the thing we've been itching to do – writing a kernel module (an LKM) to delve into some details regarding the kernel VAS.

Writing a kernel module to show information about the kernel VAS

As we have learned, the kernel VAS consists of various regions. Some are common to all architectures (arch-independent): they include the lowmem region (which contains, among other things, the uncompressed kernel image – its code and data), then we have the kernel modules region, `vmalloc/ioremap` regions, and so on.

The precise location within the kernel VAS where these regions lie, and indeed which regions may be present, is very arch (CPU)-dependent. To help understand and pin it down for any given system, let's develop a kernel module that queries and prints – in an arch-dependant fashion where required – various details regarding the kernel VAS (in fact, if asked to, it also prints some useful user space memory details as well).

Now, for you to query and print this information, you must first get familiar with some key kernel macros and globals; let's do so in the next section.

Macros and variables describing the kernel VAS layout

To write a kernel module that displays relevant kernel VAS information, we need to know how exactly to interrogate the kernel regarding these details.

In this section, we will briefly describe a few key macros and variables within the kernel representing the memory of the kernel VAS (on most architectures, *in descending order by KVA*):

- **The vector table** is a common OS data structure – it's an array of function pointers (aka a switching or jump table). It is arch-specific: ARM-32 uses it to initialize its vectors such that when a processor exception or mode change (such as an interrupt, syscall, page fault, MMU abort, and so on) occurs, the processor knows what code to run. The macro `VECTORS_BASE` is shown in the following table:

Macro or variable	Interpretation
`VECTORS_BASE`	Typically ARM-32 only; start KVA of a kernel vector table spanning 1 page

Table 7.1: Macros and variables describing the kernel VAS layout: the vector table (AArch32), optional

- **The fix map region** is a range of compile-time special or reserved virtual addresses; they are employed at boot time to fix, into the kernel VAS, required kernel elements that must have memory available for them. Typical examples include the setup of initial kernel page tables, early `ioremap` and `vmalloc` regions, and so on. Again, it's an arch-dependent region and is thus used differently on different CPUs:

Macro or variable	Interpretation
`FIXADDR_START`	Start KVA of the kernel fix map region spanning `FIXADDR_SIZE` bytes

Table 7.2: Macros and variables describing the kernel VAS layout: the fix map region

- **Kernel modules** are allocated memory at load time – for their static text and data – within a specific range in the kernel VAS. The precise location of the kernel module region varies with the architecture. On AArch32 systems, in fact, it's placed just above the user VAS, while on 64-bit systems, it's usually higher up in the kernel VAS:

Kernel modules (LKMs) region	Memory allocated from here for static code + data of LKMs
`MODULES_VADDR`	Start KVA of the kernel modules region
`MODULES_END`	End KVA of the kernel modules region; the region size is `MODULES_END - MODULES_VADDR`

Table 7.3: Macros and variables describing the kernel VAS layout: the kernel modules region

- **KASAN**: The modern kernel (4.0 onward for x86_64, 4.4 for AArch64, and 5.11 for AArch32) employs a powerful mechanism to detect and report memory issues (bugs). It's based on the user space **Address SANitizer** (**ASAN**) code base and is thus called **Kernel Address SANitizer** (**KASAN**). Its power lies in ably, via a technique called **Compile-Time Instrumentation** (**CTI**), detecting memory defects (bugs) such as **Out Of Bounds** (**OOB**) accesses (including buffer over/underflows), **Use After Free** (**UAF**) and double-free accesses. Up to 5.11, however, it works *only on 64-bit Linux* and requires a rather large **shadow memory region** (of a size that is one-eighth that of the kernel VAS, whose extents we show if it's enabled; in 5.11, though, Linus Walleij has introduced an optimized version of KASAN for the ARM-32).

- It's a kernel configuration feature (`CONFIG_KASAN`) and is typically enabled only for debug (bug hunting!) purposes (it's crucial to keep it enabled during debugging and testing!):

KASAN shadow memory region	[Optional] (only if CONFIG_KASAN is defined)
`KASAN_SHADOW_START`	Start KVA of the KASAN region
`KASAN_SHADOW_END`	End KVA of the KASAN region; the size is `KASAN_SHADOW_END - KASAN_SHADOW_START`

Table 7.4: Macros and variables describing the kernel VAS layout: the KASAN shadow memory region, optional

- The **vmemmap region** is a region within the kernel VAS that is employed when, one, the physical memory model is *sparsemem* (typically the default on modern systems), and two, when the method employed to map a **Page Frame Number** (**PFN**) to its corresponding virtual page (denoted via `struct page`) is the vmemmap one:

Vmemmap region	[Optional] (only if CONFIG_ SPARSEMEM_VMEMMAP is defined)
`VMEMMAP_START`	Start KVA of the vmemmap region
`VMEMMAP_SIZE`	Size of the kernel's vmemmap region; this macro seems to be defined only for AArch64 though...

Table 7.5: Macros and variables describing the kernel VAS layout: the vmemmap region, optional

More details can be found in the section *An introduction to physical memory models.*

- **The vmalloc region** is the kernel region from where memory for the `vmalloc()` (and friends) APIs are allocated; we will cover various memory allocation APIs in detail in the next two chapters:

The vmalloc region	For memory allocated via vmalloc() and friends
`VMALLOC_START`	Start KVA of the `vmalloc` region
`VMALLOC_END`	End KVA of the `vmalloc` region; size is `VMALLOC_END - VMALLOC_START`

Table 7.6: Macros and variables describing the kernel VAS layout: the vmalloc region

- **The lowmem region** – direct-mapped RAM (i.e., RAM mapped into the kernel VAS on a `1:1 :: physical-page-frame:kernel-logical/virtual-page` basis). It's the region where the Linux kernel maps and manages (typically) all RAM.

- Also, it's often set up as `ZONE_NORMAL` within the kernel (we will cover zones a bit later):

Lowmem region	Direct-mapped memory region
`PAGE_OFFSET`	Start KVA of the lowmem region; also represents the start of the kernel VAS/segment on some architectures and is (often) the VM split value on 32-bit.
`high_memory`	End KVA of the lowmem region, the upper bound of direct-mapped memory. In effect, this value minus `PAGE_OFFSET` is the amount of (platform) RAM on the system (careful though, this is not necessarily the case on all architectures); not to be confused with `ZONE_HIGHMEM`.

Table 7.7: Macros and variables describing the kernel VAS layout: the lowmem region

- **The highmem region** or zone is an optional region. It might exist on some 32-bit systems (typically, where the amount of RAM present is greater than the size of the kernel VAS itself). It's often set up as `ZONE_HIGHMEM` in this case (we will cover zones a bit later. Also, you can refer back to more on this highmem region in the earlier section entitled *High memory on 32-bit systems*):

Highmem region (only possible on 32-bit)	[Optional] HIGHMEM may be present on some 32-bit systems
`PKMAP_BASE`	Start KVA of the highmem region, runs until `LAST_PKMAP` pages; represents the kernel mapping of so-called high-memory pages (only possible on 32-bit)

Table 7.8: Macros and variables describing the kernel VAS layout: the highmem region, optional

- The (uncompressed) kernel image itself – its code, init, and data regions – are always present but are private symbols and, thus, unavailable to kernel modules. Thus, we don't even attempt to print them in the code of our upcoming kernel module:

Kernel (static) image	Uncompressed kernel image region (symbols not exported, thus unavailable to kernel modules)
`_text, _etext`	Start and end KVAs (respectively) of the kernel text (code) region
`__init_begin, __init_end`	Start and end KVAs (respectively) of the kernel `init` section region
`_sdata, _edata`	Start and end KVAs (respectively) of the kernel static data region
`__bss_start, __bss_stop`	Start and end KVAs (respectively) of the kernel BSS (uninitialized data) region

Table 7.9: Macros and variables describing the kernel VAS layout: the symbols describing the kernel image

- **The user VAS:** The last item, of course, is the process user VAS. It's below the kernel VAS (when ordered by descending virtual address) and is of size `TASK_SIZE` bytes. It was discussed in detail earlier in this chapter (in the *Examining the process VAS* section):

User VAS	User Virtual Address Space (VAS)
User-mode VAS of size `TASK_SIZE` bytes	(Examined in detail earlier via `procfs` and/or our `procmap` utility script); the kernel macro `TASK_SIZE` represents the size of the user VAS (in bytes).

Table 7.10: Macros and variables describing the kernel VAS layout: the user VAS

Well, that's that; we've seen several kernel macros and variables that, in effect, describe the kernel VAS. Moving on to the code of our kernel module, you'll soon see that its `init` method calls two functions (that matter):

- `show_kernelvas_info()`, which prints relevant kernel VAS details
- `show_userspace_info()`, which prints relevant user VAS details (its execution is optional, decided via a kernel parameter, and is off by default)

We will start by describing the kernel VAS function and seeing its output. Also, the way the `Makefile` is set up, it links into the object file of our kernel "library" code, `klib.c`, and generates a kernel module object called `show_kernel_vas.ko`.

Trying it out – viewing kernel VAS details

For clarity, we will show only relevant parts of the source code in this section. Do clone and use the complete code from this book's GitHub repository. Also, recall the `procmap` utility mentioned earlier; it has a kernel component, an LKM, which indeed does a similar job to this one – making kernel-level information available to user space. With it being a bit more sophisticated, we won't delve into its code here; seeing the code of the following demo kernel module `show_kernel_vas` is sufficient for now:

```
// ch7/show_kernel_vas/kernel_vas.c
[...]
static void show_kernelvas_info(void)
{
    unsigned long ram_size;

#if LINUX_VERSION_CODE >= KERNEL_VERSION(5, 0, 0)
    ram_size = totalram_pages() * PAGE_SIZE;
#else // totalram_pages() undefined on the BeagleBone running an older 4.19 
kernel..
    ram_size = totalram_pages * PAGE_SIZE;
#endif
    pr_info("PAGE_SIZE = %lu, total RAM ~= %lu MB (%lu bytes)\n",
        PAGE_SIZE, ram_size/(1024*1024), ram_size);
```

We first query the amount of physical RAM on the system; this is achieved via the `totalram_pages()` inline function in the `<linux/mm.h>` header (from kernel version 5.0). Having the amount of RAM helps us accurately calculate the direct-mapped lowmem region of the kernel VAS. We do verify whether the system is an AArch64 with a 64K page size and a 48-bit address space for each VAS (via the `VA_BITS` macro); if so, we don't handle this case (see the *caveats* mentioned at the end of this section as well). Continuing with the code:

```
    pr_info("Some Kernel Details [by decreasing address; values are
            approximate]\n"
            "+-------------------------------------------------------
            ------+\n");
#if defined(CONFIG_ARM)
  /* On ARM, the definition of VECTORS_BASE turns up only in kernels >= 4.11 */
#if LINUX_VERSION_CODE > KERNEL_VERSION(4, 11, 0)
    pr_info("|vector table: "
      " %px - %px                        | [%5zu KB]\n",
    SHOW_DELTA_K((void *)VECTORS_BASE, (void *
          )(VECTORS_BASE + PAGE_SIZE)));
#endif
#endif
```

The preceding code snippet displays the extent of the ARM (AArch32) vector table. Of course, it's conditional; the output's only emitted on an AArch32 – hence the `#if defined(CONFIG_ARM)` preprocessor directive. Also, our use of the `%px, %zu` printk format specifiers ensures the code is portable.

The `SHOW_DELTA_*()` macros used here in this demo kernel module are defined in our `convenient.h` header and are helpers that enable us to easily display the low and high values passed to it, calculate the delta (the difference) between the two quantities passed, and display it in a human-readable manner; here's the relevant code:

```
// convenient.h
[...]
/* SHOW_DELTA_*(low, hi) :
 * Show the low val, high val and the delta (hi-low) in either bytes/KB/MB/GB,
as required.
 * Inspired from raspberry pi kernel src: arch/arm/mm/init.c:MLM()
 */
#define SHOW_DELTA_b(low, hi) (low), (hi), ((hi) - (low))
#define SHOW_DELTA_K(low, hi) (low), (hi), (((hi) - (low)) >> 10)
#define SHOW_DELTA_M(low, hi) (low), (hi), (((hi) - (low)) >> 20)
#define SHOW_DELTA_G(low, hi) (low), (hi), (((hi) - (low)) >> 30)
#define SHOW_DELTA_MG(low, hi) (low), (hi), (((hi) - (low)) >> 20), (((hi) -
(low)) >> 30)
#if (BITS_PER_LONG == 64)
```

```
#define SHOW_DELTA_MGT(low, hi) (low), (hi), (((hi) - (low)) >> 20), (((hi) -
(low)) >> 30), (((hi) - (low)) >> 40)
#else // 32-bit
#define SHOW_DELTA_MGT(low, hi) (low), (hi), (((hi) - (low)) >> 20), (((hi) -
(low)) >> 30)
#endif
```

In the following code, we show the code snippet that emits messages to the kernel log (via `printk`) describing the extents of the following regions:

- Kernel module region
- (Optional) KASAN region
- The vmalloc region
- The lowmem, and a possible highmem, region

Regarding the kernel modules region, as explained in the detailed comment in the following source, we try and keep the order by descending KVAs:

```
// ch7/show_kernel_vas/kernel_vas.c
[...]
/* kernel module region
 * For the modules region, it's high in the kernel segment on typical 64-
 * bit systems, but the other way around on many 32-bit systems
 * (particularly ARM-32); so we rearrange the order in which it's shown
 * depending on the arch, thus trying to maintain a 'by descending address'
ordering. */
#if (BITS_PER_LONG == 64)
    pr_info("|module region: "
        " %px - %px | [%9zu MB]\n",
            SHOW_DELTA_M((void *)MODULES_VADDR, (void *)MODULES_END));
#endif

#ifdef CONFIG_KASAN     /* KASAN region: Kernel Address  SANitizer */
    pr_info("|KASAN shadow: "
#if (BITS_PER_LONG == 64)
        " %px - %px     | [%9zu MB = %6zu GB ~= %3zu TB]\n",
        SHOW_DELTA_MGT((void *)KASAN_SHADOW_START, (void *)KASAN_SHADOW_END)
#else  // 32-bit with KASAN enabled
        " %px - %px                     | [%9zu MB = %6zu GB]\n",
        SHOW_DELTA_MG((void *)KASAN_SHADOW_START, (void *)KASAN_SHADOW_END)
#endif
    );
#endif
```

```
    [ ... ]
  /* vmalloc region */
    pr_info("|vmalloc region: "
#if (BITS_PER_LONG == 64)
    " %px - %px     | [%9zu MB = %6zu GB ~= %3zu TB]\n",
    SHOW_DELTA_MGT((void *)VMALLOC_START, (void *)VMALLOC_END)
#else  // 32-bit
    " %px - %px                     | [%5zu MB]\n",
    SHOW_DELTA_M((void *)VMALLOC_START, (void *)VMALLOC_END)
#endif
    );
```

Notice how the code tends to be arch (CPU) and platform-dependent. Now, let's see the code to show the extents of the kernel *lowmem* region – in effect, the region where the platform RAM is direct-mapped into the kernel VAS:

```
  /* lowmem region (RAM direct-mapping)  */
    pr_info("|lowmem region: "
#if (BITS_PER_LONG == 32)
    " %px - %px                     | [%5zu MB]\n"
    "|                    ^^^^^^^^                      |\n"
    "|                   PAGE_OFFSET                    |\n",
#else
    " %px - %px     | [%9zu MB]\n"
    "|                    ^^^^^^^^^^^^^^^^              |\n"
    "|                      PAGE_OFFSET                 |\n",
#endif
SHOW_DELTA_M((void *)PAGE_OFFSET, (void *)(PAGE_OFFSET) + ram_size));
```

A look at how a possible (on 32-bit) highmem region is seen:

```
/* (possible) highmem region; may be present on some 32-bit systems */
#if defined(CONFIG_HIGHMEM) && (BITS_PER_LONG==32)
    pr_info("|HIGHMEM region:      "
        " %px - %px                     | [%5zu MB]\n",
        SHOW_DELTA_M((void *)PKMAP_BASE, (void *)((PKMAP_BASE) +
            (LAST_PKMAP * PAGE_SIZE))));
#endif
```

Finally, a quick look at the code that determines the extents of the kernel module region:

```
[ ... ]
#if (BITS_PER_LONG == 32)    /* modules region: see the comment above reg this
*/
```

```
    pr_info("|module region:       "
        " %px - %px                    | [%5zu MB]\n",
        SHOW_DELTA_M((void *)MODULES_VADDR, (void *)MODULES_END));
#endif
    pr_info(ELLPS);
}
```

Right, let's build and insert our LKM on an ARM-32 Raspberry Pi Zero W running the stock Raspberry Pi OS (here, it so happens that it too is running a 6.1-based kernel!). The following screenshot shows it being set up and then the kernel log:

```
rpi0w $ sudo rmmod show_kernel_vas 2>/dev/null ; sudo dmesg -C ; uname -r
6.1.21+
rpi0w $ sudo insmod ./show_kernel_vas.ko ; dmesg
[ 3563.737085] show_kernel_vas: inserted
[ 3563.737126] minsysinfo(): minimal platform info:
               CPU: ARM-32, little-endian; 32-bit OS.
[ 3563.737137] PAGE_SIZE = 4096, total RAM ~= 429 MB (450142208 bytes)
[ 3563.737156] Some Kernel Details [by decreasing address; values are approximate]
               +------------------------------------------------------------+
[ 3563.737165] |                        [ . . . ]                           |
               |vector table:       ffff0000 - ffff1000                     | [    4 KB]
[ 3563.737181] |                        [ . . . ]                           |
               |fixmap region:      ffc80000 - fff00000                     | [    2 MB]
[ 3563.737193] |vmalloc region:     dc800000 - ff800000                     | [  560 MB]
[ 3563.737205] |lowmem region:      c0000000 - dad4a000                     | [  429 MB]
               |                    ^^^^^^^^                                |
               |                    PAGE_OFFSET                             |
[ 3563.737221] |module region:      bf000000 - c0000000                     | [   16 MB]
[ 3563.737232] |                        [ . . . ]                           |
[ 3563.737239] +------------------------------------------------------------+
[ 3563.737245] show_kernel_vas: skipping show userspace...
rpi0w $
```

Figure 7.13: Output from the show_kernel_vas.ko LKM on a Raspberry Pi Zero W running stock Raspberry Pi OS 32-bit Linux

As can be deciphered from the value of `PAGE_OFFSET` (the KVA `0xc000 0000` in *Figure 7.13*), our Raspberry Pi kernel's VM split is configured as 3:1 (GB) (as the hexadecimal value `0xc000 0000` is 3 GB in decimal base). This VM split value is typically the default for the Raspberry Pi 32-bit OS on recent kernels.

Do note that with a 3:1 (GB) VM split, user space should span from 0 to 3 GB and kernel space the remainder. Technically, however, on AArch32 (ARM-32) systems at least, user space is slightly under 2 GB (*2 GB – 16 MB = 2,032 MB*) as this 16 MB is used as the *kernel module region* (just below `PAGE_OFFSET`); indeed, exactly this can be seen in *Figure 7.11* (the kernel module region here spans from `0xbf00 0000` to `0xc000 0000` for 16 MB). Also, as you'll soon see, the value of the `TASK_SIZE` macro – the size of the user VAS – reflects this fact as well.

We present much of this information in the following detailed diagram:

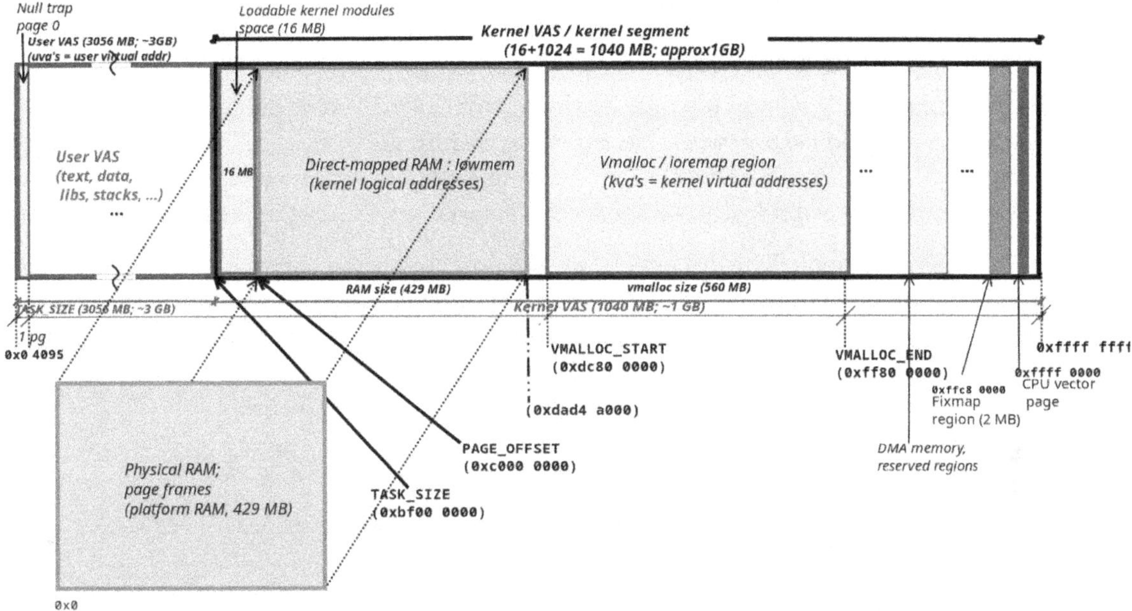

Figure 7.14: The complete VAS of a process on AArch32 (Raspberry Pi Zero W) with a 3:1 GB VM split

Do note that due to variations in differing models, the amount of usable RAM, or even the device tree, the layout shown in *Figure 7.12* may not precisely match that of the Raspberry Pi you have.

Okay, now you know how to access relevant kernel macros and variables related to the VAS, helping you understand the kernel VM layout on any Linux system!

We'll also mention a few caveats regarding our rather simplistic kernel module:

- The description of kernel macros/constants is often very arch-dependant and dependent on the kernel config as well; it's thus incomplete here – for example, recent features like the **Kernel Memory Sanitizer** (**KMSAN**), when enabled, can cause changes in the memory layout.
- We don't attempt to handle the memory layout on an ARMv8.2 or later with LPA enabled.
- Hardware I/O regions (like PCI I/O, DMA) aren't tracked or displayed.
- Due to the non-availability of floating-point calculation when running in kernel mode, at times, our module's size accuracy reduces when large values are expressed (an example of the reason why I mention this: the value 131071.999996185 GB will show up as 127 TB, not 128 TB, which of course is really the value it's very close to).
- The kernel does have similar functionality that our module exhibits, of course – the "dump tables" one; the feature can be configured via `CONFIG_PTDUMP_DEBUGFS`.
- It's still a work in progress!

Nevertheless, our module is a good start! In the following section, we will attempt to "see" (visualize) the kernel VAS – this time, via our `procmap` utility.

The kernel VAS via procmap

Okay, this is interesting: the view of the memory map layout seen in some detail in *Figure 7.14* is exactly what our aforementioned `procmap` utility provides; we introduced usage of this utility in the *The procmap process VAS visualization utility* section. As promised there, let's now see screenshots of the kernel VAS when running `procmap` (the earlier section showed screenshots of the process user VAS).

To keep in sync with the immediate discussion, we will now show screenshots of `procmap` providing a "visual" view of the kernel VAS on the very same AArch32 Raspberry Pi Zero W system (we could specify the `--only-kernel` switch to show only the kernel VAS; we don't do so here, though). As we have to run `procmap` on some process, we arbitrarily choose *systemd* PID `1`; we also use the `--verbose` option switch. However, it might initially fail, as seen in the following screenshot:

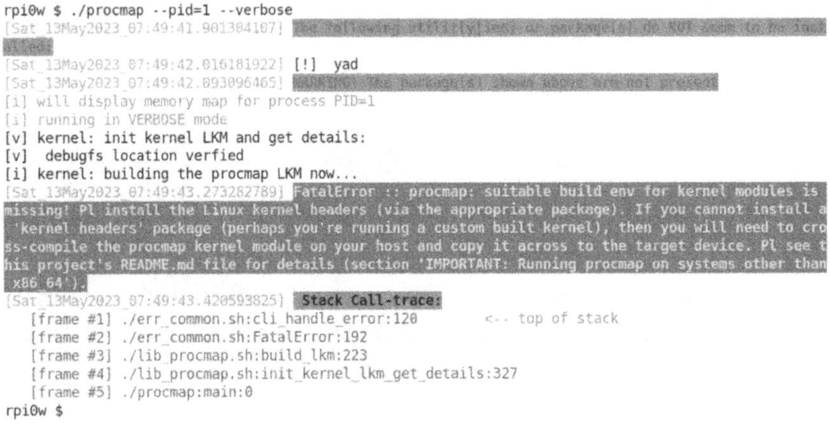

```
rpi0w $ ./procmap --pid=1 --verbose
[Sat_13May2023_07:49:41.901304107] The following utility(ies) or package(s) do NOT seem to be installed:
[Sat_13May2023_07:49:42.016181922] [!]  yad
[Sat_13May2023_07:49:42.093096465] WARNING! The package(s) shown above are not present
[i] will display memory map for process PID=1
[i] running in VERBOSE mode
[v] kernel: init kernel LKM and get details:
[v]  debugfs location verfied
[i] kernel: building the procmap LKM now...
[Sat_13May2023_07:49:43.273282789] FatalError :: procmap: suitable build env for kernel modules is missing! Pl install the Linux kernel headers (via the appropriate package). If you cannot install a 'kernel headers' package (perhaps you're running a custom built kernel), then you will need to cross-compile the procmap kernel module on your host and copy it across to the target device. Pl see this project's README.md file for details (section 'IMPORTANT: Running procmap on systems other than x86_64').
[Sat_13May2023_07:49:43.420593825] Stack Call-trace:
   [frame #1] ./err_common.sh:cli_handle_error:120        <-- top of stack
   [frame #2] ./err_common.sh:FatalError:192
   [frame #3] ./lib_procmap.sh:build_lkm:223
   [frame #4] ./lib_procmap.sh:init_kernel_lkm_get_details:327
   [frame #5] ./procmap:main:0
rpi0w $
```

Figure 7.15: Truncated screenshot showing the procmap kernel module build failing

It failed, as building the *procmap* kernel module failed; but why? I mention this possibility in the project's `README.md` file (`https://github.com/kaiwan/procmap/blob/master/README.md#procmap`):

```
[…] to build a kernel module on the target system, you will require it to have
a kernel development environment setup; this boils down to having the compiler,
make and - key here - the 'kernel headers' package installed for the kernel
version it's currently running upon. [...]
```

Here, it's as the kernel headers package for this kernel isn't available, hence the module build fails (this would also have happened if you had a custom kernel installed on the board). While you can conceivably copy the entire Raspberry Pi kernel source tree onto the device and set up the `/lib/module/<kver>/build` symbolic link, this isn't considered the right way to do so. So, what is? *Cross-compiling* the `procmap` kernel module for the Raspberry Pi from your host, of course! We have covered the details on cross-compiling the kernel itself for the Raspberry Pi here: *Chapter 3, Building the 6.x Linux Kernel from Source – Part 2*, in the *Kernel build for the Raspberry Pi* section; the discussion applies to cross-compiling kernel modules as well.

I want to stress this point: the `procmap` kernel module build on the Raspberry Pi only fails due to the lack of a Raspberry Pi-supplied kernel headers package when running a (custom) kernel. If you are happy to work with the stock (default) Raspberry Pi OS kernel (earlier called Raspbian OS), the kernel headers package is certainly installable (or already installed) and everything will just work (the package is named `raspberrypi-kernel-headers`). Similarly, on your typical x86_64 Linux distribution, the `procmap.ko` kernel module gets cleanly built and inserted at runtime. Do read the `procmap` project's `README.md` file in detail, especially the section labeled *IMPORTANT: Running procmap on systems other than x86_64* for details on how to cross-compile the `procmap` kernel module.

Once you successfully cross-compile the `procmap` kernel module on your host system, copy across the `procmap.ko` kernel module (via `scp`, perhaps) to the device and place it under the `<...>/procmap/procmap_kernel` directory; now you're ready to go!

Here, as an example, is the copied-in (or built) kernel module on the Raspberry Pi device:

```
cd <...>/procmap/
$ ls -l procmap_kernel/procmap.ko
-rw-r--r-- 1 pi pi 9100 Jan 22 17:32 procmap_kernel/procmap.ko
```

You can also run the `modinfo` utility on it to verify that it's built for ARM. Here, I run the stock Raspberry Pi OS on the board. With this in place, let's retry our `procmap` run to display the kernel VAS details:

```
rpi0w $ ./procmap --pid=1 --verbose | tee aarch32_rpi0w.txt
[i] will display memory map for process PID=1
[i] running in VERBOSE mode
[v] kernel: init kernel LKM and get details:
[v]  debugfs location verfied
[v]  LKM inserted into kernel
[v]  debugfs file present
[v] Parsing in various kernel variables as required

[v] set config for Aarch32:
Detected machine type: ARM-32, 32-bit OS
-------------------------------------------------
[v] System details detected ::
-------------------------------------------------
VECTORS_BASE = ffff0000
FIXADDR_START = ffc80000
MODULES_VADDR = bf000000
MODULES_END = c0000000
VMALLOC_START = dc800000
VMALLOC_END = ff800000
PAGE_OFFSET = c0000000
TASK_SIZE = bf000000
ARCH = Aarch32
IS_64_BIT = 0
PAGE_SIZE = 4096
KERNEL_VAS_SIZE = 1090519040
USER_VAS_SIZE = 3204448256
HIGHEST_KVA = 0xffffffff
START_KVA = bf000000
START_KVA_DEC = 3204448256
END_UVA = beffffff
END_UVA_DEC = 3204448255
START_UVA = 0x0
-------------------------------------------------
```

Figure 7.16: Truncated screenshot showing the procmap kernel module successfully inserted and various system details

The *procmap* kernel module does indeed get built and work now! As we've specified the `--verbose` option to `procmap`, you get to see its detailed progress, as well as – *quite usefully* – various kernel variables/macros/constants of interest and their current value.

Okay, let's continue capturing the screen and viewing the portion we're really after – the "visual map" of the kernel VAS on the Raspberry Pi Zero W, in descending order by KVA; the following (partial) screenshot captures (most of) the upper portion of this output from `procmap`:

```
VAS mappings:  name    [ size,perms,u:maptype,u:0xfile-offset]
+------------------  K E R N E L   V A S    end kva  ------------------+ ffffffff
|<... K sparse region ...> [  59 KB,--- ]                               |
+-----------------------------------------------------------------------+ ffff1000
|        vector table  [   4 KB,r-- ]                                   |
+-----------------------------------------------------------------------+ ffff0000  <-- VECTORS_BASE
|<... K sparse region ...> [ 960 KB,--- ]                               |
|                                                                       |
+-----------------------------------------------------------------------+ fff00000
|       fixmap region [   2.50 MB,r-- ]                                 |
|                                                                       |
|                                                                       |
+-----------------------------------------------------------------------+ ffc80000  <-- FIXADDR_START
|<... K sparse region ...> [   4.50 MB,--- ]                            |
|                                                                       |
+-----------------------------------------------------------------------+ ff800000  <-- VMALLOC_END
|      vmalloc region [ 560.00 MB,rw- ]                                  |
|                                                                       |
|                                                                       |
|                                                                       |
+-----------------------------------------------------------------------+ dc800000  <-- VMALLOC_START
|<... K sparse region ...> [  26.71 MB,--- ]                            |
|                                                                       |
|                                                                       |
|                                                                       |
+-----------------------------------------------------------------------+ dad4a000
|       lowmem region [ 429.28 MB,rwx ]                                 |
|                                                                       |
|                                                                       |
|                                                                       |
|   [----------------------------------------------------------]       | c0e29fff
|         Kernel data [   1.99 MB,... ]                                 |
|                                                                       |
|   [----------------------------------------------------------]       | c0bbffff
|         Kernel code [  11.71 MB,... ]                                 |
|                                                                       |
|                                                                       |
+-----------------------------------------------------------------------+ c0000000  <-- MODULES_END / PAGE_OFFSET
|      module region: [  16.00 MB,rwx ]                                 |
|                                                                       |
|                                                                       |
+------------------  K E R N E L   V A S  start kva  ------------------+ bf000000
+------------------        U S E R   V A S    end uva  ------------------+ beffffff
|<... Sparse Region ...> [   1.58 MB,---,-,0x0]                         |
|                                                                       |
```

Figure 7.17: Partial screenshot of our procmap utility's upper portion output showing the complete kernel VAS (when run on the Raspberry Pi Zero W with 32-bit Linux)

The complete kernel VAS – from `end_kva` (value `0xffff ffff`) right to the start of the kernel, `start_kva` (`0xbf00 0000`, which, as you can see, is the start of the kernel module region on this system) – is displayed.

Notice (on the right side in green color; hard-copy readers, please refer to the PDF with full-color diagrams here: `https://packt.link/gbp/9781803232225.pdf`) the label on the right of certain key addresses denoting what they are! For completeness, we also included in the preceding screenshot the kernel-user boundary. As the preceding output is on an ARM 32-bit system, the user VAS immediately follows the kernel VAS.

As we've learned, on a 64-bit system though, there is an (enormous!) "non-canonical" sparse unused region – a hole – between the start of the kernel VAS and the top of the user VAS. On the x86_64, it spans the vast majority of the overall VAS: 16,383.75 PB (out of a total VAS of 16,384 PB! Refer to *Figure 7.5* again, to refresh this fact).

Exercise

Install and run the `procmap` utility on your Linux system(s) and carefully study its output; this can be done on your x86_64 (VM or native system), any ARM (32 and/or 64-bit), or whichever box or VM you happen to have handy. It also works well on a BeagleBone Black embedded board with a 3:1 VM split, showing details as expected.

FYI, I also provide a solution in the form of three (large, stitched-together) screenshots of `procmap`'s output on a native x86_64 system, a BeagleBone Black (AArch32) board, and the Raspberry Pi running a 64-bit OS (AArch64) here: `https://github.com/PacktPublishing/Linux-Kernel-Programming_2E/tree/main/solutions_to_assgn/ch7`.

Studying the source code of `procmap`, and, especially relevant here, its kernel module component, will certainly help you learn more. I urge you to dig into it and even contribute; it's open source, after all!

Let's finish this section by glancing at the user segment view that our earlier demo kernel module – `ch7/show_kernel_vas` – provides.

Trying it out – the user segment

Now, let's go back to our `ch7/show_kernel_vas` LKM demo program. We have provided a kernel module parameter named `show_uservas` (defaulting to the value `0`); when set to `1`, some details regarding the process context's *user space* – more correctly, the user-mode process context that ran the module's code paths – are displayed as well. Here's the definition of the module parameter:

```
static int show_uservas;
module_param(show_uservas, int, 0660);
MODULE_PARM_DESC(show_uservas,
"Show some user space VAS details; 0 = no (default), 1 = show");
```

Right, once more on the same device (a Raspberry Pi Zero W), let's again run our `show_kernel_vas` kernel module, this time requesting it to display user space details as well (via the aforementioned parameter).

The following screenshot shows the complete output:

```
rpi0w $ uname -r ; sudo rmmod show_kernel_vas 2>/dev/null ; sudo dmesg -C
6.1.21+
rpi0w $ sudo insmod ./show_kernel_vas.ko show_uservas=1 ; dmesg
[ 7725.559741] show_kernel_vas: inserted
[ 7725.559783] minsysinfo(): minimal platform info:
               CPU: ARM-32, little-endian; 32-bit OS.
[ 7725.559794] PAGE_SIZE = 4096, total RAM ~= 429 MB (450142208 bytes)
[ 7725.559813] Some Kernel Details [by decreasing address; values are approximate]
               +------------------------------------------------------------+
[ 7725.559822] |                      [ . . . ]                             |
               |vector table:        ffff0000 - ffff1000                    | [     4 KB]
[ 7725.559837] |                      [ . . . ]                             |
               |fixmap region:       ffc80000 - fff00000                    | [     2 MB]
[ 7725.559850] |vmalloc region:      dc800000 - ff800000                    | [   560 MB]
[ 7725.559861] |lowmem region:       c0000000 - dad4a000                    | [   429 MB]
               |                     ^^^^^^^^                               |
               |                    PAGE_OFFSET                             |
[ 7725.559877] |module region:       bf000000 - c0000000                    | [    16 MB]
[ 7725.559888] |                      [ . . . ]                             |
[ 7725.559895] +------- Above this line: kernel VAS; below: user VAS --------+
               |                      [ . . . ]                             |
               |Process environment  bec7f8c8 - bec7ffeb                    | [ 1827 bytes]
               |          arguments  bec7f89d - bec7f8c8                    | [   43 bytes]
               |        stack start  bec7f790                               |
               |       heap segment  01947000 - 01968000                    | [       132 KB]
               |static data segment  00040c44 - 00041038                    | [ 1012 bytes]
               |       text segment  00010000 - 000303d8                    | [       128 KB]
               |                      [ . . . ]                             |
               +------------------------------------------------------------+
[ 7725.559935] Size of User VAS size (TASK_SIZE) = 3204448256 bytes           [  3056 GB]
                # userspace memory regions (VMAs) = 38
rpi0w $
```

Figure 7.18: Screenshot of our show_kernel_vas.ko LKM's output showing both kernel and user VAS details when running on a Raspberry Pi Zero W with the stock Raspberry Pi 32-bit Linux OS

This is useful! We now literally see a (more or less) complete memory map of the process – both the so-called "upper (canonical) half" kernel space as well as the "lower (canonical) half" user space – in one shot (yes, that's right, even though the `procmap` project shows this better and, in more detail).

As one more interesting experiment, let's use our `show_kernel_vas` module to view the overall VAS on our AArch64 Raspberry Pi 4 Model B board running the stock RPi 64-bit OS.

The screenshot (*Figure 7.19*) shows it all:

```
rpi4-64 $ cat /proc/version
Linux version 6.1.21-v8+ (dom@buildbot) (aarch64-linux-gnu-gcc-8 (Ubuntu/Linaro 8.4.0-3ubuntu1) 8.4.0, GNU ld (GNU
Binutils for Ubuntu) 2.34) #1642 SMP PREEMPT Mon Apr  3 17:24:16 BST 2023
rpi4-64 $
rpi4-64 $ sudo rmmod show_kernel_vas 2>/dev/null ; sudo dmesg -C
rpi4-64 $ sudo insmod ./show_kernel_vas.ko show_uservas=1 ; sudo dmesg
[  469.904037] show_kernel_vas: inserted
[  469.904072] minsysinfo(): minimal platform info:
               CPU: Aarch64, little-endian; 64-bit OS.
[  469.904085] PAGE_SIZE = 4096, total RAM ~= 1849 MB (1939038208 bytes)
[  469.904103] VA_BITS (CONFIG_ARM64_VA_BITS) = 39
[  469.904115] Some Kernel Details [by decreasing address; values are approximate]
               +------------------------------------------------------------+
[  469.904126] |                          [ . . . ]                         |
               |fixmap region:       ffffffdfdbf9000 - fffffffdfe000000     | [        4 MB]
[  469.904143] |module region:       ffffffc000000000 - ffffffc008000000     | [      128 MB]
[  469.904158] |                          [ . . . ]                         |
               |vmemmap region:      fffffffe00000000 - ffffffff00000000     | [     4096 MB =      4 GB ~=    0 TB]
[  469.904174] |vmalloc region:      ffffffc008000000 - fffffffdf0000000     | [   253568 MB =    247 GB ~=    0 TB]
[  469.904190] |lowmem region:       ffffff8000000000 - ffffff8073936000     | [     1849 MB]
               |                     ^^^^^^^^^^^^^^^^                       |
               |                        PAGE_OFFSET                         |
[  469.904205] |                          [ . . . ]                         |
[  469.904216] +------- Above this line: kernel VAS; below: user VAS --------+
               |                          [ . . . ]                         |
               |Process environment  0000007fcdef08c2 - 0000007fcdef0fe7     | [ 1829 bytes]
               |          arguments  0000007fcdef0897 - 0000007fcdef08c2     | [   43 bytes]
               |        stack start  0000007fcdeeff90                        |
               |       heap segment  00000055ad51d000 - 00000055ad53e000     | [       132 KB]
               |static data segment  000000556f956ca0 - 000000556f9580c0     | [ 5152 bytes]
               |       text segment  000000556f920000 - 000000556f946214     | [       152 KB]
               |                          [ . . . ]                         |
               +------------------------------------------------------------+
[  469.904248] Kernel, User VAS (TASK_SIZE) size each =     549755813888 bytes  [  512 GB]
                # userspace memory regions (VMAs) = 35
rpi4-64 $
```

Figure 7.19: Screenshot showing our module running on a standard AArch64 Raspberry Pi 4

Note how this particular memory layout pretty much matches what we showed in row 5 of the table in *Figure 7.6*

I will leave it as an exercise for you to run this kernel module and carefully study the output on your x86_64, or whichever box or VM. Do carefully go through the code as well. We printed the user space details that you see in the preceding screenshot, such as the segment start and end addresses, by dereferencing the `mm_struct` structure (the task structure member named `mm`) from `current`. Recall that `mm` is the abstraction of the user mapping of the process. A snippet of the code that does this is as follows:

```
// ch7/show_kernel_vas/kernel_vas.c
[ ... ]
static void show_userspace_info(void)
{
    pr_info("+------- Above this line: kernel VAS; below: user VAS --------+\n"
    ELLPS
    "|Process environment "
```

```
#if (BITS_PER_LONG == 64)
    " %px - %px      | [ %4zu bytes]\n"
    "|          arguments "
    " %px - %px      | [ %4zu bytes]\n"
    "|         stack start  %px                      |\n"
[ ... ]
#else // 32-bit
    " %px - %px                      | [ %4zu bytes]\n"
    "|          arguments "
    " %px - %px                      | [ %4zu bytes]\n"
    "|         stack start  %px                              |\n"
[ ... ]\n",
    SHOW_DELTA_b((void *)current->mm->env_start, (void *)current->mm->env_end),
    SHOW_DELTA_b((void *)current->mm->arg_start, (void *)current->mm->arg_end),
    (void *)current->mm->start_stack,
    [ ... ]
```

Remember the so-called null trap page at the very beginning of the user VAS? (Also, `procmap`'s output – see *Figure 7.11* – shows the null trap page). Let's see what it's for in the following section.

The null trap page

Did you notice within the preceding diagrams (*Figure 7.11* and *Figure 7.14*), at the extreme left edge (of *Figure 7.11*, albeit very small!), a single page at the very beginning of the user space, named the **null trap** page? What is it? That's easy: virtual page `0` is given no permissions (at the hardware MMU/PTE level). Thus, any access to this page, be it `r`, `w`, or `x` (read/write/execute), will result in the MMU raising what is called a *processor fault* or *exception*. This will have the processor jump to an OS handler routine (the fault handler). It runs, killing the culprit process trying to access a memory region with no permissions!

It's very interesting indeed: the OS handler mentioned previously runs in process context, and guess what `current` is: why, it's the process (or thread) that initiated this bad `NULL` pointer lookup! (Also note that it's not just the `NULL` or 0x0 address that generates this fault; any address from 0 to 4095 will.) Within the fault handler code, the `SIGSEGV` signal is delivered to the faulting process (`current`), causing it to die (via a segfault). In a nutshell, this is how the classic `NULL` pointer dereference bug is caught by the OS.

Viewing kernel documentation on the memory layout

Back to the kernel VAS; obviously, with a 64-bit VAS, the kernel VAS is *much* larger than on 32-bit. As we saw earlier, it's typically 128 TB on the x86_64. Study again the VM split table shown previously (*Figure 7.6* in the *Common VM splits section*); there, the column labeled "VM Split ..." is, of course, the VM split for different architectures. You can see how on the 64-bit Intel/AMD and AArch64 (ARM64), the numbers are much larger than for their 32-bit counterparts.

For arch-specific details, we refer you to the "official" kernel documentation on the process virtual memory layout here (the relevant kernel headers turn out to be very useful to the purpose as well):

Architecture	Process memory layout documentation (file location in kernel source tree, link to documentation, and relevant kernel header)
AArch32 (ARM32)	*File*: `Documentation/arm/memory.txt`. *Link*: `https://elixir.bootlin.com/linux/v6.1.25/source/Documentation/arm/memory.rst` *Header*: `arch/arm/include/asm/memory.h`.
AArch64 (ARM64)	*File*: `Documentation/arm64/memory.txt`. *Link*: `https://elixir.bootlin.com/linux/v6.1.25/source/Documentation/arm64/memory.rst` *Header*: `arch/arm64/include/asm/memory.h`
x86_64	*File*: `Documentation/x86/x86_64/mm.txt` *Link:* This document's readability was vastly improved quite recently; I recommend you browse through this file: `https://elixir.bootlin.com/linux/v6.1.25/source/Documentation/x86/x86_64/mm.rst` *Header*: `arch/x86/include/asm/pgtable_64_types.h`. For the x86_32: `arch/x86/include/asm/page_32_types.h`

Table 7.11: Arch (CPU)-specific official kernel documentation for process memory layout on Linux

At the risk of repetition, I urge you to try out the `ch7/show_kernel_vas` kernel module – and, even better, the `procmap` project (`https://github.com/kaiwan/procmap`) – on different Linux systems and study the output. You can then literally see the "memory map" – the complete process VAS – of any given process, which includes the kernel VAS! This understanding is critical when working with and/or debugging issues at the system layer.

Again, at the risk of overstating it, the previous two sections – covering the detailed examination of the *user and kernel VASs* – are very important indeed. Do take the time required to go over them and work on the sample code and suggested exercises. Great going!

Moving along on our journey through the Linux kernel's memory management subsystem, let's now check out another interesting topic – that of the [K]ASLR protection-via-memory-layout-randomization feature. Read on!

Randomizing the memory layout – KASLR

In infosec circles, it's a well-known fact that, with **proc filesystem** (**procfs**) and various powerful "hacking" tools at their disposal (heard of Kali Linux?), a malicious user, knowing in advance the precise location (virtual addresses) of various functions and/or globals within a process's VAS, could devise an attack to exploit and ultimately compromise a given system. (Why, even knowing the precise location of *one* well-known function or global in a given kernel could lead to an attack vector!) Thus, for security, to make it near impossible (or at least difficult) for attackers to rely on "known" virtual addresses, user space, as well as kernel space, supports **Address Space Layout Randomization** (**ASLR**) and **Kernel ASLR** (**KASLR**) techniques (often pronounced *Ass-ler*/*Kass-ler*).

The keyword here is *randomization:* this feature, when enabled, *changes the location* of portions of the process (and kernel) memory layout in terms of absolute numbers as it *offsets portions of memory* from a given base address by a random (page-aligned) quantity.

What "portions of memory" exactly are we talking about? With respect to user space mappings (we will talk about KASLR later), the starting addresses of shared libraries (their load address), `mmap()`-based allocations (you'll realize that any `malloc()` function (`/calloc()/realloc()`) allocating any memory above `MMAP_THRESHOLD` (typically 128 KB) becomes an `mmap`-based allocation, not memory off the heap), stack start, the heap, and the vDSO page; all of these can be randomized at process run (launch) time.

Hence, an attacker cannot depend on, say, a `glibc` function (such as `system()`) being mapped at a particular fixed UVA in any given process (as it used to be!); not only that, the location will vary every time the process runs! Before ASLR, and on systems where ASLR is unsupported or turned off, the location of symbols can be ascertained in advance for a given architecture and software version (procfs plus utilities like `objdump`, `readelf`, `nm`, and so on make this quite easy).

It's key to realize that [K]ASLR is merely a statistical protection. In fact, typically, not many bits are available for randomization and thus the entropy isn't very good. This implies that the page-sized offsets are not too many, even on 64-bit systems, thus leading to a possibly weakened implementation (to the experienced cracker's delight).

Let's now briefly look at a few more details regarding both user mode and kernel-mode ASLR (the latter being referred to as KASLR); the following sections cover these areas, respectively.

User memory randomization with ASLR

User-mode ASLR is usually what is meant by the term ASLR. It being enabled implies this protection is available on the user space mapping of every process. Effectively, ASLR being enabled implies that the absolute memory map of user-mode processes will vary every time they're run, and every process instance of the same program will be different in terms of the absolute user-space memory map.

ASLR has been supported on Linux for a very long time (since 2005 on 2.6.12). The kernel has a tunable pseudo file within procfs to query and set (as root) the ASLR state; here it is: `/proc/sys/kernel/randomize_va_space`.

It can have three possible values; the three values and their meaning are shown in the following table:

Tunable value	Interpretation of this value in `/proc/sys/kernel/randomize_va_space`
`0`	(User mode) ASLR is turned OFF, or can be turned off by passing the kernel parameter `norandmaps` at boot.
`1`	(User mode) ASLR is ON: `mmap()` based allocations, the stack, and the vDSO page are randomized. This also implies that shared library load locations and shared memory segments are randomized.
`2`	(User mode) ASLR is ON: all of the preceding (value `1`) *plus* the heap location are randomized (since 2.6.25); this is the OS value by default.

Table 7.12: Linux (user mode) ASLR tuning via a proc pseudo file

As noted in an earlier section, *The vsyscall page*, the vDSO page is a system call optimization, allowing some frequently issued system calls (`gettimeofday()` being a typical one) to be invoked with less overhead. If interested, you can look up more details on the man page on vDSO(7) here: `https://man7.org/linux/man-pages/man7/vdso.7.html`.

User-mode ASLR can be turned *off* at boot by passing the `norandmaps` parameter to the kernel (via the bootloader); why would one do that? It's at times useful to do so when debugging...leave it on in production!

Kernel memory layout randomization with KASLR

Similar to (user) ASLR – and more recently, from the 3.14 kernel onward – even *kernel* VAS can be randomized (to some extent) by having KASLR enabled. Here, the base location of some kernel sections (the lowmem, vmalloc, and vmemmap regions), as well as module code within the kernel VAS will be randomized by a page-aligned random offset from the base of RAM. This remains in effect for that session – that is, until a power cycle or reboot.

The kernel config for KASLR is named `CONFIG_RANDOMIZE_MEMORY`. KASLR appears to be supported on the x86[_64] and AArch64 platforms, not on AArch32.

Several kernel configuration variables exist, enabling the platform developer to enable or disable these randomization options. As an example specific to the x86, the following is quoted directly from `Documentation/x86/x86_64/mm.txt` (`https://elixir.bootlin.com/linux/v6.1.25/source/Documentation/x86/x86_64/mm.rst#L148`):

> Note that if `CONFIG_RANDOMIZE_MEMORY` is enabled, the direct mapping of all physical memory, vmalloc/ioremap space and virtual memory map are randomized. Their order is preserved but their base will be offset early at boot time.

Once configured, KASLR will remain on by default; one can control its state at boot by passing a kernel command-line parameter (via the bootloader):

- Explicitly turn it *off* by passing the `nokaslr` parameter
- Explicitly turn it *on* by passing the `kaslr` parameter

From the 5.13 kernel, a new security feature is available: `CONFIG_RANDOMIZE_KSTACK_OFFSET_DEFAULT`. Turning it on has the kernel-mode stack offsets randomized every time a system call is issued! (Do see the really well-written commit #39218ff4c625dbf2, here: `https://github.com/torvalds/linux/commit/39218ff4c625dbf2e68224024fe0acaa60bcd51a`).

Tip: You can search for a given (abbreviated) commit number from the Search box at `https://github.com/torvalds/linux`.

So, what's the current setting for [K]ASLR on your Linux system? And can we change it? Yes, of course (provided we have *root* access); the next section shows you how to do so via a Bash script.

Querying/setting KASLR status with a script

We provide a simple Bash script at `<book-source>/ch7/ASLR_check.sh`. It checks for the presence of both (user-mode) ASLR as well as KASLR, printing (color-coded) status information about them. It also allows you to change the ASLR value.

Let's give it a spin on our x86_64 Ubuntu 22.04 guest. As our script is programmed to be color-coded, we show a screenshot of its output here:

```
$ sudo ./ASLR_check.sh
[sudo] password for c2kp:
+++++++++++++++++++++++++++++++++++++++++++++++++++++++++++++++++++++++++++++++
Simple [Kernel] Address Space Layout Randomization / [K]ASLR checks:
Usage: ASLR_check.sh [ASLR_value] ; where 'ASLR_value' is one of:
 0 = turn OFF ASLR
 1 = turn ON ASLR only for stack, VDSO, shmem regions
 2 = turn ON ASLR for stack, VDSO, shmem regions and data segments [OS default]

The 'ASLR_value' parameter, setting the ASLR value, is optional; in any case,
I shall run the checks... thanks and visit again!
+++++++++++++++++++++++++++++++++++++++++++++++++++++++++++++++++++++++++++++++
[+] Checking for (usermode) ASLR support now ...
 (in /proc/sys/kernel/randomize_va_space)
 Current (usermode) ASLR setting = 2
 => (usermode) ASLR ON: mmap(2)-based allocations, stack, vDSO page,
 shlib, shmem locations and heap are randomized on startup
+++++++++++++++++++++++++++++++++++++++++++++++++++++++++++++++++++++++++++++++
[+] Checking for kernel ASLR (KASLR) support now ...
(need >= 3.14, this kernel is ver 5.15.0-43-generic)
 Kernel ASLR (KASLR) is On [default]
+++++++++++++++++++++++++++++++++++++++++++++++++++++++++++++++++++++++++++++++
ASLR quick test:
Now running this command *twice* :
 grep -E "heap|stack" /proc/self/maps

5638bad94000-5638badd6000 rw-p 00000000 00:00 0                          [heap]
7ffdaf9c8000-7ffdaf9e9000 rw-p 00000000 00:00 0                          [stack]

55b578f67000-55b578fa9000 rw-p 00000000 00:00 0                          [heap]
7ffe29154000-7ffe29175000 rw-p 00000000 00:00 0                          [stack]

With ASLR:
   enabled: the uva's (user virtual addresses) should differ in each run
  disabled: the uva's (user virtual addresses) should be the same in each run.

$
```

Figure 7.20: Screenshot showing the output when our ch7/ASLR_check.sh Bash script runs on an x86_64 Ubuntu guest

It runs, showing you that (at least on this box) both the user mode ASLR as well as KASLR configs are indeed turned on. Not only that, we write a small "test" routine to see ASLR functioning. It's very simple; it runs the following command twice:

```
grep -E "heap|stack" /proc/self/maps
```

From what you learned in the earlier *Interpreting the /proc/PID/maps output* section, you can now see, in *Figure 7.20*, that the respective UVAs for the heap and stack segments are *different in each run*, thus proving that the ASLR feature indeed works! For example, look at the starting heap UVA: in the first run, it's `0x5638 bad9 4000`, and in the second run, it's `0x55b5 78f6 7000` (bottom portion of *Figure 7.21*.

Next, let's pass the parameter `0` to the script, thus turning ASLR *off*; the following screenshot shows the (expected) output:

```
$ sudo ./ASLR_check.sh 0
+++++++++++++++++++++++++++++++++++++++++++++++++++++++++++++++++++++++++++++++
Simple [Kernel] Address Space Layout Randomization / [K]ASLR checks:
Usage: ASLR_check.sh [ASLR_value] ; where 'ASLR_value' is one of:
 0 = turn OFF ASLR
 1 = turn ON ASLR only for stack, VDSO, shmem regions
 2 = turn ON ASLR for stack, VDSO, shmem regions and data segments [OS default]

The 'ASLR_value' parameter, setting the ASLR value, is optional; in any case,
I shall run the checks... thanks and visit again!
+++++++++++++++++++++++++++++++++++++++++++++++++++++++++++++++++++++++++++++++
[+] Checking for (usermode) ASLR support now ...
 (in /proc/sys/kernel/randomize_va_space)
 Current (usermode) ASLR setting = 2
 => (usermode) ASLR ON: mmap(2)-based allocations, stack, vDSO page,
 shlib, shmem locations and heap are randomized on startup
+++++++++++++++++++++++++++++++++++++++++++++++++++++++++++++++++++++++++++++++
[+] Checking for kernel ASLR (KASLR) support now ...
(need >= 3.14, this kernel is ver 5.15.0-43-generic)
 Kernel ASLR (KASLR) is On [default]
+++++++++++++++++++++++++++++++++++++++++++++++++++++++++++++++++++++++++++++++
[+] Setting (usermode) ASLR value to "0" now...
ASLR setting now is: 0
 => (usermode) ASLR is currently OFF
+++++++++++++++++++++++++++++++++++++++++++++++++++++++++++++++++++++++++++++++
ASLR quick test:
Now running this command *twice* :
 grep -E "heap|stack" /proc/self/maps

555555582000-5555555d4000 rw-p 00000000 00:00 0                        [heap]
7ffffffde000-7ffffffff000 rw-p 00000000 00:00 0                        [stack]

555555582000-5555555d4000 rw-p 00000000 00:00 0                        [heap]
7ffffffde000-7ffffffff000 rw-p 00000000 00:00 0                        [stack]

With ASLR:
  enabled: the uva's (user virtual addresses) should differ in each run
 disabled: the uva's (user virtual addresses) should be the same in each run.

$
```

Figure 7.21: Screenshot showing how ASLR is turned off (via our ch7/ASLR_check.sh script on an x86_64 Ubuntu guest)

This time, we can see that ASLR was on by default, but we turned it off. This is clearly highlighted in (red) bold font via the line " => *(usermode) ASLR is currently OFF*" in the preceding screenshot. (Do remember to turn it on again.) Also, as expected, as it's now off, the UVAs of both the heap and stack (respectively) remain the same in both test runs, which is insecure. I will leave it to you to browse through and understand the source code of the script.

Note: To take advantage of ASLR, applications must be compiled with the `-fPIE` and `-pie` GCC flags (**PIE** stands for **Position Independent Executable**).

Both ASLR and KASLR protect against some types of attack vectors, the return-to-libc and **Return-Oriented Programming** (**ROP**) ones being the typical cases. However, and unfortunately, white and black hat security being the cat-and-mouse game it is, defeating [K]ASLR and similar methodologies is something advanced exploits do quite well. As an example, the well-known `setarch` utility has the `--addr-no-randomize` option to, guess what, turn off ASLR. Refer to this chapter's *Further reading* section (under the *Linux kernel security* heading) for more details.

While on the topic of security, as an added-value tip, many useful tools exist to carry out vulnerability checks on your system. Check out the following:

- The `checksec.sh` script (`http://www.trapkit.de/tools/checksec.html`) displays various "hardening" measures and their current status (for both individual files and processes): RELRO, stack canary, NX-enabled, PIE, RPATH, RUNPATH, presence of symbols, and compiler fortification.
- grsecurity's PaX suite.
- The `hardening-check` script (an alternative to checksec).
- The `kconfig-hardened-check` Perl script (`https://github.com/a13xp0p0v/kconfig-hardened-check`) checks (and suggests) kernel config options for security against some predefined checklists.
- Several others: Lynis, `linuxprivchecker.py`, memory, and so on.

So, the next time you see differing kernel or user virtual addresses on multiple runs or sessions, you will know it's probably due to the [K]ASLR protection feature (turning [K]ASLR off for debugging is fairly common). Now, let's complete this chapter by moving on to an exploration of how the Linux kernel organizes and works with physical memory.

Understanding physical memory organization

Now that we have examined the *virtual memory* view for both user and kernel VASs in quite some detail, let's turn to the topic of physical memory organization on the Linux OS.

Physical RAM organization

The Linux kernel, at boot, organizes and partitions physical RAM into a tree-like hierarchy consisting of *nodes, zones,* and *page frames* (page frames are physical pages of RAM) (see *Figure 7.22* and *Figure 7.23*). Do note that further organization via physical *memory models* is also performed at early boot and is a related discussion; we will throw some light on this in the *An introduction to physical memory models* section.

Nodes are divided into zones, and zones consist of page frames. It's essentially a tree-like hierarchy, simplistically and conceptually depicted as a three-level tree-like hierarchy like this:

- Node(s) ← Level 1
    - Zone(s) ← Level 2
        - Page frames ← Level 3

A node is a metadata structure that abstracts a physical bank of RAM; this RAM itself is associated with one or more processor (CPU) cores. At the hardware level, the microprocessors are connected to the RAM controller chip(s); any memory controller chip, and thus any RAM, can be reached from any CPU core, across an *interconnect*. Now, obviously, being able to reach the RAM physically nearest the core on which a thread is allocating or using memory will lead to performance enhancement. This very idea is leveraged by hardware and OSes that support the so-called NUMA model (the meaning is explained shortly).

Nodes and NUMA

Essentially, *nodes* are data structures used to denote and abstract a physical RAM module on the system motherboard and its associated controller chipset. Yes, we're talking actual *hardware* here being abstracted via software metadata. (Note that the term "node," as used in this context, isn't the same as when used to denote a single hardware computer on a network, perhaps). It's always associated with a physical socket (or collection of processor cores) on the system motherboard. Two types of hierarchies exist:

- **Non-Uniform Memory Access (NUMA) systems:** Where the particular CPU core on which a kernel allocation request occurs does matter (memory is treated *non*-uniformly), leading to performance improvements
- **Uniform Memory Access (UMA) systems:** Where the particular CPU core on which a kernel allocation request occurs doesn't matter (memory is treated uniformly)

True NUMA systems are those whose hardware is always multicore – implying two or more CPU cores, thus it's always **Symmetric Multi-Processor (SMP)** as well – *and* there must be two or more physical banks of RAM, each of which is associated with a CPU (or CPUs). In other words, NUMA systems will always have two or more nodes, whereas UMA systems will have exactly one node (FYI, the data structure that abstracts a node is called `pg_data_t` and is defined here: `include/linux/mmzone.h:pg_data_t` as a typedef structure).

Why all this complexity, you may wonder? Well, it's – what else – all about performance! NUMA systems (they typically tend to be rather expensive server-class machines and supercomputers) and the OSes they run (Linux/Unix/Windows Server, typically) are designed in such a way that when a process (or thread) on a particular CPU core wants to perform a kernel memory allocation, the software guarantees that it does so with high performance by taking the required memory (RAM) *from the node closest to the core* (hence the NUMA moniker!). No such benefits accrue to UMA systems (your typical embedded systems, smartphones, laptops, and desktops), nor do they matter. Enterprise, data center class servers, and supercomputer systems nowadays can have hundreds of processors and terabytes, even a few petabytes, of RAM, a node, and there will be 2 or more nodes. These are almost always architected as NUMA systems.

With the way that Linux is designed, though – and this is a key point – even **regular UMA systems are treated as NUMA systems by the kernel** (well, pseudo-NUMA). This is to avoid, at any cost, changing the code base, thus forking Linux (as you'll know, the very same Linux kernel code base is used to power any and all types of Linux systems, from tiny, embedded ones to supercomputers). They – UMA systems – will have *exactly one node*, so that's a quick way to check whether the system is NUMA or UMA: if there are two or more nodes and it has multiple CPU cores, it's a true NUMA system; only one node and/or only one CPU core, and it's a "fake NUMA" or pseudo-NUMA box. How can you check the number of nodes? The `numactl` utility is one way (try doing `numactl --hardware`). There are other ways to check (via *procfs* itself); hang on a bit, we'll get there... (FYI, checking the number of CPU cores the OS sees is easy; you can use `nproc`, `lscpu`, and/or `cat /proc/cpuinfo`).

So, a simpler way to visualize this: on a NUMA box, one or more CPU cores are associated with a bank (a hardware module) of physical RAM; this is termed a node, and MUMA systems will always have two or more nodes. Thus, a NUMA system is also always a **SMP** one, but an SMP box may be a NUMA or UMA system.

An example of a NUMA server processor

To make this discussion a bit more practical, let's briefly visualize the micro-architecture of an actual server system – one running the AMD Epyc/Ryzen/Threadripper (and the older Bulldozer) CPUs. It contains the following hardware (see *Figure 7.22*):

- A total of 32 CPU cores (as seen by the OS) within two physical *sockets* (P#0 and P#1) on the motherboard. Each socket consists of a package of 8x2 CPU cores (8x2, as there are actually 8 physical cores, each of which is hyperthreaded; the OS sees each hyperthreaded core as a usable core of course, thus yielding a total of 16 CPU cores per socket).
- A total of 32 GB of RAM split up into *four physical banks* of 8 GB each.

Clearly, here, this system is multicore (SMP) *and* has two or more RAM banks, thus qualifying as a true NUMA system. Thus, the Linux NUMA-aware memory management code, upon detecting this topography at boot, will set up *four NUMA nodes* to represent it. We won't delve into the processor's various (L1/L2/L3/etc.) caches here; see the *Tip* box after the following diagram for a way to see all of this. Furthermore, systems like this are termed **cache-coherent NUMA (ccNUMA)**, as they maintain cache coherency with help from the hardware; you'll learn more about caches and cache coherency in *Chapter 13, Kernel Synchronization – Part 2*, in the *Understanding CPU caching basics, cache effects, and false sharing* section.

The following conceptual diagram shows an approximation of the four tree-like hierarchies – one for each node – formed on some AMD server systems running the Linux OS. *Figure 7.22* conceptually shows the *nodes/zones/page frames* per physical RAM bank on the system coupled to different CPU cores:

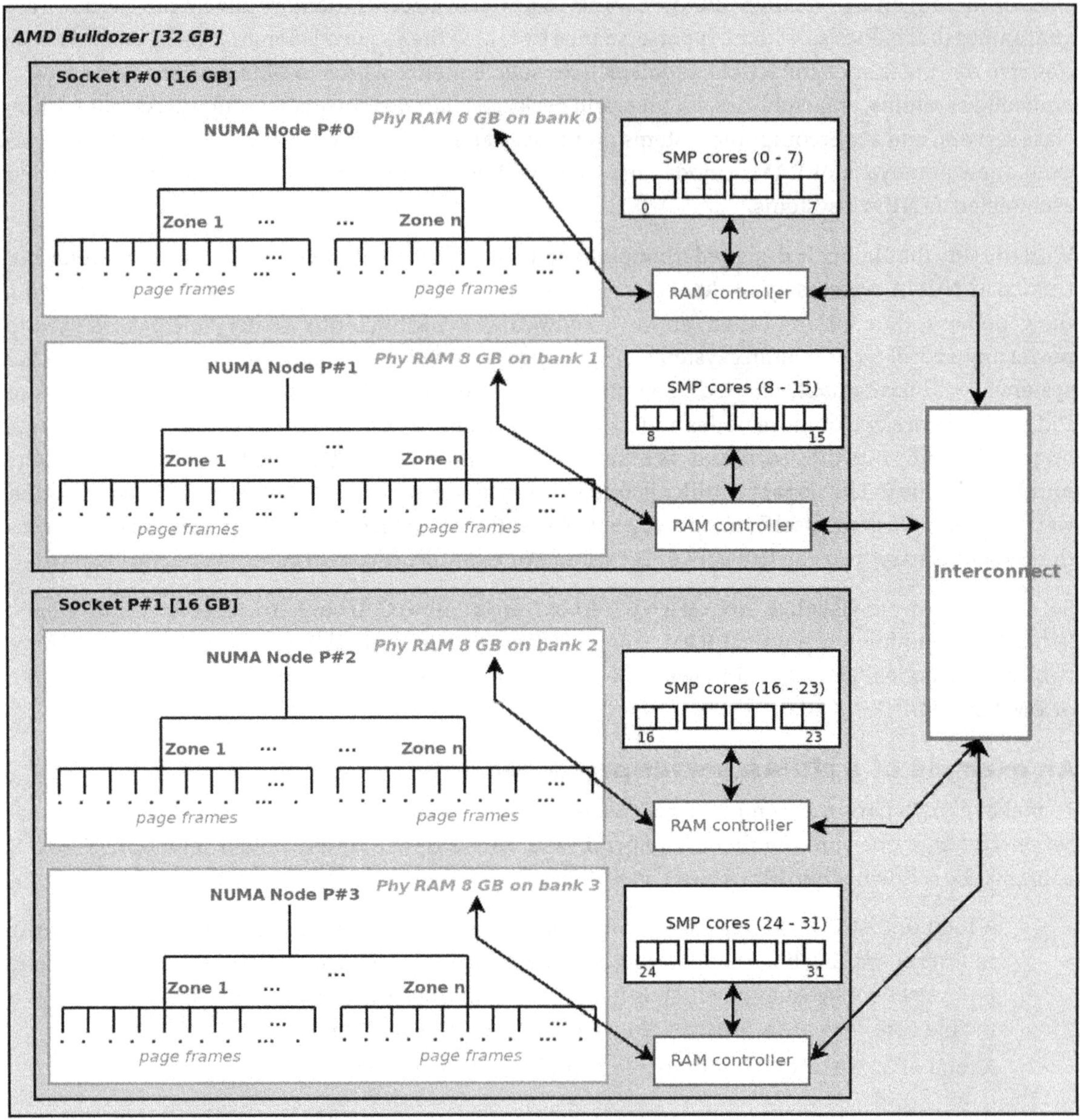

Figure 7.22: An approximate conceptual view of an AMD server: physical memory hierarchy on Linux

TIP: Use the powerful `lstopo` utility (*list topography*, and its associated `hwloc-*` – hardware locality – utilities) to graphically view the hardware (CPU) topology of your system! On Ubuntu, install it with `sudo apt install hwloc`. FYI, the hardware topography graphic of the previously mentioned AMD server system can be generated by `lstopo` and is available here: `https://en.wikipedia.org/wiki/CPU_cache#/media/File:Hwloc.png`.

To reassert the key point here: for performance (here, with respect to *Figure 7.22*), a thread running some kernel or driver code in process context on, say, CPU #18 requests the kernel for some RAM. The kernel's **memory management** (**MM**) layer, being NUMA-aware, will have the request serviced (as priority) from any free RAM page frames in any zone on NUMA node #2 (that is, from physical RAM bank #2) as it's "closest" to the processor core that the request was issued upon. Just in case there are no free page frames available in any zone within NUMA node #2 (unlikely), the kernel has an intelligent fallback system. It might now go across the interconnect and request RAM page frames from another *node:zone* (worry not, we cover these aspects in more detail in the following chapter, *8, Kernel Memory Allocation for Module Authors – Part 1*, in the *Understanding and using the kernel page allocator (or BSA)* section).

Zones within a node

Zones can be thought of as Linux's way of smoothing out and dealing with hardware quirks; these tend to proliferate on the x86, where Linux "grew up" of course. They also deal with a few software difficulties (look up `ZONE_HIGHMEM` on the now mostly legacy 32-bit x86 architecture; we discussed this concept in an earlier section, *High memory on 32-bit systems*).

Zones form the second level of the hierarchy; they always belong to a certain node (0, 1, 2, ...) and consist of *page frames* – physical pages of RAM. More technically, a range of **Page Frame Numbers** (**PFNs**) are allocated to each zone within a node:

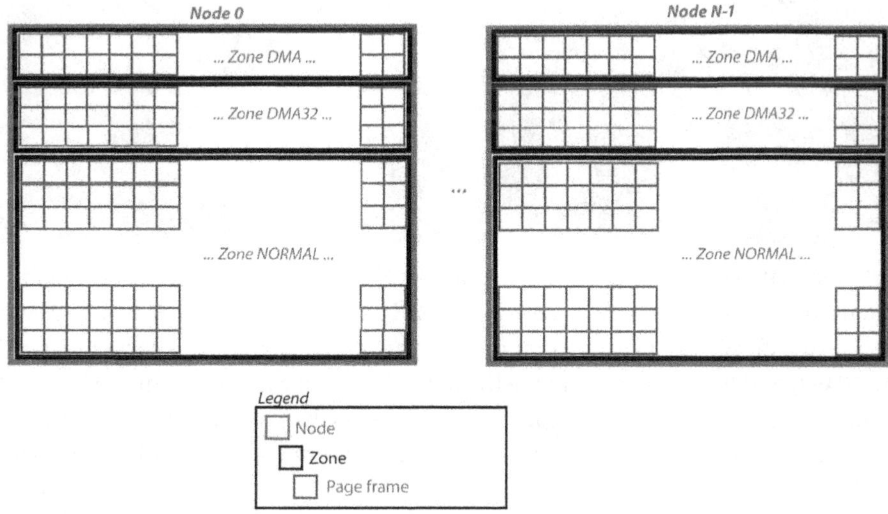

Figure 7.23: Another (conceptual and generic) view of the physical memory hierarchy on Linux – nodes, zones, and page frames

(An aside: a bit more on how PFNs are tracked by Linux can be found in the section *An introduction to physical memory models.*)

Figure 7.23 depicts a conceptual generic Linux system with *N nodes* (numbered from `0` to `N-1`), with each node consisting of, for example, three *zones*, and each zone made up of physical pages of RAM – *page frames*. The number (and name) of zones per node is dynamically determined by the kernel at boot. You can check out this physical memory hierarchy on a Linux system by delving under *procfs*. In the following code, we peek into a native Linux x86_64 system with 16 GB of RAM:

```
$ cat /proc/buddyinfo
Node 0, zone   DMA          3      2     4    3    3     1    0    0   1   1   3
Node 0, zone   DMA32    31306  10918  1373  942  505   196   48   16   4   0   0
Node 0, zone   Normal   49135   7455  1917  535  237    89   19    3   0   0   0
$
```

The leftmost column reveals that we have exactly one node across the entire system: `Node 0`. This tells us we're actually on a *UMA system*, though of course, the Linux OS will treat it as a (pseudo/fake) NUMA system. As can be seen, this single node, `Node 0`, is split into three zones, labeled `DMA`, `DMA32`, and `Normal`; each zone, of course, consists of page frames. For now, ignore the numbers on the right; we will get to their meaning in the following chapter.

Another way to notice how Linux "fakes" a NUMA node on UMA systems is visible from the kernel log. We run the following command on the same native x86_64 system with 16 GB of RAM. For readability, I replaced the first few columns showing the timestamp and hostname with ellipses:

```
$ journalctl -b -k --no-pager | grep -A7 "NUMA"
 <...>: No NUMA configuration found
 <...>: Faking a node at [mem 0x0000000000000000-0x00000004427fffff]
 <...>: NODE_DATA(0) allocated [mem 0x4427d5000-0x4427fffff]
 <...>: Zone ranges:
   <...>: DMA      [mem 0x0000000000001000-0x0000000000ffffff]
 <...>:   DMA32    [mem 0x0000000001000000-0x00000000ffffffff]
 <...>:   Normal   [mem 0x0000000100000000-0x00000004427fffff]
 <...>:   Device   empty
 $
```

We can clearly see that, as the system is detected as not NUMA (thus, UMA), the kernel fakes a NUMA node. The extents of the node are the total amount of RAM on the system (here, `0x0-0x00000004427fffff`, which is indeed 16 GB). We can also see that on this particular system, the kernel instantiates three zones – `DMA`, `DMA32`, and `Normal` – to organize the available physical page frames of RAM. This is fine and ties in with the `/proc/buddyinfo` output we saw previously. The data structure representing the *zone* on Linux is defined here: `include/linux/mmzone.h:struct zone`. We will have occasion to visit it later in the book.

Now, to better understand how the Linux kernel organizes RAM, let's start at the very beginning – boot time.

Direct-mapped RAM and address translation

At boot, the Linux kernel "maps" all (usable) system RAM (aka *platform RAM*) directly into the kernel VAS (we learned about this in the *Examining the kernel VAS* section; see *Figure 7.12* and *Figure 7.14* again if you wish). So, we have the following:

- Physical page frame `0` maps to kernel virtual page `0`.
- Physical page frame `1` maps to kernel virtual page `1`.
- Physical page frame `2` maps to kernel virtual page `2`, and so on.

Thus, we call this a 1:1 or direct mapping, identity-mapped RAM, or linear addresses. A key point is that all *these kernel virtual pages are at a fixed offset from their physical counterparts* (and, as already mentioned, these kernel addresses are pedantically referred to as kernel logical addresses). The fixed offset is the `PAGE_OFFSET` value (here let's assume it's the value `0xc000 0000`).

So, think of this. On a 32-bit system with a 3:1 (GB) VM split, physical address `0x0` is equal to kernel logical address `0xc000 0000` (`PAGE_OFFSET`). As already mentioned, the (pedantic) terminology *kernel logical address* is applied to kernel addresses that are at a fixed offset from their physical counterparts. Thus, direct-mapped RAM maps to kernel logical addresses. This region of direct-mapped memory is often referred to as the *low-memory* (or simply, **lowmem**) region within the kernel VAS.

We have already shown an almost identical diagram earlier, *Figure 7.12*. In the following figure, to emphasize the points just made, it's slightly modified to actually show how the first three (physical) page frames of RAM map to the first three kernel virtual pages (in the lowmem region of the kernel VAS):

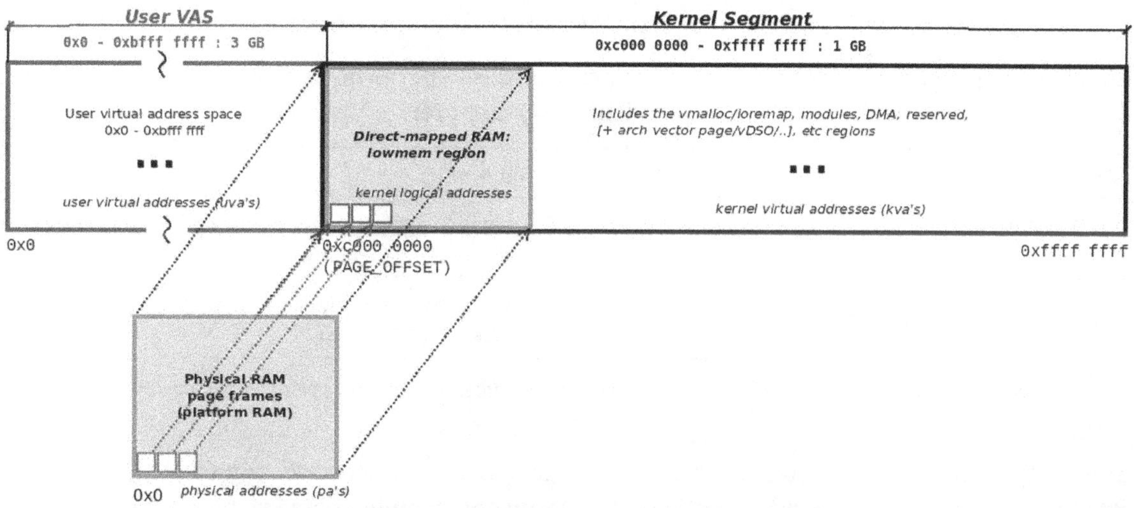

Figure 7.24: Focus on the direct-mapped RAM – the lowmem region – on 32-bit Linux with a 3:1 (GB) VM split

As an example, *Figure 7.24* shows a direct mapping of platform RAM to the kernel VAS on a 32-bit system with a 3:1 (GB) VM split. The point where physical RAM address `0x0` maps into the kernel is the `PAGE_OFFSET` kernel macro (in the preceding figure, it's kernel logical address `0xc000 0000`).

Notice how *Figure 7.24* also shows the *user VAS* on the left side, ranging from `0x0` to `PAGE_OFFSET-1` (of size `TASK_SIZE` bytes). We have already covered details on the remainder of the kernel VAS in the earlier *Examining the kernel VAS* section.

Understanding this mapping of physical-to-virtual pages might well tempt you into reaching these seemingly logical conclusions:

- Given a KVA, to calculate the corresponding **Physical Address** (PA) – that is, to perform a KVA-to-PA calculation – simply do this:

  ```
  pa = kva - PAGE_OFFSET
  ```

- Conversely, given a PA, to calculate the corresponding KVA – that is, to perform a PA-to-KVA calculation – simply do this:

  ```
  kva = pa + PAGE_OFFSET
  ```

Do refer to *Figure 7.24* again. The direct mapping of RAM to the kernel VAS (starting at `PAGE_OFFSET`) certainly predicates this conclusion. So, it is correct. But hang on, please pay careful attention here: **these address translation calculations work only for direct-mapped or linear addresses** – in other words, KVAs (technically, the kernel logical addresses) – **within the kernel's lowmem region, nothing else!** For all UVAs, and all KVAs *besides* the lowmem region (which includes module addresses, `vmalloc`/`ioremap` (MMIO) addresses, KASAN addresses, the (possible) highmem region addresses, DMA memory regions, and so on), it does *not* work! This fact is also further reinforced by the realization that physical RAM on a system isn't always contiguous or linear; it can have holes! Again, this discussion is deferred to the upcoming section *An introduction to physical memory models.*

As you will anticipate, the kernel does indeed provide APIs to perform these address conversions; of course, their implementation is arch-dependent. Here they are:

Kernel API	What it does
`phys_addr_t virt_to_phys(volatile void *address)`	Converts the given virtual address to its physical counterpart (return value)
`void *phys_to_virt(phys_addr_t address)`	Converts the given physical address to a virtual address (return value)

Table 7.13: Converting between physical to kernel virtual addresses (and vice versa) within the lowmem region

The `virt_to_phys()` API for the x86 has a comment above it clearly advocating that this API (and its ilk) are **not to be used by driver authors;** for clarity and completeness, we have reproduced the comment in the kernel source here (`https://elixir.bootlin.com/linux/v6.1.25/source/arch/x86/include/asm/io.h#L118`):

```
// arch/x86/include/asm/io.h
/**
 *  virt_to_phys    -   map virtual addresses to physical
```

```
 * @address: address to remap
 *
 * The returned physical address is the physical (CPU) mapping for
 * the memory address given. It is only valid to use this function on
 * addresses directly mapped or allocated via kmalloc.
 *
 * This function does not give bus mappings for DMA transfers. In
 * almost all conceivable cases a device driver should not be using
 * this function
 */
static inline phys_addr_t virt_to_phys(volatile void *address)
```

The preceding comment mentions the (very common) `kmalloc()` API. Worry not, it's covered in depth in the following two chapters. Of course, a similar comment to the preceding is in place for the `phys_to_virt()` API as well.

So, who – sparingly – uses these address conversion APIs (and the like)? The kernel internal *MM* code, of course! As a demo, we do use them in at least a couple of places in this book: in the following chapter, in an LKM called `ch8/lowlevel_mem` (well actually, its usage is within a function in our "kernel library" code, `klib.c`). Also FYI, the powerful `crash` utility can indeed translate any given virtual address to a physical address via its `vtop` (virtual-to-physical) command (and vice versa, via its `ptov` command!).

Moving along, another key point to note: by mapping all physical RAM into itself, do not get misled into thinking that the kernel is *reserving* RAM for itself. It isn't; it's merely *mapping* all the available RAM, thus making it available for allocation to anyone who wants it – core kernel code, kernel threads, device drivers, or user space applications. This is part of the job of the OS; it is the system resource manager, after all.

Of course, a certain portion of RAM will be taken up (allocated) – by the static kernel code, data, kernel page table, and so on – at boot, no doubt. Furthermore, using the kernel's kdump and/or **Contiguous Memory Allocator (CMA)** functionalities can entail reserving larger specified amounts of RAM at boot for them. Typically, though, you should realize that the amount of RAM used directly by the kernel itself is quite small. As an example, on my guest VM with 1 GB RAM, the kernel code, data, and BSS typically take up a combined total of about 25 MB of RAM. All kernel memory comes to about 100 MB, whereas user space memory usage is in the region of 550 MB! It's almost always user space that is the memory hogger.

TIP: Try using the `smem` utility with the `--system -p` option switches to see a summary of memory usage as percentages (also, use the `--realmem=` switch to pass the actual amount of RAM on the system).

Back to the point: we know that kernel page tables are set up early in the boot process. So, by the time applications start up, *the kernel has all RAM mapped and available*, ready for allocation! Thus, we understand that while the kernel *direct-maps* page frames into its VAS, user mode processes are not so lucky – they can only *indirectly map* page frames via paging tables set up by the OS (at process creation – `fork()` – time) on a per-process basis. It's also interesting to realize that memory mapping via the powerful `mmap()` system call can provide the illusion of "direct mapping" files or anonymous pages into the user VAS (under the hood, its all page table manipulation).

A few additional points to note:

- For performance, allocated kernel memory pages can *never be swapped*, even if they aren't in use.
- Sometimes, you might think, it's quite obvious that user space memory pages map to (physical) page frames (assuming the page is resident) via the paging tables set up by the OS on a per-process basis.

  Yes, but what about kernel memory pages? Please be very clear on this point: all kernel pages also map to page frames via the kernel "master" paging table (named `swapper_pg_dir`). *Kernel memory, too, is virtualized, just as user space memory is.*

  In this regard, for you, the interested reader, check out a Q&A I initiated on Stack Overflow: *How exactly do kernel virtual addresses get translated to physical RAM?* at `http://stackoverflow.com/questions/36639607/how-exactly-do-kernel-virtual-addresses-get-translated-to-physical-ram`.
- Several memory optimization techniques have been baked into the Linux kernel (well, many are configuration options); among them are **Transparent Huge Pages** (**THPs**) and, critical for cloud/virtualization workloads, **Kernel Samepage Merging** (**KSM**, aka memory de-duplication). I refer you to the *Further reading* section of this chapter for more information.

Now that we've covered the first two levels of the physical memory hierarchy (nodes and zones), let's delve into the organization of the third and last one: page frames!

An introduction to physical memory models

Any way you cut it, physical memory is a precious resource. Add to this the fact that memory organization on modern systems can be complex hierarchies, with large *holes* (or *sparse regions*) scattered in between the memory space being quite common (the previous sections glossed over this). Not only that, each NUMA node on a server-type system needs its own set of memory management metadata. Hardware too has become more demanding – the ability to hot-plug (and remove) memory banks (and CPUs), the desire for setting up page-level mapping within some types of persistent storage devices, and so on... Thus, the Linux kernel community has devised abstractions to better mimic and thus manage physical memory. These abstractions are in the form of a **memory model**; in fact, to date, three such models have been proposed and implemented: ***flatmem***, ***discontigmem***, and ***sparsemem***. Realistically though, the *sparsemem* model is the one enjoying wide deployment, especially with modern 64-bit systems with large-ish amounts of RAM; the *discontigmem* model has been deprecated and the *flatmem* one still clings on for dear life (servicing 32-bit systems with small amounts of RAM).

In all models, a basic notion is the ability to track *every single physical page of RAM* on the system. The metadata structure that tracks this is `struct page` (`https://elixir.bootlin.com/linux/v6.1.25/source/include/linux/mm_types.h#L73`); it's commonly called the *page descriptor*. (Track every single RAM page with a *page* structure? Yes indeed, and that's why it's kept small, just 64 bytes; nevertheless, it does eat into memory, especially when the amount of RAM is large.) It has metadata within it regarding the page it tracks – what it's being used for (or that it's currently free), and various flag values, including the mapping.

Back to the memory models. Linux requires each architecture to manage its memory using a model – either *flatmem* or *sparsemem*. The arch-specific code in early boot sets this up (for example, `sparse_init()` sets up the *sparsemem* model). A few characteristics true of every memory model follow:

- As physical RAM tends to be consecutive page frames, possibly broken up by holes, the models typically implement one or more arrays of *struct page* within *units*; *sparsemem* calls a unit a *section*. Each physical frame of RAM is denoted by a **Page Frame Number** (**PFN**), effectively the index to an array of *struct page* objects. A 1:1 mapping always exists between a PFN and a *struct page*.
- This mapping requires helper functions to convert between the two: the helpers `page_to_pfn()` and `pfn_to_page()` will minimally be defined for each model.

Let's delve a bit more into the sparsemem memory model.

Understanding the sparsemem[-vmemmap] memory model in brief

The *sparsemem* model is effectively the versatile one, the one most often used in practice. It can abstract and support required modern features like the hot-plugging and removal of memory banks, persistent device memory mapping, deferred init of memory, and so on.

This model can implement the mandatory helper APIs (`page_to_pfn()` and `pfn_to_page()`) in two ways: via "classic sparse" or the "*sparse vmemmap*" approach. When using the latter (the common approach), a *vmemmap* pointer points to the base of an array of *struct page* objects (more precisely, it's set to `(struct page *)VMEMMAP_START`. For a more visual look, I suggest you look up the diagrams (and read the article): *Reducing page structures for huge pages, Jon Corbet, LWN, Dec 2020* at `https://lwn.net/Articles/839737/`. Now you'll better understand the *vmemmap region* – the metadata described originating from this pointer – mentioned in the earlier *Macros and variables describing the kernel VAS layout* section.

With the *sparsemem-vmemmap* model being the preferred one for 64-bit Linux these days, you'll often find the kernel configs `CONFIG_SPARSEMEM_VMEMMAP=y` and `CONFIG_SPARSEMEM_VMEMMAP_ENABLE=y` set on such systems (modern x86_64 *always* uses the *sparsemem* model and AArch64 platforms seem to adhere to this as well. FYI, the common Android **Generic Kernel Image** (**GKI**) kernel for AArch64 has these config options enabled as well by default, demonstrating that it too uses this model).

Two macros are required to be defined for the *sparsemem* model:

- `SECTION_SIZE_BITS`: Number of physical address bits to cover the maximum amount of memory in a (physically contiguous) section; the size of each memory section will be 2 raised to this power.

- `MAX_PHYSMEM_BITS`: Maximum number of bits in a physical address; in effect, the maximum amount of RAM supported. This value is the same as the `CONFIG_ARM64_PA_BITS` kernel configurable on the AArch64.

These macros are typically defined within the `arch/<arch>/include/asm/sparsemem.h` header. A tiny module to reveal these numbers, as well as provide a little more info on the *sparsemem* model, is here: `ch7/sparsemem_show`. The essential code is as follows:

```
// ch7/sparsemem_show/sparsemem_show.c
#include <linux/mmzone.h>
[ ... ]
#ifdef CONFIG_SPARSEMEM_VMEMMAP
    pr_info("VMEMMAP_START = 0x%016lx\n"
    "SECTION_SIZE_BITS = %u, so size of each section=2^%u = %u bytes = %u MB\n"
    "max # of sections=%lu\n"
    "MAX_PHYSMEM_BITS=%u; so max supported physical addr space (RAM): %lu GB =
%lu TB\n",
    VMEMMAP_START,
    SECTION_SIZE_BITS, SECTION_SIZE_BITS, (1 << SECTION_SIZE_BITS),
    (1 << SECTION_SIZE_BITS) >> 20, NR_MEM_SECTIONS, MAX_PHYSMEM_BITS,
    (1UL << MAX_PHYSMEM_BITS) >> 30, (1UL << MAX_PHYSMEM_BITS) >> 40);
#else
    pr_info("SPARSEMEM_VMEMMAP not supported\n");
#endif
```

A summary table showing the values obtained for the sparsemem-related macros via this module on different 64-bit platforms is shown:

Attributes / Platforms	AArch64: stock Raspberry Pi OS	AArch64: custom Yocto build (Poky, 4.0.4), with CONFIG_ARM64_VA_BITS=52	x86_64 VM or native Ubuntu 22.04 LTS
SECTION_SIZE_BITS	27	29	27
Size of each sparsemem section: **2^ SECTION_SIZE_BITS**	128 MB	512 MB	128 MB
MAX_PHYSMEM_BITS	48	52	46
Max physical address space (max RAM, in effect): **2^ MAX_PHYSMEM_BITS**	256 TB	4,096 TB = 4 PB	64 TB

Table 7.14: Summary of values obtained with regard to some of the sparsemem macros on different architectures (platforms) running 64-bit Linux

Notice how the third column here matches up with row #7 (the last AArch64 row) in *Figure 7.6*. I don't show entries for the 32-bit platforms as they simply don't support the *sparsemem[-vmemmap]* model (and instead employ the much simpler *flatmem* one). For more detail on these topics, refer to the links within the *Further reading* document.

Alright, with this coverage of some aspects of physical RAM management behind us, we complete this chapter. Good going; excellent progress!

Summary

In this chapter, we delved – in quite some depth – into the big topic of kernel memory management in a level of detail sufficient for a kernel module or device driver author like you; also, there's more to come! A key piece of the puzzle – the VM split and how it's achieved on various architectures running the Linux OS – served as a starting point.

We then moved into a deep examination of both regions of this split: first, user space (the user mode process VAS) and then the kernel VAS (or kernel segment). Here, we covered many details and tools/utilities on how to examine it (including via the quite powerful `procmap` utility). We built a demo kernel module that can literally generate a pretty complete memory map of the kernel and the calling process. User and kernel memory layout randomization technology ([K]ASLR) was also briefly discussed. We closed the chapter by looking at the physical organization of RAM within the Linux OS, including the memory models in use (especially the *sparsemem* one).

The information and the concepts learned within this chapter are *very useful*, not only for designing and writing better kernel/device driver code but also when you encounter system-level issues and bugs; it makes you better able to debug complex scenarios.

This chapter has been a long and, indeed, critical one; great job on completing it! Next, in the following two chapters, you will move on to learning about key and practical aspects of how exactly to allocate (and of course deallocate) kernel memory efficiently, along with related important concepts behind this common activity. On, on!

Questions

As we conclude, here is a list of questions for you to test your knowledge regarding this chapter's material: `https://github.com/PacktPublishing/Linux-Kernel-Programming_2E/blob/main/questions/ch7_qs_assignments.txt`. You will find some of the questions answered in the book's GitHub repo: `https://github.com/PacktPublishing/Linux-Kernel-Programming_2E/tree/master/solutions_to_assgn`.

Further reading

To help you delve deeper into the subject with useful materials, we provide a rather detailed list of online references and links (and at times, even books) in a *Further reading* document in this book's GitHub repository. The *Further reading* document is available here: `https://github.com/PacktPublishing/Linux-Kernel-Programming_2E/blob/master/Further_Reading.md`.

Learn more on Discord

To join the Discord community for this book – where you can share feedback, ask questions to the author, and learn about new releases – follow the QR code below:

`https://packt.link/SecNet`

8

Kernel Memory Allocation for Module Authors – Part 1

In the previous two chapters, one on the kernel's internal aspects and architecture and the other on the essentials of memory management internals, we covered key aspects that serve to provide the required background information for this and the following chapter. In this and the next chapter, we will get down to the actual allocation and freeing of kernel memory by various means. We will demonstrate this via kernel modules that you can test and tweak, elaborate on the whys and hows of allocation, and provide many real-world tips and tricks to enable a kernel or driver developer like you to gain maximum efficiency when working with memory within your kernel module.

In this chapter, we will cover the kernel's two primary memory allocators – the **Page Allocator** (**PA**) (aka the **Buddy System Allocator** (**BSA**)) and the **slab allocator**. We will delve into the nitty-gritty of working with their APIs within kernel modules. We will go well beyond simply seeing how to use the APIs, clearly demonstrating why memory efficiency isn't optimal in many cases, and how to overcome these situations. *Chapter 9, Kernel Memory Allocation for Module Authors – Part 2,* will continue our coverage of the kernel memory allocators, delving into a few more advanced areas.

In this chapter, we will cover the following topics:

- Introducing kernel memory allocators
- Understanding and using the kernel page allocator (or BSA)
- Understanding and using the kernel slab allocator
- Size limitations of the kmalloc API
- Slab allocator – a few additional details
- Caveats when using the slab allocator

Technical requirements

I assume that you have gone through *Online Chapter, Kernel Workspace Setup*, and have appropriately prepared a guest **Virtual Machine** (**VM**) running Ubuntu 22.04 LTS (or a later stable release) and installed all the required packages. If not, I highly recommend you to do this first.

To get the most out of this book, I strongly recommend first setting up the workspace environment, including cloning this book's GitHub repository (`https://github.com/PacktPublishing/Linux-Kernel-Programming_2E`) for the code, and working on it in a hands-on fashion.

The topics covered here assume you have at least basic familiarity with the process **Virtual Address Space (VAS)** and virtual memory basics (including physical memory layout concepts like UMA and NUMA); we covered these areas in *Chapter 6, Kernel Internals Essentials – Processes and Threads*, and *Chapter 7, Memory Management Internals – Essentials*. Ensuring you understand them before going into the following material is important.

Introducing kernel memory allocators

Dynamically allocating, and subsequently freeing, kernel memory – both physical and virtual – is the key topic area here. The Linux kernel, like any other OS, requires a sturdy algorithm and implementation to perform this really key task. The primary (de)allocator engine in the Linux OS is referred to as the PA, or the BSA. Internally, it uses a so-called **buddy system algorithm** to efficiently organize and parcel out free chunks of system RAM. You will find more on the algorithm in the *Understanding and using the kernel page allocator (or BSA)* section of this chapter.

In this chapter and throughout this book, when we use the notation *(de)allocate*, please read it as both words: *allocate* and *deallocate*.

Of course, being imperfect, the page allocator is not the only or always the best way to obtain and subsequently release system memory. Other technologies exist within the Linux kernel to do so. High on the list of these is the kernel's **slab allocator** or **slab cache** system (we use the word *slab* here as the generic name for this type of allocator as it originated with this name; in practice, though, the internal implementation of the modern slab allocator used by the Linux kernel is called **SLUB** (the unqueued slab allocator); more on this later).

Think of it this way: the slab allocator solves some issues and optimizes performance with the page allocator. What issues exactly? Patience, you'll soon see... For now, though, it's important to understand that the *only way in which to actually (de)allocate physical memory is via the page allocator*. In effect, the page allocator is the primary 'engine' for memory (de)allocation on the Linux OS!

To avoid confusion and repetition, we will from now on refer to this primary allocation engine as the **page allocator.** You will understand that it's also known as the BSA (derived from the name of the algorithm that drives it).

Thus, the slab allocator is layered upon (or above) the page allocator. Various core kernel subsystems, as well as non-core code within the kernel, such as modules and device drivers, can allocate (and deallocate) memory either directly via the page allocator or indirectly via the slab allocator. The following diagram illustrates this:

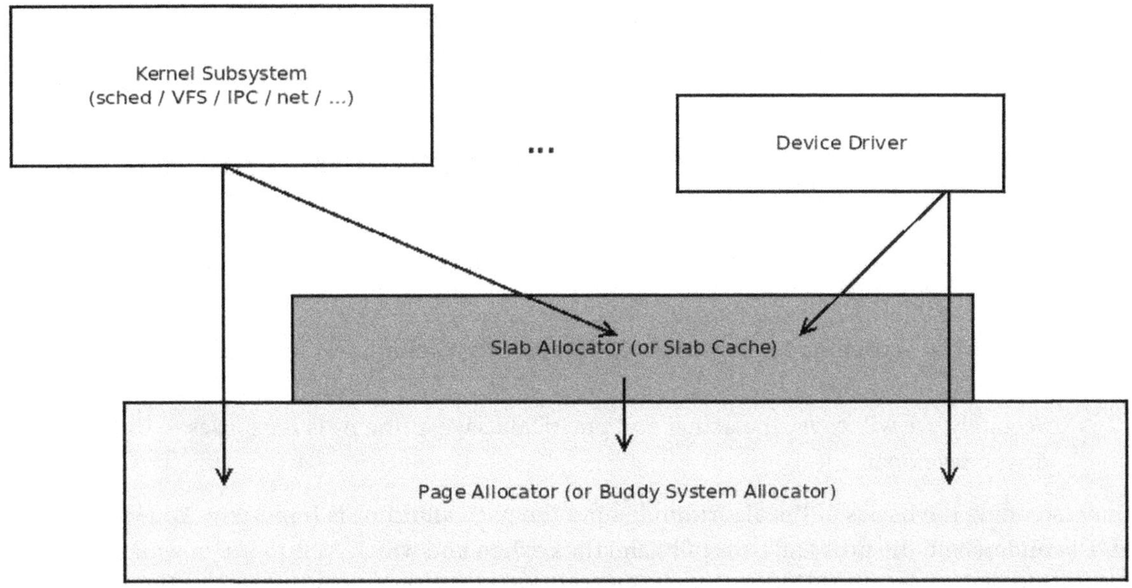

Figure 8.1: Linux's page allocator engine with the slab allocator layered above it

A few things to be clear about at the outset are:

- The Linux kernel and all of its core components and subsystems (excluding the memory management subsystem itself and perhaps a few very arch-specific users) ultimately use the page allocator (or BSA) for memory (de)allocation. This includes non-core stuff, such as kernel modules and device drivers.
- The slab and page allocators reside completely in kernel (virtual) address space and are not directly accessible from user space.
- The page frames (RAM) from where the page allocator gets memory is within the kernel's *lowmem* region, or the direct-mapped RAM region of the kernel segment (we covered the kernel segment in detail in the previous chapter)
- The slab allocator is ultimately a user of the page allocator, and thus gets its memory from there itself (which again implies that the memory ultimately comes from the kernel lowmem region)
- *User space* dynamic memory allocation with the familiar `malloc` family of APIs does *not* directly map to the preceding layers (that is, calling `malloc()` in user space does *not* directly result in a call to the page or slab allocator). It does so indirectly. How exactly? You will learn how, patience! (This key coverage is found in two sections of *Chapter 9, Kernel Memory Allocation for Module Authors – Part 2,* within the ones involving *demand paging*; look out for it as you cover that chapter!)
- Also, to be clear, Linux kernel memory is non-swappable. It can never be swapped out to disk; this was decided in the early Linux days to keep performance high. User space memory pages are always swappable by default, though this can be changed by the system programmer via the `mlock()`/`mlockall()` system calls.

Now, fasten your seatbelts! With this very basic understanding, let's begin the journey of learning (the basics of) how the Linux kernel's memory allocators work and, more importantly, how to work well with them.

Understanding and using the kernel page allocator (or BSA)

In this section, you will learn about two aspects of the Linux kernel's primary (de)allocator engine, the page allocator (or BSA):

- First, we will cover the fundamentals of the algorithm behind this software (called the "buddy system").
- Secondly, we will cover the actual and practical usage of the APIs it exposes to the kernel or driver developer.

Understanding the basics of the algorithm behind the page allocator is important. You will then be able to understand the pros and cons of it, and thus, when and which APIs to use in which situation. Let's begin with its inner workings. Again, remember that the scope of this book with regard to the internal memory management details is limited. We shall cover it to a depth deemed sufficient for a typical module/driver author and no more.

The fundamental workings of the page allocator

We will break up this discussion into a few relevant parts. Let's begin with how the kernel's page allocator tracks free physical page frames via its *freelist* data structures.

Understanding the organization of the page allocator freelist

The key to the page allocator (buddy system) algorithm is its primary internal metadata structure. It's called the **buddy system freelist** and consists of an array of pointers to (the oh-so-common!) doubly linked circular lists. The index of this array of pointers is called the **order** of the list – it's the power to which to raise 2 to. The array length is from `0` to `MAX_ORDER-1`. The value of `MAX_ORDER` is arch-dependent; on the x86 and ARM platforms, it's 11, whereas on a large-ish system such as the Itanium, it's 17. Thus, on the x86 and ARM, the order (indices) range from 2^0 to 2^{10}; that is, from 1 to 1,024. What does that mean? Do read on...

Each doubly linked circular list *points to free and physically contiguous page frames of size 2^{order} pages*. Thus, assuming a page size of 4 KB, we end up with eleven lists (0-10) with the following size characteristics:

- Order 0: 2^0 = 1 page = 4 KB chunks
- Order 1: 2^1 = 2 pages = 8 KB chunks
- Order 2: 2^2 = 4 pages = 16 KB chunks

- Order 3: 2^3 = 8 pages = 32 KB chunks
- [... and so on ...]
- Order 10: 2^{10} = 1024 pages = 1024*4 KB = 4 MB chunks

A diagram will help! The following is a simplified conceptual illustration of (a single instance of) the page allocator freelist:

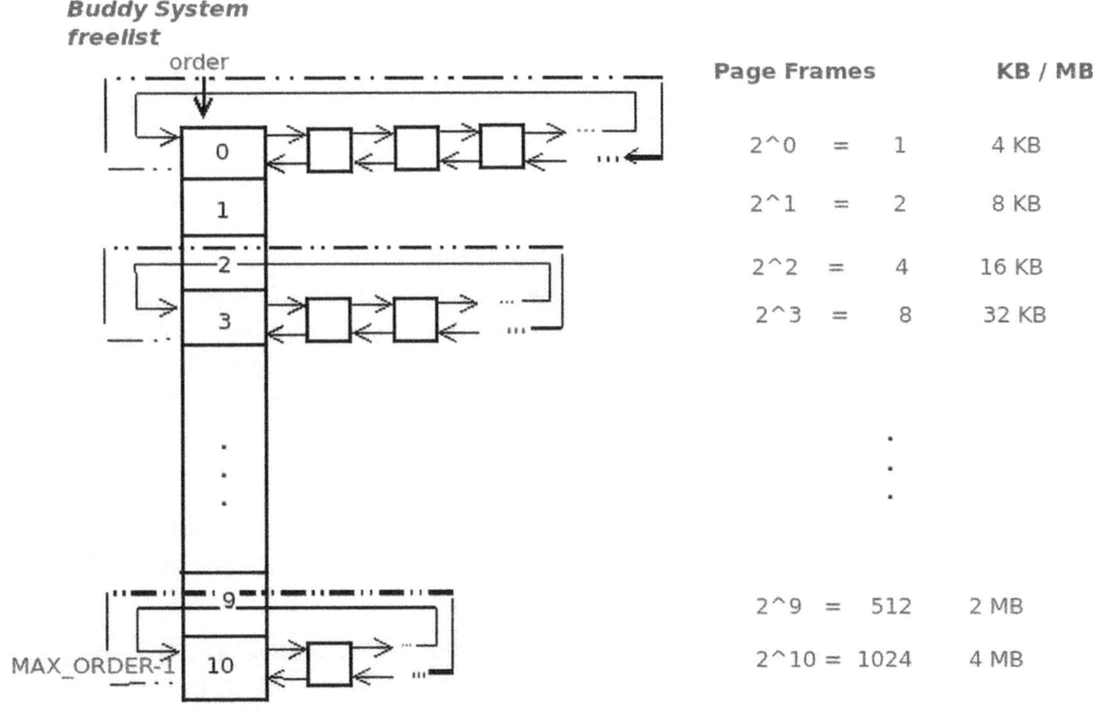

Figure 8.2: Buddy system/page allocator freelist on a system with a page size of 4 KB and MAX_ORDER of 11

In *Figure 8.2*, each memory "chunk" is represented by a square box (to keep it simple, we use the same size for all chunks in our diagram). Internally, of course, these aren't the actual memory pages; rather, the boxes represent metadata structures (*struct page*) that point to physical memory frames. (Recall, we mentioned the importance of struct page in *Chapter 7*, *Memory Management Internals – Essentials*, in the *An introduction to physical memory models* section).

On the right side of *Figure 8.2*, we show the size of each individual physically contiguous free memory chunk that could possibly be enqueued on the list to the left.

The kernel gives us a convenient (summarized) view into the current state of the page allocator via the proc filesystem (via the /proc/buddyinfo pseudofile); here's an example from our Ubuntu VM with 1 GB RAM:

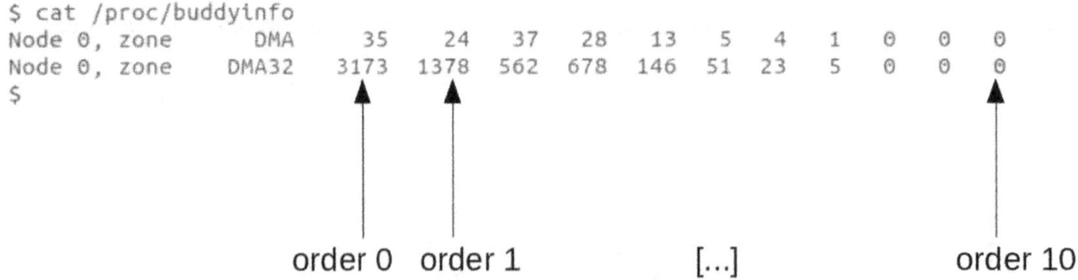

Figure 8.3: Annotated screenshot of sample /proc/buddyinfo output

As you can tell from *Figure 8.3*, our VM is a pseudo-NUMA box with one node (`Node 0`) and two zones (`DMA` and `DMA32`). (We covered what NUMA is, and associated topics, in *Chapter 7, Memory Management Internals – Essentials,* under the *Physical RAM organization* section). The numbers following `zone XXX` are the number of free (physically contiguous!) page frames on the order 0, order 1, order 2, ..., right up to the order `MAX_ORDER-1` (here, *11 – 1 = 10*) list. So, let's take a couple of examples from the preceding output (*Figure 8.3*):

- There are 35 single-page free chunks of RAM in the order `0` list for node `0`, zone DMA.
- In node `0`, zone DMA32, order `3`, the number shown in *Figure 8.3* is 678; now, take $2^{order} = 2^3 =$ *8 page frames = 32 KB* (assuming a page size of 4 KB); this implies that there currently are 678 32-KB physically contiguous free chunks of RAM on that list.

It is important to note that **each chunk is guaranteed to be physically contiguous RAM in and of itself.** Also, notice that the size of the memory chunks on a given order is always double that of the previous order (and half that of the next one). This mathematical property is going to be the case, of course, as they're all powers of 2.

Note that `MAX_ORDER` can (and does) vary with the architecture. On regular x86 and ARM systems, it's `11`, yielding the largest possible chunk size of 4 MB of physically contiguous RAM on order 10 of the freelist(s) (as the last order will be 2^{10} = 1024 page frames = 1024 * 4KB = 4096 KB = 4 MB). As an alternate example, on some high-end enterprise servers running the Itanium (IA-64) processor, `MAX_ORDER` can be as high as `17` (implying a largest chunk size on order (17-1), thus of 2^{16} = 65,536 pages = 512 MB chunks of physically contiguous RAM on order 16 of the freelists, for a 4 KB page size). The IA-64 MMU supports up to eight page sizes ranging from a mere 4 KB right up to 256 MB. As another example, with a page size of 16 MB, the order 16 list could potentially have physically contiguous RAM chunks of size *65,536 * 16 MB = 1 TB* each!

Another key point: the kernel keeps **multiple PA/BSA freelists – one for every node:zone that is present on the system!** This lends itself to a natural way to allocate memory on a NUMA system.

The following diagram shows how the kernel instantiates multiple freelists – *one per node:zone present on the system* (diagram credit: *Professional Linux Kernel Architecture,* Mauerer, Wrox Press, Oct 2008):

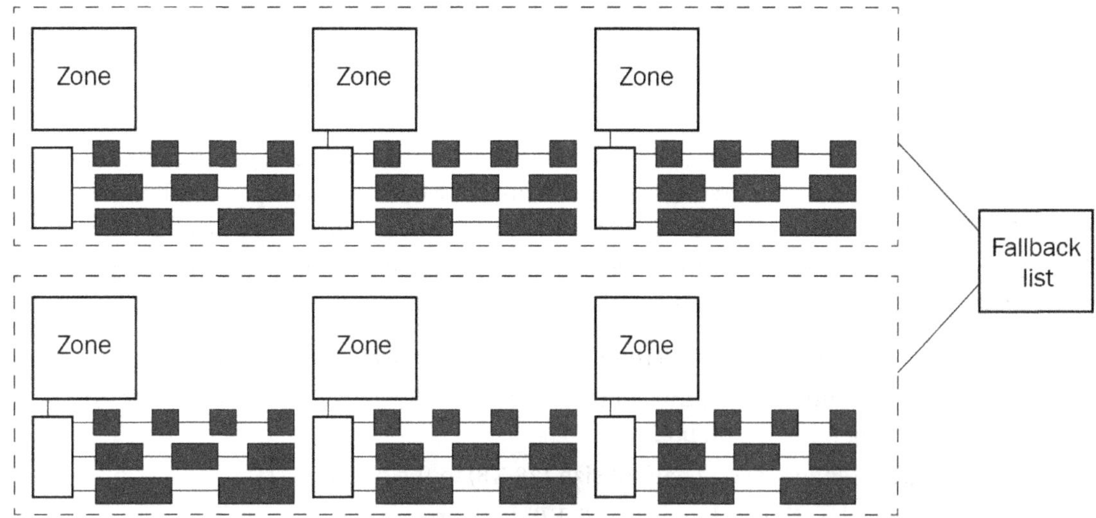

Figure 8.4: Page allocator (BSA) "freelists," one per node:zone on the system; Diagram credit: Professional Linux Kernel Architecture, Mauerer, Wrox Press, Oct 2008

Furthermore (as can be seen in *Figure 8.4*), when the kernel is called upon to allocate RAM via the page allocator, it picks the optimal freelist to allocate memory from – the one associated with the node upon which the thread asking the request is running (recall the NUMA architecture from the previous chapter). If the zone(s) on this node is out of memory or cannot allocate it for whatever reason, the kernel then uses a fallback list to figure out which freelist to attempt to allocate memory from. (In reality, the real picture is even more complex; we provide some more details in the *Page allocator internals – a few more details* section).

Let's now understand (in a conceptual way) how all of this works to (de)allocate memory pages!

The workings of the page allocator

The actual (de)allocation strategy can be explained by using a simple example. Let's say a device driver requests 128 KB of memory. To fulfill this request, the page allocator (BSA) algorithm will do this (simplified and conceptually):

1. The algorithm expresses the amount to be allocated (128 KB here) in pages. Thus, here (assuming a page size of 4 KB), it's *128K/4K = 32 pages.*
2. Next, it determines to what power 2 must be raised to get 32. That's *$log_2 32$ (or ln 2),* which is 5 (as 2^5 is 32). Thus the order 5 list will have free memory chunks of precisely this required size.
3. It now checks the list on order 5 of the appropriate (closest) *node:zone* page allocator freelist. If a memory chunk is available (it will be of size *2^5 pages = 128 KB*), dequeue it from the list on order 5, update the list, and allocate it to the requester. Job done! Return to caller.

Why do we say of the appropriate (closest) *node:zone* page allocator freelist? Does that mean there's more than one of them? Yes, indeed! We repeat: the reality is that there will be several PA freelist data structures, one each per *node:zone* on the system. (Also see more details in the upcoming *Page allocator internals – a few more details* section.)

4. If not even one memory chunk is available on the order 5 list (that is, if it's null), then check the list on the next order; that is, the order 6 linked list (if it's not empty, it will have 2^6 *pages = 256 KB* memory chunks enqueued on it, each chunk being double the size of what we want).
5. If the order 6 list is non-null, then take (dequeue) a chunk of memory from it (which will be 256 KB in size, *double* what's required), and do the following:
    - Update the (order 6) list to reflect the fact that one chunk is now removed.
    - Cut the chunk in half, thus obtaining two 128 KB halves or **buddies**! (Please see the following information box.)
    - Migrate (enqueue) one half (of size 128 KB) to the order 5 list.
    - Allocate the other half (of size 128 KB) to the requester.
    - Job done! Return to caller.
6. If the order 6 list is also empty, then it repeats the preceding process with the order 7 list, and so on (recursively), until it succeeds.
7. If all the remaining higher-order lists are empty (null), it will fail the request (this would be very unlikely).

We can cut or slice a memory chunk in half because every chunk on the list is guaranteed to be *physically* contiguous memory. Once cut, we have two halves; each is called a **buddy block**, hence the name of this algorithm. Pedantically, it's called the binary buddy system as we use power-of-2-sized memory chunks. More correctly, a *buddy block* is defined as a block that is of the same size and physically adjacent to another.

You will understand that the preceding description is conceptual. The actual code implementation is certainly more complex and optimized. By the way, the code – the *"heart" of the zoned buddy allocator*, as its comment literally says, is here: `mm/page_alloc.c:__alloc_pages()` (FYI, here, on 6.1.25: `https://elixir.bootlin.com/linux/v6.1.25/source/mm/page_alloc.c#L5523`). Being beyond the scope of this book, we won't attempt to delve into the code-level details of the allocator.

Working through a few scenarios

Now that we have the basics of the algorithm, let's consider a few scenarios: first, a simple straightforward case, and after that, a couple of more complex cases.

The simplest case

Let's say that a kernel-space device driver (or some core code) requests 128 KB and receives a memory chunk from the order 5 list of one of the freelist data structures. At some later point in time, it will (should/must) necessarily free the memory chunk by employing one of the page allocator-free APIs. Now, this "free" API's algorithm calculates - via its order (and thus size) - that the just-freed chunk belongs on the order 5 list; thus, it enqueues it there. (You'll soon see that a parameter to both the alloc and free APIs is the order of the list.)

A more complex case

Now, let's say that, unlike the previous simple case, when the device driver requests 128 KB, the order 5 list is null; thus, as per the page allocator algorithm, we go to the list on the next order, 6, and check it. Let's say it's non-null; the algorithm now dequeues an order 6 chunk of size 256 KB, and *splits (or cuts) it in half*. So, we now have two halves that are physically adjacent and equally sized - ***buddy blocks***. Now, one buddy block (of size 128 KB) goes to the requester, and the remaining buddy block (again, of size 128 KB) is enqueued on to the order 5 list.

The really interesting property of the buddy system is what happens when the requester (the driver), at some later point in time, frees the memory chunk. As expected, the algorithm calculates (via its order) that the just-freed chunk belongs on the order 5 list. But before blindly enqueuing it there, **it looks for its buddy block**, and in this case, it (possibly) finds it! It now merges the two buddy blocks into a single larger block (of size 256 KB) and places (enqueues) the merged block on the *order 6* list. This is fantastic - it has even helped to defragment memory!

The downfall case

Let's make it interesting now by *not* using a convenient rounded power-of-2 size as the allocation requirement. This time let's say that the device driver requests a memory chunk of size 132 KB. What will the buddy system allocator do? As, of course, it cannot allocate less memory than requested, it allocates more - you guessed it (see *Figure 8.2*) - the next appropriately-sized (equal to or more than the request) available memory chunk is on order 7 of the freelist, of size 256 KB. But the consumer (the driver) is only going to see and use the first 132 KB of the 256 KB chunk allocated to it. The remaining (124 KB) is wasted (think about it, that's close to 50% wastage!). This is called **internal fragmentation (or wastage)** and is the critical failing of the binary buddy system!

You will learn, though, that there is indeed a mitigation to this: a patch was contributed to deal with similar scenarios (via the `alloc_pages_exact()` / `free_pages_exact()` APIs). We will cover the APIs to use the page allocator shortly.

Page allocator internals – a few more details

In this book, we do not intend to delve into code-level detail on the internals of the page allocator. Having said that, here's the thing: in terms of data structures, the `zone` structure contains an array of `free_area` structures, the so-called page allocator 'freelists'. This makes sense; as you've learned, there can be (and usually are) multiple page allocator freelists on the system, one per node:zone; let's look them up at the code level:

```
// include/linux/mmzone.h
struct zone {
    [ ... ]
    /* free areas of different sizes */
    struct free_area free_area[MAX_ORDER];
    [ ... ]
};
```

The `free_area` structure is the implementation of the doubly-linked circular lists (of free memory page frames within that node:zone) along with the number of page frames that are currently free within it:

```
struct free_area {
    struct list_head free_list[MIGRATE_TYPES];
    unsigned long nr_free;
};
```

Why does the `free_area` structure contain an array of linked lists and not just one list? Without delving into the details, we'll mention that the kernel layout for the buddy system freelists is more complex than let on until now: from the 2.6.24 kernel, each freelist we have seen is further broken up into multiple (up to 6) individual freelists to cater to different *page migration types* (referring to whether the kernel can literally migrate or move around the page to reduce fragmentation effects. There are five possible migration types (all of which may not be enabled): unmovable, movable, reclaimable, CMA, and isolate). This was required to deal with complications when trying to keep memory (RAM) defragmented.

Besides, as mentioned earlier, these freelists exist per *node:zone* on the system. So, for example, on an actual NUMA system with 4 nodes and 3 zones per node, there will be 12 (4 x 3) freelists. Not just that, we now realize that each `free_area` structure actually consists of (up to) 6 freelists, one per migration type. Thus, on such a system, a total of up to *6* x *12* = *72* freelist data structures would exist system-wide!

If you are interested, dig into the details; the output of `/proc/buddyinfo` is merely a nice summary view of the state of the buddy system freelists (as *Figure 8.3* shows). For a more detailed and realistic view (of the type mentioned previously, showing *all* the freelists), look up `/proc/pagetypeinfo` (this requires root access); it shows *all* the freelists (broken up into page migration types as well). Note, though, that all six migration types need not necessarily be defined on a system.

The design of the page allocator (buddy system) algorithm is one of the *best-fit* classes. It confers the major benefit of helping defragment physical memory as the system runs. The pros of the page allocator (buddy system) algorithm are as follows:

- Helps defragment memory (external fragmentation is mitigated)
- Guarantees the allocation of a physically contiguous memory chunk
- Guarantees CPU cacheline-aligned memory blocks
- Fast (well, fast enough; the algorithmic time complexity is *O(log n)*)

On the other hand, by far the biggest downside is that *internal fragmentation or wastage* can be much too high.

Okay, great! We have covered a good deal of background material on the internal workings of the page or buddy system allocator. Time to get our hands dirty; let's now dive into understanding and using the page allocator APIs to allocate and free kernel memory.

Learning how to use the page allocator APIs

The Linux kernel provides (exposes to the core and modules) a set of APIs to allocate and deallocate memory (RAM) via the page allocator. These are often referred to as the *low-level* (de)allocator routines. The following table summarizes the page allocation APIs; you'll notice that in all the APIs or macros that have two parameters, the first parameter is called the GFP flags or bitmask (its named `gfp_mask`); we shall explain it in detail shortly, please ignore it for now. The second parameter is *order* – the order of the freelist, that is, the amount of memory to allocate is 2^{order} page frames.

In the following discussions, we quite often use the term 'kernel logical address'. As stated in *Chapter 7, Memory Management Internals – Essentials*, in the *Examining the kernel segment* section, the distinction between the 'logical' and 'virtual' address is quite pedantic. Kernel logical address is meant to denote an address within the kernel *lowmem* region; all other kernel addresses are termed **kernel virtual addresses** (**KVAs**). For all practical purposes, you can treat them as the same.

All the following API prototypes can be found in `include/linux/gfp.h`:

API or macro name	Comments	API signature or macro
`__get_free_page()`	Allocates exactly one page frame. The allocated memory will have random content; it's a macro, a wrapper around the `__get_free_pages()` API. The return value is a pointer to the just-allocated memory's kernel logical address.	`#define __get_free_page(gfp_mask) \ __get_free_pages((gfp_mask), 0)`

`__get_free_pages()`	Allocates 2^{order} physically contiguous page frames. The allocated memory will have random content; the return value is a pointer to the just-allocated memory's kernel logical address.	`unsigned long __get_free_pages(gfp_t gfp_mask, unsigned int order);`
`get_zeroed_page()`	Allocates exactly one page frame; its contents are set to ASCII zero (`NULL`; that is, it's zeroed out); the return value is a pointer to the just-allocated memory's kernel logical address.	`unsigned long get_zeroed_page(gfp_t gfp_mask);`
`alloc_page()`	Allocates exactly one page frame. The allocated memory will have random content; it's a macro (wrapper) over the `alloc_pages()` API; the return value is a pointer to the just-allocated memory's `page` metadata structure; it can be converted into a kernel logical address via the `page_address()` function.	`#define alloc_page(gfp_mask) \ alloc_pages(gfp_mask, 0)`
`alloc_pages()`	Allocates 2^{order} physically contiguous page frames. The allocated memory will have random content; the return value is a pointer to the start of the just-allocated memory's `page` metadata structure; it can be converted into a kernel logical address via the `page_address()` function.	`struct page *` `alloc_pages(gfp_t gfp_mask, unsigned int order);`

Table 8.1: Low-level (BSA/page) allocator – popular exported allocation APIs

All the preceding APIs are exported (via the `EXPORT_SYMBOL()` macro), and hence available to kernel module and device driver developers; worry not, you will soon see a kernel module that demonstrates using them.

As mentioned before, the Linux kernel considers it worthwhile to maintain a (small) metadata structure to track every single page frame of RAM. It's called the `page` structure. The point here is to be careful: unlike the usual semantics of returning a pointer (a virtual address) to the start of the newly allocated memory chunk, notice how both the `alloc_page()` and `alloc_pages()` APIs mentioned previously (*Table 8.1*) return a pointer to the start of the newly allocated memory's page structure, not the memory chunk itself (as the other APIs do). You must obtain the actual pointer to the start of the newly allocated memory by invoking the `page_address()` API on the page structure address that is returned. In any case, the example code in the *Writing a kernel module to demo using the page allocator APIs* section of this chapter will illustrate the usage of all the preceding APIs.

Before we can make use of the page allocator APIs mentioned here, though, it's imperative to understand at least the basics regarding the **Get Free Page** (**GFP**) flags, which are the topic of the section that follows.

Dealing with the GFP flags

You will notice that the first parameter to all the previous allocator APIs (or macros) is `gfp_t gfp_mask`. What does this mean? Essentially, these are the GFP - Get Free Page - flags; they affect the behavior of the kernel when allocating memory; we shall come to the details shortly. These are flags (there are several of them) used by the kernel's internal memory management code layers. For all practical purposes, for the typical kernel module (or device driver) developer, just two GFP flags are crucial (as mentioned before, the rest are for internal usage). They are as follows:

- `GFP_KERNEL`
- `GFP_ATOMIC`

Deciding which of these to use when performing memory allocation via the page allocator APIs is important; a key rule to always remember is the following:

If the code is running in process context and it is safe to sleep, use the GFP_KERNEL flag. If it is unsafe to sleep (typically, when in any type of atomic or interrupt context), you must use the GFP_ATOMIC flag.

Following the preceding rule is critical. Getting this wrong can result in the entire machine freezing, kernel crashes, and/or random bad stuff happening. So, what exactly do the statements *safe/unsafe to sleep* really mean? For this and more, we defer to the *The GFP flags – digging deeper* section in this chapter, that follows. It is important though, so I definitely recommend you read it.

Linux Driver Verification (**LDV**) project: back in *Online Chapter, Kernel Workspace Setup*, in the *The LDV – Linux Driver Verification – project* section, we mentioned that this project has useful "rules" with respect to various programming aspects of Linux modules (drivers, mostly) as well as the core kernel.

Regarding our current topic, here's one of the rules, a negative one, implying that you cannot do this: *"Using a blocking memory allocation when spinlock is held"* (`http://linuxtesting.org/ldv/online?action=show_rule&rule_id=0043`). When holding a spinlock, you're not allowed to do anything that might block; this includes kernel-space memory allocations with anything but `GFP_ATOMIC`. Thus, very important, you *must* use the `GFP_ATOMIC` flag when performing a memory allocation in any kind of atomic or non-blocking context, like when holding a spinlock (you will learn that this isn't the case with the mutex lock; you are allowed to perform blocking activities while holding a mutex). Violating this rule leads to instability and even raises the possibility of (an implicit) deadlock. The LDV page mentions a device driver that was violating this very rule and the subsequent fix (`https://git.kernel.org/pub/scm/linux/kernel/git/torvalds/linux.git/commit/?id=5b0691508aa99d309101a49b4b084dc16b3d7019`). Take a look: the patch clearly shows (in the context of the `kzalloc()` API, which we shall soon cover) the `GFP_KERNEL` flag being replaced with the `GFP_ATOMIC` flag.

Another commonly used GFP flag is `__GFP_ZERO`. Its usage implies to the kernel that you want zeroed-out memory pages. It's often bitwise-ORed with `GFP_KERNEL` or `GFP_ATOMIC` flags to return memory initialized to zero (a good programming practice in general!).

The kernel developers have taken the trouble to document the GFP flags in detail. Look up `include/linux/gfp_types.h`. Within it, there's a long and detailed comment; it's headed *DOC: Useful GFP flag combinations* (on 6.1.25, here: `https://elixir.bootlin.com/linux/v6.1.25/source/include/linux/gfp_types.h#L263`).

For now, and so that we get off the ground quickly, just understand that using the Linux kernel's memory allocation APIs with the `GFP_KERNEL` flag is indeed the common case for kernel-internal allocations.

Freeing pages with the page allocator

The flip side of allocating memory is freeing it, of course. We all understand the basic rule: *free memory that you (dynamically) allocated, so as to prevent leakage.* Memory leaks within the kernel are definitely not something you'd like to contribute to. For the page allocator APIs shown in *Table 8.1,* here are the corresponding free APIs:

API or macro name	Comment	API signature or macro
`free_page()`	Free the (single) page that was allocated via the `__get_free_page()`, `get_zeroed_page()`, or `alloc_page()` APIs; it's a simple wrapper over the `free_pages()` API.	`#define free_page(addr) __free_pages((addr), 0)`
`free_pages()`	Free multiple pages that were allocated via the `__get_free_pages()` or `alloc_pages()` APIs (it's a wrapper over `__free_pages()`).	`void free_pages(unsigned long addr, unsigned int order)`
`__free_pages()`	(*Same as the preceding row, plus*) it's the underlying routine where the work gets done; also, note that the first parameter is a pointer to the `page` metadata structure.	`void __free_pages(struct page *page, unsigned int order)`

Table 8.2: Common free page(s) APIs to use with the page allocator

You can see that the actual underlying API in the preceding functions is `free_pages()`, which itself is just a wrapper over the `mm/page_alloc.c:__free_pages()` code. The first parameter to the `free_pages()` API is the pointer to the start of the memory chunk being freed; this, of course, being the return value from the allocation routine. However, the first parameter to the underlying API, `__free_pages()`, is the pointer to the *page* metadata structure of the start of the memory chunk being freed.

Generally speaking, unless you really know what you are doing, you're advised to invoke the `foo()` wrapper routine and not its internal `__foo()` routine. One reason to do so is simply correctness (perhaps the wrapper uses some necessary synchronization mechanism – like a lock – prior to invoking the underlying routine). Another reason to do so is validity checking (which helps code remain robust and secure). Often, the `__foo()` routines bypass validity checks in favor of speed.

Let's now move on to some useful guidelines to follow when allocating or freeing memory within the kernel.

A few guidelines to observe when (de)allocating kernel memory

As all experienced C/C++ application developers know, allocating and subsequently freeing memory is a rich source of bugs! This is primarily because C is an unmanaged language, as far as memory is concerned; hence, you can quite easily hit all sorts of memory bugs. These include the well-known memory leakage, buffer overflows/underflows for both read/write, double-free, and **Use After Free** (**UAF**) bugs.

Unfortunately, it's no different in kernel space; it's just that the consequences are (much) worse! Be extra careful! Please do take care to ensure the following:

- Favor routines that initialize the just-allocated memory to zero.
- Think about and use the appropriate GFP flag(s) when performing an allocation; more on this in the *The GFP flags – digging deeper* section, but briefly, note the following:
    - When in process context where it's safe to sleep, use `GFP_KERNEL`.
    - When in an atomic context, such as when processing a (hardware) interrupt or holding a spinlock, use `GFP_ATOMIC`.
- When using the page allocator (as we're doing now), try as much as possible to keep the allocation size as a rounded power-of-2 pages (again, the rationale behind this and ways to mitigate this – when you don't require so much memory, the typical case – are covered in detail in the coming sections of this chapter).
- You only ever attempt to free memory that you allocated earlier; needless to say (but we'll say it!), don't miss freeing it, don't use the wrong pointer, and don't double-free it.
- Don't attempt to access already-freed memory (this results in the UAF bug).
- Keep the original memory chunk's pointer safe from reuse, manipulation (`ptr ++` or something similar), and corruption, so that you can correctly free it when done.
- Check (and recheck!) the parameters passed to APIs. Is a pointer to the previously allocated block required, or rather a pointer to its underlying page structure?

Finding it difficult and/or worried about issues in production? Don't forget, you have help via the kernel's (as well as third-party) static and dynamic analysis tooling! Make sure to learn how to use powerful static analysis tools found within the kernel (Coccinelle, `sparse`, and others, such as `cppcheck` or `smatch`). For dynamic analysis, learn how to install and use **KASAN** (the **Kernel Address Sanitizer**). The *Linux Kernel Debugging* book covers this and more...

Also recall the 'better' Makefile template I provided in *Chapter 5, Writing Your First Kernel Module – Part 2*, in the *A better Makefile template* section. It contains targets that use several of these tools; please do use it!

Alright, now that we've covered both the (common) allocation and free APIs of the page allocator, it's time to put this learning to use. Let's write some code!

Writing a kernel module to demo using the page allocator APIs

Let's now get hands-on with the low-level page allocator and free APIs that we've learned about so far. In this section, we will show relevant code snippets (not every single line), followed by an explanation where warranted, from our demo kernel module (`ch8/lowlevel_mem/lowlevel_mem.c`).

In the primary worker routine, `bsa_alloc()`, of our small LKM, we highlight some code comments that show what we are trying to achieve. A few points to note:

1. First, we do something very interesting: we use our small kernel "library" function `klib.c:show_phy_pages()` to literally show you how physical RAM page frames are identity mapped to kernel virtual pages in the kernel lowmem region (The exact working of the `show_phy_pages()` routine is discussed very shortly):

```
// ch8/lowlevel_mem/lowlevel_mem.c
[...]
static int bsa_alloc(void)
{
    int stat = -ENOMEM;
    u64 numpg2alloc = 0;
    const struct page *pg_ptr1;

    /* 0. Show the identity mapping: physical RAM page frames to kernel
virtual addresses, from PAGE_OFFSET for 5 pages */
pr_info("0. Show identity mapping: RAM page frames : kernel virtual pages
:: 1:1\n" "(PAGE_SIZE = %ld bytes)\n", PAGE_SIZE);
  /* Show the virt, phy addr and PFNs (page frame numbers). ... */
    show_phy_pages((void *)PAGE_OFFSET, 5 * PAGE_SIZE, 1);
```

2. Second, we allocate one page of memory via the underlying `__get_free_page()` page allocator API (that we saw previously in *Table 8.1*):

```
/* 1. Allocate one page with the __get_free_page() API */
gptr1 = (void *)__get_free_page(GFP_KERNEL);
if (!gptr1) {
        pr_warn("__get_free_page() failed!\n");
        /* As per convention, we emit a printk above saying that the
         * allocation failed. In practice it isn't required; the kernel
         * will definitely emit many warning printk's if a memory alloc
         * request ever fails! Thus, we do this only once (here; could
         * also use the WARN_ONCE()); from now on we don't pedantically
         * print any error message on a memory allocation request
         * failing. */
        goto out1;
}
pr_info("#.     BSA/PA API      Amt alloc'ed         KVA\n"); // header
pr_info("1.  __get_free_page()      1 page     %px\n", gptr1);
```

We emit a couple of `printk` functions here, showing a header line and then showing the details of a particular allocation made (**KVA**, of course, is **Kernel Virtual Address**). Recall from the previous chapter that the KVAs seen are page allocator memory that lies very much in the direct-mapped RAM or lowmem region of the kernel segment/VAS.

Now, for security, we should consistently, and only, use the `%pK` format specifier when printing kernel addresses so that a hashed value and not the real virtual address shows up in the kernel logs. However, here, in order to show you the actual KVA, we simply use the `%px` format specifier (which, like `%pK`, is portable as well; for security, please don't use the `%px` format specifier in production!).

Next, notice the detailed comment in the error code path of the first `__get_free_page()` API (in the preceding code snippet). It mentions the fact that you don't really have to print an out-of-memory error or warning messages. (Curious? To find out why, visit `https://lkml.org/lkml/2014/6/10/382`.)

In this example module (as with several earlier ones and more to follow), we code our printk (or `pr_<foo>()` macro) instances for portability by using appropriate printk format specifiers (such as `%zd`, `%zu`, `%pK`, `%px`, and `%pa`).

3. Now, let's move on to our second memory allocation using the page allocator; see the following code snippet:

```
/*2. Allocate 2^bsa_alloc_order pages with the __get_free_pages() API */
numpg2alloc = powerof(2, bsa_alloc_order); /* returns 2^bsa_alloc_order
*/
gptr2 = (void *)__get_free_pages(GFP_KERNEL|__GFP_ZERO, bsa_alloc_order);
if (!gptr2) {
        /* no error/warning printk now; see above comment */
```

```
        goto out2;
    }
    pr_info("2. __get_free_pages()  2^%d page(s)  %px\n",
            bsa_alloc_order, gptr2);
```

In the preceding code snippet, we allocate 2^3 – that is, 8 – pages of memory via the page allocator's `__get_free_pages()` API (as the default value of our module parameter, `bsa_alloc_order`, is 3).

An aside: notice that we use the `GFP_KERNEL|__GFP_ZERO` GFP flags to ensure that the allocated memory is zeroed out, a best practice. Then again, zeroing out large memory chunks can result in a slight performance hit.

Now, we ask ourselves the question: is there a way to verify that the memory is physically contiguous (as promised)? It turns out that yes, we can actually retrieve and print out the physical address of the start of each allocated page frame and retrieve its **Page Frame Number (PFN)** as well.

The **PFN** is a simple concept: it's just the index or page number – for example, the PFN of physical address 8192 is 2 (8192/4096). As we showed previously how to (and importantly, when you can) translate kernel virtual addresses to their physical counterparts earlier (and vice versa; this coverage is in *Chapter 7, Memory Management Internals – Essentials*, in the *Direct-mapped RAM and address translation* section), we won't repeat it here.

To do this work of translating virtual addresses to physical addresses and checking for contiguity, we write a small "library" function, which is kept in a separate C file in the root of this book's GitHub source tree, `klib.c`. Our intent is to modify our kernel module's `Makefile` to link in the code of this library file as well! (Doing this properly was covered back in *Chapter 5, Writing Your First Kernel Module – Part 2*, in the *Performing library emulation via multiple source files* section.) Here's our invocation of our library routine (just as was done in the beginning portion of our `bsa_alloc()` function):

```
show_phy_pages(gptr2, numpg2alloc * PAGE_SIZE, 1);
```

The following is the code of our library routine (in the `<booksrc>/klib.c` source file; again, for clarity, we won't show the entire code here):

```
// klib.c
[...]
int loops = len/PAGE_SIZE, i;
/* show_phy_pages() - show the virtual, physical addresses and PFNs of
the memory range provided on a per-page basis.
 * ! NOTE   NOTE   NOTE !
```

```
 * The starting kernel address MUST be a 'linear' address, i.e., an
adrress within the 'lowmem' direct-mapped region of the kernel segment,
else this will NOT work and can possibly crash the system.
 * @kaddr: the starting kernel virtual address
 * @len: length of the memory piece (bytes)
 * @contiguity_check: if True, check for physical contiguity of pages
 * 'Walk' the virtually contiguous 'array' of pages one by one (that is,
page by page), printing the virt and physical address (and PFN- page
frame number). This way, we can see if the memory really is *physically*
contiguous or not.
 */
void show_phy_pages(const void *kaddr, size_t len, bool contiguity_check)
{
    [...]
    if (len % PAGE_SIZE)
        loops++;
    for (i = 0; i < loops; i++) {
        pa = virt_to_phys(vaddr+(i*PAGE_SIZE));
        pfn = PHYS_PFN(pa);

        if (!!contiguity_check) {
        /* what's with the 'if !!(<cond>) ...' ??
         * a 'C' trick: ensures that the if condition always
         * evaluates to a boolean - either 0 or 1. Cool, huh. */
            if (i && pfn != prev_pfn + 1)
             pr_notice(" *** physical NON-contiguity detected ***\n");
        }
        [...]
        pr_info("%05d 0x%px %pa %ld\n", i, vaddr+(i*PAGE_SIZE), &pa,
pfn);
        if (!!contiguity_check)
            prev_pfn = pfn;
    }
}
```

Study the preceding function. We walk through our given memory range, (virtual) page by (virtual) page, obtaining the physical address and PFN, which we then emit via `printk` (notice how we use the %pa format specifier to portably print a physical address – it requires it to be passed by reference though). Not only that, if the third parameter to our `klib.c:show_phy_pages()` function, `contiguity_check`, is `1`, we check whether the PFNs are just a single digit apart, thus checking that the pages are indeed physically contiguous or not. (By the way, the simple `powerof()` function that we make use of is also within our library code).

4. Now, let's continue with the kernel module code:

```
/* 3. Allocate and init one page with the get_zeroed_page() API */
gptr3 = (void *)get_zeroed_page(GFP_KERNEL);
if (!gptr3)
        goto out3;
[...]
pr_info("3.  get_zeroed_page()   1 page      %px\n", gptr3);
```

As seen in the preceding snippet, we allocate a single page of memory but ensure it is zeroed out by employing the PA `get_zeroed_page()` API. `pr_info()` shows the actual KVAs (using either the `%px` or `%pK` format specifiers has the addresses printed in a port-able fashion as well, irrespective of whether you're running a 32- or 64-bit system).

5. Next, we allocate one page with the `alloc_page()` API. Careful here! It does *not* return the pointer to the allocated page, but rather the pointer to the metadata structure page representing the allocated page; here's the function signature: `struct page * alloc_page(gfp_mask)`. Thus, we must use the `page_address()` helper to convert it into a kernel logical (or virtual) address:

```
/* 4. Allocate one page with the alloc_page() API. */
pg_ptr1 = alloc_page(GFP_KERNEL);
if (!pg_ptr1)
      goto out4;
gptr4 = page_address(pg_ptr1);
pr_info("4.       alloc_page()   1 page      %px\n"
        " (struct page addr = %px)\n", (void *)gptr4, pg_ptr1);
```

In the preceding code snippet, we allocate one page of memory via the `alloc_page()` PA API. As explained, we need to convert the page metadata structure returned by it into a KVA (or kernel logical address) via the `page_address()` API.

6. Next, allocate *2^3 = 8 pages* with the `alloc_pages()` API. The same warning as the preceding code snippet applies here too:

```
/* 5. Allocate 2^3 = 8 pages with the alloc_pages() API. */
gptr5 = page_address(alloc_pages(GFP_KERNEL, 3));
if (!gptr5)
      goto out5;
pr_info("5.      alloc_pages()  %lld pages     %px\n",
        powerof(2, 5), (void *)gptr5);
```

In the preceding code snippet, we combine `alloc_pages()` wrapped within the `page_address()` API to allocate *2^3 = 8* pages of memory!

Interestingly, we use several local `goto` statements in the code (have a glance at the code in the repo). Looking carefully at it, you will notice that it actually keeps the error handling code paths clean and logical! This is indeed part of the Linux kernel coding style guidelines.

Usage of the (often pedantically controversial) `goto` statement in Linux is documented clearly right here: `https://elixir.bootlin.com/linux/v6.1.25/source/Documentation/process/coding-style.rst#L526`. I urge you to check it out! Once you understand the usage pattern, you'll find that it helps reduce the all-too-typical memory leakage (and similar) cleanup errors!

7. Finally, in the cleanup method, prior to being removed from kernel memory, we free up all the memory chunks we just allocated in the cleanup code of the kernel module.
8. In order to link our library `klib` code with our `lowlevel_mem_lkm` kernel module, the `Makefile` changes to have the following (recall that we learned about compiling multiple source files into a single kernel module in *Chapter 5*, *Writing Your First Kernel Module – Part 2*, in the *Performing library emulation via multiple source files* section):

```
FNAME_C := lowlevel_mem
[ ... ]
PWD              := $(shell pwd)
obj-m            += ${FNAME_C}_lkm.o
lowlevel_mem_lkm-objs := ${FNAME_C}.o ../../klib.o
```

Phew! That was a fair bit to cover. Do ensure you understand the code, and then read on to see it in action.

Deploying our lowlevel_mem_lkm kernel module

Okay, time to see our `ch8/lowlevel_mem` kernel module in action! Let's build and deploy it on both a Raspberry Pi 4 (running the default Raspberry Pi OS built for 32-bit) and on an x86_64 VM (running Ubuntu 22.04 LTS).

Trying it out on a 32-bit Raspberry Pi

On the Raspberry Pi 4 Model B (here, running a stock Raspberry Pi 32-bit kernel version 5.15.76-v7l+), we build and then `insmod` our `lowlevel_mem_lkm` kernel module (this portion isn't seen in *Figure 8.5*). We then do an `lsmod` followed by removing (unloading) it with `rmmod`, and then showing the kernel log via `dmesg`.

The following screenshot shows the output:

```
rpi 4 $ lsmod |grep lowlevel_mem_lkm
lowlevel_mem_lkm        16384  0
rpi 4 $ sudo rmmod lowlevel_mem_lkm ; sudo dmesg
[  754.543365] lowlevel_mem_lkm:bsa_alloc(): 0. Show identity mapping: RAM page frames : ke
rnel virtual pages :: 1:1
                (PAGE_SIZE = 4096 bytes)
[  754.543378] lowlevel_mem_lkm:bsa_alloc(): [--------- show_phy_pages() output follows:
[  754.543384]  start kaddr c0000000, len 20480, contiguity_check is on
[  754.543391] -pg#-  ----va----     --------pa--------      --PFN--
[  754.543395] 00000  0xc0000000     0x0000000000000000            0
[  754.543402] 00001  0xc0001000     0x0000000000001000            1
[  754.543408] 00002  0xc0002000     0x0000000000002000            2
[  754.543414] 00003  0xc0003000     0x0000000000003000            3
[  754.543419] 00004  0xc0004000     0x0000000000004000            4
[  754.543425] lowlevel_mem_lkm:bsa_alloc():  --------- show_phy_pages() output done]
[  754.543431] lowlevel_mem_lkm:bsa_alloc(): #.      BSA/PA API      Amt alloc'ed         KVA
[  754.543436] lowlevel_mem_lkm:bsa_alloc(): 1.    __get_free_page()       1 page     c4f8b000
[  754.543453] lowlevel_mem_lkm:bsa_alloc(): 2.  __get_free_pages()   2^3 page(s)  c4e90000
[  754.543459] lowlevel_mem_lkm:bsa_alloc(): [--------- show_phy_pages() output follows:
[  754.543464]  start kaddr c4e90000, len 32768, contiguity_check is on
[  754.543470] -pg#-  ----va----     --------pa--------      --PFN--
[  754.543474] 00000  0xc4e90000     0x000000004e90000        20112
[  754.543480] 00001  0xc4e91000     0x000000004e91000        20113
[  754.543486] 00002  0xc4e92000     0x000000004e92000        20114
[  754.543491] 00003  0xc4e93000     0x000000004e93000        20115
[  754.543497] 00004  0xc4e94000     0x000000004e94000        20116
[  754.543502] 00005  0xc4e95000     0x000000004e95000        20117
[  754.543508] 00006  0xc4e96000     0x000000004e96000        20118
[  754.543513] 00007  0xc4e97000     0x000000004e97000        20119
[  754.543518] lowlevel_mem_lkm:bsa_alloc():  --------- show_phy_pages() output done]
[  754.543525] lowlevel_mem_lkm:bsa_alloc(): #.      BSA/PA API      Amt alloc'ed         KVA
[  754.543530] lowlevel_mem_lkm:bsa_alloc(): 3.  get_zeroed_page()     1 page         c4f8c000
[  754.543537] lowlevel_mem_lkm:bsa_alloc(): 4.       alloc_page()     1 page         c4f93000
                (struct page addr = d9ab30ac)
[  754.543547] lowlevel_mem_lkm:bsa_alloc(): 5.      alloc_pages()   32 pages       c5400000
[  929.591289] lowlevel_mem_lkm:lowlevel_mem_exit(): free-ing up the prev allocated BSA/PA
memory chunks...
[  929.591328] lowlevel_mem_lkm:lowlevel_mem_exit(): removed
rpi 4 $ 
```

physically contiguous memory pages

physically contiguous memory pages

Figure 8.5: The lowlevel_mem_lkm kernel module's output on a Raspberry Pi 4 Model B

Check it out! In step 0 of the output in *Figure 8.5* our `show_phy_pages()` 'library' routine clearly shows that KVA `0xc000 0000` has pa (physical address) `0x0`, KVA `0xc000 1000` has pa `0x1000`, and so on, for five pages (along with the PFN on the extreme right); you can literally see the 1:1 identity mapping of physical RAM page frames to kernel virtual pages (in the lowmem region of the kernel segment)!

Next, after showing a header line, the initial memory allocation with the `__get_free_page()` API goes through as expected. More interesting is our case 2. Here (middle-to-lower portion of *Figure 8.5*), we can clearly see that the physical address and PFN of each allocated page (from 0 to 7, for a total of 8 pages, via the `__get_free_pages()` API, our point #3 in the code narrative) is consecutive, (again) showing that the memory pages allocated are indeed physically contiguous!

Trying it out on an x86_64 VM

We build and run the same module on an x86_64 VM running Ubuntu 22.04 LTS (running a custom 6.1 kernel). The following screenshot shows the output:

```
$ uname -r
6.1.11-lkp-kernel
$ sudo rmmod lowlevel_mem_lkm ; sudo dmesg -C
[sudo] password for c2kp:
$ sudo insmod ./lowlevel_mem_lkm.ko ; sudo dmesg
[30002.831039] lowlevel_mem_lkm:bsa_alloc(): 0. Show identity mapping: RAM page frames : kernel virtual pages :: 1:1
               (PAGE_SIZE = 4096 bytes)
[30002.831056] lowlevel_mem_lkm:bsa_alloc(): [--------- show_phy_pages() output follows:
[30002.831057]  start kaddr ffff934cc0000000, len 20480, contiguity_check is on
[30002.831058] -pg#-   --------va--------   --------pa--------   ---PFN---
[30002.831058] 00000  0xffff934cc0000000   0x0000000000000000         0
[30002.831059] 00001  0xffff934cc0001000   0x0000000000001000         1
[30002.831060] 00002  0xffff934cc0002000   0x0000000000002000         2
[30002.831061] 00003  0xffff934cc0003000   0x0000000000003000         3
[30002.831062] 00004  0xffff934cc0004000   0x0000000000004000         4
[30002.831063] lowlevel_mem_lkm:bsa_alloc():  --------- show_phy_pages() output done]
[30002.831064] lowlevel_mem_lkm:bsa_alloc(): #.    BSA/PA API     Amt alloc'ed       KVA
[30002.831064] lowlevel_mem_lkm:bsa_alloc(): 1.   __get_free_page()     1 page     ffff934d284a0000
[30002.831066] lowlevel_mem_lkm:bsa_alloc(): 2.  __get_free_pages()  2^3 page(s)  ffff934cd95b8000
[30002.831067] lowlevel_mem_lkm:bsa_alloc(): [--------- show_phy_pages() output follows:
[30002.831068]  start kaddr ffff934cd95b8000, len 32768, contiguity_check is on
[30002.831069] -pg#-   --------va--------   --------pa--------   ---PFN---
[30002.831069] 00000  0xffff934cd95b8000   0x00000000195b8000    103864
[30002.831070] 00001  0xffff934cd95b9000   0x00000000195b9000    103865
[30002.831071] 00002  0xffff934cd95ba000   0x00000000195ba000    103866
[30002.831072] 00003  0xffff934cd95bb000   0x00000000195bb000    103867
[30002.831072] 00004  0xffff934cd95bc000   0x00000000195bc000    103868
[30002.831073] 00005  0xffff934cd95bd000   0x00000000195bd000    103869
[30002.831074] 00006  0xffff934cd95be000   0x00000000195be000    103870
[30002.831074] 00007  0xffff934cd95bf000   0x00000000195bf000    103871
[30002.831075] lowlevel_mem_lkm:bsa_alloc():  --------- show_phy_pages() output done]
[30002.831076] lowlevel_mem_lkm:bsa_alloc(): #.    BSA/PA API     Amt alloc'ed       KVA
[30002.831076] lowlevel_mem_lkm:bsa_alloc(): 3.  get_zeroed_page()   1 page      ffff934cc48b9000
[30002.831077] lowlevel_mem_lkm:bsa_alloc(): 4.       alloc_page()   1 page      ffff934cf65da000
               (struct page addr = ffffc63500d97680)
[30002.831083] lowlevel_mem_lkm:bsa_alloc(): 5.      alloc_pages()  32 pages     ffff934d328c0000
$
```

Figure 8.6: The lowlevel_mem_lkm kernel module's output on a x86_64 VM running Ubuntu 22.04 LTS

This time, with the `PAGE_OFFSET` value being a 64-bit quantity (the value here happens to be `0xffff934cc0000000`; realize it can certainly change due to KASLR!), you can again clearly see the identity mapping of physical RAM frames to kernel virtual addresses (for 5 pages). Let's take a moment and look carefully at the kernel logical addresses returned by the page allocator APIs. In this particular case, in *Figure 8.6*, you can see that they are all in the range `0xffff 934c c000 xxxx`. All these very much fall within the kernel's direct-mapped physical memory (RAM is mapped here into the kernel segment), aka the *lowmem* region.

If all this seems new and strange to you, please refer to *Chapter 7, Memory Management Internals – Essentials,* particularly the *Examining the kernel segment* and *Direct-mapped RAM and address translation* sections.

The reality is that the page allocator memory (in effect the buddy system free lists) directly maps the (free) physical RAM within the direct-mapped or lowmem region of the kernel VAS. Thus, it obviously returns memory from this region. The specific address range used is of course very arch-specific.

Though it is tempting to claim that you're now done with the page allocator and its APIs, the reality is that this is (as usual) not quite the case. Do read on to see why – it's really important to understand these aspects.

The page allocator and internal fragmentation (wastage)

Though all looks good and innocent on the surface, I urge you to delve a bit deeper. Just under the surface, a massive (unpleasant!) surprise might await you: the blissfully unaware kernel/driver developer. The APIs we covered previously regarding the page allocator (see *Table 8.1*) have the dubious distinction of being able to internally fragment – in simpler terms, waste – very significant portions of kernel memory!

To understand why this is the case, you must understand at least the basics of the page allocator algorithm and its freelist data structures. The section titled *The fundamental workings of the page allocator* covered this (just in case you haven't read it, please do so).

In the *Working through a few scenarios* section, you have seen that when we make an allocation request of convenient, perfectly rounded powers-of-two-size pages (note, pages, not bytes), it goes very smoothly. However, when this isn't the case – let's say the driver requests 132 KB of memory – then we end up with a major issue: the internal fragmentation or wastage is very high (as seen, this by default results in an allocation of 256 KB, thus incurring a wastage of 256 - 132 = 124 KB). This is a serious downside and must be addressed. We will see how, in two ways, in fact. Do read on!

The 'exact' page allocator API pair

Realizing the vast potential for memory wastage within the default page allocator (or BSA), a developer from Freescale Semiconductor (see the information box) contributed a patch to the kernel page allocator that extends the PA API, adding a couple of new ones.

In the `2.6.27-rc1` series, on 24 July 2008, Timur Tabi submitted a patch to mitigate the page allocator wastage issue. Here's the relevant commit: `https://github.com/torvalds/linux/commit/2be0ffe2b29bd31d3debd0877797892ff2d91f4c`.

Using these APIs leads to more efficient allocations for large-ish chunks (multiple pages) of memory **with far less wastage.** The new (well, it was new back in 2008, at least) pair of APIs to allocate and free memory are as follows:

```
#include <linux/gfp.h>
void *alloc_pages_exact(size_t size, gfp_t gfp_mask);
void free_pages_exact(void *virt, size_t size);
```

The first parameter to the `alloc_pages_exact()` API, `size`, is in bytes: obviously, it's the number of bytes to allocate; the second is the "usual" GFP flags value discussed earlier (in the *Dealing with the GFP flags* section; `GFP_KERNEL` for the typical might-sleep process context cases, and `GFP_ATOMIC` for the can't-sleep interrupt or atomic context cases).

Note that the memory allocated by this API is still guaranteed to be physically contiguous. Also, the amount that can be allocated at a time (with one function call) is limited by `MAX_ORDER`; in fact, this is true of all the other regular page allocation APIs that we have seen so far. We will discuss a lot more about this aspect in the upcoming *Size limitations of the kmalloc API* section. There, you'll realize that the discussion is in fact not limited to the slab cache but to the page allocator as well!

The `free_pages_exact()` API must only be used to free memory allocated by its counterpart, `alloc_pages_exact()`. Also, note that the first parameter to the "free" routine is the value returned by the matching `alloc` routine (the pointer to the allocated memory chunk).

The implementation of `alloc_pages_exact()` is simple and clever: it first allocates the entire memory chunk requested "as usual" via the `__get_free_pages()` API. Then, it loops over pages – from the end of the memory region specified (from the nearest page boundary) for use to the amount of actually allocated memory (which is typically far greater) – freeing up those unnecessary memory pages! So, in our example, if you allocate 132 KB via the `alloc_pages_exact()` API, it will actually first internally allocate 256 KB via `__get_free_pages()` (because it must; that's the algorithm), but will then free up the unused memory from 132 KB to 256 KB!

Another example of the beauty of open source! A demo of using these APIs can be found in the module code here: `ch8/page_exact_loop`; we will leave it to you to try it out.

Before we began this section, we mentioned that there were two ways in which the wastage issue of the page allocator can be addressed. One is by using (more efficient than the other PA APIs) the `alloc_pages_exact()` and `free_pages_exact()` API pair, as we just learned; the other – in fact, the common approach – is by using a different layer altogether to allocate memory – the *slab allocator*. We will soon cover it; until then, hang in there. Next, let's cover more details, *crucial to understand*, on the (typical) GFP flags and how you, the kernel module or driver author, are expected to use them.

The GFP flags – digging deeper

With regard to our discussions on the low-level page allocator APIs, the first parameter to every function is the so-called GFP mask. When discussing the APIs and their usage, we mentioned a *key rule*.

If your kernel/driver code is running in *process context and it is safe to sleep,* use the `GFP_KERNEL` flag. If it is currently *unsafe to sleep* (typically, when in any type of atomic, including interrupt, context, or when holding a spinlock), you *must* use the `GFP_ATOMIC` flag.

We will elaborate on this in the following sections.

Never sleep in interrupt or atomic contexts

Just what does the phrase *safe to sleep* actually mean? To answer this, think of blocking calls (APIs): a *blocking call* is one where the calling process (or thread) is put into a sleep state because it is waiting on something, an *event*, to occur and the event it's waiting on hasn't occurred yet. Thus, it waits – it "sleeps". When, at some future point in time, the event it is waiting on occurs or arrives, it is *woken up* by the kernel (or device driver) and proceeds forward.

One example of a user-space blocking API includes the `sleep()` C library API. Say you do this: `sleep(5);`. Here, the event being waited upon is the elapse of a certain amount of time (the parameter to the function, in seconds). So, the process context that called this function goes to sleep, and is woken up once the event occurs – the elapse of 5 seconds of time! Another example is the `read()` system call and its variants, where the event being waited on is (typically) storage or network data becoming available. With the `wait4()` system call, the event being waited on is the death or stoppage/continuing of a child process, and so on.

So, any function that might possibly block can end up spending some time asleep; note that while asleep, it's certainly off the CPU run queues, and in a wait queue; in effect, while sleeping, it isn't even a candidate to be scheduled! Invoking this *possibly blocking* functionality when in kernel mode (which, of course, is the mode we are in when working on kernel modules) is only allowed when in process context and, within this, when it's safe to sleep (holding certain locks, like a spinlock, makes it unsafe to sleep, whereas holding a mutex lock implies it's okay to sleep or block). It is a bug to invoke a blocking call of any sort in a context where it is unsafe to sleep, such **as when in an interrupt or atomic context**. Think of this as a golden rule. Violating this, that is, sleeping in an atomic context – is wrong, is buggy, and must *never* happen.

You might wonder, *how can I know in advance whether my code is running in an atomic (or interrupt) or in process context*? In one way, the kernel helps us out: when configuring the kernel (recall `make menuconfig` from *Chapter 2, Building the 6.x Linux Kernel from Source – Part 1*), under the **Kernel Hacking / Lock Debugging** menu, there is a Boolean tunable called `"Sleep inside atomic section checking"`. Turn it on! (The config option is named `CONFIG_DEBUG_ATOMIC_SLEEP`; you can always grep your kernel config file for it. Again, in *Chapter 5, Writing Your First Kernel Module – Part 2*, under the *Configuring a "debug" kernel* section, this is something you should definitely turn on). Furthermore, we shall come to a few macros (like `in_task()`) that allow you to figure out whether you're executing code in process context or not.

Another way to think of this situation is by considering this: how exactly do you put a process or thread to sleep? The short answer is, by having it invoke the scheduling code – the `schedule()` function. Thus, by implication of what we have just learned (as a corollary), `schedule()` must only be called from within a context where it's safe to sleep; process context usually is safe, interrupt context never is (a lot of detail on CPU scheduling is the content of both *Chapter 10* and *Chapter 11*).

This is really important to keep in mind! (We briefly covered what process and interrupt context are in *Chapter 4, Writing Your First Kernel Module - Part 1*, in the *Process and interrupt contexts* section. Moreover, as covered there, you can always employ the `in_task()` macro to determine whether the code is currently running in a process or interrupt context). Similarly, you can use the `in_atomic()` macro; if the code is an *atomic context* – where it must typically run to completion without interruption – it returns `True`; otherwise, `False`. Note well: you can be in process context but be atomic at the same time – for example, when holding certain kinds of locks (spinlocks; we will, of course, cover this in the final two chapters on *synchronization*); the converse cannot happen (i.e., in interrupt context, the code is *always* atomic).

A quick note on other internal-usage GFP flags

Besides the GFP flags we, as module/driver authors, are focused upon – the `GFP_KERNEL` and `GFP_ATOMIC` ones – the kernel has several other rather exotic `[__]GFP_*` flags that are used internally; several are for the express purpose of reclaiming memory. These include (but are not limited to) `__GFP_IO`, `__GFP_FS`, `__GFP_DIRECT_RECLAIM`, `__GFP_KSWAPD_RECLAIM`, `__GFP_RECLAIM`, and `__GFP_NORETRY`.

Internally, the kernel uses these GFP flags to specify precisely the kind of expected internal behavior when performing a memory allocation. To help make this clear, consider the following (typical) examples.

If memory runs short, the kernel can free up some memory by first writing it to persistent storage (into, typically, the swap partition or file), marking its location in swap, and then freeing it. Now, let's say a filesystem component makes a memory request; as RAM is in short supply, the kernel tries to free some memory by asking to write out some 'older' memory pages (possibly using a **Least Frequently Used** (**LFU**) algorithm). But, consider this: what if the same filesystem code paths are involved in writing out these memory pages and those code paths require a memory allocation just then? This could result in twisted recursive behavior, and even in deadlock! Thus, the kernel, when performing a memory allocation, in this case, does *not* use a GFP flag named `__GFP_IO` that specifies whether, while performing the allocation, physical I/O can be started (or not). Similarly, there's a flag named `__GFP_FS` specifying whether or not the kernel's able to call down to the low-level filesystem code paths. The `GFP_NOWAIT` flag specifies semantics that tell the kernel not to wait for direct memory reclamation, not to start physical I/O, and not to use any low-level filesystem callbacks.

When the kernel is in a situation where the code path must remain atomic (it can't sleep or block in any manner) and the allocation must succeed or fail immediately (preferably the former), the `GFP_ATOMIC` flag's the ticket. This, of course, is the case with atomic code paths, like any kind of interrupt context or when holding a spinlock (note, though, that even this isn't considered okay for handling **Non-Maskable Interrupts** (**NMIs**)).

Viewing the flags mentioned at the code level: `include/linux/gfp_types.h`

```
#define GFP_ATOMIC  (__GFP_HIGH|__GFP_ATOMIC|__GFP_KSWAPD_RECLAIM)
#define GFP_KERNEL  (__GFP_RECLAIM | __GFP_IO | __GFP_FS)
```

So, with the GFP_KERNEL flag, we can now see that the alloc API can actively reclaim memory, and start physical and/or filesystem I/O, thus increasing the chances of obtaining the requested memory.

As already mentioned, the gfp_types.h header has detailed comments on the meaning of each of these flags; do look it up.

A quick aside on the **Linux Driver Verification (LDV)** project: back in *Online Chapter, Kernel Workspace Setup*, we mentioned that this project has useful "rules" with respect to various programming aspects of Linux modules (drivers, mostly) as well as the core kernel.

With regard to our current topic, here's one of the rules, a negative one, implying that you cannot do this: *Not disabling IO during memory allocation while holding a USB device lock* (http://linuxtesting.org/ldv/online?action=show_rule&rule_id=0077). Some quick background: when you specify the GFP_KERNEL flag, it implicitly means (among other things) that the kernel can start an I/O (input/output; reads/writes) operation to reclaim memory. The trouble is, at times this can be problematic and should not be done; to get over this, you're expected use the GFP_NOIO flag as part of the GFP bitmask when allocating kernel memory.

That's precisely the case that this LDV 'rule' is referring to: in the particular situation encountered, between the usb_lock_device() and usb_unlock_device() APIs, the GFP_KERNEL flag shouldn't be used and the GFP_NOIO flag should be used instead. (You can see several instances of this flag being used in this code: drivers/usb/core/message.c.) The LDV page mentions the fact that a couple of USB-related driver source files were fixed to adhere to this rule.

Page allocator – pros and cons

Let's conclude this section on the page allocator/BSA by quickly summarizing its pros and cons.

Page allocator/BSA – pros:

- It's fast; uses the kernel's identity-mapped RAM (obtained and mapped at boot); pages are in the lowmem region and thus pre-mapped; no page table setup is required.
- Guarantees memory chunks that are physically contiguous and CPU cacheline-aligned.
- Actually helps to defragment RAM by merging 'buddy' blocks (a buddy block is a block that's of the same size and physically contiguous).

Page allocator/BSA – cons:

- Primary issue: internal fragmentation or wastage can be (too) high.
- The granularity of an allocation request is a page; so (assuming 4K pages), asking for 128 bytes gets you 4,096 (this is mitigated in many cases by using the superior alloc_pages_exact()/free_pages_exact() API pair (but not in this case!).
- There is an upper limit to how much can be allocated with a single API call; practically, assuming MAX_ORDER=11 and page size of 4K, it's 4 MB.

Alright, now that you're armed with a good amount of detail on the page allocator (it is, after all, the internal "engine" of RAM (de)allocation!), its APIs, and how to use them, let's move on to a very important topic – the motivation(s) behind the slab allocator, its APIs, and how to use them.

Understanding and using the kernel slab allocator

As seen in the first section of this chapter, *Introducing kernel memory allocators*, the *slab allocator* or *slab cache* is layered above the page allocator (or BSA; refer back to *Figure 8.1*). The slab allocator justifies its very existence with two primary ideas or purposes:

- **Object caching:** Here, it serves as a cache of common "objects"; these are the frequently allocated data structures within the Linux kernel. This idea – which we'll of course expand upon – has the slab cache allocate (and subsequently freeing) these objects on demand, resulting in higher performance.
- Mitigate the high wastage (internal fragmentation) of the page allocator by providing small, conveniently sized caches, of various sizes that are typically **fragments of a page**.

Let's now examine these ideas in a more detailed manner.

The object caching idea

Okay, we begin with the first of these design ideas – the notion of a cache of common objects. A long time ago, a SunOS developer, Jeff Bonwick, noticed that certain kernel objects – data structures, typically – were allocated and deallocated frequently within the OS. He thus had the idea of pre-allocating them in a cache of sorts. This evolved into what we call the slab cache.

Thus, on the Linux OS as well, the kernel (as part of boot time initialization) pre-allocates a fairly large number of objects into several slab caches. The reason: performance! When core kernel code (or a device driver) requires memory for one of these objects, it directly requests the slab allocator. If cached, the allocation is almost immediate (the converse being true as well, at deallocation). You might wonder, *is all this really necessary*? Indeed, it is!

A good example of high performance being required is within the critical code paths of the network and block I/O subsystems. Precisely for this reason, several network and block I/O data structures (the network stack's socket buffer, `sk_buff`, the block layer's `biovec`, and, of course, the core `task_struct` data structures or objects, being a few good examples) are *auto-cached* (*pre-allocated*) by the kernel within the slab caches. Similarly, filesystem metadata structures (such as the `inode` and `dentry` structures, and so on), the memory descriptor (`struct mm_struct`), and several more are *pre-allocated* on slab caches. Can we see these cached objects? Yes, just a bit further down, we will do precisely this (via `/proc/slabinfo`).

The other reason that the slab (or, more correctly now, the SLUB) allocator has far superior performance is simply that when traditional heap-based allocators allocate and deallocate memory often, they end up creating "holes" in memory (fragmentation). Because the slab objects are allocated once (at boot) into the caches, and freed back there (thus not really "freed up"), performance remains high. Of course, the modern kernel has the intelligence to, in a graceful manner, start freeing up the slab caches when memory pressure – the demand for precious RAM – gets too high.

Again, there are typically gaps between theory and practice; at times, when RAM runs very low, the kernel responds in a rather heavy-handed fashion, bringing in its heavy gun – the **Out-Of-Memory** (**OOM**) killer! (Worry not, we discuss it in depth in *Chapter 9, Kernel Memory Allocation for Module Authors – Part 2*, in the *Stayin' alive – the OOM killer* section). Let's be patient, and leave that discussion until then.

The current state of the slab caches – the object caches, the number of objects in a cache, the number in use, the size of each object, and so on – can be looked up in several ways: raw views via the `proc` and `sysfs` filesystems, or a more human-readable view via various frontend utilities, such as `smem`, `vmstat`, `slabtop` (along with the eBPF `slabratetop`), and (the kernel-supplied app) `slabinfo`.

A few FAQs regarding (slab) memory usage and their answers

To help understand things, the typical FAQ format can prove useful. Before jumping into it, a quick note on how to use the powerful `slabinfo` utility: within your kernel source tree, simply switch to the `tools/vm` directory, and type **`make`**; the `slabinfo` utility (along with others) will get built. Running it requires root access. Once it's built, for convenience, I usually create a soft link to `slabinfo` within a directory that's in the default PATH (with something like this):

```
sudo ln -s <path/to/>linux-6.1.25/tools/vm/slabinfo /sbin
```

What slab caches are currently in use by the kernel?

Running `sudo vmstat -m` reveals all of them; try it (note that some caches can be empty).

The following is some example output:

```
$ sudo vmstat -m | grep -C3 task_struct
files_cache                 644    644    704     23
signal_cache               1165   1484   1152     28
sighand_cache               750    825   2112     15
task_struct                2210   2356   6592      4
cred_jar                   4770   6111    192     21
anon_vma_chain           147015 153792     64     64
anon_vma                  98484 101517    104     39
$
```

(Spot the task structure in the output!) The man page on `vmstat` reveals the precise meaning of each column (from left to right):

```
FIELD DESCRIPTION FOR SLAB MODE
       cache: Cache name
       num: Number of currently active objects
       total: Total number of available objects
       size: Size of each object  (bytes)
       pages: Number of pages with at least one active object
```

How can I see the overall system-wide memory usage?

There are several ways; one is via the smem utility:

```
$ smem --kernel /home/c2kp/kernels/linux-6.1.25/vmlinux -wk --realmem=3G
Area                          Used      Cache   Noncache
firmware/hardware            47.6M          0      47.6M
kernel image                 37.4M          0      37.4M
kernel dynamic memory         1.2G       1.1G     156.1M
userspace memory              1.0G     378.6M     649.6M
free memory                 684.3M     684.3M          0
```

Running smem with --kernel (give the path to the uncompressed vmlinux image; this is why we're using our x86_64 VM with our custom 6.1 kernel – we know where the vmlinux file is), -w (show whole system), -k (show abbreviated units), and --realmem (specifies the actual amount of RAM on the system) reveal this.

Here, we can see that within 'kernel dynamic memory' (highlighted), 1.1 GB of 1.2 GB being used is cached! For what?

A major user of cache memory (which is simply page frames, RAM) is the kernel's *page cache*; it's essentially RAM that the kernel uses to cache every single piece of filesystem I/O (a bit more on this follows shortly). The free utility shows the approximate cache usage as 1.5 GB here (the column named buff/cache); of this, most is indeed taken up by the kernel page cache:

```
$ free -h
              total        used        free      shared  buff/cache   available
Mem:          2.9Gi       796Mi       693Mi       7.0Mi       1.5Gi       2.0Gi
Swap:         8.9Gi          0B       8.9Gi
```

Don't confuse the system's page cache usage with that of slab caches... page cache memory usage is typically much higher, accounting for significant amounts of RAM. What about the amount taken by the kernel's slab caches?

How much RAM exactly is taken up by the slab caches?

The slabinfo utility can easily answer this FAQ; we run it with the -T (totals) option switch:

```
$ sudo ./slabinfo -T |grep -A3 "^% PartObj"
% PartObj          4%                0%              54%                4%
Memory           1.2M              4.0K            31.0M            144.7M
Used             1.1M              4.0K            30.2M            138.5M
Loss            52.0K                 0             1.2M              6.1M
```

Here, a total of almost 145 MB of RAM is being currently used by the slab caches, of which 138.5 MB is in active use (see the row entitled Used) and 6.1 MB is 'lost', in effect, wasted (see the row entitled Loss); more on this follows.

Another common way to look this up:

```
$ grep "^Slab" /proc/meminfo
Slab:              148172 kB
```

As an important aside, learning to use and interpret the myriad `slabinfo` options (as well as the `slab[rate]top` utilities) can in general be very helpful when learning and/or debugging user and kernel memory issues. My earlier book, *Linux Kernel Debugging*, covers these topics in more detail. As a sample, the next FAQ is from there.

Of the many slab caches that are currently allocated (and have some data content), which takes up the most kernel memory?

This is easily answered, again, by the `slabinfo` utility: one way is to run it with the `-B` switch, to display the space taken in bytes, allowing you to easily sort on this column. Even simpler, the `-S` option switch has `slabinfo` sort the slab caches by size (largest first) with nice, human-readable size units displayed.

As is now hopefully apparent, significant amounts of memory (RAM) can be used by caches maintained by the kernel (among them are the page cache, dcache, icache, slab caches, and so on)! This, in fact, is a common feature on Linux that puzzles those new to it: the kernel can and will use RAM for caching, thus greatly improving performance. It is, of course, designed to intelligently throttle down the amount of memory used for caching as memory pressure increases. As mentioned, a significant percentage of memory can go toward caching (especially the page cache; it's used to cache the content of files as I/O is performed upon them). This is fine, as long as memory pressure is low.

Look up `/proc/meminfo` for fine-granularity detail on system memory usage; the fields displayed are numerous. The man page on `proc(5)` describes them under the `/proc/meminfo` section.

Now that you understand the motivation behind the slab allocator (there's more on this too), let's dive into learning how to use the APIs it exposes for both the core kernel and module/driver authors.

Learning how to use the slab allocator APIs

You may have noticed that, so far, we haven't explained the second "design idea" behind the slab allocator (or slab cache), namely, mitigate the high wastage (internal fragmentation) of the page allocator by providing small, conveniently sized caches, typically, fragments of a page. We will soon see what exactly this means in a practical fashion, along with the kernel slab allocator APIs.

Allocating slab memory

Though several APIs to perform memory allocation and freeing exist within the slab layer, there are just a couple of really key ones, with the rest falling into a "convenience or helper" functions category (which we will of course mention later). The key slab allocation APIs for the kernel module or device driver author are `kmalloc()` and `kzalloc()` (the latter being preferred, as will soon be explained); their signatures are as follows:

```
#include <linux/slab.h>
void *kmalloc(size_t size, gfp_t flags);
void *kzalloc(size_t size, gfp_t flags);
```

Be sure to include the `<linux/slab.h>` header file when using any slab allocator APIs.

The `kmalloc()` and `kzalloc()` routines tend to be the **most frequently used APIs for memory allocation** within the kernel. A quick check – we're not aiming to be perfectly precise here – with the very useful `cscope` code browsing utility on the 6.1.25 Linux kernel source tree reveals the (approximate) frequency of usage: `kmalloc()` is called close to 4,800 times and `kzalloc()` is called a little over 13,000 times!

Both the `kmalloc()` and `kzalloc()` functions accept two parameters:

- The first parameter to pass is the size of the memory chunk required in bytes.
- The second parameter is the type of memory to allocate, specified via the now-familiar GFP flags. (We covered this topic in the *Dealing with the GFP flags* and *The GFP flags – digging deeper* sections. If you're not familiar with them, I suggest you read those sections first.)

To mitigate the risk of **Integer Overflow** (**IoF**) bugs, you should avoid dynamically calculating the size of memory to allocate (the first parameter to these APIs). The kernel documentation warns us precisely about this:

`https://www.kernel.org/doc/html/latest/process/deprecated.html#open-coded-arithmetic-in-allocator-arguments`. In general, always avoid using deprecated stuff in the kernel; they're even documented here: *Deprecated Interfaces, Language Features, Attributes, and Conventions*: `https://www.kernel.org/doc/html/latest/process/deprecated.html#deprecated-interfaces-language-features-attributes-and-conventions`.

Right, back to the APIs; upon successful allocation, the return value is a pointer, the *kernel logical address* (remember, it's still a virtual address, a KVA, *not* a physical address) of the start of the memory chunk (or slab) just allocated. Indeed, you'll notice that but for the second parameter, the `kmalloc()` and `kzalloc()` APIs closely resemble their user space counterpart, the all-too-familiar glibc `malloc()` (and friends) APIs. Don't get the wrong idea, though: *internally, they're completely different*. The `malloc()` API returns a user space virtual address (a UVA) and, as mentioned earlier, there is *no direct correlation* between the user-mode `malloc()` and the kernel-mode `kmalloc()/kzalloc()` APIs (we'll typically abbreviate this key pair of APIs as `k{m|z}alloc()`). So, no, a call to `malloc()` does not result in an immediate call to `k{m|z}alloc()`; more on this later!

Next, it's important to understand that the memory returned by these slab allocator APIs **is guaranteed to be physically contiguous**. Furthermore, another key benefit is that the return address is guaranteed to be on a CPU cacheline boundary; that is, it will be hardware **cacheline-aligned**. Both these points are important performance-enhancing benefits.

Every CPU reads and writes data (from and to CPU caches <-> RAM) in an atomic unit called the **CPU cacheline**. The size of the cacheline varies with the CPU. You can usually look this up with the `getconf` utility – for example, try doing `getconf -a|grep CACHE_LINESIZE` (note that the unit is *bytes*, not bits). On modern CPUs, the cachelines for instructions and data are often separated out (as are the CPU caches themselves). A typical CPU cacheline size is 64 bytes. We cover (a lot) more on this topic in *Chapter 13, Kernel Synchronization – Part 2,* in the *Understanding CPU caching basics, cache effects, and false sharing* section. You can also refer to the *Further reading* section for more.

The content of a memory chunk immediately after allocation by `kmalloc()` is random (again, like the `malloc()`). Indeed, the reason why the `kzalloc()` API is the preferred and recommended API to use is that it *sets to zero* the allocated memory, preventing several classes of bugs (such as the common **Uninitialized Memory Reads** (**UMR**) one). Some may argue that the initialization of the memory slab takes some time, thus reducing performance. Our counter argument is that unless the memory allocation code is in an extremely time-critical code path (which, you could reasonably argue, is not good design in the first place, but sometimes can't be helped), you should, as a best practice, *initialize your memory upon allocation*. A whole slew of memory bugs and security side effects can thereby be avoided.

Many parts of the Linux kernel core code certainly use the slab layer for memory. Within these, there are time critical code paths – good examples can be found within the network and block I/O subsystems. For maximizing performance, the slab (actually SLUB) layer code has been written to be lockless (via a lock-free technology called **per-CPU variables**). We cover lock-free kernel synchronization in *Chapter 13, Kernel Synchronization – Part 2,* in the *Lock-free programming with per-CPU and RCU* section. Also, see more on the performance challenges and implementation details of the slab layer in the *Further reading* section.

Freeing slab memory

Of course, you must free the slab memory you allocated at some point in the future (thus not leaking memory); the `kfree()` routine serves this purpose.

Using kfree() to free slab memory

Analogous to the user space `free()` API, `kfree()` takes a single parameter – the pointer to the memory chunk to free. It must be a valid kernel logical (or virtual) address (a KVA) and must have been initialized by, that is, it should be the return value of one of the slab layer APIs (`k{m|z}alloc()` or one of its helpers). Its API signature is simple:

```
void kfree(const void *);
```

Just as with `free()`, there is no return value. As mentioned before, take care to ensure that the parameter to `kfree()` is the precise value returned by `k{m|z}alloc()`. Passing an incorrect value will result in memory corruption, ultimately leading to an unstable system.

There are a few additional points to note. Let's assume we have allocated some slab memory with `kzalloc()`:

```
static char *kptr = kzalloc(1024, GFP_KERNEL);
```

Later, after usage, we would like to free it, so we do the following:

```
if (kptr)
    kfree(kptr);
```

This code – checking that the value of `kptr` is not `NULL` before freeing it – is unnecessary; just perform `kfree(kptr);` and it's done.

Another example of incorrect code (pseudo-code) is shown as follows:

```
static char *kptr = NULL;
while (<some-condition-is-true>) {
       if (!kptr)
            kptr = kzalloc(num, GFP_KERNEL);
       [... work on the slab memory ...]
       kfree(kptr);
}
```

Interesting: here, from the second loop iteration onward, the programmer has assumed that the `kptr` pointer variable will be set to `NULL` upon being freed! This is definitely not the case (it would have been quite a nice semantic to have though; also, the same argument applies to the "usual" user space library APIs). Thus, here, we hit a dangerous bug: on the loop's second iteration, the `if` condition will likely turn out to be false, thus skipping the allocation. Then, we hit the `kfree()`, which, of course, will now corrupt memory (due to a double-free bug)!

We provide a demo of this very case in the LKM here: `ch8/slab2_buggy`, and leave it to you to study and try out.

Using kfree_sensitive() (previously kzfree()) to free slab memory

With regard to initializing memory buffers after (or during) allocation, just as we mentioned with regard to allocations, the same holds true for freeing memory. You should realize that the `kfree()` API merely returns the just-freed slab to its corresponding cache, leaving the internal memory content intact! Thus, just prior to freeing up your memory chunk, a (slightly pedantic) best practice is to wipe out (overwrite) the memory content. This is especially true for security reasons (such as in the case of an info-leak where a malicious attacker could conceivably scan freed memory for secrets). The Linux kernel provides the `kfree_sensitive()` API for this express purpose (the signature is identical to that of `kfree()`).

The API was (mis)named `kzfree()` in earlier kernels; from 5.9, this has been fixed, renaming it to `kfree_sensitive()`. The rationale is explained in the commit here: `https://github.com/torvalds/linux/commit/453431a54934d917153c65211b2dabf45562ca88`.

Careful! In order to overwrite "secrets," a simple `memset()` of the target buffer might just not work. Why not? The compiler might optimize away the code (as the buffer is no longer to be used). David Wheeler, in his excellent work *Secure Programming HOWTO* (`https://dwheeler.com/secure-programs/`), mentions this very fact and provides a solution: *"One approach that seems to work on all platforms is to write your own implementation of memset with internal "volatilization" of the first argument."* (This code is based on a workaround proposed by Michael Howard):

```
void *guaranteed_memset(void *v,int c,size_t n)
{ volatile char *p=v; while (n--) *p++=c; return v; }
```

Then place this definition into an external file to force the function to be external (define the function in a corresponding `.h` file, and `#include` the file in the callers, as is usual). This approach appears to be safe at any optimization level (even if the function gets inlined).

The kernel's `kfree_sensitive()` API should work just fine. Take care when doing similar stuff in user space.

Data structures – a few design tips

Using the slab APIs for memory allocation in kernel space is highly recommended. For one, it guarantees both physically contiguous as well as (hardware) cacheline-aligned memory. This is good for performance; in addition, let's check out a few quick tips that can yield big returns.

CPU caching can provide tremendous performance gains. Thus, especially for time-critical code, take care to design your data structures for best performance:

- Keep the most important (frequently accessed, "hot") members together and at the top of the structure. To see why, imagine there are five important members (of a total size of say, 56 bytes) in your data structure and you keep them all together and at the top of the structure. Say the CPU cacheline size is 64 bytes. Now, when your code accesses any one of these five important members (for anything, read or write), *all five members will be fetched into the CPU cache(s) as the CPU's memory read/writes work in an atomic unit of CPU cacheline size*; this optimizes performance (as working on the cache memory silicon is typically multiple times faster than working on RAM).
- Try and align the structure members such that a single member does not "fall off a cacheline." Usually, the compiler helps in this regard, and you can even use compiler attributes to explicitly specify this.

- Accessing memory sequentially results in high performance due to effective CPU caching. However, we can't seriously push the case for making all our data structures arrays! Experienced designers and developers know that using linked lists is extremely common. But doesn't that actually hurt performance? Well, yes, to some extent. Thus, a suggestion: when using linked lists, keep the "node" of the list as a large data structure (with "hot" members at the top and together). This way, we try and maximize the best of both cases as the large structure is essentially an array. (Think about it, the list of task structures that we saw in *Chapter 6, Kernel Internals Essentials – Processes and Threads*, the *task list* is a perfect real-world example of a linked list with large data structures as nodes).
- See more techniques on leveraging CPU caches in the *Further reading* section.

The upcoming section deals with a key aspect: we learn exactly which slab caches the kernel uses when allocating (slab) memory via the popular `k{m|z}alloc()` APIs.

The actual slab caches in use for kmalloc

We'll take a quick deviation – very important, though – before trying out a kernel module using the basic slab APIs. It's important to understand where exactly the memory allocated by the `k{m|z}alloc()` APIs is coming from. Well, it's from the slab caches, yes, but which ones exactly? A quick `grep` on the output of `sudo vmstat -m` reveals this for us (the following screenshot is on our x86_64 Ubuntu 22.04 LTS guest running (one of) our custom 6.1 kernels. Also, we deliberately filter out the `kmalloc-rcl-*` caches here; they're explained very shortly):

```
$ uname -r
6.1.11-lkp-kernel
$ sudo vmstat -m | head -n1
Cache                         Num  Total   Size  Pages
$ sudo vmstat -m | grep --color="auto" "^kmalloc-[0-9].."
kmalloc-8k                    144    144   8192      4
kmalloc-4k                   1185   1200   4096      8
kmalloc-2k                   1072   1072   2048     16
kmalloc-1k                   1160   1232   1024     16
kmalloc-512                  2439   2464    512     16
kmalloc-256                  2616   2672    256     16
kmalloc-192                  3024   3024    192     21
kmalloc-128                  1545   1664    128     32
kmalloc-96                   2124   2688     96     42
kmalloc-64                   7323   7552     64     64
kmalloc-32                   4354   4480     32    128
kmalloc-16                   6886   6912     16    256
kmalloc-8                    5632   5632      8    512
$
```

Figure 8.7: Screenshot of sudo vmstat -m showing the kmalloc-n slab caches

That's very interesting! The kernel has a slew of dedicated slab caches for generic `kmalloc` memory of varying sizes, *ranging from 8,192 bytes down to a mere 8 bytes!* This has us realize something critical – with the page allocator, if we had requested, say, 12 bytes of memory, it would have ended up giving us a whole page (4 KB) – the wastage is just too high. **Here, with the slab allocator, an allocation request for 12 bytes ends up actually allocating just 16 bytes** (from the second-to-last cache seen in *Figure 8.7*)! **Fantastic.** It should be quite clear why it works this way: the kernel slab allocator uses a *best-fit* approach. Plus, if we request an amount (12 bytes) that's in between two slab caches (the `kmalloc-8` one and the `kmalloc-16` one), obviously, the kernel will pick the greater of the two (as we can always allocate more than what was asked, never less). Thus, it actually allocates a 16 byte 'slab' while letting us think it's 12 bytes in size.

Also, note the following:

- Upon calling `kfree[_sensitive]()`, the memory is freed back into the appropriate slab cache.
- The precise sizing of the slab caches for `kmalloc` varies with the architecture. On our Raspberry Pi system (an ARM CPU, of course), the generic memory `kmalloc-N` caches ranged from 64 bytes to 8,192 bytes.
- The preceding screenshot also reveals a clue. Very often, the demand is for small-to-tiny fragments of memory. As an example, in the preceding screenshot, the column labeled `Num` represents the *Number of currently active objects*; the maximum number is from the 8- and 16-byte `kmalloc` slab caches (of course, this may not always be the case). *Quick tip:* use the `slabtop` utility (you'll need to run it as root): the rows toward the top reveal the current frequently used slab caches.

Linux keeps evolving, of course. As of the 5.0 mainline kernel, there is a newly introduced `kmalloc` cache type, called the **reclaimable cache** (the naming format is `kmalloc-rcl-N`). Thus, performing a `grep` as done previously on a 5.x kernel or later will also reveal these caches:

```
$ sudo vmstat -m | grep --color=auto "^kmalloc"
kmalloc-rcl-8k              0       0    8192      4
kmalloc-rcl-4k              0       0    4096      8
kmalloc-rcl-2k              0       0    2048     16
[...]
kmalloc-8k                144     144    8192      4
kmalloc-4k               1185    1200    4096      8
kmalloc-2k               1072    1072    2048     16
[...]
```

The new `kmalloc-rcl-N` caches help internally with more efficiencies (to reclaim pages under pressure and as an anti-fragmentation measure). However, a module author like you need not be concerned with these details. (The commit for this work can be viewed here: `https://github.com/torvalds/linux/commit/1291523f2c1d631fea34102fd241fb54a4e8f7a0`.)

`vmstat -m` is essentially a wrapper over the kernel's `/sys/kernel/slab` content (more on this follows). Deep internal details of the slab caches can be seen using utilities such as `slabinfo`, `slabtop`, as well as the powerful `crash` utility (on a "live" system, the relevant crash command is `kmem -s` (or `kmem -S`)).

Right! Time to again get hands-on with some code to demonstrate the usage of the slab allocator APIs!

Writing a kernel module to use the basic slab APIs

In the following code snippet, look at the demo kernel module code (found at `ch8/slab1/`). In the initialization code, we merely perform a couple of slab layer allocations (via the `kmalloc()` and `kzalloc()` APIs), print some information, and free the buffers in the cleanup code path (of course, the full source code is accessible at this book's GitHub repository). Let's look at the relevant parts of the code step by step.

At the start of the `init` code of this kernel module, we initialize a global pointer (`gkptr`) by allocating 1,024 bytes to it (*C programmers, remember: pointers have no memory!*) via the `kmalloc()` slab allocation API. Notice that, as we're certainly running in process context here and it is thus "safe to sleep," we use the `GFP_KERNEL` flag for the second parameter (just in case you want to refer back for more information, the earlier section titled *The GFP flags – digging deeper* has it covered):

```
// ch8/slab1/slab1.c
#define pr_fmt(fmt) "%s:%s(): " fmt, KBUILD_MODNAME, __func__
[...]
#include <linux/slab.h>
[...]
static char *gkptr;
struct myctx {
    u32 iarr[100];
    u64 uarr[100];
    char uname[128], passwd[16], config[16];
};
static struct myctx *ctx;

static int __init slab1_init(void)
{
    /* 1. Allocate slab memory for 1 KB using the kmalloc() */
    gkptr = kmalloc(1024, GFP_KERNEL);
    if (!gkptr) {
        WARN_ONCE(1, "kmalloc() failed!\n");
        /* As mentioned earlier, there is really no need to print an
         * error msg when a memory alloc fails; the situation "shouldn't"
         * typically occur, and if it does, the kernel will emit a chain
```

```
         * of messages in any case. Here, we use the WARN_ONCE()
         * macro pedantically, and as this is a 'learning' program..
         * In practice, simply returning an appropriate -ve errno (-ENOMEM
         * here) is enough. */
        goto out_fail1;
    }
    pr_info("kmalloc() succeeds, (actual KVA) ret value = %px\n", gkptr);

    /* Commit 091cb09 (5.4 kernel) has the print_hex_dump_bytes() perform
     * printing at log level KERN_DEBUG; in effect, it only shows up if
     * -DDEBUG (or you use the dynamic debug facility). So, I've set DEBUG on
in the Makefile */
    print_hex_dump_bytes("gkptr before memset: ", DUMP_PREFIX_OFFSET, gkptr,
32);
    memset(gkptr, 'm', 1024);
    print_hex_dump_bytes(" gkptr after memset: ", DUMP_PREFIX_OFFSET, gkptr,
32);
```

In the preceding code, also notice that we use the `print_hex_dump_bytes()` kernel convenience routine as a convenient way to dump the buffer memory in a human-readable format. Its signature is:

```
void print_hex_dump_bytes(const char *prefix_str, int prefix_type,
    const void *buf, size_t len);
```

Its parameters are as follows: `prefix_str` is any string you would like to prefix to each line of the hex dump; `prefix_type` is one of `DUMP_PREFIX_OFFSET`, `DUMP_PREFIX_ADDRESS`, or `DUMP_PREFIX_NONE`; `buf` is the source buffer to hex-dump; and `len` is the number of bytes to dump. Note that (as shown in the comment), from Linux 5.4, it only emits `printk` when the symbol `DEBUG` is defined (or when using the kernel's powerful dynamic debug facility, a topic covered in depth in the *Linux Kernel Debugging* book).

Quick Tip: When given a Git Commit ID (hash) of the Linux kernel project (it's often an abbreviated one, like the one in the above code comment – 091cb09), you can easily look it up on a Git-based repo via the `git blame` command. Alternatively, for the Linux kernel, you can navigate to `https://github.com/torvalds/linux` and search the repo for via the commit ID; it will yield the resulting commit. (In this example, here it is: `https://github.com/torvalds/linux/commit/091cb0994edd20d67521094ac9c6ec9804058d9a`).

Up next is a typical strategy – *a best practice* – followed by many device drivers: they keep all their required or 'context' information in a single data structure, often termed the *driver context* (or the driver's *private*) structure. We mimic this by declaring a (silly/sample) data structure called `myctx`, as well as a global pointer to it called `ctx` (the structure and pointer definition is in the preceding code block). Its allocation is serviced by the `kzalloc()` API:

```
    /* 2. Allocate memory for and initialize our 'context' structure */
    ctx = kzalloc(sizeof(struct myctx), GFP_KERNEL);
```

```
    if (!ctx)
        goto out_fail2;
    pr_info("context struct alloc'ed and initialized (actual KVA ret = %px)\n",
ctx);
    print_hex_dump_bytes("ctx: ", DUMP_PREFIX_OFFSET, ctx, 32);
    return 0;        /* success */
out_fail2:
    kfree(gkptr);
out_fail1:
    return -ENOMEM;
}
```

Here, we allocate and initialize `ctx` to the size of the `myctx` data structure via the useful `kzalloc()` wrapper API. The subsequent hexdump will show (provided `DEBUG` is defined!) that it is indeed initialized to all zeroes (for readability, we will only "dump" the first 32 bytes).

Do notice how we handle the error paths using `goto`; this has already been mentioned a few times earlier in this book, so we won't repeat ourselves here. Finally, in the cleanup code of the kernel module, we `kfree()` both buffers, preventing any memory leakage:

```
static void __exit slab1_exit(void)
{
    kfree(ctx);
    kfree(gkptr);
    pr_info("freed slab memory, removed\n");
}
```

A screenshot of a sample run on my Raspberry Pi 4 follows. I used our `../../lkm` convenience script to build, load the module, and run `dmesg`:

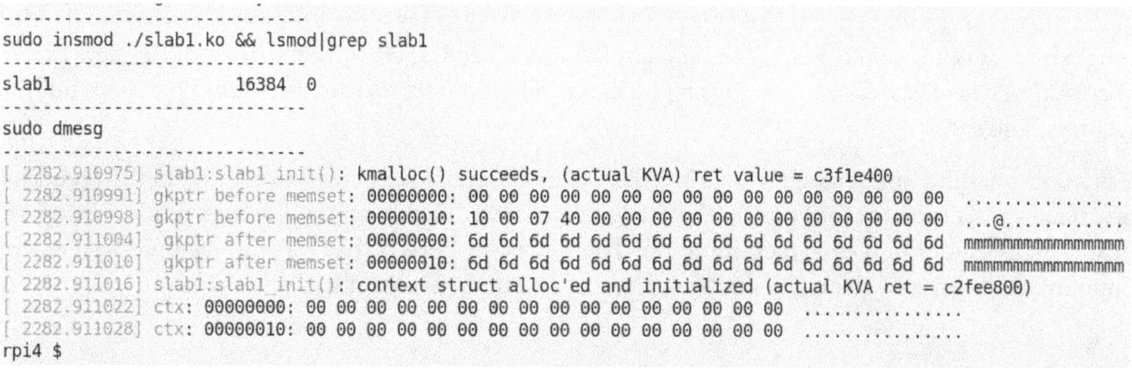

```
------------------------------
sudo insmod ./slab1.ko && lsmod|grep slab1
------------------------------
slab1                  16384  0
------------------------------
sudo dmesg
------------------------------
[ 2282.910975] slab1:slab1_init(): kmalloc() succeeds, (actual KVA) ret value = c3f1e400
[ 2282.910991] gkptr before memset: 00000000: 00 00 00 00 00 00 00 00 00 00 00 00 00 00 00 00  ................
[ 2282.910998] gkptr before memset: 00000010: 10 00 07 40 00 00 00 00 00 00 00 00 00 00 00 00  ...@............
[ 2282.911004]  gkptr after memset: 00000000: 6d 6d 6d 6d 6d 6d 6d 6d 6d 6d 6d 6d 6d 6d 6d 6d  mmmmmmmmmmmmmmmm
[ 2282.911010]  gkptr after memset: 00000010: 6d 6d 6d 6d 6d 6d 6d 6d 6d 6d 6d 6d 6d 6d 6d 6d  mmmmmmmmmmmmmmmm
[ 2282.911016] slab1:slab1_init(): context struct alloc'ed and initialized (actual KVA ret = c2fee800)
[ 2282.911022] ctx: 00000000: 00 00 00 00 00 00 00 00 00 00 00 00 00 00 00 00  ................
[ 2282.911028] ctx: 00000010: 00 00 00 00 00 00 00 00 00 00 00 00 00 00 00 00  ................
rpi4 $
```

Figure 8.8: Partial screenshot of our slab1.ko kernel module in action on a Raspberry Pi 4

Okay, now that you have a grip on the basics of using the common slab allocator APIs, `kmalloc()`, `kzalloc()`, and `kfree[_sensitive]()`, let's go further.

In the next section, we will dive deep into a really key concern – the reality of size limitations on the memory you can obtain via the slab (and page) allocators. Read on!

Size limitations of the kmalloc API

One of the key advantages of both the page and slab allocators is that the memory chunk they provide upon allocation is not only virtually contiguous (obviously) but is also guaranteed to be *physically contiguous memory*. Now that is a big deal and will certainly help performance.

But (there's always a *but*, isn't there!), precisely because of this guarantee, it becomes impossible to serve up *any* given large (memory) size when performing an allocation. In other words, there must be a limit to the amount of memory you can obtain from the slab allocator with a single call to our dear `k{m|z}alloc()` APIs. What is the limit? (This is indeed a really frequently asked question).

Firstly, you should understand that, technically, this limit is determined by two factors:

- One, the system page size (determined by the `PAGE_SIZE` macro).
- Two, the number of "orders" (determined by the `MAX_ORDER` macro); that is, the number of lists in the page allocator (or BSA) freelist data structures (see *Figure 8.2*).

With a standard 4-KB page size and a (typical) MAX_ORDER value of 11, the maximum amount of memory that can be allocated with a single `kmalloc()` or `kzalloc()` API call is 4 MB. This is the case on both the x86_64 and ARM (32 and 64-bit) architectures.

You might wonder, *how exactly is this 4-MB limit arrived at*? Think about it: once a slab allocation request exceeds the maximum slab cache size that the kernel provides (often 8 KB), the kernel simply passes the request down to the page allocator. The page allocator's maximum allocable size is determined by `MAX_ORDER`. With it set to `11`, the maximum allocable buffer size is $2^{(MAX\_ORDER-1)} = 2^{10}$ *pages = 1024 pages = 1024 * PAGE_SIZE bytes = 1024 * 4 KB = 4 MB*!

Testing the limits – memory allocation with a single call

A key thing for developers (and everyone else, for that matter) is to **be empirical** in your work! The English word *empirical* means based on what is experienced or seen, rather than on theory. *This is a critical rule to always follow – do not simply assume things or take them at face value.* Try them out for yourself and see.

Let's do something interesting: let's write a kernel module that allocates memory from the (generic) slab caches (via the `kmalloc()` API). We will do so in a loop, allocating – and subsequently freeing – a (calculated) amount on each loop iteration. The key point here is that we will keep increasing the amount allocated by a given *step size* (we'll even modularize this quantity).

The loop terminates when `kmalloc()` fails; this way, we can test just how much memory we can actually allocate with a single call to `kmalloc()` (you'll realize, of course, that `kzalloc()`, being a simple wrapper over `kmalloc()`, faces precisely the same limits).

In the following code snippet, we show the relevant code. The `test_maxallocsz()` function is called from the `init` code of the kernel module:

```
// ch8/slab3_maxsize/slab3_maxsize.c
[...]
static int stepsz = 204800;  // 200 Kb
module_param(stepsz, int, 0644);
MODULE_PARM_DESC(stepsz,
"Amount to increase allocation by on each loop iteration (default=200 KB");

static int test_maxallocsz(void)
{
  size_t size2alloc = 0;
  void *p;

  while (1) {
      p = kmalloc(size2alloc, GFP_KERNEL);
      if (!p) {
          pr_alert("kmalloc fail, size2alloc=%zu\n", size2alloc);
   // WARN_ONCE(1, "kmalloc fail, size2alloc=%zu\n", size2alloc);
          return -ENOMEM;
      }
      pr_info("kmalloc(%7zu) = 0x%px\n", size2alloc, p);
      kfree(p);
      size2alloc += stepsz;
  }
  return 0;
}
```

Notice how our `printk` functions use the `%zu` format specifier for the `size_t` (essentially an unsigned integer) variable? `%zu` is a portability aid; it makes the variable format correct for both 32- and 64-bit systems!

Let's run this kernel module on our Raspberry Pi 4 device running the stock Raspberry Pi OS. Almost immediately, upon `insmod`, you will see an error message, `Cannot allocate memory`, printed by the `insmod` process; the following (truncated) screenshot shows what has occured:

```
[ 4667.413286] slab3_maxsize: inserted
[ 4667.413298] kmalloc(      0) = 0x00000010
[ 4667.413312] kmalloc( 204800) = 0xc5580000
[ 4667.413322] kmalloc( 409600) = 0xc5580000
[ 4667.413335] kmalloc( 614400) = 0xc5600000
[ 4667.413346] kmalloc( 819200) = 0xc5600000
[ 4667.413357] kmalloc(1024000) = 0xc5600000
[ 4667.413372] kmalloc(1228800) = 0xc5600000
[ 4667.413386] kmalloc(1433600) = 0xc5600000
[ 4667.413400] kmalloc(1638400) = 0xc5600000
[ 4667.413413] kmalloc(1843200) = 0xc5600000
[ 4667.413427] kmalloc(2048000) = 0xc5600000
[ 4667.413456] kmalloc(2252800) = 0xc5800000
[ 4667.413475] kmalloc(2457600) = 0xc5800000
[ 4667.413495] kmalloc(2662400) = 0xc5800000
[ 4667.413514] kmalloc(2867200) = 0xc5800000
[ 4667.413534] kmalloc(3072000) = 0xc5800000
[ 4667.413553] kmalloc(3276800) = 0xc5800000
[ 4667.413573] kmalloc(3481600) = 0xc5800000
[ 4667.413592] kmalloc(3686400) = 0xc5800000
[ 4667.413612] kmalloc(3891200) = 0xc5800000
[ 4667.413631] kmalloc(4096000) = 0xc5800000
[ 4667.413644] ------------[ cut here ]------------
[ 4667.413649] WARNING: CPU: 2 PID: 8162 at mm/page_alloc.c:5418 __alloc_pages+0x914/0x1138
[ 4667.413667] Modules linked in: slab3_maxsize(O+) slab1(O) cmac algif_hash aes_arm_bs crypto_
simd cryptd algif_skcipher af_alg bnep hci_uart btbcm bluetooth ecdh_generic ecc 8021q garp stp
 llc brcmfmac brcmutil cfg80211 vc4 snd_soc_hdmi_codec v3d cec gpu_sched rfkill drm_kms_helper
raspberrypi_hwmon snd_soc_core i2c_brcmstb i2c_bcm2835 bcm2835_codec(C) rpivid_hevc(C) bcm2835_
isp(C) bcm2835_v4l2(C) v4l2_mem2mem bcm2835_mmal_vchiq(C) videobuf2_dma_contig snd_bcm2835(C) v
ideobuf2_vmalloc videobuf2_memops videobuf2_v4l2 videobuf2_common videodev snd_compress snd_pcm
_dmaengine snd_pcm vc_sm_cma(C) mc snd_timer snd syscopyarea uio_pdrv_genirq sysfillrect nvmem_
rmem sysimgblt fb_sys_fops uio drm i2c_dev hello(O) fuse drm_panel_orientation_quirks backlight
 ip_tables x_tables ipv6 [last unloaded: slab1]
[ 4667.413970] CPU: 2 PID: 8162 Comm: insmod Tainted: G         C O      5.15.76-v7l+ #1597
[ 4667.413976] Hardware name: BCM2711
[ 4667.413979] Backtrace:
[ 4667.413984] [<c0bd7354>] (dump_backtrace) from [<c0bd75a0>] (show_stack+0x20/0x24)
[ 4667.413997]  r7:0000152a r6:c0e3f708 r5:00000000 r4:60000013
[ 4667.414000] [<c0bd7580>] (show_stack) from [<c0bdbcb0>] (dump_stack_lvl+0x70/0x94)
[ 4667.414008] [<c0bdbc40>] (dump_stack_lvl) from [<c0bdbcec>] (dump_stack+0x18/0x1c)
[ 4667.414017]  r7:0000152a r6:00000009 r5:c0427f88 r4:c0e53968
[ 4667.414020] [<c0bdbcd4>] (dump_stack) from [<c02226c0>] (__warn+0xfc/0x114)
[ 4667.414029] [<c02225c4>] (__warn) from [<c0bd7c60>] (warn_slowpath_fmt+0x70/0xd8)
[ 4667.414037]  r7:0000152a r6:c0e53968 r5:c1205048 r4:00000000
[ 4667.414040] [<c0bd7bf4>] (warn_slowpath_fmt) from [<c0427f88>] (__alloc_pages+0x914/0x1138)
[ 4667.414050]  r9:0000000b r8:c043ed74 r7:00000cc0 r6:0041a000 r5:0000000b r4:00000000
[ 4667.414053] [<c0427674>] (__alloc_pages) from [<c03f6cc4>] (kmalloc_order+0x48/0xc0)
[ 4667.414063]  r10:0041a000 r9:0000000b r8:c043ed74 r7:00000cc0 r6:0041a000 r5:0000000b
[ 4667.414066]  r4:0041a000
[ 4667.414069] [<c03f6c7c>] (kmalloc_order) from [<c03f6d68>] (kmalloc_order_trace+0x2c/0xc4)
```

Figure 8.9: The first insmod of our slab3_maxsize.ko kernel module on a Raspberry Pi 4 running a stock kernel

This is expected! Think about it, the `init` function of our kernel module code has indeed failed (with failure code `-ENOMEM`) after all. Don't get thrown by this; looking up the kernel log reveals what transpired (*Figure 8.9*).

The fact is that on the very first test run of this kernel module, you will find that at the place where `kmalloc()` fails, the kernel dumps some diagnostic information, including a pretty lengthy kernel stack trace. This is due to kernel code paths invoking the `WARN()` macro when a kernel memory allocation failed, as, in the normal case, this simply shouldn't happen. Look again at the bottom portion of *Figure 8.9*; the stack trace includes the `__alloc_pages()` function; in the earlier *The workings of the page allocator* section, we literally pointed out that this function is documented as being the *'heart' of the zoned buddy allocator!* It being here in the stack (call or back) trace shows that it failed; hence, we get the result we see.

(A nagging thing, perhaps: why are the return pointers from the `kmalloc()`, the KVAs – as seen in *Figure 8.8* – often the same? Easy: it's as we allocated and then almost immediately release (free) the memory chunk, often leading to it being available and thus reused.)

So, our slab memory allocations worked, up to a point. To clearly see the failure point, simply scroll down in the kernel log (`dmesg`) display. The following screenshot shows this:

```
[ 4667.414053] [<c0427674>] (__alloc_pages) from [<c03f6cc4>] (kmalloc_order+0x48/0xc0)
[ 4667.414063]  r10:0041a000 r9:0000000b r8:c043ed74 r7:00000cc0 r6:0041a000 r5:0000000b
[ 4667.414066]  r4:0041a000
[ 4667.414069] [<c03f6c7c>] (kmalloc_order) from [<c03f6d68>] (kmalloc_order_trace+0x2c/0xc4)
[ 4667.414077]  r7:00000cc0 r6:0041a000 r5:c5800000 r4:0041a000
[ 4667.414080] [<c03f6d3c>] (kmalloc_order_trace) from [<c043ed74>] (__kmalloc+0x48c/0x4fc)
[ 4667.414091]  r10:0041a000 r9:00000cc0 r8:00000000 r7:bf318094 r6:bf319000 r5:c5800000
[ 4667.414094]  r4:0041a000
[ 4667.414097] [<c043e8e8>] (__kmalloc) from [<bf087048>] (slab3_maxsize_init+0x48/0x1000 [slab
3_maxsize])
[ 4667.414111]  r10:c1205048 r9:c3f52c48 r8:00000000 r7:bf318094 r6:bf319000 r5:c5800000
[ 4667.414114]  r4:0041a000
[ 4667.414116] [<bf087000>] (slab3_maxsize_init [slab3_maxsize]) from [<c02021a4>] (do_one_init
call+0x50/0x244)
[ 4667.414128]  r7:00000002 r6:bf087000 r5:c1205048 r4:bf319040
[ 4667.414131] [<c0202154>] (do_one_initcall) from [<c02d18e8>] (do_init_module+0x54/0x23c)
[ 4667.414141]  r8:bf319040 r7:00000002 r6:c3406f40 r5:00000002 r4:bf319040
[ 4667.414144] [<c02d1894>] (do_init_module) from [<c02d4148>] (load_module+0x24f8/0x294c)
[ 4667.414153]  r6:c3f52c00 r5:00000002 r4:c3f0df30
[ 4667.414156] [<c02d1c50>] (load_module) from [<c02d4818>] (sys_finit_module+0xc8/0xfc)
[ 4667.414166]  r10:0000017b r9:c3f0c000 r8:c0200244 r7:00000003 r6:0002de04 r5:00000000
[ 4667.414169]  r4:c1205048
[ 4667.414172] [<c02d4750>] (sys_finit_module) from [<c0200040>] (ret_fast_syscall+0x0/0x1c)
[ 4667.414180] Exception stack(0xc3f0dfa8 to 0xc3f0dff0)
[ 4667.414184] dfa0:                   00000000 00000002 00000003 0002de04 00000000 b6f28074
[ 4667.414189] dfc0: 00000000 00000002 9d2c7000 0000017b 01d98d68 00000002 be9eb7d4 00000000
[ 4667.414193] dfe0: be9eb600 be9eb5f0 00023bc0 b6c15580
[ 4667.414197]  r7:0000017b r6:9d2c7000 r5:00000002 r4:00000000
[ 4667.414201] ---[ end trace 439cc7506ad82102 ]---
[ 4667.414206] kmalloc fail, size2alloc=4300800
rpi4 $
```

Figure 8.10: Partial screenshot showing the lower part of the dmesg output (of our ch8/slab3_maxsize.ko kernel module) on a Rasperry Pi 4

Aha, look at the last line of output (*Figure 8.10*): `kmalloc()` fails when attempting to allocate above 4 MB (here, at 4,300,800 bytes), precisely as expected; until then, it succeeds.

Also, now the beginning of the stack backtrace can be seen; read it bottom-up of course and it's all crystal clear – you can literally see our `kmalloc()` ending up in `__alloc_pages()`, which, because it failed, led to this turn of events.

As an interesting aside, notice that we have (quite deliberately) performed the very first allocation in the loop with size `0`; it does not fail (but that doesn't mean you use it):

`kmalloc(0, GFP_xxx);` returns the zero pointer; on x86[_64], it's the value `16` or `0x10` (see `include/linux/slab.h` for details). In effect, it's an invalid virtual address living in the page `0` `NULL` pointer trap. Accessing it will, of course, lead to a page fault (originating from the MMU), a bug.

Similarly, attempting `kfree(NULL);` or `kfree()` on the zero pointer results in `kfree()` becoming a no-op.

Hang on, though – there's an extremely important point to note here: in the section titled *The actual slab caches in use for kmalloc*, we saw that the slab caches that are used to allocate memory to the caller are the *kmalloc-N* slab caches, where *N* ranges **from 64 to 8,192 bytes** (on the Raspberry Pi, and thus the ARM[64] processor for this discussion). Also, FYI, you can perform the following quickly to verify this:

```
sudo vmstat -m | grep -v "\-rcl\-" | grep --color=auto "^kmalloc-[0-9]"
```

But clearly, in the preceding kernel module code example, we have allocated via `kmalloc()` much larger quantities of memory (right from 0 bytes through 8 KB and then right up to 4 MB). Under the hood, the way it really works is that the `kmalloc()` API only uses the `kmalloc-N` slab caches for memory allocations less than or equal to 8,192 bytes (if available); any allocation request for larger memory chunks is then passed to the underlying page (or buddy system) allocator (see *Figure 8.1*)!

Now, recall what we learned: the page allocator uses the buddy system freelists (on a per node:zone basis) and the maximum size of memory chunks enqueued on the freelists are $2^{(MAX\_ORDER-1)} = 2^{10}$ *pages*, which, given a page size of 4 KB and `MAX_ORDER` of `11`, is 4 MB. This neatly ties in with our theoretical discussions.

So, there we have it: both in theory and in practice, you can now see that (again, given a page size of 4 KB and `MAX_ORDER` of `11`), the maximum size of memory that can be allocated via a single call to `kmalloc()` or `kzalloc()` is 4 MB.

Checking via the /proc/buddyinfo pseudofile

It's really important to realize that although we figured out that 4 MB of RAM is the maximum we can get at one shot, it definitely doesn't mean that you will always get that amount when you request it. No sir, of course not; it completely depends upon the amount of free memory present within the particular freelist at the time of the memory request. Think about it: what if you were running on a Linux system that has been up for several days (or weeks)? The likelihood of finding physically contiguous 4 MB chunks of free RAM is quite low (again, this depends upon the amount of RAM on the system and its workload).

As a rule of thumb, if the preceding experiment did not yield a maximum allocation of what we have deemed to be the maximum size (that is, 4 MB), why not try it on a freshly booted guest system? Now, the chances of having physically contiguous 4 MB chunks of free RAM are a lot better.

Unsure about this? Let's get empirical again and look up the content of `/proc/buddyinfo` – both on an in-use and a freshly booted system – to figure out whether the memory chunks are available. In the following command-line snippet, on my in-use native x86_64 Ubuntu system (up for over 7 days, with 32 GB RAM), we look it up, highlighting the order 10 quantity of free chunks – i.e., 4 MB free chunks of RAM – in the `Normal` zone in bold (the output wraps):

```
$ grep -w "Normal" /proc/buddyinfo
Node 0, zone   Normal  23653   5314   1284    747    233     67     16     17
18      0      0
```

As we learned earlier (in the *Understanding the organization of the page allocator freelist* section), the numbers seen in the preceding code block are in the sequence order `0` to `MAX_ORDER-1` (typically, *0* to *11 – 1, so 0 to 10*), and they represent the number of 2^{order} contiguous free page frames in that order.

In the preceding output, we can see that, for the *Normal* zone, we do *not* have free blocks on the order `10` list (that is, the 4 MB chunks; it's zero). On a freshly booted Linux system, the chances are high that we will. In the following output, on the same system that's just been rebooted, we see that there are 6,571 (!) chunks of free, physically contiguous 4 MB RAM available in node `0` in the *Normal* zone:

```
$ grep -w "Normal" /proc/buddyinfo
Node 0, zone   Normal   3859   6778   6131   5025   3810   2274   1130    382
78      9  6571
```

You'll realize that, as system uptime increases, chances of "cutting" the higher sized chunks into smaller pieces as required, goes up... Thus we find that memory tends to percolate from the higher to lower orders as time marches on.

Of course, there's more to explore. In the following section, we cover more on using the slab allocator – the resource-managed API alternatives, additional slab helper APIs that are available, and a note on cgroups and memory in modern Linux kernels.

Slab allocator – a few additional details

A few more key points remain to be explored. First, we'll divulge some information on using the kernel's *resource-managed* versions of the memory allocator APIs, followed by a few additionally available slab helper routines within the kernel, and then have a brief look at cgroups and memory. We recommend you go through these sections as well. Please, do read on!

Using the kernel's resource-managed memory allocation APIs

Especially useful when working on device drivers, the kernel provides a few managed APIs for memory allocation. These are formally referred to as the **device resource-managed or devres APIs** (the link to the official kernel documentation on this is `https://www.kernel.org/doc/html/latest/driver-api/driver-model/devres.html`). All these APIs are prefixed with `devm_`; though there are several of them, we will focus on only one common use case here – that of using the resource-managed versions in place of the usual memory allocation APIs, the `k{m|z}alloc()` ones. They are as follows:

```
#include <linux/device.h>
```

```
void *devm_kmalloc(struct device *dev, size_t size, gfp_t gfp);
void *devm_kzalloc(struct device *dev, size_t size, gfp_t gfp);
```

The reason why these resource-managed APIs are useful is that there is *no need for the developer to explicitly free the memory allocated by them*. The kernel resource management framework guarantees that it will automatically free the memory buffer upon driver detach, or if it's a kernel module, when the module is removed (or the device is detached, whichever occurs first). This feature immediately enhances code robustness. Why? Simple, we're all human and make mistakes. Leaking memory (especially on error code paths) is indeed a pretty common bug!

A few relevant points regarding the usage of these APIs:

- A key point – please do **not** attempt to blindly replace all `k{m|z}alloc()` instances with the corresponding `devm_k{m|z}alloc()`! These resource-managed allocations are designed to be used only in the `init` and/or `probe()` methods of a device driver (all drivers that work with the kernel's unified device model will typically supply the `probe()` and `remove()` (or `disconnect()`) methods. We will not delve further into these aspects here).
- `devm_kzalloc()` is usually preferred as it initializes the buffer as well. Internally (as with `kzalloc()`), it is merely a thin wrapper over the `devm_kmalloc()` API.
- The second and third parameters are the usual ones, as with the `k{m|z}alloc()` APIs – the number of bytes to allocate and the GFP flags to use. The first parameter, though, is a pointer to `struct device`. Quite obviously, it represents the *device* that your driver is driving (every driver will have access to this structure).
- As the memory allocated by these APIs is auto-freed (on driver detach or module removal), you don't have to do anything (to free the memory). It can, though, be freed via the `devm_kfree()` API. Your doing this manually, however, is usually an indication that the managed APIs are the wrong ones to use.
- Licensing: The managed APIs are exported (and thus available) only to modules licensed under the GPL.

Additional slab helper APIs

There are several helper slab allocator APIs, friends of the `k{m|z}alloc()` API family. These include the `kcalloc()` and `kmalloc_array()` APIs for allocating memory for an array, as well as `krealloc()`, whose behavior is analogous to `realloc()`, the familiar user space API. (Under the hood, `kmalloc_array()` calls the `check_mul_overflow()` macro, which performs multiplication with overflow checking! Neat.)

In conjunction with allocating memory for an array of elements, the `array_size()` and `struct_size()` kernel helper routines can be very helpful. In particular, `struct_size()` has been heavily used to prevent – and indeed fix – many integer overflow (IoF, and related) bugs when allocating an array of structures, a common task indeed. As a quick example, here's a small code snippet from `net/bluetooth/mgmt.c`:

```
rp = kmalloc(struct_size(rp, addr, i), GFP_KERNEL);
if (!rp) {
      err = -ENOMEM; [...]
```

FYI, it's worth browsing through the `include/linux/overflow.h` kernel header file.

(Keep in mind the already covered `kfree_sensitive()` API as well.) The resource-managed versions of these APIs are also available: `devm_kcalloc()` and `devm_kmalloc_array()`.

Control groups and memory

The Linux kernel supports a very sophisticated resource management system called **cgroups** (**control groups**), which, in a nutshell, is used to hierarchically organize processes and perform resource management.

We shall cover a lot more on cgroups, with examples that leverage the cgroups v2 CPU controller, in *Chapter 11, The CPU Scheduler – Part 2,* in the section titled *An introduction to cgroups.* If the following paragraphs here don't make too much sense to you now, I recommend you first read up on cgroups (in the chapter and section just mentioned) and then come back here.

Among the several resource controllers within the cgroups framework is one for controlling memory bandwidth. By carefully configuring it, the sysadmin can effectively regulate the distribution of memory on the system. Memory protection is possible, both as (what is called) hard and best-effort protection via certain `memcg` (memory cgroup) pseudofiles (particularly, the `memory.min` and `memory.low` files). In a similar fashion, within a cgroup, the `memory.high` and `memory.max` pseudofiles are the main mechanism to control the memory usage of a cgroup. Of course, as there is a lot more to it than is mentioned here, I refer you to the kernel documentation on the new cgroups (v2) here: `https://www.kernel.org/doc/html/latest/admin-guide/cgroup-v2.html`.

Right, now that you have learned how to better use the slab allocator APIs, let's dive a bit deeper still. The reality is that there are still a few important caveats regarding the size of the memory chunks allocated by the slab allocator APIs. Do read on to find out what they are!

Caveats when using the slab allocator

We will split up this discussion into three parts. We will first re-examine some necessary background (which we covered earlier), then actually flesh out the problem that we're getting at with two use cases – the first being very simple; the second being a more real-world case of the issue at hand.

Background details and conclusions

So far, you have learned some key points:

- The *page* (or *buddy system*) *allocator* allocates pages to the caller in powers of 2 pages; in other words, the **granularity of an allocation request is a page (typically 4K)**. The power to raise 2 is called the *order*; it typically ranges from `0` to `10` (on both x86[_64] and ARM[_64], assuming a page size of 4K and `MAX_ORDER` of 11).
- This is fine, except when it's not. When the amount of memory requested is very small, or just over a certain threshold, the *wastage* (or internal fragmentation) can be huge.

- In the day-to-day operation of the OS and its drivers, **requests for fragments of a page ((much) less than 4,096 bytes) are very common.** Thus, the *slab allocator, layered upon the page allocator* (see *Figure 8.1*) is designed with object caches, as well as small generic memory caches, to efficiently fulfill requests for small amounts of memory.
- The page allocator guarantees physically contiguous page and hardware cacheline-aligned memory.
- The slab allocator too guarantees physically contiguous and hardware cacheline-aligned memory.

So, fantastic – this leads us to conclude that when the amount of memory required is large-ish and a perfect (or close) power of 2, use the page allocator. When it's quite small (less than a page), use the slab allocator. Indeed, the kernel source code of `kmalloc()` has a comment that neatly sums up how the `kmalloc()` API should be used (reproduced in bold font as follows):

```
// include/linux/slab.h
[...]
 * kmalloc - allocate memory
 * @size: how many bytes of memory are required.
 * @flags: the type of memory to allocate.
 * kmalloc is the normal method of allocating memory
 * for objects smaller than page size in the kernel.
```

Sounds great, but there is still a problem! To see it, let's learn how to use another useful slab API, `ksize()`. Its signature is as follows:

```
size_t ksize(const void *);
```

The parameter to `ksize()` is a pointer to an existing slab cache (it must be a valid one). In other words, it's the return address from one of the slab allocator APIs (typically one of `[devm_]k{m|z}alloc()`). The return value of `ksize()` is the *actual number of bytes allocated.*

Okay, now that you know what `ksize()` is for, let's use it in a more practical fashion, first with a simple use case and then with a better one!

Testing slab allocation with ksize() – case 1

To understand what we're getting at, consider a small example. (For readability, we will not show the essential validity checks. Also, as this is a tiny code snippet, we haven't provided it as a kernel module in the book's code base):

```
struct mysmallctx {
    int tx, rx;
    char passwd[8], config[4];
} *ctx;

pr_info("sizeof struct mysmallctx = %zd bytes\n", sizeof(struct mysmallctx));
```

```
ctx = kzalloc(sizeof(struct mysmallctx), GFP_KERNEL);
pr_info("(By now, the context structure is allocated and initialized to
zero)\n"
        "*actual* size allocated = %zu bytes\n", ksize(ctx));
```

The resulting output on my x86_64 Ubuntu guest system is as follows:

```
$ dmesg
[...]
sizeof struct mysmallctx = 20 bytes
(By now, the context structure is allocated and initialized to zero)
*actual* size allocated = 32 bytes
```

So, we attempted to allocate 20 bytes with `kzalloc()`, but actually obtained 32 bytes, thus incurring a wastage of 12 bytes, or 60%! This is expected. Recall the `kmalloc-N` slab caches – on x86, there is one for 16 bytes and another for 32 bytes (among the many others). So, when we ask for an amount in between the two, we obviously get memory from the higher of the two (the allocators essentially use a version of the best-fit approach). (Incidentally, and FYI, on our ARM-based Raspberry Pi system, the smallest slab cache for `kmalloc` is 64 bytes, so, of course, there we will get 64 bytes when we ask for 20 bytes.)

Note that the `ksize()` API works only on allocated slab memory; you cannot use it on the return value from any of the page allocator APIs (which we saw in the *Learning how to use the page allocator APIs* section).

Now for the second, and more interesting, use case.

Testing slab allocation with ksize() – case 2

Okay, now, let's extend our earlier kernel module (`ch8/slab3_maxsize`) to

`ch8/slab4_actualsize`. Here, we will perform the same loop, allocating memory with the `kmalloc()` API and freeing it as before, but this time, we will also document the *actual* amount of memory allocated to us in each loop iteration by the slab layer, by invoking the `ksize()` API:

```
// ch8/slab4_actualsize/slab4_actualsize.c
static int test_maxallocsz(void)
{
   /* This time, initialize size2alloc to 100, as otherwise we'll get a
     * divide error! */
    size_t size2alloc = 100, actual_alloced;
    void *p;

    pr_info("kmalloc(      n) :  Actual : Wastage : Waste %%\n");
    while (1) {
        p = kmalloc(size2alloc, GFP_KERNEL);
```

```
        if (!p) {
          pr_alert("kmalloc fail, size2alloc=%zu\n", size2alloc); // pedantic
          return -ENOMEM;
        }
        actual_alloced = ksize(p);
        /* Print the size2alloc, the amount actually allocated,
         * the delta between the two, and the percentage of waste
         * (integer arithmetic, of course :-)  */
        pr_info("kmalloc(%7zu) : %7zu : %7zu : %3zu%%\n",
            size2alloc, actual_alloced, (actual_alloced - size2alloc),
            (((actual_alloced - size2alloc) * 100) / size2alloc));
        kfree(p);
        size2alloc += stepsz;
    }
    return 0;
}
```

The output of this kernel module is indeed interesting to scan! In the following figure, we show a partial screenshot of the output I got on my x86_64 Ubuntu 22.04 LTS guest running our custom-built 6.1 kernel (loaded via our convenience `lkm` script):

```
------------------------------
sudo insmod ./slab4_actualsize.ko && lsmod|grep slab4_actualsize
------------------------------
insmod: ERROR: could not insert module ./slab4_actualsize.ko: Cannot allocate memory
 ^--[FAILED]
------------------------------
sudo dmesg
------------------------------
[ 3948.215217] slab4_actualsize: inserted
[ 3948.215220] kmalloc(      n) :  Actual : Wastage : Waste %
[ 3948.215220] kmalloc(    100) :     128 :      28 :  28%
[ 3948.215232] kmalloc( 204900) :  262144 :   57244 :  27%
[ 3948.215249] kmalloc( 409700) :  524288 :  114588 :  27%
[ 3948.215513] kmalloc( 614500) : 1048576 :  434076 :  70%
[ 3948.215537] kmalloc( 819300) : 1048576 :  229276 :  27%
[ 3948.215559] kmalloc(1024100) : 1048576 :   24476 :   2%
[ 3948.215617] kmalloc(1228900) : 2097152 :  868252 :  70%
[ 3948.215660] kmalloc(1433700) : 2097152 :  663452 :  46%
[ 3948.215700] kmalloc(1638500) : 2097152 :  458652 :  27%
[ 3948.215741] kmalloc(1843300) : 2097152 :  253852 :  13%
[ 3948.215782] kmalloc(2048100) : 2097152 :   49052 :   2%
[ 3948.216309] kmalloc(2252900) : 4194304 : 1941404 :  86%
[ 3948.216396] kmalloc(2457700) : 4194304 : 1736604 :  70%
[ 3948.216478] kmalloc(2662500) : 4194304 : 1531804 :  57%
[ 3948.216614] kmalloc(2867300) : 4194304 : 1327004 :  46%
[ 3948.216710] kmalloc(3072100) : 4194304 : 1122204 :  36%
[ 3948.216792] kmalloc(3276900) : 4194304 :  917404 :  27%
[ 3948.216874] kmalloc(3481700) : 4194304 :  712604 :  20%
[ 3948.216956] kmalloc(3686500) : 4194304 :  507804 :  13%
[ 3948.217038] kmalloc(3891300) : 4194304 :  303004 :   7%
[ 3948.217120] kmalloc(4096100) : 4194304 :   98204 :   2%
[ 3948.217126] ------------[ cut here ]------------
[ 3948.217127] WARNING: CPU: 2 PID: 124052 at mm/page_alloc.c:5534 __alloc_pages+0x2
2a/0x1270
[ 3948.217135] Modules linked in: slab4_actualsize(OE+) tls drm_ttm_helper ttm drm_k
ms_helper syscopyarea sysfillrect sysimgblt fb_sys_fops vboxsf(OE) binfmt_misc snd_i
ntel8x0 snd_ac97_codec ac97_bus snd_pcm intel_rapl_msr snd_seq joydev snd_timer snd_
seq_device intel_rapl_common crct10dif_pclmul crc32_pclmul ghash_clmulni_intel aesni
_intel snd crypto_simd input_leds cryptd video rapl wmi serio_raw soundcore vboxgues
t(OE) mac_hid drm sch_fq_codel msr parport_pc ppdev lp parport ramoops pstore_blk re
ed_solomon efi_pstore pstore_zone ip_tables x_tables autofs4 hid_generic usbhid hid
psmouse e1000 ahci i2c_piix4 libahci pata_acpi
[ 3948.217165] CPU: 2 PID: 124052 Comm: insmod Tainted: G           OE      6.1.11-l
kp-kernel #1
[ 3948.217167] Hardware name: innotek GmbH VirtualBox/VirtualBox, BIOS VirtualBox 12
/01/2006
[ 3948.217168] RIP: 0010:__alloc_pages+0x22a/0x1270
```

Figure 8.11: Partial screenshot of our ch8/slab4_actualsize/slab4_actualsize.ko kernel module in action

The module's `printk` output can be seen clearly in the upper and middle portions of *Figure 8.11*. The remainder of the screenshot is some of the diagnostic information from the kernel – this is emitted as a kernel-space memory allocation request failed. All this kernel diagnostic information is a result of the first invocation of the kernel calling the `WARN_ONCE()` macro, as the underlying page allocator code, `mm/page_alloc.c:__alloc_pages()` – the "heart" of the buddy system allocator, as it's known – failed! This should typically never occur, hence the diagnostics (the details on the kernel diagnostic is beyond this book's scope, so we will leave this aside. Having said that, we do examine the kernel stack backtrace to some extent in the coming chapters. Furthermore, details like this are covered in depth in the *Linux Kernel Debugging* book).

By the way, in this module, we don't even code the module exit (cleanup) routine as we know it's going to return failure in the `init` code path itself and thus we'll never need to `rmmod` it.

Interpreting the output from case 2

Look closely at the preceding screenshot (*Figure 8.11*; here, we will simply ignore the kernel diagnostics emitted by the `WARN_ONCE()` macro, which got invoked because a kernel-level memory allocation failed!). Though quite self-explanatory, the key portion of *Figure 8.11*'s output has the timestamp from `dmesg` (we ignore it), followed by four columns, as its header line shows:

- `kmalloc(n)`: The number of bytes requested by `kmalloc()` (where `n` is the required amount).
- `Actual`: The actual number of bytes allocated by the slab allocator (revealed via `ksize()`).
- `Wastage`: The wastage (bytes); the difference between the actual and required bytes.
- `Waste %`: The wastage as a percentage.

As an example, in the second allocation (highlighted in *Figure 8.11* by a rectangle), we requested 204,900 bytes, but actually obtained 262,144 bytes (256 KB). This makes sense, as this is the precise size of one of the page allocator lists on a buddy system freelist (it's *order 6*, as 2^6 = *64 pages = 64 x 4 = 256 KB*; see *Figure 8.2*). Hence, the delta, the wastage really, is *262,144 – 204,900 = 57,244 bytes*, which, when expressed as a percentage, is 27%.

It works like this: the closer the requested (or required) size gets to the kernel's available (or actual) size, the less the wastage will be; the converse is true as well.

Let's look at another example from the preceding output (the snipped output is reproduced as follows for clarity):

```
[ … ]
[ 3948.215700] kmalloc(1638500) : 2097152 :  458652 :  27%
[ 3948.215741] kmalloc(1843300) : 2097152 :  253852 :  13%
[ 3948.215782] kmalloc(2048100) : 2097152 :   49052 :   2%
[ 3948.216309] kmalloc(2252900) : 4194304 : 1941404 :  86%
[ 3948.216396] kmalloc(2457700) : 4194304 : 1736604 :  70%
[ 3948.216478] kmalloc(2662500) : 4194304 : 1531804 :  57%
[ … ]
```

From the preceding output, you can see that when `kmalloc()` requests 1,638,500 bytes (a little over 1.5 MB), it actually gets 2,097,152 bytes (exactly 2 MB), and the wastage is 27%. The wastage then successively *reduces as we get closer to an allocation "boundary" or threshold* (the actual size of the kernel's slab cache or page allocator memory chunk) as it were: to 13%, then down to just 2%. But look: with the next allocation, when we *cross that threshold,* asking for *just over* 2 MB (2,252,900 bytes), we get 4 MB, *a wastage of 86%*! Then, the wastage again drops as we move closer to the 4-MB memory size...

This is important! You might think you're being very efficient by virtue of the mere use of the slab allocator APIs, but the slab layer invokes the page allocator when the amount of memory requested is above the maximum size that the slab layer can provide (typically, 8 KB, which is the case in our preceding experiments). Thus, **the page allocator, suffering from its usual wastage issues, ends up allocating far more memory than you require, or indeed ever use.** What a waste!

The moral of the story: *check and recheck your code that allocates memory with the slab APIs.* Run trials on it using `ksize()` to figure out how much memory is actually being allocated, not how much you think is being allocated.

There are no shortcuts. Well, there is one:

If you require less than a page of memory (a very, very typical use case), just use the slab APIs.

If you require more than 1 page of memory, the preceding discussion comes into play. Another thing: using the `alloc_pages_exact()/free_pages_exact()` APIs (covered in the *The 'exact' page allocator API pair* section) should help reduce wastage as well (of course this implies you've decided to use these APIs and not the slab ones; it's not the typical case).

Even better, you can check the actual – and more accurate – size of slab memory allocated via the `/sys/kernel/slab/<slab-name>/slab_size` pseudofile (we shall come across this in the following chapter, in the *Extracting useful information regarding slab caches* section)

Graphing it

As an interesting aside, we can use the well-known `gnuplot` utility to plot a graph from the previously gathered data. We must minimally modify the kernel module to only output what we'd like to graph: the required (or requested) amount of memory to allocate (x axis), and the percentage of waste that occurred at runtime (y axis). You can find the code of our slightly modified kernel module in the book's GitHub repository here: `ch8/slab4_actualsz_wstg_plot` (`https://github.com/PacktPublishing/Linux-Kernel-Programming_2E/tree/main/ch8/slab4_actualsz_wstg_plot`).

We also reduce the 'step size' to 20 KB (from 200K) to make the graph more granular. So, we build and insert this kernel module, "massage" the kernel log, saving the data in an appropriate column-wise format as required by `gnuplot` (in a file called `plotdata.txt`). The good news: all of this, including generating the plot (via `gnuplot`) is automated via our `plot_graph.sh` helper script; simply run it and see! (To understand the workings of the very useful gnuplot utility, refer to the *Further reading* section of this chapter.)

We also provide a helpful `Readme` file in the GitHub repo within this directory to guide you. Lo and behold, the plot:

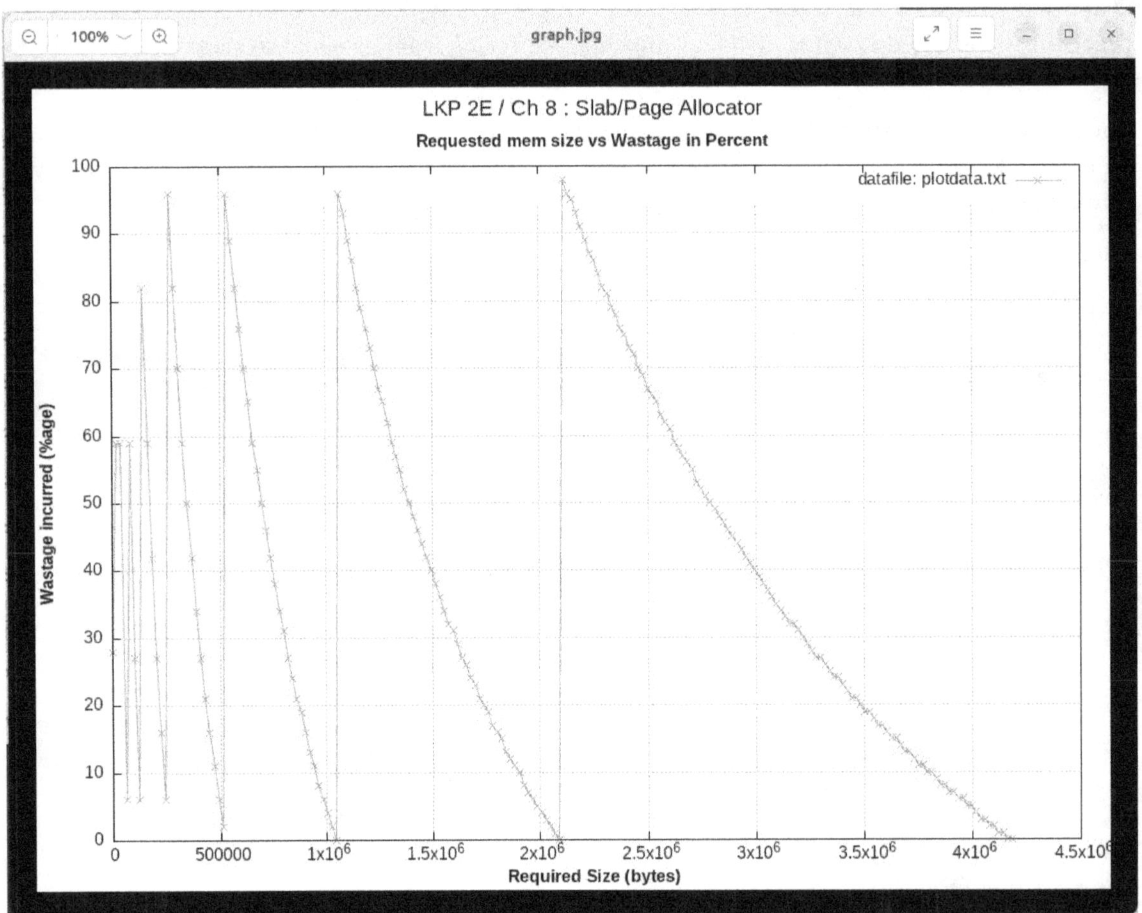

Figure 8.12: A gnuplot generated graph showing the size requested by kmalloc() (x-axis, bytes) versus the wastage incurred (y-axis, as a percentage)

This "saw-tooth"-shaped graph helps visualize what you just learned. The closer a `kmalloc()` (or `kzalloc()`, or indeed *any* page allocator API) allocation request size is to any of the kernel's predefined freelist sizes, the less wastage there is. But the moment this threshold is crossed, the wastage zooms up – spikes! – to close to 100% (as seen by the almost perfectly vertical lines in the preceding graph).

Finding internal fragmentation (wastage) within the kernel

The previous section demonstrated how internal fragmentation – wastage – can occur when allocating memory via the two now familiar slab APIs (the `k{m|z}alloc()` pair). The wastage seen there, though, tends to naturally be more so when the amount to be allocated goes well beyond a page.

Not to say that the wastage doesn't increase even for smaller numbers – the absolute amount wasted is less, though the percentage of wastage is consistent and can be pretty high when you ask for amounts of memory just over a given 'threshold' (the typical example: requesting 63 bytes gets you 64, whereas requesting 65 bytes gets you 128 (the `ksize()` API can be used to verify this)).

Here, we concern ourselves with exactly how much wastage is incurred for memory allocation requests of less than a page in size, the typical quantities requested. We'll show you two ways to quantify it: one, an easy approach, the second a bit more cumbersome (made easier with a custom script) but very valuable. Let's begin with the first approach.

The easy way with slabinfo

The `slabinfo` utility, which we learned to build in the *A few FAQs regarding (slab) memory usage and their answers* section earlier in this chapter, can be very helpful. An easy way to lookup wastage or 'loss' is simply to run `slabinfo` (as root) with the `-L` option switch (`Sort by loss`; the first command is used to simply show the help for the `-L` option switch):

```
# sudo slabinfo -h |grep "^-L"
-L|--Loss                  Sort by loss
# slabinfo -L | head -n5
Name             Objects  Objsize      Loss Slabs/Part/Cpu  O/S O %Fr %Ef  Flg
dentry            148162      192     14.9M      5297/20/0   28 1   0  65  PaZFU
kmalloc-4k          1153     4096     14.2M       579/4/0     2 3   0  24  PZFU
buffer_head       132952      104     13.4M      6648/4/0    20 0   0  50  PaZFU
lsm_inode_cache 134919         24     13.0M      3988/53/0   34 0   1  19  PZFU
ext4_inode_cache 58224      1184      7.4M       2330/2/0    25 3   0  90  PaZFU
```

The key column here, of course, is the one labeled `Loss`; it specifies the wastage the given slab cache (first column) has incurred so far; the output is also sorted by the loss (highest first). The second and third columns specify the number of objects in the slab and the size of each object respectively; thus, their product is the total amount of space taken by that slab cache. FYI, O/S stands for **Objects/Slab**, O is the order of the allocation, the next two columns specify the amount of free memory and effective memory usage within the cache in percentage terms, and the last column is the slab flags (here, they're populated as I passed them when booting; this aspect is covered very shortly).

The `slabinfo -L` output has the advantage of showing the loss (wastage) incurred by all slab caches (not just the `kmalloc-N` ones) but the downside of only showing the total wastage; contrast this with the following approach.

Even better, you can check the actual - and more accurate - size of slab memory allocated via the `/sys/kernel/slab/<slab-name>/slab_size` pseudofile (we shall come across this in the following chapter, in the *Extracting useful information regarding slab caches* section)

More details with alloc_traces and a custom script

As will be fully explained in the following chapter, the kernel can sometimes run into a difficult situation where primary memory (RAM) *and* swap are both exhausted; in this dire circumstance, the kernel has no choice but to invoke the (dreaded?) OOM killer!

This is a kernel component that literally kills the key culprit, the process that is the memory hogger (it kills its descendants as well! Talk about being ruthless...).

Conceivably, on a system with limited memory (think embedded, or otherwise), if the slab caches suffer from a lot of wastage (loss), it's entirely possible that the OOM killer will be invoked, at times during system boot itself while the kernel's initializing. This very situation is what led to a developer posting a patch (which got merged recently into the **6.1 kernel**); here's the link to the commit:

`https://github.com/torvalds/linux/commit/6edf2576a6cc46460c164831517A36064eb8109c`. This patch sets up a means for one to track all of the `kmalloc-N` slab cache's wastage (and other info), and importantly, even shows the call stack leading up to the allocation request! This lets us see the details of the allocation request, and thus helps us debug such situations...

The crux of this as it relates to our discussion: once you turn on certain SLUB debug flags, the wastage info gets integrated into a new pseudofile (named `alloc_traces`) under debugfs here: `/sys/kernel/debug/slab/kmalloc-*`; let's check it out (here, we assume that `CONFIG_SLUB_DEBUG=y`; it typically is):

1. Reboot your Linux box, and at the bootloader, pass along this kernel command-line parameter (in addition to the usual ones): `slub_debug=FZPU`.
    a. You'll find the list of all kernel parameters documented here (very valuable): `https://www.kernel.org/doc/html/latest/admin-guide/kernel-parameters.html`.
    b. The kernel SLUB layer typically has the full debugging technology built in but off (disabled) by default (as it can affect performance). To turn it on, certain SLUB debug flags are provided, which are passed via the `slub_debug=<debug-flags>` kernel parameter. The detailed info on the SLUB debug flags we employ here (`FZPU`) is covered in the *Linux Kernel Debugging* book; we shan't repeat it here. You can also look it up via the official kernel doc on it: `https://docs.kernel.org/mm/slub.html`.
2. Once within a root shell, look up the pseudofiles for a given slab (SLUB) cache's allocation and free tracking; as an example, let's look it up for the `kmalloc-64` slab cache:

```
# ls /sys/kernel/debug/slab/kmalloc-64/
alloc_traces  free_traces
```

Right, got it, now let's look at the `alloc_traces` file's current content:

```
# cat /sys/kernel/debug/slab/kmalloc-64/alloc_traces
975 populate_error_injection_list+0x8d/0x110 waste=15600/16
age=6363993/6363993/6363994 pid=1 cpus=0
        __kmem_cache_alloc_node+0x279/0x2b0
        kmalloc_trace+0x2a/0xa0
        populate_error_injection_list+0x8d/0x110
        init_error_injection+0x1b/0x75
        do_one_initcall+0x49/0x210
        kernel_init_freeable+0x27b/0x2e8
        kernel_init+0x1b/0x160
        ret_from_fork+0x22/0x30

    403 get_mountpoint+0xab/0x190 waste=9672/24 age=6360787/6361173/6363388
pid=332-1199 cpus=0-1,3-5
        __kmem_cache_alloc_node+0x279/0x2b0
        kmalloc_trace+0x2a/0xa0
        get_mountpoint+0xab/0x190
        lock_mount+0x53/0x100
        path_mount+0x535/0xb80
        __x64_sys_mount+0x10c/0x150
        do_syscall_64+0x5c/0x90
        entry_SYSCALL_64_after_hwframe+0x63/0xcd
[ ... ]
```

The output is *sorted by maximum wastage*; the first two are seen above. Here's how to interpret them (let's do so for the first trace):

```
975 populate_error_injection_list+0x8d/0x110 waste=15600/16
age=6363993/6363993/6363994 pid=1 cpus=0
```

The `populate_error_injection_list()` function (at an offset of `0x8d` bytes into the function, whose length is `0x110` bytes) made 975 requests for memory of 64 bytes each (as this is the `kmalloc-64` cache, after all), wasting 16 bytes each time, thus wasting a total of 975 * 16 = 15,600 bytes. The stack trace (read it bottom-up) shows the history, the way it got here (great for debugging!).

Let's delve deeper: the function's here: `lib/error-inject.c:populate_error_injection_list()`; looking up its code (ver 6.1.25) reveals a single memory allocation via the slab layer here:

```
struct ei_entry *ent;
[…]
ent = kmalloc(sizeof(*ent), GFP_KERNEL);
if (!ent)
            break;
[…]
```

Now, the structure instance being allocated – `struct ei_entry` – has a size of 46+2 = 48 bytes (on 64-bit; check by looking it up, keeping in mind that padding can occur due to alignment requirements (there are 2 bytes of padding here)). Thus, each allocation is serviced by, guess who, the `kmalloc-64` slab cache (best-fit) and thus wastes exactly 64 - 48 = 16 bytes. This may seem acceptable, but, apparently, it happened (until now) 975 times, thus ending up wasting 15,600 bytes! Even this isn't too bad *if the memory is freed up quickly*; what if it isn't? Ah, then we have problems (including the kind of trigger-happy OOM killer fellow) ... at least we can now effectively debug!

Exercise: Interpret the second trace instance seen above.

Do note carefully – the wastage tracking is only for the `kmalloc-*` slab caches (as these are the ones from where (slab) memory requests will typically be satisfied).

Running a script to see the SLUB wastage or loss

To ease the burden of looking up the `/sys/kernel/debug/slab/kmalloc-*/alloc_traces` pseudofile, I wrote a quick custom script (`ch8/wastage_kmalloc_slabs.sh`); it shows only the top line (not the stack trace) of allocation instances where wastage has occurred, sorted by highest wastage first, for all `kmalloc-*` caches (it essentially iterates over the output of this command: `grep -r -H -w waste /sys/kernel/debug/slab/kmalloc-[1-9]*/alloc_traces`). Not only that, it separates the slab allocations performed by the kernel (internal) routines and by kernel modules. To keep the output human-readable, it shows only the top ten 'wasters'; the output 'reports' containing the full details are pointed to. Do go through its code and try it out.

Note: To try this script, you must be running a recent 6.1 kernel (or later) and boot the system passing `slub_debug=FZPU` on the kernel command line.

Here's a sample run (for a change, this one's on my native x86_64 Fedora 39 system running a recent 6.6 kernel; I've truncated a portion of the right side so as to obtain a readable screenshot):

```
ch8 $ sudo ./waste_kmalloc_slabs.sh
[sudo] password for c2kp:
waste_kmalloc_slabs.sh: gathering data, please be patient ...
...........................
======== Wastage (highest-to-lowest with duplicate lines eliminated) ========
--------------- kernel internal ----------------
   Top 10 wasters (in desc order). (To see all, lookup the full report here: kint.waste)
/sys/kernel/debug/slab/kmalloc-2k/alloc_traces: 229 bpf_prog_alloc_no_stats+0x74/0x130 waste=230832/1008 age=1
/sys/kernel/debug/slab/kmalloc-2k/alloc_traces: 51 cgroup_mkdir+0xde/0x410 waste=51816/1016 age=36775/2707513/
/sys/kernel/debug/slab/kmalloc-2k/alloc_traces: 42 sk_prot_alloc+0x97/0x110 waste=39984/952 age=28/2635537/278
/sys/kernel/debug/slab/kmalloc-2k/alloc_traces: 29 bpf_prog_alloc_no_stats+0x74/0x130 waste=29232/1008 age=127
/sys/kernel/debug/slab/kmalloc-2k/alloc_traces: 39 acpi_add_single_object+0x43/0x6b0 waste=25272/648 age=27892
/sys/kernel/debug/slab/kmalloc-1k/alloc_traces: 55 find_css_set+0x1ad/0x670 waste=23760/432 age=36762/2680068/
/sys/kernel/debug/slab/kmalloc-512/alloc_traces: 102 pids_css_alloc+0x16/0x50 waste=22848/224 age=36783/272597
/sys/kernel/debug/slab/kmalloc-1k/alloc_traces: 63 tty_register_device_attr+0x9f/0x200 waste=18648/296 age=278
/sys/kernel/debug/slab/kmalloc-2k/alloc_traces: 23 alloc_super.isra.0+0x22/0x2b0 waste=16192/704 age=1280832/2
/sys/kernel/debug/slab/kmalloc-4k/alloc_traces: 8 bpf_prog_store_orig_filter+0x52/0x80 waste=13248/1656 age=27
-none-

--------------- kernel modules ----------------
   Top 10 wasters (in desc order). (To see all, lookup the full report here: kmods.waste)
-none-
ch8 $
```

Figure 8.13: Partial screenshot showing our ch8/wastage_kmalloc_slabs.sh script executing; focus on the waste=total/each portion of the the highlighted output

The script (a wrapper over the recent kernel's `alloc_traces` pseudofile) essentially iterates over each `kmalloc-N` slab and shows only those allocations that resulted in positive wastage (and the precise amount of wastage incurred); it can take a while though... In this particular run, there's no memory wastage by modules.

The interpretation remains the same; also, modules might show up (with their name within square brackets). The `waste=x/y` is the critical thing; `x` is the total number of bytes wasted, `y` is the number of bytes wasted each time; we sort the output from highest to lowest `x`, i.e., the total bytes wasted.

This script thus gives us a quick way to spot if our kernel or a module code's is wasting precious kernel memory!

(Notice the 'age=...' fields? We'll cover what this means in the next chapter, so keep a keen eye out).

So, with this, we've covered a significant amount of stuff. The next section very briefly highlights the pros and cons of the SLUB layer.

Slab layer - pros and cons

Right, let's now summarize the slab (SLUB) allocator pros and cons in a table:

Slab allocator: pros	Slab allocator: cons
Fast – uses the kernel's identity-mapped RAM (obtained and mapped at boot); pages are in the lowmem region and thus pre-mapped; no page table setup is required (just as with the page allocator/BSA).	There is an upper limit to how much can be allocated with a single API call; practically, assuming MAX_ORDER=11 and a page size of 4K, it's 4 MB.
Guarantees that memory chunks are physically contiguous and CPU cacheline-aligned.	Security: by default, freed memory is not cleared, which could result in info-leakage scenarios. Can build the kernel with `CONFIG_PAGE_POISONING` or use `kfree_sensitive()` (can result in a slight performance impact).
The ability to allocate fragments of a page (typically from as small as 8 bytes (on the x86) up to 8K).	Have to carefully check allocations (with `ksize()` or slab tools or `alloc_traces`) to ensure too much wastage doesn't occur.

Table 8.3: Summary of the pros and cons of the slab allocator (k{m|z}alloc() / k[z]free())

We finish with a quick mention of different internal implementations of the slab allocator.

Slab layer – a word on its implementations within the kernel

We end this section and this chapter with a quick word on the actual slab layer implementations (yes, there are several) within the kernel. Let's check it out!

There are at least three different mutually exclusive kernel-level implementations of the slab allocator; only one of them can be in use at runtime. The one to be used at runtime is selected at the time of configuring the kernel (you learned this procedure in detail in *Chapter 2, Building the 6.x Linux Kernel from Source – Part 1*). The relevant kernel configuration options are as follows:

- `CONFIG_SLAB`
- `CONFIG_SLUB`
- `CONFIG_SLOB`

The first (`SLAB`) is the early, well-supported, but quite under-optimized one. The second implementation, `SLUB`, aka the *Unqueued Allocator*, is a major improvement on the first in terms of memory efficiency, performance, and better diagnostics and is the one selected by default. The `SLOB` allocator (aka the *Simple Allocator*) is a drastic simplification and, as per the kernel config help, "does not perform well on large systems." The official kernel documentation has a useful page on SLUB here: *Short users guide for SLUB* (`https://www.kernel.org/doc/html/latest/mm/slub.html`; much of the content here is to do with interpreting and analyzing the output of `slabinfo`, something I cover in some detail in the *Linux Kernel Debugging* book).

Summary

In this chapter, you learned – to a good level of detail – how both the page (or buddy system) and the slab allocators work. Recall that the actual "engine" of allocating (and freeing) RAM within the kernel is ultimately the *page (or buddy system) allocator*, with the slab allocator being layered on top of it to provide optimization for typical less-than-a-page-in-size allocation requests, and to efficiently allocate several well-known kernel data structures ('objects').

You learned how to efficiently use the APIs exposed by both the page and slab allocators, with several demo kernel modules to help show this in a hands-on manner. A good deal of focus was (quite rightly) given to the real issue of the developer issuing a (slab) memory request for a certain *N* number of bytes, but you learned that it can be very sub-optimal, with the kernel actually allocating much more (the wastage can climb very close to 100%)! You now know how to check for and mitigate these cases. Well done!

The next chapter covers more on optimal allocation strategies, as well as some more advanced topics on kernel memory allocation, including the creation of custom slab caches, using the kernel's `vmalloc` allocation interfaces, what the *OOM killer* is all about, and more. So, first ensure you've understood the content of this chapter and worked on the kernel modules and assignments (as follows). Then, let's get you on to the next one!

Questions

As we conclude, here is a list of questions for you to test your knowledge regarding this chapter's material: `https://github.com/PacktPublishing/Linux-Kernel-Programming_2E/blob/main/questions/ch8_qs_assignments.txt`. You will find some of the questions answered in the book's GitHub repo: `https://github.com/PacktPublishing/Linux-Kernel-Programming_2E/tree/master/solutions_to_assgn`.

Further reading

To help you delve deeper into the subject with useful materials, we provide a rather detailed list of online references and links (and at times, even books) in a *Further reading* document in this book's GitHub repository. The *Further reading* document is available here: `https://github.com/PacktPublishing/Linux-Kernel-Programming_2E/blob/master/Further_Reading.md`.

Leave a review!

Kernel Memory Allocation for Module Authors – Part 2

The previous chapter covered the basics (and a lot more!) of using the available APIs for memory allocation via both the **page allocator** (**PA**) or **Buddy System Allocator** (**BSA**) and the slab allocators within the kernel. In this chapter, we will delve further into this large and interesting topic. We will cover the creation of custom slab caches, the `vmalloc` interfaces, and, very importantly, given the wealth of choice, which APIs to use in which situation. We shall then delve into some key kernel internal details regarding memory reclamation, the dreaded **Out of Memory** (**OOM**) killer, and demand paging.

These areas tend to be important to understand when working with kernel modules, especially with device drivers. A Linux system project's sudden crash with merely a `Killed` message on the console requires some explanation, yes!? The OOM killer's likely the sweet chap behind this...

Briefly, within this chapter, these are the main areas covered:

- Creating a custom slab cache
- Debugging kernel memory issues – a quick mention
- Understanding and using the kernel vmalloc() API
- Memory allocation in the kernel – which APIs to use when
- Memory reclaim – a key kernel housekeeping task
- Stayin' alive – the OOM killer

Technical requirements

I assume that you have gone through *Online Chapter, Kernel Workspace Setup*, and have appropriately prepared a guest VM (or a native system) running Ubuntu 22.04 LTS (or a later stable release) and installed all the required packages. If not, I highly recommend you do this first.

Also, the last section of this chapter deliberately runs a *very* memory-intensive app, so intensive that the kernel will take some drastic action (killing off some process(es))! I highly recommend you try out stuff like this on a safe, isolated system, preferably a Linux test VM (with no important data on it).

To get the most out of this book, I strongly recommend you first set up the workspace environment, including cloning this book's GitHub repository for the code, and work on it in a hands-on fashion. The GitHub repository can be found at `https://github.com/PacktPublishing/Linux-Kernel-Programming_2E`.

Creating a custom slab cache

As explained in detail in the previous chapter, a key design concept behind slab caches is the powerful idea of object caching. By caching frequently used objects – data structures, really – the memory allocation/free work for those objects are much quicker and thus overall performance receives a boost.

So, think about this: what if we're writing a driver and within it, we notice that a certain data structure (an object) is very frequently allocated and freed? Normally, we would use the usual `kzalloc()` (or `kmalloc()`) followed by the `kfree()` APIs to allocate and free this object. Some good news: the Linux kernel sufficiently exposes the slab layer API to us, the module/driver authors, allowing us to create *custom slab caches*. In this section, you'll learn how you can leverage this powerful feature.

Creating and using a custom slab cache within a kernel module

In this section, we will create, use, and subsequently destroy a custom slab cache. At a broad level, we'll be performing the following three steps:

1. Creating a custom slab cache of a given size with the `kmem_cache_create()` API. This is often done as part of the init code path of the kernel module (or within the probe method of a driver).
2. Using the slab cache. Here we will do the following:
   - Issue the `kmem_cache_alloc()` API to allocate a single instance of the custom object(s) within your slab cache.
   - Use the object.
   - Free it back to the custom cache with the `kmem_cache_free()` API.
3. Destroying the custom slab cache when done with `kmem_cache_destroy()`. This is often done as a part of the cleanup code path of the kernel module (or within the remove/detach/disconnect method of a driver).

Let's explore each of these APIs in a bit of detail. We'll start with the creation of a custom (slab) cache.

Step 1 – creating a custom slab cache

First, of course, let's learn how to create the custom slab cache. The signature of the `kmem_cache_create()` kernel API is as follows:

```
#include <linux/slab.h>
struct kmem_cache *kmem_cache_create(const char *name,
        unsigned int size, unsigned int align, slab_flags_t flags,
        void (*ctor)(void *));
```

The first parameter is the name of the cache, as will be revealed by proc (and hence by other wrapper utilities over proc, such as vmstat, slabtop, and so on). It usually matches the name of the data structure or object being cached (but does not have to).

The second parameter, size, is a key one here – it's the size in bytes for each object within the new cache. Based on this object size (using a best-fit algorithm), the kernel's slab layer constructs a cache of objects. The actual size of each object within the cache could be (slightly) larger than what's requested due to three reasons:

- We can always provide more, but never less, than the memory size requested.
- Some space for metadata (housekeeping information) is required.
- The kernel is limited in being able to provide a cache of the exact size required. It uses the memory of the closest possible (equal to or larger) matching size from the available slab caches (recall what you learned in *Chapter 8, Kernel Memory Allocation for Module Authors – Part 1*, in the *Caveats when using the slab allocator* section, where we clearly saw that more memory (sometimes a lot!) could be used).

Recall from *Chapter 8, Kernel Memory Allocation for Module Authors – Part 1*, that the `ksize()` API can be used to query the actual size of the allocated object. There is another API with which we can query the size of the individual objects within the new slab cache: `unsigned int kmem_cache_size(struct kmem_cache *s);` you'll see this being used shortly.

The third parameter, align, is the *alignment* required for the objects within the cache. If unimportant, just pass it as 0. Quite often, though, there are very particular alignment requirements, for example, to ensure that the object is aligned to the size of a word on the machine (32 or 64 bits). To do so, pass the value as sizeof(long) (the unit for this parameter is bytes, not bits).

The fourth parameter, flags, can either be 0 (implying no special behavior) or the bitwise-OR operator of the following flag values. For clarity, we have directly reproduced the information on the following flags from the comments within the source file, mm/slab_common.c:

```
// mm/slab_common.c
[...]
 * The flags are
 *
 * %SLAB_POISON - Poison the slab with a known test pattern (a5a5a5a5)
 * to catch references to uninitialized memory.
 *
 * %SLAB_RED_ZONE - Insert `Red` zones around the allocated memory to
 * check for buffer overruns.
 *
 * %SLAB_HWCACHE_ALIGN - Align the objects in this cache to a hardware
 * cache line. This can be beneficial if you're counting cycles as
 * closely as davem.
[...]
```

Let's quickly check out these flags:

- The first of them, `SLAB_POISON`, provides slab "poisoning," that is, initializing the cache memory to a previously known value (`0xa5a5a5a5`). Doing this can help during debug situations (where we peek at raw memory to see what's going on and this memory pattern catches the eye).
- The second flag, `SLAB_RED_ZONE`, is interesting, inserting red zones (analogous to guard pages) around the allocated buffer. This is a common way of checking for the unfortunately pretty common **OOB** (**Out Of Bounds**) or buffer overflow/underflow errors. It's almost always used in a debug context (typically during development/test).
- The third possible flag, `SLAB_HWCACHE_ALIGN`, is commonly used and is recommended for performance. It guarantees that all the cache objects are aligned to the hardware (CPU) cache line size. The usage of this flag is precisely why the memory allocated via the popular `k{m|z}alloc()` APIs is aligned to the hardware (CPU) cache line. (We cover some details on CPU caching concepts in *Chapter 13, Kernel Synchronization – Part 2*.)

Finally, the fifth parameter to `kmem_cache_create()` is very interesting too: a function pointer, `void (*ctor)(void *);`. This is modeled as a constructor function (as in object orientation and OOP languages). It conveniently allows you to initialize the slab object allocated from the custom slab cache the moment it's allocated! As one example of this feature in action within the kernel, see the code of the **Linux Security Module** (**LSM**) called `integrity` (here `security/integrity/iint.c:integrity_iintcache_init()`); it invokes the following:

```
iint_cache = kmem_cache_create("iint_cache", sizeof(struct integrity_iint_
cache), 0, SLAB_PANIC, init_once);
```

The `init_once()` function – the constructor – initializes the cached object instance (which was just allocated). The constructor function is called whenever new pages are allocated by this cache.

Though it may seem counter-intuitive, the fact is that the modern Linux kernel is quite object-oriented in design terms. The code, of course, is mostly plain old C, a traditional procedural language. Nevertheless, a vast number of architecture implementations within the kernel (the driver model being a big one) are quite object-oriented in design: method dispatch via virtual function pointer tables, the so-called "strategy" design pattern, and so on. See a two-part article on LWN depicting this in some detail here: *Object-oriented design patterns in the kernel, part 1*, June 2011 (`https://lwn.net/Articles/444910/`).

The return value from the `kmem_cache_create()` API is a pointer to the newly created custom slab cache on success, and `NULL` on failure. This pointer is usually kept global, as you will require access to it in order to actually allocate objects from it (our next step).

It's important to understand that the `kmem_cache_create()` API can only be called from process context. A fair bit of kernel code (including many drivers) creates and uses its own custom slab caches. For example, in the 6.1 Linux kernel, there are over 430 instances of this API being invoked.

All right, now that you have a custom (slab) cache available, how exactly do you use it to allocate memory objects? Read on; the next section covers precisely this.

Step 2 – using our custom slab cache's memory

So, we created a custom slab cache; internally, the kernel will have pre-allocated several objects of your custom structure within it, ready for you to use! To use it, you must issue the `kmem_cache_alloc()` API. Its job is that given the pointer to a slab cache, it returns a single instance of an object on that slab cache (in fact, this is how the `k{m|z}alloc()` APIs work under the hood!). Its signature is as follows (of course, remember to always include the `<linux/slab.h>` header for all slab-based APIs):

```
void *kmem_cache_alloc(struct kmem_cache *s, gfp_t gfpflags);
```

Let's look at its two parameters:

- The first parameter to `kmem_cache_alloc()` is the pointer to the (custom) cache that we created in the previous step (the pointer being the return value from the `kmem_cache_create()` API).
- The second parameter is the usual GFP flags to pass along (remember the essential rule for driver/module authors: use `GFP_KERNEL` for normal process-context allocations when it's safe to sleep, else `GFP_ATOMIC` when in any kind of atomic context).

As with the now-familiar `k{m|z}alloc()` APIs, the return value is a pointer to the newly allocated memory chunk - a kernel virtual address.

Use the newly allocated memory object, and when done, do not forget to free it with the following:

```
void kmem_cache_free(struct kmem_cache *s, void *x);
```

Here, take note of the following with respect to the `kmem_cache_free()` API:

- The first parameter to `kmem_cache_free()` is, again, the pointer to the (custom) slab cache that you created in the previous step (the return value from `kmem_cache_create()`).
- The second parameter is the pointer to the memory object you wish to free - the object instance that you were just allocated/given with `kmem_cache_alloc()` - and thus have it return to the cache specified by the first parameter!

Similar to the `k[z]free()` APIs, there is no return value.

Step 3 – destroying our custom cache

When completely done (often in the cleanup or exit code path of the kernel module, or your driver's `remove` method), you must destroy the custom slab cache that you created earlier using the following API:

```
void kmem_cache_destroy(struct kmem_cache *);
```

The parameter, of course, is the pointer to the (custom) cache that you created in the first step (the return value from the `kmem_cache_create()` API).

Now that you have understood the procedure and its related APIs, let's get hands-on with a kernel module that creates its own custom slab cache, uses it, and then destroys it.

Custom slab – a demo kernel module

Time to get our hands dirty with some code! Let's look at a simple demonstration of using the preceding APIs to create our very own custom slab cache. As usual, we show only relevant code snippets here. I urge you to clone the book's GitHub repository, browse the full code, and try it out yourself! You can find the code for this file in `ch9/slab_custom/slab_custom.c`.

Let's get started:

```
// ch9/slab_custom/slab_custom.c
#define OURCACHENAME   "our_ctx"
/* Our 'demo' structure, one that (we imagine) is often allocated and freed;
 * hence, we create a custom slab cache to hold pre-allocated 'instances'
 * of it... Size of one structure instance: 328 bytes.
 */
struct myctx {
    u32 iarr[10];         // 40 bytes; total=40
    u64 uarr[10];         // 80 bytes; total=120
    s8 uname[128], passwd[16], config[64]; // 128+16+64=208 bytes;
                                           // total=328
};
static struct kmem_cache *gctx_cachep;
```

In the preceding code, we declared a (global) pointer (`gctx_cachep`) to the to-be-created custom slab cache – which will hold objects; namely, our fictional often-allocated-and-freed data structure, `myctx`.

In the following, see the code that creates the custom slab cache:

```
static int create_our_cache(void)
{
    int ret = 0;
    void *ctor_fn = NULL;

    if (use_ctor == 1)
        ctor_fn = our_ctor;
    pr_info("sizeof our ctx structure is %zu bytes\n"
            " using custom constructor routine? %s\n",
            sizeof(struct myctx), use_ctor==1?"yes":"no");

  /* Create a new slab cache:
   * kmem_cache_create(const char *name, unsigned int size, unsigned int
     align, slab_flags_t flags, void (*ctor)(void *));
   * When our constructor's enabled (the default), this call will trigger it.
*/
    gctx_cachep = kmem_cache_create(OURCACHENAME, // name of our cache
```

```
        sizeof(struct myctx), // (min) size of each object
        sizeof(long),                // alignment
        SLAB_POISON |          /* use slab poison values (explained soon) */
        SLAB_RED_ZONE |    /* good for catching buffer under|over-flow bugs */
        SLAB_HWCACHE_ALIGN, /* good for performance */
        ctor_fn);                    // ctor: here, on by default

    if (!gctx_cachep) {
        [...]
        if (IS_ERR(gctx_cachep))
            ret = PTR_ERR(gctx_cachep);
    }
    return ret;
}
```

Hey, that's interesting: notice that our cache creation API supplies a constructor function to help initialize any newly allocated object; here it is:

```
/* The parameter to the constructor is the pointer to the just-allocated memory
 * 'object' from our custom slab cache. Here, in our 'constructor' routine,
 * we initialize the just-allocated memory object.
 */
static void our_ctor(void *new)
{
    struct myctx *ctx = new;
    struct task_struct *p = current;

    /* TIPS:
     * 1. To see how exactly we got here, insert this call: dump_stack(); (read
its output bottom-up ignoring call frames that begin with '?')
     * 2. Count the number of times the below printk's emitted; its, in effect,
the number of objects cached by the kernel within this slab cache. */
    pr_info("in ctor: just alloced mem object is @ 0x%px\n", ctx);  /* %pK in
production */
    memset(ctx, 0, sizeof(struct myctx));

    /* As a demo, we init the 'config' field of our structure to some
     * (arbitrary) 'accounting' values from our task_struct
     */
    snprintf_lkp(ctx->config, 6 * sizeof(u64) + 5, "%d.%d,%ld.%ld,%ld,%ld",
        p->tgid, p->pid, p->nvcsw, p->nivcsw, p->min_flt, p->maj_flt);
}
```

The comments in the preceding code are self-explanatory; do read them. The constructor routine, if set up (depending on the value of our `use_ctor` module parameter; it's `1` by default), will be auto-invoked by the kernel whenever a new memory object is allocated to our cache. (Also, as we employ our safer `snprintf_lkp()` wrapper, which is defined in our `klib.c` library file, we have our Makefile link to the `klib.o` file as well.)

Next, within the init code path, we call our `use_our_cache()` function. It allocates an instance of our `myctx` object via the `kmem_cache_alloc()` API, and, if our custom constructor routine is enabled, *it runs*, initializing the object. We then dump its memory to show that it was indeed initialized as coded, freeing it when done (for brevity, we'll leave out showing the error code paths):

```
    obj = kmem_cache_alloc(gctx_cachep, GFP_KERNEL);
    pr_info("Our cache object size is %u bytes; ksize=%lu\n",
            kmem_cache_size(gctx_cachep), ksize(obj));
    print_hex_dump_bytes("obj: ", DUMP_PREFIX_OFFSET, obj, sizeof(struct
myctx));
    kmem_cache_free(gctx_cachep, obj);
```

Finally, in the exit code path, we destroy our custom slab cache:

```
kmem_cache_destroy(gctx_cachep);
```

The following output from a sample run helps us understand how it works. *Figure 9.1* is a partial screenshot showing the output on our x86_64 Ubuntu 22.04 LTS guest running our custom Linux 6.1 kernel:

```
[32377.321242] slab_custom:slab_custom_init(): inserted
[32377.321244] slab_custom:create_our_cache(): sizeof our ctx structure is 328 bytes
               using custom constructor routine? yes
[32377.321263] slab_custom:our_ctor(): in ctor: just alloced mem object is @ 0xffff9878b3c6eac0
[32377.321265] slab_custom:our_ctor(): in ctor: just alloced mem object is @ 0xffff9878b3c6f8c0
[32377.321266] slab_custom:our_ctor(): in ctor: just alloced mem object is @ 0xffff9878b3c6f000
[32377.321267] slab_custom:our_ctor(): in ctor: just alloced mem object is @ 0xffff9878b3c6fa80
[32377.321268] slab_custom:our_ctor(): in ctor: just alloced mem object is @ 0xffff9878b3c6f1c0
[32377.321269] slab_custom:our_ctor(): in ctor: just alloced mem object is @ 0xffff9878b3c6fc40
[32377.321270] slab_custom:our_ctor(): in ctor: just alloced mem object is @ 0xffff9878b3c6ec80
[32377.321271] slab_custom:our_ctor(): in ctor: just alloced mem object is @ 0xffff9878b3c6e580
[32377.321272] slab_custom:our_ctor(): in ctor: just alloced mem object is @ 0xffff9878b3c6f380
[32377.321273] slab_custom:our_ctor(): in ctor: just alloced mem object is @ 0xffff9878b3c6e200
[32377.321274] slab_custom:our_ctor(): in ctor: just alloced mem object is @ 0xffff9878b3c6e740
[32377.321275] slab_custom:our_ctor(): in ctor: just alloced mem object is @ 0xffff9878b3c6fe00
[32377.321276] slab_custom:our_ctor(): in ctor: just alloced mem object is @ 0xffff9878b3c6ee40
[32377.321277] slab_custom:our_ctor(): in ctor: just alloced mem object is @ 0xffff9878b3c6f700
[32377.321278] slab_custom:our_ctor(): in ctor: just alloced mem object is @ 0xffff9878b3c6e3c0
[32377.321279] slab_custom:our_ctor(): in ctor: just alloced mem object is @ 0xffff9878b3c6e900
[32377.321280] slab_custom:our_ctor(): in ctor: just alloced mem object is @ 0xffff9878b3c6f540
[32377.321281] slab_custom:our_ctor(): in ctor: just alloced mem object is @ 0xffff9878b3c6e040
[32377.321282] slab_custom:use_our_cache(): Our cache object (@ ffff9878b3c6eac0, actual=ffff9878b3c6eac0)
 size is 328 bytes; actual ksize=328
$
```

Figure 9.1: Some output from our slab_custom kernel module on an x86_64 VM

Great! Hang on, though, there are a couple of key points to take note of here:

- As our constructor routine is enabled by default (the value of our `use_ctor` module parameter is 1), it runs whenever a new object instance is allocated by the kernel slab layer to our new cache. Here, we performed just a single `kmem_cache_alloc()`, yet our constructor routine has run 18 times, implying that the kernel's slab code (pre)allocated 18 objects to our brand-new cache! Of course, this number can vary.
- As seen in the preceding screenshot, the *size* of each object is seemingly 328 bytes, which matches the actual structure size (as shown by these three APIs: `sizeof()`, `kmem_cache_size()`, and `ksize()`). However, again, **this is not true!** The actual size of the object as allocated by the kernel is larger; we can see this via `vmstat`:

```
$ sudo vmstat -m | head -n1
Cache                         Num  Total  Size  Pages
$ sudo vmstat -m | grep our_ctx
our_ctx                         0     18   448     18
```

As highlighted in the preceding output block, the actual size of each allocated object is not 328 bytes but 448 bytes (thus, the wastage is 120 bytes per object allocated, close to 27%!). Just as we saw earlier, this is important for you to realize, and indeed check for.

FYI, you can always check out the man page of `vmstat(8)` for the precise meaning of each column.

Right, let's now move on to seeing how we can extract some internal details regarding the slab caches.

Extracting useful information regarding slab caches

Continuing the discussion from the previous section, if the `ksize()` API returns the size of the allocated object as the same number (328 here) as the `sizeof()` operator does, then how exactly are we to figure out that there's indeed some wastage incurred? Using `vmstat` is one easy way, via the slab statistics maintained by the kernel.

Within the sysfs pseudo filesystem, in the `/sys/kernel/slab` directory, there will exist a (pseudo) directory for every slab cache currently on the system! Looking up its content reveals useful internal details regarding the cache in question (you will require root access). As an example, once our previous `slab_custom` kernel module is inserted into the kernel, run this command (as root):

```
# grep . /sys/kernel/slab/our_ctx/*
```

Recall that we named our custom slab cache `our_ctx`.

Tip: For (pseudo) files with a single value within them (very common with sysfs), to see the filename followed by its current content, you can leverage this `grep` command style (as we do just below): `grep . <path/to/files>/*`.

Using grep this way is an efficient way to quickly reveal the value of each (pseudo) file within the directory; a screenshot of the output I got when doing so follows:

```
# grep . /sys/kernel/slab/our_ctx/*
/sys/kernel/slab/our_ctx/aliases:0
/sys/kernel/slab/our_ctx/align:64
/sys/kernel/slab/our_ctx/cache_dma:0
/sys/kernel/slab/our_ctx/cpu_partial:0
/sys/kernel/slab/our_ctx/cpu_slabs:0
/sys/kernel/slab/our_ctx/ctor:our_ctor+0x0/0x91 [slab_custom]
/sys/kernel/slab/our_ctx/destroy_by_rcu:0
/sys/kernel/slab/our_ctx/hwcache_align:1
/sys/kernel/slab/our_ctx/min_partial:5
/sys/kernel/slab/our_ctx/objects:0
/sys/kernel/slab/our_ctx/object_size:328
/sys/kernel/slab/our_ctx/objects_partial:0
/sys/kernel/slab/our_ctx/objs_per_slab:18
/sys/kernel/slab/our_ctx/order:1
/sys/kernel/slab/our_ctx/partial:1 N0=1
/sys/kernel/slab/our_ctx/poison:1
/sys/kernel/slab/our_ctx/reclaim_account:0
/sys/kernel/slab/our_ctx/red_zone:1
/sys/kernel/slab/our_ctx/remote_node_defrag_ratio:100
/sys/kernel/slab/our_ctx/sanity_checks:0
/sys/kernel/slab/our_ctx/skip_kfence:0
/sys/kernel/slab/our_ctx/slabs:1 N0=1
/sys/kernel/slab/our_ctx/slabs_cpu_partial:0(0)
/sys/kernel/slab/our_ctx/slab_size:448
/sys/kernel/slab/our_ctx/store_user:0
/sys/kernel/slab/our_ctx/total_objects:18 N0=18
/sys/kernel/slab/our_ctx/trace:0
/sys/kernel/slab/our_ctx/usersize:0
#
```

Figure 9.2: Screenshot showing how one can quickly extract useful info from any existing kernel slab cache (with relevant values highlighted here)

Do study *Figure 9.2*; we get all relevant details regarding our slab cache! To "zoom" in on the specific values of interest, simply do this:

```
# grep . /sys/kernel/slab/our_ctx/* | grep -E "object_size|slab_size"
/sys/kernel/slab/our_ctx/object_size:328
/sys/kernel/slab/our_ctx/slab_size:448
```

So, given the name of a slab cache, writing a small script to extract ("grok") these values and then calculate the wastage is an exercise I'll leave to you, dear reader! We'll round off the discussion on creating and using custom slab caches with understanding the slab shrinker interface.

Understanding slab shrinkers

Caches are good for performance. Visualize reading the content of a large file from disk as opposed to reading its content from RAM. There's no question that the RAM-based I/O is much faster! As can be imagined, the Linux kernel leverages these ideas and thus maintains several caches – the page cache, dentry cache, inode cache, slab caches, and so on. These caches indeed greatly help performance, but think about it; they're not a mandatory requirement.

Thus, when memory pressure reaches high levels (implying that too much memory is in use and too little is free), the Linux kernel has mechanisms to intelligently free up caches; this constitutes a part of what's called **memory reclamation** – it's a continuous ongoing process. Kernel threads (typically named `kswapd*`) reclaim memory as part of their housekeeping chores (more on this in the *Memory reclaim – a key kernel housekeeping task* and *Stayin' alive – the OOM killer* sections).

In the case of the slab cache(s), the fact is that some kernel subsystems and drivers create their own custom slab caches, as we covered earlier in this chapter. To integrate well and cooperate with the kernel, best practice demands that your custom slab cache code registers a *shrinker interface*. When this is done, and when memory pressure gets high enough, the kernel might well invoke several slab shrinker callbacks, which are expected to ease the memory pressure by freeing up (shrinking) slab objects.

The API to register a shrinker function with the kernel is the `register_shrinker()` API; this is its signature:

```
int register_shrinker(struct shrinker *shrinker, const char *fmt, …);
```

The first parameter is a pointer to a `shrinker` structure. This structure contains (among other housekeeping members) function pointers to two callback routines:

- The first routine, `count_objects()`, merely counts and returns the number of objects that would be freed (when it is actually invoked). If it returns `0`, this implies that the number of freeable memory objects cannot be determined now, or that we should not even attempt to free any right now. Also, it returns `SHRINK_EMPTY` if there aren't any objects to free in the first place.
- The second routine, scan_objects(), is invoked only if the first callback routine returns a non-zero value (besides SHRINK_EMPTY). This is the routine that, when invoked, actually frees up, or shrinks, the slab cache in question. It returns the actual number of objects freed up in this reclaim cycle or SHRINK_STOP if the reclaim attempt could not progress (due to possible deadlocks).

FYI, the `register_shrinker()` API was updated recently, in the 6.0 kernel, to include an identifying name printf-format-style string (commit e33c267: `https://github.com/torvalds/linux/commit/e33c267ab70de4249d22d7eab1cc7d68a889bac2`). Further, if `CONFIG_SHRINKER_DEBUG` is enabled (set to `y`), the content of `/sys/kernel/debug/shrinker` shows all registered slab shrinkers. Only a few components within the 6.1 kernel seem to be making use of the shrinker interface; as one example, the code in `drivers/md/dm-bufio.c` (a way to perform cached I/O on devices) registers a shrinker interface and then, when its `scan_objects` callback is invoked, it uses a workqueue function (`drivers/md/dm-bufio.c:shrink_work()`) to try and evict buffers as required.

We'll now wrap up the discussion on the slab layer with a quick summary of the pros and cons of using this layer for memory (de)allocation. This is very important for you as a kernel/driver author and something to be keenly aware of!

Summarizing the slab allocator pros and cons

In this section, we will very briefly summarize things you have already learned by now. This is intended as a way for you to quickly look up and recollect these key points!

The **pros** of using the slab allocator (or slab cache) APIs to allocate and free kernel memory are as follows:

- (Very) fast (as it uses pre-cached memory objects).
- Allocation of a physically contiguous memory chunk is guaranteed.
- Hardware (CPU) cacheline-aligned memory is guaranteed when the `SLAB_HWCACHE_ALIGN` flag is specified when creating the cache. This is the case for `kmalloc()`, `kzalloc()`, and so on.
- You can create your own custom slab cache for (frequently alloc-ed/freed) objects.

The **cons** of using the slab allocator (or slab cache) APIs are the following:

- A limited amount of memory can be allocated at a time; typically, just 8 KB directly via the slab interfaces, or up to 4 MB indirectly via the page allocator on most current platforms (the precise upper limits depend upon the architecture and system page size; these are the values for both x86[_64] and ARM platforms with a page size of 4,096 bytes and a `MAX_ORDER` value of 11).
- Using the `k{m|z}alloc()` APIs incorrectly: asking for too much memory, or asking for a memory size just over a threshold value (discussed in detail in *Chapter 8*, *Kernel Memory Allocation for Module Authors – Part 1*, under the *Size limitations of the kmalloc* API section), can certainly lead to internal fragmentation (wastage). It's designed to be optimized for common cases only, really – for allocations of size less than one page.

Now, let's move on to – merely mentioning – another key aspect for the kernel/driver developer to understand: effectively debugging when things go wrong with respect to memory allocations/freeing, particularly within the slab layer.

Debugging kernel memory issues – a quick mention

Memory corruption is unfortunately a very common root cause of bugs, and being able to debug them is a key skill. Well, unfortunately (and with apologies), we don't cover kernel memory debugging in this book due to two primary reasons:

1. This isn't a book on kernel debugging.
2. Topics on kernel debugging have been covered in depth in my recent (up to date as of the 5.10 LTS kernel) **Linux Kernel Debugging** (**LKD**) book, including two whole chapters of very detailed coverage on debugging kernel memory issues.

Nevertheless, in this book, I consider it my duty to at least mention the various tools and approaches one typically employs when debugging kernel memory issues. You would do well to gain familiarity with the powerful dynamic (runtime) analysis frameworks/tools that are mentioned here:

- The Sanitizer toolset:
    - **KASAN** (the **Kernel Address Sanitizer**): Available for x86_64 and AArch64, 4.x kernels onward
    - **KMSAN** (the **Kernel Memory Sanitizer**): The kernel equivalent of the user-space **MSAN (Memory Sanitizer)** tooling, which can catch the deadly **UMR (Uninitialized Memory Reads)** bug (recent, 6.1 onward)
    - **UBSAN (Undefined Behavior Sanitizer)**: Catches many types of arithmetic and memory-related UB, including the UMR bug (4.5 onward)
- **SLUB debug** techniques:
    - Relies on `CONFIG_SLUB_DEBUG=y` and passing various flags at boot to the kernel to enable different kinds of SLUB memory debug (the flags include `F:sanity checks, Z:red zoning, P:poisoning, U:user tracking`, and so on... FYI, we employed SLUB debugging and its flags a bit in *Chapter 8, Kernel Memory Allocation for Module Authors – Part 1*, in the *More details with alloc_traces and a custom script* section).
    - Learn how to use the powerful *slabinfo* utility.
- **kmemleak** (to catch memory leaks, though KASAN seems superior).
- The eBPF tool `memleak[-bpfcc]` can be used to track kernel-level allocations that haven't been freed yet. It's powerful – it even shows the amount of memory allocated as well as the call trace leading up to the allocation! (Try it: `sudo memleak-bpfcc -s 5`). Well, as of now it isn't completely stable, a case of **Your Mileage May Vary (YMMV)**...
- **KFENCE**: A statistical approach to catching memory bugs; must run for long periods, but with low enough overhead to even run in production! (In fact, if supported on the arch being compiled for, and either SLAB or SLUB is enabled, it's typically turned on by default).
- kmemcheck (note, though, that kmemcheck was removed in Linux 4.15).

There's more (quoting verbatim from the *LKD* book):

You can catch kernel memory bugs indirectly with the following:

- Static analysis tools: *checkpatch.pl, sparse, smatch, Coccinelle, cppcheck, flawfinder*, and *GCC*
- Tracing techniques
- K[ret]probes instrumentation
- Post-mortem analysis tooling (logs, Oops analysis, kdump/crash, and [K]GDB)

Okay. I'll summarize this information with specifics on the tools you can use in the following table.

Type of memory bug or defect	Tool(s)/techniques to detect it
Uninitialized Memory Reads (UMR)	Compiler (warnings) [1], static analysis
Out-of-bounds (**OOB**) memory accesses: read/write underflow/overflow defects on compile-time and dynamic memory (including the stack)	KASAN **[2]**, SLUB debug
Use-After-Free (**UAF**) or dangling pointer defects (aka **Use-After-Scope** (**UAS**) defects)	KASAN, SLUB debug
Use-After-Return (**UAR**) aka UAS defects	Compiler (warnings), static analysis
Double-free	Vanilla kernel [3], SLUB debug, KASAN
Memory leakage	kmemleak

Table 9.1: A summary of tools (and techniques) you can use to detect kernel memory issues

A few notes to match the numbers in square brackets in the second column follow:

- [1]: Modern GCC/Clang compilers definitely emit a warning for UMR, with recent ones even being able to auto-initialize local variables (if so configured).
- [2]: KASAN catches (almost!) all the OOB/UAF/UAS/double-free/leakage defects - wonderful! The SLUB debug approach can catch a couple of these, but not all. Vanilla kernels don't seem to catch any.
- [3]: By vanilla kernel, I mean that this defect was caught on a regular distro kernel (with no special config set for memory checking).

(At the risk of over-selling, I'd urge you to also read the *Linux Kernel Debugging* book). Okay, let's now move on to learning about another important kernel memory region, the vmalloc region and it's allocation API, `vmalloc()` (and friends).

Understanding and using the kernel vmalloc() API

As we learned in the previous chapter, ultimately there is just one engine for memory allocation within the kernel - the page (or buddy system) allocator. Layered on top is the slab allocator (or slab cache) machinery. In addition to these two layers, there is another completely virtual address space within the kernel's virtual address space from where virtual pages can be allocated at will - this is called the **kernel vmalloc region.**

Within the kernel segment or VAS is the *vmalloc* address space, aka the *vmalloc* region, extending from `VMALLOC_START` to `VMALLOC_END-1` (the precise addresses and the space available are arch-dependant; incidentally, we covered all this in some detail in *Chapter 7, Memory Management Internals - Essentials*, under the *Examining the kernel segment* section). It's a completely virtual region to begin with, that is, its virtual pages are initially not mapped to any physical page frames.

For a quick refresher, revisit the diagram of the user and kernel segments – in effect, the complete VAS – by re-examining, f.e., *Figure 7.12*. You will find this in *Chapter 7, Memory Management Internals – Essentials*, under the *Examining the kernel VAS* section.

In this book, our purpose is not to delve into the gory internal details regarding the kernel's vmalloc region. Instead, we present enough information for you, the module or driver author, to use this region for the purpose of allocating this type of memory at runtime.

Learning to use the vmalloc family of APIs

You can allocate virtual memory (in kernel space, of course) from the kernel's vmalloc region – you can think of it as a global "pool" of allocatable virtual memory pages within the kernel – using the `vmalloc()` API:

```
#include <linux/vmalloc.h>
void *vmalloc(unsigned long size);
```

Some key points to note about *vmalloc*:

- The `vmalloc()` API allocates contiguous virtual memory to the caller. There is no guarantee that the allocated region will be physically contiguous; it may or may not be (the larger the allocation, the less chance that it's physically contiguous).
- The content of the virtual pages allocated is, in theory, random; in practice, it appears to be arch-dependent (the x86_64, at least, seems to zero out the memory region); of course (at the risk of a slight performance hit), you're recommended to ensure memory zeroing out by employing the `vzalloc()` wrapper API (it's signature is identical to that of the `vmalloc()`).
- The `vmalloc()` (and friends) API must only ever be invoked from a safe-to-sleep (or non-atomic) process context (as it might cause the caller to sleep; thus, calling it in process context but while holding a spinlock violates this rule; don't do it).
- The return value of `v{m|z}alloc()` is the kernel virtual address (or KVA) (within the kernel's vmalloc region) on success or `NULL` on failure; though unlikely to fail, it's your job as a good developer to check for the failure case.
- The start of the vmalloc memory region just allocated is guaranteed to be on a page boundary (in other words, the allocation is always page-aligned). This tells us something: the `v{m|z}alloc()` APIs are meant to be used for large-ish memory allocations, when the fact that the allocated region may not be physically contiguous is okay.
- The actual allocated memory (from the page allocator) might well be larger than what's requested (as again, it internally allocates sufficient pages to cover the size requested).

It will strike you that this API seems very similar to the familiar user space `malloc()`. Indeed it is at first glance, except that, of course, it's a kernel space allocation (and again, as with the `kmalloc()` family, remember that there is no direct correlation between the user and kernel allocation APIs).

This being the case, how is vmalloc() helpful to us module or driver authors? When you require a large, virtually contiguous buffer of a size greater than the slab APIs (that is, k{m|z}alloc() and friends) can provide – recall that it's typically 4 MB with a single allocation on both ARM and x86[_64] – then you should use vmalloc()!

FYI, the kernel uses vmalloc() for various reasons. Some of them are as follows:

- Allocating space for the (static) memory of kernel modules when they are loaded into the kernel (deep within kernel/module/main.c:load_module()).
- If CONFIG_VMAP_STACK is defined (it typically is these days), then vmalloc() is used for the allocation of the kernel-mode stack of every thread at birth (in kernel/fork.c:alloc_thread_stack_node()).
- Internally, while servicing an operation called the ioremap() (and friends).
- Within the Linux socket filter (bpf) code paths, and so on.

For convenience, the kernel provides the vzalloc() wrapper API (analogous to kzalloc()) to allocate and zero out the memory region – a good coding practice, no doubt, but one that might hurt time-critical code paths slightly:

```
void *vzalloc(unsigned long size);
```

Once you are done with using the allocated virtual buffer, you must of course free it:

```
void vfree(const void *addr);
```

As expected, the parameter to vfree() is the return address from v{m|z}alloc() (or even the underlying __vmalloc() API that these invoke). Passing NULL to vfree() causes it to just harmlessly return.

Trying out vmalloc()

In the following code snippet, we show some sample code from our ch9/vmalloc_demo kernel module. As usual, I urge you to clone the book's GitHub repository and try it out yourself (for brevity, we don't show the whole of the source code in the following snippet; we show our primary vmalloc_try() function invoked by the module's init code):

```
static int vmalloc_try(void)
{
    if (!(vptr_rndm = vmalloc(10000))) {
        pr_warn("vmalloc failed\n");
        goto err_out1;
    }
    pr_info("1. vmalloc(): vptr_rndm = 0x%pK (actual=0x%px)\n",
            vptr_rndm, vptr_rndm);
    print_hex_dump_bytes(" content: ", DUMP_PREFIX_NONE, vptr_rndm, DISP_
BYTES);
```

The `vmalloc()` API in the preceding code block allocates a contiguous kernel virtual memory region of (at least) 10,000 bytes; in reality, the memory allocated will be page-aligned! We then employ the kernel's `print_hex_dump_bytes()` helper routine to dump the first 16 bytes of this region.

Moving on, the following code employs the `vzalloc()` API to again allocate another contiguous kernel virtual memory region of (at least) 10,000 bytes (again, it will be page-aligned memory, though); this time, the memory contents are set to zeroes (performing the memory dump via the `print_hex_dump_bytes()` helper will prove this):

```
/* 2. vzalloc(); memory contents are set to zeroes */
if (!(vptr_init = vzalloc(10000))) {
    pr_warn("%s: vzalloc failed\n", OURMODNAME);
    goto err_out2;
}
pr_info("2. vzalloc(): vptr_init = 0x%pK (actual=0x%px)\n",
        vptr_init, (TYPECST)vptr_init);
print_hex_dump_bytes(" content: ", DUMP_PREFIX_NONE, vptr_init,
           DISP_BYTES);
```

A couple of points regarding the just-seen code:

- One, notice the error handling with the goto statement (at the target labels of multiple goto instances, where we use `vfree()` or whatever's appropriate to free up previously allocated memory buffers as required), typical of kernel code.
- Two, for now, please ignore the `kvmalloc()`, `kcalloc()`, and `__vmalloc()` friend routines; we'll cover them in the *Friends of vmalloc()* section. Let's continue with the code:

```
/* 3. kvmalloc(): allocate 'kvn' bytes with the kvmalloc(); if kvn is
 * large (enough), this will become a vmalloc() under the hood, else
 * it falls back to a kmalloc() */
 if (!(kv = kvmalloc(kvn, GFP_KERNEL))) {
     pr_warn("kvmalloc failed\n");
     goto err_out3;
 }
 [...]

 /* 4. kcalloc(): allocate an array of 1000 64-bit quantities and zero
out the memory */
 if (!(kvarr = kcalloc(1000, sizeof(u64), GFP_KERNEL))) {
     pr_warn("kvmalloc_array failed\n");
     goto err_out4;
 }
 [...]
 /* 5. __vmalloc(): <seen later> */
```

```
    [...]
    return 0;
err_out5:
  vfree(kvarr);
err_out4:
    vfree(kv);
err_out3:
    vfree(vptr_init);
err_out2:
    vfree(vptr_rndm);
err_out1:
    return -ENOMEM;
}
```

The reality is that from the 5.8.0 kernel onward, the `vmalloc()` API is a wrapper over the underlying `__vmalloc_node()` function (the "node" implies that it's NUMA-aware code, that's all). `__vmalloc()` takes just two parameters - the size to allocate and the GFP flags bitmask.

On kernels prior to 5.8.0, though, the exported wrapper `__vmalloc()` API takes an additional parameter, a **page protection bitmask**, via which the protections of the pages allocated can be set! We can leverage this to, say, allocate pages that are read-only (which is precisely what we do in our demo). We cover more on this aspect in the upcoming *Specifying memory protections* section.

Now, check out the code that uses the `__vmalloc()` API on kernels below version 5.8.0:

```
[ ... ]
/* 5. __vmalloc(): allocate some 42 pages and set protections to RO */
#undef WR2ROMEM_BUG
    /* #define WR2ROMEM_BUG */
[ ... ]
    /* In 5.8.0, commit 88dca4c 'mm: remove the pgprot argument to __vmalloc'
has removed the pgprot arg from the __vmalloc(). So, only attempt this when
we're on kernels < 5.8.0 */
#if LINUX_VERSION_CODE < KERNEL_VERSION(5, 8, 0)
    vrx = __vmalloc(42 * PAGE_SIZE, GFP_KERNEL, PAGE_KERNEL_RO);
    if (!vrx) {
        pr_warn("__vmalloc failed\n");
        goto err_out5;
    }
    pr_info("5. __vmalloc():        vrx = 0x%pK (actual=0x%px)\n", vrx, vrx);

    /* Try reading the memory, should be fine */
    print_hex_dump_bytes(" content: ", DUMP_PREFIX_NONE, vrx, DISP_BYTES);
[ ... ]
```

We also have a bit of an experiment going on (though it's enabled only for the older < 5.8 kernels): if the macro WR2ROMEM_BUG is defined (it's not by default), we attempt to write into the read-only vmalloc region we allocated! This, of course, will cause a kernel bug, an *Oops*; here's the relevant code snippet:

```
 ifdef WR2ROMEM_BUG
    /* Try writing to the RO memory! We should, of course, find that the kernel
crashes (emits an Oops!) */
    pr_info("6. Attempting to now write into a kernel vmalloc-ed region that's
RO!\n");
    *(u64 *) (vrx + 4) = 0xba;
#endif /* WR2ROMEM_BUG */
```

Try this out when running on an older, < 5.8, kernel, and be prepared for a nice kernel Oops!

The cleanup code path of our kernel module is straightforward; we of course free all the previously allocated memory regions:

```
static void __exit vmalloc_demo_exit(void)
{
#if LINUX_VERSION_CODE < KERNEL_VERSION(5, 8, 0)
    vfree(vrx);
#endif
    kvfree(kvarr);
    kvfree(kv);
    vfree(vptr_init);
    vfree(vptr_rndm);
    pr_info("removed\n");
}
```

We'll leave it to you to try out and verify this demo kernel module.

Now, let's delve briefly into another key and *really interesting* aspect – how exactly does a user-space malloc() memory allocation become physical memory? Do read on to find out!

A brief note on user-mode memory allocations and demand paging

Without delving into deep detail regarding the internal workings of vmalloc() or the user-space malloc() APIs, we'll nevertheless cover some crucial points that a competent kernel/driver developer like you must understand.

In a user-space app (within a process or thread), let's say we do this:

```
p = malloc(10000);
```

Let's assume it succeeds. It's a very common misconception to think that now, 10,000 bytes of physical RAM are available at address **p** onward. Guess what: on modern OSs like Linux, this simply isn't necessarily true (at least immediately)! We don't blame you for thinking this; it's just how modern OSs work.

It's quite enlightening to clearly understand how memory allocation really works for the popular user-space `malloc()` API and friends – it's all via **demand paging**! Meaning, the successful return of these APIs does not mean anything in terms of *physical* memory allocation. When a user-space `malloc()` returns success, all that has really happened so far is that a virtual memory region has been reserved; no physical memory has actually been allocated yet! The actual allocation of a physical page frame only happens on a per-page basis as and when the virtual page is accessed (for anything: reading, writing, or execution).

But how does this happen internally? The answer: it's simply a part, a side effect, of the usual memory management code translating a virtual address to its corresponding physical counterpart (how else would the machine work?). We covered this topic in brief in *Chapter 7, Memory Management Internals – Essentials*, in the *Getting from virtual to physical addresses* section.

So, to quickly summarize: whenever the kernel (running in process or interrupt context), or a user-space process or thread, *accesses a virtual address*, this virtual address is sent – via the CPU – to, and is interpreted by, the **Memory Management Unit** (**MMU**), which is a part of the silicon on the CPU core. Hardware optimizations are checked: is the code/data already within the CPU internal L1/L2/... caches; if so, work on it and the job's done. If not, do we have a **Translation Lookaside Buffer** (**TLB**) hit? If so, the memory translation (virtual-to-physical address) is already available; if not (we have a TLB-miss – expensive!), and the MMU will now *walk* the paging tables of the process (which are in kernel memory), effectively translating the virtual address and thus obtaining the *physical address*. It then caches the translation in the TLB(s) and places the physical address on the address bus, and the CPU goes on its merry way.

But, think about this: **what if the MMU cannot translate** the given virtual address to a physical address? This can happen for several reasons; some are as follows:

- The virtual address doesn't (yet) have a mapping to a physical address.
- The virtual address is simply incorrect (a bug).

The first case is in fact what will happen here with our `malloc()` example – when the thread "touches" any memory in any manner (reads/writes/executes) within the newly allocated virtual page, the virtual address is sent to the MMU. Now, think, the MMU fails to translate it to physical address as, naturally, there *is no mapping, no physical* page frame (yet), **only a virtual page**. At this point, the MMU essentially gives up as it cannot handle this situation (it's silicon, after all, poor fellow). Instead, it hands off control to the OS, ***invoking its page fault handler code*** – an exception or fault handler routine that runs in the current process's context – in other words, in the context of `current`. The page fault handler resolves the situation; in our case, with `malloc()`, once it determines that this is indeed a legal access, it requests the page allocator (or BSA) for a single physical page frame (at order `0`) and maps it to the virtual page (updating the paging table of the process).

Thus, we can now see that the virtual page was actually allocated physically only on demand, only when it was accessed and not before. This approach to physical memory allocation is termed ***demand paging*** (or *lazy allocation*, and all modern OSs essentially do the same thing)!

Also, as a corollary: if a user-space process or thread performs an `malloc()` and then a `memset()` on the memory region, the physical frames will be allocated the moment each virtual page is written to! This also applies to APIs like `calloc()`; thus, in a real-time system, we prefer the developer to use APIs like this to pre-fault in the required memory frames before hitting the time-critical code paths.

It's equally important to realize that this demand paging (or lazy allocation) OS technique is *not the case for kernel memory allocations* carried out via the page (buddy system), the slab allocator, and even the vmalloc memory region.

Thus, **when kernel memory is allocated via kernel APIs, understand that actual physical page frames are allocated immediately**. (On Linux, it's all very fast because, recall, by the time the system is up, the buddy system allocator per-node and per-zone freelists have already mapped all system physical RAM into the kernel's *lowmem* region and can therefore use it at will.) `vmalloc()` (and friends) too ultimately uses the page allocator to satisfy memory allocation requests; it's just that the virtual pages come from a designated "vmalloc region" within the kernel VAS.

You can go ahead and refer to *Figure 9.9*, which quite nicely maps all of this out; for now, ignore the portion on the OOM killer. We will of course cover it in a later section.

Recall what we did in an earlier program, `ch8/lowlevel_mem`; there, we used our `show_phy_pages()` library routine to display the virtual address, the physical address, and the **Page Frame Number (PFN)** for a given memory range, thereby verifying that the low-level page allocator routines really do allocate physically contiguous memory chunks. Now, you might think, why not call this same function in this `ch9/vmalloc_demo` kernel module? If the PFNs of the allocated (virtual) pages are not consecutive, we again prove that, indeed, it's only virtually contiguous. It sounds tempting to try, but it doesn't work! Why? Simply because, as stated earlier (in *Chapter 8, Kernel Memory Allocation for Module Authors – Part 1*), we should not attempt to translate from virtual to physical any addresses other than direct-mapped (identity-mapped/*lowmem* region) ones – the ones the page or slab allocators supply. It just doesn't work with addresses within the vmalloc region. However (and as we've learned), it's still important to understand that, under the hood, though the vmalloc-ed memory isn't necessarily physically contiguous, it has indeed been physically allocated immediately (and is not demand-paged as user-space memory allocations are).

A few more points on `vmalloc()` and some associated information follow; do read on.

Friends of vmalloc()

A few `vmalloc()`-related APIs, "friends" of `vmalloc()`, in an FAQ style, follow.

Is this vmalloc-ed (or module region) memory?

To figure out whether a given KVA (kernel virtual address) was allocated by one of the `vmalloc()` family, a quick helper function is available:

```
mm/vmalloc.c:bool is_vmalloc_addr(const void *x)
```

The parameter, of course, is the address to check; if it was allocated via `vmalloc()` (or friends), it returns `True`, else `False`. Internally, it checks whether the address passed is within the kernel's vmalloc region. (This routine was merged in the 5.6 kernel.)

On a related note, the kernel provides another helper routine:

```
mm/vmalloc.c:int is_vmalloc_or_module_addr(const void *x)
```

Clearly, it returns `True` if the parameter (the kernel virtual address) passed is within either the kernel module region or the kernel vmalloc region. If neither, it returns `False`.

Unsure which API to use? Try kvmalloc()

In many cases, the precise API (or memory layer) used to perform a memory allocation does not really matter to the caller. So, a pattern of usage that emerged in a lot of in-kernel (and module) code paths went something like the following pseudocode:

```
kptr = kmalloc(n);
if (!kptr) {
    kptr = vmalloc(n);
    if (unlikely(!kptr))
        <... failed, cleanup ...>
}
<ok, continue with kptr>
```

The cleaner alternative to this kind of code is the interesting `kvmalloc()` API. Internally, it attempts to allocate the requested `n` bytes of memory like this:

1. First, try the allocation via the more efficient `kmalloc()`.
2. If it succeeds, fine, we have quickly obtained physically contiguous memory and our job's done; return to caller.
3. If not, fall back to allocating the memory request via the slower but surer `vmalloc()` API (thus obtaining virtually contiguous memory) and then return to the caller.

Its signature is as follows:

```
#include <linux/mm.h>
void *kvmalloc(size_t size, gfp_t flags);
```

(Remember to include the header file `<linux/mm.h>`.) Note that for the invoked `vmalloc()` to go through (if it comes to that), only the `GFP_KERNEL` flag must be supplied. As usual, the return value is a pointer (a kernel virtual address) to the allocated memory, or `NULL` on failure.

Free the memory obtained with kvfree:

```
void kvfree(const void *addr);
```

Here, the parameter, of course, is the return address from kvmalloc().

Similarly, and analogous to the {k|v}zalloc() APIs, we also have the kvzalloc() API, which, of course, zeroes the memory content. I'd suggest you use it in preference to the kvmalloc() API (with the usual caveat: it's safer but a bit slower).

Further, you can use the kvmalloc_array() API to allocate virtual contiguous memory for an array of items. It allocates n elements of size bytes each. Its implementation is shown as follows:

```
// include/linux/slab.h
static inline __alloc_size(1, 2) void *kvmalloc_array(size_t n, size_t
size, gfp_t flags)
{
        size_t bytes;
        if (unlikely(check_mul_overflow(n, size, &bytes)))
                return NULL;
        return kvmalloc(bytes, flags);
}
```

A key point here: notice how a validity check for the dangerous **integer overflow** (IoF) bug is made (via the check_mul_overflow() routine); that's important and interesting. Do write robust code by performing similar validity checks in your code where required.

This check for IoF is key; but why exactly? Think about this: when allocating memory for an array of objects, we often write the code something like this:

```
ptr = kmalloc(num*sizeof(struct mystruct));
```

Well, so? What if the multiplication causes an overflow? In fact, it often occurs that, due to overflow and the *size* parameter's data type being unsigned, the result is a number *smaller* than intended; this can be leveraged by malicious hackers to try and incorporate a BoF (buffer overflow) attack vector. The results can be disastrous, a security-critical bug waiting to strike at an inopportune moment! (See the *Further reading* section for more on arithmetic overflow.)

Next, the kvcalloc() API is functionally equivalent to the calloc() user-space API, and is just a simple wrapper over the kvmalloc_array() API:

```
void *kvcalloc(size_t n, size_t size, gfp_t flags);
```

We also mention that for code requiring *NUMA awareness* (we covered NUMA and associated topics in *Chapter 7, Memory Management Internals – Essentials,* in the *Physical RAM organization* section), the following APIs are available, with which we can specify the particular NUMA node to allocate the memory from as a parameter (this being the point to NUMA systems; do see the information box that follows shortly):

```
void *kvmalloc_node(size_t size, gfp_t flags, int node);
```

Similarly, we have the `kzalloc_node()` API as well, which sets the memory content to zero.

In fact, generally, most of the kernel-space memory APIs we have seen ultimately boil down to one *that takes a NUMA node as a parameter*. For example, take the call chain for one of the primary page allocator APIs, `__get_free_page()`:

```
__get_free_page() -> __get_free_pages() -> alloc_pages()
 ->  mm/page_alloc.c:struct page *__alloc_pages(gfp_t gfp, unsigned int order,
int preferred_nid, nodemask_t *nodemask)
(Its comment says: "This is the 'heart' of the zoned buddy allocator.").
```

Of course, you must free the memory you take; for the preceding `kv*()` APIs (and the `kcalloc()` API), free the memory obtained with `kvfree()`.

Another internal detail worth knowing about, and a reason the `k[v|z]malloc[_array]()` APIs are useful: with a regular `kmalloc()`, the kernel will indefinitely retry allocating the memory requested if it's small enough (this number currently being defined as `CONFIG_PAGE_ALLOC_COSTLY_ORDER`, which is 3, implying 8 pages or less); this can actually hurt performance! With the `kvmalloc()` API, this indefinite retrying is not done (this behavior is specified via the GFP flags `__GFP_NORETRY|__GFP_NOWARN`), thus speeding things up. This LWN article goes into detail regarding the rather weird indefinite-retry semantics of the slab allocator: *The "too small to fail" memory-allocation rule, Jon Corbet, December 2014* (`https://lwn.net/Articles/627419/`).

Regarding the `vmalloc_demo` kernel module we saw earlier in this section, take a quick look at the code again (`ch9/vmalloc_demo/vmalloc_demo.c`). We use the `kvmalloc()` as well as `kcalloc()` APIs (*steps 3* and *4* in the comments). Let's run it on an AArch64 Raspberry Pi 4 (running a recent 6.1 64-bit kernel) system and see the output:

```
$ journalctl -b -o short-monotonic |head -n5
-- Journal begins at Tue 2023-02-21 09:38:04 IST, ends at Tue 2023-04-11 10:11:20 IST. --
[    0.000000] rpi kernel: Booting Linux on physical CPU 0x0000000000 [0x410fd083]
[    0.000000] rpi kernel: Linux version 6.1.21-v8+ (dom@buildbot) (aarch64-linux-gnu-gcc-8 (Ubuntu/Linaro 8.4.0-3ubu
ntu1) 8.4.0, GNU ld (GNU Binutils for Ubuntu) 2.34) #1642 SMP PREEMPT Mon Apr  3 17:24:16 BST 2023
[    0.000000] rpi kernel: random: crng init done
[    0.000000] rpi kernel: Machine model: Raspberry Pi 4 Model B Rev 1.4
$ sudo dmesg -C
$ sudo insmod ./vmalloc_demo.ko ; dmesg
[ 5925.362001] vmalloc_demo:vmalloc_demo_init(): inserted
[ 5925.362056] vmalloc_demo:vmalloc_try(): 1. vmalloc():   vptr_rndm = 0x00000000ce8d7fec (actual=0xffffffc008057000)
[ 5925.362083]   content: 01 00 00 00 00 00 00 00 ff ff ff ff ff ff ff ff  ................
[ 5925.362118] vmalloc_demo:vmalloc_try(): 2. vzalloc():   vptr_init = 0x00000000c1d3f714 (actual=0xffffffc00805f000)
[ 5925.362135]   content: 00 00 00 00 00 00 00 00 00 00 00 00 00 00 00 00  ................
[ 5925.362268] vmalloc_demo:vmalloc_try(): 3. kvmalloc() :        kv = 0x00000000bd7bc75a (actual=0xffffffc009800000)
      (for 5242880 bytes)
[ 5925.362289]   content: 01 00 00 00 00 00 00 00 00 00 00 00 00 00 00 00  ................
[ 5925.362307] vmalloc_demo:vmalloc_try(): 4. kcalloc() :      kvarr = 0x00000000ede691e5 (actual=0xffffff8048a04000)
[ 5925.362323]   content: 00 00 00 00 00 00 00 00 00 00 00 00 00 00 00 00  ................
[ 5925.362334] vmalloc_demo:vmalloc_try(): 5. >= 5.8.0 : __vmalloc(): no page prot param; can use __vmalloc_node_rang
e() but it's not exported.. so, simply skip this case
$
```

Figure 9.3: Output on loading our vmalloc_demo.ko kernel module

We can see from *Figure 9.3* that we're running on a Raspberry Pi 4 with a recent 6.1.21-v8+ stock kernel. Further, look at the "actual" return (kernel virtual) addresses from the APIs in the preceding output – note that they all belong within the kernel's *vmalloc region.*

Given the memory address (the KVA), how can one check whether the allocation is from the kernel's vmalloc range? There are several ways:

- Directly look up the `/proc/vmallocinfo` pseudofile.
- Use the *procmap* utility; it will show all kernel regions with their start and end address values (https://github.com/kaiwan/procmap).
- Another (programmatic way) is via the already-mentioned `is_vmalloc_addr()` API.

Notice the return address of `kvmalloc()` (step 3 in *Figure 9.3*); let's search for it under `proc`:

```
$ sudo grep "^0x00000000bd7bc75a" /proc/vmallocinfo
0x00000000bd7bc75a-0x0000000035088340 6295552 0xffffffdf7fdca170 pages=1536
vmalloc vpages
```

There it is! We can clearly see how using the `kvmalloc()` API for a large quantity of memory (5 MB) resulted in the `vmalloc()` API being internally invoked (the `kmalloc()` API would have failed and would not have emitted a warning, nor retried) and thus, as you can see, it under `/proc/vmallocinfo`.

To interpret the preceding fields of /proc/vmallocinfo, refer to the kernel documentation here: https://www.kernel.org/doc/Documentation/filesystems/proc.txt.

Exercise: A little something for you to try out - in our ch9/vmalloc_demo kernel module, change the amount of memory to be allocated via kvmalloc() by passing kvnum=<# bytes to alloc> as a module parameter.

Miscellaneous helpers – vmalloc_exec() and vmalloc_user()

FYI, up to the 5.8.0 version, the kernel provided an internal helper API, vmalloc_exec() – it's (again) a wrapper over the vmalloc() API and is used to allocate a virtually contiguous memory region that has execute permissions set upon it. An interesting user is the kernel module allocation code path (kernel/module.c:module_alloc()); the space for the kernel module's (executable section) memory is allocated via this routine. This routine isn't exported, though. From 5.8.0 (see commit 7a0e27b), vmalloc_exec() has been removed (merged into its only caller, the module_alloc() function). It's now a wrapper over the recent way to specify memory protections, the __vmalloc_node_range() API(which isn't exported and thus unavailable to modules). The other helper routine we mention is vmalloc_user(); it's (yet again) a wrapper over the __vmalloc_node_range() API and is used to allocate a zeroed-out virtually contiguous memory region suitable for mapping into user VAS. This routine is exported; it's used, for example, by several device drivers as well as the kernel's perf events ring buffer.

Specifying memory protections

As we've already seen from the *Trying out vmalloc()* section, up until the 5.8.0 kernel, the __vmalloc() API could be easily leveraged to specify the particular protections (for example, read-only) of a (vmalloc-ed) memory region. But, after 5.8.0, only the underlying kernel routine __vmalloc_node_range() can do this; so, what's the problem? It isn't exported (whereas __vmalloc() is). Practically speaking, this implies that only kernel-internal code (and in-tree drivers) can leverage such facilities.

Of course, with sufficient effort, one can overcome this seeming limitation: instead of using __vmalloc() or the __vmalloc_node_range() APIs, we might avail of some alternate means, perhaps memory mapping some kernel memory into user space via an mmap() method, and using the mprotect() system call from a user-space app to set up appropriate protections (or, better, even setting up protections through well-known and tested LSM frameworks, such as SELinux, AppArmor, Integrity, and so on).

In user-space applications, performing a similar memory protection setting upon an arbitrary memory region can be done via the mprotect() system call; do look up its man page for usage details (it even kindly provides example code! Link: https://man7.org/linux/man-pages/man2/mprotect.2.html).

Hang on, *why* make memory read-only? Specifying memory protections at allocation time to, say, read-only may appear to be a pretty useless thing to do: how would you then initialize that memory to some meaningful content? Well, think about it – **guard pages** are the perfect use case for this scenario (like the red zone regions that the SLUB layer keeps when in debug mode).

Okay, something new-ish: from the 5.13 kernel, a performance feature regarding `vmalloc()` was merged, allowing the possibility to **employ huge pages** when `vmalloc()` is called (the commit is entitled *mm/vmalloc: hugepage vmalloc mappings* commit # 121e6f3).

Huge pages are MMU-supported pages of a large size (typically much larger than the usual 4K page we're all so used to and love). For example, the x86_64 uses huge pages of size 2 MB for the kernel region when possible. Using huge pages has the beneficial effect of significantly reducing TLB pressure (and thus improving the TLB hit rate, as each huge page of size 2 MB provides a 2 M / 4 K = 512X improvement!).

So, is there a way to figure out whether a `vmalloc()` we've performed is using huge pages? Yes, indeed: the `bool is_vm_area_hugepages(const void *addr)` inline function returns a Boolean. It, however, works via `find_vm_area()`, which isn't exported; hence we can't use it in an out-of-tree module. Oh well. Also, on smaller systems (embedded), using huge pages can result in significant wastage (due to internal fragmentation); the kernel provides a parameter you can pass at boot, `nohugevmalloc`, to turn off this feature.

We conclude this section with a quick comparison between the typical kernel memory allocator APIs: `kmalloc()` and `vmalloc()`.

The kmalloc() and vmalloc() APIs – a quick comparison

A quick comparison between the `kmalloc()` (or `kzalloc()`) and `vmalloc()` (or `vzalloc()`) APIs is presented in the following table:

Characteristic	kmalloc() / kzalloc() APIs	vmalloc() / vzalloc() APIs
Memory allocated is	Physically contiguous	Virtually contiguous (no guarantee on physical contiguity)
Memory alignment	Aligned to hardware (CPU) cache line	Page-aligned
Minimum granularity	Arch-dependent; as low as 8 bytes on x86[_64]	One page
Performance	Faster (physical RAM allocated) for small memory allocations (the typical case); ideal for allocations < 1 page	Slower but can service large (virtual) allocations (even over 4 MB; size depends on the size of the arch-specific vmalloc region)
Size limitation	Limited in size (to typically 4 MB)	Can be very large (and is arch-dependant; the kernel vmalloc region's size can even be several terabytes on 64-bit systems, though much less on 32-bit)

Suitability	Suitable for almost all use cases where performance matters, and the memory required is small, including DMA (still, use the DMA API); can work in atomic/interrupt contexts	Suitable where large software (virtually) contiguous buffers are required; slower, cannot be used in atomic/interrupt contexts
Debugging (via JTAG)	Easy and direct (wrt debugging memory via hardware JTAG (or the like)) as this memory is direct-mapped	Indirect; the JTAG debugger, if supported, will have to "walk" the paging tables to translate from a virtual to a physical address

Table 9.2: Comparison between the kmalloc() and vmalloc() families of APIs

This does not imply that one is superior to the other. Their usage depends upon the circumstance at hand. This leads us on to our next – indeed very important – topic: how do you decide which memory allocation API to use when? Making the right decision is critical for the best possible system performance and stability – do read on to find out how to make that choice!

Memory allocation in the kernel – which APIs to use when

A really quick summary of what we have learned so far: the kernel's underlying engine for memory allocation (and freeing) is called the page (or buddy system) allocator. Ultimately, every single memory allocation (and subsequent free) goes through this layer. It has its share of problems, though, the chief one being internal fragmentation or wastage (due to its minimum granularity being a page or multiple pages). Thus, we have the slab allocator (or slab cache) layered above it, providing the power of object caching and caching fragments of a page (helping alleviate the page allocator's wastage issues). Also, don't forget that you can create your own custom slab caches. Further, the kernel has a `vmalloc` region and APIs to allocate large virtual memory swathes from within it.

With this information in mind, let's move along. To understand which API to use when, let's first look at the kernel memory allocation API set.

Visualizing the kernel memory allocation API set

The following conceptual diagram shows us the Linux kernel's memory allocation layers as well as the prominent APIs within them; do note the following:

- Here we only show the typically/often-used APIs exposed by the kernel to module/driver authors (with the exception being the one that ultimately performs all the allocation work under the hood: the `__alloc_pages()` API right at the bottom! See *Figure 9.4*).
- For brevity, we haven't shown the corresponding memory-freeing APIs.

The following is a diagram showing several of the (exposed to module/driver authors) kernel memory allocation APIs:

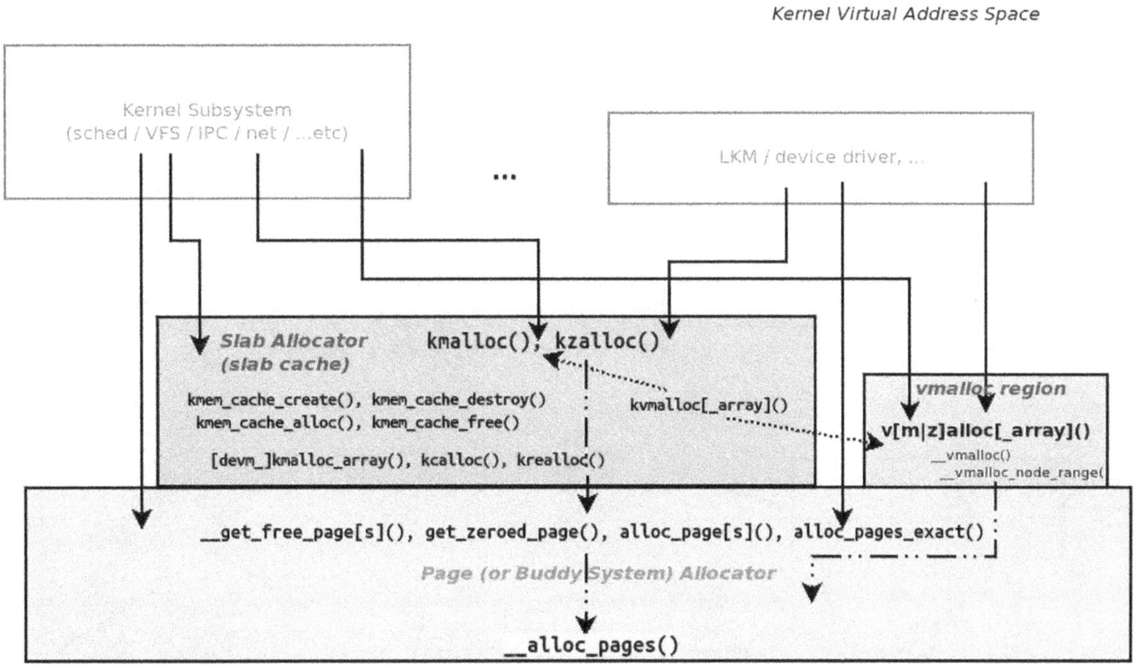

Figure 9.4: Conceptual diagram showing the kernel's memory allocation API set (mostly for module/driver authors)

(*Figure 9.4* also shows that anyone in kernel space can use (most of) these APIs - core code, any subsystem code, device drivers, modules, and so on; its understood that out-of-tree modules can use only exported routines). Now that you have seen the wealth of (exposed) memory allocation APIs available, the following sections delve into helping you make the right decision as to which to use under what circumstances.

Selecting an appropriate API for kernel memory allocation

With this wide choice of memory allocation APIs, how do we choose which to use when? Though we have already talked about this very case in this chapter as well as the previous one, we'll again summarize it as it's very important. Broadly speaking, there are two ways to look at it – the API to use depends upon the following:

- The amount of memory required
- The type of memory required

We will illustrate both cases in this section.

First, to decide which API to use by the type, amount, and contiguity of the memory to be allocated, scan through the following flowchart (starting in the upper-right corner from the label *Start here*):

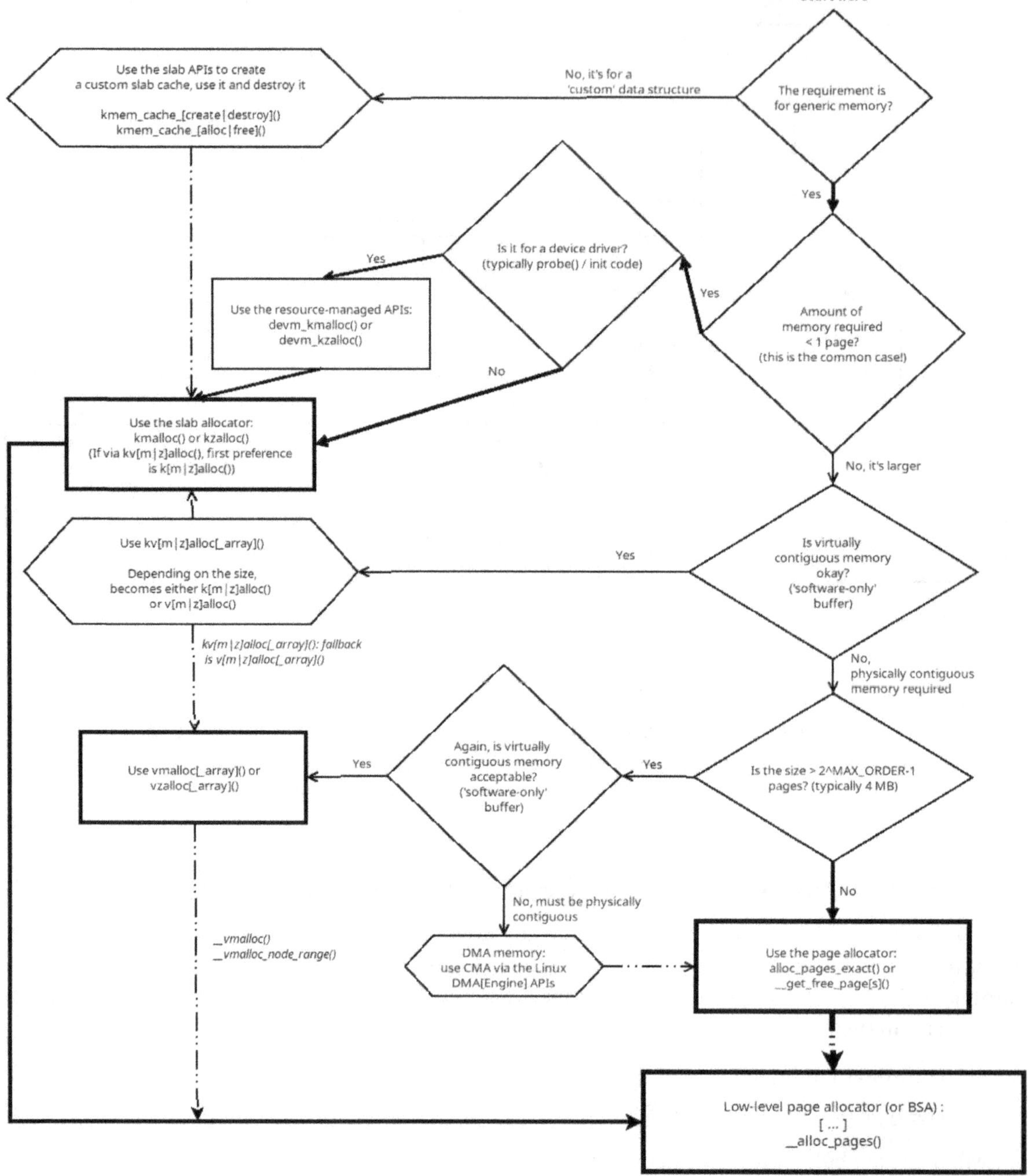

Figure 9.5: Decision flowchart for which kernel memory allocation API(s) to use for a module/driver

Of course, it's not trivial; not only that, but I'd also like to remind you to recall the detailed discussions we covered earlier in the previous chapter, including the GFP flags to use (and the *do not sleep in atomic context* rule); in effect, the following key points:

- When in any atomic ("can't sleep") context, including interrupt contexts, ensure you only use the `GFP_ATOMIC` flag.
- Else, when in process context, you decide whether to use the `GFP_ATOMIC` or `GFP_KERNEL` flag; use `GFP_KERNEL` when it's safe to sleep.
- Then, as covered in *Chapter 8, Kernel Memory Allocation for Module Authors – Part 1*, in the *Caveats when using the slab allocator* section: when using the `k{m|z}alloc()` API and friends, make sure to check the actual allocated memory size with `ksize()`. Even better, check the actual size allocated via the /sys/kernel/slab/<slab-name>/slab_size pseudofile (as we learned in the *Extracting useful information regarding slab caches* section).

Next, to decide which API to use by the type of memory to be allocated, scan through the following table:

<table>
<tr><th>Type of memory required</th><th>Allocation method</th><th>APIs to use</th></tr>
<tr><td>Kernel modules, typical case: regular usage for small amounts (less than one page), physically contiguous</td><td>Slab allocator</td><td>kmalloc(), kzalloc(), kcalloc(), krealloc()</td></tr>
<tr><td>Device drivers: regular/typical usage for small amounts (< 1 page), physically contiguous; recommended for drivers in their probe() or init methods (no explicit free required!)</td><td>Resource-managed APIs</td><td>devm_kmalloc() and devm_kzalloc()</td></tr>
<tr><td>Physically contiguous, general-purpose usage</td><td>Page allocator</td><td>__get_free_page[s](), get_zeroed_page(), and alloc_page[s][_exact]()</td></tr>
<tr><td>Physically contiguous, for Direct Memory Access (DMA)</td><td>Purpose-built DMA API layer, with CMA (or slab/page allocator)</td><td>(not covered here: dma_alloc_coherent(), dma_map_[single|sg](), Linux DMA Engine APIs, and so on)</td></tr>
<tr><td>Virtually contiguous (for large software-only buffers)</td><td>Indirect via page allocator</td><td>vmalloc() and vzalloc()</td></tr>
<tr><td>Virtually or physically contiguous, when unsure of runtime size</td><td>Either slab or vmalloc region</td><td>kvmalloc[_array]()</td></tr>
<tr><td>Custom data structures (objects)</td><td>Creates and uses a custom slab cache</td><td>kmem_cache_{create|destroy}() and kmem_cache_{alloc|free}()</td></tr>
</table>

Table 9.3: Guidelines for deciding on which memory allocation API to use depending on the situation at hand

Of course, there is some overlap with this table and the flowchart in *Figure 9.5*.

As a generic rule of thumb, your first choice should be the slab allocator APIs, that is, via `[devm_]kzalloc()` or `[devm_]kmalloc()`; these are the most efficient for typical allocations, those less than a page in size.

Also, recall that when unsure of the runtime size required, you could use the `kvmalloc()` API. Again, if the size required happens to be a perfectly rounded power-of-2 number of pages (2^0, 2^1, ..., $2^{MAX\_ORDER-1}$ *pages*), then using the low-level page allocator APIs will be optimal.

A word on DMA and CMA

On the topic of DMA, though its study and usage are beyond the scope of this book, I would nevertheless like to mention that Linux has a purpose-built set of APIs for DMA christened the *DMA Engine*. Driver authors performing DMA operations are very much expected to use these APIs and *not* directly use the slab or page allocator APIs (subtle hardware issues do turn up if you do so).

Further, several years back, Samsung engineers successfully merged a patch into the mainline kernel called the **Contiguous Memory Allocator** (**CMA**). Essentially, it allows the allocation of *large, physically contiguous memory* chunks (of sizes over the typical 4 MB limit!). This is required for DMA on some memory-hungry devices (want to stream that ultra-HD quality movie on a big-screen tablet or TV?). The cool thing is that the CMA code is transparently built into the DMA Engine and DMA APIs. Thus, as usual, driver authors performing DMA operations should just stick to using the Linux DMA Engine layer.

Also, realize that our discussion has mostly been with regard to the typical kernel module or device driver author. Within the OS itself, the demand for single pages tends to be quite high (due to the OS servicing demand paging via the page fault handler, for servicing what are called *minor* or *"good" faults*). Thus, under the hood, the memory management subsystem tends to issue the `__get_free_page[s]()` APIs quite frequently. Also, to service the memory demand for the *page cache* (and other internal caches), the page allocator plays an important role.

All right, well done; with this you have (almost!) completed our two chapters of coverage on the various kernel memory allocation layers and APIs focused on module/driver authors! Let's finish off this large topic with a couple of remaining important areas – the Linux kernel's memory reclamation procedures and the (fairly controversial) OOM killer; do read on!

Memory reclaim – a key kernel housekeeping task

As you will be aware, the kernel tries, for optimal performance, to keep the working set of memory pages as high up as possible in the memory pyramid (or hierarchy).

The so-called memory pyramid (or memory hierarchy) on a system consists of (in order, from smallest size but fastest speed to largest size but slowest speed): CPU registers, CPU caches (LI, L2, L3, ...), RAM, and swap (raw disk/flash/SSD partition). In the following discussion, we ignore CPU registers as their size is minuscule.

In a modern processor, as code executes and data is worked upon, the processor uses its hardware caches (L1, L2, and so on) to hold the current working set of pages within its multilevel CPU instruction and data caches. But of course, CPU cache memory is very limited, thus it will soon run out, causing the memory to spill over into the next hierarchical level – RAM. On modern systems, even many embedded ones, there's quite a bit of RAM; still, it does run short at times. When this occurs (simplifying it), the OS spills over the memory pages that can no longer fit in RAM into a raw disk partition – *swap*. Thus, the system continues to work well, albeit at a significant performance cost once swap is (often) used (of course, we're referring to user space pages being possibly swapped here; as you know, kernel pages are simply non-swappable. The, again, it's typically user space that's the resource hogger).

Zone watermarks and kswapd

The Linux kernel, in an effort to ensure that a given minimum amount of free memory pages are always available within RAM, **continually performs background page reclamation work** – indeed, you can think of this as routine housekeeping. Who, within the kernel, performs this work?

(Hey, what exactly do we mean by the term "to reclaim" memory? Freeing up the memory page, in effect, making it available to the system.)

The `kswapd` kernel thread(s) are continually monitoring memory usage on the system and invoke a page reclaim mechanism when they sense that memory is running low.

This page reclamation work is done on a per-*node:zone* basis. The kernel uses so-called *watermark levels* – min, low, and high – per *node:zone* to determine when to reclaim memory pages in an intelligent fashion. You can always look up `/proc/zoneinfo` to see the current watermark levels; note that the unit of watermark levels is *pages*.

A quick example (with truncated output) on my native x86_64 laptop running Ubuntu 22.04 is as follows:

```
$ sudo grep -A2 -E "^Node |min" /proc/zoneinfo
Node 0, zone      DMA
[ ... ]
--
Node 0, zone   Normal
  pages free     75060
        boost    0
        min      16188
        low      23952
        high     31716
[ ... ]
```

Here, the situation's quite comfortable; within our node 0, zone Normal region, the "high" watermark is at 31,716 pages, and currently, 75,060 pages are free. If the number of free pages in the zone is somewhere between the *low* and *min* watermarks, or less than that, we have a problem. Thus, here, no problems. Also, as we mentioned earlier, caches are typically the first victims of page reclamation and are (intelligently) shrunk down as memory pressure increases.

So, what if memory pressure does increase? The kernel's **page reclamation algorithm** – so-called **PFA** or simply "**reclaim**" – loosely (and typically) encodes these steps:

- If *number of low (zone watermark) pages < free pages < high pages*

  (In other words, if the number of free pages is between the high and low zone watermarks), the kernel begins to evict some caches and performs "gentle" swapping; this should free sufficient pages to have the "pages free" number climb above "high" and all's well.

- If this doesn't work and *number of min (zone watermark) pages < free pages < low pages*

  (IOW, if the number of free pages is between the low and min zone watermarks), the kernel continues to evict caches (to a greater extent) and performs aggressive swapping; this should help...

- If it doesn't and the number of free pages continues to fall and *number of free pages < min (zone watermark) pages*

  (In other words, if the number of free pages goes below the *min* zone watermark), the kernel is now "alarmed"; it now works hard to aggressively reclaim memory pages by doing the following:

  - It aggressively continues to evict (drop) caches.
  - It continues to perform aggressive swapping.
  - It denies any new memory requests on that zone.

Quite recently, a couple of new features related to page reclaim have been merged; we'll briefly introduce them now.

The new multi-generational LRU (MGLRU) lists feature

The Linux kernel has traditionally figured out which pages to reclaim (evict, in effect) via a typical **Least Recently Used** (**LRU**) algorithm; pages that haven't been accessed recently are unlikely to be required in the near future and vice versa. So, traditionally, (but simplifying it), the kernel has maintained a pair of LRU lists: active and inactive LRU lists. This seems reasonable, but experience has shown that it doesn't always work well and can in fact contribute to slowing the system down! (For example, sequentially reading a large file is a common operation in many scenarios, so pages that have just been read go onto the active list. But if your app(s) will not need those pages again; now it's very sub-optimal). Also, the way anonymous pages (those that aren't file backed) are scanned at runtime using a reverse-mapping (*rmap*) approach (see this link for details: `http://lastweek.io/notes/rmap/`) is considered expensive.

Merged as recently as in the 6.1 kernel, the new **multi-generational LRU** (aka **MGLRU**) feature has essentially these characteristics:

- There now exists a whole bunch of LRU lists between the active and inactive ones to provide much finer granularity in picking a page to reclaim; each is called a *generation*.

  They're organized by *page age*, from the youngest generation (generation 0, which of course holds the most recently used pages, the active ones) to the oldest generation (generation N-1, which holds the LRU pages). At reclaim time, older-generation pages are the obvious candidates for eviction.
- The number of generations N (and thus lists) is set within the kernel code; Android-based systems typically have the number as 4, and high-end cloud/enterprise servers might use a much higher value. In the default codebase, we have (in `include/linux/mmzone.h`):

```
 /* MAX_NR_GENS is set to 4 so that the multi-gen LRU can support twice
the
  * number of categories of the active/inactive LRU when keeping track of
  * accesses through page tables. [ … ]
  */
#define MIN_NR_GENS              2U
#define MAX_NR_GENS             4U
```

- It does not use the costly rmap approach to finding anonymous pages; instead, it scans the process PTEs themselves, looking for the recent bit flag being set or cleared (and, as an optimization, for only those threads that have run since the last scan). It's thus definitely faster.
- It uses less CPU ("demonstrates 51% less CPU usage from kswapd").
- It uses a little more memory ("~500 bytes per-memcg and per-node and ~50 bytes per-process memory overhead").

The multi-generational LRU itself is a kernel-configurable feature (`CONFIG_LRU_GEN`); as of the time of writing, it's still considered an opt-in feature and isn't tightly meshed into the memory management subsystem (the traditional code using the pair of active/inactive lists is still present, though unused when MGLRU is activated). On my Ubuntu 22.04 LTS distro kernel, it's not configured by default. Even if configured, it may be off by default and can be turned on via a tunable under debugfs (`/sys/kernel/debug/lru_gen`). (More recently, Fedora 39 running the 6.6-based distro kernel, and Ubuntu 23.04 running a 6.5-based one, enables MGLRU.) When configured, various sysctls (or tunables) appear under `/sys/kernel/mm/lru_gen` and a couple in debugfs.

Trying it out – seeing histogram data from MGLRU

MGLRU generates data via callbacks to the following pseudofiles: `/sys/kernel/debug/lru_gen` and `/sys/kernel/mm/lru_gen`. Here, let's check out the first one, the simpler case: reading from the `/sys/kernel/debug/lru_gen` psuedofile yields the data, essentially, a histogram of the number of pages that were accessed at a different point in time. This is shown for each memcg (memory control group; FYI, we cover control groups in *Chapter 11, The CPU Scheduler – Part 2*) and for each NUMA node.

The output format is as follows:

```
memcg  memcg_id  memcg_path
   node  node_id
         min_gen_nr  age_in_ms  nr_anon_pages  nr_file_pages
         ...
         max_gen_nr  age_in_ms  nr_anon_pages  nr_file_pages
```

Control groups are essentially a means to elegantly apply resource constraints (where the resource can be CPU/memory/IO/network/... bandwidth) on a group of processes, i.e., on a control group (a *cgroup*). So, groups of processes – *cgroups* – with memory bandwidth constraints fall under "memory cgroups" or, simply, *memcgs*. As one example, they are very useful on cloud workloads, ensuring boundaries that VMs must adhere to in terms of memory (as well as CPU and IO) usage! You'll find a bit more information about cgroups in the last portion of this chapter as well.

So, let's look at and learn how to interpret MGLRU data! A quick sampling of the data generated by MGLRU (on my x86_64 Ubuntu guest) is as follows:

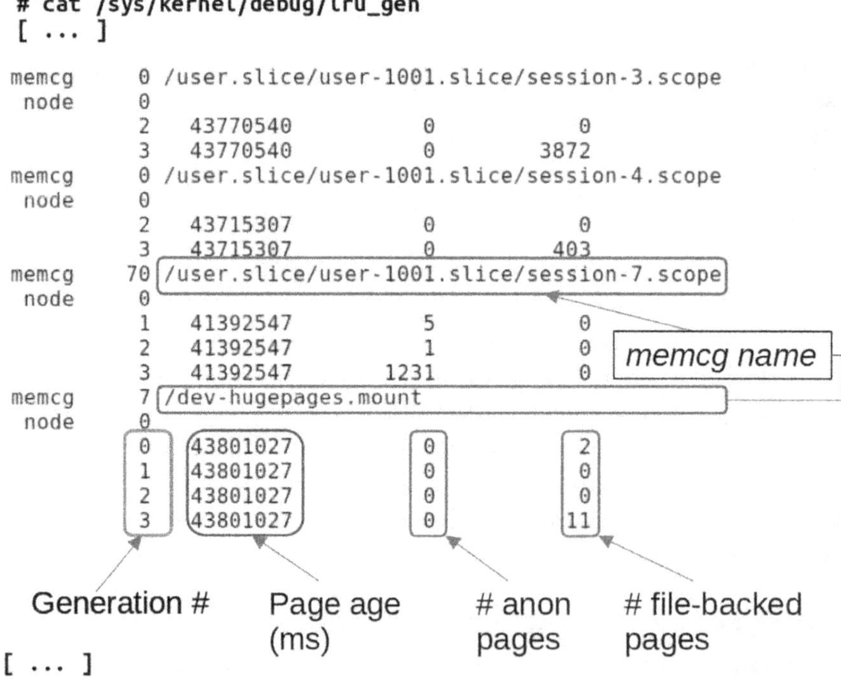

Figure 9.6: An annotated screenshot of a portion of the MGLRU output

As we're on a pseudo-NUMA system, it's always memcgs with node #0... More interesting are the *generation numbers*, which, here, can range from 0 to 3, implying a total of four generations (as `MAX_NR_GENS = 4`), with 0 being the youngest and 3 the oldest.

It's followed by the "age" of the page (in ms) and the number of pages accessed (in that time interval) in that generation – for both anonymous pages followed by file-backed pages. So now, we can literally see how the kernel tracks page usage in one sense and can thus make better decisions when it comes to page reclamation.

The following material makes use of the *systemd-cgls* utility; if it isn't clear, don't worry – we cover all of this (systemd slices and scope, and so on) in *Chapter 11, The CPU Scheduler – Part 2*, in the *Systemd and cgroups* section.

So, looking at *Figure 9.6*, you may well wonder: which processes belong to a given named memcg (for example, to the memcg named `/user.slice/user-1001.slice/session-7.scope`)? A quick way is to run the `systemd-cgls` command; it will show the entire cgroup hierarchy (tree) from which you can pick up the relevant cgroup(s). *Tip*: Passing the `-u <cgroup-name>` option switch can help narrow it down; for example, on my system, I looked up the processes belonging to the `<...>/session-7.scope` memcg (which happens to be for the user with UID 1001) like this:

```
$ systemd-cgls -u session-7.scope
Unit session-7.scope (/user.slice/user-1001.slice/session-7.scope):
├─2598 sshd: c2kp [priv]
├─2712 sshd: c2kp@pts/4
└─2713 -bash
```

So, we can now see that this particular memcg - memory cgroup - whose processes pages are being tracked by MGLRU, has three processes within it: 2 sshd and 1 Bash process.

To delve deeper into MGLRU, I refer you to the official kernel documentation on it:

- *Multi-Gen LRU*: Provides documentation on quick start, enabling/disabling it, the various tunables present, and their meaning: `https://www.kernel.org/doc/html/v6.1/admin-guide/mm/multigen_lru.html`
- *MGLRU design doc*: `https://www.kernel.org/doc/html/v6.1/mm/multigen_lru.html`

A quick introduction to DAMON – the Data Access Monitoring feature

The purpose of **DAMON** – the **Data Access MONitor** – is to capture and analyze memory access patterns of user-space processes, helping developers to gain insight into them and thus be able to optimize them. This feature was merged in the 5.15 kernel.

The kernel component of DAMON essentially divides the target process into equally sized regions that it monitors for data access. It's clever: if a region is looking to be a hotspot, it divides it further. The number of accesses made to each region is provided as a histogram. In effect, we have a *producer-consumer* architecture: the DAMON kernel component is the producer (of data access patterns of a given user-space process).

The consumer can be a user-mode app or the kernel itself; by analyzing the data access patterns to the workload, they can even request changes to the memory region (to optimize access!), effected via calls to the well-known `madvise()` system call. (To do so, DAMON sets up a way to specify "schemes"; details can be found in the official kernel docs here: `https://www.kernel.org/doc/html/v6.1/admin-guide/mm/damon/usage.html#schemes-n`.)

To deal with the very large VAS on 64-bit Linux, DAMON begins by only monitoring data accesses to the text, heap, and stack segments (or mappings).

Now, to make it suitable for production use (which is the intent), monitoring every single page in each of these mappings would be prohibitively expensive. So (kind of like the **KCSAN** (**Kernel Concurrency SANitizer**) does), it instead uses a sampling-based or statistical approach, picking a random page in each of these mappings and recording accesses made to it, in the hope that they represent the typical data access pattern to that region. It uses intelligence to "zoom" in on hotspots and further sub-divide them into more regions, thus getting a better sampling.

To use DAMON, several interfaces are made available:

- Via a DAMON user-mode tool named *damo* (`https://github.com/awslabs/damo`).
- Via sysfs (under `/sys/kernel/mm/damon/admin/`); these are documented here: `https://www.kernel.org/doc/html/v6.1/admin-guide/mm/damon/usage.html#sysfs-interface`.
- Via a debugfs interface; this, however, was used initially and is now considered deprecated. Use the sysfs ones instead.
- Via kernel APIs (for kernel devs; `https://www.kernel.org/doc/html/v6.1/mm/damon/api.html`).

It's typically easiest to use the *damo* front-end. Let's do so!

Running a memory workload and visualizing it with DAMON's damo front-end

A sample run with DAMON's *damo* front-end can help us visualize the data access patterns (on a given process) that it superbly extracts. We'll use *masim*, a "simulator for intensive memory access," to provide a process that accesses memory (intensively) in a given pattern (via a config file specified to it. In fact, the `masim` and `damo` apps are written by none other than the lead author of the DAMON patches, SeongJae Park). To do so, follow these steps:

1. Install `masim` (the workload):

```
git clone https://github.com/sjp38/masim
cd masim
make
```

2. Install `damo`: `git clone https://github.com/awslabs/damo` (the `damo` Python script is there)
3. Start the memory-intensive workload process (`masim`; or, you could start your own workload) and then (immediately) start recording the memory accesses (which are internally being captured by the DAMON kernel component!) via `damo`, writing the recording into a file. Right, let's start the workload process, `masim` (we're currently within the `damo` tool's directory):

```
$ ../masim/masim ../masim/configs/stairs_30secs.cfg &

# Now, ASAP, let's start the recorder process, damo:
$ sudo ./damo record -o damon.data $(pidof masim)
initial phase:                    102,870 accesses/msec, 5001 msecs run
Press Ctrl+C to stop
phase 0:                    107,374 accesses/msec, 2500 msecs run
phase 1:                    101,921 accesses/msec, 2500 msecs run
phase 2:                    102,666 accesses/msec, 2501 msecs run
[ … ]
phase 9:                    104,962 accesses/msec, 2500 msecs run
[ perf record: Woken up 2 times to write data ]
[ perf record: Captured and wrote 1.198 MB damon.data (2920 samples) ]
[1]+  Done                    ../masim/masim ../masim/configs/
stairs_30secs.cfg
$
```

Here, I am using a workload called `stairs_30secs.cfg`; you can use any other one, or your own workload.

4. Visualize the data accesses via *damo's* heatmap feature:

```
sudo ./damo report heats --heatmap stdout
```

```
damo $ sudo ./damo report heats --heatmap stdout
# access_frequency:  0  1  2  3  4  5  6  7  8  9
# x-axis: space (140702942609408-140703053238208: 105.504 MiB)
# y-axis: time (56317015659000-56341983181000: 24.968 s)
# resolution: 80x40 (1.319 MiB and 624.188 ms for each character)
damo $ _
```

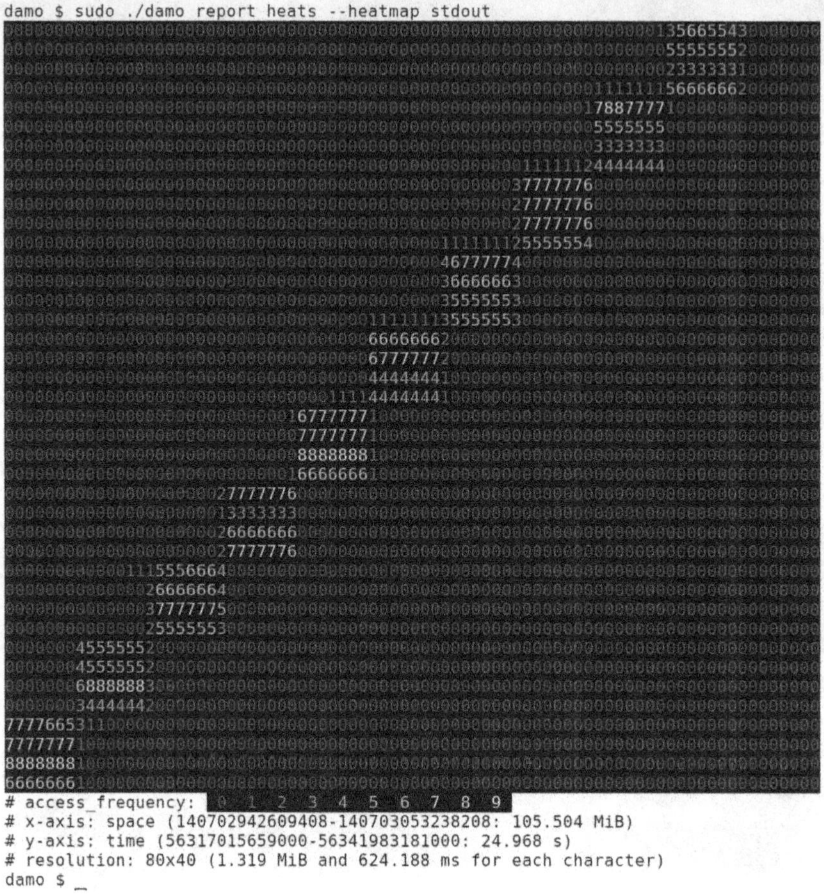

Figure 9.7: Screenshot of running a "stairs" memory workload (via masim) and DAMON's damo front-end showing a visualization of the memory access patterns (as a "heatmap")

An aside: damo requires the perf utility to be installed and working. Now, perf will be installed and ready to use if you've followed the install instructions from *Online Chapter, Kernel Workspace Setup*, and you're running the stock/distro kernel, not a custom one. The issue is that perf is tightly coupled with the kernel it runs upon; hence, if you're running a custom kernel (like our 6.1.25 one that we built back in *Chapters 3* and *4*), then you'll need to manually build perf from within its kernel source tree. In turn, perf requires a Python module named setuptools, which in turn needs pip (yes, a descent into dependency hell). So, if needed, on a custom kernel system, you can do this: sudo apt install python3-pip.

After this, switch to your custom kernel source tree and build perf; for example:

```
cd ~/kernels/linux-6.1.25/tools/perf ; make
```

(Don't forget to add the perf location to the PATH).

Now, perf, and thus damo, should work.

How does DAMON actually help? Several interesting case studies dealing with how profiling and optimization were affected by leveraging DAMON have been written up by SeongJae Park; do refer to the *Further reading* section for links on this and more. We also provide many useful links there on MGLRU (including good articles by LWN) and DAMON.

So, to conclude, all of this reclaim technology should certainly help the kernel reclaim sufficient memory frames when needed such that it's comfy once again! But can it *always* do enough? If not (hint: sometimes that's the case!), heaven forbid, the kernel's dreaded OOM killer may pay a visit!

Stayin' alive – the OOM killer

Now that we've covered background details regarding kernel memory management, particularly the reclaiming of free memory, you're well placed to understand what the **Out of Memory** (**OOM**) killer kernel component is, how to work with it, even how to deliberately invoke it and, to an extent, control it.

Let's revisit a key point, that of memory (RAM and swap) running short. Let's play devil's advocate: what if RAM runs low and all this memory reclamation work (which we just covered in the previous section) simply doesn't help, and memory pressure keeps increasing to the point where the complete memory pyramid is exhausted, where a kernel allocation of even a few pages fails (or infinitely retries, which, frankly, is just as useless, perhaps worse)? In other words, what if all CPU caches, RAM, *and* swap are (almost completely) full!? Well, most systems just die at this point (actually, they don't technically die; they just become so painfully slow that it appears to us humans that they're permanently hung).

Furthermore, realize that on many embedded systems, there's no swap partition (or file), thus enhancing the likelihood of completely running out of memory!

The Linux kernel, though, being Linux, tends to get aggressive in these situations; it now invokes a component aptly named the **OOM killer**. The OOM killer's job is to identify and summarily kill the memory-hogger process and its descendants by sending them the fatal `SIGKILL` signal; thus, it could end up killing a whole bunch of processes.

Quoting directly from the kernel docs: "The OOM killer selects a task to sacrifice for the sake of the overall system health. The selected task is killed in a hope that after it exits enough memory will be freed to continue normal operation."

As you might imagine, it has had its fair share of controversy. Early versions of the OOM killer have been (quite rightly) criticized (occasionally, it killed the wrong chap!).

My extremely able technical reviewer, ChiThanh Hoang, says this regarding the OOM killer: "... "fairly" controversial is an understatement. Most of time people see OOM kills but don't know how to deal with it, especially on embedded systems where there is no swap."

There's some hope, though; recent versions of the kernel use superior OOM killer victim-id heuristics that do work better.

You can find more information on the improved OOM killer work (the kick-in strategy and the OOM reaper thread) in this LWN article (December 2015): *Toward more predictable and reliable out-of-memory handling, Jon Corbet*: `https://lwn.net/Articles/668126/`.

Deliberately invoking the OOM killer

Rather than dwell on theory and internal implementation, we'd rather live a little, testing our friend, the OOM killer, provoking the kernel enough to have it unleash her! To do so, of course, we shall have to put enormous memory pressure on the system. If we succeed in this, the kernel will quite happily join battle, unleashing its weapon – the OOM killer, which, once invoked, will identify and kill some process (or processes).

Quite obviously, I highly recommend you try out stuff like this on a safe, isolated system, preferably a test Linux VM (with no important data on it).

Invoking the OOM killer via Magic SysRq

The kernel provides an interesting feature dubbed *Magic SysRq*: essentially, certain keyboard key combinations (or accelerators) result in a callback to some kernel code. For example (and assuming it's enabled), pressing the `Alt-SysRq-b` key combination on an x86[_64] system results in a cold reboot! Take care – don't just type anything; do read the relevant documentation here: `https://www.kernel.org/doc/html/latest/admin-guide/sysrq.html`

Let's try out some interesting things; to see the *Magic SysRq* state, we run the following on our x86_64 Ubuntu 22.04 Linux VM:

```
$ cat /proc/sys/kernel/sysrq
176
```

This reveals the current state of the Magic SysRq bitmask; being positive, it shows that the feature is partially enabled (the kernel documentation mentioned at the start of this section provides the details). To be able to invoke the OOM killer via the SysRq feature, we need to first enable it fully. To do so, run the following command:

```
sudo sh -c "echo 1 > /proc/sys/kernel/sysrq"
```

Right, *now* we can use Magic SysRq to invoke the OOM killer!

Careful! Invoking the OOM killer, via Magic SysRq or otherwise, will cause some process – typically the heavyweight one(s) – to unconditionally die!

How? As root, just run the following command:

```
# echo f > /proc/sysrq-trigger
```

Look up the kernel log to see whether anything interesting occurred! A detailed interpretation of the kernel log when the OOM killer strikes is covered in detail in the upcoming *Demand paging and OOM* section; for now, we're just looking at getting it invoked.

Invoking the OOM killer with a crazy allocator program

Here's a more hands-on and interesting way by which you can (most probably) invite the OOM killer in: let's write a simple user-space C program that behaves as a "crazy allocator," performing (typically) tens of thousands of memory allocations, writing something to each page, and – of course! – never freeing up the memory pages allocated, thus putting tremendous pressure on memory resources.

Though we show you the code here, we put off running it – and triggering the OOM killer – for a bit. This is because the whole thing becomes much clearer once we understand the kernel's VM overcommit settings and what they imply, which is the topic of the following main section. So, I request your patience...

As usual, we show only the most relevant parts of the source code in the following snippet; please refer to and clone the book's GitHub repo for the full code. Remember, this is a user-mode app, not a kernel module:

```
// ch9/oom_killer_try/oom_killer_try.c
#define BLK      (getpagesize()*2)
static int force_page_fault = 0;
int main(int argc, char **argv)
{
    char *p;
    int i = 0, j = 1, stepval = 5000, verbose = 0;
    [...]

    do {
        p = (char *)malloc(BLK);
        if (!p) {
            fprintf(stderr, "%s: loop #%d: malloc failure.\n",
                  argv[0], i);
            break;
        }

        if (force_page_fault) {
           p[2048] &= 0xde; // write something into a byte of the 1st page
           p[6143] |= 0xad; // write something into a byte of the 2nd page
        }
        if (!(i % stepval)) { // every 'stepval' (5000) iterations..
           if (!verbose) {
              if (!(j%5)) printf(". ");
              else printf(".");
              fflush(stdout);
              j++;
             [...]
           }
        i++;
    } while (p && (i < atoi(argv[1])));
```

Hang on, though; before we can have fun with this program and fully understand what goes on under the hood when it executes, we require some background information; let's get to it!

Understanding the three VM overcommit_memory policies

Here's an interesting point: the Linux kernel follows a **VM overcommit** policy, deliberately over-committing memory (to a certain extent). To understand this, let's look up the current VM *overcommit memory* setting:

```
$ cat /proc/sys/vm/overcommit_memory
0
```

This setting, 0, is the default value. The three permissible values (settable only by root) are as follows:

- `0`: Allow memory overcommitting using a heuristic algorithm; this is the *default* (corresponds to the macro `OVERCOMMIT_GUESS`).
- `1`: Always overcommit; in other words, never refuse any `malloc()`; this is useful for some types of scientific apps that use sparse memory (corresponds to the macro `OVERCOMMIT_ALWAYS`).
- `2`: Never overcommit (corresponds to the macro `OVERCOMMIT_NEVER`), aka the "strict accounting" policy.

  The following notes are direct quotes from the official kernel documentation on this value 2, `https://www.kernel.org/doc/html/v6.1/mm/overcommit-accounting.html`:

  "Don't overcommit. The total address space commit for the system is not permitted to exceed swap plus a configurable amount (default is 50%) of physical RAM. Depending on the amount you use, in most situations this means a process will not be killed while accessing pages but will receive errors on memory allocation as appropriate. Useful for applications that want to guarantee their memory allocations will be available in the future without having to initialize every page."

Now, the overcommit extent is determined by the *overcommit ratio*:

```
$ cat /proc/sys/vm/overcommit_ratio
50
```

Worry not, we'll soon get to what exactly these overcommit values mean practically. Also, we often refer to these numbers via the expression `vm.overcommit_memory`; it has the same meaning as the value of the sysctl `/proc/sys/vm/overcommit_memory`. This is as the `sysctl` utility (and the `/etc/sysctl.conf` configuration file) express a kernel system tunable – a "sysctl" – in this manner – foo.bar, where "foo" is a subsystem under `/proc/sys` – one of fs, kernel, net, or vm, and "bar" is the tunable in question within that subsystem. So, for the VM overcommit tunables, let's look up the defaults:

```
$ sudo sysctl -a |grep "vm\.overcommit"
vm.overcommit_kbytes = 0
vm.overcommit_memory = 0
vm.overcommit_ratio = 50
```

Great, it's as expected. We'll examine what exactly these VM overcommit values mean, via both code-based and empirical (running it!) viewpoints.

VM overcommit from the viewpoint of the __vm_enough_memory() code

This section's a bit of a deviation, but important overall; do follow along...

When a process attempts to allocate (virtual) memory pages, the kernel doesn't really interfere by default; if the process requests too much, the allocation either fails or, in the worst case, the kernel OOM machinery takes over.

For the **Linux Security Modules** (**LSM**) mechanism, though, the kernel does provide a helper routine to check whether a process can allocate the number of pages being requested, taking into consideration the three VM overcommit policies (the `vm.overcommit_memory` value). Here is the function: `mm/util.c:__vm_enough_memory()`.

Firstly, the three VM overcommit policies (with values 0, 1, and 2, respectively) are defined as macros here:

```
// include/uapi/linux/mman.h
...
#define OVERCOMMIT_GUESS        0          << this is the default >>
#define OVERCOMMIT_ALWAYS       1
#define OVERCOMMIT_NEVER        2
```

A quick glance at the `__vm_enough_memory()` function is interesting (we've added some annotations `<< like this >>` to help make it clear):

```
// mm/util.c
int __vm_enough_memory(struct mm_struct *mm, long pages, int cap_sys_admin)
{
[ … ]
    /*
     * Sometimes we want to use more memory than we have
     */
    if (sysctl_overcommit_memory == OVERCOMMIT_ALWAYS)
        return 0;

   << this is the default >>
    if (sysctl_overcommit_memory == OVERCOMMIT_GUESS) {
        if (pages > totalram_pages() + total_swap_pages)
            goto error;
        return 0;
    }
   << thus, this is the OVERCOMMIT_NEVER case >>
   allowed = vm_commit_limit();
    [ … ]
```

```
error:
    pr_warn_ratelimited("%s: pid: %d, comm: %s, no enough memory for the
allocation\n", __func__, current->pid, current->comm);
    vm_unacct_memory(pages);
    return -ENOMEM;
}
```

Glance at the code; clearly, for the VM overcommit memory values of 1 and 2, the logic is simple:

- `OVERCOMMIT_ALWAYS (1)`: Simply let the process go ahead (it's on his head!)
- `OVERCOMMIT_GUESS (0)`: The default – flag an error if the number of pages requested is more than the total available on the system (RAM + swap), else allow the allocation to go ahead. It *could* later result in issues, but that's okay – the kernel's OOM machinery will handle the worst case, if it comes to it. Do look at this *if* condition carefully (repeated here for clarity: *if (pages > totalram_pages() + total_swap_pages)*): if a single very large allocation request (in effect, a spike) is attempted, this code fails it; however, even if we request many tens of thousands of *small* allocations (as we indeed do in our earlier experiment (ch9/oom_killer_try/oom_killer_try.c), requesting just 8 KB at a time), one at a time, it tries to go through for as long as it can... In effect, the kernel won't bother us until one of these conditions occurs:
    - Mode 2 (`OVERCOMMIT_NEVER` policy) only: the kernel's CommitLimit value is breached (see the next section *A note on the Commit* limits in proc*).
    - The process user virtual address space becomes full (quite feasible on the relatively small VAS on 32-bit (2 or 3 GB), unlikely on 64-bit).
    - The entire memory hierarchy pyramid has been exhausted (all CPU caches + RAM + swap memory); then, the OOM killer jumps in...

It's important to realize that this `__vm_enough_memory()` function is *not* called every time a user-mode process attempts to allocate virtual pages; it's only invoked on-demand by some kernel LSMs, on occasion. Thus, here, it's all quite academic, merely helping us gain some insight into the underlying machinery, that's all.

A note on the Commit* limits in proc

Within the `/proc/meminfo` pseudofile content, you'll find the so-called *Commit** VM limits. This is on my AArch64 Raspberry Pi 4 with close to 2 GB RAM, 100 MB swap, and running a recent 6.1 stock kernel:

```
# grep Commit /proc/meminfo
CommitLimit:      1049192 kB
Committed_AS:      203344 kB
```

How do we interpret these values? The man page on `proc(5)` reveals the answer:

```
  CommitLimit %lu (since Linux 2.6.10)
                   This  is  the  total  amount  of memory currently
available to be allocated on the system, expressed in kilobytes.  This limit is
```

```
 adhered to only if strict overcommit  accounting  is  enabled  (mode  2  in /
proc/sys/vm/overcommit_memory). ......
  Committed_AS %lu
                    The  amount of memory presently allocated on the system.
The committed memory is a sum of all of the memory which has been allocated by
processes, even if it has not been "used" by them as  of  yet. A  process which
allocates 1 GB of memory (using malloc(3) or similar), but touches only 300 MB
of that memory will show up as using only 300 MB of memory even if it has the
address space allocated for the entire 1 GB. ...
```

Okay, let's now practically understand things by checking out two cases of the VM overcommit memory tunable while running our "crazy allocator" code on each of them (on our Raspberry Pi 64-bit board) in the following sections.

Case 1: vm.overcommit_memory == 0 (the default, OVERCOMMIT_GUESS)

This *is* the default VM overcommit setting; thus:

- The sysctl `vm.overcommit_memory` is set to `0` (not 2).
- With this, the `vm.overcommit_ratio` and `vm.overcommit_kbytes` don't matter.

From the kernel documentation: *"The overcommit amount can be set via vm.overcommit_ratio (percentage) or vm.overcommit_kbytes (absolute value). These only have an effect when vm.overcommit_memory is set to 2."*

With the `vm.overcommit_memory` tunable set to `0`, the default, there is no formula to calculate the total overcommitted memory we're "allowed" to swallow before being checked. The macro's name gives it away – `OVERCOMMIT_GUESS`: it is a very heuristic approach. In effect, our memory hogger process will be allowed to consume many, many *virtual* pages of memory before either (a bit of repetition here):

- It runs out of **Virtual Address Space** (**VAS**); on older 32-bit systems, this is a distinct possibility, very much less so on 64-bit (the one we're typically running upon, as the VAS is vast there).
- It truly exhausts all memory (in the entire hierarchy, the memory pyramid) – CPU caches, RAM, and swap are all used up; in spite of efforts to reclaim memory frames, this can fail in the face of enormous memory pressure (as is indeed going to be applied here!). Now there's no choice: the kernel invokes the OOM killer.

So, on our AArch64 Raspberry Pi 4, with `vm.overcommit_memory == 0` and 1,849 MB of RAM, no "huge pages or `total_huge_TLB [1]`" and 100 MB of swap, let's be empirical, run our "crazy allocator" program, and see what happens!

[1] *Huge pages* are a kernel feature allowing the kernel to employ larger-sized memory pages; it's typically employed on mid-to-large systems, including the x86_64 by default (but not by default on embedded devices). Here, it implies the amount of memory reserved for huge pages.

Right, finally, let's run our custom "crazy allocator" app (`ch9/oom_killer_try/`; we saw it in the *Invoking the OOM killer with a crazy allocator program* section). We run it with an enormous number of two-page allocations attempted – 500,000 – that aren't ever freed:

```
$ ./oom_killer_try 500000 0
./oom_killer_try: PID 874 (verbose mode: off)
..... ..... ..... ..... ..... ..... ..... ..... ..... ..... ..... ..... .....
..... ..... ..... ..... ..... .... Killed$
```

Ah, see the `Killed` message? The OOM killer has indeed struck! (As the memory pressure gets very high, the slow down on my embedded machine is very, very palpable!)

So, here:

- See the periods (`.`) in the output block? They're very deliberate `printf()`'s! Here, we got 18 groups of 5 periods plus 4 more, which is 18*5 + 4 = 94 "periods" before it was killed.
- Each period emitted by our code represents 5,000 memory allocations - `malloc()'s` - , each is an allocation of 2 pages (8 KB)); so, we have allocated a total of:

94 * 5000 * 8192 ~= 3,672 MB ~= 3.5 GB of virtual memory!

...and then we (quite obviously) hit the second condition just mentioned, we completely ran out of all memory on the system, thus forcing the kernel's hand – the OOM killer stepped in and terminated us. The kernel log will indeed reveal this to be the case!

```
$ dmesg | tail -n1
[ ...] Out of memory: Killed process 874 (oom_killer_try) total-vm:3725008kB,
anon-rss:1779856kB, file-rss:0kB, shmem-rss:0kB, UID:1000 pgtables:7320kB oom_
score_adj:0
```

We shall interpret the kernel log details when OOM occurs in more detail in an upcoming section.

Intercepting the OOM killer via systemd-oomd

Sometimes, depending on how the system's configured, you may *not* get a kernel-level OOM killer invocation, though you expected to! How come? Well, it's the usual culprit – the powerful (and invasive) user-space `systemd` framework in action again!

Systemd includes a component named `systemd-oomd`, a user-space daemon process. If so configured, its job is to monitor processes going overboard on memory usage. It does this via the really powerful cgroups (v2) kernel infrastructure (we will cover this in some detail in *Chapter 11, The CPU Scheduler–Part 2*). Also, the relevant cgroup(s) need to have memory accounting enabled and the kernel needs to support the somewhat new-ish **Pressure Stall Information** (**PSI**) feature. See the man page on `systemd-oomd.service(8)` for more details.

Very briefly, **PSI** is a kernel feature (4.20 and later) that allows insight into situations caused by severe resource (CPU, memory, and I/O) crunches. As the official kernel docs on PSI states: "Having an accurate measure of productivity losses caused by resource scarcity aids users in sizing workloads to hardware--or provisioning hardware according to workload demand." The kernel documentation on PSI resides here: `https://docs.kernel.org/accounting/psi.html`.

You can think of `systemd-oomd` as a kind of "memory monitoring agent" between user-space apps (processes) and the Linux kernel (a user-space OOM killer!), thus allowing it to kill processes that exceed their memory usage limits. **This can be seen as an advantage in some situations as it avoids sending the kernel into a possible OOM kill situation.** As can be expected, its usage again is controversial; Ubuntu 22.04 does seem to ship with it enabled by default. Other distros (like Arch Linux) disable it by default. PSI made its way into the 4.20 kernel; this effectively implies the same for the systemd-based OOM killer. (For more on this, see the *Further reading* section of this chapter.)

As an example (it's in fact from *Chapter 11, The CPU Scheduler – Part 2*), you can try out this systemd-based service unit, which deliberately causes memory to be stressed; we do so by running this deliberate "lowram" service – `ch11/cgroups/cpu_constrain/systemd_svcunit/svc3_primes_lowram.service` – under the aegis of systemd via our `setup_service` wrapper script (located in the same folder). Our service unit invokes the well-known `stress-ng` program like this:

```
stress-ng --oomable --pathological --vm 1 --vm-bytes 100% --vm-populate
--malloc-bytes 100G --malloc-touch --malloc 12
```

This – very deliberately – can cause mayhem with system memory by putting tremendous stress on it, so as to try and invoke the OOM killer, our dubious objective here, of course.

Somewhat ironically, something to note, though: triggering the OOM killer isn't always as straightforward (as our `oom_killer` app!); in order to invoke the OOM killer quicker (which is what we want here), it's advisable to:

- Disable swap first with `sudo swapoff -a` and then test
- Have the system *physically* allocate the memory page(s) by touching them (via a write or execute on them; we learned about this in the *A brief note on user-mode memory allocations and demand paging* section, and we shall cover more on this in the upcoming *Demand paging and OOM* section). (An aside: why physically allocate memory only on page *write* or *execute*? Why not on the *read* as well? The upcoming *The optimized (unmapped) read* section explains it; you can ignore it for now.)

See this link as well: `https://forum.linuxfoundation.org/discussion/858099/lab-13-1-invoking-the-oom-killer`. Again, *Chapter 11, The CPU Scheduler–Part 2*, has more details on cgroups and systemd.

Case 2: vm.overcommit_memory == 2 (VM overcommit turned off, OVERCOMMIT_NEVER) and vm.overcommit_ratio == 50

Firstly, remember, this is *not* the default.

- `vm.overcommit_memory` is now set to 2, i.e., VM overcommit memory *is off*; aka "strict accounting" mode (Setting it's easy: `sudo sh -c "echo 2 > /proc/sys/vm/overcommit_memory"`.)
- `vm.overcommit_ratio` is 50, effectively a *percentage* value (the default).

With the `vm.overcommit_memory` tunable set to 2, there is a formula to calculate the total (possibly overcommitted) available/allowed memory, as follows:

Total allowed memory = (total_RAM - total_huge_TLB) * (overcommit_ratio / 100) + total_swap

Again, note that this formula only applies when `vm.overcommit_memory == 2`.

So, on our AArch64 Raspberry Pi 4, with `vm.overcommit_memory == 2`, with 1,849 MB of RAM, no huge pages, the `vm.overcommit_ratio` percentage as 50, and 100 MB of swap, the above formula yields the following (expressed in units of megabytes, MB):

```
Total allowed memory (MB) = (1849 - 0) * (50/100) + 100 = 924.5 + 100 = 1024.5
MB = 1,049,088 KB ~= 1 GB.
```

This figure also ties in almost perfectly with the value of the commit limit revealed via proc (it is, after all, derived via the same formula):

```
$ grep CommitLimit /proc/meminfo
CommitLimit:     1049192 kB
```

Right, let's run the app like the last time (with an enormous number of two-page allocations attempted – 500,000 – that aren't ever freed) and see what happens this time:

```
# ./oom_killer_try 500000 0
./oom_killer_try: PID 929 (verbose mode: off)
..... ..... ..... ..... ../oom_killer_try: loop #103137: malloc failure.
```

Did you notice!? This time it didn't die due to the OOM killer. No, it exited due to the `malloc()` API (on loop iteration 103,137) failing! Strict accounting in action!

This time, there are a total of only 21 periods in the process' output, implying that a total of:

```
21 * 5000 * 8192 = 860,160,000 bytes ~= 820 MB
```

of memory was allocated before the `malloc()` request failed. This is in line with, and (kind of) close to, the expected theoretical maximum of 1 GB that we calculated and that the `CommitLimit` is set to.

But why is it less? Why did the process get killed at 820 MB and not at 1 GB? Ah, the actual code allows for some slippage (reserving some memory for root-owned processes and keeping some memory in reserve in any case). Indeed, this time the kernel log reveals the message:

```
__vm_enough_memory: pid: 929, comm: oom_killer_try, no enough memory for the
allocation
```

...which of course is emitted from the `mm/util.c:__vm_enough_memory()` function we saw earlier (in the *VM overcommit from the viewpoint of the __vm_enough_memory() code* section).

Right. Moving along, as we know, the VM memory overcommit is, after all, tunable (by root), including the ability to switch it off (by setting `vm.overcommit_memory` to the value 2). At first glance, it may appear tempting to do so, turning off the kernel's "memory overcommit" feature. Do pause, though, and think it through; leaving the VM overcommit at the kernel defaults is best on most workloads.

For example, setting the `vm.overcommit_memory` value to 2 on my x86_64 Ubuntu guest VM (with 2 GB RAM) caused the effective available memory to change to just 2 GB! The typical memory usage, especially with the GUI running, far exceeded this, causing the system to be unable to even log in the user in GUI mode!

This is an important point to ponder; the following links help throw more light on the subject:

- Linux kernel documentation: `https://www.kernel.org/doc/html/v6.1/mm/overcommit-accounting.html`.
- *What are the disadvantages of disabling memory overcommit in Linux?*: `https://www.quora.com/What-are-the-disadvantages-of-disabling-memory-overcommit-in-Linux`.
- Chris Siebenmann's blog articles on Linux memory overcommit (and Linux in general) are very good and deal with real-world production servers: `https://utcc.utoronto.ca/~cks/space/blog/__IndexChron` (search for overcommit here!).

Also, do see the *Further reading* section of this chapter for more useful links.

Real-world experience shows that there's no single hard-and-fast rule regarding the best approach to determine which overcommit mode to employ in production: take the time to run your test suite under intense (memory) load and see what works best for your project.

We're not done with the OOM killer or even our crazy allocator program yet! The next section is very interesting; be sure to study it!

Demand paging and OOM

Recall the really important fact we learned earlier in this chapter, in the *A brief note on user-mode memory allocations and demand paging* section: because of the demand paging (or lazy allocation) policy that the OS uses, when a user-space memory page is allocated by `malloc()` (and friends), it only actually causes virtual memory space to be reserved in a region of the process user VAS; *no physical memory is allocated at this time.*

Only when you perform some action on any byte(s) of the virtual page, "touching" it in any manner – performing a read, write, or execute – does the MMU raise a page fault (internally accounted as a *minor fault or 'good' fault)* and the OS's page fault handler runs as a result. If it deems that this memory access is legal, it has the kernel's page allocator machinery a single physical frame! (which of course is mapped to the process).

In our simple `oom_killer_try` app, we manipulate this very idea via its third parameter, `force_page_fault`: when set to `1`, we emulate precisely this situation – forcing a page fault – by writing something, anything really, into a byte – any byte – of each of the two pages allocated per loop iteration (peek at the code again if you need to: `https://github.com/PacktPublishing/Linux-Kernel-Programming_2E/blob/main/ch9/oom_killer_try/oom_killer_try.c`).

So, now that you know this, let's re-run our shy and retiring app (I am very tempted to put in a smiley here!) with the third parameter, `force_page_fault`, set to `1`, to force page faults!

Okay, before running our app, let's check out the state of system memory on our trusty AArch64 Raspberry Pi 4 machine:

```
$ free -h
               total        used        free      shared  buff/cache
available
Mem:           1.8Gi        74Mi        27Mi       1.0Mi     1.7Gi          1.7Gi
Swap:           99Mi          0B        99Mi
```

Notice how, as the system has been performing a lot of file I/O (file reads and writes), a very significant percentage of RAM (1.7GB of the 1.8GB available!) is taken up by the page cache (the column labeled *buff/cache*; it's mostly *page cache* data).

You'll understand by now that this is fine and normal; as memory pressure increases, the kernel's page reclamation machinery swings into action, freeing up caches as required. The last column, *available*, hints at precisely this fact (the *available* column has wrapped around; its value is 1.7Gi, showing us that the majority of RAM's actually available when really required).

With the VM overcommit memory setting on the default (`0; OVERCOMMIT_GUESS`), let's run our app again, this time with the `force_page_fault` parameter set to 1, and, as usual, trying to allocate way too much memory:

```
$ cat /proc/sys/vm/overcommit_memory
0
$ ./oom_killer_try 500000 1
./oom_killer_try: PID 812 (verbose mode: off)
..... ..... ..... ..... ..... ..... ..... ..... ..... Killed
$
```

As expected, the OOM killer is triggered, jumps in, and kills our app. The kernel log reveals this clearly:

```
$ dmesg |tail -n1
[  750.748864] Out of memory: Killed process 812 (oom_killer_try) total-
vm:1791976kB, anon-rss:1711176kB, file-rss:4kB, shmem-rss:0kB, UID:1000
pgtables:3536kB oom_score_adj:0
$
```

This time (too), you can feel the system struggle as it fights for memory. This time, it runs out of memory much sooner (than we saw in the *Case 1 :: vm.overcommit_memory == 0 (the default, OVERCOMMIT_GUESS)* section) as actual physical memory was allocated.

From the preceding output, we see in this particular case:

- Exactly 9 x 5 = 45 periods, with each period representing 5,000 malloc()s of 8 KB each.
- Thus, the amount of *physical* memory allocated by the process (before it was ignominiously killed) was 45 * 5,000 * 8192 ~= 1,758 MB.
- We say *physical memory* because we passed the *force_page_fault parameter* as 1, causing – forcing – demand paging to kick in!

Apparently, at this point, one of two things occurred:

- The system ran completely out of memory – CPU caches, RAM, and swap – thus failing to allocate a page and inviting the OOM killer in.
- The calculated (artificial) kernel VM commit limit was exceeded.

We can easily look up this kernel VM commit value (again on the Raspberry Pi 4 where I ran this):

```
$ grep CommitLimit /proc/meminfo
CommitLimit:       1049192 kB
```

This works out to just over (almost exactly) a gigabyte (well under our calculation of 1,758 MB). So here, it's very likely that all the system memory ran out (as our dear app ate it all up!) and the first case came into play. Also notice how, before running our program, the amount of memory used by the larger system caches (the page and buffer caches) is significant (this does, of course, depend on the amount of file IO exercised; a system with higher uptime typically uses a significant percentage of RAM for the page cache).

The column entitled `buff/cache` in the output of the `free` utility shows this. Before running our crazy allocator app, recall that 1.7 GB of the 1.8 GB of RAM available was being used for the page cache. Once our crazy allocator program runs, though, it applies so much memory pressure that the OS's reclamation machinery jumps into action, quickly reclaiming memory frames from all caches. Inevitably (as we refused to free any memory), the OOM killer jumps in and kills us, causing large amounts of memory to be reclaimed (freed up). Now let's again run the `free` program:

```
$ free -h
                total        used         free        shared    buff/cache
available
```

```
Mem:          1.8Gi       93Mi       1.6Gi     0.0Ki     98Mi       1.7Gi
Swap:          99Mi       58Mi        41Mi
$
```

Wow, look at that! The free memory and the cache usage right after the OOM killer cleans up are 1.6 GB and 98 MB, respectively, as expected. The cache usage right now is very low; it will increase as the system runs.)

Next, looking up the kernel log reveals that indeed, the OOM killer has paid us a visit! Note that the following partial screenshot of the kernel log following the OOM killer shows only the kernel-mode stack dump - of our oom_killer_try process context of course - on the AArch64 Raspberry Pi 4 running the stock 6.1 kernel:

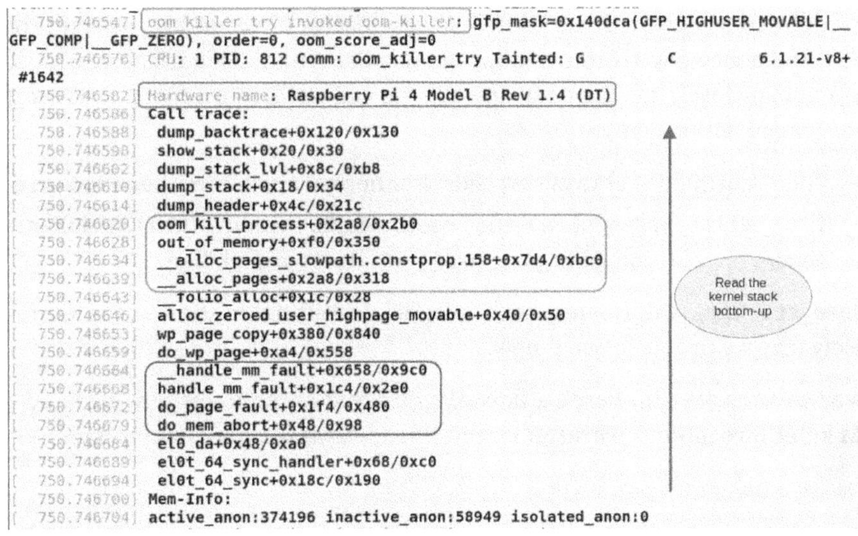

Figure 9.8: The kernel log after the OOM killer, showing the kernel call stack (of our oom_killer_try process context)

Examine the kernel-mode stack in *Figure 9.8* carefully, reading it in a bottom-up fashion (ignoring any frames that start with ?); clearly, the sequence of events reveals itself:

- A *page fault* occurred; you can see the call frames: `do_mem_abort()` (on ARM[64], one type of page fault – like this one – is called the *Data Abort* exception, hence the function name `do_mem_abort()`) `-->` `do_page_fault()` `-->` `handle_mm_fault()` `-->` [...].
    - Why did this page fault occur? We've discussed this: when a fresh page is allocated by our process (via `malloc()`), only the virtual page exists; when accessed in any manner (here we wrote into it), the virtual address is sent via the CPU to the MMU for translation. When the MMU fails to translate the virtual address to a physical address (as the physical page frame doesn't exist yet), it raises a *Data Abort exception* on the ARM[64], a page fault. (By the way, the `el0_*()` call frames at the bottom indicate that the fault was taken at Exception Level 0 (EL0) – user mode on AArch64 processors.)

- The OS runs the fault-handling code (in the process context of the faulting process, i.e., our `oom_killer_try` process; in effect, `current` runs the fault-handling code (and all that follows...)).
- The fault-handling code proceeds to figure out what actually occurred (it has a big complex flow); here, it figures that we have a minor fault, a *good fault*. It thus begins to invoke page allocator code to allocate a page frame!
- This ultimately becomes the code deep inside the page allocator: `__alloc_pages()`! It's literally the heart of the zoned buddy system (or page) allocator – the engine of memory allocation within the kernel, as we've seen (refer back to *Figure 9.4*). It's key to understand that all these steps, up to now, are completely normal, expected behavior. ***It's how physical memory is allocated to user-space tasks, after all!*** It's demand-paging in action.
- But (there's always a but, right?) *now*, because of the intense memory pressure cooker situation we deliberately created, the kernel's page allocator (the `__alloc_pages()` routine) simply couldn't find a free page frame to allocate; **it fails**. What!? This triggers, who else, the OOM killer! You can clearly see the code flow via the call stack (*Figure 9.8*): `__alloc_pages() --> __alloc_pages_slowpath()`.
    - This – normally – always succeeds! But here, we've simply been too greedy...
    - The `__alloc_pages_slowpath()` routine fails, invoking the OOM killer code: `out_of_memory() --> oom_kill_process()`! It's how we got killed.
- The `dump_stack()` ... call frames are the diagnostic routines to show us the kernel-mode stack; precisely why we're able to see it now...

To make this as clear as we can, here's a flow diagram that accentuates how demand paging – and possible OOM killer invocation – works. It is simplified; we've skipped many details.

Nevertheless, it does show how this key process works:

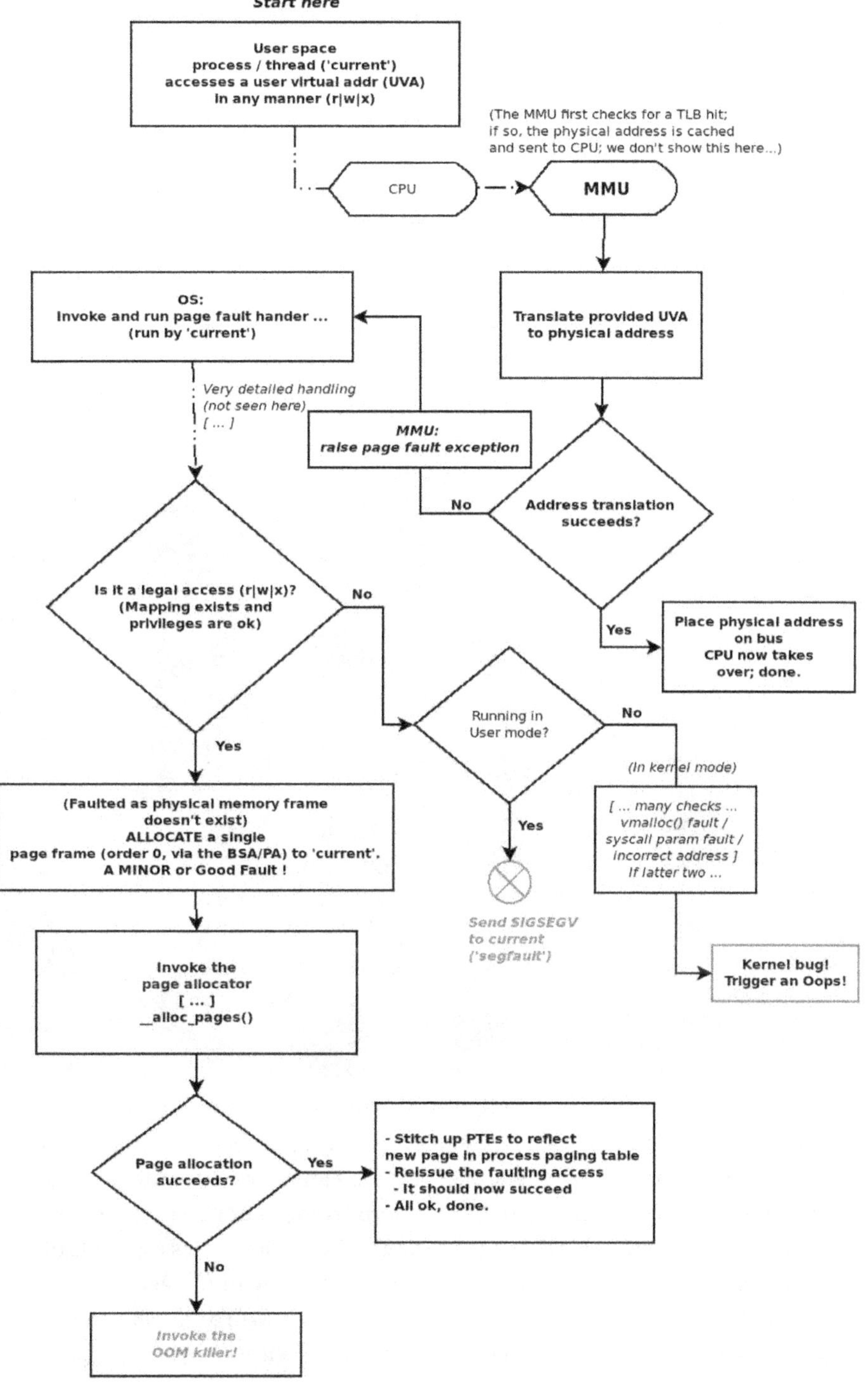

Figure 9.9: A (simplified) view on how demand paging – and possible OOM killer invocation – works

So, there we are. At this point, it might be interesting, perhaps, to look back at *Chapter 7*, *Memory-Management Internals – Essentials, Figure 7.3* and *Figure 7.4*, to see the continuity with *Figure 9.9* here, to see, more, how address translation works and could – though rarely – end up in an OOM killer or kernel bug (an Oops!) situation.

Back to the kernel log with regard to the OOM killer's fatal attack; toward the latter part of its diagnostic dump, the kernel tries hard to justify its reason for killing a given process(es). It shows statistics regarding the system memory state, including that of the page allocator per-node, per-zone freelists, and a table of all threads alive at the time of the OOM kill and their memory usage (and various other statistics). These statistics are displayed due to the `sysctl /proc/sys/vm/oom_dump_tasks` being on (1) by default. Here's a sampling (in the following output, we have eliminated the leftmost timestamp column of `dmesg` to make the data more readable):

```
$ sudo dmesg
[ ... ]
Tasks state (memory values in pages):
[  pid  ]  uid tgid total_vm rss pgtables_bytes swapents oom_score_adj name
[    144] 0    144  12280     1   90112          231      -250          systemd-
journal
[    177] 0    177  5414      1   53248          621      -1000         systemd-
udevd
[    299] 103 299  22027     0   69632          194       0            systemd-
timesyn
-- snip --
[    728] 1000 728 2024      1   49152          420       0            bash
[    812] 1000 812 447994 427795 3620864        19725     0            oom_
killer_try
oom-kill:constraint=CONSTRAINT_NONE,nodemask=(null),cpuset=/,mems_
allowed=0,global_oom,task_memcg=/,task=oom_killer_try,pid=812,uid=1000
Out of memory: Killed process 812 (oom_killer_try) total-vm:1791976kB, anon-
rss:1711176kB, file-rss:4kB, shmem-rss:0kB, UID:1000 pgtables:3536kB oom_score_
adj:0
$
```

In the preceding output block, we have highlighted in bold the `rss` (*resident set size*) column (the fifth column) as it's a good indication of physical memory usage by the process in question (careful, the unit of *rss* is *pages*). Clearly, our `oom_killer_try` process is using an enormous amount of physical memory (here, it happens to be over 1,671 MB!). Further, it's important to notice that it's not just application (user-space) memory that plays a role in determining the OOM victim; it's all kinds of memory usage that can cumulatively sum up to the maximum memory usage by the hogger process. Here, for our oom_killer_try process, notice how the `pgtables_bytes` (over 3 MB) and `swapents` (number of swap entries) column values are very high as well! (Recall, on x86_64 with 4-level paging, each mapped page costs us 16 KB in paging table metadata! It all adds up.)

Modern kernels (4.6 onward) at times use a specialized `oom_reaper` kernel thread to perform the work of reaping (killing; it's the "grim reaper"!) the victim process (though not the case here; at such times, the last line of the preceding output will show that this kernel thread "reaped" the victim process). Interestingly, the Linux kernel's OOM can be also thought of as a (last) defense against fork bombs and similar **(Distributed) Denial of Service** ((D)DoS) attacks.

The optimized (unmapped) read

Well, there's always more to understand: as usual, the reality is nuanced... Earlier, when beginning the discussion on demand paging and OOM, we said, and I quote: *Only when you perform some action on any byte(s) of the virtual page, "touching" it in any manner – performing a read, write, or execute – does the MMU raise a page fault (internally accounted as a minor fault) and the OS's page fault handler runs as a result.*

This is mostly correct; it's just that **reads on pages that don't exist yet, on unmapped memory**, are treated differently. How? The kernel is clever: when a virtual unmapped memory page is read from – imagine that instead of writing to memory in our `oom_killer_try` app, we just read instead – we know that it must get allocated to a clean zero-filled physical memory page frame. This approach, however, would have the kernel service every page read by allocating a page frame that must be initialized to zeroes; that's expensive! the kernel has a far superior performance-benefiting approach instead. It merely *maps* every new virtual page that's freshly read from to an underlying *single* kernel frame that's populated with zeroes! This way, even if you perform ten thousand (or ten million) reads on ten thousand (or ten million) virtual pages, the kernel ends up physically mapping just one page frame to all of them.

In effect, the kernel does not have to service every (minor) page fault on new pages by allocating a frame – no, just the ones that are targets of a *write or execute*! A detail, perhaps, but a very interesting one, implying that one can allocate enormous amounts of virtual memory if it's sparse (literally meaning, empty) data; there are a few real-world apps that indeed do this (webkit, Address Sanitizer, and KASAN). (Again, *offlinemark* does a wonderful job of detailing this in this blog article: *What they don't tell you about demand paging in school*, Oct 2020: `https://offlinemark.com/2020/10/14/demand-paging/`. Be sure to check it out.)

Understanding the OOM score

In order to speed up the discovery of who the top memory-hogging process is at crunch time (when the OOM killer is invoked), the kernel assigns and maintains an *OOM score* on a per-process basis (you can always look up the value in the `/proc/<pid>/oom_score` pseudofile).

Moreover, the OOM score has a range; it's from `0` to `1000`:

- An OOM score of `0` implies that the process is not using any memory available to it.
- An OOM score of `1000` implies the process is using 100 percent of the memory available to it.

Obviously, the process with the highest OOM score "wins." Its reward – it's selected as the victim to be killed by the OOM killer (talk about dry humor). Not so fast, though: the kernel has heuristics to protect important tasks. The baked-in heuristics imply that the OOM killer will not select as its victim any root-owned process, any kernel thread, or any task that has a hardware device open.

What if we would like to ensure that a certain process will *never be killed* by the OOM killer? It's quite possible to do so, though it does require root access. The kernel provides a tunable, `/proc/<pid>/oom_score_adj`, an OOM "adjustment" value (with the default being `0`). The **net OOM score** is the sum of the `oom_score` value and the adjustment value:

```
net_oom_score = oom_score + oom_score_adj
```

Thus, setting the `oom_score_adj` value of a process to `1000` pretty much guarantees that it will be killed, whereas setting it to `-1000` has exactly the opposite effect – it will never be selected as a victim.

A quick way to query (and even set) a process's OOM score, as well as its OOM adjustment value, is via the `choom(1)` utility. For example, to query the OOM score and OOM adjustment value of the `systemd` process, just do `choom -p 1`. We did the obvious thing – wrote a script, a simple wrapper over `choom` – to query the OOM score of all processes currently alive on the system (it can be found here: `ch9/query_process_oom.sh`; do try it out on your box). *Quick tip*: The (10) processes with the highest OOM score on the system can quickly be seen with this command line (the third column is the net OOM score):

```
./query_process_oom.sh | sort -k3n | tail
```

Closing thoughts on the OOM killer and cgroups

So, now that we've covered the OOM killer, what do you think? A rational response, I'd venture, would be: *don't let the situation that triggers the OOM killer occur in the first place!* Perhaps easier said than done, granted, but care should be taken; it is important to think this way (don't throw the baby out with the bath water, right?).

As Andries Brouwer put it, in a tongue-in-cheek remark (back in Sept 2004: `https://lwn.net/Articles/104185/`), the OOM killer algorithm is perhaps akin to chucking passengers out of an aircraft when it runs too low on fuel.

Then again, some production environments differ in individual server system setup; no one rule may be best. On some setups, having the OOM killer invoked and killing off the misbehaving app is preferable to having innocent smaller apps fail because they can't get memory. It's the usual thing: *there is no silver bullet*. Test and arrive at the best approach for your particular setup.

Cgroups and memory bandwidth – a note

Something we've mentioned quite often is **control groups (cgroups)**; worry not, quite some detail on this key, powerful, and implicit kernel infrastructure is covered in *Chapter 11, The CPU Scheduler – Part 2*, in the section entitled *An introduction to cgroups* (it's more than an intro).

Cgroups (now in their v2 implementation) allow one to apply resource constraints as required, whether the resource is the CPU, memory (RAM), IO, maximum number of PIDs (processes/threads) allowed, and so on, via so-called resource controllers. The *memory* resource controller is obviously used to apply constraints to a group of processes within a cgroup on their combined memory usage. To understand it, and make use of and thus leverage it, I suggest you do two things:

1. Read the content on cgroups in *Chapter 11, The CPU Scheduler – Part 2*, in the section entitled *An introduction to cgroups*.
2. In parallel, refer to the official kernel documentation on it, particularly for the memory controller; it's available here: `https://docs.kernel.org/admin-guide/cgroup-v2.html#memory-interface-files`. You'll see all the possible memory controller interface files and what exactly they mean.

   Further, the brief section entitled *Usage Guidelines* (`https://docs.kernel.org/admin-guide/cgroup-v2.html#usage-guidelines`) is useful to read.

Next, as our later coverage on cgroups (v2) is a bit biased toward the topic of constraining CPU bandwidth, we'll quickly mention a few points regarding the memory controller here:

- Constraining the memory usage of processes within a cgroup is not exactly the same as that of constraining CPU usage. Think about this: CPU time is indeed infinite in the real sense of the term; memory simply isn't. Thus, we can set different types of limits on memory usage (via the cgroup's `memory.min|low|high|max` pseudofiles; refer to the kernel documentation link just mentioned for details).
- The 4.19 kernel introduced a tiny implementation of a cgroup-aware OOM killer component; this allows – to quote – "an ability to kill a cgroup as a single unit and so guarantee the integrity of the workload." The interface is the `memory.oom.group` pseudofile; when set and a process within a cgroup triggers the OOM killer, it kills all the processes within that cgroup (or none at all).
- Internals: The kernel incorporates a slab memory controller (*kmemcg*), which essentially replicated slab memory internals for every existing memory cgroup (`memcg`). This led to major under-utilization of the slab memory from within these memcgs (as they didn't share slab memory; each had its own slabs), which in turn led to serious issues – significant slab memory wastage, in effect. This led to the creation of a new slab memory controller (by Roman Gushchin and others) from 5.9 Linux. The slab memory savings range from 35% to 50% (on both desktop and server-class systems).
- A perhaps easier way to control cgroups memory bandwidth allocation is to do so via the interfaces made available by ***systemd***. Similar to the kernel memory cgroup (*memcg*) interfaces, it provides *MemoryMin, MemoryLow, MemoryHigh,* and *MemoryMax* interfaces for the processes within a given systemd unit (all set to a quantity, expressed in bytes/KB/MB/GB/TB units). Alternatively, you can specify memory as a percentage of total RAM for them. Even swap memory usage can be constrained via the *MemorySwapMax* and *MemoryZSwapMax* interfaces.

  The *MemoryHigh* interface is typically the one recommended to be specified; to quote the documentation on its usage (from the man page here: `https://man7.org/linux/man-pages/man5/systemd.resource-control.5.html` "Specify the throttling limit on memory usage of the executed processes in this unit. Memory usage may go above the limit if unavoidable, but the processes are heavily slowed down and memory is taken away aggressively in such cases. This is the main mechanism to control memory usage of a unit."

The *MemoryMax* interface allows specifying a non-negotiable upper (or absolute) limit on memory usage by the unit's processes; if memory usage crosses this threshold, the OOM killer is invoked.

There are a few caveats to setting up the systemd-based memcg settings; be sure to carefully read the documentation.

With this, we conclude this section and indeed this chapter.

Summary

In this chapter, we continued where we left off in the previous chapter. We covered, in a good amount of detail, how you can create and use your own custom slab caches (useful when your driver or module very frequently allocates and frees a certain data structure). We then provided an overview of available kernel debug techniques for debugging memory issues. Next, we learned about and used the kernel `vmalloc()` API (and friends). With the wealth of memory APIs available, how do you select which one to use in a given situation? We covered this important concern with a useful *decision chart* and table. We then delved into an understanding of the kernel page reclaim procedures; this discussion covered the zone watermarks, the kswapd kernel thread(s), the new MG-LRU lists, and the DAMON data access monitoring technology.

We then went into what exactly the kernel's dreaded *OOM killer* (and the `systemd-oomd` daemon) component is and how to work with it.

As I have mentioned before, sufficiently deep knowledge of Linux memory management internals and exported API set will go a long way in helping you as a kernel module and/or device driver author. The reality, as we well know, is that a significant amount of time is spent by developers on troubleshooting and debugging code; the intricate knowledge and skills gained in these last few chapters on kernel internals and memory management will help you better navigate these mazes.

This completes the explicit coverage of Linux kernel memory management in this book. Though we have covered many areas, we have also left out or only skimmed over some of them. The fact is that Linux memory management is a huge and complex topic, well worth understanding for the purposes of learning, writing more efficient code, and debugging complex situations. Learning the (basic) usage of the powerful `crash` utility (used to look deep within the kernel, via either a live session or a kernel dumpfile), and then re-looking at this and the previous chapter's content armed with this knowledge, is indeed a powerful way to learn!

So, great job on having covered Linux memory management! The next two chapters will have you learning about another core OS topic – how *CPU (task) scheduling* is performed on the Linux OS. Take a breather, work on the following assignments and questions, and browse through the *Further reading* materials that capture your interest. Then, revitalized, jump into the next exciting area with me!

Questions

As we conclude, here is a list of questions for you to test your knowledge regarding this chapter's material: `https://github.com/PacktPublishing/Linux-Kernel-Programming_2E/blob/main/questions/ch9_qs_assignments.txt`. You will find some of the questions answered in the book's GitHub repo: `https://github.com/PacktPublishing/Linux-Kernel-Programming_2E/tree/master/solutions_to_assgn`.

Further reading

To help you delve deeper into the subject with useful materials, we provide a rather detailed list of online references and links (and, at times, even books) in a *Further reading* document in this book's GitHub repository. The *Further reading* document is available here: `https://github.com/PacktPublishing/Linux-Kernel-Programming/blob/master/Further_Reading.md`.

Learn more on Discord

To join the Discord community for this book – where you can share feedback, ask questions to the author, and learn about new releases – follow the QR code below:

`https://packt.link/SecNet`

10

The CPU Scheduler – Part 1

The previous two chapters delved pretty deep into aspects of Linux memory management, with a focus on how exactly you, as a kernel/driver developer, can efficiently dynamically allocate and deallocate kernel memory (besides the APIs, we covered interesting stuff like MGLRU, DAMON, and the OOM killer!).

In this chapter and the next, you will dive into the details regarding a key OS topic – that is, CPU (or task) scheduling on the Linux OS. I will try and keep the learning both conceptual and hands-on, by asking (and answering) typical questions and performing common tasks related to scheduling. Understanding how CPU scheduling works at the level of the OS is not only important from a kernel (and driver) developer viewpoint, it will also automatically make you a better system architect, even for user space applications.

We shall begin by covering essential background material; this will include the notion of the **Kernel Schedulable Entity** (**KSE**) on Linux, as well as the POSIX scheduling policies that Linux implements. We will then move on to using tools – `perf` and others – to visualize the flow of control as the OS runs tasks on CPUs and switches between them. This can prove very useful when you need to profile apps as well!

After that, we will dive deeper into the details of how exactly CPU scheduling works on Linux (within kernel space), covering modular scheduling classes, including **Completely Fair Scheduling** (**CFS**), the running of the core schedule function, the "who" and "when" of it, and so on. Along the way, we will also cover how you can programmatically and dynamically query the scheduling policy and priority of any thread on the system.

Something to point out at the onset: as you know, the CPUs on a system are a resource shared among the users, processes, and threads that execute upon them. The CPU (or task) scheduler is of course a way to arbitrate access to this precious resource. Still, CPUs are but one resource; memory, IO devices, networking, and more are also shared resources. The Linux kernel has an advanced, deeply ingrained framework for a more highly abstracted, practically important resource-sharing feature: **cgroups (control groups)**! (In fact, our very first diagram, *Figure 10.1*, alludes to it!). A detailed internal understanding of CPU scheduling must also include the key role that cgroups can play as well; worry not – we shall cover some key details on cgroups in the following chapter.

In this chapter, we will cover the following areas:

- Learning about the CPU scheduling internals – part 1 – essential background
- Visualizing the flow
- Learning about the CPU scheduling internals – part 2
- Querying a given thread's scheduling policy and priority
- Learning about the CPU scheduling internals – part 3

Now, let's get started with this interesting topic!

Technical requirements

I assume that you have gone through *Online Chapter, Kernel Workspace Setup,* and have appropriately prepared a guest **Virtual Machine** (**VM**) running Ubuntu 22.04 LTS (or a later stable release) and installed all the required packages. If not, I highly recommend you do this first.

To get the most out of this book, I strongly recommend you first set up the workspace environment, including cloning this book's GitHub repository for the code and working on it in a hands-on fashion. The repository can be found here: `https://github.com/PacktPublishing/Linux-Kernel-Programming_2E`.

Learning about the CPU scheduling internals – part 1 – essential background

Let's take a quick look at the essential background information we require to understand CPU scheduling on Linux.

Note that, in this book, we do not intend to cover material that competent (user space) system programmers on Linux should already be well aware of; this includes basics such as process (or thread) states, basic information on what real time is, the POSIX scheduling policies, and so on. This (and more) has been covered in some detail in my earlier book, *Hands-On System Programming with Linux,* published by Packt in October 2018. Nevertheless, we do touch upon some of the basics here.

What is the KSE on Linux?

As you learned in *Chapter 6, Kernel Internals Essentials – Processes and Threads,* in the *Organizing processes, threads, and their stacks – user and kernel space* section, every (user mode) *thread* alive on the system is bestowed with a task structure (`struct task_struct`) and both a user-mode and a kernel-mode stack. As we've learned, kernel threads do have a task structure but only a kernel-mode stack.

Every OS needs to perform task scheduling of course. Modern operating systems – Linux/Unix/Windows/Mac – have the notion of processes and threads. Now, the key question to ask is: when task scheduling is performed, *what "object" does it act upon?* In other words, what is the **Kernel Schedulable Entity**, the **KSE**? On Linux, **the KSE is a thread**, not a process (of course, every process contains a minimum of one thread). Thus, *the thread is the granularity level at which scheduling is performed.*

An example – along with a diagram (*Figure 10.1*) – will help explain this: if we have a hypothetical situation where we have just one CPU core and three user space processes alive (P1, P2, and P3), consisting of one, two and five threads each respectively, plus a few kernel threads (say, three of them), then we have a total of (1+2+5+3), which equals 11 threads.

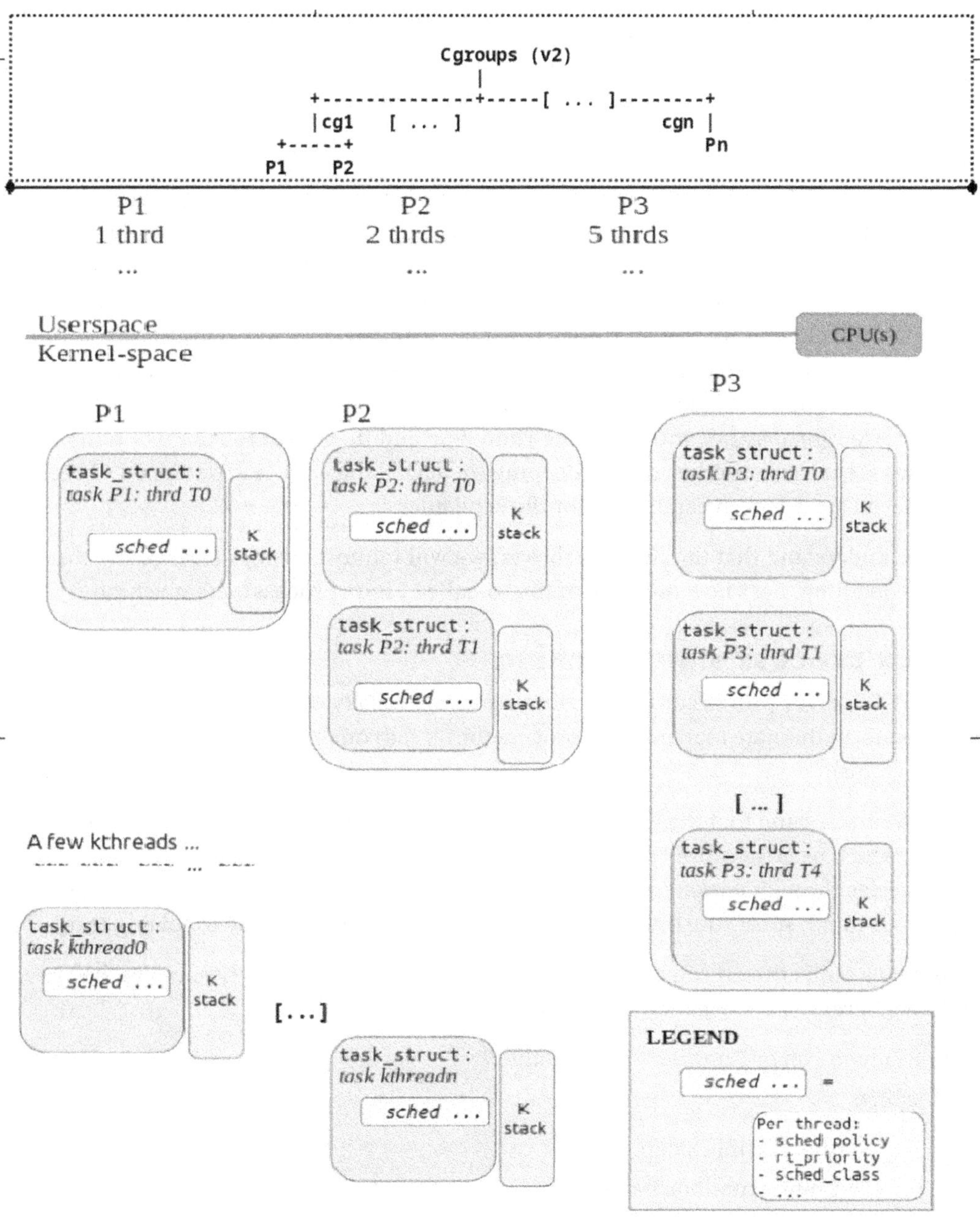

Figure 10.1: A conceptual diagram showing the kernel layout of a few processes and their threads, stacks, and kthreads within the kernel (for now, ignore the "cgroups" abstraction at the very top; we will get to it in the following chapter)

Okay, I'm asking you to ignore the elephant in the room. The very top portion of *Figure 10.1* shows an abstracted, highly conceptual view of the cgroups kernel framework, a framework that is now deeply embedded inside the very fabric of the kernel. Here, what I'm asking you to visualize - without worrying at all about the "why/how/who/when/..." questions - is that this cgroups layer, if you will, is pretty much ever-present in modern Linux, and it certainly can have an impact on CPU scheduling. For now, just be aware it exists; details - and trying it out! - are covered in the following chapter.

With respect to *Figure 10.1*, each thread - except for the kernel threads - has a user and kernel stack and a task structure (the kernel threads only have kernel stacks and task structures; all of this was thoroughly explained in *Chapter 6, Kernel Internals Essentials - Processes and Threads*, in the *Organizing processes, threads, and their stacks - user and kernel space* section). Now, if all these 11 threads are in a runnable state, that is, they're ready-to-run, then they all compete for the single processor core. (Although it's unlikely that they're all runnable simultaneously, let's just consider it for the sake of discussion). It's key to understand that we now have 11 *threads* in competition for the CPU resource, not three processes and three kernel threads. A more realistic scenario is, of the 11 threads alive, perhaps four are *runnable*, i.e., they're in the runqueue and they want to run, while the remaining seven threads are in various other states (sleeping (in a blocking call) or are stopped (frozen)); these remaining seven are not even candidates for the scheduler.

Now that we understand that the KSE is a thread, we will (almost) always refer to the thread in the context of scheduling. Let's now move on to the so-called Linux process state machine.

The Linux process state machine

On the Linux OS, every process or thread runs through a variety of definite states, and by encoding these, we can form the state machine of a process (or thread) on the Linux OS (do refer to *Figure 10.2* while reading this).

Since we now understand that the KSE on the Linux OS is a thread and not a process, we shall ignore convention, which uses the word "process," and instead use the word "thread" when describing the entity that cycles through various states of the state machine. (If it's more comfortable, you could always, in your mind, substitute the word 'process' in lieu of 'thread' in the following text.)

The states that a Linux thread can cycle through are as follows (the `ps` utility encodes the state via the letters shown here):

- R: Ready-to-run or Running (or "runnable")
- Sleeping:
    - S: Interruptible Sleep
    - D: Uninterruptible Sleep
- T: Stopped (or suspended/frozen)

- Z: Zombie (or defunct)
- X: Dead

When a thread is newly created either via the `fork()` or `clone()` system calls, or the `pthread_create()` API, and once the OS determines that the thread is fully born, it informs the scheduler of its existence by putting the thread into a runnable (R) state. A thread in the R state is either actually running on a CPU core or is in the ready-to-run state. What we need to understand is that in both cases, the thread is enqueued on a data structure within the OS called a **runqueue**. *Linux maintains one runqueue per CPU core on the system* (actually, the reality is even more nuanced; we will get to that soon enough). The threads in the runqueue are the valid candidates to run on a given CPU core; no thread can possibly run unless it is enqueued on an OS runqueue. Linux does not explicitly distinguish between the ready-to-run and actually running state; it merely marks the thread in either state as R. Of course, as we learned in *Chapter 6, Kernel Internals Essentials – Processes and Threads*, it's a member within the task structure (`unsigned int __state;`) that is set to an appropriate value to mark the thread being in a particular state.

The following diagram represents the Linux state machine for any process or thread:

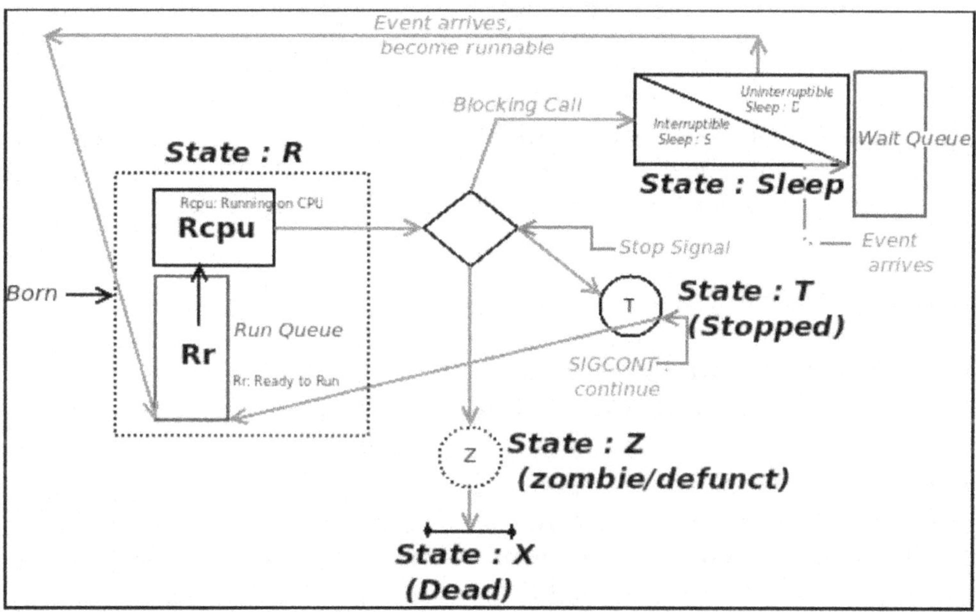

Figure 10.2: The Linux process/thread state machine

The preceding diagram shows transitions between states via (red) arrows. Do note that for clarity, some transitions (for example, a thread can be killed while asleep or stopped) are not explicitly shown in the preceding diagram. (FYI, a process that dies but isn't waited for by its parent ends up in a half-dead, half-alive state, termed a *zombie* or defunct process; it's meant to be a transient state. The rule to be followed to prevent zombies remaining on the system forever, is that every `fork()` requires a corresponding `wait*()` system call. (Linux, as usual, has a neat trick: killing the parent of a zombie(s), has the zombie(s) 'reaped' (die) as well.)

We know that every thread alive (both user and kernel) has its own task structure. Thus, at the level of the code, the task structure member named `__state` (it was earlier named `state`) holds the "state" of the task – the thread – in the so-called "state machine." When runnable, it will be set to the value `TASK_RUNNING` (represented by the "R" overall, and internally by the `Rr` and `Rcpu` states in *Figure 10.2*).

A **wait queue** is a data structure where tasks that are in a sleep state are enqueued – that is, they're waiting for an event (in effect, they land up here within a blocking call). Linux has two possibilities for a sleeping thread; the thread can be in an:

- *Interruptible sleep*: `__state = TASK_INTERRUPTIBLE`: it "sleeps," waiting on a certain event; however, any signal delivered to the process/thread will awaken it and run the signal handler (represented by the letter 'S' in *Figure 10.2*)
- *Uninterruptible sleep*: `__state = TASK_UNINTERRUPTIBLE`: it "sleeps," waiting on a certain event; any/all signal(s) delivered to the process/thread will have no effect on it (represented by the letter 'D' in *Figure 10.2*).

When the event it was waiting upon arises, the OS will issue it a wakeup call, causing it to become runnable (dequeued from its wait queue and enqueued in a runqueue) again. Note that the thread won't run instantaneously; it will become runnable (`Rr` in *Figure 10.2*) and a candidate for the scheduler; soon enough, it will get a chance and actually run on the CPU (`Rcpu`).

A common misconception is to think that the OS maintains one run queue and one wait queue. The Linux kernel maintains one run queue per CPU. Wait queues are often created and used by device drivers (as well as the kernel); thus, there can be any number of them.

With these basics out of the way, let's get going!

The POSIX scheduling policies

Think of the term **scheduling policy** as being roughly equivalent to **scheduling algorithm**. It's important to realize that the Linux kernel does not have just one policy that implements CPU scheduling; the fact is that the POSIX standard specifies a minimum of three scheduling policies (algorithms, in effect) that a POSIX-compliant OS must adhere to. Linux goes above and beyond, implementing these three as well as more, with a powerful design called *scheduling classes* (covered in the *Understanding modular scheduling classes* section later in this chapter).

For now, let's just briefly summarize the POSIX scheduling policies and what effect they have in the following table; before that though, it's key to understand that **every thread alive (both user and kernel space) is associated with exactly one of these scheduling policies** at any given point in time (it can be changed at runtime). Here's the table:

Scheduling policy	Key points	Priority scale
`SCHED_OTHER` or `SCHED_NORMAL`	Always the default; threads with this policy are non-real-time; they are internally implemented as a **Completely Fair Scheduling** (**CFS**) class (seen in the upcoming *The working of the Completely Fair Scheduling (CFS) class in brief* section). The motivation behind this scheduling policy is fairness (as in, "be fair to all runnable threads, and avoid starving anyone of CPU time") and overall throughput.	Threads with this policy have a real-time priority of `0`. The non-real-time priority is called the *nice value*; it ranges from -20 to +19 (a lower number implies superior priority) with a base (initial) value of 0
`SCHED_RR`	The motivation behind this scheduling policy is to provide a (soft) real-time policy that's moderately aggressive. Threads belonging to this `sched` class have a finite timeslice (typically defaulting to 100 ms). A `SCHED_RR` thread will yield the processor IFF (if and only if): • It blocks on I/O (goes to sleep). • It stops or dies. • A higher-priority real-time thread becomes runnable (which will preempt this one). • Its timeslice expires.	(Soft) real-time: 1 to 99 (a higher number implies superior priority)
`SCHED_FIFO`	The motivation behind this scheduling policy is to provide a (soft) real-time policy that's (by comparison to `SCHED_RR`, very) aggressive. A `SCHED_FIFO` thread will yield the processor IFF: • It blocks on I/O (goes to sleep). • It stops or dies. • A higher-priority real-time thread becomes runnable (which will preempt this one). It has, in effect, infinite timeslice.	(The same as `SCHED_RR`)
`SCHED_BATCH`	The motivation behind this scheduling policy is one that's suitable for low-priority non-interactive batch jobs, with less preemption.	Nice value range (-20 to +19)

SCHED_IDLE	Special case: typically, the PID `0` kernel thread (traditionally called the `swapper`; in reality, it's the *per CPU idle thread*) uses this policy. It's always guaranteed to be the lowest-priority thread on the system and only runs when no other thread wants the CPU.	The lowest priority of all (think of it as being of even lower priority than nice value +19)

Table 10.1 : The Linux (POSIX-compliant) scheduling policies and their meaning in brief

It's important to note that when we say real-time in the preceding table, we mean soft (or at best, firm) real-time and not hard real-time, as in a **Real-Time Operating System (RTOS)**. Linux is a **general-purpose OS (GPOS)**, not an RTOS. Having said that, you can convert vanilla Linux into a true hard real-time RTOS by applying an external patch series (called RTL, supported by the Linux Foundation); you'll learn how to do precisely this in the following chapter, in the *Converting mainline Linux into an RTOS* section.

Study the table carefully; notice that a `SCHED_FIFO` thread has, in effect, infinite timeslice and can thus run on a CPU core for as long as it wishes to; it gets preempted (taken off the CPU) when one of the preceding mentioned conditions comes true. Also, other key differences between `SCHED_FIFO` and `SCHED_RR` include:

- While a `SCHED_FIFO` thread effectively has infinite timeslice, a `SCHED_RR` thread has a finite timeslice (tunable via the `/proc/sys/kernel/sched_rr_timeslice_ms` sysctl; 100 ms by default)
- `SCHED_RR` threads at the same priority level are scheduled in a round robin fashion (thus allowing other `SCHED_RR` threads to get a share of the processor). With `SCHED_FIFO` threads, this isn't the case – the one preempted will go back to being the next task to run (thus denying others at the same priority level the processor). Thus, when employing `SCHED_FIFO`, you should avoid keeping threads at the same priority level (we cover more on this in the following chapter, in the *Setting the policy and priority within the kernel – on a kernel thread* section).

At this point, it's important to understand that on an OS such as Linux, the reality is that hardware (and software) **interrupts** are always superior to and will always preempt even (kernel or user space) `SCHED_FIFO` threads! (Do refer back to *Figure 6.1* in *Chapter 6, Kernel Internals Essentials – Processes and Threads*, to see this. If you'd like to delve into the topic of hardware interrupts in detail, please see the (free!) *Linux Kernel Programming, Part 2* book, *Chapter 4, Handling Hardware Interrupts*). For our discussion here, we will ignore interrupts for the time being. Now, let's look into per-thread priority values in more detail.

Thread priorities

The priority scaling for threads is simple (the following is from lower to higher priority; refer to *Figure 10.3*):

- Non-real-time threads (`SCHED_OTHER`) have a real-time priority of `0`; this ensures that they cannot even compete with real-time threads; they're not even on the same playing field! Then, how would one distinguish, priority-wise, between all the non-real-time threads?

Easy – they use an (old UNIX-style) priority value called the **nice value**, which ranges from -20 to +19, with -20 being the highest priority, 0 the base or default, and +19 the lowest. (Appropriate interfaces on both the command line and APIs are present to query and set them; we'll deal with these in the upcoming *Querying a given thread's scheduling policy and priority* section as well as in *Chapter 11, The CPU Scheduler – Part 2*.(Something to take into account: many distros enable a kernel feature called **autogroups** (`CONFIG_SCHED_AUTOGROUP`); it helps speed up interactive response time for the foreground terminal process(es) and their threads. When enabled, traditional notions of nice value aren't really used (under the hood, cgroups is leveraged). See more on this in the *Further reading* section.)

- Real-time threads (having a policy of `SCHED_FIFO` or `SCHED_RR`) have a real-time priority scale from 1 to 99, 1 being the least and 99 being the highest priority. Think of it this way: on a non-preemptible kernel with one CPU, a `SCHED_FIFO` priority 99 thread spinning in an unbreakable infinite loop will effectively hang the machine! (Of course, even this too can be preempted by interrupts – both hard and soft; see *Chapter 6, Kernel Internals Essentials – Processes and Threads, Figure 6.1*.) *Figure 10.2* reveals the thread priority scale on Linux:

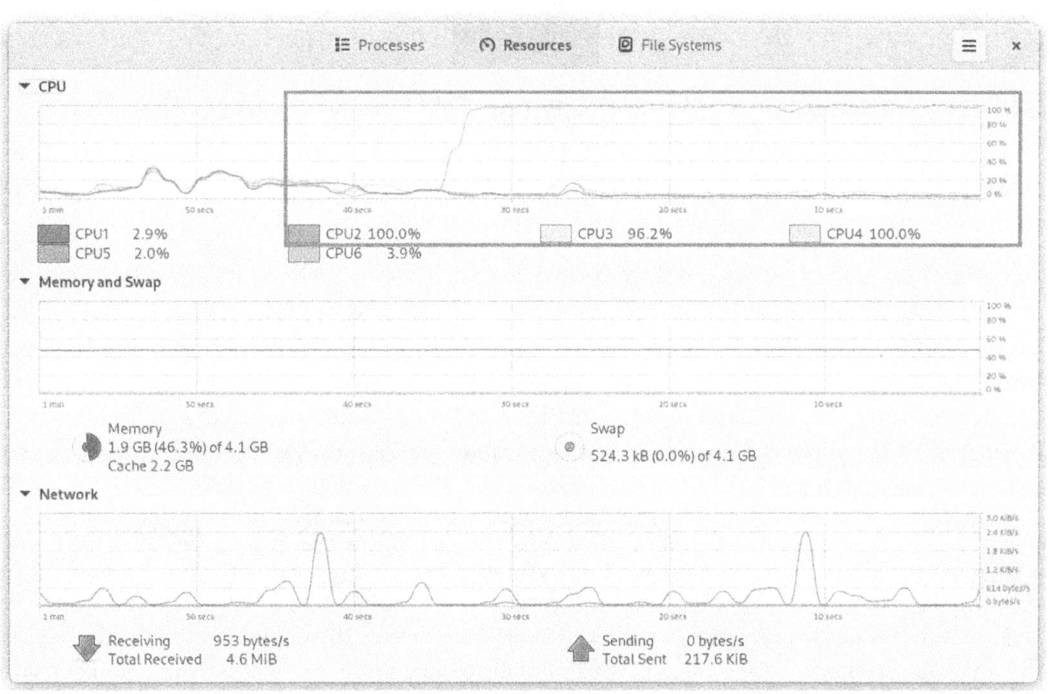

Figure 10.3: Thread priority scale for SCHED_OTHER, SCHED_RR / SCHED_FIFO policies

The scheduling policy is (loosely) specified via a member within the task structure, the *scheduling class*. Also, the thread policy and priorities (both the static nice value and real-time priority) are members of the task structure (as *Figure 10.1* clearly shows). Note that the scheduling class that a thread belongs to is exclusive; **a thread can only belong to one scheduling class** at a given point in time (worry not – we'll cover scheduling classes in some detail later, in the upcoming *Learning about the CPU scheduling internals – part 2* section).

Also, you should realize that on a modern Linux kernel, there are other scheduling classes (*stop-sched* and *deadline*) that are superior (in priority) to the FIFO/RR ones we've been seeing (again, more on this follows). Now that you have an idea of the basics, let's move on to something interesting: how we can *visualize* the flow of threads as they execute. Read on!

Visualizing the flow

Multicore systems have led to processes – well, threads really (both user - and kernel-space ones) – *executing concurrently* on different processors. This is useful for gaining higher throughput and thus performance, but it also causes synchronization headaches when they work with shared writable data (we shall deal in depth with the really important topic of kernel synchronization in this book's last two chapters).

So, for example, on a hardware platform with, say, six processor cores, we can expect processes (threads) to execute in parallel on them; this is nothing new. Is there a way, though, to actually *see* which processes or threads are executing on which CPU core – that is, a way to visualize a processor timeline? It turns out there are indeed a few ways to do so. In the following sections, we will look at a couple of interesting ways: with the `gnome-system-monitor` GUI program, `perf`, as well as other possibilities.

Using the gnome-system-monitor GUI to visualize the flow

The GNOME project provides a superb GUI to view and monitor system activity, appropriate for laptop, desktop, and server class systems (indeed, for any system powerful enough to run the GNOME GUI environment): the `gnome-system-monitor` application.

To quickly test using it to see the flow of work across a few CPU cores, let's first run a few processes concurrently. For our simple test case here, a process that continually executes on the CPU – i.e., a CPU-bound one – is called for. A good candidate's the simple utility named `yes`, which merely prints the character `y` continually to stdout (try it out!). So, let's say we run it three times, in the background. To make this experiment meaningful, we'd like to *affine* (tie) each process to a particular CPU core on the system. The `nproc` command reveals the number of cores; they're numbered starting at 0. On my x86_64 Fedora 38 VM:

```
$ nproc
6
```

Right, I have six CPU cores here (numbered 0–5). We leverage the useful `taskset` utility (it belongs to the *util-linux* package, which we specified installing back in *Chapter 1*). Running `taskset` with the `-c` option allows us to specify which CPU core(s) the command is to run upon; thus, if we do this

```
taskset -c 2 yes >/dev/null
```

it has `yes` run upon CPU core #2 only (we redirect stdout to the null device so that we don't have to see `y` continually)!

So, let's quickly rig up a test case: we'll run the `yes` utility three times (in the background of course), with each process instance on a different CPU core – for example, like this:

```
taskset -c 1 yes >/dev/null &
taskset -c 2 yes >/dev/null &
taskset -c 3 yes >/dev/null &
```

(Of course, you must have a system with at least four cores for the above to work (as core numbering starts at 0).) While they're alive and running (like heck), run the `gnome-system-monitor` application (we assume you're on a system with it installed, in GUI mode). The screenshot shows the typical output:

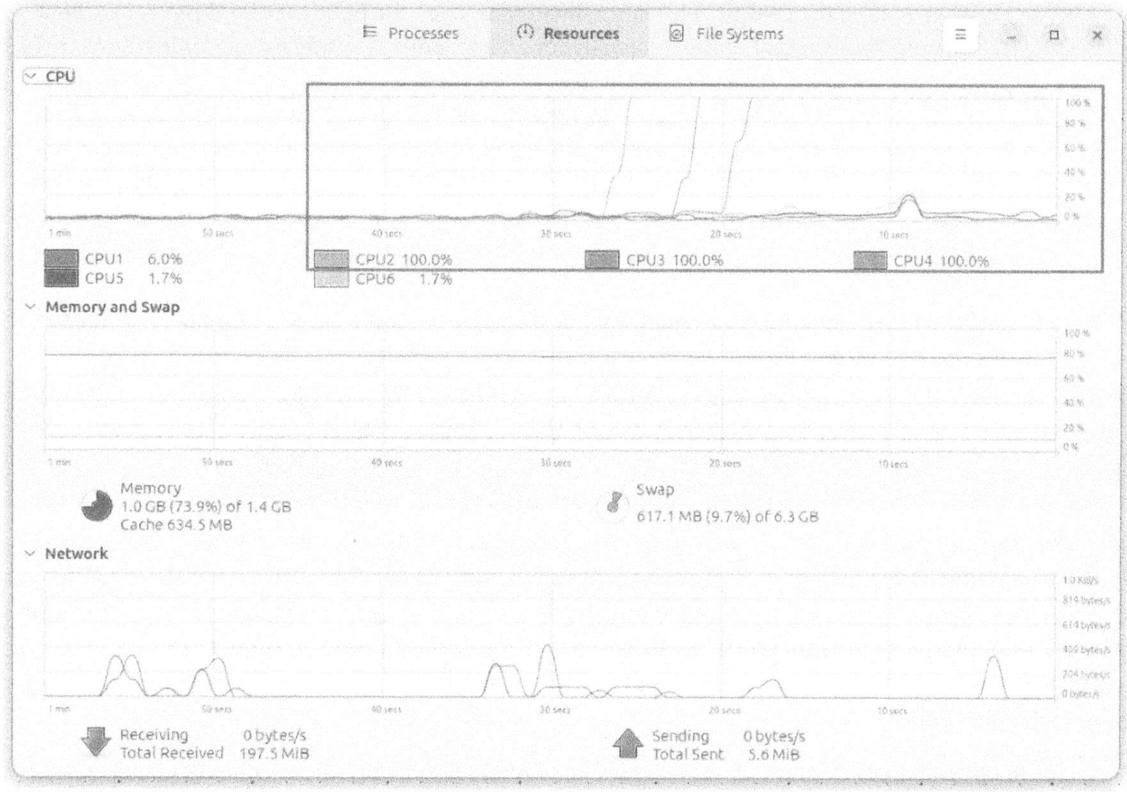

Figure 10.4: Screenshot showing the gnome-system-monitor GUI app; our three "yes" processes can be seen hammering away at 100% CPU usage on their three designated CPU cores (CPU2, CPU3, and CPU4)

Notice that the `gnome-system-monitor` GUI application numbers the CPU cores starting from 1 (not 0). Now, doing

```
pkill yes
```

will have all three `yes` process instances terminate. So, in *Figure 10.4* (in the upper **CPU** pane), you can literally see the concurrency, and the parallelism, while they execute on different CPU cores. (Also, as you can see, besides the **Resources** tab, it also has **Processes and File Systems** tab views).

Right, let's now move on to using the powerful perf utility to see the flow of processes/threads on CPU cores!

Using perf to visualize the flow

Linux, with its vast arsenal of developer and **Quality Assurance** (**QA**) tools, has a really powerful one in perf. In a nutshell, the perf toolset is one of the ways to perform CPU profiling on a Linux box. (Besides a few tips, we do not cover perf in any detail in this book.)

Akin to the venerable top (and the newer htop) utility, to get a "thousand-foot view" of what's eating the CPU (in a lot more detail than top), the perf set of utilities is excellent. Do note, though, that, quite unusually for an app, perf is tightly coupled with the kernel that it runs upon. It's important that (on Ubuntu at least), to get perf, you install the linux-tools-$(uname -r) package. Also, the perf-related distribution packages will not be available for the custom 6.1 kernel we have built; so, when using perf, I suggest you boot your guest VM with one of the standard (or distro) kernels, install the linux-tools-$(uname -r) package, and then try using perf. (Of course, you can always manually build perf from within the kernel source tree, under the tools/perf/ folder.)

With perf installed and running, you could try out these perf top-related commands (refer to the man page or tutorials for details):

```
sudo perf top
sudo perf top --sort comm,dso
sudo perf top -r 90 --sort pid,comm,dso,symbol
```

The aforementioned perf top variants help us not only get a birds-eye view of what's eating the CPU but also see what code paths they're executing, even allowing us to zoom into each task to see further details (something traditional tooling typically doesn't let us do). By the way, comm implies the name of the command/process, and **dso** is an abbreviation for **dynamic shared object**. The man page on perf(1) provides the details; use the man perf-<foo> notation – for example, man perf-top – to get help with perf top.

But back to our main point: one way to use perf is to obtain a clear picture of exactly what task(s) are running on what CPU core; this is done via the sched map sub-command in perf. First, though, you must record events using perf, which can be done both system-wide as well as for a specific process. To record events, run the following command:

```
sudo perf sched record [command]
```

With no parameter following the record keyword, it records events system-wide; passing a parameter makes it record events for only that command process, and its descendants. Terminate the recording session with the SIGINT signal (^C). This will generate a binary data file named perf.data by default; we'll soon see how it can be intuitively interpreted.

Trying it out – the command-line approach

Firstly, of course, lets run our CPU exerciser process - yes - concurrently, each instance affined to different CPU cores. To make the job easy, I've set up a couple of simple wrapper scripts in this folder: `ch10/concurrent_exercise/`. Let's do a sample run and see, from within your copy of this book's GitHub repo (available at `https://github.com/PacktPublishing/Linux-Kernel-Programming_2E`):

```
cd ch10/concurrent_exercise
```

As the VM we're running upon (an x86_64 Fedora 38) is configured with six CPU cores, we'll instruct the script to spawn off six yes processes, each of which will be affixed to one CPU core:

```
$ ./concurrency 6
concurrency: spawning a 'yes' process on ...
... CPU core #0 now ...
... CPU core #1 now ...
... CPU core #2 now ...
... CPU core #3 now ...
... CPU core #4 now ...
... CPU core #5 now ...
```

Now (in another Terminal window), start recording (sampling, CPU profiling) with `perf` (note that we're performing system-wide recording; that's fine):

```
sudo perf sched record
```

(As an aside, viewing the **Resources** tab of the `gnome-system-monitor` app is interesting at this point, though it can skew the 'benchmark'.)

Next, after letting the processes run for a bit (half a minute should be more than enough), we shut down the concurrent processes and get `perf` to give us a detailed map of CPU usage, as seen during the recording:

```
$ pkill yes
Terminated
Terminated
Terminated
Terminated
Terminated
Terminated
```

Press ^C in the Terminal where the `perf record` command is running to have it stop recording. Then, do this to generate a report:

```
$ sudo perf sched map > mymap.txt
```

Done. Let's look up the results!

The report file, `mymap.txt` here, is formatted as follows (as the columns aren't explicitly labelled, I've inserted an annotation at the top of *Figure 10.5*):

- From the left, each column represents a CPU core, starting with 0 (please ignore the line numbers at the extreme left).
- After the **CPU core** columns (there will be six here as we have six CPUs), comes a **Timestamp** column (with the system uptime expressed in the `sec.us` format).
- Then comes the **Legend** column. Here, each thread that executes is assigned a two-character name; in our run, as a good example (see *Figure 10.5*), `A0` is the `kthread` named `migration/0:18`, `B0` is one of our yes threads (`yes:6428`), `C0` is another `kthread migration/1:23`, and so on.

```
           CPU Core #
      0   1   2  3   4   5        Timestamp            Legend
      |   |   |  |   |   |
1    *A0                        2986.732221 secs A0 => migration/0:18
2    *B0                        2986.732226 secs B0 => yes:6164
3     B0 *C0                    2986.733960 secs C0 => migration/1:23
4     B0 *D0                    2986.733966 secs D0 => yes:6145
5     B0  D0 *E0                2986.747286 secs E0 => migration/2:29
6     B0  D0 *F0                2986.747291 secs F0 => yes:6144
7     B0  D0  F0 *G0            2986.748225 secs G0 => migration/3:35
8     B0  D0  F0 *H0            2986.748230 secs H0 => yes:6153
9     B0  D0  F0  H0 *I0        2986.749260 secs I0 => migration/4:41
10    B0  D0  F0  H0 *J0        2986.749266 secs J0 => yes:6165
11    B0  D0  F0  H0  J0 *K0    2986.750322 secs K0 => yes:6148
12    B0  D0 *L0  H0  J0  K0    2986.758874 secs L0 => VBoxClient:3004
13    B0  D0 *F0  H0  J0  K0    2986.758883 secs
14    B0  D0 *L0  H0  J0  K0    2986.787236 secs
15    B0  D0 *F0  H0  J0  K0    2986.787255 secs
16    B0 *M0  F0  H0  J0  K0    2986.791238 secs M0 => gnome-system-mo:4725
17   *N0  M0  F0  H0  J0  K0    2986.793457 secs N0 => gnome-shell:2280
```

Figure 10.5: Screenshot of the upper portion of the map file captured via perf sched map (with an annotation inserted at the very top) and our yes processes highlighted

So, we can see that our six yes processes (which are single-threaded of course), are the ones labeled B0, D0, F0, H0, J0, and K0. To make the visualization easier, I opened the map file in vim and searched to match them with a regular expression, like this:

```
/B0..D0..F0..H0..J0..K0
```

It's worked! You can see them highlighted in both *Figure 10.5* and *Figure 10.6*. Now, the yes processes – here, B0, D0, F0, H0, J0, and K0 – all being in one line **indicates that all six were running concurrently, in parallel, on CPU cores 0 to 5, respectively** (as row numbers 11, 13, and 15 in *Figure 10.5* clearly show)! Nice.

```
 B0  D0  F0  H0 *H1  K0     2986.919273 secs H1 => kworker/4:3-eve:996
 B0  D0  F0  H0 *J0  K0     2986.919277 secs
 B0  D0  F0  H0  J0 *Y0     2986.919279 secs
 B0  D0  F0  H0  J0 *K0     2986.919283 secs
 B0  D0 *L0  H0  J0  K0     2986.923245 secs
 B0  D0 *F0  H0  J0  K0     2986.923266 secs
 B0  D0 *I1  H0  J0  K0     2986.927245 secs I1 => systemd-oomd:1150
 B0  D0 *F0  H0  J0  K0     2986.927914 secs
 B0  D0  F0 *X0  J0  K0     2986.928214 secs
 B0  D0  F0 *H0  J0  K0     2986.928217 secs
 B0  D0  F0 *X0  J0  K0     2986.932184 secs
 B0  D0  F0 *H0  J0  K0     2986.932186 secs
 B0  D0  F0 *X0  J0  K0     2986.936189 secs
 B0  D0  F0 *H0  J0  K0     2986.936191 secs
 B0 *J1  F0  H0  J0  K0     2986.943242 secs J1 => gmain:2291
*K1  J1  F0  H0  J0  K0     2986.943252 secs K1 => gmain:2505
```

Figure 10.6: Screenshot of another portion of the perf map file, showing our six yes processes running concurrently on the six CPU cores on the box

(FYI, an asterisk indicates that a change of context, typically a context-switch on that core, occurred.) It's instructive to see how other threads also run at different points; obviously, it's not just our **yes-men** that want to run!

Again, you don't have to do exactly what we did here to test this; any sufficiently CPU-bound application can be executed (benchmarking programs like `stress-ng`, the kernel build itself, and so on are decent candidates), and the `perf` commands we showed can be run to capture and report details.

We've just used `perf` to get a command-line (or console) based "view" of threads executing on our CPU cores; now, let's leverage `perf` to show the same thing graphically!

Trying it out – the graphical approach

The steps to carry out here remain identical to those in the previous section except for the last command (the `sudo perf sched map > mymap.txt` command); instead, we generate a graphical representation of CPU scheduling with the following `perf` command:

```
sudo perf timechart -i ./perf.data
```

This command generates a **Scalable Vector Graphics** (**SVG**) file named `output.svg` by default! It can be viewed with vector drawing utilities (such as Inkscape, or via the `display` command in ImageMagick) or simply within a web browser. It can be useful to study the time chart; try it out. However, do note that the vector images can be quite large and, therefore, take a while to open.

FYI, it is possible to configure perf to work with non-root access. The SVG was generated using the same perf.data file from our run in the previous section and is seen within a browser:

Figure 10.7: Partial screenshot of the SVG file generated by perf using the data from the previous run

You can zoom in and out of this SVG file, studying the scheduling and CPU events that are recorded by default by perf. Clearly, our six concurrent yes processes are hammering away on the six CPU cores (the blue color represents the **Running** state, as can be seen in the legend, the first line). Other processes (mostly asleep) show up below them.

The following figure is another partial screenshot when zoomed in 400% to the top-left region of the preceding screenshot; again, it's pretty clear that the yes processes are the ones on the CPU:

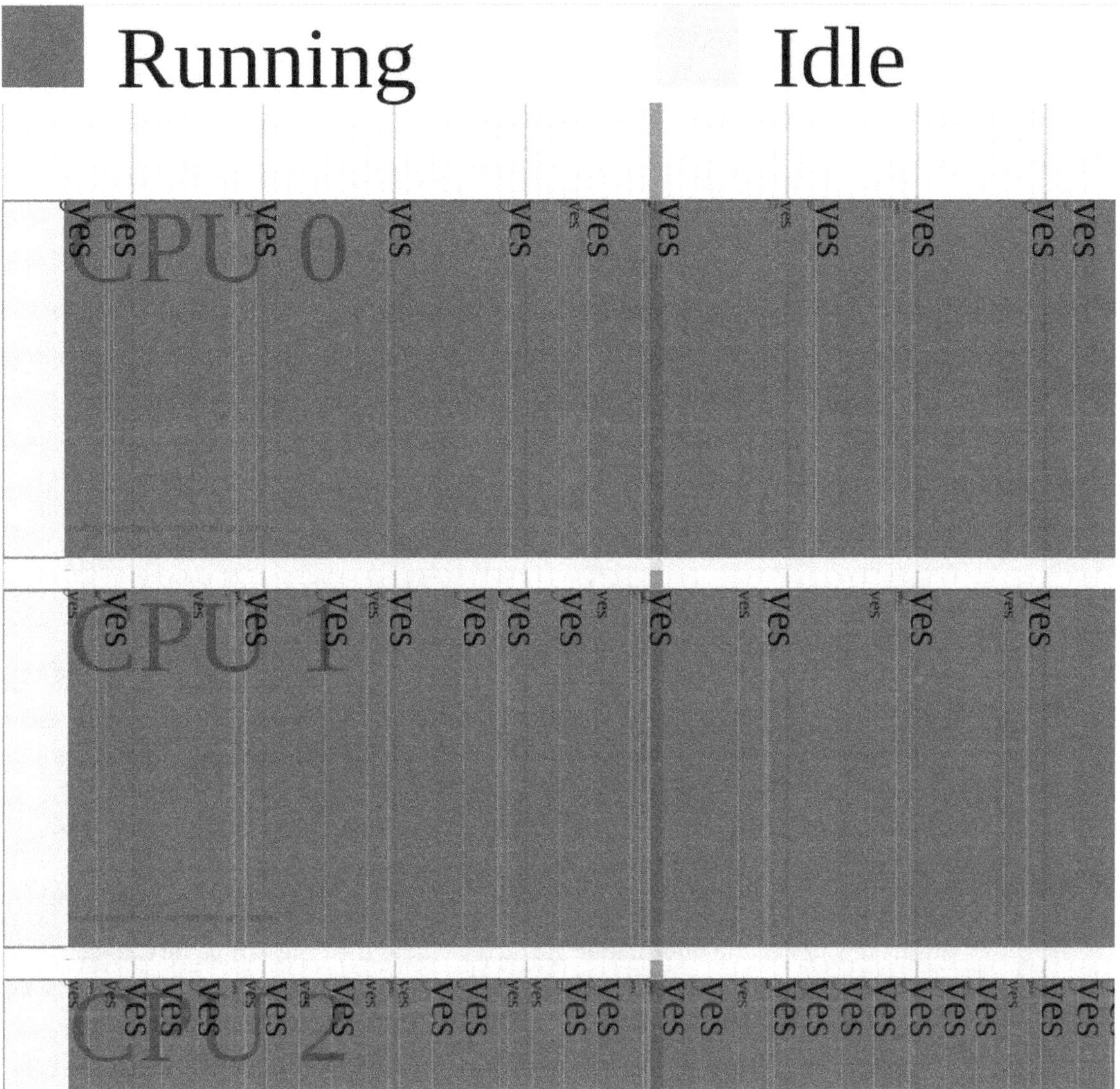

Figure 10.8: Partial screenshot of perf timechart's SVG file, when zoomed in 400% to the top-left region

What else? By using the `-I` option switch to `perf` timechart record, you can request that only system-wide disk I/O (and network, apparently) events be recorded. This could be especially useful as, often, real performance bottlenecks are caused by I/O activity (not CPU). The `man` page on `perf-timechart(1)` details further useful options.

You can convert perf's binary `perf.data` record file into the popular **Common Trace Format** (**CTF**) file format, using `perf data convert --all --to-ctf <dir>`, where the last argument is the directory where the CTF file(s) get stored. Why is this useful? CTF is the native data format used by powerful GUI visualizers and analyzer tools such as Trace Compass. (I find that this works on Fedora, whereas Ubuntu seems to have some nagging issues.)

Visualizing the flow via alternate approaches

There are, of course, alternate ways to visualize what's running on each processor. A quick overview follows.

The fact is several of these visualization tools are covered in a lot more detail in the *Linux Kernel Debugging* book, in *Chapter 9, Tracing the Kernel Flow*. Hence, we don't repeat the same information here but do provide a quick overview.

Let's first mention a few **console-based** visualization tools:

- **perf-tools**: Brendan Gregg has a very useful series of scripts that perform a lot of the hard work required when monitoring production Linux systems using `perf`; do take a look at them here: `https://github.com/brendangregg/perf-tools` (some distributions include them as a package called `perf-tools[-unstable]`).
- **Ftrace**: a very powerful kernel tracer system built into the fabric of the Linux kernel (of course, it has to be enabled; most distros do enable Ftrace by default):
    - Raw ftrace can be used to trace (almost) every single function, as it executes deep inside the kernel; the information gleaned includes the timestamp, the CPU core that it's executing upon, the context that's performing the execution, the interrupt state, and the function name.
    - A front-end to raw Ftrace is the superb **trace-cmd** utility; it provides essentially the same information as raw Ftrace while being much simpler to use. It can also be configured to show the parameters being passed to kernel functions. (I have built a simple front-end to it, named *trccmd* (`https://github.com/kaiwan/trccmd`); do give it a spin.)
- Some simple **Bash scripting** can show what's executing on a given core (via a simple wrapper over `ps`). In the following snippet, we show sample Bash functions; for example, the following `c0()` function shows what is currently executing on CPU core `#0`, while `c1()` does the same for core `#1` (this does imply that you're using the regular GNU `ps` and not an abbreviated variant like `busybox ps`):

```
# Show thread(s) running on cpu core 'n' - func c'n'()
function c0()
{
 ps -eLF | awk '{ if($5==0) print $0}'
```

```
}
function c1()
{
 ps -eLF | awk '{ if($5==1) print $0}'
}
```

- **LTTng**: The **Linux Trace Toolkit – next generation** (**LTTng**) is a powerful and popular open-source tracing system for the Linux kernel as well as user space apps and libraries. Its original version (LTT) dates to 2005, and LTTng is actively maintained. It has made a name for itself in helping track down performance and debug issues on multicore parallel and real-time systems.

A few GUI visualization tools are here:

- **KernelShark**: A good GUI front-end to the data generated by raw Ftrace and by `trace-cmd`. (For raw Ftrace, you'd have to first do a `trace-cmd extract` to get the data into a file format recognized by it.)
- **TraceCompass**: An excellent GUI front-end to visualize the traces generated by LTTng; in fact, it works with the popular CTF format, so any traces in this format can be imported and thus visualized by it.

Do give these alternatives a try!

Right, now that you've learned how you can visualize the flow of processes/threads on CPU cores by various means, let's move our focus back to the kernel internals side for a while.

Learning about the CPU scheduling internals – part 2

This section delves into kernel CPU scheduling internals in some detail, the emphasis being on the core aspect of modern design, modular scheduling classes.

Understanding modular scheduling classes

Ingo Molnar, a key kernel developer (along with others), redesigned the internal structure of the kernel scheduler, introducing a new approach called **scheduling classes** (this was back in October 2007 with the release of the 2.6.23 kernel).

As a side note, the word `class` in `scheduler classes` isn't a coincidence; many Linux kernel features are intrinsically, and quite naturally, designed with an **object-oriented** nature. The C language, of course, does not allow us to express this directly in code (hence the preponderance of structures with both data and function pointer members, emulating a class). Nevertheless, the design is very often object-oriented (as you shall quite clearly see with the driver model in the *Linux Kernel Programming – Part 2* book). Please see the *Further reading* section of this chapter for more details on this aspect.

A layer of abstraction was introduced into the core scheduling code, the function `kernel/sched/core.c:schedule()`. This layer in `schedule()` is generically called the scheduling classes and is modular in design.

Note that the word *modular* here implies that scheduler classes can be added or removed from the in-tree kernel code; it has nothing to do with the **Loadable Kernel Module** (**LKM**) framework.

The basic idea is this: the Linux kernel (as of 6.1, and, as of this writing, the latest 6.7 kernel in fact), contains five scheduling classes, each of which is associated with a priority level. When the core scheduler code – the `schedule()` function (itself a thin wrapper over `__schedule()`) – is invoked, *it iterates over each of the classes in a predefined priority order, asking each one if it has a thread that's ready to run* (how exactly, we shall soon see). Of the five scheduling classes present, it's guaranteed that one of them will answer in the affirmative and pick a candidate thread to run next; as soon as this occurs, the core scheduling code context-switches to that lucky thread (skipping any remaining scheduling classes), and the job is done. The following flow diagram encapsulates this design:

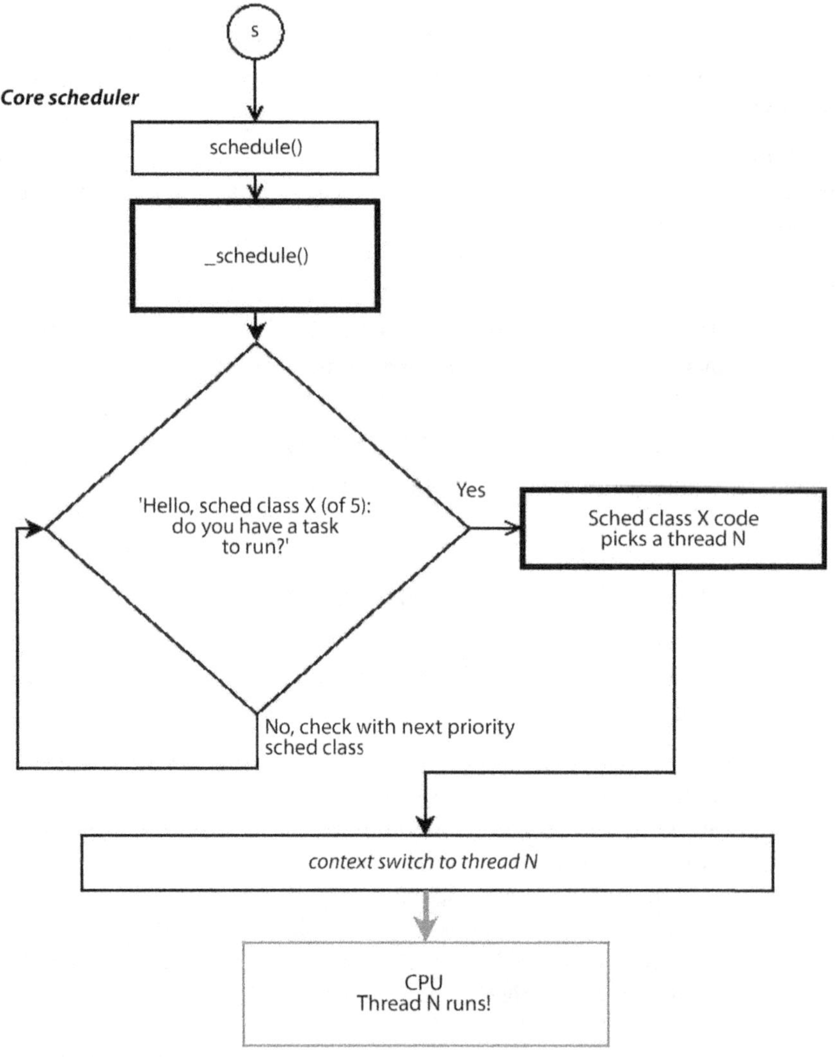

Figure 10.9: Flow diagram of the core scheduling code paths, showing the overall modular scheduler class architecture (this diagram's a key one!)

You, being alert, have spotted the connector labeled **S** at the very top of *Figure 10.9*; what does it imply? It's there to answer a key question: *"Who calls, and when exactly, is the core scheduler function, schedule(), called?"* (This is answered in detail in an upcoming section, *Learning about the CPU scheduling internals - part 3*; relax – we'll get there!)

As of the 6.1 Linux kernel (and, as of this writing, the latest 6.7 kernel in fact), these are the **five scheduler classes** within the kernel, listed in priority order, with the highest priority shown first:

Scheduling policy	**Scheduler class**	**sched_class data structure name**	**Defined in (via the DEFINE_SCHED_CLASS() macro)**
-	Stop-task / Stop-sched (`__sched_class_highest`)	`stop_sched_class`	`kernel/sched/stop_task.c`
`SCHED_DEADLINE`	(Early) Deadline First	`dl_sched_class`	`kernel/sched/deadline.c`
`SCHED_FIFO` / `SCHED_RR`	RT (real-time)	`rt_sched_class`	`kernel/sched/rt.c`
`SCHED_OTHER` (or `SCHED_NORMAL`): the default	CFS	`fair_sched_class`	`kernel/sched/fair.c`
`SCHED_IDLE`	Idle (`__sched_class_lowest`)	`idle_sched_class`	`kernel/sched/idle.c`

Table 10.2: The five modular scheduling classes

So, there we have the five modular scheduler classes – stop-task, deadline, (soft) real-time (RT), fair (CFS), and idle – in highest to lowest priority order. The data structures that abstract these scheduler classes, `struct sched_class`, are strung together on a singly linked list, which the core scheduling code iterates over. (We'll come to what the `sched_class` structure is later; ignore it for now.)

Every thread is associated with its unique task structure (`struct task_struct`); within the task structure (as we saw in *Chapter 6*, you can look it up here: `https://elixir.bootlin.com/linux/v6.1.25/source/include/linux/sched.h#L737`), the following applies:

- The member `struct sched_class *` holds the pointer to the modular scheduling class that the thread belongs to; it's exclusive – a thread can only belong to one scheduling class at any given point (it will be one of those mentioned in *Table 10.2*, third column). By default, it will point to CFS (`fair_sched_class`).
- The `policy` member specifies the scheduling policy that the thread adheres to (it will be one of those mentioned in *Table 10.2*, first column). It too is exclusive – a thread can only adhere to one scheduling policy at any given point in time (it can be changed though).
- The thread *priority* value(s); the members incorporating them are `prio`, `static_prio`, `normal_prio`, and `rt_priority`.

- Both the scheduling policy and priority are dynamic and can be queried and set programmatically (or via utilities; you will soon see this).

FYI, threads belonging to the `stop-sched` class are very few; this is because it's an extreme priority level. When a stop-sched thread gains the processor, the kernel turns off execution (as well as anything locking related, interrupts, and everything else) on all other cores on the system. Thus, the `stop-sched` class thread executes - on a core where all interrupts and kernel preemptions are masked - literally alone, with absolutely no chance of being preempted. Who needs this level of priority and non-preemptability? Well, one example is the Ftrace kernel tracing subsystem; another is live kernel patching. Also, the next priority sched class – **Deadline** – is meant for real-time tasks that have an associated deadline they must meet (in classic RTOS-like fashion). As `stop-sched` and `deadline` threads tend to be rare, we focus on the RT, and – mostly – fair (CFS) threads; the "fair class" threads are the ones typically alive and running.

So, knowing this, you will now realize that all threads that adhere to either the `SCHED_FIFO` or `SCHED_RR` scheduling policy map to the `rt_sched_class` (via the `sched_class` member within their task structures), all threads that are `SCHED_OTHER` (or `SCHED_NORMAL`) map to the `fair_sched_class`, and the CPU idle threads (`swapper/n`, where `n` is the CPU number starting from `0`) always map to the `idle_sched_class` scheduling class.

When the kernel needs to schedule, this is the essential call sequence:

```
schedule() --> __schedule() --> pick_next_task()
```

The actual iteration over the preceding scheduling classes occurs here; see the (partial) code of `pick_next_task()`, as follows: https://elixir.bootlin.com/linux/v6.1.25/source/kernel/sched/core.c#L5850:

```
static inline struct task_struct *
__pick_next_task(struct rq *rq, struct task_struct *prev, struct rq_flags *rf)
{
        const struct sched_class *class;
        struct task_struct *p;

        /*
         * Optimization: [ ... ] */
    [ ... ]
        put_prev_task_balance(rq, prev, rf);
        for_each_class(class) {
                p = class->pick_next_task(rq);
                if (p)
                        return p;
        }
        BUG(); /* The idle class should always have a runnable task. */
}
```

(The reality is that a good amount of optimization goes into the task picking code logic; for clarity, we will skip it and focus upon the canonical case.) The preceding `for_each_class()` macro sets up a `for` loop to iterate over all scheduling classes. Its implementation is as follows:

kernel/sched/sched.h

```
#define for_class_range(class, _from, _to) \
        for (class = (_from); class < (_to); class++)

#define for_each_class(class) \
        for_class_range(class, __sched_class_highest, __sched_class_lowest)
```

You can see from the preceding code snippets that it results in each class, from `__sched_class_highest` to `__sched_class_lowest`, being "asked" via the `class->pick_next_task()` "method" who to schedule next (as *Figure 10.9* conceptually shows). Now, it's left to the scheduling class code to determine whether it has any candidates that want to execute. How? That's simple actually; it merely looks up its **runqueue** data structure.

Now, this is a key point: *the kernel maintains one runqueue for every processor core and for every scheduling class*! So, if we have a system with, say, six CPU cores, then we will have *6 cores * 5 sched classes = 30 runqueues*! (There is an exception: on **Uniprocessor** (**UP**) systems, the `stop-sched` class does not exist.) Runqueues are implemented per scheduling class; here's, for example, the runqueue for CFS: `struct cfs_rq` (`https://elixir.bootlin.com/linux/v6.1.25/source/kernel/sched/sched.h#L550`); similarly, for the RT class, it's `struct rt_rq`, and so on. The following diagram is an attempt to present this information:

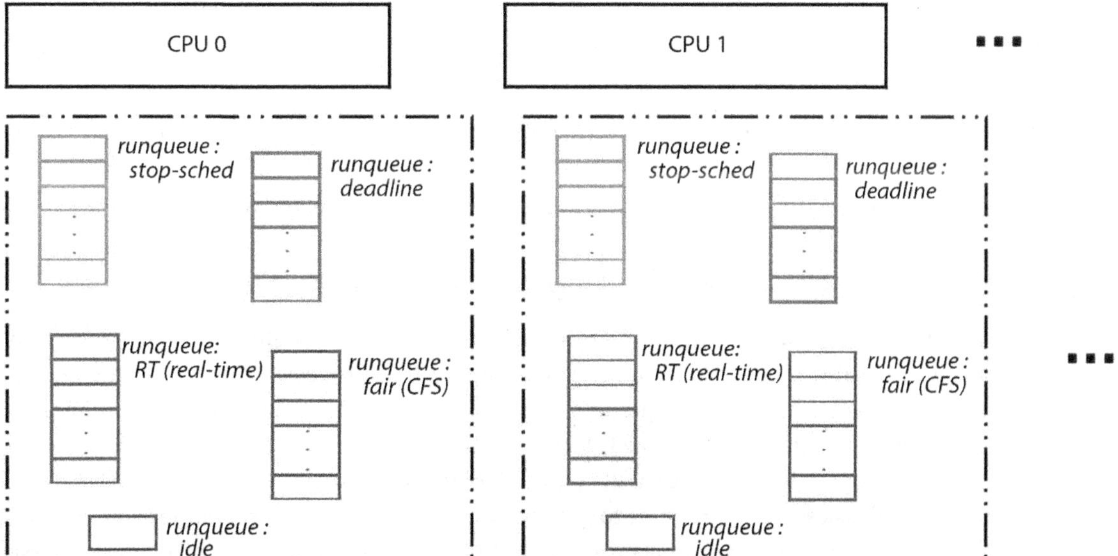

Figure 10.10: There is 1 runqueue per CPU core and per scheduling class

Please note that in the preceding diagram, the way I show the runqueues makes them perhaps appear to be arrays. That isn't the intention at all; it's merely a conceptual diagram. The actual runqueue data structure used depends on the scheduling class (the class code implements the runqueue after all). It could be an array of linked lists (as with the real-time class), a tree – a **red-black (rb) tree,** in fact – with the fair class, and so on.

A conceptual example to help understand scheduling classes

To help better understand the scheduler class model, we will devise an example: let's say, on a **Symmetric Multi Processor** (SMP) or multicore system, we have 100 threads alive (in both user and kernel space). Among them, we have a few competing for the CPUs; that is, they are in the ready-to-run (Rr) state, implying they are runnable and thus enqueued on runqueue data structures (see *Figure 10.2*, the state machine). Let's say the runnable threads fall into various scheduling classes as follows:

- One **stop-sched** (SS) class thread, S1
- Two **Deadline** (DL) class threads, D1 and D2
- Two **Real Time** (RT) class threads, RT1 and RT2
- Three **Completely Fair Scheduling** (CFS) (or fair) class threads, F1, F2, and F3
- One idle class thread I1

Now let's imagine that, to begin with, thread F2 is on a processor core, happily executing code. At some point, the kernel wishes to context switch to some other task on that CPU. (What triggers this? You shall soon see.) On the scheduling code path, the kernel code ultimately ends up in the core scheduling code `kernel/sched/core.c:void schedule(void)` kernel routine (again, code-level details follow later). What's important to understand for now is that the `pick_next_task()` routine, invoked by `schedule()`, which becomes `__schedule()`, iterates over the linked list of scheduler classes, asking each whether it has a candidate to run (again, refer to *Figure 10.9*). Its code path (conceptually, of course) looks something like this:

1. Core scheduler code (`__schedule()`): *"Hey, SS (stop-sched class), do you have any threads that want to run?"*
2. SS class code: Iterates over *its* runqueue and does find a runnable thread (S1); it thus replies: *"Yes, I do; it's thread S1."*
3. Core scheduler code (`__schedule()`): *"Okay, great; let's context switch to S1."*

And the job is done (for this scheduling round or epoch, at least). But let's change the scenario a bit: what if there is no runnable thread S1 on the SS runqueue for that processor (or it has gone to sleep, or is stopped, or it's on another CPU's runqueue)? Then, SS will say *"no,"* and the next most important scheduling class, *Deadline (DL)*, will be asked. If it has potential candidate threads that want to run (D1 and D2, in our example), its class code will run its algorithm to identify which of D1 or D2 should run, return that thread task structure pointer to `__schedule()`, and the kernel scheduler will faithfully context switch to it. This process continues for the *Real-time (RT)* and *Fair (CFS)* scheduling classes. (A picture is worth a thousand words, right? See *Figure 10.11.*)

It's likely that (on your typical moderately loaded Linux system), there will be no *SS*, *DL*, or *RT* candidate threads that want to run on the CPU in question, and there often will be at least one fair (CFS) thread that will want to run.

Hence, the competition is typically between the fair (CFS) runnable threads; the one picked by the fair class implementation (CFS) will be the one that is context-switched to.

If there are really no threads that want to run (no *SS/DL/RT/CFS* class thread wants to run), it implies that **the system is presently idle** (lazy chap). Now, the `Idle` class is asked whether it wants to run; *it always says yes!* This makes sense; after all, it is the CPU idle thread's job to run on the processor when no one else needs/wants to. Hence, in such a case, the kernel switches context to the idle thread, typically labelled `swapper/n`, where `n` is the CPU number that it's executing upon (starting from `0`; and yes, I know what you're probably thinking: *why is it called "swapper"?...* it's just old Unix history coming back to haunt us – nothing else).

Also, note that the `swapper/n` (CPU idle) kernel thread does not show up in the `ps` listing, even though it's always present (recall the code we demonstrated in *Chapter 6, Kernel Internals Essentials – Processes and Threads*, here: `ch6/foreach/thrd_showall/thrd_showall.c` (`https://github.com/PacktPublishing/Linux-Kernel-Programming_2E/blob/main/ch6/foreach/thrd_showall/thrd_showall.c`) There, we wrote a `disp_idle_thread()` routine to show some details of the CPU idle thread, as even the kernel's `do_each_thread() { ... } while_each_thread()` loop that we employed there does not show the idle thread.

The following diagram neatly sums up the way the core scheduling code invokes the scheduling classes in priority order, context-switching to the ultimately selected "next" thread:

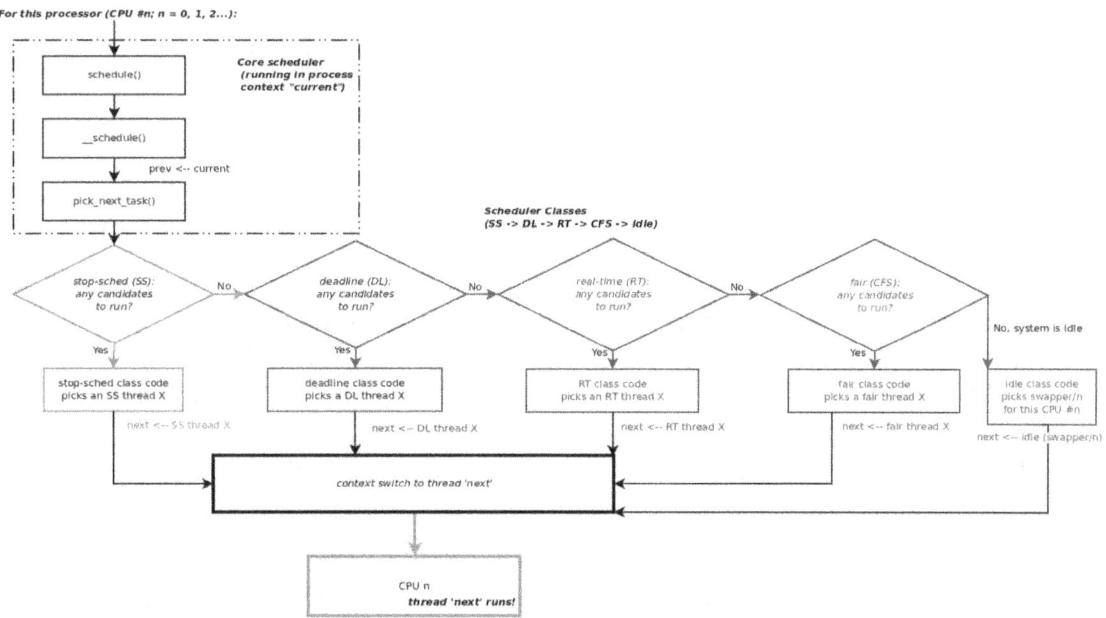

Figure 10.11: Core scheduler code iterating over every scheduling class to pick the task that will run next

Asking the scheduling class

How exactly does the core scheduler code (in `pick_next_task()`) *ask* the scheduling classes whether they have any threads that want to run? We have already seen this, but I feel it's worthwhile repeating the following code fragment for clarity (called mostly from `__schedule()` and also from the thread migration code path):

kernel/sched/core.c

```
static inline struct task_struct *
__pick_next_task(struct rq *rq, struct task_struct *prev, struct rq_flags *rf)
{
        const struct sched_class *class;
        struct task_struct *p;

        /*
         * Optimization: [ ... ] */
     [ ... ]
        put_prev_task_balance(rq, prev, rf);
        for_each_class(class) {
                p = class->pick_next_task(rq);
                if (p)
                        return p;
        }
    [ … ]
```

Notice the object orientation in action: the `class->pick_next_task(rq)` code, for all practical purposes, invokes a method, `pick_next_task()`, of the scheduling class `class`! The return value, conveniently and deliberately, is the pointer to the task structure of the picked task, which the code can now context-switch to. As can also be seen, `pick_next_task()` returning NULL implies that the current class does not have any candidate to schedule; so we move onto the next class and ask it. The loop will always terminate as, if nothing else, the idle class will always return a non-null candidate to schedule, the idle thread for that core.

The preceding code and paragraph implies, of course, that there is a pre-populated `class` structure per scheduling class, embodying what we really mean by the scheduling class. Indeed, this is the case: it contains all possible operations, as well as useful hooks, that you might require in a scheduling class. It's (surprisingly) called the `sched_class` structure:

`https://elixir.bootlin.com/linux/v6.1.25/source/kernel/sched/sched.h#L2147`

```
struct sched_class {
    [ ... ]
        void (*enqueue_task) (struct rq *rq, struct task_struct *p, int flags);
        void (*dequeue_task) (struct rq *rq, struct task_struct *p, int flags);
    [ ... ]
```

```
        struct task_struct *(*pick_next_task)(struct rq *rq);
        void (*task_tick)(struct rq *rq, struct task_struct *p, int queued);
        void (*task_fork)(struct task_struct *p);
    [ ... ]
};
```

(There are many more members of this structure than we've shown here; do look it up in the kernel code.) As should be obvious by now, each scheduling class instantiates this structure, appropriately populating it with methods (via function pointers, of course). The core scheduling code, iterating over the linked list of scheduling classes (as well as elsewhere in the kernel), invokes – as long as it's not `NULL` – the methods and hook functions as required (again, *Figure 10.9* and *Figure 10.11* remind us of this).

As an example, let's consider how the fair scheduling class (CFS) implements its scheduling class:

`https://elixir.bootlin.com/linux/v6.1.25/source/kernel/sched/fair.c#L12365`

```
DEFINE_SCHED_CLASS(fair) = {
        .enqueue_task               = enqueue_task_fair,
        .dequeue_task               = dequeue_task_fair,
         [ ... ]
        .pick_next_task             = __pick_next_task_fair,
         [ ... ]
        .task_tick                  = task_tick_fair,
        .task_fork                  = task_fork_fair,
        .prio_changed               = prio_changed_fair,
        [ ... ]
};
```

(Earlier, the structure was directly defined; from 5.11, the `DEFINE_SCHED_CLASS()` macro's used to do so. To see why, read the commit: `https://github.com/torvalds/linux/commit/43c31ac0e665d942fcaba83a725a8b1aeeb7adf0`.)

So, now you see it: the code used by the fair sched class to pick the next task to run (when asked by the core scheduler) is the function `__pick_next_task_fair()` (a thin wrapper over `pick_next_task_fair()`). FYI, the `task_tick` and `task_fork` members are good examples of scheduling class *hooks*; these functions will be invoked by the scheduler core on every timer tick (that is, each timer interrupt, which fires – in theory, at least – `CONFIG_HZ` times a second) and when a thread belonging to this scheduling class forks, respectively.

An interesting in-depth Linux kernel project, perhaps: create your own scheduling class with its particular methods and hooks, implementing its internal scheduling algorithm(s). Link all the bits and pieces as required (into the scheduling classes linked list, inserted at the desired priority, and so on) and test! Now, you'll see why they're called modular scheduling classes.

Great – now that you understand the architecture behind how the modern modular CPU scheduler works, let's take a brief look at the algorithm behind CFS, perhaps the most used scheduling class on generic Linux.

The workings of the Completely Fair Scheduling (CFS) class in brief

Since version 2.6.23 (back in 2007), CFS has been the de facto kernel CPU scheduling code for regular threads; most threads on Linux tend to belong to the `SCHED_OTHER` policy by default, which is driven by CFS. The driver behind the CFS algorithm is to provide fairness and overall throughput.

It's implementation, in a nutshell, is this: the kernel keeps track of the actual CPU runtime (at nanosecond granularity) of every runnable CFS (`SCHED_OTHER / SCHED_NORMAL`) thread; the thread with the smallest runtime is the thread that most deserves to run and will be awarded the processor on the next scheduling cycle or epoch (the word "epoch" denotes "the beginning of a period in the history of someone or something"). Conversely, threads that continually hammer on the processor will accumulate a large amount of runtime and will thus be penalized (it's quite karmic, really)!

We divide this discussion into two sections: the first briefly looks at the CFS concepts of *vruntime* and its internal rb-tree runqueue, and the second looks at how CFS dynamic timeslices work. Let's, of course, begin with the first.

A note on the CFS vruntime value and its runqueue

Without delving into too many details regarding the internals of the CFS implementation, embedded within the task structure is another data structure, `struct sched_entity`, which contains within it an unsigned 64-bit value called `vruntime` (or **virtual runtime**) (`https://elixir.bootlin.com/linux/v6.1.25/source/include/linux/sched.h#L556`). This is, at a simplistic level, the monotonic counter that keeps track of the amount of time, in nanoseconds, that the thread has accumulated (run) on the processor.

In practice, in implementation, a lot of code-level tweaks, checks and balances are required. For example, often, the kernel will reset the `vruntime` value to `0`, triggering another scheduling epoch. As well, there are various tunables (or sysctls) under `/proc/sys/kernel/sched_*`, helping to better fine-tune the CPU scheduler behavior (some of them are documented here: `https://www.kernel.org/doc/html/latest/admin-guide/sysctl/kernel.html`).

How CFS picks the next task to run is encapsulated in the `kernel/sched/fair.c:pick_next_task_fair()` function. In theory, the way CFS works is simplicity itself: enqueue all runnable tasks (for that CPU) onto the CFS runqueue, which is an rb-tree (a type of self-balancing binary search tree), in such a manner that the task that has spent the least amount of time on the processor is the leftmost leaf node on the tree, with succeeding (leaf) nodes to the right representing the next task to run, and then the one after that.

In effect, scanning the tree from left to right gives *a timeline of future task execution*. How is this assured? By using the aforementioned `vruntime` value as the key via which tasks are enqueued onto the rb-tree!

Why is it called *vruntime* and not just *runtime*? It's because the vruntime member's value isn't simply the time spent on the processor by the thread; it's more nuanced: it takes the thread's priority – the nice value – into account when calculating this all-important number (after all, the thread's "position" in the CFS rb-tree run queue is based on this vruntime quantity). So here's what's done: the lower the nice value (the higher the priority), the more the vruntime value is scaled down, thus having it enqueued more to the left; the higher the nice value (the lower the priority), the more the vruntime value is scaled up, thus having it enqueued more to the right. (This approach to task scheduling is broadly called a *weighted fair queuing* scheduler.)

When the core scheduler needs to schedule, and it asks CFS "*do you have any threads that want to run?*", the CFS class code – which we've already mentioned (the `pick_next_task_fair()` function) – runs and *simply picks the leftmost leaf node on the tree*, returning the pointer to the task structure pointer embedded there. Think about it – it is, by definition, the task with the lowest `vruntime` value, so effectively, the one that has run the least! (Traversing a tree is an *O(log n)* time-complexity algorithm, but due to some code optimization and clever caching of the leftmost leaf node, the implementation effectively has a very desirable *O(1)* time complexity.) Of course, the actual code is a lot more complex than is let on here; it requires several checks and balances. We won't delve into all the gory details here.

We refer those of you that are interested in learning more about CFS to the kernel documentation on the topic, at `https://www.kernel.org/doc/html/v6.1/scheduler/sched-design-CFS.html`.

Also, the kernel contains several tunables (sysctls) under `/proc/sys/kernel/sched_*` that have a direct impact on scheduling. Notes on these and how to use them can be found here: *Analyzing the impact of sysctl scheduler tunables*, IBM LTC: `https://events.static.linuxfound.org/slides/2011/linuxcon/lcna2011_rajan.pdf`, and an excellent real-world use case can be found in the article at `https://www.scylladb.com/2016/06/10/read-latency-and-scylla-jmx-process/`.

Further, several of these kernel-level scheduling tunables have more recently moved to debugfs here: `/sys/kernel/debug/sched/*`. You will therefore, of course, need root access to see/change them.

A note on the CFS scheduling period and timeslice

Did you notice? Unlike traditional OS schedulers, with CFS, timeslices are *dynamic!* Tasks that sip on the CPU (the so-called IO-bound ones) automatically accumulate less *vruntime* and thus migrate toward the left of the CFS rb-tree. The opposite happens for CPU-bound tasks; they (by having bigger *vruntime* values) move toward the right, lessening their chance of gaining the CPU quickly.

In the scheduler context, the *scheduling period* (sometimes confusingly called "scheduling latency"), refers to the time for which a complete scheduling cycle (or *epoch*) runs; within this time period, *the OS guarantees that every thread will be given a chance to run on the CPU.* So what is it? The default value is in the `sysctl /sys/kernel/debug/sched/latency_ns` (with a typical default of 24 ms on Ubuntu 22.04 and 18 ms on Fedora 38/39). Further, at runtime, it can change – it itself is dynamic (more on this soon follows) – and is calculated as follows:

```
Scheduling Period length = min_granularity_ns * nr_running ;
```

Here, `nr_running` is the number of tasks currently runnable (this is on a per-runqueue basis). These scheduler tunables, in recent kernels, are found in `debugfs`; let's look up some values on x86_64 Ubuntu 22.04 LTS. (The `grep . <files-spec>` syntax used here is a fantastic way to both list the (single - line) tunable and see its current value! Of course, the `grep -v` syntax is used to negate some matches.)

```
# grep -E . /sys/kernel/debug/sched/* 2>/dev/null | grep -v -E  "^\/sys\/
kernel\/debug\/sched\/debug\:|features"
/sys/kernel/debug/sched/idle_min_granularity_ns:750000
/sys/kernel/debug/sched/latency_ns:24000000
/sys/kernel/debug/sched/latency_warn_ms:100
/sys/kernel/debug/sched/latency_warn_once:1
/sys/kernel/debug/sched/migration_cost_ns:500000
/sys/kernel/debug/sched/min_granularity_ns:3000000
/sys/kernel/debug/sched/nr_migrate:32
/sys/kernel/debug/sched/preempt:none (voluntary) full
/sys/kernel/debug/sched/tunable_scaling:1
/sys/kernel/debug/sched/verbose:N
/sys/kernel/debug/sched/wakeup_granularity_ns:4000000
```

The above `grep` command line also filters out the tunables named `debug` and `features`. Okay, here's the meaning of some relevant tunables:

- `latency_ns`: This is the current calculated "period" value (in nanoseconds (ns)); so, being 24,000,000 ns, the default scheduling period here is 24 ms, i.e., *each thread is guaranteed it gets a chance to run at least once every 24 ms!*
- `min_granularity_ns`: This is the minimum required "distance" between nodes on the CFS rb-tree; in effect, *this is the minimum timeslice guaranteed for each CFS thread when it runs (default: 3 ms).*

Here are a couple of points to note before moving on:

- The defaults can and do vary across systems (kernels)! On my x86_64 Fedora 38 VM, the period and minimum granularity (the minimum timeslice) is 18 ms and 2.25 ms, respectively. Check what it is on your system.
- The above scheduler tunables are created in the function `sched_init_debug()` (called during system initialization here: `https://elixir.bootlin.com/linux/v6.1.25/source/kernel/sched/debug.c#L299`).

Think about it – the number of currently runnable tasks (`nr_running`) directly influences the period (and thus the timeslice as well). So, given these facts:

- The equation *Scheduling Period length = min_granularity_ns * nr_running* (as we saw)
- The `min_granularity_ns` is, in effect, the CFS (dynamic) timeslice value
- Given a default `min_granularity_ns` value of 3 ms, and a period (`latency_ns`) of 24 ms (as is the case on my Ubuntu 22.04 (and 23.10) box), to see how this works, we provide the calculations of the effective timeslice and scheduling period for a set of sample values, as shown in the following table.

Number of runnable threads in the CFS runqueue (nr_running)	**Effective timeslice (ms) = Scheduling Latency (ms) / nr_running.** **Is it less than the minimal granularity or timeslice (min_granularity_ns)?**	**Scheduling period (epoch duration) = min_granularity_ns * nr_running (ms)**
3	24/3 = 8 ms // okay	3*3 = 9 ms
6	24/6 = 4 ms // okay	6*3 = 18 ms
8	24/8 = 3 ms // minimum allowed! (as per `min_granularity_ns`)	8*3 = 24 ms
12	24/12 = 2 ms /* less than min_granularity_ns; => untenable! recalculate period */	12 *3 = 36 ms

Table 10.3: Effective CFS task timeslice based on the number of runnable threads

The effective timeslice is calculated and seen in the second column; as long as its value is within the `min_granularity_ns` value (3 ms), all's well. However, as can be seen, when the number of runnable threads gets sufficiently large, the situation obviously becomes untenable (the last row of *Table 10.3*); now, the scheduler, instead of giving up, cleverly and dynamically adjusts the *scheduling period*, recalculating it! (Well, that's one way to put it; the cunning kernel goes: "Did I promise you a chance on the CPU at least once in every 24 ms period? No, no, you're mistaken; I meant once in every 36 ms period...".)

More technically, the check done is this (pseudocode):

```
effective_timeslice (ms) = latency_ns (ms) / nr_running
if effective_timeslice < min_granularity_ns
    recalculate the scheduling period
```

Put another way:

```
if nr_running > (latency_ns / min_granularity_ns)
   recalculate the scheduling period
```

(The terminology can be confusing; remember, `latency_ns` is the scheduling period.) Check out the *Further reading* section for more on CFS and these scheduler-related tunables. Let's now move on to the (brief) topic of looking up task scheduling statistics maintained by the kernel (if so configured).

Scheduling statistics

Can you see the state of scheduling, both at system and process granularity? Indeed, when configured (requires `CONFIG_SCHEDSTATS=y`), the kernel makes available these pseudofiles under `proc`:

- `/proc/schedstat` : System-wide scheduling statistics. These have been around a long time; they display per-CPU scheduling statistics (as well as domain - CPU / micro-architecture hierarchy - information on SMP). The information includes the number of times the scheduler (and the `try_to_wake_up()` function) was called (on that CPU core), the cumulative sum of all time spent running/ waiting by tasks on the core, and the number of timeslices on that core. It's best to refer to the official kernel documentation on it: `https://elixir.bootlin.com/linux/v6.1.25/source/Documentation/scheduler/sched-stats.rst`.
- `/proc/PID/schedstat` and `/proc/PID/sched` : Process/thread-level granularity:
    - The `/proc/PID/schedstat` content consists of three whitespace-delimited numbers, all with regard to the thread with the process ID PID:
        - The time spent on the CPU (nanoseconds)
        - The time spent waiting on a runqueue (nanoseconds)
        - The number of timeslices run on this CPU
    - The `/proc/PID/sched` content is pretty detailed; it includes many fields of the task's `sched_entity` structure (shown as `se.foo`; this includes the task `vruntime` (in nanoseconds) as `se.vruntime`!), context-switching statistics, the scheduling policy and priority, as well as some NUMA statistics.

Quick tip: Try this to keep a "watch" on the fields as they change. As an interesting experiment, run yes in the background (`yes >/dev/null &`), note its PID, and then do this:

```
watch -d -n1 'cat /proc/<PID_of_yes>/sched'
```

Break off with `^C`.

The Linux kernel's always evolving (that's really its secret sauce!); task (CPU) scheduling hasn't been spared. From the 6.6 kernel (October 2023), a new scheduler replaces the venerable CFS for the fair class; it's named **EEVDF** – the **Earliest Eligible Virtual Deadline First** scheduler (Peter Zijlstra is one of its key developers). Interestingly, the overall approach is similar to that of CFS – both use a virtual time-based or weighted fair queuing approach. EEVDF addresses limitations of CFS and brings certain advantages; among these, it addresses the tight latency requirements that some threads have (enabling them to run in shorter bursts but more often on the processor, with lower latency); it's also considered to be a cleaner implementation (getting rid of several heuristics that CFS relies upon). As we focus on the 6.1 (S)LTS kernel, this edition doesn't go into further details; as always, see the *Further reading* section for more on EEVDF.

Now, let's move on to learning how to query the scheduling policy and priority of any given thread.

Querying a given thread's scheduling policy and priority

In this section, you'll learn how to query the scheduling policy and priority of any given thread on the system via the command line. (But what about programmatically querying and *setting* the same? We defer that discussion to *Chapter 11, The CPU Scheduler – Part 2,* in the *Querying and setting a thread's scheduling policy and priority* section.)

We learned that, on Linux, the thread is the KSE; it's what gets scheduled and runs on the processor. Also, Linux has several choices for the scheduling policy (or algorithm) to use. Both the scheduling policy and priority is assigned on a per-thread basis, with the default policy always being `SCHED_OTHER` and the default real-time priority being `0` (in other words, it's a non-real-time thread; see *Table 10.1*).

On a given Linux system, we can always see all processes alive (via a simple `ps -A`), or, with GNU `ps`, even *every thread alive* (one way is with `ps -LA`). However, this does not reveal a key fact: what scheduling policy and priority these tasks run under. How do we query that?

This turns out to be pretty simple: on the shell, the `chrt` utility is admirably suited to query and set a given process' scheduling policy and/or priority. Issuing `chrt` with the `-p` option switch and providing the PID as a parameter has it display both the scheduling policy as well as the real-time priority of the task in question; for example, let's query this for the `init` process (or systemd) PID `1`:

```
$ chrt -p 1
pid 1's current scheduling policy: SCHED_OTHER
pid 1's current scheduling priority: 0
```

It's quite clear: the PID 1 process (typically, its *systemd*, of course) has the `SCHED_OTHER` scheduling policy (which CFS drives) and a real-time priority of 0 (as it isn't a real-time task; see *Figure 10.3* to be reminded of the priority scale). As usual, the `man` page on `chrt(1)` provides all the option switches and their usage; do take a peek at it.

In the following (partial) screenshot, we show a run of a simple Bash script (`ch10/query_task_sched.sh`, a wrapper over `chrt`, essentially) that iterates over, queries, and displays the scheduling policy and priority (both nice and (soft)real-time) of all threads currently alive (on my x86_64 Fedora 38 VM):

```
$ ./query_task_sched.sh
  PID        TID            Name                           Sched Policy  Prio *RT   Nice  CPU-affinity-mask
    1          1                               systemd     SCHED_OTHER    0           0                  3f
    2          2                              kthreadd     SCHED_OTHER    0           0                  3f
    3          3                                rcu_gp     SCHED_OTHER    0         -20                  3f
    4          4                            rcu_par_gp     SCHED_OTHER    0         -20                  3f
    5          5                           slub_flushwq    SCHED_OTHER    0         -20                  3f
    6          6                                 netns     SCHED_OTHER    0         -20                  3f
    8          8              kworker/0:0H-events_highpri  SCHED_OTHER    0         -20                  1
   10         10                          mm_percpu_wq     SCHED_OTHER    0         -20                  3f
   12         12                      rcu_tasks_kthread    SCHED_OTHER    0           0                  3f
   13         13                 rcu_tasks_rude_kthread    SCHED_OTHER    0           0                  3f
   14         14                rcu_tasks_trace_kthread    SCHED_OTHER    0           0                  3f
   15         15                           ksoftirqd/0     SCHED_OTHER    0           0                  1
   16         16                           rcu_preempt     SCHED_OTHER    0           0                  3f
   17         17                           migration/0      SCHED_FIFO   99   ***      -                  1
   18         18                         idle_inject/0      SCHED_FIFO   50     *      -                  1
   20         20                               cpuhp/0     SCHED_OTHER    0           0                  1
   21         21                               cpuhp/1     SCHED_OTHER    0           0                  2
   22         22                         idle_inject/1      SCHED_FIFO   50     *      -                  2
   23         23                           migration/1      SCHED_FIFO   99   ***      -                  2
   24         24                           ksoftirqd/1     SCHED_OTHER    0           0                  2
   25         25                 kworker/1:0-mm_percpu_wq  SCHED_OTHER    0           0                  2
   26         26              kworker/1:0H-events_highpri  SCHED_OTHER    0         -20                  2
   27         27                               cpuhp/2     SCHED_OTHER    0           0                  4
   28         28                         idle_inject/2      SCHED_FIFO   50     *      -                  4
   29         29                           migration/2      SCHED_FIFO   99   ***      -                  4
   30         30                           ksoftirqd/2     SCHED_OTHER    0           0                  4
   32         32              kworker/2:0H-events_highpri  SCHED_OTHER    0         -20                  4
   33         33                               cpuhp/3     SCHED_OTHER    0           0                  8
   34         34                         idle_inject/3      SCHED_FIFO   50     *      -                  8
   35         35                           migration/3      SCHED_FIFO   99   ***      -                  8
   36         36                           ksoftirqd/3     SCHED_OTHER    0           0                  8
   38         38              kworker/3:0H-events_highpri  SCHED_OTHER    0         -20                  8
   39         39                               cpuhp/4     SCHED_OTHER    0           0                 10
   40         40                         idle_inject/4      SCHED_FIFO   50     *      -                 10
```

Figure 10.12: (Partial) screenshot of our ch10/query_task_sched.sh Bash script in action

A few things to notice:

- In our script, by using GNU `ps`, via invocation of the `ps -LA` command, we're able to capture all the threads that are alive on the system; their PIDs and TIDs are displayed. As you learned in *Chapter 6, Kernel Internals Essentials – Processes and Threads*, the PID is the user space equivalent of the kernel TGID, and the TID is the user space equivalent of the kernel PID. We can thus conclude the following:

    - If the PID and TID match, it – the thread seen in that row (the third column has its name) – is the `main()` thread of the process.
    - If the PID and TID match and the PID shows up only once, it's a single-threaded process.
    - If the PID repeats (the leftmost column) with varying TIDs (the second column), the process is multithreaded and those (the TIDs) are the children (more accurately, the worker) threads of the process. Our script shows this by indenting the TID number a bit to the right (in the truncated screenshot of *Figure 10.12*, you can't see it, as this occurs later).

- Notice how the vast majority of threads on a typical Linux box (desktop, server, or even embedded), will tend to be non-real-time (belonging to the policy `SCHED_OTHER`). A few (soft) real-time threads (SCHED_FIFO/SCHED_RR) may show up as well. **Deadline** (**DL**) and **Stop-Sched** (**SS**) threads are rare indeed (though of course, things like this tend to be project specific).
- Do notice the following observations regarding the real-time threads that showed up in the preceding output:
    - Our script highlights any real-time threads (one with **Sched Policy**, the fourth column, as `SCHED_FIFO` or `SCHED_RR`) by displaying an asterisk in the (sixth) column, labeled `*RT`.
    - Moreover, any soft real-time threads with a real-time priority of 99 (the maximum possible value) will have three asterisks in this same column (these tend to be specialized kernel threads).
- The `SCHED_RESET_ON_FORK` flag, when Boolean ORed with the scheduling policy, has the effect of disallowing any children (via `fork()`) from inheriting a privileged scheduling policy (a security measure).
- The penultimate column is the thread's "nice" value (recall, the range is -20 to +19, with -20 being the best priority and 19 the worst, with a default value of 0). Note that the nice value's only valid for threads with the scheduling policy `SCHED_OTHER` (as well as the batch and idle ones).
- The far-right column shows the CPU affinity mask (in hexadecimal), again a per-thread attribute! This implies that you can set the CPU cores that the thread can possibly be scheduled upon (we cover this aspect in detail in the following chapter). As an example, the mask value 0x3f in binary is 0011 1111, implying that this thread can run on any of the cores set to 1, and thus here, on any core (as there are a total of six cores on the system).

  Quick quiz: what does the CPU affinity mask value 0x8 imply?

- Changing the scheduling policy and/or priority of a thread can be performed with `chrt`; however, you should realize that this is a sensitive operation requiring root privileges (or, nowadays, the preferred mechanism would be via the capabilities model, with `CAP_SYS_NICE` being the required capability bit in question).

We will leave it to you to examine the code of the script (`ch10/query_task_sched.sh`). Also, be aware that performance and shell scripting do not really go together (so don't expect much in terms of performance here). Think about it – every external command issued within a shell script (and we have several here, such as `awk`, `grep`, and `cut`) involves a `fork-exec-wait` semantic and context-switching. Also, these all execute within a loop.

The `tuna(8)` program, a powerful GUI-based (or console-mode) system monitoring, tuning, and profile management tool, can be used to both query and set various attributes; this includes process-/thread-level scheduling policy/priority and the CPU affinity mask, as well as IRQ affinity. It's a well-designed GUI; do check it out! (Installing tuna: `https://tuna.readthedocs.io/en/stable/installation.html`).

Also, the `schedtool` utility is somewhat similar to `chrt` and can be used to query and set any or all task scheduling parameters on given threads; the man page on schedtool(8) covers its usage.

You might wonder, will the (few) threads with the `SCHED_FIFO` policy and a real-time priority of 99 always hog the system's processors? No, not really; the reality is that these threads are typically asleep most of the time. When the kernel does require them to perform some work, it wakes them up. Now, precisely due to their real-time policy and very high priority, it's pretty much guaranteed that they will more or less immediately get a CPU core and execute for as long as is required (going back to sleep once the work is done). The key point is, when they require the processor, they will get it (somewhat akin to an RTOS, but without the iron-clad guarantees, low scheduling latency, and determinism that an RTOS delivers).

How exactly does the `chrt` utility query (and set) the real-time scheduling policy/priority? Ah, that should be obvious: as they reside within the task structure in the kernel **Virtual Address Space** (**VAS**), the `chrt` process must issue a system call. There are several system call variations that perform these tasks: the one used by `chrt` to query the scheduling policy and priority is `sched_getattr()`, and the `sched_setattr()` system call is used to set the scheduling policy and priority. (Be sure to look up the `man` page on `sched(7)` for details on these and more scheduler-related system calls.) A quick `strace` run on `chrt` will indeed verify this!

```
$ strace chrt -p 1
[ … ]
sched_getattr(1, {size=56, sched_policy=SCHED_OTHER, sched_flags=0, sched_
nice=0, sched_priority=0, sched_runtime=0, sched_deadline=0, sched_period=0,
sched_util_min=0, sched_util_max=0}, 56, 0) = 0
newfstatat(1, "", {st_mode=S_IFCHR|0620, st_rdev=makedev(0x88, 0), ...}, AT_
EMPTY_PATH) = 0
write(1, "pid 1's current scheduling polic"..., 47pid 1's current scheduling
policy: SCHED_OTHER
) = 47
write(1, "pid 1's current scheduling prior"..., 39pid 1's current scheduling
priority: 0
) = 39
[ … ]
```

Now that you have the practical knowledge to query a thread's scheduling policy/priority and CPU affinity mask, it's time to dig a bit deeper. In the following section, we delve further into the internal workings of Linux's CPU scheduler.

We shall understand the basics of user and kernel preemption, figuring out who exactly runs the code of the kernel task scheduler and when it runs. Curious? I hope so! Read on!

Learning about the CPU scheduling internals – part 3

Let's begin by exploring the topic of preemption.

Preemptible kernel

Please visualize this hypothetical situation: you're running on a system with just one CPU core. An analog clock app is running on the GUI along with a C program, `a.out`, whose one line of code is (groan) `while (1);`. So, what do you think: will the CPU - hogger *while 1* process indefinitely hog the CPU, thus causing the GUI clock app to stop ticking (will its second hand stop moving altogether)?

A little thought (and experimentation) will reveal that, indeed, the GUI clock app keeps ticking in spite of the naughty CPU hogger app! This is really the whole point of having an OS-level scheduler: it can, and does, *preempt (kick out!)* the CPU-hogging user space process. (We briefly discussed the CFS algorithm previously; CFS will cause the aggressive CPU-bound process to accumulate a huge `vruntime` value and thus move more to the right on its rb-tree runqueue, thus penalizing itself of the processor!) All modern OSes support this type of preemption – it's called **user-mode preemption**.

But now consider this: what if you write a kernel module that performs the same `while(1)` infinite loop on a single core system? This poses a problem: the system will now simply hang. Why? Because, by default, most OS kernels are non-preemptible; that is, they cannot preempt themselves!

Besides this trivial example, real-world cases definitely exist where, if and when the kernel's non-preemptible, it can have a negative impact on (real-time) thread scheduling. Think about it: what if your high - priority thread is on a runqueue (and thus runnable) and needs to run urgently, but the kernel's stuck processing some long winding loop in a non-preemptible section of kernel (or kernel module/driver) code? Alternatively, a hardware interrupt occurs but can't be serviced, due to interrupts being masked in, again, a long - running non-preemptible section of code... Locking plays a role here (as you'll learn in a lot more detail in the last two chapters); in fact, for many years, the kernel had a notorious, very coarse-grained (and recursive) lock nicknamed the **Big Kernel Lock** (**BKL**); when held, it kept the kernel in a non-preemptible state for long-ish periods of time (with interrupts masked), thus wreaking havoc with performance and latency response. It was (finally!) completely removed in the 2.6.39 kernel.

So, the question arises: can, when required, the kernel preempt itself? Well, guess what: for many years now, Linux has provided a build-time configuration option *to make the kernel preemptible*; it's called `CONFIG_PREEMPT`. (Actually, this is merely evolution toward the long-term goal of cutting down latencies and improving the kernel and scheduler response. A large body of this work came from earlier, and some ongoing, efforts: the **Low Latency** (**LowLat**) patches, (the old) RTLinux work, and particularly, the **RTL** (**Real-Time Linux**) project (where Linux runs as a true RTOS! We will cover a bit on setting up RTL in the following chapter).)

Once this `CONFIG_PREEMPT` kernel config option is enabled and the kernel is built and booted into, we run on a preemptible kernel – where, at most points, the OS has the ability to preempt itself!

To check out this option, within the `make menuconfig` UI, navigate to **General Setup | Preemption Model.**

Actually, there's more to it; by default, modern Linux kernels have three kernel config options available as far as kernel preemption goes:

Preemption type	Characteristics
`CONFIG_PREEMPT_NONE`	Traditional model, geared toward high overall throughput (meant for servers). Though occasional long(er) delays can occur, this is the option to select to "maximize the raw processing power of the kernel, irrespective of scheduling latencies." (Here, kernel preemption's limited to only where explicit `cond_resched()` calls are invoked.)
`CONFIG_PREEMPT_VOLUNTARY`	Preemptible kernel (meant for desktop/laptop); more explicit preemption opportunity points within the OS (those points where `cond_resched()` and `might_sleep()` calls are invoked). Leads to lower scheduling latencies and better app response, at the cost of slightly lower throughput. Often the default for distros.
`CONFIG_PREEMPT`	Aka the LowLat kernel; (almost) the entire kernel is preemptible (meant for a desktop/laptop/powerful embedded system). Selecting this option implies that involuntary preemption of even kernel code paths is now possible (any/all sections of kernel code where preemption isn't explicitly disabled are now preemptible). Yields even lower scheduling latencies (a tens to low hundreds microseconds range on average) at the cost of slightly lower throughput and slight runtime overhead. Select this if you are building a kernel for a desktop or embedded system with latency requirements in the milliseconds range

Table 10.4: Linux OS kernel preemption – the available configuration options

The `kernel/Kconfig.preempt` kbuild configuration file contains the relevant menu entries for the preemptible kernel options. (As you will see in the following chapter, when building Linux as an RTOS, a fourth choice for kernel preemption, named `CONFIG_PREEMPT_RT`, appears.)

Wouldn't it be nice if we could dynamically select the kernel-preemption behavior at boot? The following section shows you how!

The dynamic preemptible kernel feature

From the 5.12 kernel, a new *dynamic preemptible* kernel feature (`CONFIG_PREEMPT_DYNAMIC`) was introduced. Having this feature enabled allows you to tune the kernel's preemption behavior (or mode) at boot by passing a kernel parameter (via the bootloader of course); it's named `preempt=`.

The values that it can be set to are as follows (refer *Table 10.4*):

- `preempt=none`: The kernel preemption behavior is the same as `CONFIG_PREEMPT_NONE`.
- `preempt=voluntary`: The kernel preemption behavior is close to that of `CONFIG_PREEMPT_VOLUNTARY`.
- `preempt=full` [default]: The kernel preemption behavior is the same as that of `CONFIG_PREEMPT`.

A typical intended use case is to allow distributions to ship one kernel binary image – built with `CONFIG_PREEMPT`, with (almost) full kernel preemption baked in – but be able to allow the end user to modify the kernel preemption mode at boot, thus allowing the user to select whether to run it as a typical server class system (with `preempt=none`), desktop (`preempt=voluntary`), or with full preemption enabled. This way, the distro (or product) doesn't have to ship and maintain different kernel images to serve differing use cases.

The kernel config is named `CONFIG_PREEMPT_DYNAMIC`; setting it to `y` enables the feature. (To do so within the `make menuconfig` UI, go to **General Setup | Preemption behaviour defined on boot**. It's a Boolean; turn it on, rebuild, and reboot, passing along the `preempt=<value>` parameter.)

FYI, `uname -a` as well as `/proc/version` will show this feature, if enabled (for example, I enabled it on a custom 6.1 kernel (on my Fedora VM)):

```
$ uname -a
Linux fedora 6.1.25-onfc38 #4 SMP PREEMPT_DYNAMIC Wed Jul 26 21:49:07 IST 2023
x86_64 GNU/Linux
```

Suggested exercise: Enable `CONFIG_PREEMPT_DYNAMIC` on your 6.1 LTS custom kernel; build it, and then boot with different values for the `preempt=` kernel parameter (and verify that it's been passed along by looking up `/proc/cmdline`).

Right, with that done, we will get back to some more details on the internal aspects. Let's very briefly summarize some key points we've learned in the preceding sections. You learned that the core kernel scheduling code is anchored within the `void schedule(void)` function, a thin wrapper over the worker function, `__schedule()`, and that it iterates over the modular scheduler classes in priority order, ending up with a thread picked (by one of the underlying scheduling class's code) to be context-switched to. All of this is fine; a couple of key questions now are: *who* exactly calls this "task scheduling" core code path, and *when* exactly is it run? (Precisely what was alluded to by the connector labelled "S" at the very top of *Figure 10.9*!) The following sections attempt to answer these questions; on, on!

Who runs the scheduler code?

A subtle yet key misconception regarding how scheduling works is unfortunately held by many: we imagine that some kind of kernel thread (or some such entity) called the "scheduler" is present, which periodically runs and schedules tasks. This is just plain wrong; in a monolithic OS such as Linux, scheduling is carried out by the process contexts themselves, the regular threads that run on the CPU in kernel mode!

In fact, the scheduling code is always run by the process context that is currently executing the code of the kernel – in other words, by `current` (we covered what exactly `current` is back in *Chapter 6, Kernel Internals Essentials – Processes and Threads,* in the *Accessing the task structure with current* section).

This may also be an appropriate time to remind you of what we shall call one of the *golden rules* of the Linux kernel: ***the scheduler must never ever run in any kind of atomic context (which includes interrupt context)***.

In other words, atomic (literally, indivisible) and/or interrupt context code must be guaranteed to be non-blocking and atomic – it must run to completion without interruption. This is why you cannot call `schedule()`, as it doesn't make any sense. (How can a code path be atomic when we put the caller to sleep? It can't, so don't call it.) Alternatively, as another example, calling `kmalloc()` with the `GFP_KERNEL` flag in any atomic context is wrong – as the `GFP_KERNEL` flag says it might block! But with the `GFP_ATOMIC` flag, it's all right, as that instructs the kernel memory management code to never block.

It's also important to realize that, in a preemptible kernel, kernel preemption is disabled while the schedule code paths run; this makes sense as well.

A quick summary of the key points learned here: the core task scheduling code on Linux – the code of `[__]schedule()` and all that it invokes – is run in process context by the process (thread, really) that's going to kick itself off the CPU by ultimately context-switching to some other thread! And this thread is what? Why, it's `current`, of course! (Relook at *Figure 10.9* and *Figure 10.11*; the code paths there – except the last box at the bottom where the "next" thread runs on the core – will be run by `current`). Also, there's a "rule": never call `schedule()` in any kind of atomic context (which includes interrupt context).

But *when* does this occur, IOW, when exactly is `schedule()` invoked? Read on to find out...

I'll understand if you find the following sections a bit dense on first reading... that's okay. If you prefer, you could skim the details for now (and perhaps peek at the *CPU scheduler entry points – a summary* section to digest the summary). Come back to the details when you're ready.

When does schedule() run?

The job of the OS scheduler is to arbitrate access to the processor (CPU) resources, sharing it between competing entities (threads) that want to use it. But what if the system is busy, with many threads continually competing for and acquiring and then relinquishing the processor(s)? More correctly, what we're really getting at is this: in order to ensure fair sharing of the CPU resources between tasks, you must ensure that the policeman in the picture, the scheduler itself, runs periodically on the processor. Sounds good, but how exactly can you ensure that?

Here's a (seemingly) logical way to go about it: have the OS hook into the timer chip's interrupt at boot and, as part of the "housekeeping" done in the interrupt handler routine when the timer interrupt fires, invoke the scheduler! Now, (a bit simplistically) in contrast to what we've learned before, the timer interrupt "fires" `CONFIG_HZ` times a second.

So if we call `schedule()` here, it gets a chance to run `CONFIG_HZ` times a second (which is often set to the value 250 on x86_64 Ubuntu, and 1,000 on x86_64 Fedora)! Hang on, though, we learned a golden rule in *Chapter 8, Kernel Memory Allocation for Module Authors – Part 1*, in the *Never sleep in interrupt or atomic contexts* section: *you cannot invoke the scheduler in any kind of atomic (including interrupt) context* (as was again just mentioned). So, going by this rule, we simply cannot invoke the scheduler code path within the timer interrupt (doing so would immediately lead to a kernel bug).

So what does the OS do? We'll get to it, but first, we need to cover some basics on a structure called `thread_info`.

Minimally understanding the thread_info structure

To clearly understand the following points, you'll need to know the basics regarding an arch-dependent per-thread data structure named `thread_info`. This structure is small, containing a few key "hotspot" members (that used to originally be within the task structure). The reason for this `thread_info` structure is the usual one – performance; looking up `thread_info` content is much faster than looking up the content of a large task structure, as it's much smaller and is typically designed to fit within a single CPU cacheline. (Well, on both the x86_64 and AArch64 at least, this is the case: its size is just 24 bytes. Also, this structure's leveraged in the calculation of `current` in the AArch32.)

The `thread_info` structure has a colorful placement history. In early Linux, it was a separate entity altogether; from 2.6 Linux, for 32-bit platforms (x86-32 and AArch32 at least), it lived within the kernel-mode stack of each thread. On more recent Linux versions (from 4.0 or 4.4) and for some arches, when `CONFIG_THREAD_INFO_IN_TASK=y`, it has become a member of the task structure itself (mostly due to security issues).

You can see the definition of `thread_info` for the x86[_64] at `https://elixir.bootlin.com/linux/v6.1.25/source/arch/x86/include/asm/thread_info.h#L56` and for the AArch64 at `https://elixir.bootlin.com/linux/v6.1.25/source/arch/arm64/include/asm/thread_info.h#L24`.

For our immediate purposes, just one of the members of `thread_info` is meaningful – a bitmask: `unsigned long flags`. As you can guess, it's a bitmask of various flag values (all in the style `TIF_<FOO>`, where `TIF` is an abbreviation for 'thread_info flag'). You'll find all the `TIF_*` flag macros defined here in the code base: `https://elixir.bootlin.com/linux/v6.1.25/source/arch/x86/include/asm/thread_info.h#L76`. Among the many flags that exist, the key one in our discussion on task scheduling is `TIF_NEED_RESCHED`. As will shortly be explained, if set, it implies the kernel "needs to reschedule ASAP"; if cleared, it doesn't need to reschedule.

Another topic that tends to (occasionally) come up in these discussions is that of hardware interrupts and their handling (`hardirqs` and `softirqs`) on Linux. Again, the (free!) *Linux Kernel Programming, Part 2* book, in *Chapter 4, Handling Hardware Interrupts*, covers these areas in depth; do check it out if you need to.

Right, now that you have understood these basics with regard to the `thread_info` structure, let's jump back to our discussions! The "when" of the scheduling code path (`schedule()`) and everything it calls is divided into two parts: one, when is the `TIF_NEED_RESCHED` bit set?

And, two, when is the `TIF_NEED_RESCHED` bit checked? Let's see:

- *When is the "need-resched" (`TIF_NEED_RESCHED`) bit set?*

  The `thread_info.flags:TIF_NEED_RESCHED` bit (this isn't meant to be C code; it's just to show the "need resched" bit conceptually), when set, is effectively akin to a red flag informing the kernel that (re)scheduling must be performed ASAP (in effect, `current` must be preempted now!). This flag can be set in the following contexts:

  - *(Timer) Interrupt housekeeping:* On every timer interrupt (technically, in the timer `softirq` code path), scheduler-related "housekeeping" is performed; among the work done here, a key question is: does `current` need to be preempted? If yes, set the `TIF_NEED_RESCHED` bit. (Caution: only the bit's set; `schedule()` isn't called here, as that's not allowed.)
  - *Task awakening*: When a task's awoken, it's enqueued in an appropriate runqueue; now, if it's determined that it must preempt `current`, set the `thread_info.flags:TIF_NEED_RESCHED` bit.

- *When is the "need-resched" (`TIF_NEED_RESCHED`) bit checked?*

  Scheduling Opportunity – Process context recognition: Have `current` check for the `thread_info.flags:TIF_NEED_RESCHED` bit being set at certain well-designed process context "opportunity points"; if the bit's set, invoke the schedule code path; otherwise, continue as usual.

Did you notice? In both the *(Timer) Interrupt housekeeping* and the *Task awakening* cases, even if we run in interrupt context, nothing goes wrong. This is because we do not invoke `schedule()` here and now; we merely set the `TIF_NEED_RESCHED` flag to inform the kernel that we need to reschedule ASAP, and that a reschedule must occur at the next available "opportunity point"! We expand on this discussion in the following few sections.

The timer interrupt housekeeping – setting TIF_NEED_RESCHED

Within the timer interrupt (in the code of `kernel/sched/core.c:scheduler_tick()`, wherein interrupts are disabled on the local core), the kernel performs the meta-work (the housekeeping) necessary to keep scheduling running smoothly; this involves the constant updating of the per - CPU runqueues as appropriate, task load balancing work, and so on. Please be aware that the actual `schedule()` function is *never called here*. At best, the scheduling class hook function (for the process context `current` that was interrupted), `sched_class:task_tick()`, if non-null, is invoked (refer *Figure 10.13*). For example, for any thread belonging to the fair (CFS) class, the update of the `vruntime` member (the virtual runtime), and the (priority-biased) time spent on the processor by the task is done here, in the hook function named `task_tick_fair()`.

More technically, all this work described in the preceding paragraph, the code of `scheduler_tick()`, runs within the timer interrupt softirq code path, `TIMER_SOFTIRQ`.

Now (besides the other sched-related housekeeping), while within this timer interrupt (softirq) context, we must decide: does `current` needs to be preempted or not? This key decision is made by checking the following conditions and preempting current if any of them are true:

- Has `current` exceeded its time quantum? Furthermore, is it exceeded by a large-enough threshold to its neighbor? This is precisely what the tunable (or sysctl) `/sys/kernel/debug/sched/min_granularity_ns` holds; it's the effective CFS timeslice. The defaults vary (it's 3 ms on x86_64 Ubuntu 22.04, and 2.25 ms on Fedora 38/39. As a quick reminder, we covered information on this, the effective CFS timeslice, in the *A note on the CFS scheduling period and timeslice* section).
- Does a newly born and run, or recently woken, task (on this CFS runqueue) have superior priority to `current`?
- Does the CFS rb-tree have a task with a lower `vruntime` than `current` (in other words, is `current` no longer the leftmost leaf node on this tree)?

(Thinking about it, the first and third points are quite alike; if the time quantum is exceeded, it's very likely the `vruntime` value would have increased to the extent that it's no longer the left-most leaf node in the runqueue.)

Let's say the kernel code paths just described have decided that a new task deserves the CPU more; then what? Does it call `schedule()`? No! As just explained above, we can't call `schedule()` in interrupt context. The kernel now merely "marks" the fact that we need to reschedule as soon as is feasible, at the "next scheduling opportunity point," by setting a "global" flag – the `thread_info->flags`'s `TIF_NEED_RESCHED` bit. (The reason we put the word global within quotes is that it's not really a kernel-wide global; it's actually simply a bit within the `current` instance's `thread_info->flags` bitmask, named `TIF_NEED_RESCHED`. Why? It's faster to access the bit that way rather than via a global!) Furthermore, it bears keeping in mind that if a newborn or recently woken task needs to be scheduled, it is placed into an appropriate runqueue, and it isn't run just yet; it will run in the near future when the next scheduling opportunity arrives.

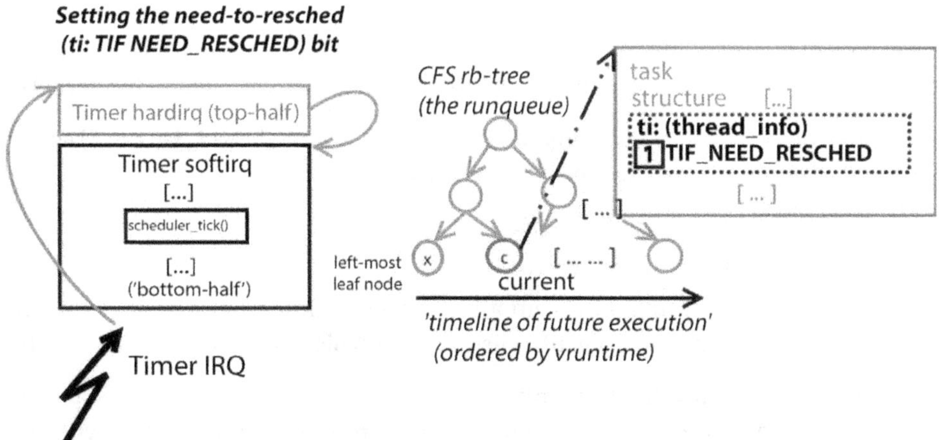

Figure 10.13: A conceptual diagram showing the setting of the thread_info->flags.TIF_NEED_RESCHED bit within the context of the timer softirq

Figure 10.13 shows this in a conceptual fashion. When the timer interrupt (or IRQ) fires, the fast-and-small so-called top-half or hardirq handler portion runs, very quickly performing its work; when done, the kernel invokes the timer bottom-half or softirq handler (the left portion of *Figure 10.13*). Here, a portion of work is the timer housekeeping tasks, as already described. We show the CFS rb-tree conceptually, with node **x** representing the left-most leaf node and **c** being the node representing `current` (the middle portion of *Figure 10.13*). So, in this diagram, `current` isn't the left-most leaf node any longer (implying that its `vruntime` value is higher than that of **x**). The code in the timer (scheduler-related) housekeeping code paths detects this case and sets the `TIF_NEED_RESCHED` bit (`ti` is the abbreviation for `thread_info`, which, in modern Linux, is within the task structure, as the right portion of *Figure 10.13* shows).

The other place where similar checks are made is when a task is **awoken**. When this occurs, it's dequeued from the wait queue it was on and enqueued in a CPU runqueue. Now, if it's found that the recently woken task (on that core:runqueue) must preempt `current` (due to it having superior priority, less `vruntime`, or whatever), the `TIF_NEED_RESCHED` bit is set. (We discuss this case in more detail in the upcoming *CPU scheduler entry points – a summary* section.)

It's also worth emphasizing that, in the typical (likely) case, when running the timer softirq code paths, there will typically be no need to preempt `current`, and thus there is no need to set the `thread_info.flags:TIF_NEED_RESCHED` bit; it will remain clear. If set, scheduler activation will occur soon, but when exactly? Do read on...

The process context part – checking TIF_NEED_RESCHED

One side of the coin, the just-described timer interrupt (softirq) portion of the scheduling housekeeping work, where it possibly **sets** the `thread_info:TIF_NEED_RESCHED` bit to "signal" to the kernel that the schedule code should be called as soon as possible, is continually carried out, as the system runs.

The other side of the coin is **checking** or recognizing whether this bit is set and invoking `schedule()` if so; this latter portion – the invoking of `schedule()` – is, note carefully:

- performed in process context only, and
- only at certain specific "opportunity points" that are sprinkled throughout the kernel code paths

The following are the typical so-called **opportunity points** where `thread_info->flags.TIF_NEED_RESCHED` is checked (often via the `need_resched()` helper):

- On return from the system call code path.
- On return from the interrupt code path.
- In general, any switch from non-preemptible to preemptible mode within the kernel is an opportunity point (when `preempt_enable()`is called). A typical one is when a spinlock is unlocked.

So, think about it: every time any thread running in user space issues a **system call**, that thread is (context-) switched to kernel mode and then runs code within the kernel, with kernel privilege (this is the monolithic kernel design). Of course, system calls are finite in length; when done, there is a well-known return path that they will follow in order to switch back to user mode and continue execution there.

On this return path in the kernel, a scheduling opportunity point is introduced: a check is done to see whether the `TIF_NEED_RESCHED` bit (within the `thread_info` structure's `flags` member) is set. If yes, the scheduler is activated, by having the process context call `schedule()`.

The other places where `schedule()` can be invoked are:

- Any explicit (or implicit) call to `schedule()` (e.g. when a blocking call's issued)
- Any calls to `cond_resched*()` functions can result in `schedule()` being invoked.

Obviously, an explicit or implicit invocation of `schedule()` while in the process context will trigger it being run. Further, the kernel provides a few "conditional schedule" APIs (like `cond_resched()`), allowing, say, a driver to check: *am I eating up too much CPU? If so, yield...* They lead to `schedule()` being invoked as and when required (in effect, only if the `current->ti->flags.TIF_NEED_RESCHED` bit is set).

Regarding the return-from-kernel-to-user-mode code path, here's the flow that activates the scheduling code path:

```
// include/linux/entry-common.h
#define EXIT_TO_USER_MODE_WORK                              \
    (_TIF_SIGPENDING | _TIF_NOTIFY_RESUME | _TIF_UPROBE |          \
     _TIF_NEED_RESCHED | _TIF_PATCH_PENDING | _TIF_NOTIFY_SIGNAL |   \
     ARCH_EXIT_TO_USER_MODE_WORK)
```

Notice that it's not just for scheduling that the kernel can prepare to switch back to user mode; it includes several other things that it must take care of (in effect, this is an opportunity point for even these other things – like handling pending signals, uprobes, kernel live patching, and more). If any of the above bits are set, the `EXIT_TO_USER_MODE_WORK` macro returns `True`; the kernel then sets up for the exit-to-user-mode and calls `schedule()`; here's the code view:

```
kernel/entry/common.c
static void exit_to_user_mode_prepare(struct pt_regs *regs)
{
  …
  ti_work = read_thread_flags();
  if (unlikely(ti_work & EXIT_TO_USER_MODE_WORK))
          ti_work = exit_to_user_mode_loop(regs, ti_work);
[ … ]
static unsigned long exit_to_user_mode_loop(struct pt_regs *regs,
                                unsigned long ti_work)
{
    /*
     * Before returning to user space ensure that all pending work
     * items have been completed.
     */
    while (ti_work & EXIT_TO_USER_MODE_WORK) {
```

```
    local_irq_enable_exit_to_user(ti_work);

    if (ti_work & _TIF_NEED_RESCHED)
        schedule();
    [ … ]
    if (ti_work & (_TIF_SIGPENDING | _TIF_NOTIFY_SIGNAL))
        arch_do_signal_or_restart(regs);
    [ … ]
```

Similarly, the same can occur after handling an (any) **hardware interrupt** (and any associated softirq handlers that needed to be run). While handling any hardware interrupt, the kernel is of course set to be non-preemptible; but once the handling's done, it's reset back to be preemptible. *Here lies an opportunity point:* the kernel checks the `TIF_NEED_RESCHED` bit, and if it is set, `schedule()` is invoked. (By the way, the same opportunity arises when a spinlock is unlocked, as then, the kernel is set to be preemptible!)

The following diagram (*Figure 10.14*) attempts to conceptually show this work in action. The top portion (points 1A to 5A; read them sequentially to follow what occurs) shows the checking of the `TIF_NEED_RESCHED` bit in the "return from system call" code path opportunity point; the bottom portion (points 1B to 3B) shows the "hardware interrupt (hardirq) return path opportunity point":

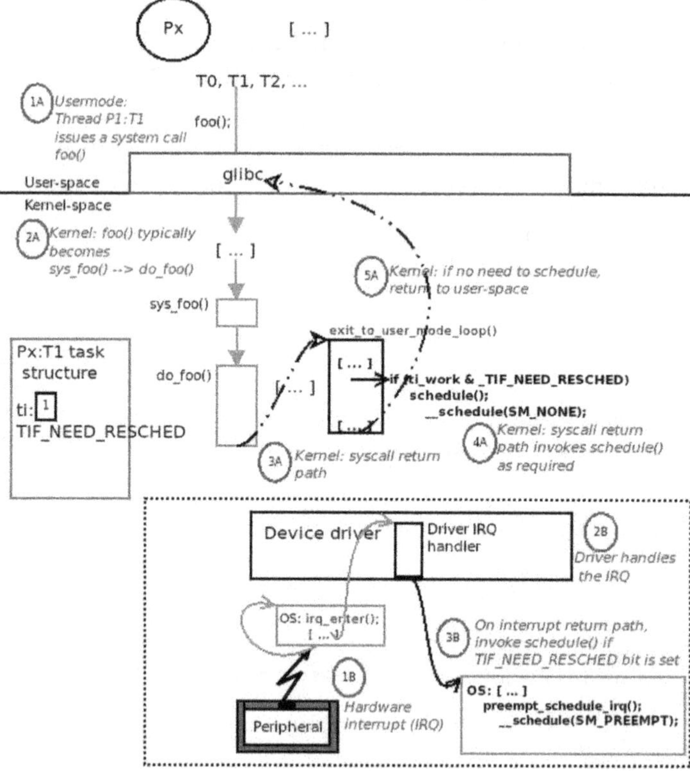

Figure 10.14: A conceptual diagram showing the checking of the TIF_NEED_RESCHED bit within the context of both the process-driven system call return path and the IRQ return path opportunity points

CPU scheduler entry points – a summary

Here, we summarize the preceding discussions on the setting and recognition of the `TIF_NEED_RESCHED` bit; in effect, we precisely answer the questions of when exactly `schedule()` is entered, or when exactly scheduling is activated (relooking at *Figure 10.9, Figure 10.13,* and *Figure 10.14* can be useful while reading this); so, `schedule()` is invoked (or the core scheduler activated):

- When any *explicit blocking call* is made by a process/thread: this makes sense. Blocking APIs end up calling `schedule()`, as the caller is required to wait on some event (when that event arises, the kernel (or driver that put it to sleep), will awaken it).
- *User-mode preemption from timer housekeeping*: if any of the three conditions described in the *The timer interrupt housekeeping – setting TIF_NEED_RESCHED* section come true; let's quickly restate them (see *Figure 10.13*):
    - Has `current` exceeded its time quantum (it's effective timeslice) by a sufficient amount (the value `min_granularity_ns`)?
    - Does a newborn, or recently woken, task have superior priority to `current` (the task currently executing the kernel code paths)?
    - Does the CFS rb-tree have a task with a lower `vruntime` than `current` (in other words, is `current` no longer the left-most leaf node on this tree)?
- *Timer (softirq) housekeeping:* check if the `TIF_NEED_RESCHED` bit needs to be set (the check is carried out in the timer softirq context function `scheduler_tick()`; see *Figure 10.13*). Note that the `TIF_NEED_RESCHED` bit can be set here but `schedule()` will never be invoked here.
- A summary of all so-called "opportunity points": here, we check whether the `TIF_NEED_RESCHED` bit is set. The following are the "opportunity points" where the check's carried out in the process context (see *Figure 10.14*):
    - On the system call return path
    - On the interrupt return path
- Other points where scheduling's activated (a bit of repetition here):
    - any call to `schedule()`
    - any `cond_resched*()` calls that can result in `schedule()` being invoked
    - (in general, any switch from non-preemptible to preemptible mode within the kernel).
- ***Wakeups*** occurring: this scenario has a task added to a runqueue. Must this task preempt `current`? If yes, set `TIF_NEED_RESCHED`, and invoke `schedule()` at the next opportunity point:
    - In a preemptible kernel: at any kernel preemption point (in effect, anytime `preempt_enable()` is called – for example, a spinlock being unlocked):
    - in any system call, when returning
    - in any hardware interrupt, when returning
- In a non-preemptible kernel, at the next:
    - return from system call to user space

- return from interrupt to user space
- any cond_resched*() calls
- any explicit call to schedule()

The detailed comments present just before the core kernel scheduling function kernel/sched/core.c:__schedule() are well worth reading through; they specify all the possible entry points to the kernel CPU scheduler (as we've just learned). For the 6.1 kernel code base, you'll find them here; do take a look: https://elixir.bootlin.com/linux/v6.1.25/source/kernel/sched/core.c#L6396.

The core scheduler code in brief

As stated earlier, always keep in mind that the scheduler code is being run in process context by the process (the thread, really) that's going to kick itself off the CPU by ultimately context-switching to some other thread! And this thread is what? Why, it's current, of course!

The __schedule() function has (among others) two local variables, pointers to struct task_struct named prev and next. The pointer named prev is set to rq->curr, which is nothing but current! The pointer named next will be set to the task that's going to be context-switched to and that's going to run next! So you see, current runs the scheduler code, calling itself the previous task, performing the work of running the *pick-next-task* for that modular scheduling class (the scheduling algorithm in effect), and then kicking itself off the processor by context-switching to next! Again, it's worth relooking at *Figure 10.9*, realizing that the code paths there – except for the last box at the bottom where the "next" thread runs on the core – are run by none other than current!

A few snippets from the core scheduling code follow (with annotations << like this >> reiterating the points we have discussed): https://elixir.bootlin.com/linux/v6.1.25/source/kernel/sched/core.c#L6435.

```
static void __sched notrace __schedule(unsigned int sched_mode)
{
        struct task_struct *prev, *next;
        unsigned long *switch_count;
        unsigned long prev_state;
        struct rq_flags rf;
        struct rq *rq;
        int cpu;

        cpu = smp_processor_id();
        rq = cpu_rq(cpu);
        prev = rq->curr;          << this is current ! >>
    [ ... ]
        next = pick_next_task(rq, prev, &rf);
            << Here we 'pick' the task to run next in an 'object-
                oriented' manner, as discussed earlier in detail... >>
        clear_tsk_need_resched(prev);
```

```
        clear_preempt_need_resched();
    [ ... ]
        if (likely(prev != next)) {    << switching to another thread's likely…
>>
         [ ... ]
           /* Also unlocks the rq: */
            rq = context_switch(rq, prev, next, &rf);
    [ ... ]
}
```

A quick word on the actual context switch follows.

The context switch

To finish this chapter, let's take a quick look at the (CPU scheduler) context switch. The job of the context switch (in the context of the CPU/task scheduler) is quite obvious: before simply switching to the next task, the OS must save the state of the previous, that is, the currently executing task (in other words, the state of `current`). The task structure holds another structure embedded within it to store/retrieve the *thread's hardware context*; it's the member `struct thread_struct thread` (on the x86, it's always the very last member of the task structure).

In Linux, an inline function, `kernel/sched/core.c:context_switch()`, performs the job of context-switching; you will by now realize of course that its code is run by `current`, which is considered to be the "previous" task. It performs the work of switching from the `prev` task (from `current`) to the `next` task, the winner of this scheduling round or preemption battle. This context switch is essentially performed in two, very arch-specific, stages:

- **The memory (MM) switch:** Switch an arch-specific CPU register to point to the memory descriptor structure (`struct mm_struct`) of `next`. On the x86[_64], this register is called **CR3** (**Control Register 3**); on ARM (AArch32), it's called the **TTBR0** (**Translation Table Base Register 0**) register. Why? Because it's from the `mm_struct`, where the kernel can "see" the entire memory picture of the process, including, importantly, the pointer to the base of its paging tables; setting this pointer correctly will thus make the MMU refer to `next`'s paging tables when it performs address translation as `next` executes!
- **The actual CPU switch:** Switch from `prev` to `next` by saving the stack and CPU register state of `prev` and restoring the stack and CPU register state of `next` to the processor; this is done within the `switch_to()` macro. This will have `next` resume processing precisely where it left off...

A more detailed implementation of the context switch is not something we shall cover here; do check out the *Further reading* section for more resources.

Finally, something very interesting: the kernel provides a way **to isolate** a given set of processors from disturbances – in effect, *from the effect of scheduler classes and SMP load balancing!* This is achieved by specifying a kernel parameter, `isolcpus=[flag-list,]<cpu-list>`.

The `flag-list` defaults to `domain`; when specified – and as default – all CPU cores in the `cpu-list` specified are isolated from "the general SMP balancing and scheduling algorithms." (This tends to be useful for some types of real-time applications.)

Hang on a moment though: the `isolcpus=` kernel parameter is now considered deprecated; in its place, you're advised to leverage the more powerful and flexible *cpusets* cgroups (v2) controller instead. Worry not – we shall cover many details on cgroups in the coming chapter.

Summary

In this chapter, you learned about several areas and facets of the versatile Linux kernel's CPU (or task) scheduler. Firstly, you saw how the actual KSE is a thread and not a process. We also learned that, in a monolithic OS like Linux, there is no "scheduler" thread within the kernel: scheduling is performed by a process context thread – the `current` thread – running the scheduling code paths, itself context-switching to the next task when done (thereby kicking itself off the processor – of course, the timer interrupt softirq housekeeping code paths have a key role to play in scheduler-related housekeeping as well!).

We then gained an appreciation of the available scheduling policies that the OS implements. Next, you understood that to support multiple CPUs in a superbly scalable fashion, the kernel powerfully mirrors this with a design that employs one runqueue per CPU core per modular scheduling class (again, the cgroups framework implies there's more to this, as we'll learn later). How to query any given thread's scheduling policy and priority, and more in-depth details on the internal implementation of the CPU scheduler, were then covered. We focused on how the modern scheduler leverages the modular scheduling class design, who exactly runs the actual scheduler code (`current`!) and when, and ended with a brief note on the internal code implementation of the core scheduling code and the context switch.

The next chapter continues on this very interesting journey, gaining more insight into and details about the inner workings of the kernel-level CPU scheduler. I suggest you first fully digest this chapter's content, work on the questions and exercises given, and then move on to the next chapter. Great going! Let's schedule the next session soon!

Questions

As we conclude, here is a list of questions for you to test your knowledge regarding this chapter's material: `https://github.com/PacktPublishing/Linux-Kernel-Programming_2E/blob/main/questions/ch10_qs_assignments.txt`. You will find some of the questions answered in the book's GitHub repo: `https://github.com/PacktPublishing/Linux-Kernel-Programming_2E/tree/main/solutions_to_assgn`.

Further reading

To help you delve deeper into the subject with useful materials, we provide a rather detailed list of online references and links (and, at times, even books) in a *Further reading* document in this book's GitHub repository, available here: `https://github.com/PacktPublishing/Linux-Kernel-Programming_2E/blob/main/Further_Reading.md`.

Learn more on Discord

To join the Discord community for this book – where you can share feedback, ask questions to the author, and learn about new releases – follow the QR code below:

`https://packt.link/SecNet`

11

The CPU Scheduler – Part 2

In this, our second chapter on the Linux kernel CPU (or task) scheduler, we continue our coverage from the previous chapter. In the preceding chapter, we covered several key areas regarding the workings (and visualization) of the CPU scheduler on the Linux OS. This included topics on what exactly the **Kernel Schedulable Entity** (**KSE**) on Linux is (it's the thread!), the POSIX scheduling policies that Linux implements, using `perf` (and other tools) to see the thread/scheduler flow, and how the design of the modern scheduler is based upon modular scheduling classes. We also covered how to query any thread's scheduling policy and priority (using a couple of command-line utilities), and finished by delving a lot deeper into the internal workings of the OS task scheduler.

With this background in place, we're now ready to explore more on the CPU scheduler on Linux; in this chapter, we shall cover the following areas:

- Understanding, querying, and setting the CPU affinity mask
- Querying and setting a thread's scheduling policy and priority
- An introduction to cgroups
- Running Linux as an RTOS – an introduction
- Miscellaneous scheduling related topics

We do expect that you've read (or have the equivalent knowledge of) *Chapter 10, The CPU Scheduler – Part 1*, before tackling this one.

Technical requirements

I assume you have gone through the complete chapter on *Kernel Workspace Setup*, that has been uploaded online, and have appropriately prepared a guest **Virtual Machine** (**VM**) running Ubuntu 22.04 LTS (or a later stable release, or a recent Fedora distro) and installed all the required packages. If not, I highly recommend you do this first.

To get the most out of this book, I strongly recommend you first set up the workspace environment, including cloning this book's GitHub repository for the code, and work on it in a hands-on fashion. The repository can be found here: `https://github.com/PacktPublishing/Linux-Kernel-Programming_2E`.

Understanding, querying, and setting the CPU affinity mask

The task structure – the root data structure for the thread (or task), containing several dozen thread attributes – has a few attributes directly pertaining to scheduling: the priority (the **nice** as well as the **Real-Time** (**RT**) priority values), the scheduling class structure pointer, the runqueue the thread is on (if any), and so on. (FYI, we covered generic details on the task structure back in *Chapter 6, Kernel Internals Essentials – Processes and Threads*).

Among these is an important member, the **CPU affinity bitmask** (the actual structure member is `cpumask_t *cpus_ptr`. FYI, until the 5.3 kernel, it was a member named `cpus_allowed`; this was changed in this commit: `https://github.com/torvalds/linux/commit/3bd3706251ee8ab67e69d9340ac2abdca217e733`). This bit mask is just that: a bit mask of the CPU cores that the thread (represented by that task structure) is allowed to run on. A simple visualization helps; on a system with 8 CPU cores, this is what the typical CPU affinity bitmask would (conceptually) look like:

7	6	5	4	3	2	1	0	←CPU core #
0	0	1	1	1	1	1	1	←Affinity bit

In the example above, each cell represents a CPU core; the top row denotes the CPU core number, and the cells in the row below show a sample value for it: a bottom row cell can be set to either 0 or 1, representing whether the thread can run on the corresponding CPU core or not. So, here, with a CPU bitmask value of `0x3f` (converting the binary number 0011 1111 to hexadecimal), it implies that this thread can be scheduled on CPU cores 0 to 5 but never on cores 6 and 7.

By default, all the CPU affinity mask bits are set; thus, by default, a thread can run on any core; this makes sense. For example, on a box with (the OS seeing) 8 CPU cores, the default CPU affinity bitmask for each thread alive would be binary `1111 1111` (`0xff` hexadecimal).

As this CPU affinity bitmask member's held within the task structure, this tells us that the *CPU affinity bitmask is a per-thread quantity*; this makes sense too – the KSE on Linux is a thread, after all. At run-time, the scheduler decides which core the thread will actually run upon. In fact, think about it, it's implicit: by design, each CPU core has a runqueue associated with it. Every runnable thread will be on a single CPU runqueue; it's thus eligible to run, and by default, runs on the CPU that its runqueue represents. Of course, the scheduler has a *load balancer* component that can migrate threads to other CPU cores (runqueues, really) as the need arises (kernel threads called `migration/n` assist in this task when required, where `n` is the core number).

The kernel does expose APIs to user space (system calls, of course, `sched_{s,g}etaffinity(2)` and their `pthread` wrapper library APIs), which allows an application to *affine*, or associate, a thread (or multiple threads) to particular CPU cores as it sees fit (by the same logic, we can do this within the kernel as well for any given kernel thread). For example, setting the CPU affinity mask to `1000 0001` binary, which equals `0x81` in hexadecimal, implies that the thread can execute *only* upon CPU cores 7 and 0 (remember, core counting starts from 0).

A key point: though you can manipulate the CPU affinity mask for a given thread(s), *the recommendation is to avoid doing so*; the kernel scheduler subsystem understands the CPU topography (or domains) in detail and can best load-balance the system.

Having said that, explicitly setting the CPU affinity mask of a thread can be beneficial due to the following reasons:

- Frequent cache invalidation (and thus unpleasant cache "bouncing") can be greatly reduced by ensuring a thread always runs on the same CPU core. (*Chapter 13, Kernel Synchronization – Part 2*, has coverage on CPU caching in more detail).
- Thread migration costs between cores can be effectively eliminated.
- To fulfill *CPU reservation* – a strategy to bestow the core(s) exclusively to one thread by guaranteeing all other threads are explicitly not allowed to execute upon that core.

The first two are useful in some corner cases; the third one, CPU reservation, tends to be a technique used in some time-critical, real-time systems where the cost of doing so is justified (FYI, this was achieved via the `isolcpus=` kernel parameter; these days, it's considered deprecated and you're to use the *cpusets cgroup* controller instead).

Now that you understand the theory behind it, let's actually write a user space C program to query and/or set the CPU affinity mask of any given thread.

Querying and setting a thread's CPU affinity mask

As a demonstration, we provide a small user space C program to query and set a user space process's (thread, really) CPU affinity mask. Querying the CPU affinity mask is achieved with the `sched_getaffinity()` system call, and setting it is done with its counterpart, the `sched_setaffinity()` system call:

```
#define _GNU_SOURCE
#include <sched.h>

int sched_getaffinity(pid_t pid, size_t cpusetsize,
                        cpu_set_t *mask);
int sched_setaffinity(pid_t pid, size_t cpusetsize,
                        const cpu_set_t *mask);
```

A specialized data type called `cpu_set_t` is what is used to represent the CPU affinity bitmask (the third parameter). It's quite sophisticated: its size is dynamically allocated based on the number of CPU cores seen on the system. This CPU mask (of type `cpu_set_t`) must first be initialized to zero; the `CPU_ZERO()` macro achieves this (several similar helper macros exist; do refer to the man page on `CPU_SET(3)`). The second parameter in both the preceding system calls is the size of the CPU set (we simply use the `sizeof` operator to obtain it). The first parameter is the **Process ID** (**PID**) of the process or thread whose CPU affinity mask you'd like to query or set.

To understand this better, it's instructive to see a sample run of our code (on GitHub, here: https://github.com/PacktPublishing/Linux-Kernel-Programming_2E/tree/main/ch11/cpu_affinity). Here, we run it on a native Linux system with 12 CPU cores:

```
$ nproc
12
$ make
gcc -Wall -O3 userspc_cpuaffinity.c -o userspc_cpuaffinity
gcc -g -Wall -O0 userspc_cpuaffinity.c -o userspc_cpuaffinity_dbg
$
$ ./userspc_cpuaffinity
Detected 12 CPU cores [for this process ./userspc_cpuaffinity:237363]
CPU affinity mask for PID 237363:
 237363 pts/2    00:00:00 userspc_cpuaffi
        +---+---+---+---+---+---+---+---+---+---+---+---+
core#   | 11| 10|  9|  8|  7|  6|  5|  4|  3|  2|  1|  0|
        +---+---+---+---+---+---+---+---+---+---+---+---+
cpumask|  1|  1|  1|  1|  1|  1|  1|  1|  1|  1|  1|  1|
        +---+---+---+---+---+---+---+---+---+---+---+---+
$
```

Figure 11.1: Our demo user space app showing the CPU affinity mask of the calling process

Here, we have run the app with no parameters. In this mode, it queries the CPU affinity mask of itself (meaning, of the userspc_cpuaffinity calling process). We print out the bits of the bitmask: as you can clearly see in the preceding screenshot (*Figure 11.1*), it's binary 1111 1111 1111 (which is equivalent to 0xfff), implying that, by default, the process is eligible to run on any of the 12 CPU cores available on the system!

The app internally detects the number of CPU cores available by running the nproc utility via the useful popen() library API. Do note, though, that the value returned by nproc is the number of CPU cores available to the calling process; it may be less than the actual number of (online and offline) CPU cores, though it's usually the same. The number of available cores can be changed in a few ways, the proper way being via the cgroup cpuset resource controller (we cover information on control groups (cgroups) later in this chapter).

The querying code is as follows (the source file is ch11/cpu_affinity/userspc_cpuaffinity.c):

```
static int query_cpu_affinity(pid_t pid)
{
    cpu_set_t cpumask;

    CPU_ZERO(&cpumask);
    if (sched_getaffinity(pid, sizeof(cpu_set_t), &cpumask) < 0) {
        perror("sched_getaffinity() failed");
        return -1;
    }
    disp_cpumask(pid, &cpumask, numcores);
```

```
    return 0;
}
```

Our `disp_cpumask()` function draws the bitmask (we leave it to you to check it out).

If additional parameters are passed to this program – the PID of the process (or thread) as the first parameter, and a CPU bitmask (in hexadecimal) as the second parameter – we then attempt to *set* the CPU affinity mask of that process (or thread) to the value passed. Of course, changing the CPU affinity bitmask requires you to own the process or have root privileges (more correctly, to have the `CAP_SYS_NICE` capability).

Here's a quick demo: in *Figure 11.2*, `nproc` shows us the number of CPU cores (12); then, we run our app to query and set our (`bash`) shell process's CPU affinity mask. On a laptop with 12 cores, let's say that the affinity mask of `bash` is `0xfff` (binary `1111 1111 1111`) to begin with, as expected; here, we change it to `0xdae` (binary `1101 1010 1110`) and query it again to verify the change:

```
$ nproc
12
$ ps
    PID TTY          TIME CMD
   6397 pts/2    00:00:00 bash
  35507 pts/2    00:00:13 retext
  35514 pts/2    00:00:00 python3
  35515 pts/2    00:00:00 python3
 126098 pts/2    00:00:27 gitg
 126289 pts/2    00:00:00 git
 237565 pts/2    00:00:00 ps
$
$ ./userspc_cpuaffinity 6397 0xdae
Detected 12 CPU cores [for this process ./userspc_cpuaffinity:237571]
CPU affinity mask for PID 6397:
   6397 pts/2    00:00:00 bash
         +---+---+---+---+---+---+---+---+---+---+---+---+
core#    | 11| 10|  9|  8|  7|  6|  5|  4|  3|  2|  1|  0|
         +---+---+---+---+---+---+---+---+---+---+---+---+
cpumask|   1|  1|  1|  1|  1|  1|  1|  1|  1|  1|  1|  1|
         +---+---+---+---+---+---+---+---+---+---+---+---+

Setting CPU affinity mask for PID 6397 now...
CPU affinity mask for PID 6397:
   6397 pts/2    00:00:00 bash
         +---+---+---+---+---+---+---+---+---+---+---+---+
core#    | 11| 10|  9|  8|  7|  6|  5|  4|  3|  2|  1|  0|
         +---+---+---+---+---+---+---+---+---+---+---+---+
cpumask|   1|  1|  0|  1|  1|  0|  1|  0|  1|  1|  1|  0|
         +---+---+---+---+---+---+---+---+---+---+---+---+
$
$ nproc
8
$
```

Figure 11.2: Our demo app queries and then sets the CPU affinity mask of bash to 0xdae

Okay, this is interesting. To begin with, the app correctly detects the number of CPU cores available to it as 12. It then queries the (default) CPU affinity mask of the bash process (as we pass its PID as the first parameter); it shows up, as `0xfff`, as expected. Then, as we've also passed a second parameter – the bitmask to now set (`0xdae`) – it does so, setting the CPU affinity mask of bash to `0xdae`. Now, as the terminal window we're on is this very same bash process, running `nproc` again shows the value as 8, not 12! That's indeed correct: the bash process now has only eight CPU cores available to it. (This is because we don't revert the CPU affinity mask to its original value on exit.)

Here's the relevant code to set the CPU affinity mask:

```
// ch11/cpu_affinity/userspc_cpuaffinity.c
static int set_cpu_affinity(pid_t pid, unsigned long bitmask)
{
    cpu_set_t cpumask;
    int i;

    printf("\nSetting CPU affinity mask for PID %d now...\n", pid);
    CPU_ZERO(&cpumask);

    /* Iterate over the given bitmask, setting CPU bits as required */
    for (i=0; i<sizeof(unsigned long)*8; i++) {
        /* printf("bit %d: %d\n", i, (bitmask >> i) & 1); */
        if ((bitmask >> i) & 1)
            CPU_SET(i, &cpumask);
    }

    if (sched_setaffinity(pid, sizeof(cpu_set_t), &cpumask) < 0) {
        perror("sched_setaffinity() failed");
        return -1;
    }
    disp_cpumask(pid, &cpumask, numcores);
    return 0;
}
```

In the preceding code snippet, you can see we first set up the `cpu_set_t` bitmask appropriately (by looping over each bit; as you'll know, the expression `(bitmask >> i) & 1` tests the ith bit for being 1) and then employs the `sched_setaffinity()` system call to set the new CPU affinity mask on the given `pid`.

It's important – and quite correct – to note that while anyone can always query any task's CPU affinity mask, you can't set it unless you own the task, have root access, or have `CAP_SYS_NICE` capability.

Using taskset to perform CPU affinity

Akin to how (in the preceding chapter) we used the convenient user space utility program, `chrt`, to get (or set) a process's (or thread's) scheduling policy and/or priority, you can use the user space `taskset` utility to get and/or set a given process's (or thread's) CPU affinity mask. A couple of quick examples follow; note that these examples were run on an x86_64 Linux VM with 6 CPU cores:

Use `taskset` to query the CPU affinity mask of systemd (PID 1):

```
$ taskset -p 1
pid 1's current affinity mask: 3f
$
```

Think about it: 0x3f is 0011 1111 in binary, which represents (all) 6 CPU cores being enabled for the process/thread in question (here, systemd).

Now, as an example, let's run the compiler under the aegis of `taskset`, using it to ensure that GCC – and its descendants (the assembler and linker processes) – run only on the first two CPU cores; the first parameter to `taskset` is the CPU affinity bitmask (`03` is binary `0011`):

```
$ taskset 03 gcc userspc_cpuaffinity.c -o userspc_cpuaffinity -Wall -O3
```

Done. Do look up the man page on `taskset(1)` for complete usage details. (FYI, as mentioned in the previous chapter, the `schedtool(8)` utility too can be used to get/set the CPU affinity bitmask of a given thread/process.)

Setting the CPU affinity mask on a kernel thread

As a pretty interesting example, if we want to demonstrate a synchronization technique called **per-CPU variables** (as we shall indeed learn about and do in *Chapter 13, Kernel Synchronization – Part 2*, in the *Per-CPU – an example kernel module* section), we are required to create two kernel threads (kthreads) and guarantee that each of them runs on a separate CPU core. To do so, we must of course explicitly set the CPU affinity mask of each kernel thread to be different and non-overlapping (to keep it simple, let's set the first kthread's affinity mask to `0`, and the second kthread's mask to `1`, in order to guarantee they execute on only CPU cores `0` and `1`, respectively). But there's a problem...the following section explains it.

Hacking the availability of non-exported symbols

The thing is, setting CPU affinity from within a module nowadays is unfortunately not a clean job – quite a *hack*, to be honest; we show it here but it's definitely *not* recommended for production purposes.

The reason is that the API within the kernel that we require to set the CPU affinity bitmask – `sched_setaffinity()` – is present but isn't exported. As we learned from our earlier chapters on writing modules, an out-of-tree kernel module (like ours) can only use exported functions (and data). So what do we do?

The "usual" approach that module developers used for many years (and indeed, that I used in the first edition of this book!) is to employ the convenience routine `kallsyms_lookup_name()` to look up any given symbol within the kernel and obtain its (kernel virtual) address.

Armed with this, any decent C programmer can treat the address as a function pointer and invoke it at will, thus effectively overcoming the restriction that only exported functions can be called from an out-of-tree module! (A neat hack! Experienced kernel folk, though, will almost certainly groan at this.)

True, but from kernel version 5.7, the community decided it was time to stop this (silly) abuse and simply unexported the `kallsyms_lookup_name()` (and the similar `kallsyms_on_each_symbol()`) functions! (The short form commit ID is `0bd476e6c671`, have a look.) So, now what? Well, we can always look up any kernel symbol via the `/proc/kallsyms` pseudofile, as long as we have root access (that's the security). Furthermore, with kernel address space layout randomization (KASLR) enabled (it typically is in modern kernels), this value changes on every boot, and thus can't be hardcoded (also good for security). So, we write a small wrapper script to do this (it's here: `ch13/3_lockfree/percpu/run`; yes, the code's from *Chapter 13, Kernel Synchronization – Part 2*) and pass the address (of the `sched_setaffinity()` routine looked up via `/proc/kallsyms`) to the module as a parameter (`ch13/3_lockfree/percpu/percpu_var.c`), which then, treating it as a function pointer, manages to call it. Whew!

Here's the signature of the `sched_setaffinity()` function:

```
long sched_setaffinity(pid_t pid, const struct cpumask *new_mask);
```

A brief snippet of the relevant code – where we use the passed the (via the module parameter named `func_ptr`) `sched_setaffinity()` function pointer to set the CPU affinity mask as desired – follows:

```
    // ch13/3_lockfree/percpu/percpu_var.c
    [ … ]
    static unsigned long func_ptr;
    module_param(func_ptr, ulong, 0);
    unsigned long (*schedsa_ptr)(pid_t, const struct cpumask *) = NULL;
    [ … ]
    // set up the function pointer
    schedsa_ptr = (unsigned long (*)(pid_t pid, const struct cpumask *in_mask))
func_ptr;
    [ … ]
          /* pr_info("setting cpu mask to cpu #%u now...\n", cpu); */
    cpumask_clear(&mask);
    cpumask_set_cpu(cpu, &mask); // 1st param is the CPU number, not bitmask
    /* !HACK! sched_setaffinity() is NOT exported, we can't call it
        * sched_setaffinity(0, &mask);   // 0 => on self
     * so we invoke it via it's function pointer
     */
    ret = (*schedsa_ptr)(0, &mask);   // 0 => on self
    [ … ]
```

While unconventional and controversial, it does work, but please avoid hacks like this in production.

Now that you know how to get/set a (kernel) thread's CPU affinity mask, let's move on to the next logical step: how to programatically get/set a thread's scheduling policy and priority! The next section delves into the details.

Querying and setting a thread's scheduling policy and priority

In *Chapter 10, The CPU Scheduler – Part 1*, in the *Thread Priorities* section, you learned how to query the scheduling policy and priority of any given thread via the `chrt` utility (we also demonstrated a simple Bash script to do so). There, we mentioned the fact that `chrt` internally invokes the `sched_getattr()` system call in order to query these attributes.

Very similarly, setting the scheduling policy and priority can be performed either by using the `chrt` utility (making it simple to do so within a script, for example), or programmatically within a (user space) C application with the `sched_setattr()` system call. In addition, the kernel exposes other APIs: `sched_{g,s}etscheduler()` and its `pthread` library wrapper APIs, `pthread_{g,s}etschedparam()` (as these are all user space APIs, we leave it to you to browse through their man pages to get the details and try them out for yourself).

Setting the policy and priority within the kernel – on a kernel thread

As you know by now, the kernel is most certainly neither a process nor a thread. Having said that, the Linux kernel is certainly multithreaded-capable and does contain threads, the so-called **kernel threads** (or **kthreads**). Like their user space counterparts, kernel threads can be created as required (from within the core kernel, a device driver, or a kernel module; the kernel exposes APIs for this purpose).

They *are* schedulable entities (KSEs!) and, of course, each of them has a task structure and a kernel-mode stack; thus, just as with regular threads, they compete for the CPU resource and their scheduling policy and priority can be programmatically queried or set as required.

If you'd like to learn more about and use kthreads, might I suggest the (free e-book!) companion guide *Linux Kernel Programming – Part 2*, particularly *Chapter 5, Working with Kernel Timers, Threads, and Work Queues*.

Regarding kthreads, how does one interpret their (quite weird) names (f.e., `'[kworker/u12:2]'`)? This link provides the answer: `https://unix.stackexchange.com/a/152865`

So, to the point at hand: in user space, the modern preferred system call to query and set a thread's scheduling attributes is `sched_getattr()` and `sched_setattr()`, respectively. In earlier days, it used to be the `sched_{g|s}et_scheduler()` pair of system calls. Now, the `sched_{g|s}etattr()` system calls receive a pointer to `struct sched_attr`, which holds all possibly required details; see the man page (`https://man7.org/linux/man-pages/man2/sched_setattr.2.html`).

So, going the modern way, one would assume we'll use the kernel implementation of these system calls to perform similar work within the kernel. Not so fast; the kernel community deems that the old(er) design – allowing user (app) and module developers to happily invoke these APIs with a policy such as `SCHED_FIFO` and use any (real-time) priority they think is correct – is fundamentally broken.

Why? As we can easily land up with situations like this: two or more SCHED_FIFO threads with the same priority, and/or the usage of "random" priority values – with no careful thought behind them – being chosen. These can cause chaos in terms of CPU scheduling, and thus resource management. Thus, what's been done from the 5.9 kernel follows (allow me to quote directly from the commit, as it's really the best way to get the message across); here's a portion of the commit, https://github.com/torvalds/linux/commit/7318d4cc14c8c8a5dde2b0b72ea50fd2545f0b7a:

```
…
Therefore it doesn't make sense to expose the priority field; the
kernel is fundamentally incapable of setting a sensible value, it
needs systems knowledge that it doesn't have.

Take away sched_setschedule() / sched_setattr() from modules and
replace them with:

 - sched_set_fifo(p); create a FIFO task (at prio 50)
 - sched_set_fifo_low(p); create a task higher than NORMAL,
       which ends up being a FIFO task at prio 1.
 - sched_set_normal(p, nice); (re)set the task to normal

This stops the proliferation of randomly chosen, and irrelevant, FIFO
priorities that dont't really mean anything anyway.

The system administrator/integrator, whoever has insight into the
actual system design and requirements (userspace) can set-up
appropriate priorities if and when needed. ...
```

Ah; so, now, dear module authors, *we're* to use *these* APIs – sched_set_fifo(), sched_set_fifo_low(), and sched_set_normal() – when setting up (SCHED_)FIFO tasks (threads) within the kernel. As the above commit mentioned, we trust the administrator and/or user space developer(s) to program *user apps* and provide them with correct and meaningful real-time priority values as required; the kernel (or modules) aren't supposed to know or question such decisions – it merely carries them out (again, an example of the *provide mechanism, not policy* design guideline in action).

The first two APIs:

- are wrappers around the sched_setscheduler_nocheck() function within the kernel
- set the thread's scheduling policy to SCHED_FIFO
- set the thread's (real-time) priority to MAX_RT_PRIO/2 (i.e. 50) and 1, respectively

The sched_set_normal():

- is a wrapper over the sched_setattr_nocheck()

- sets the thread's scheduling policy to `SCHED_NORMAL` (same as `SCHED_OTHER`, meaning non real-time, Completely Fair Scheduler (CFS) driven)
- sets the thread's `nice` value to the second parameter

Here, the `*_nocheck()` notation is meant to imply that the kernel doesn't even bother to check whether the process context that's running these APIs has sufficient privileges; it goes through regardless. (See the comment here: `https://elixir.bootlin.com/linux/v6.1.25/source/kernel/sched/core.c#L7742`.)

Furthermore, these three APIs are **GNU Public License** (GPL) exported, implying they can only be employed by modules licensed under the GNU GPL.

A real-world example – threaded interrupt handlers

One example of the kernel's usage of kernel threads is when the kernel (quite commonly) uses threaded interrupts (work queues are another example). Here, the kernel must create a dedicated kernel thread with the `SCHED_FIFO` (soft) real-time scheduling policy and a real-time priority value of `50` (halfway between), to correctly handle what's called **threaded interrupts**. Let's see the relevant code paths, `https://elixir.bootlin.com/linux/v6.1.25/source/kernel/irq/manage.c#L1448`:

```
static int
setup_irq_thread(struct irqaction *new, unsigned int irq, bool secondary)
{
	struct task_struct *t;

	if (!secondary) {
		t = kthread_create(irq_thread, new, "irq/%d-%s", irq,
				   new->name);
	} else {
		t = kthread_create(irq_thread, new, "irq/%d-s-%s", irq,
				   new->name);
	}
	[ … ]
```

The `kthread_create()` macro takes care of creating the kernel thread (the kthread). Now, the `irq_thread()` kernel-only API (invoked via the `kthread_create()` macro as the *thread function*) will, as part of its code path, set the scheduling policy and priority appropriately, `https://elixir.bootlin.com/linux/v6.1.25/source/kernel/irq/manage.c#L1286`:

```
/* Interrupt handler thread */
static int irq_thread(void *data)
{
	struct callback_head on_exit_work;
	struct irqaction *action = data;
```

```
[ ... ]
sched_set_fifo(current);
[ ... ]
```

Aha! Notice the call to `sched_set_fifo()`; as we saw, it sets up the kthread (the calling thread, referenced by `current`) to use the policy `SCHED_FIFO` and a (real-time) priority of `50`. Done.

Curious why `SCHED_FIFO` and priority `50` for IRQ (interrupt) threads? In fact, why use threaded IRQ handlers at all? (FYI, nowadays, the norm is for the majority of drivers to use threaded handlers.) Learn all about it in this book's companion volume *Linux Kernel Programming - Part 2* (free e-book!), in *Chapter 4, Handling Hardware Interrupts*.

Right, now that you understand to a good extent how CPU (or task) scheduling works at the level of the OS, we'll move on to yet another quite compelling discussion – that of cgroups; read on!

An introduction to cgroups

In the hazy past, the kernel community struggled mightily with a rather vexing issue: though scheduling algorithms and their implementations – the early 2.6.0 O(1) scheduler, and a little later (with 2.6.23), the **Completely Fair Scheduler** (**CFS**) – promised, well, completely fair scheduling, it really wasn't "completely fair" in any meaningful sense of the term!

Think about this for a moment: let's say you are logged in to a Linux server along with nine other people. Everything else being equal, it is likely that processor time is (more or less) fairly shared between all ten of you; of course, you will understand that it's not really people that run on the processor(s) and eat memory, it's processes and threads that do so on their behalf.

For now, at least, let's assume it's (mostly) fairly shared. But, what if you, one of the ten users logged in, write a user space program that, in a loop, indiscriminately spawns off several new threads, each of which performs a lot of CPU-intensive work (and perhaps as an added bonus, allocates large swathes of memory as well) in each loop iteration!? The CPU bandwidth allocation (even via CFS) *is no longer fair* in any real sense of the term; your account will effectively hog the CPUs (and perhaps other system resources, such as memory and I/O, as well)!

A generic solution that precisely and effectively managed CPU (and other resource) bandwidth was required, throttling (checking, not allowing) more of that resource to be consumed when the specified limits were reached. Many proposed patches were discussed and discarded; ultimately, engineers from Google, IBM, and others obliged with a patch set that put the modern-day **control groups** (**cgroups**) solution into the Linux kernel (back in version 2.6.24, October 2007. The original ideas and implementation were by Paul Menage and Rohit Seth at Google in 2006.). In a nutshell, cgroups is a kernel feature that allows the system administrator (or anyone with root access) to be able to elegantly perform bandwidth allocation and fine-grained resource management on the various resources or *controllers* (as they are called in the cgroup lexicon) on a system. Do note: using cgroups, it's not just the processors (CPU bandwidth), but also memory and block I/O bandwidth (and more) that can be carefully partitioned or allocated and monitored as required by your project or product.

So, with the example we began this topic with – ten users on a Linux system – if all the processes were placed within the same that cgroup's *and* the cgroups CPU controller was enabled for it, it would then, in the face of CPU contention, really result in fair CPU shares to each process! Or, you, as the sysad, can do more sophisticated things: you could divvy up the system into several cgroups – one for, say, building a project (a Yocto build, say), one for the web browsers, one for virtual machines, and so on – and then fine-tune and allocate resources (CPU, memory, and I/O) to each cgroup as desired! Indeed, it's what pretty much all modern distros do automatically, courtesy of the powerful *systemd* framework (more on this follows); it's what embedded Linux usually does as well, including Android.

So, hey, you're interested now! How do you enable this cgroups feature? Simple – it's a kernel feature you enable (or disable) at quite a fine granularity in the usual way: by configuring the kernel. The relevant menu (via the convenient `make menuconfig` user interface) is **General setup | Control Group support**. Try this: `grep` your kernel config file for `CGROUP`; then if required, tweak your kernel config, rebuild, reboot with the new kernel, and test. (We covered kernel configuration in detail back in *Chapter 2, Building the 6.x Linux Kernel from Source – Part 1*, and the kernel build and install in *Chapter 3, Building the 6.x Linux Kernel from Source – Part 2*).

Good news: cgroups is enabled by default on any (recent enough) Linux system that runs the *systemd* init framework. As mentioned just now, you can query the cgroup controllers enabled by `grep`-ping your kernel config file, and modify the config as desired; on desktop and server class systems, this typically shouldn't be required.

From its initiation in 2.6.24, cgroups, like all other kernel features, continually evolve. Fairly recently, a point was reached where sufficiently improved cgroup features became incompatible with the old, resulting in a new cgroup design and release, one christened **cgroups v2** (or simply cgroups2 – Tejun Heo is the maintainer); this was declared production-ready in the 4.5 kernel series (with the older one now referred to as cgroups v1 or as the legacy cgroups implementation). Note that, as of the time of this writing, both can and do exist together, with some limitations; many applications and frameworks still use the older cgroups v1 and are yet to migrate to v2. This is changing, though; soon, if not already, cgroups2 will become the de facto one to use, so plan on using it.

In this coverage, we shall focus almost exclusively on using the modern version, cgroups v2. The best documentation is the official kernel one, available here (for kernel 6.1): `https://www.kernel.org/doc/html/v6.1/admin-guide/cgroup-v2.html`. (FYI, the docs for the latest kernel version are always available as well, here: `https://docs.kernel.org/admin-guide/cgroup-v2.html`.

A detailed rationale of why to use cgroups v2 as opposed to cgroups v1 can be found within the kernel documentation here: `https://www.kernel.org/doc/html/latest/admin-guide/cgroup-v2.html#issues-with-v1-and-rationales-for-v2`.

Cgroup controllers

A cgroup controller is the underlying kernel component that's responsible for distributing a given resource (like CPU cycles, memory and I/O bandwidth, and so on) within and through a cgroup hierarchy: a cgroup and its descendants. You can think of it as a "resource limiter" of sorts for a given cgroup hierarchy.

The man page on `cgroups(7)` describes in some detail the interfaces and various available **(resource) controllers** (or *subsystems*, as they are sometimes referred to). The cgroups v2 controllers typically available are as follows (*Table 11.1* shows cgroups v2 stuff; the original cgroups v1 implementations for many of the controllers date back to 2.6.24):

Cgroups v2 controller name	What it controls (or constrains or regulates) when enabled	Since kernel version
cpu	CPU bandwidth (cycles)	4.15
cpuset	CPU affinity and memory-node placement (especially useful for large NUMA systems)	5.0
memory	Memory (RAM) usage	4.5
io	Distribution of I/O resources	4.5
pids	Hard limit on the number of processes in the cgroup	4.5
devices	Creation of and access to device files (only via cgroup BPF programs)	4.15
rdma	Distribution and accounting of **remote direct memory access (RDMA)** resources	4.11
hugetlb	Limits the HugeTLB (huge pages) usage per cgroup	5.6
misc	Various; see `https://www.kernel.org/doc/html/v6.1/admin-guide/cgroup-v2.html#misc`	5.13

Table 11.1: A summary of available cgroups v2 controllers on modern Linux systems

We refer interested readers to said official kernel docs and the man pages for details; as an example, the PIDS controller is very useful in preventing fork bombs by allowing you to limit the number of processes that can be forked off from that cgroup or its descendants. (Fork bombs are a silly but nevertheless deadly DoS attack where the `fork()` system call is issued typically within an infinite loop!)

Next, very importantly, how are the kernel cgroups made visible to (exposed), or interfaced with, user space? Ah, in the usual manner on Linux: control groups are exposed via a purpose-built synthetic or pseudo filesystem! It's the `cgroup` filesystem, typically mounted at `/sys/fs/cgroup`. Well, with cgroups v2, the filesystem type is now called `cgroup2` (you can simply do **mount | grep cgroup** to see this). There are plenty of interesting goodies to explore within it; it's what we do as we make progress...

Let's begin with this: how can I find which controllers are enabled for my system (kernel, really)? It's easy:

```
$ cat /sys/fs/cgroup/cgroup.controllers
cpuset cpu io memory hugetlb pids rdma misc
```

Clearly, it shows a space-separated list of available controllers (I ran this on my x86_64 Fedora 38 VM. Also, note that using `/proc/cgroups` to peek at the controllers is cgroups v1 compatible only; don't rely on it for cgroups v2.). The exact controllers you see here depend on how the kernel is configured.

In cgroups v2, all controllers are mounted in a single hierarchy (or tree). This is unlike cgroups v1, where multiple controllers could be mounted under multiple hierarchies or groups. The modern init framework, *systemd*, is a user of both the v1 and v2 cgroups. In fact, it's `systemd` that auto-mounts the cgroups v2 filesystem during startup (at `/sys/fs/cgroup/`).

Exploring the cgroups v2 hierarchy

Looking under the cgroups (v2) pseudo filesystem mount point – always `/sys/fs/cgroup` by default – can make you stare in wonder at all the pseudo files (and folders) within (go ahead, peek at *Figure 11.3*); this section will explore many of its more interesting and useful nooks and crannies!

Let's begin by confirming where the cgroups v2 hierarchy is mounted:

```
$ mount | grep cgroup2
cgroup2 on /sys/fs/cgroup type cgroup2
(rw,nosuid,nodev,noexec,relatime,nsdelegate,memory_recursiveprot)
```

Clearly, it's as anticipated, at `/sys/fs/cgroup`. (Curious about the various mount options that follow in parentheses? They're documented here: `https://www.kernel.org/doc/html/v6.1/admin-guide/cgroup-v2.html#mounting`).

Just in case you're running an older distro (like Ubuntu 18.04 or so, as we did in this book's first edition), it's possible you won't find any controllers present in `cgroup2`. This is the case in the presence of *mixed* cgroups, v1 and v2. To exclusively make use of the later version (as we expect here) – and thus have all configured controllers visible – you must first disable cgroups v1 by passing this kernel command-line parameter at boot: `cgroup_no_v1=all` (recall, all available kernel parameters can be conveniently seen at `https://www.kernel.org/doc/html/latest/admin-guide/kernel-parameters.html`. Then reboot and re-check. With a later distro (like Ubuntu 22.04 or Fedora 38), you need not do this.

Now let's start exploring it!

For this session, I am working on an x86_64 Fedora 38 VM, where I built and booted into a custom 6.1.25 kernel. Let's look at the overall scene to begin with:

```
$ uname -r
6.1.25-onfc38
$
$ mount |grep cgroup2
cgroup2 on /sys/fs/cgroup type cgroup2 (rw,nosuid,nodev,noexec,relatime,nsdelegate,memory_recursiveprot)
$
$ ls /sys/fs/cgroup
cgroup.controllers      cpuset.cpus.effective  io.prio.class      proc-sys-fs-binfmt_misc.mount/
cgroup.max.depth        cpuset.mems.effective  io.stat            sys-fs-fuse-connections.mount/
cgroup.max.descendants  cpu.stat               irq.pressure       sys-kernel-config.mount/
cgroup.pressure         dev-hugepages.mount/   machine.slice/     sys-kernel-debug.mount/
cgroup.procs            dev-mqueue.mount/      memory.numa_stat   sys-kernel-tracing.mount/
cgroup.stat             init.scope/            memory.pressure    system.slice/
cgroup.subtree_control  io.cost.model          memory.reclaim     user.slice/
cgroup.threads          io.cost.qos            memory.stat
cpu.pressure            io.pressure            misc.capacity
$
```

Figure 11.3: The root of the cgroups v2 hierarchy

Under the root cgroup location – `/sys/fs/cgroup` – you can see several files and folders (needless to say, these are all volatile pseudofile objects; they're mounted in RAM via sysfs). First off:

- The "regular" files seen – like `cgroup.controllers`, `cpu.pressure`, and so on – are cgroup2 interface files.

  These are further subdivided into *core and controller interfaces*; all the `cgroup.*` files are core interface files, the `cpu.*` ones are interface files for the CPU controller, `memory.*` for the memory controller, and so on.

- The *folders* seen represent – at last! – the control groups or *cgroups*! Among the many seen, you'll find that not all are constrained. You might wonder who created them; the short answer (for the ones there by default at least) is *systemd*; more on this soon follows.

Enabling or disabling controllers

Let's check out a key core interface file, `cgroup.controllers`. The previous section briefly mentioned this. Its content is the list of available controllers for the cgroup; for the root cgroup, it's the controllers that the kernel has been configured with. As we saw earlier, by default for a modern distro, it's usually these: `cpuset cpu io memory hugetlb pids rdma misc`.

Careful here: a controller showing up in this list does *not* necessarily mean it's enabled within the cgroup hierarchy; in fact, the default is that none are! Enabling a controller indicates that its constraints on target resource distribution will take effect on the immediate children. To enable a controller, you write the string +*<controller-name>* to the `cgroup.subtree_control` pseudofile (and conversely, -*<controller-name>* to it to disable it). So, for example, to enable the CPU and I/O controllers but disable the memory controller on the current cgroup (and thus have it take effect on its descendants, the hierarchy that flows under it), do (as root):

```
echo "+cpu +io -memory" > cgroup.subtree_control
```

So, now we know the `cgroup.subtree_control` file yields a space-separated list of controllers enabled to control resource distribution from this cgroup to its children.

The kernel docs state it like this `https://docs.kernel.org/admin-guide/cgroup-v2.html`:

> *"Top-down Constraint*
>
> *Resources are distributed top-down and a cgroup can further distribute a resource only if the resource has been distributed to it from the parent. This means that all non-root "cgroup.subtree_control" files can only contain controllers which are enabled in the parent's "cgroup.subtree_control" file. A controller can be enabled only if the parent has the controller enabled and a controller can't be disabled if one or more children have it enabled."*

Take a moment to digest that statement. A couple more useful core cgroup interface files – that are only present within a cgroup folder (hierarchy) – to be aware of follow:

- `cgroup.events`: Read-only; can have these values:
    - `populated`: 0 or 1. If 1, the cgroup or its descendants contains live processes, else it's 0. Thus, *only if the value of* `populated` *is 1 is the cgroup worth delving into!* Else, it's an empty or unpopulated cgroup (our cgroups v2 "explorer" script, which we shall soon see, will show this).
    - `frozen`: 0 or 1. If 1, the cgroup is frozen, else it's 0 (freezing a cgroup is akin to placing all its processes – and all its descendant cgroups and their processes as well! – in the "freezer," implying that they remain in a *stopped* state until thawed).

On a system where systemd is being used as the init manager (this is typical these days and we more or less assume it), there will always be a cgroup named `init.scope` (scope? worry not, we cover all of this). It consists of just the init process PID 1 (systemd). Let's look up this init cgroup's `cgroup.events` file:

```
$ cat /sys/fs/cgroup/init.scope/cgroup.events
populated 1
frozen 0
```

As it is populated, one could delve further into it. Next, within a cgroup hierarchy, you'll also find this cgroup interface file:

- `cgroup.kill`: Write-only; writing 1 here has the cgroup tree and all its descendants die – the processes within the target cgroup tree will be sent the `SIGKILL` signal.

A few more cgroup files follow in the coming sections. Again, worry not, all core interfaces are documented here: `https://www.kernel.org/doc/html/v6.1/admin-guide/cgroup-v2.html#core-interface-files`.

Kernel documentation on cgroups v2

What we've covered so far are the very basics; as you'll appreciate, explaining cgroups in complete detail is simply too much to cover in this chapter and book; more importantly, it's the job of a book to explain concepts and show examples helping you to learn. It's completely pointless to simply repeat the details within the kernel docs; with regard to cgroups v2, the resource models, the interface files (for both the core as well as for each controller), the rules to work with them, namespace management, and so forth are covered excruciatingly well within the official 6.1 series kernel documentation for cgroups2 at: `https://www.kernel.org/doc/html/v6.1/admin-guide/cgroup-v2.html#control-group-v2`. And, for the latest kernel version, here: `https://docs.kernel.org/admin-guide/cgroup-v2.html#control-group-v2`.

So, it is important to learn how to efficiently peruse the kernel docs (we talked about this briefly back in *Online Chapter, Kernel Workspace Setup*, in the *Locating and using the Linux kernel documentation* section). A screenshot (hopefully) helps you remember that these details are very much in the kernel documentation, just waiting for you to refer to them when required:

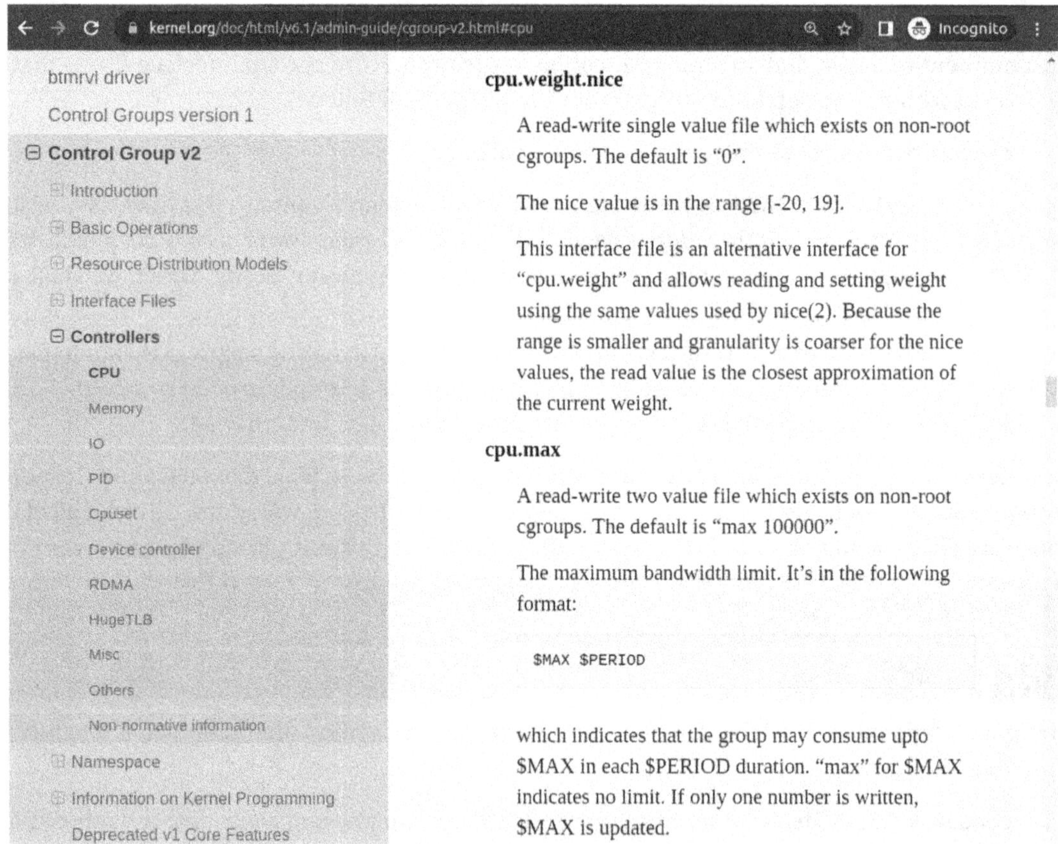

Figure 11.4: Partial screenshot of the kernel documentation for Cgroups v2 / Controllers / CPU

At the risk of repetition, I urge you to browse through the relevant docs when working with cgroups v2.

Let's move along and look within!

The cgroups within the hierarchy

Back to the cgroups within the root of the cgroups tree (or hierarchy, see *Figure 11.3*). As mentioned, the folders seen within there represent cgroups. For now, let's hold off on the question of who (and how) they're created; suffice it to say that they've been set up via systemd at boot.

Let's take a simple cgroup to begin with, represented by the folder `init.scope`. Taking a look within, we find that the kernel pre-populated it with a lot of (pseudo) files and interfaces; here is a truncated view:

```
$ cd /sys/fs/cgroup/
$ ls init.scope/
cgroup.controllers cgroup.threads io.latency     memory.high memory.swap.current
cgroup.events      cgroup.type    io.max         memory.low  memory.swap.events
cgroup.freeze      cpu.idle       io.pressure    memory.max  memory.swap.high
cgroup.kill        cpu.max        io.prio.class  memory.min  memory.swap.max
[ … ]
cgroup.subtree_control           io.bfq.weight memory.events.local  memory.stat
pids.peak
$
```

It's not all strange; we just spoke about a couple of key core interface files (in the previous section) – the `cgroup.controllers` and `cgroup.subtree_control` ones – and what they mean.

For now, let's investigate a few of the more interesting files here:

```
$ cat init.scope/cgroup.procs
1
```

Here, `cgroup.procs` shows the list of PIDs of processes that belong to this cgroup. It makes sense – systemd has set up an `init.scope` cgroup that contains just the init process, PID 1 – in fact, it's systemd itself. (Analogously, the `<cgroup-name>/cgroup.threads` pseudofile holds the PIDs of all *threads* belonging to the cgroup.)

To *migrate* a process to a given cgroup, write its PID to the target cgroup's `cgroup.procs` file. The writer requires the appropriate permission to do so; root, of course, will work, but even non-root can write provided permission matches check out. (Migrating a process to another cgroup is akin to a cut-paste operation; it's implicitly removed from the source cgroup. The changes can take a bit of time to propagate, though.)

What if you forget what this interface file represents or how exactly to manipulate it? Simple: **look up the kernel documentation.** The relevant entry is here, of course: `https://www.kernel.org/doc/html/v6.1/admin-guide/cgroup-v2.html#cpu`. (The previous section was about precisely this.)

Let's look up some key interfaces with regard to the CPU controller within the `init.scope` cgroup (recall, we're currently in the `/sys/fs/cgroup` folder):

```
$ cat init.scope/cpu.max  init.scope/cpu.weight  init.scope/cpu.weight.nice
max 100000
100
0
```

We can see their values (what do they mean? Patience, we're getting there: it's covered in the *Trying it out – constraining the CPU resource via cgroups v2* section!).

Now, using `cat` is fine, but see how much easier – and nicer! – it is to employ a quick `grep` hack when the content of the files is just a single line (or so), this way: `grep . <file-spec>` ; so here, using this technique, we look up all CPU controller interfaces and their values like this:

```
$ pwd
/sys/fs/cgroup
$
$ alias grep
alias grep='grep --color=always'
$
$ grep . init.scope/cpu.*
init.scope/cpu.idle:0
init.scope/cpu.max:max 100000
init.scope/cpu.max.burst:0
init.scope/cpu.pressure:some avg10=0.00 avg60=0.00 avg300=0.00 total=215428
init.scope/cpu.pressure:full avg10=0.00 avg60=0.00 avg300=0.00 total=181823
init.scope/cpu.stat:usage_usec 3895248
init.scope/cpu.stat:user_usec 1225449
init.scope/cpu.stat:system_usec 2669799
init.scope/cpu.stat:core_sched.force_idle_usec 0
init.scope/cpu.stat:nr_periods 0
init.scope/cpu.stat:nr_throttled 0
init.scope/cpu.stat:throttled_usec 0
init.scope/cpu.stat:nr_bursts 0
init.scope/cpu.stat:burst_usec 0
init.scope/cpu.weight:100
init.scope/cpu.weight.nice:0
$
```

Figure 11.5: Screenshot conveniently showing all the CPU controller interface files and their current value under the init.scope cgroup

So now, I hope you realize that being able to see the wealth of interface (and other) pseudo files, and having the kernel docs at hand to understand them, empowers you to really start understanding cgroups (as with anything, really).

You should also realize that cgroups can be *nested*; a cgroup can contain cgroups, which themselves contain cgroups. To see this, just look under a cgroup – a folder under the root one – looking for more folders.

Let's take another cgroup (which, again, systemd creates at boot) – `system.slice` – and look under it for sub-folders, thus *its* cgroups. A quick way to do so is like this:

```
$ find /sys/fs/cgroup/system.slice/ -maxdepth 1 -type d
/sys/fs/cgroup/system.slice/
/sys/fs/cgroup/system.slice/system-dbus\x2d:1.15\x2dorg.freedesktop.problems.
slice
/sys/fs/cgroup/system.slice/abrt-journal-core.service
/sys/fs/cgroup/system.slice/system-systemd\x2dfsck.slice
/sys/fs/cgroup/system.slice/sysroot.mount
/sys/fs/cgroup/system.slice/low-memory-monitor.service
[ … ]
/sys/fs/cgroup/system.slice/abrt-oops.service
/sys/fs/cgroup/system.slice/var-lib-nfs-rpc_pipefs.mount
$
```

(Though I kept the depth parameter to a max of `1`, this command revealed 55 cgroups on my box!) So, under each of these cgroups (represented by folders), we see similar (mirrored) content – their pseudo files representing core and controller interfaces – in effect limiting system resources, and possibly more cgroups (folders)! The design is deliberately recursive in nature.

Don't forget though: the resource-constraining feature of a cgroup *only* comes into play if the `cgroup.subtree_control` file within it, or in a parent cgroup, has the relevant controllers enabled!

So now, let's begin with a small step: on a systemd-based Linux (which pretty much all modern distros are), we'll use the `systemd-cgls` utility (`cgls` = cgroup list) to "see" the cgroups under the `init.scope` one; running the utility with a parameter – the cgroup hierarchy to look within – has it show precisely that:

```
$ systemd-cgls /sys/fs/cgroup/init.scope
Directory /sys/fs/cgroup/init.scope:
└─1 /usr/lib/systemd/systemd --switched-root --system -deserialize=32
```

It reveals *every process running under the cgroup*, along with its command line! This output was on a Fedora 38 VM; on Ubuntu, the output's a tad more readable:

```
$ systemd-cgls /sys/fs/cgroup/init.scope
Directory /sys/fs/cgroup/init.scope:
└─1 /sbin/init
```

Great; now lets look at the `system.slice` cgroup (on Fedora 38):

```
$ systemd-cgls /sys/fs/cgroup/system.slice
Directory /sys/fs/cgroup/system.slice:
```

```
├─abrt-journal-core.service
│ └─1386 /usr/bin/abrt-dump-journal-core -D -T -f -e
├─bolt.service
│ └─1425 /usr/libexec/boltd
├─low-memory-monitor.service
│ └─1296 /usr/libexec/low-memory-monitor
├─systemd-udevd.service …
│ └─udev
│   ├─   688 /usr/lib/systemd/systemd-udevd
│   ├─610731 (udev-worker)
│   ├─610732 (udev-worker)
[ … ]
```

It shows several "services" (processes – most of which began life under the aegis of systemd) and their sub-processes (descendants); nice. Now let's look at the big picture, *all cgroups*, by running the `systemd-cgls` utility with no parameters, which then reveals the entire cgroups hierarchy (under `/sys/fs/cgroup` by default; the `--no-pager` parameter is to simply not "page" the output with `less`):

```
$ systemd-cgls --no-pager
Working directory /sys/fs/cgroup:
├─user.slice (#1483)
│ → user.invocation_id: cd8fcd26621e4d6f9bb00aaa2be35ef7
│ └─user-1000.slice (#4982)
│   → user.invocation_id: 69fd059af5484948a57c6590527e6168
│   ├─session-10.scope (#14609)
│   │ ├─1283732 sshd: kaiwan [priv]
│   │ ├─1283739 sshd: kaiwan@pts/3
│   │ ├─1283749 -bash
│   │ ├─1324666 /usr/libexec/git-core/git credential-cache--daemon /home/kaiwan/.cache/git/credential/socket
│   │ └─1376512 systemd-cgls --no-pager
│   ├─session-9.scope (#14478)
│   │ ├─1283542 sshd: kaiwan [priv]
│   │ ├─1283547 sshd: kaiwan@pts/4
│   │ └─1283555 -bash
│   ├─user@1000.service … (#5124)
│   │ → user.delegate: 1
│   │ → user.invocation_id: 124eed885f8f44d5ae5a655dabc76caa
│   │ ├─session.slice (#5408)
│   │ │ ├─gvfs-goa-volume-monitor.service (#6547)
```

Figure 11.6: Partial screenshot of the cgroup hierarchy on the system

It's too big, of course, to show the entire thing here; do try it out on your system and study it. (Doing `systemd-cgls -h` reveals the help screen; see the man page for details.)

Systemd and cgroups

The manual management of cgroups can be a daunting task; the powerful **systemd** init framework that most workstation distros, enterprise and data center servers, and even embedded Linux run by default comes to the rescue. As we've already begun to notice, an interesting facet of systemd is that it takes over the role of creating and managing cgroups at boot, thus automatically leveraging their power for the benefit of users and their apps. (Of course, the more you learn about how it does so, the more you can tweak it to suit your project.)

Also, realize that there exist several tools to **help you visualize the defined cgroups** (and slices/scopes) on the system; they include `ps`, and some from the systemd project itself – `systemd-cgls`, `systemctl`, and `systemd-cgtop`. (We'll soon see our very own cgroups visualizer script as well!)

Slice and scope

As *Figure 11.6* quite clearly reveals, systemd has the intelligence to auto-construct cgroups, logically grouping processes together. To do so, it defines and employs artifacts – a *slice* and *scope*. A **slice** is used to represent all processes belonging to a particular user, or, it could represent a "bunch" of applications (processes) whose resources are managed via that unit. (In *Figure 11.6*, for my user account with a UID value of 1000, clearly, my slice is called `user-1000.slice`).

A **scope** represents further logical splicing or splitting of the slice (quite a tongue-twister, no?); as a good example, all processes running within a terminal window are typically grouped by systemd into a *session-<number>.scope* cgroup (the term is prefixed with the word "session" as a *session* represents the processes that are spawned and managed within a Terminal window)! Again, in *Figure 11.6*, you can clearly see that a Terminal window, represented by the scope named `session-9.scope`, is one that's been set up via sshd and has a Bash shell (whose PID is 1283555 and teletypewriter (tty) device is pts/4), is being represented or organized as a descendant of the user slice, and within it, in a "session" type scope.

Furthermore, systemd organizes the hierarchy with (boot-time) scope and service units being assigned to an appropriate slice (internally getting their own cgroups). As mentioned, every user that logs in to the system will also be seen as a "slice" under the generic `user.slice` node in the tree and the apps they run will show up under that cgroup, of course (again, you can literally see my user slice in *Figure 11.6;* it shows up as `user-1000.slice`, and under this hierarchy comes the "scope" units). Quoting from the systemd docs:

> *"The (slice) name consists of a dash-separated series of names, which describes the path to the slice from the root slice. The root slice is named* `-.slice`*. Example:* `foo-bar.slice` *is a slice that is located within* `foo.slice`*, which in turn is located in the root slice* `-.slice`*. ..."*

(You can see the root slice `-.slice` as the first slice in *Figure 11.7*).

What if you'd like to change the defaults, the manner in which systemd sets up cgroups? There are broadly **three ways** to do so: first, by manually editing the service unit file(s); second, by using the `systemctl set-property` sub-command to perform the edits; and third, by employing what's known as drop-in files within the systemd directory structure. As a quick example:

```
$ cat /usr/lib/systemd/system/user@.service
[ … ]
[Unit]
Description=User Manager for UID %i
```

```
Documentation=man:user@.service(5)
After=user-runtime-dir@%i.service dbus.service systemd-oomd.service
Requires=user-runtime-dir@%i.service
IgnoreOnIsolate=yes

[Service]
User=%i
PAMName=systemd-user
Type=notify-reload
ExecStart=/usr/lib/systemd/systemd --user
Slice=user-%i.slice
KillMode=mixed
Delegate=pids memory cpu
TasksMax=infinity
[ … ]
```

Do look up the *systemd* portion in the *Further reading* section for this chapter; there are plenty of links to good materials that will guide you with the details, along with examples.

Visualizing cgroups (and slices and scopes) with systemctl and systemd-cgtop

We've already seen how to use `systemd-cgls` to scan the cgroup hierarchy. Another means to view it (under the aegis of *systemd*) is via the `systemctl` app and the systemd unit `type`. systemd defines several unit types: `service`, `mount`, `swap`, `socket`, `target`, `device`, `automount`, `timer`, `path`, `slice`, and `scope`. Of these, only the last two are relevant to us here, so let's look at them via the `systemctl` command:

```
$ systemctl -t slice --all --no-pager
  UNIT                                              LOAD   ACTIVE SUB    DESCRIPTION
  -.slice                                           loaded active active Root Slice
  machine.slice                                     loaded active active Virtual Machine and Container Slice
  system-dbus\x2d:1.15\x2dorg.freedesktop.problem… loaded active active Slice /system/dbus-:1.15-org.freedesktop.problems
  system-getty.slice                                loaded active active Slice /system/getty
  system-modprobe.slice                             loaded active active Slice /system/modprobe
  system-sshd\x2dkeygen.slice                       loaded active active Slice /system/sshd-keygen
  system-systemd\x2dcryptsetup.slice                loaded active active Cryptsetup Units Slice
  system-systemd\x2dfsck.slice                      loaded active active Slice /system/systemd-fsck
  system-systemd\x2dzram\x2dsetup.slice             loaded active active Slice /system/systemd-zram-setup
  system.slice                                      loaded active active System Slice
  user-1000.slice                                   loaded active active User Slice of UID 1000
  user.slice                                        loaded active active User and Session Slice

LOAD   = Reflects whether the unit definition was properly loaded.
ACTIVE = The high-level unit activation state, i.e. generalization of SUB.
SUB    = The low-level unit activation state, values depend on unit type.
12 loaded units listed.
To show all installed unit files use 'systemctl list-unit-files'.
$
$
$ systemctl -t scope --all --no-pager
  UNIT               LOAD   ACTIVE SUB     DESCRIPTION
  init.scope         loaded active running System and Service Manager
  session-1.scope    loaded active running Session 1 of User kaiwan
  session-10.scope   loaded active running Session 10 of User kaiwan
  session-11.scope   loaded active running Session 11 of User kaiwan
  session-9.scope    loaded active running Session 9 of User kaiwan
```

Figure 11.7: Screenshot showing how to view the systemd-generated cgroup hierarchy at the slice and scope levels

Now that you have a grip on the basics of cgroups2, here's a partial screenshot of the `cgroups(7)` man page highlighting some more key points; do study it:

```
CGROUPS VERSION 2
       In  cgroups  v2, all mounted controllers reside in a single unified hierarchy.  While (differ-
       ent) controllers may be simultaneously mounted under the v1 and v2 hierarchies, it is not pos-
       sible to mount the same controller simultaneously under both the v1 and the v2 hierarchies.

       The  new behaviors in cgroups v2 are summarized here, and in some cases elaborated in the fol-
       lowing subsections.

       •  Cgroups v2 provides a unified hierarchy against which all controllers are mounted.

       •  "Internal" processes are not permitted.  With the exception of the root  cgroup,  processes
          may  reside only in leaf nodes (cgroups that do not themselves contain child cgroups).  The
          details are somewhat more subtle than this, and are described below.

       •  Active cgroups must be specified via the files cgroup.controllers  and  cgroup.subtree_con-
          trol.

       •  The  tasks  file has been removed.  In addition, the cgroup.clone_children file that is em-
          ployed by the cpuset controller has been removed.

       •  An improved mechanism for notification of empty cgroups is provided  by  the  cgroup.events
          file.

       For  more  changes,  see the Documentation/admin-guide/cgroup-v2.rst file in the kernel source
       (or Documentation/cgroup-v2.txt in Linux 4.17 and earlier).

       Some of the new behaviors listed above saw subsequent modification with the addition in  Linux
       4.14 of "thread mode" (described below).
```

Figure 11.8: Partial screenshot of the cgroups.7 man page

Is there a way to see which cgroup a given process belongs to, if any? Of course: one way is to see them all and filter for the process of interest. Viewing the process listing via ps with cgroup info is easy; simply tack on the `-o cgroup` option; here is an example (the `f` option has the benefit of showing parent-child relationships):

```
$ ps fw -eo pid,user,cgroup,args
    PID USER     CGROUP                             COMMAND
      2 root     -                                  [kthreadd]
      3 root     -                                   \_ [rcu_gp]
      4 root     -                                   \_ [rcu_par_gp]
      5 root     -                                   \_ [slub_flushwq]
      [ … ]
```

Notice how kernel threads don't belong to any cgroup (makes sense, as they're part of the kernel). Okay, here's more of the above ps output showing a few processes that are attached (by systemd) to the `init.scope` and `system.slice` cgroups (of course, the `init.scope` cgroup has just one process within it – systemd itself):

```
1377834 root     -                                  \_ [kworker/1:1-ata_
sff]
      1 root     0::/init.scope                    /usr/lib/systemd/systemd
--switched-root --system [...]
   1096 root     0::/system.slice/systemd-jo /usr/lib/systemd/systemd-journald
   1109 root     0::/system.slice/systemd-ud /usr/lib/systemd/systemd-udevd
   1228 systemd+ 0::/system.slice/systemd-oo /usr/lib/systemd/systemd-oomd
   1229 systemd+ 0::/system.slice/systemd-re /usr/lib/systemd/systemd-resolved
    [ … ]
```

(The cgroup names do get truncated here.) And here – interesting! – is this ps process itself:

```
[ … ]
      1392 root     0::/system.slice/sshd.servi sshd: /usr/sbin/sshd -D
[listener] 0 of 10-100 startups
1283542 root     0::/user.slice/user-1000.sl  \_ sshd: kaiwan [priv]
1283547 kaiwan   0::/user.slice/user-1000.sl  |   \_ sshd: kaiwan@pts/4
1283555 kaiwan   0::/user.slice/user-1000.sl  |       \_ -bash
1283732 root     0::/user.slice/user-1000.sl  \_ sshd: kaiwan [priv]
1283739 kaiwan   0::/user.slice/user-1000.sl  |   \_ sshd: kaiwan@pts/3
1283749 kaiwan   0::/user.slice/user-1000.sl  |       \_ -bash
1377837 kaiwan   0::/user.slice/user-1000.sl  |           \_ ps fw -eo
pid,user,cgroup,args
[ … ]
```

Notice how it belongs to the `user.slice/user-1000.slice` cgroup (also, I did an `ssh` to my Fedora VM; hence my ps process is within the `sshd … / bash` hierarchy).

Glance back at *Figure 11.7* to see all the slices. Something to question: if you dig a bit, you'll often find that no processes seem to belong to several of the slices seen (like `machine.slice`, `system-dbus...`, `system-getty`, `system-modprobe`, and so on); how come? Simple: the cgroups do exist (created by systemd at boot) but are currently *unpopulated* – there are no processes within them, which is okay.

So, when the user spawns off a new process (an app, say), does it appear by default in a cgroup? If so, which one? Ah, these are important questions. The answer's roughly like this: with systemd managing the system (typically the case), it will ensure it places every new process into an appropriate slice and scope, and thus cgroup. So, when I run, say, `vim`, to edit a file, this is what happens.

Before running `vim`:

```
$ ps fw -eo pid,user,cgroup,args | grep "[v]im"
$
```

Quick tip: using the regular expression as `"[v]im"` and not `"vim"` is very helpful; it has `grep` avoid showing itself in the output!

After running `vim`:

```
$ ps fw -eo pid,user,cgroup,args | grep "[v]im"
1378003 kaiwan   0::/user.slice/user-1000.sl                    \_ vim cgroupsv2_
explore
$
```

Aha! As expected with systemd, our dear `vim` process has become part of my user slice! It's thus tracked and managed, and any resource allocation constraints that apply to the `user-1000.slice` (and ancestors) slice(s) will apply to it as well.

Back to the question: what if it's a box not running systemd? Then there is no cgroups management... it's left to you. You can set it up so that processes are placed into cgroups (we do cover the basics of manually creating and managing a cgroup with a CPU controller in the upcoming *The manual way – a cgroups v2 CPU controller* section). Be aware that there is tooling to help with **cgroup management** besides the big guy, systemd (see `systemd-run(1)`); a good example is the *cg** tool suite (via the cgroup-tools/libcgroup package, which installs tools like `cgcreate`, `cgexec`, and `cgclassify`).

Getting a quick brief on kernel namespaces

A quick but useful deviation: it's interesting to realize that the whole shebang with regard to **containers** – a powerful, industry-standard, and de facto way to manage application deployment – is essentially based on two key technologies within the Linux kernel: cgroups and namespaces. You can think of containers as essentially lightweight VMs (to some extent); the majority of container technologies in use today (Docker, LXC, Kubernetes, and others) are, at heart, a marriage of these two baked-in Linux kernel technologies: cgroups and namespaces.

A **kernel namespace** is a critically important notion and structure for the whole container idea's implementation (the structure in the kernel is `struct nsproxy`). With namespaces, the kernel can partition its resources in a way that a set of processes in one namespace sees certain values and a set of processes in another namespace sees certain other values. Why is this required? Take, as an example, two containers; to have clean isolation, each must see processes with PID 1, 2, and so on. Similarly, each must have, perhaps, its own domain name and hostname, its own set of mounts whose content is unique to that container (`/proc` for one), network interfaces that are unique to each, and so on. The kernel can maintain many namespaces; by default, they're all optional, so the kernel always maintains the notion of a *<FOO> global namespace* for each of them (where *FOO* is the name of the namespace, like mount, PID, and so on):

Namespace	What it provides
Mount	Each mount namespace has its own filesystem layout (thus the contents of `/proc` in mount ns1 are different from its content in mount ns2!)
PID	Process isolation (thus each namespace can have a PID 1 process within it)
Network	Network isolation
UTS	Domain name and hostname isolation
IPC	IPC resources (shared memory, message queues, and semaphores) can be isolated
User	User ID isolation (allowing processes in different namespaces to have the same UID/GID)

Table 11.2: Kernel namespaces

On a related note: as you're probably aware, the `clone()` system call is used under the hood to create threads on Linux (`pthread_create()` calls it). Among its many flags – used to inform the kernel on how to create a custom process, or in other words, a thread – are flags labeled `CLONE_NEW*` (for example, `CLONE_NEWPID`, `CLONE_NEWNS`, `CLONE_NEWNET`, and so on). These are the ones that have the kernel create the process in a new namespace. The other namespace-related system calls include the `setns()`, `unshare()`, and `ioctl_ns()`; do check out their man pages for more.

(Again, the *Further reading* section for this chapter has links to more on kernel namespaces and container tech.) Okay, back to our cgroups discussions!

Using systemd-cgtop

Another way to both visualize the cgroups hierarchy *and* simultaneously observe at runtime which cgroups – and the slices/services within them – are using the lion's share of resources is via the very useful `systemd-cgtop` tool (in effect, the equivalent of the venerable *top* utility for systemd cgroups)!

By default, in the output of `systemd-cgtop`, cgroups are ordered by CPU load. The help screen is useful:

```
$ systemd-cgtop -h
```

```
systemd-cgtop [OPTIONS...] [CGROUP]

Show top control groups by their resource usage.

  -h --help                        Show this help
     --version                     Show package version
  -p --order=path                  Order by path
  -t --order=tasks                 Order by number of tasks/processes
  -c --order=cpu                   Order by CPU load (default)
  -m --order=memory                Order by memory load
  -i --order=io                    Order by IO load
  -r --raw                         Provide raw (not human-readable) numbers
     --cpu=percentage              Show CPU usage as percentage (default)
     --cpu=time                    Show CPU usage as time
  -P                               Count userspace processes instead of tasks
(excl. kernel)
  -k                               Count all processes instead of tasks (incl.
kernel)
     --recursive=BOOL        Sum up process count recursively
  -d --delay=DELAY           Delay between updates
  -n --iterations=N          Run for N iterations before exiting
  -1                         Shortcut for --iterations=1
  -b --batch                 Run in batch mode, accepting no input
     --depth=DEPTH           Maximum traversal depth (default: 3)
  -M --machine=              Show container

See the systemd-cgtop(1) man page for details.
```

You can also interactively switch the sorted-on field; the man page explains this:

```
...
(type) p, t, c, m, i
           Sort the control groups by path, number of tasks, CPU load, memory
usage, or I/O load, respectively. This setting may also be controlled using the
--order= command line switch.
…
```

I suggest you run the utility on your Linux system and try out the option switches.

Careful, though: merely running the `systemd-cgtop` utility isn't enough; (as stated before) if resource accounting isn't enabled on a cgroup(s), it can't tell anything about its resource usage. The `systemd-cgtop` man page states this key point as follows:

```
... unless "CPUAccounting=1", "MemoryAccounting=1" and "BlockIOAccounting=1"
are enabled for the services in question, no resource accounting will be
```

```
available for system services and the data shown by systemd-cgtop will be
incomplete.
```

Our cgroups v2 explorer script

With the existing cgroups visualization tooling, one issue is this: we can't immediately see if a cgroup is populated or empty; further, even if it is populated, we can't straight away see *which controllers* within it are enabled (or disabled). Knowing these is key to understanding the cgroup tree on the system. Our Bash script `cgroupsv2_explore` (`https://github.com/PacktPublishing/Linux-Kernel-Programming_2E/blob/main/ch11/cgroups/cgroupsv2_explore`) attempts to remedy the situation by showing the following things (and more):

- Given a starting cgroup as a parameter, it recursively iterates over all nested cgroups; if none is specified, it simply begins at the root of the cgroup tree (`/sys/fs/cgroup`) and thus scans the entire tree.
- For each cgroup it parses, it first checks: if it isn't populated (if it has no live processes within it), skip to the next cgroup, else display some content regarding it, such as:
    - Sub-controllers (in effect, the content of `cgroup.subtree_control` pseudofile for that cgroup! Refer back to our brief on sub-controllers...)
    - The cgroup type (domain/domain threaded/threaded/...)
    - The frozen state (0/1, no/yes)
    - Processes belonging to this cgroup: by default, it shows the number of processes (within parentheses) and then the list of PIDs; if the `-p` option is passed to this script, it displays the processes within it (via `ps`)
    - Threads belonging to this cgroup: by default, it shows the number of threads (within parentheses) and then the list of PIDs; if the `-t` option is passed to this script, it displays the threads within it (via `ps`)
    - Data pertaining to a few controllers within the cgroup; as of now (it's still evolving!):
        - CPU
        - Memory

Furthermore, there are option switches our script accepts:

- `-d`: Controls the depth to which to scan the tree
- `-v`: Displays in verbose mode
- `-p` / `-t`: Shows the processes and/or threads belonging to each cgroup (as already noted)

The `cgroupv2_explore` script's *help screen* reveals all of this in one stroke:

```
$ ./cgroupsv2_explore -h
Usage: cgroupsv2_explore [-v -p -t] [-d depth] [CGROUP]
This script recursively shows the cgroups metadata from the specified
CGROUP (the last parameter), down through it's hierarchy (tree).
If no particular CGROUP's specified, it shows the entire system CGROUP
hierarchy (typically the content of /sys/fs/cgroup). It assumes we're
running on a Cgroups v2 supported system.

All parameters are optional, and can be used in any combination (except
for CGROUP; it must be the last one).

-v : run in verbose mode
      Note! It's _very_ verbose, showing verbatim the content of all interface files for
            the various controllers. On the plus side, it's nice colorful output! (provided
            your terminal supports it).
      Off by default.

-p : show the name(s) of the processes belonging to the cgroup
     Note that this option can increase processing time.
-t : show the name(s) of the *threads* belonging to the cgroup
     Note that this option can increase processing time.

-d depth : a positive integer that affects the depth to which the Cgroups v2 hierarchy
is shown. The 'depth' value can be:
 1       => show only a very top-level overview of the hierarchy
 2,3,... => show to 2,3,... level(s) of the cgroup v2 hierarchy, whatever's specified.
 Must immediately follow the -d (f.e. pass as '-d2' and not as '-d 2').

 Practically speaking, most distros (i tested on Ubuntu/Fedora) will max out at 6 or 7
 levels of depth. (On mainstream distros, systemd typically sets up the Cgroup v2 hierarchy
 at boot).
 Default : shows _all_ levels of the specified CGROUP v2 hierarchy.

CGROUP : Path to any cgroup; for example: /sys/fs/cgroup/system.slice/wpa_supplicant.service
 (Tip: you can first use systemd-cgls, or this script, with no particular CGROUP parameter,
  to list all cgroups currently defined in the system).
 This Must be the last parameter.
$
```

Figure 11.9: Screenshot of our cgroupsv2_explore bash script showing its help screen

Right. Let's just run it with a depth of 1 (thus having it report only the first level folders under the cgroup root) to keep the output minimized (the screenshot's truncated for brevity):

```
$ ./cgroupsv2_explore -d1
cgroupsv2_explore: settings: depth=1, verbose=0, show-processes=0, show-threads=0

==================== cgroups v2 hierarchy ====================
<< Recursively from /sys/fs/cgroup >>

/sys/fs/cgroup
  /sys/fs/cgroup/dev-hugepages.mount                : unpopulated (no live processes)
  /sys/fs/cgroup/dev-mqueue.mount                   : unpopulated (no live processes)

<<--------------- /sys/fs/cgroup/init.scope                     ----------------------------
    (Sub)Controllers                                : -none-    [1]
    cg type                                         : domain
    cg frozen?                                      : 0         [2]
    Process PIDs                                    : (    1) : 1
    Thread PIDs                                     : (    1) : 1
    irq.pressure                                    : full avg10=0.00 avg60=0.00 avg300=0.00 total=41024    [3]
    CPU                [4]
      cpu.weight                                    : 100
      cpu.weight.nice                               : 0
      cpu.max                                       : max 100000
      cpu.pressure                                  : some avg10=0.00 avg60=0.00 avg300=0.00 total=126442
full avg10=0.00 avg60=0.00 avg300=0.00 total=68194
    MEMORY             [5]
      mem.current                                   : 79228928 (75.55 MB)
      mem.min                                       : 0
      mem.low                                       : 0 (0 B)
      mem.high                                      : max ()
    cg stat                                         : nr_descendants 0 nr_dying_descendants 0
--------------------------------------------------------------------------------------------->>

  /sys/fs/cgroup/machine.slice                      : unpopulated (no live processes)
  /sys/fs/cgroup/proc-sys-fs-binfmt_misc.mount      : unpopulated (no live processes)
  /sys/fs/cgroup/sys-fs-fuse-connections.mount      : unpopulated (no live processes)
  /sys/fs/cgroup/sys-kernel-config.mount            : unpopulated (no live processes)
  /sys/fs/cgroup/sys-kernel-debug.mount             : unpopulated (no live processes)
  /sys/fs/cgroup/sys-kernel-tracing.mount           : unpopulated (no live processes)

<<--------------- /sys/fs/cgroup/system.slice                   ----------------------------
    (Sub)Controllers                                : memory pids
    cg type                                         : domain
    cg frozen?                                      : 0         [2]
    Process PIDs                                    : (    0) : - (Has 56 descendants)
    Thread PIDs                                     : (    0) : -
```

Figure 11.10: Truncated screenshot of our cgroupsv2_explore bash script showing the output when run at a depth setting of 1

The output format corresponds to what was just described. Notice here (*Figure 11.10*) how the `/sys/fs/cgroup/init.scope` cgroup has no sub-controllers, implying that no resource constraints are actually being imposed! But the `/sys/fs/cgroup/system.slice` cgroup does have the `memory` and `pids` sub-controllers enabled, thus the system will keep those resources under the constraints specified therein. You can also see how here (*Figure 11.10*), several cgroups – starting from `machine.slice` and up to the `sys-kernel-tracing.mount` one (among others) – are empty, *unpopulated* (with no live processes within).

You can run the script passing it a particular cgroup; it will recursively show that cgroup's metadata as well as all its descendant cgroups, if any. We try out this feature in the next section when we create our own cgroup! For now, as an exercise, do try out the script yourself on various cgroups.

Trying it out – constraining the CPU resource via cgroups v2

As we're focused on CPU (task) scheduling here, we now take the time to look at two ways in which we can specify resource constraints on CPU usage by process(es) within a cgroup:

- The easy way, by leveraging *systemd* (with a service that starts at boot).
- The manual way, by creating and managing a cgroup (v2) via a Bash script that imposes resource constraints on CPU usage for a demo app.

Let's now get started with the first way.

This discussion does assume you have some basic knowledge with regard to systemd and setting up service units; if not, do refer to resources within the *Further reading* section (you'll find a few systemd resources here: `https://github.com/PacktPublishing/Linux-Kernel-Programming_2E/blob/main/Further_Reading.md#chapter-4-writing-your-first-kernel-module-lkms-part-1---further-reading`) and familiarize yourself with these basics.

Leveraging systemd to set up CPU resource constraints on a service

Systemd can appear to act as an abstraction layer above cgroups, allowing the sys admin to quite easily set resource constraints on cgroups. Systemd lets the administrator (or root users) define services that it auto-launches at boot, subjecting the process(es) to many attributes and/or constraints that you specify in the **service unit** file – metadata that defines what's to be run and how exactly it is to be run. A systemd service unit file is typically named `<foo>.service` (you'll usually find several of the system-defined ones under `/etc/systemd/system`). It's a plain ASCII text file divided into a few sections: `[Unit]`, `[Service]`, `[Install]`, and so on (do learn the details via the resources provided in the *Further reading* section just mentioned). For example, the pathname to the app (service) to run is specified via the `ExecStart=` directive in the `[Service]` section. (You'll very shortly see a demo of a few simple service units.)

Among various specifiers within the `[Service]` section, you can specify the `CPUQuota=` directive; it can be set to the percentage of processor cycles (CPU time) to allocate at maximum to the service! Similarly, the `AllowedCPUs=n` sets the maximum number of CPU cores the service can employ while it executes. Analogously, memory, process, I/O, and even network accounting constraints can be applied to the service! To see these, and more, do look up the man page on `systemd.resource-control(5)`.

Also, recall what we've learned regarding thread scheduling policies and priorities (refer to *Chapter 10, The CPU Scheduler – Part 1*, in the *The POSIX Scheduling Policies* section). With systemd, you can set the nice value, CPU scheduling policy, priority, and CPU affinity mask very easily on a service unit by employing appropriate directives.

The man page on `systemd.exec(5)` (`https://man7.org/linux/man-pages/man5/systemd.exec.5.html`) shows this clearly; a brief quote from it via a partial screenshot follows:

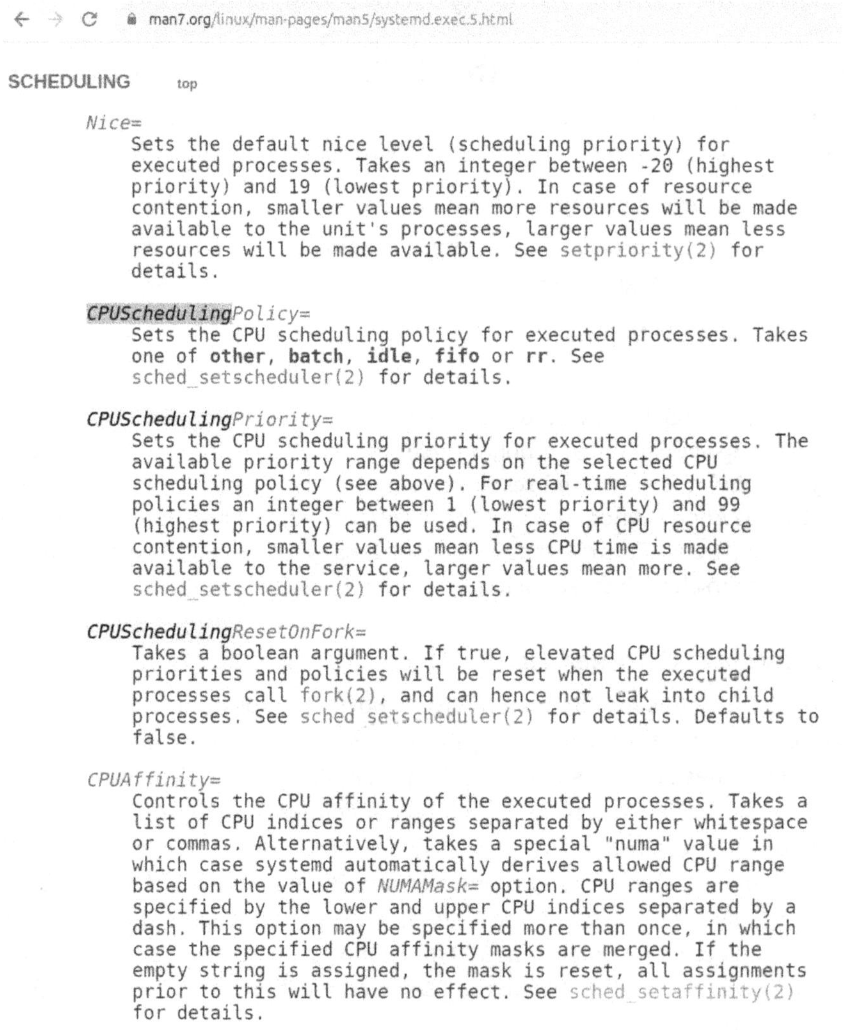

man7.org/linux/man-pages/man5/systemd.exec.5.html

SCHEDULING top

```
Nice=
    Sets the default nice level (scheduling priority) for
    executed processes. Takes an integer between -20 (highest
    priority) and 19 (lowest priority). In case of resource
    contention, smaller values mean more resources will be made
    available to the unit's processes, larger values mean less
    resources will be made available. See setpriority(2) for
    details.

CPUSchedulingPolicy=
    Sets the CPU scheduling policy for executed processes. Takes
    one of other, batch, idle, fifo or rr. See
    sched_setscheduler(2) for details.

CPUSchedulingPriority=
    Sets the CPU scheduling priority for executed processes. The
    available priority range depends on the selected CPU
    scheduling policy (see above). For real-time scheduling
    policies an integer between 1 (lowest priority) and 99
    (highest priority) can be used. In case of CPU resource
    contention, smaller values mean less CPU time is made
    available to the service, larger values mean more. See
    sched_setscheduler(2) for details.

CPUSchedulingResetOnFork=
    Takes a boolean argument. If true, elevated CPU scheduling
    priorities and policies will be reset when the executed
    processes call fork(2), and can hence not leak into child
    processes. See sched_setscheduler(2) for details. Defaults to
    false.

CPUAffinity=
    Controls the CPU affinity of the executed processes. Takes a
    list of CPU indices or ranges separated by either whitespace
    or commas. Alternatively, takes a special "numa" value in
    which case systemd automatically derives allowed CPU range
    based on the value of NUMAMask= option. CPU ranges are
    specified by the lower and upper CPU indices separated by a
    dash. This option may be specified more than once, in which
    case the specified CPU affinity masks are merged. If the
    empty string is assigned, the mask is reset, all assignments
    prior to this will have no effect. See sched_setaffinity(2)
    for details.
```

Figure 11.11: Partial screenshot of the systemd.exec(5) online man page showing a few relevant CPU scheduler-related attributes that can be easily set on a systemd service unit

As a demo, we have written a C program to generate prime numbers (from 2 to a specified maximum). It's designed to accept two parameters:

- The maximum number to which to generate primes up to
- The maximum amount of time it can take to do so (in seconds; an alarm's used to have it commit suicide (kill itself) once this time has elapsed)

The idea is this: when run with no (CPU) resource constraints, it will be able to generate quite a lot of prime numbers in the allotted time period.

But, when run under severe CPU constraints – set up via systemd using cgroups v2 kernel technology! – you'll find it can generate (relatively) few primes!

Here, we shan't delve into the nitty-gritty of setting up systemd service units (nor of the prime number generator program); do browse through links provided in the *Further reading* section on systemd. We shall leave it to you to go through and try out the prime number generator code in detail; it's here: `ch11/cgroups/cpu_constrain/primegen/`; also, the systemd service units and a couple of supporting scripts are at `ch11/cgroups/cpu_constrain/systemd_svcunit/` (link: `https://github.com/PacktPublishing/Linux-Kernel-Programming_2E/tree/main/ch11/cgroups/cpu_constrain/systemd_svcunit`).

A setup Bash script – `setup_service` – sets things up so that the service is installed and run immediately (of course, we can set it to run every time the system boots but this is a much easier way to test it and, of course, we don't want to trouble your system by running a useless prime number generator program every time you reboot!).

Without further ado, here's the service unit for running our prime number generator program without any constraints; in fact, we've given it a boost by – so easily! – setting the CPU scheduling policy to `SCHED_FIFO` and the priority to (a pretty arbitrary) value of 83 (recall, the range is 1 to 99):

```
$ cd <book_src>/ch11/cgroups/cpu_constrain/systemd_svcunit
$ ls
run_primegen    svc1_primes_normal.service  svc3_primes_lowram.service
setup_service    svc2_primes_lowcpu.service
$ cat svc1_primes_normal.service
# svc1_primes_normal.service
# NORMAL version, no artificial CPU (or other) constraints applied
[Unit]
Description=My test prime numbers generator app to launch at boot (normal
version)
[ … ]
After=mount.target

[Service]
# run_primegen: the script that launches the app.
# Our setup_service script ensures it copies all required files to this
location.
# UPDATE the /path/to/executable if required.
ExecStart=/usr/local/bin/systemd_svcunit_demo/run_primegen

# Optional: Apply 'better' cpu sched settings for this process
CPUSchedulingPolicy=fifo
CPUSchedulingPriority=83
```

```
# (Below) So that the child process - primegen - runs with these sched
settings!
# (well it's anyway the default)
CPUSchedulingResetOnFork=false
# Nice value applies for only the default 'other/normal' cpu sched policy
#Nice=-20

[ … ]
# UPDATE to your preference
[Install]
WantedBy=graphical.target
#WantedBy=multi-user.target
$
```

A tiny bash script, `run_primegen`, is executed, which, in turn, invokes the prime number generator program, essentially like this:

```
/usr/local/bin/systemd_svcunit_demo/primegen 100000 3
```

Thus, it will have our simple prime number generator process attempt to generate primes up to a maximum of 100,000 for a maximum of 3 seconds.

Attempt 1.1 – Executing the prime number generator under systemd with no resource constraints, SCHED_FIFO policy, and rtprio 83

Right, let's simply have systemd execute the script. We execute our setup script passing the service unit file as a parameter, which has systemd execute the `run_primegen` program (see the source for details; the output below is from my x86_64 Fedora 38 guest running our custom 6.1.25 kernel):

```
kaiwan@fedora:~/lkp2e/ch11/cgroups/cpu_constrain/systemd_svcunit
$ ./setup_service svc1_primes_normal.service
[sudo] password for kaiwan:
make: Nothing to be done for 'all'.
setup_service: enable and run the "svc1_primes_normal.service" service unit NOW
Created symlink /etc/systemd/system/graphical.target.wants/svc1_primes_normal.service → /usr/lib/syst
emd/system/svc1_primes_normal.service.
setup_service: asked to disable program on boot, disabling...
Removed "/etc/systemd/system/graphical.target.wants/svc1_primes_normal.service".
$
$ systemctl status svc1_primes_normal.service --no-pager -l
○ svc1_primes_normal.service - My test prime numbers generator app to launch at boot (normal version)
     Loaded: loaded (/usr/lib/systemd/system/svc1_primes_normal.service; disabled; preset: disabled)
    Drop-In: /usr/lib/systemd/system/service.d
             └─10-timeout-abort.conf
     Active: inactive (dead)

Aug 27 17:19:50 fedora run_primegen[52503]:  98473,  98479,  98491,  98507,  98519,  98533,  98543,
98561,  98563,  98573,  98597,  98621,  98627,  98639,  98641,  98663,
Aug 27 17:19:50 fedora run_primegen[52503]:  98669,  98689,  98711,  98713,  98717,  98729,  98731,
98737,  98773,  98779,  98801,  98807,  98809,  98837,  98849,  98867,
Aug 27 17:19:50 fedora run_primegen[52503]:  98869,  98873,  98887,  98893,  98897,  98899,  98909,
98911,  98927,  98929,  98939,  98947,  98953,  98963,  98981,  98993,
Aug 27 17:19:50 fedora run_primegen[52503]:  98999,  99013,  99017,  99023,  99041,  99053,  99079,
99083,  99089,  99103,  99109,  99119,  99131,  99133,  99137,  99139,
Aug 27 17:19:50 fedora run_primegen[52503]:  99149,  99173,  99181,  99191,  99223,  99233,  99241,
99251,  99257,  99259,  99277,  99289,  99317,  99347,  99349,  99367,
Aug 27 17:19:50 fedora run_primegen[52503]:  99371,  99377,  99391,  99397,  99401,  99409,  99431,
99439,  99469,  99487,  99497,  99523,  99527,  99529,  99551,  99559,
Aug 27 17:19:50 fedora run_primegen[52503]:  99563,  99571,  99577,  99581,  99607,  99611,  99623,
99643,  99661,  99667,  99679,  99689,  99707,  99709,  99713,  99719,
Aug 27 17:19:50 fedora run_primegen[52503]:  99721,  99733,  99761,  99767,  99787,  99793,  99809,
99817,  99823,  99829,  99833,  99839,  99859,  99871,  99877,  99881,
Aug 27 17:19:50 fedora run_primegen[52503]:  99901,  99907,  99923,  99929,  99961,  99971,  99989,
99991,
Aug 27 17:19:50 fedora systemd[1]: svc1_primes_normal.service: Deactivated successfully.
$
$
$ 
```

Figure 11.12: Screenshot showing our systemd service unit has set up and executed the prime number generator program with no resource constraints

All right! The `systemctl status <service.unit>` command shows its status and whatever output it generated (*Figure 11.12*; usefully, systemd auto-saves all `stdout`, `stderr`, and kernel `printk` output to the logs).

In this particular run, it managed to generate prime numbers from 2 to 99,991 within the 3 seconds it had, with no resource (CPU) constraints, and running as `SCHED_FIFO` with a high priority. (You'll realize, of course, that the number of primes generated can vary quite a bit depending on the hardware system. *Quick tip*: to see the full output, simply run `journalctl -b`.)

By the way, our script deliberately disables the service after running it once; you can change this by changing, in the `setup_service` script, the variable `KEEP_PROGRAM_ENABLED_ON_BOOT` to the value `1`.

As *Figure 11.12* shows, run

```
systemctl status svc1_primes_normal.service --no-pager -l
```

to see the status of the service.

Furthermore, and quite often useful, the `systemctl show <service-unit-name>` command shows *all* settings for that service unit; let's do this, `grep`-ping for CPU-related ones:

```
$ systemctl show svc1_primes_normal.service |grep CPU
CPUUsageNSec=[not set]
CPUAccounting=yes
CPUWeight=[not set]
StartupCPUWeight=[not set]
CPUShares=[not set]
StartupCPUShares=[not set]
CPUQuotaPerSecUSec=infinity
CPUQuotaPeriodUSec=infinity
LimitCPU=infinity
LimitCPUSoft=infinity
CPUSchedulingPolicy=1
CPUSchedulingPriority=83
CPUAffinityFromNUMA=no
CPUSchedulingResetOnFork=no
$
```

Figure 11.13: Screenshot shows the CPU-related settings within the service unit

We've highlighted a couple of the CPU settings we explicitly specified in the service unit file `svc1_primes_normal.service`. The others are set to defaults, of course (by the way, the settings named `LimitCPU*` are to specify the (older-style) resource limits for processes within the service unit).

So, keep in mind that this run – with no CPU constraints at all, a `SCHED_FIFO` policy, and an RT priority of 83 (and within about 3 seconds) – yielded around 99,991 primes on my system.

Attempt 1.2 – Executing the prime number generator under systemd with constraints on the CPU resource, SCHED_OTHER, and rtprio 0

Now we run the same prime number generation program on the very same system, but this time, *with some definite CPU constraints* specified via systemd. The service unit file used now is this one: `ch11/cgroups/cpu_constrain/systemd_svcunit/svc2_primes_lowcpu.service`. Pretty much everything is the same as the first one (that we just saw in the previous section) except this:

```
#--- Apply CPU constraints ---
```

```
CPUQuota=10%
AllowedCPUs=1
```

We also remove the lines `CPUSchedulingPolicy=fifo` and `CPUSchedulingPriority=83`, thus keeping the process to defaults: a scheduling policy of `SCHED_OTHER` and a real-time priority of 0, and, of course, we've constrained it to use just 10% CPU bandwidth and just 1 core! Let's run it and then check the status:

```
$ ./setup_service svc2_primes_lowcpu.service
[ … ]
$ systemctl status svc2_primes_lowcpu.service --no-pager -l
o svc2_primes_lowcpu.service - My test prime numbers generator app to launch at boot (CPU constrained version)
     Loaded: loaded (/usr/lib/systemd/system/svc2_primes_lowcpu.service; disabled; preset: disabled)
    Drop-In: /usr/lib/systemd/system/service.d
             └─10-timeout-abort.conf
     Active: inactive (dead)

Aug 27 17:26:27 fedora run_primegen[52673]:  30347,  30367,  30389,  30391,  30403,  30427,  30431,  30449,  30467,  30469,  30491,  30493,  30497,  30509,  30517,  30529,
[ … ]
Aug 27 17:26:28 fedora run_primegen[52673]:  31531,  31541, primegen.c:buzz()
Aug 27 17:26:28 fedora run_primegen[52672]: Terminated
Aug 27 17:26:28 fedora systemd[1]: svc2_primes_lowcpu.service: Deactivated successfully.
$
```

Aha! This time, the program – under the constraint of only 10% CPU bandwidth (quota), allowed to run on exactly 1 core, `SCHED_OTHER`, and rtprio 0 – yields primes up to only the number 31,541 in the 3 seconds it's (internally) allowed to run for, as compared to well over 99 thousand primes in the first "normal" case, showing that the first case generated close to 70% more primes than the second, proving the efficacy of systemd cgroups control. So there we are.

As a demo to illustrate how **memory limits** can be set on a cgroup, we have provided another sample service unit: `svc3_primes_lowram.service`. Within it, the cgroup memory limits are specified via the `MemoryHigh` and `MemoryMax` systemd settings (do look up the `systemd.resource-control` man page, specifically the section called *Memory Accounting and Control* for details). Our service deliberately has the *stress-ng* program allocate huge amounts of memory, thus breaching the specified limits and causing the **Out of Memory** (**OOM**) killer (or the `systemd-oomd` process, if configured) to kill off the cgroup task(s). (We covered the OOM killer in some detail in *Chapter 9, Kernel Memory Allocation for Module Authors – Part 2*, in the *Stayin' alive – the OOM killer* section). Be careful when you run it; we definitely recommend doing so on a test VM.

The manual way – a cgroups v2 CPU controller

Let's try, no, let's *do* something interesting (*Do, or do not. There is no try -Yoda.)*. We shall now manually create a new cgroup under the cgroups v2 hierarchy on the system. We'll then set up a CPU controller for it and set a user-specified upper limit on how much CPU bandwidth processes within the cgroup can actually make use of! We'll then run our prime number generator program within it to see how it's affected by the constraints we've set.

Here, we outline the steps you will typically take to do this (all of these steps require you to be running with root access):

1. Ensure your kernel supports cgroups v2; we expect you're running on a 4.5 or later kernel, with cgroups v2 support enabled.

How can I check I am running cgroups v2?

Simple: run `mount | grep cgroup`; the output must contain the substring `type cgroup2`.

2. Create a cgroup within the hierarchy (typically within `/sys/fs/cgroup/`). This is achieved by simply creating a directory with the required cgroup name under the cgroup v2 hierarchy; for example, to create a sub-group called `test_group`, do this:

```
mkdir /sys/fs/cgroup/test_group
```

3. Add a `cpu` controller to the new cgroup; this is achieved by doing this (as root):

```
echo "+cpu" > /sys/fs/cgroup/test_group/cgroup.subtree_control
```

Recall that without controllers, no resource constraints are being applied to the cgroup (and its descendants; reread the *Enabling or disabling controllers* section, especially the *Top-down constraint* paragraph we mentioned there if you wish).

4. The juicy bit is here: set up the maximum allowable CPU bandwidth for the processes that will belong to this cgroup. This is effected by writing two integers into the `<cgroups-v2-mount-point>/<our-cgroup>/cpu.max` (pseudo) file. For clarity, the explanation of this file, as per the kernel documentation (`https://docs.kernel.org/admin-guide/cgroup-v2.html#cpu-interface-files`), is reproduced here:

```
cpu.max
A read-write two value file which exists on non-root cgroups. The default
is "max 100000".
The maximum bandwidth limit. It's in the following format:
$MAX $PERIOD
which indicates that the group may consume up to $MAX in each $PERIOD
duration. "max" for $MAX indicates no limit. If only one number is
written, $MAX is updated.
```

In effect, all processes in the cgroup will be collectively allowed to run for `$MAX` out of a period of `$PERIOD` microseconds; so, for example, with `MAX = 300,000` and `PERIOD = 1,000,000`, we're effectively allowing all processes within the sub-control group to run for 0.3 seconds out of a period of 1 second! In other words, with 30% CPU bandwidth or utilization. As the kernel doc says, by default, the MAX is the same as the PERIOD thus implying 100% CPU utilization by default.

5. Insert a process (or several) into the new cgroup; this is achieved by writing their PIDs into the `<cgroups-v2-mount-point>/<our-cgroup>/cgroup.procs` pseudofile.

That's it; *the processes under the new cgroup will now perform their work under the CPU bandwidth constraint imposed (if any);* when done, they will die as usual... you can remove (or delete) the cgroup with a simple `rmdir <cgroups-v2-mount-point>/<our-cgroup>`.

For convenience, a Bash script that carries out the preceding steps is available here: `ch11/cgroups/cpu_constrain/cpu_manual/cgv2_cpu_ctrl.sh` (`https://github.com/PacktPublishing/Linux-Kernel-Programming_2E/blob/main/ch11/cgroups/cpu_constrain/cpu_manual/cgv2_cpu_ctrl.sh`). Do check it out!

To make it interesting, this script allows you to pass the maximum allowed CPU bandwidth as a parameter – the `$MAX` value discussed in *step 4*; in effect, it's a way to specify the maximum allowed CPU bandwidth to all processes within the cgroup (we'll see this shortly). Now, within the script, after the new cgroup has been created and set up (with the maximum allowed CPU bandwidth you specify), we run our prime number generator program like this:

```
primegen 1000000 5
```

Everything's similar to the previous case with systemd, except that we ask it to generate more primes and allow it more time, 5 seconds, to run; this is to give it a chance to do some work, as the script has to then query and write it's PID into the `<cgroups-v2-mount-point>/<our-cgroup>/cgroup.procs` pseudofile.

A couple of test runs here will help you understand it's working:

```
$ sudo ./cgv2_cpu_ctrl.sh
[+] Checking for cgroup v2 kernel support
cgv2_cpu_ctrl.sh: detected cgroup2 fs here: /sys/fs/cgroup

Usage: cgv2_cpu_ctrl.sh max-to-utilize(us) [run-cgroupsv2_explore-script]
max-to-utilize : REQUIRED: This value (microseconds) is the max amount of
  time the processes in the control group we create will be allowed to utilize
the
  CPU; it's relative to the period, which is set to the value 1000000.
  So, f.e., passing the value 300,000 (out of 1,000,000) implies a max CPU
utiltization
  of 0.3 seconds out of 1 second (i.e., 30% utilization).
```

```
  The valid range for the $MAX value is [1000-1000000].
run-cgroupsv2_explore-script : OPTIONAL: OFF by default.
  Passing 1 here has the script invoke our cgroupsv2_explore bash script,
passing
  the new cgroup as the one to show details of.
$
```

You're expected to run it as root and to pass, as the first parameter, the `$MAX` value (the usage screen seen previously quite clearly explains it, including displaying the valid range – the microseconds value).

Right, here we run our Bash script with the parameter `800000`, implying a CPU bandwidth of 800,000 out of a period of 1,000,000; in effect, a quite high CPU utilization of 0.8 seconds out of every 1 second on the CPU (i.e., 80% CPU utilization):

```
$ sudo ./cgv2_cpu_ctrl.sh 800000
[+] Checking for cgroup v2 kernel support
cgv2_cpu_ctrl.sh: detected cgroup2 fs here: /sys/fs/cgroup
[+] Creating a cgroup here: /sys/fs/cgroup/test_group
[+] Adding a 'cpu' controller to it's cgroups v2 subtree_control file

***
Now allowing 800000 out of a period of 1000000 to all processes in this cgroup,
i.e., 80.000% !
***

[+] Launch the prime number generator process now ...
../primegen/primegen 1000000 5 &

  2,  3,      5,      7,      11,      13,      17,      19,      23,      29,
31,      37,      41,      43,      47,      53,
[ … ]
   3071 pts/1    00:00:00 primegen
[+] Insert the 3071 process into our new CPU ctrl cgroup
   227,     229,     233,     239,     241,     251,     257,     263,     269,     271,
277,     281,     283,     293,     307,     311,
cat /sys/fs/cgroup/test_group/cgroup.procs
3071
[ … ] 7541,    7547,    7549,    7559,    7561,

.............. sleep for 6 s, allowing the program to execute ...............

  7573,    7577,    7583,    7589,    7591,    7603,    7607,    7621,    7639,    7643,
7649,    7669,    7673,    7681,    7687,    7691,
```

```
[ … ]

 41143,  41149,  41161,  41177,  41179,  41183,  41189,  41201,  41203,  41213,
41221,  41227,  41231,  41233,  41243,  41257,
primegen.c:buzz()
    [+] Removing our (cpu) cgroup
$
```

Study our script's output; you can see that it does its job. After verifying cgroup v2 support, it creates a cgroup called `test_group` under `/sys/fs/cgroup/` and adds a `cpu` controller to it. It then proceeds to set up the maximum allowed CPU bandwidth to 800,000 out of a period of 1,000,000 (i.e., 80% CPU utilization) maximum, as this is the value we passed. It then launches our prime number generator app (with appropriate parameters), and – key of course – adds its PID to the cgroup. In our test run here, with 80% CPU bandwidth available to it, it happens to generate primes up to the number 41,257 (within the 5 seconds allotted; of course, this number can and very likely will vary on different systems). The script then cleans up, deleting the cgroup.

However, to really appreciate the cgroups effect, let's run our script again (study the following output in *Figure 11.14*), but this time, with a maximum CPU bandwidth of just 1,000 (the `$MAX` value) – **in effect, a max CPU utilization of just 0.1%!**:

```
$ sudo ./cgv2_cpu_ctrl.sh 1000
[+] Checking for cgroup v2 kernel support
cgv2_cpu_ctrl.sh: detected cgroup2 fs here: /sys/fs/cgroup
[+] Creating a cgroup here: /sys/fs/cgroup/test_group
[+] Adding a 'cpu' controller to it's cgroups v2 subtree_control file

***
Now allowing 1000 out of a period of 1000000 to all processes in this cgroup, i.e., .100% !
***

[+] Launch the prime number generator process now ...
../primegen/primegen 1000000 5 &

  2,  3,      5,       7,      11,      13,      17,      19,      23,      29,      31,      37,      41,
     43,      47,      53,
    59,      61,      67,      71,      73,      79,      83,      89,      97,     101,     103,     107,
   109,     113,     127,     131,
   3181 pts/1    00:00:00 primegen
[+] Insert the 3181 process into our new CPU ctrl cgroup
   137,     139,     149,     151,     157,     163,     167,     173,     179,     181,     191,     193,
   197,     199,     211,     223,
   227,     229,     233,     239,     241,     251,     257,     263,     269,     271,     277,     281,
   283,     293,     307,     311,
   313,     317,     331,     337,     347,     349,     353,     359,     367,     373,     379,     383,
   389,     397,     401,     409,
   419,     421,     431,     433,     439,     443,     449,     457,     461,     463,     467,     479,
   487,     491,     499,     503,
   509,     521,     523,     541,     547,     557,     563,     569,     571,     577,     587,     593,
   599,     601,     607,     613,
cat /sys/fs/cgroup/test_group/cgroup.procs
3181

............... sleep for 6 s, allowing the program to execute ...............

   617,     619,     631,     641,     643,     647,     653,     659, primegen.c:buzz()
[+] Removing our (cpu) cgroup
$
```

Figure 11.14: Screenshot showing how our Bash script creates a cgroup, sets it up to allocate only 0.1% CPU bandwidth, and then adds and runs our prime number generator program within it

What a difference! This time – with a measly 0.1% CPU bandwidth allocation – our poor prime number generator process could emit prime numbers from 2 to just 659 (as opposed to up to the prime 41,257 with 80% CPU bandwidth). This clearly proves the efficacy of the cgroups v2 CPU controller.

One more thing: let's now rerun the script the same way, except that we'll **pass the second parameter as 1**; under the hood, this enables our `cgroupsv2_explore` bash script – with the first parameter set to `-p` to show the processes within it, and the second parameter as the new cgroup! – to execute, thus revealing some details about it. In the truncated screenshot (*Figure 11.15*), we show and highlight only its output:

```
.............. sleep for 6 s, allowing the program to execute ...............
cgroupsv2_explore: settings: depth=full, verbose=0, show-processes=1, show-threads=0

==================== cgroups v2 hierarchy ====================
<< Recursively from /sys/fs/cgroup/test_group >>

<<---------------  /sys/fs/cgroup/test_group  --------------------------
    (Sub)Controllers                                  : cpu
    cg type                                           : domain threaded
    cg frozen?                                        : 0          [2]
    Process PIDs                                      : (    1) : 3220
UID          PID    PPID  C STIME TTY           TIME CMD
root        3220    3202  6 19:49 pts/1     00:00:00 ../primegen/primegen 1000000 5
    Thread PIDs                                       : (    1) : 3220
    irq.pressure                                      : full avg10=0.00 avg60=0.00 avg300=0.00 total=49     [3]
    CPU                     [4]
      cpu.weight                                        : 100
      cpu.weight.nice                                   : 0
      cpu.max                                           : 1000 1000000
      cpu.pressure                                      : some avg10=0.00 avg60=0.00 avg300=0.00 total=1135724
full avg10=0.00 avg60=0.00 avg300=0.00 total=1135724
    MEMORY                  [5]
      mem.current                                     : 0 (0 B)
      mem.min                                         : 0
      mem.low                                         : 0 (0 B)
      mem.high                                        : max ()
    cg stat                                           : nr_descendants 0 nr_dying_descendants 0
------------------------------------------------------------------------------------------->>

[2] See cgroup.freeze (and cgroup.events) under https://docs.kernel.org/admin-guide/cgroup-v2.html#core-
interface-files
[3] See cgroup.pressure, irq.pressure under https://docs.kernel.org/admin-guide/cgroup-v2.html#core-inte
rface-files ; plus https://docs.kernel.org/accounting/psi.html#psi
[4] cpu: see https://docs.kernel.org/admin-guide/cgroup-v2.html#cpu-interface-files
[5] memory: see https://www.kernel.org/doc/html/latest/admin-guide/cgroup-v2.html#memory
Parsed a total of 1 (v2) CGROUPs (0 were empty / unpopulated).
   313,     317,     331,     337,     347,     349,     353, primegen.c:buzz()
[+] Removing our (cpu) cgroup
$
```

Figure 11.15: Truncated screenshot of the output of our cgv2_cpu_ctrl.sh script – this time, with its second parameter set to 1, thus allowing our (earlier) cgroupsv2_explore script to display some details regarding the new cgroup

Fantastic; it's all there as expected.

On a related note, one can affect the *weightage* of CPU distributed to processes within a cgroup by fiddling with the `cpu.weight` pseudofile. It gets a default value of 100 for all tasks within the cgroup.

As such, it simply means that all of them will receive an equal share of the CPU resource; the actual value assigned doesn't really matter. It's only when you vary the weight between processes (tasks) that it matters.

A key point to realize with the CPU resource controller (and others in general): the constraints you specify only have meaning and are applied when the resource in question - here, the CPU cores - are *saturated*; otherwise, processes are free to (happily?) use as much CPU as they'd like to. Makes sense!

So, now, finally, I hope you can see why, in the prior chapter's *Figure 10.1*, at the very top of it, the "Cgroups v2" layer was displayed. We can now quite clearly see how cgroups are deeply embedded into the very fabric of the Linux kernel; thus, they can be programmed to affect the allocation (or bandwidth/utilization) of various resources, including the CPU, and thus realize how much this affects CPU scheduling.

Simple exercise: run the `ch11/cgroups/cpu_constrain/cpu_manual/cgv2_cpu_ctrl.sh` Bash script on your system, having it allocate different amounts of CPU bandwidth utilization and enabling it to show output from our earlier `cgroupsv2_explore` script. Study the results.

With that, we complete our coverage of a really powerful and useful kernel feature: cgroups (v2). Let's move on to the final section of this chapter: an introduction to learning what a **Real-Time Operating System** (**RTOS**) is, and how it's possible to convert regular Linux into one by applying appropriate patches and configuring it.

Running Linux as an RTOS – an introduction

Mainline or vanilla Linux (the kernel you download from `https://kernel.org`, or even a typical Linux Git kernel tree) is decidedly *not* an **RTOS**; it's a **General Purpose Operating System** (**GPOS**; as is Windows, macOS, and Unix). In an RTOS, where hard **real-time** (RT) characteristics come into play, not only must the software obtain the correct result but there are *deadlines* associated with doing so; it must guarantee it meets these deadlines, every single time.

One can very broadly categorize an OS based on its RT characteristics in this manner (see *Figure 11.16*); at the left extreme are the non-RT OSs, and at the right extreme is the RTOS:

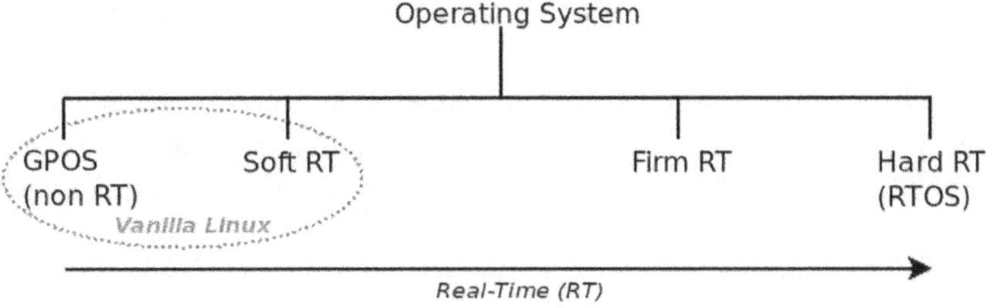

Figure 11.16: Categorizing an OS on the RT scale

The mainline or "vanilla" Linux OS, though not an RTOS, does a tremendous job performance-wise without even breaking a sweat. It easily qualifies as being a soft real-time OS: one where deadlines are met most of the time, on a "best effort" basis (it's sometimes said to meet the "five-9s" qualifier, as it meets its deadlines 99.999% of the time!). Nevertheless, true hard real-time domains (for example, many types of military operations, transport, robotics, telecom, factory floor automation, stock exchanges, medical electronics, and so on) *require* hard real-time guarantees and thus require an RTOS. Thus, for these domains, plain vanilla Linux, a GPOS, just won't cut it.

A key point in this context is that of **determinism:** an oft-missed point regarding real time is that the software response time to an (external) event need not always be really fast (responding, say, within a few microseconds). It may be a lot slower (in the range of, say, tens of milliseconds); by itself, that isn't what really matters in an RTOS. What does matter is that the system is reliable and predictable, working in the same consistent manner and always guaranteeing the deadline is met; such systems are said to have *deterministic response* and this is a key characteristic of real-time systems.

For example, the time taken to respond to a scheduling request should be consistent, predictable, and not bounce all over the place. The variance from the required time (or baseline) is often referred to as the **jitter**; an RTOS works to keep the jitter tiny, even negligible. In a GPOS, this is often impossible and isn't even a design goal in the first place! Thus, in such non-RT systems, the jitter can vary tremendously, at one point being low and very high at the next.

Overall, the ability to maintain a stable, even, predictable response with minimal jitter – even in the face of extreme workload pressures – is termed *determinism* and is the hallmark of an RTOS. To provide such a deterministic response, its algorithms must, as far as possible, be designed to correspond to *O(1)* (big-Oh 1) algorithmic time complexity.

Another goal of an RT system is that of reducing latencies and delays. Actually, that statement's not quite accurate: **reducing the *maximum or worst-case latency* experienced to an acceptable level is the goal**; the (ironic) reality is that the minimum and average latencies can be – and often are – worse than on non-RT systems.

Thomas Gleixner, along with community support, has worked toward the goal of converting the regular (or plain vanilla) non-RT Linux kernel into a hard RTOS for a long while now. He and his collaborators succeeded a long while back: ever since the 2.6.18 kernel (released in September 2006), there have been out-of-tree patches that convert the Linux kernel into an RTOS! These patches can be found, for many versions of the kernel, here: `https://mirrors.edge.kernel.org/pub/linux/kernel/projects/rt/`. The older name for this project was "preemptible real-time" or simply, `PREEMPT_RT`. Later (from October 2015, kernel version 4.1 onward), the **Linux Foundation** (**LF**) took over stewardship of this project – a very positive step! – and renamed it the **Real-Time Linux** (**RTL**) Collaborative Project (`https://wiki.linuxfoundation.org/realtime/rtl/start`), or simply, RTL (don't confuse this project with co-kernel approaches such as Xenomai or RTAI, or the older and now-defunct attempt called RTLinux).

An FAQ, of course, is "Why aren't these patches to convert Linux into a hard real-time OS in mainline itself?" Well, it turns out that:

- Much of the RTL work has indeed been merged into the mainline kernel; this includes important areas such as the scheduling subsystem, mutexes and spinlocks, lockdep, threaded interrupts, PI (priority inheritance), tracing, and so on. In fact, an ongoing primary goal of RTL is to get itself merged into mainline; as of this writing, it's (very) close!
- Traditionally, Linus Torvalds deems that Linux, being primarily designed and architected as a GPOS, should not have highly invasive features that only an RTOS really requires; so, though patches do get merged in, it's a slow deliberated process.

We have included several interesting articles and references to RTL (and hard real-time in general) in the *Further reading* section of this chapter; do take a look.

What you're going to do next is interesting indeed: we present a brief on how you can patch the mainline 6.1 LTS kernel with the (currently still) out-of-tree RTL patches, configure it, build it, and boot it; you will thus end up running an RTOS – *Real-Time Linux or RTL*! We shall do this on our x86_64 Linux VM (or native system).

Pointers to building RTL for the mainline 6.x kernel (on x86_64)

Here, instead of covering in detail how to patch a vanilla Linux kernel, configure it, and build it as a real-time kernel (as RTL), we simply point you to a brief tutorial on the superb RTL Wiki site: *HOWTO setup Linux with PREEMPT_RT properly:* `https://wiki.linuxfoundation.org/realtime/documentation/howto/applications/preemptrt_setup`.

As a matter of fact, the first edition of this book went into an in-depth explanation of patching, configuring, and building RTL; do refer to it for these details if you wish.

This short tutorial – a portion of the large and very useful RTL Wiki site – explains things:

- Download the RTL patch set for a given stable kernel (from `https://mirrors.edge.kernel.org/pub/linux/kernel/projects/rt/`; as of this writing, the tutorial just mentioned uses the RTL patches for the 4.4 LTS kernel):
    - FYI, the patch set often shows up in two forms. The first is a single patch file named `patch-<kver>-rt<xy>.tar.xz` (as well as a gzip one, and a `sign` file); it's easier to download and extract this one – the entire RTL patch is in this one file. The other form is a patch series (multiple patch files) in a file named `patches-<kver>-rt<xy>.tar.xz`; with this approach, when extracted, the RTL patch files are in a folder named `patches/`. There can be several patches (which must be applied in order – see the series file; for example, with the `patches-6.1.75-rt23.tar.xz` file, it expands into a series of 63 patch files).

- Extract the patch(es) and patch the kernel source tree
- Configure the kernel, setting it up for full kernel preemption, for RTL (`CONFIG_PREEMPT_RT=y`); see the *tips* that follow
- Build it in the usual manner (which we covered in-depth in *Chapter 2, Building the 6.x Linux Kernel from Source - Part 1*, and *Chapter 3, Building the 6.x Linux Kernel from Source - Part 2*, in any case)

A few quick tips regarding kernel configuration for RTL: To configure building Linux as RTL, within the `make menuconfig` UI, navigate to **General Setup | Preemption Model.** Assuming you've downloaded and applied the (out-of-tree) RTL patch(es), you should see 4 choices here (in place of the usual 3). The fourth is the one we want: Fully Preemptible Kernel (Real-Time). Select it (`CONFIG_PREEMPT_RT=y`). However, you might find only the usual first 3 choices; this is as the fourth one depends upon `CONFIG_EXPERT` being on (y):

```
Depends on: <choice> && EXPERT [=y] && ARCH_SUPPORTS_RT [=y]
```

So, if, in spite of applying the RTL patch set, the fourth RTL option is not seen, first, turn on **General Setup | Configure standard kernel features (expert users)**; the fourth RTL choice should then be visible and you can enable it.

Also, it's interesting to note that the *dynamic preemption* model is turned off when `PREEMPT_RT` (RTL, in effect) is turned on. We covered this in the previous chapter in the section *The dynamic preemptible kernel feature.*

As of this writing, the supported patche to 6.1 LTS were for the 6.1.77[-rt24], 6.1.75[-rt23], 6.1.67[-rt20], 6.1.46[-rt14] and 6.1.38[-rt13] kernels (here: `https://mirrors.edge.kernel.org/pub/linux/kernel/projects/rt/6.1/`); at the time of writing this material, I selected the 6.1.46 one (note that the particular RTL patches available against a given base kernel can and do change as things evolve).

Once built, perform the usual remaining steps (on the x86_64: `sudo make modules_install && sudo make install`), then reboot. From the bootloader menu, be sure to select your cool new RTL kernel! Once on a shell, verify that it is indeed the real-time kernel. Here's some output I got:

```
$ uname -a
Linux fedora 6.1.46-rt14-rc1 #1 SMP PREEMPT_RT [ ... ] x86_64 GNU/Linux
```

Interestingly, the presence of the `/sys/kernel/realtime` (pseudo)file, and its value being 1, indicates it's a real-time kernel, RTL:

```
$ cat /sys/kernel/realtime
1
```

So great, we're now – as promised – running Linux in its RTL avatar, a hard real-time operating system, an RTOS!

Do note carefully: merely running the OS as a real-time system is nowhere near sufficient when deploying a real-time project; the *applications too* need to be carefully designed and written – especially their time-critical regions – with real-time guidelines in mind. Again, the RTL Wiki site provides many pointers on the same; be sure to check them out.

Miscellaneous scheduling related topics

We close this chapter with a couple of miscellaneous yet useful topic mentions.

A few small (kernel space) routines to check out

Here are a few kernel routines you may find useful (we leave it to you to look up the details and sample usage within the kernel):

- `rt_prio()`: Given the priority as a parameter, returns a Boolean to indicate whether it's a real-time task or not.
- `rt_task()`: Based on the priority value of the task, given the task structure pointer as a parameter, returns a Boolean to indicate whether it's a real-time task or not (a wrapper over `rt_prio()`).
- `task_is_realtime()`: Similar, but based on the scheduling policy of the task. Given the task structure pointer as a parameter, returns a Boolean to indicate whether it's a real-time task or not.

As the above routines all fall within a header (`include/linux/sched/rt.h`), simply include it and directly use them within your module.

The ghOSt OS

These days, there's a steady rise of whole new technology domains with regard to hardware and software; take, for example, NUMA with chiplets, processor accelerators (GPUs, TPUs, DPUs), AMD CCX, AWS Nitro, and so on. Not only that, but cloud-based workloads have special needs; for example, microsecond (scheduling) latencies are required for certain high-speed networking/telecom projects. These greatly stress – if not overwhelm – the capabilities of existing OS schedulers.

Sure, we could build or employ a data plane OS or write several custom scheduling classes to address these concerns, but testing, refining, and deploying them at scale (think datacenter workloads) is hard to do quickly, correctly and efficiently.

These and similar scenarios are giving rise to an OS and user space abstraction of sorts with regard to scheduling; Google has proposed a relatively new system, christened **ghOSt** – providing *fast and flexible user-space delegation of Linux scheduling* – to address just such needs and scenarios. It's aimed at both the research community and at datacenters in production. It consists of both kernel and user space code; the intent is that the kernel code remains largely stable (essentially, it's a custom scheduling class). The user space portion is where policy decisions are made (via regular processes called *agents* that communicate with the ghOSt kernel code); thus, it can and does change fast, and is kept open source (`https://github.com/google/ghost-userspace`).

The overall idea seems to be to provide *"policies for a range of scheduling objectives, from µs-scale latency, to throughput, to energy efficiency…."* The kernel implementation uses eBPF to help accelerate code paths. As of this writing, it seems to be in an experimental development phase and is not considered to be an officially supported Google product. The *Further reading* section of this chapter has more on this topic area.

Summary

In this, our second chapter on CPU (or task) scheduling on the Linux OS, you have learned several key things. Among them, you learned how to programmatically query and set any (user or kernel) thread's CPU affinity mask; this naturally led to how you can programmatically query and set any thread's scheduling policy and priority.

The whole notion of being "completely fair" (via the CFS implementation) was brought into question, and some light (quite a bit!) was shed on the elegant solution called cgroups (v2), which is now deeply embedded into the Linux kernel. We covered how *systemd* helps auto-integrate cgroups into modern distros, servers and even embedded systems, automatically and dynamically creating and maintaining various cgroups (via its slice and scope artifacts). You even learned how to leverage the cgroups v2 CPU controller to allocate CPU bandwidth (or utilization) as desired to processes in a sub-group, both via systemd unit files and manually.

We then briefly delved into what an RTOS is and found that though vanilla Linux is a GPOS, an RTL (out-of-tree) patch set very much exists, which, once applied and the kernel is configured and built, has you running Linux as a true hard real-time system, an RTOS. (We do note that carefully designing and writing your RT apps is equally important for true RT systems.) We ended with a quick mention of a few kernel scheduler-related routines and Google's research project, the ghOSt OS.

That's quite a bit! Take the time to properly understand the material and work on it in a hands-on fashion. Once done, do join me on a really important aspect; let's together adventure into the intricacies of kernel synchronization in this book's final two chapters!

Questions

As we conclude, here is a list of questions for you to test your knowledge regarding this chapter's material: `https://github.com/PacktPublishing/Linux-Kernel-Programming_2E/blob/main/questions/ch11_qs_assignments.txt`. You will find some of the questions answered in the book's GitHub repo: `https://github.com/PacktPublishing/Linux-Kernel-Programming_2E/tree/main/solutions_to_assgn`.

Further reading

To help you delve deeper into the subject with useful materials, we provide a rather detailed list of online references and links (and at times, even books) in a *Further reading* document in this book's GitHub repository. The *Further reading* document is available here: `https://github.com/PacktPublishing/Linux-Kernel-Programming_2E/blob/main/Further_Reading.md`.

Leave a review!

Enjoying this book? Help readers like you by leaving an Amazon review. Scan the QR code below for a 20% discount code.

**Limited Offer*

12

Kernel Synchronization – Part 1

With the previous chapter and the one preceding it (*Chapters 11* and *10*, respectively), you learned a good deal about CPU (or task) scheduling on the Linux OS. In this chapter and the following one, we shall dive into the – at times necessarily complex – topic of kernel synchronization.

As any developer familiar with programming in a multithreaded environment is well aware, there is a need for **synchronization** whenever two or more threads (code paths in general) may work upon a shared writable data item. Without synchronization (or mutual exclusion) in accessing the shared data, they can race; that is, the outcome cannot be predicted. This is called a *data race*. (In fact, data races can even occur when multiple single-threaded processes work on any kind of shared memory object, or where interrupts are a possibility.) Pure code itself is never an issue as its permissions are read+execute (`r-x`); reading and executing code simultaneously on multiple CPU cores is not only perfectly fine and safe but actively encouraged (as it results in better throughput, and that is why multithreading is a good idea). However, the moment you're working on shared writable data is the moment you need to start being very careful (it's why multithreading and concurrency tend to be complex topics!).

The discussions around concurrency and its control via synchronization are varied, especially with as rich and complex a piece of software as the Linux kernel (and related drivers/modules), which is what we're dealing with in this book. Thus, for convenience, we will split this large topic into two chapters, this one and the next.

In this chapter, we will cover the following topics:

- Critical sections, exclusive execution, and atomicity
- Concurrency concerns within the Linux kernel
- Mutex or spinlock? Which to use when
- Using the mutex lock
- Using the spinlock
- Locking and interrupts
- Locking – common mistakes and guidelines

Technical requirements

I assume you have gone through *Online Chapter, Kernel Workspace Setup*, and have appropriately prepared a guest **Virtual Machine** (**VM**) running Ubuntu 22.04 LTS (or a later stable release, or a recent Fedora distro) and installed all the required packages. If not, I highly recommend you do this first.

To get the most out of this book, I strongly recommend you first set up the workspace environment, including cloning this book's GitHub repository for the code, and work on it in a hands-on fashion. The repository can be found here: `https://github.com/PacktPublishing/Linux-Kernel-Programming_2E`.

In this and the following chapter, we refer to small (and simple) device driver examples now and then. If you're not familiar with the basic Linux driver concepts, though not a required prerequisite, may I again suggest the *Linux Kernel Programming - Part 2 - Char Device Drivers and Kernel Synchronization* book; this is explained in the first chapter itself: *Chapter 1, Writing a Simple misc Character Device Driver* (this *Part 2* book is in fact considered the companion volume to this book; also, the e-book is freely downloadable).

With this done, fasten your seatbelts, and let's get started on this important and interesting topic area!

Critical sections, exclusive execution, and atomicity

Imagine you're writing software for a multicore system (well, nowadays, it's typical that you will work on multicore systems, even on most embedded projects). As we mentioned in the introduction, running multiple code paths in parallel is not only safe but also desirable (why spend those dollars otherwise, right?). On the other hand, concurrent (parallel and simultaneous) code paths within which **shared writable data** (also known as **shared state**) is accessed in any manner is where you are required to guarantee that, at any given point in time, only one thread can work on that data at a time! This is key. Why? Think about it: if you allow multiple concurrent code paths to work in parallel on shared writable data, you're asking for trouble: **data corruption** (a "data race") can occur as a result. The following section, after covering some key points, will clearly illustrate the *data race* concept with a few pseudocode examples (also, you could peek at *Figure 12.6* if you'd like).

What is a critical section?

The points that follow are very important; please read them carefully.

A **critical section** is a code path that must fulfill these two conditions:

- **Condition one: The code path is possibly concurrent**, that is, there exists a possibility that it can run in parallel.

  AND

- **Condition two:** That it works upon (reads and/or writes) **shared writable data** (aka shared state).

Thus, critical sections, by definition, require protection from parallelism.

In other words, a critical section is a piece of code that must run exclusively, even, at times, atomically.

- By exclusively, we're implying that at any given point in time, exactly one thread is running the code of the critical section; that is, it runs alone (serialized, as opposed to parallelized). This is required for data safety reasons.
- The word *atomic* implies something indivisible; in this context, it means the ability to *run to completion without interruption.*

If two or more threads can execute a critical section's code concurrently, it is a **bug** or a defect; this is often referred to as a **race condition** or **data race.**

Identifying and protecting critical sections from simultaneous execution (from data races) is an implicit requirement for correct software that you – the designer/architect/developer – must ensure. Learning how to protect critical sections is (relatively) easy; correctly *identifying* every critical section is a skill you must master. We'll try and help with that via the "exercises" that follow.

Exercise 1. In the following user-mode Pthreads (pseudo) code snippet, where you can assume the function code can be run in parallel (by multiple threads), does the code region between time *t1* and *t2* (shown in comments) constitute a critical section?

```
void qux(int factor)
{
    int nok;
    [ … ]
    //------------------------ t1
    nok += 10*PI;
    //------------------------ t2
    printf("..."); [ … ]
}
```

Solution 1. Okay, I'll help you with this first exercise! Ask yourself, what exactly constitutes a critical section? It's when these two conditions are fulfilled: the code path can possibly execute in parallel and works upon shared writable data. So, now, does the code in question (the line between *t1* and *t2*) fulfill these two pre-conditions? Well, it can run in parallel (as explicitly stated), but it does not work on shared writable data (the variable nok isn't shared; it's a local variable, thus each thread that executes this code path will get a copy of the variable on its stack). So, the answer here is clearly no, it isn't a critical section. In other words, it can run in parallel – any number of instances – without any kind of explicit protection.

Onto another couple of small exercises.

Exercises 2 and 3. In the following kernel module (pseudo) code snippets, where you can assume the code can run in parallel (by multiple user-mode threads switching into kernel space to these code paths, running them in process context), do the code regions between time *t1* and *t2* (shown in comments) constitute critical sections?

```
static struct quux_drvctx {
        ...
} *mydrv;

write_quux()      /*  driver
write method    */
{
    [ … ]
    //-------------------- t1
    mydrv->sensor2 = 1;
    mydrv->hw = hw_zoom;
    //-------------------- t2
            [ … ]
}
```

```
static int glob;
static __init my_kmod_init(void) /*
init method */
{
        [ … ]
        //-------------------- t1
    glob += 14;
    pr_info(...);
        //-------------------- t2
        [ … ]
}
[ … ]
module_init(my_kmod_init);
module_exit(my_kmod_cleanup);
```

(Solutions can be found at the end of this chapter in the *Solutions(s)* section; please try solving them for yourself before you check the solution.)

Let's now revisit the key notion of *atomicity:* an atomic operation is one that is indivisible. On any modern processor, two operations are generally considered to **always be atomic**; that is, they will run to completion without interruption:

- The execution of a single machine language instruction.

- Reads or writes to an aligned primitive data type that is within the processor's word size (typically 32 or 64 bits); so, reading or writing a 32-bit or 64-bit integer on a 64-bit system is guaranteed to be atomic. Threads reading that variable will never see an in-between, *torn*, or *dirty* result; they will either see the old or the new value. On the other hand, a 32-bit processor operating on a 64-bit item is not guaranteed to be atomic and *could* result in a torn or dirty read (or write).

A word to the wise: it pays to tread very carefully here! It's been shown that with modern highly optimizing compilers on modern hardware processors, even this "truth" – that loads/stores (meaning reads/writes) to aligned primitive data types within the processor's word size are always atomic – may not hold true! Compilers can now employ load-/store-tearing techniques, and more; read more in this excellent article: *Who's afraid of a big bad optimizing compiler?* LWN, July 2019: `https://lwn.net/Articles/793253/` (plus more in the *Further reading* section of this chapter).

So, if you have some lines of code that work upon shared (global or static) writable data, they cannot – in the absence of an explicit synchronization mechanism – be guaranteed to run exclusively. Note that at times, running the critical section's code *atomically*, as well as exclusively, is required, but not all the time; let's delve further into this aspect.

When the code of the critical section is running in a safe-to-sleep possibly blocking *process context* (such as typical file operations on a driver via a user app (open, read, write, ioctl, mmap, and so on), or the execution path of a kernel thread or workqueue), it might well be acceptable to not have the critical section being truly atomic, but it does need to be exclusive. However, when its code is running in a non-blocking *atomic context* (such as within a hardware interrupt: a hardirq, tasklet, or softirq), *it must run atomically as well as exclusively* (we shall cover these points in more detail in the *Mutex or spinlock? Which to use when* section).

A conceptual example will help clarify things. Let's say that three threads (from user-space app(s)) issue the `open()` and `read()` system calls, thus working on your driver (implemented as a kernel module or within the kernel) more or less simultaneously on a multicore system. (Recall, Linux is a *monolithic kernel*; when a process or thread issues a system call, it switches to kernel mode and runs the appropriate kernel/driver code paths in process context.)

Without any intervention, they may well end up running the critical section's code in parallel, thus working on the shared writable data in parallel (a data race!) and very likely corrupting it! For now, let's look at a conceptual diagram to see how non-exclusive execution within a critical section's code path is wrong (we won't even talk about atomicity here):

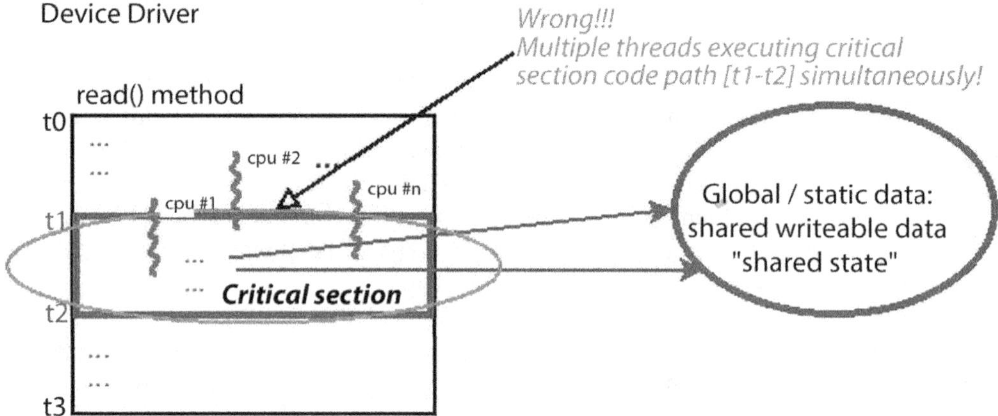

Figure 12.1: A conceptual diagram showing how a critical section code path is violated by having more than one thread running within it simultaneously, a data race

As shown in the preceding diagram, in your device driver, within its (say) read method, you're having it run some code to perform its job (reading some data from the hardware). Let's take a more in-depth look at this diagram *in terms of data accesses being made* at different points in time:

- From time `t0` to `t1`: None or only local variable data is accessed. This is concurrent-safe, with no protection required, and can run in parallel (since each thread has its own private stack).
- From time `t1` to `t2`: Global/static shared writable data is accessed. This is *not* automatically concurrent-safe; it's **a critical section** (as *Figure 12.1* shows that it definitely fulfills the two conditions for it). Thus, it must be **protected** from concurrent access. It must run exclusively (alone, exactly one thread at a time, and serialized) and, perhaps, atomically.
- From time `t2` to `t3`: None or only local variable data is accessed. This is concurrent-safe, with no protection required, and can run in parallel (since each thread has its own private stack).

In this book, with the material covered so far and your own knowledge, we assume that you are by now aware of the need to synchronize critical sections; we will not discuss this particular point any further. Those of you who are interested in learning more may refer to my earlier book, *Hands-On System Programming with Linux* (Packt, October 2018), which covers these points in detail (especially *Chapter 15, Multithreading with Pthreads Part II – Synchronization*).

So, knowing this, we can now restate the notion of a critical section while also mentioning when the situation arises. *A critical section* is code that must run as follows:

- **(Always) Exclusively**: Alone (serialized)

- **(When in an atomic context) Atomically:** Indivisibly, to completion, without interruption

In the next section, we'll look at a classic scenario – the increment of a global integer.

A classic case – the global i ++

Think of this classic example: a global integer `i` is being incremented within a concurrent code path, one within which multiple threads of execution can simultaneously execute. A naive understanding of computer hardware and software will lead you to believe that this operation is obviously atomic. However, the reality is that modern hardware and software (the compiler and OS) are much more sophisticated than you may imagine, thus causing all kinds of invisible (to the app developer) performance-driven optimizations.

We won't attempt to delve into too much detail here, but the reality is that modern processors are extremely complex: among the many technologies they employ toward better performance, a few are superscalar and super-pipelined execution in order to execute multiple independent instructions and several parts of various instructions in parallel (respectively), performing on-the-fly instruction and/or memory reordering, caching memory in complex hierarchical on-CPU caches, load/store tearing, and so on! We will delve into some of these details in *Chapter 13, Kernel Synchronization – Part 2*, in the *Understanding CPU caching basics, cache effects and false sharing* and *Introducing memory barriers* sections. Several papers (and books) worth delving into on these fascinating topics can be found in the *Further reading* section.

All of this makes for a situation that's more complex than it appears to be at first glance. Let's continue with the classic `i ++` in a possibly concurrent code path:

```
static int i = 5;
[ ... ]
foo()
{
    [ ... ]
    i ++;       /* Is this safe? Yes, if this code path is truly exclusive or
atomic...
                 * Alternately, if this code path's not exclusive, is the i ++
truly atomic?? */
}
```

So, is this increment safe? The short answer is no, you must protect it. Why? *It's a critical section* – we're accessing (reading/writing) shared writable data in a possibly concurrent code path! The longer answer is that it really depends on:

a. Whether the code of function foo() is guaranteed to run exclusively.
b. Whether the increment operation here is truly atomic (indivisible); if it is, then `i ++` poses no danger in the presence of parallelism – if not, it does!

So, assuming the execution of the code path surrounding the `i ++` is non-exclusive – that is, other threads can execute this code path `foo()` in parallel – to have it work correctly, we then require the `i ++` operation to be truly atomic. Being a simple high-level language operation, it may "appear" at first glance to be atomic, but is it really? How do we know whether this is the case or not? Two things determine this:

- The processor's **Instruction Set Architecture** (**ISA**), which determines (among several things related to the processor at a low level) the machine instructions that execute at runtime when this operation runs
- The compiler, which of course converts the high-level language source to assembly (and then the assembler generates the machine code that ultimately executes on the processor)

If the ISA has within it the facility to employ a *single* machine instruction to perform an integer increment, *and* the compiler has the intelligence and occasion to use it, *then* it's truly atomic – it's safe and doesn't require explicit protection, typically, locking. Otherwise, it's not safe and requires locking! So, on your machine and with your current compiler, how do we know?

Try this out: Navigate your browser to this wonderful compiler explorer website: `https://godbolt.org/`. Select C as the programming language and then, in the left pane, declare the global `i` integer and increment it within a function. Compile in the right pane with an appropriate compiler and compiler options. You'll see the actual machine code generated for the C high-level `i ++;` statement. If it's indeed a single machine instruction, then it will be safe and atomic; if not, it will require locking. Typically, on CISC-based machines like the `x86[_64]`, compiler optimization levels of 2 and above do make the code atomic, but this isn't always the case with RISC machines (like ARM-based ones).

By and large, you will find that you can't really tell; in effect, you *cannot* afford to assume things – you will have to assume that the `i ++` operation is unsafe, that is, non-atomic, by default and protect it! This can be seen in the following screenshot:

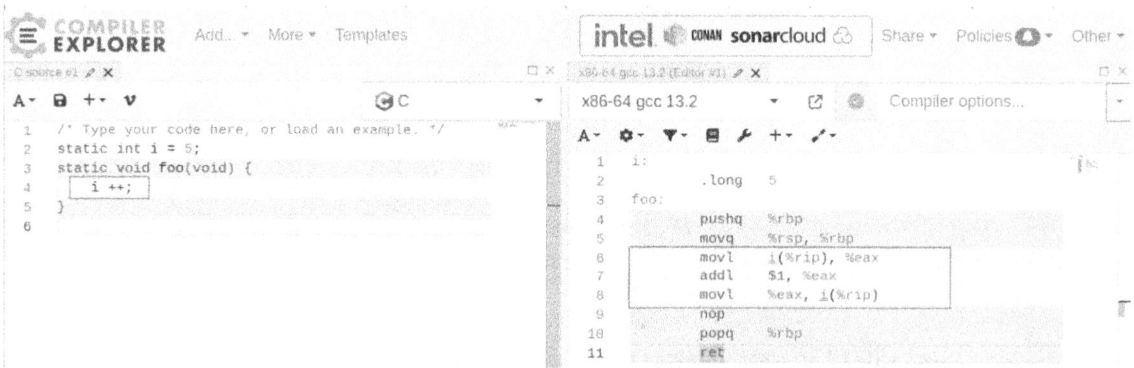

Figure 12.2: Even with the latest stable gcc version but no optimization, the x86_64 gcc produces multiple machine instructions for the i ++

The preceding screenshot clearly shows this: the light/yellow background regions in the left- and right-hand panes are the C source for our famous `i ++;` statement and the corresponding assembly generated by the compiler, respectively (based on the `x86_64` ISA and the particular compiler's optimization level). (FYI, in *Figure 12.2*, I changed the assembly syntax from Intel to the more generalized AT&T one by deselecting **Intel asm syntax** from the Setting gear wheel in the right pane). Typically, with no optimization, `i ++` becomes *three machine instructions*. This is exactly what we expect: it corresponds to the *fetch or load of* `i` (memory to register), the *increment*, and the *store* (register to memory)! Now (in this case at least, without optimization) this is *not* atomic; thus it's entirely possible that, after one of the machine instructions executes, the control unit interferes and switches the instruction stream to a different point. This could even result in another process or thread being context-switched in (unless you used locking around it)!

The good news is that with a quick `-O2` in the **Compiler options...** window, `i ++` becomes just one machine instruction – truly atomic! (It doesn't always show though.) However, we can't predict these things in advance; one day, your code may execute on a fairly low-end ARM (RISC) system, increasing the chance that multiple machine instructions are required for `i ++`. You might by now be wondering – to fix this, using something like a mutex (or even a spinlock) lock for an `i ++` seems pretty sub-optimal; you'd be right. We shall cover an optimized locking technology specifically for integer operations in *Chapter 13, Kernel Synchronization – Part 2*, in the *Using the atomic_t and refcount_t interfaces* section.

Modern languages provide native atomic operators; for C/C++, it's fairly recent (from 2011); the ISO C++11 and ISO C11 standards provide ready-made and built-in atomic variables for this. A little googling will quickly reveal them to you. Modern glibc also makes use of them. As an example, if you've worked with signaling in user space, you will know to use the `volatile sig_atomic_t` data type to safely access and/or update an integer atomically within a signal handler. What about in the kernel? In the next chapter, you'll learn about the Linux kernel's solution to this key issue.

Do realize early: **the Linux kernel is, of course, a concurrent environment**; multiple threads of execution run in parallel on multiple CPU cores. Not only that, but even on uni-processor (UP/single CPU) systems, the presence of hardware interrupts, traps, faults, exceptions, and software signals can cause data integrity issues (data races). Needless to say, protecting against concurrency at required points in the code path, at critical sections, is easier said than done; identifying and protecting critical sections using technologies such as locking – as well as other synchronization primitives and technologies – is absolutely essential, which is why this is the core subject matter of this chapter and the next.

Concepts – the lock

We require synchronization because of the fact that, without any intervention, threads can concurrently execute critical sections where shared writable data (shared state) is being worked upon. To defeat concurrency in these critical sections, we need to get rid of parallelism; we need to *serialize* the flow of code in the critical section.

To force a code path to become serialized, a common technique is to use a **lock.** Essentially, a lock works by guaranteeing that precisely one thread of execution can "take" or own the lock at any given point in time; once taken, only this thread can move forward - we term it the "winner." These notions are expanded on shortly. Thus, using a lock to protect a critical section in your code will give you what we're after - running the critical section's code exclusively (and also perhaps atomically; more on this to come). Check out this diagram (*Figure 12.3*):

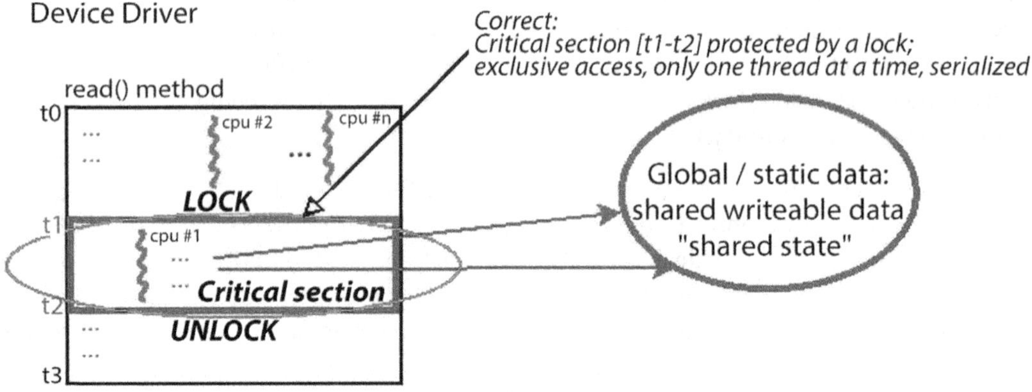

Figure 12.3: A conceptual diagram showing how a critical section code path is honored, given exclusivity, by using a lock

Figure 12.3 shows one way to fix the situation mentioned previously (*Figure 12.1*): using a lock to protect the critical section! How does the lock (and unlock) work, conceptually?

The basic premise of a lock is that whenever there is contention for it - that is, when multiple competing threads (say, `n` threads) attempt to acquire the lock (via the conceptual `LOCK` operation) - **exactly one thread will succeed**. This thread is deemed the "winner," the "owner," of the lock. It sees the *lock* API as a non-blocking call and thus continues to run happily - and exclusively! - while executing the code of the critical section (the critical section is effectively the code between the *lock* and the *unlock* operations!).

What happens to the `n-1` "loser" threads? They (perhaps) see the lock API as a blocking call; they, to all practical effect, wait. Wait for what? The `UNLOCK` operation, of course, which is (must be) performed by the owner of the lock (the "winner" thread) when it's done with the work of the critical section! Once unlocked, the remaining `n-1` threads now compete for the next "winner" slot; of course, exactly one of them will win and proceed forward. In the interim, the `n-2` losers will now wait upon the (new) winner's *unlock*. This repeats until all `n` threads (finally and sequentially) acquire the lock.

Now, locking works of course, but - and this should be quite intuitive - it results in (pretty steep!) **overhead, as it defeats parallelism and serializes** the execution flow! To help you visualize this situation, think of a funnel, with the narrow stem being the critical section where only one thread can fit at a time. All other threads get choked, waiting on the unlock; locking creates bottlenecks. The following figure is meant to show this analogy:

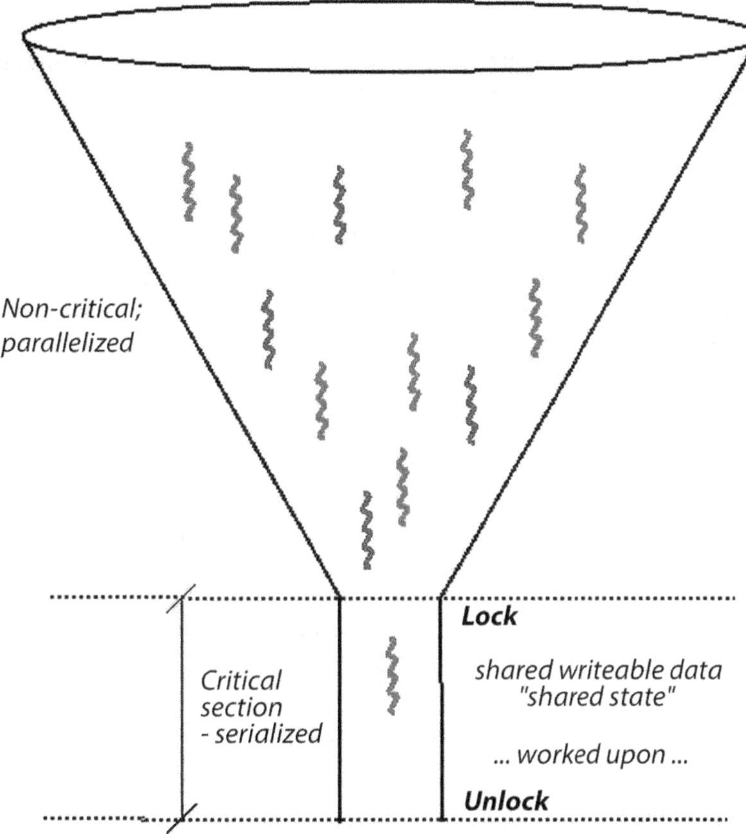

Figure 12.4: A lock creates a bottleneck, analogous to a physical funnel

Another oft-mentioned physical analog to locking is a highway with several lanes merging into one very busy – and choked with traffic – lane, a poorly designed toll booth, perhaps. Again, parallelism – cars (threads) driving in parallel with other cars in different lanes (CPUs) – is lost, and serialized behavior is required; cars are forced to queue one behind the other.

Thus, it is imperative that we, as software architects, try and design our products/projects with *minimal locking*. While completely eliminating global variables is not practically possible in most real-world projects, optimizing and minimizing their usage is required. We shall cover more regarding this, including some very interesting *lockless* programming techniques, later.

Another really key point: a newbie programmer might naively assume that performing reads on a shared writable data object is perfectly safe and thus requires no explicit protection; (with the exception of an aligned primitive data type that is within the size of the processor's bus) this is untrue. This situation can lead to what's called **dirty or torn reads**, a situation where possibly stale and/or inconsistent data is being read even as another writer thread is simultaneously writing.

The moral's clear: you need to protect all accesses, read and write, to shared writable data in possibly concurrent code paths, in effect, in all critical sections.

Since we're on the topic of atomicity, as we just learned, on a typical modern microprocessor, the only things guaranteed to be atomic are a single machine language instruction or (as we learned, possibly) a read/write to an aligned primitive data type within the processor bus's width. So, how can we mark a few lines of C code so that they're truly atomic? In user space, this isn't even possible (we can come close, but cannot guarantee atomicity).

How do you "come close" to atomicity in user-space apps? You can always construct a user thread to employ the `SCHED_FIFO` task scheduling policy and a **real-time** (**RT**) priority of `99`. This way, when it wants to run, pretty much nothing besides hardware interrupts/exceptions can preempt it. (The old audio subsystem implementation heavily relied on this semantic.)

In kernel space, we can write code that's truly atomic. How, exactly? The short answer is that we use **spinlocks** to do so! We'll learn about spinlocks in more detail shortly.

Critical sections – a summary of key points

Let's summarize some key points regarding critical sections. It's really important to go over these carefully, keep these handy, and ensure you use them in practice:

- A **critical section** is a possibly concurrent code path, one that can execute in parallel, and that works upon (reads and/or writes) shared writable data (aka shared state).
- Because it works on shared writable data, the critical section:
    - Requires protection from parallelism and concurrency (that is, it must run alone/serialized/in a mutually exclusive fashion).
    - When running in an atomic non-blocking context (which includes any kind of interrupt context), you must guarantee it runs atomically: indivisibly, to completion, without interruption.

Once protected, you can safely access your shared state until you "unlock."

- Every critical section in the codebase must be identified and protected:
    - Identifying critical sections is critical! Carefully review your code and make sure you don't miss them. (Any global or static variable is a typical red flag; but it's not only these, any kind of shared state – hardware registers, mailboxes, and so on – in a possibly concurrent code path can be a critical section.)
    - Protecting them can be achieved via various technologies; one very common technique is *locking* (which we'll shortly get into in detail). There are also *atomic operators* and *lock-free* programming techniques, which we'll look at in the next chapter.

    - A common mistake is only protecting critical sections that *write* to shared writable data; you must also protect critical sections that *read* shared writable data. Otherwise, you risk a **torn or dirty read!** To help make this key point clear, visualize an unsigned 64-bit data item being read and written on a 32-bit system; in such a case, the operation can't be atomic (two load/store operations are required per read/write). Thus, what if, while one thread's reading the value of the data item, it's being simultaneously written to by another thread!? The writer thread would have taken a "lock" of some sort on the access but because you thought reading is safe, the lock isn't taken by the reader thread; due to an unfortunate timing coincidence, you can end up performing a partial/torn/dirty read. We will learn how to overcome these issues by using various locking techniques in the coming sections now and in the next chapter.
    - Another deadly mistake is not using the same (correct) lock to protect a given data item. For example, if you're using lock A to protect a global data structure X, then you must always use lock A whenever it's accessed; using lock B doesn't help. This can be more difficult than it seems at first, as large projects (like the Linux kernel) can have tens of thousands of locks!
    - Failing to protect critical sections leads to a **data race**, a situation where the outcome – the actual value of the shared data being read/written – is "racy," which means it varies, depending on runtime circumstances and timing. This is a defect, a bug (a bug that, once in "the field," is extremely difficult to see, reproduce, determine its root cause, and fix. We will cover some very powerful stuff to help you with this in the next chapter, in the *Lock debugging within the kernel* section; be sure to read it then!).
- **Exceptions**: You are safe (implicitly, without explicit protection) in the following situations:
    - When you are working on local variables. They're allocated on the private stack of the thread (or, in interrupt context, on the local IRQ stack) and are thus, by definition, safe.
    - When you are working on shared writable data in code that cannot possibly run in another context; that is, it's serialized by nature. In our context, the *init* and *cleanup* methods of a kernel module qualify (they run exactly once, serially, on `insmod` (or `modprobe`) and `rmmod` only).
    - When you are working on shared data that is truly constant and read-only (don't let C's `const` keyword fool you, though!).
- Locking is inherently complex; you must carefully think, design, and implement your locking schema while avoiding *deadlocks*. We'll cover this in more detail in the *Locking – common mistakes and guidelines* section.

Data races – a more formal definition

A more formal approach to the important notions of memory consistency in the presence of multiple concurrent load/store operations is provided by a memory (consistency) model; it "models" the system memory behavior predicting what values may result via load (read from memory) operations when code executes on that system. The Linux kernel has just such a model; it's called the **Linux-Kernel Memory Model (LKMM)**.

Delving into its details isn't really required here, but does make for some fascinating reading; do look up the `explanation.txt` document in the official kernel documentation here: `https://elixir.bootlin.com/linux/v6.1.25/source/tools/memory-model/Documentation/explanation.txt`.

(I'd definitely recommend you read at least the first three sections: *INTRODUCTION, BACKGROUND*, and *A SIMPLE EXAMPLE*. The notion of *memory ordering* is also covered in the LKMM.)

The LKMM provides two kinds of access to memory:

- **Plain access**: These are the typical accesses made to memory via C language statements, for example, `i ++` or `y = 42 - x;`.
- **Marked access**: These are "special" accesses; they're designed and implemented to implicitly *guarantee atomicity*:
    - Loads (reads) via the `READ_ONCE()` macro (e.g.., `READ_ONCE(x);`)
    - Stores (writes) via the `WRITE_ONCE()` macro (e.g., `WRITE_ONCE(y, 42-x);`)
    - (In fact, it's not just these macros that guarantee atomicity; the `atomic_*()`, `refcount_*()`, `smp_load_acquire()`, and similar macros fall into the class of marked access.)

Okay, now to the key point here: what exactly, in the view of the LKMM, constitutes a data race?

From the same `explanation.txt` document, the section *PLAIN ACCESSES AND DATA RACES* (here: `https://elixir.bootlin.com/linux/v6.1.25/source/tools/memory-model/Documentation/explanation.txt#L1977`) goes into the details. It's best to quote directly a bit from this document:

```
A "data race" occurs when there are two memory accesses such that:

1.      they access the same location,

2.      at least one of them is a store,

3.      at least one of them is plain,

4.      they occur on different CPUs (or in different threads on the same CPU),
and

5.      they execute concurrently.

In the literature, two accesses are said to "conflict" if they satisfy 1
and 2 above.  We'll go a little farther and say that two accesses are "race
candidates" if they satisfy 1 - 4.  Thus, whether or not two race candidates
actually do race in a given execution depends on
```

```
whether they are concurrent. [ … ]
```

This is wonderful! A clear and formal definition, via the LKMM, of precisely what constitutes a data race.

You'd think this knowledge should be leveraged to write a tool to detect data races, right? This is precisely what the **Kernel Concurrency Sanitizer** (**KCSAN**) does! (Marco Elver is the current maintainer of KCSAN.) In fact, the *Linux Kernel Debugging* book in *Chapter 8, Lock Debugging*, covers using KCSAN in some detail (with examples). I'd definitely recommend you learn how to use KCSAN.

Marco Elver, in a presentation at the Linux Plumbers Conference, August 2020, entitled *Data-race detection in the Linux kernel* (link: `https://lpc.events/event/7/contributions/647/attachments/549/972/LPC2020-KCSAN.pdf`), explains data races as defined by the LKMM with this slide:

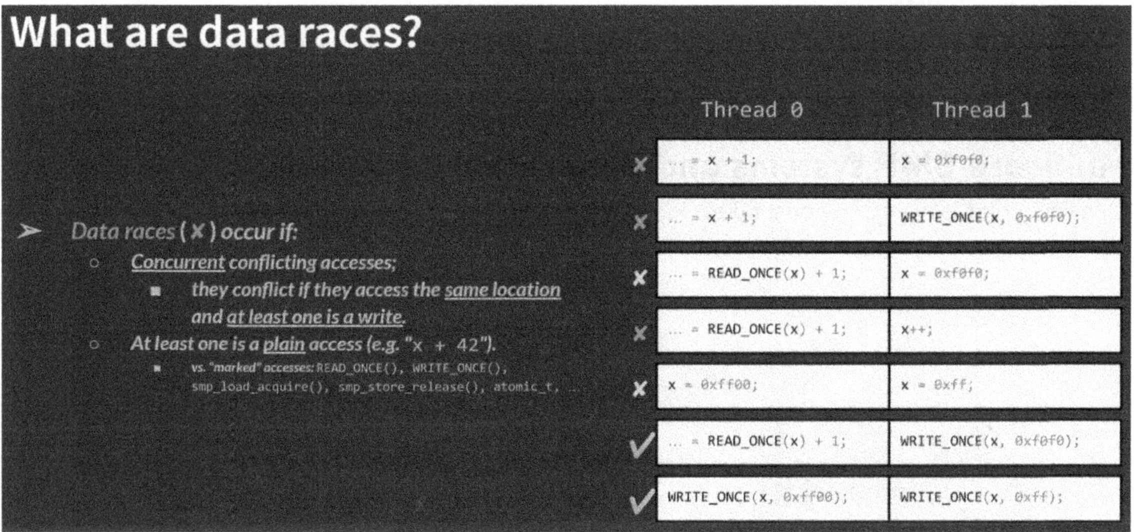

Figure 12.5: The "What are data races?" slide. Credit: Marco Elver

As you can see, on the right of *Figure 12.5* are examples of two threads running concurrently (on different CPU cores) working on the same memory object (same location, shared writable data). The red cross to the left indicates a data race, the green tick mark implies it's fine (the in-betweens – rows 3 and 5 – are a "maybe"; they race when interpreted "strictly").

This might well have you thinking: "*Why not always use marked accesses and never have data races?*" Please do NOT do this. Marked accesses are meant to be used internally by the kernel code and/or when you know there's a potential data race but don't care too much about it (a typical example is network driver statistics code that increments a counter without explicit locking primitives; a data race here is deemed to not matter too much). The key point here though, especially for module/driver authors, is this: using marked accesses actually *prevents* tools like KCSAN from catching data races. It's important to continue to use plain C accesses for most of the cases.

Also, careful – though marked accesses guarantee atomic loads and stores, they do not guarantee memory ordering; this must be done via memory barriers (the following chapter throws some light on this). Now, we will move on to another key area – identifying concurrency and critical sections when working with the Linux OS.

Concurrency concerns within the Linux kernel

Recognizing critical sections within a piece of kernel code is of critical importance; how can you protect it if you can't even see it? The following are a few guidelines to help you, as a budding kernel/driver developer, recognize where concurrency concerns – and thus critical sections – may arise:

- The presence of **Symmetric Multi-Processor (SMP)** systems (`CONFIG_SMP=y`)
- The presence of a preemptible kernel (`CONFIG_PREEMPTION=y`)
- Blocking I/O
- Hardware interrupts (on both SMP and/or UP systems)

These are critical points to understand, and we will discuss each in this section.

Multicore SMP systems and data races

This first point is pretty obvious; take a look at the pseudocode shown in *Figure 12.6*:

```
      driver_read_method()
      {
t1        [time t1] : < do work|w1() >
t2        [time t2] : <... iterate ...
critical                   ... over ...
section                    ... global ('shared-writable') array ...
t3        [time t3] :      ... of structures ... >
          [...]
time  }
```

Figure 12.6: Pseudocode – a critical section (time t2 to t3) within a (fictional) driver's read method; it's potentially buggy as there's no locking

It's a similar situation to what we showed in *Figure 12.1* and *Figure 12.3*. As shown, from time `t2` to time `t3`, the function can run in parallel and it's working on some global shared writable data, thus making this a critical section.

Now, visualize a system with, say, four CPU cores (an SMP/multicore system); two user-space processes, P1 (running on, say, CPU 0) and P2 (running on, say, CPU 2), can concurrently open the device file and simultaneously issue a `read()` system call. Now, both processes will be concurrently executing the driver's read "method," thus simultaneously working on shared writable data! As we know, the code between `t2` and `t3` is a critical section, and since we violate the fundamental exclusivity rule – critical sections must be executed by only a single thread at any point in time – we can very well end up corrupting the data, the application, or worse.

In other words, this approach can lead to a **data race**; depending on delicate timing coincidences, we may or may not generate an error (a bug). This very uncertainty – the delicate timing coincidence – is what makes finding and fixing errors like this extremely difficult (it can escape your testing effort).

This aphorism is all too true, unfortunately: *Testing can detect the presence of errors, not their absence.* Adding to this, you're worse off if your testing fails to catch data races (and bugs), allowing them free rein in the field.

You might argue that since your product is a small embedded system running on one CPU core (UP), this discussion regarding controlling concurrency (often, via locking) does not apply to you. We beg to differ: pretty much all modern products, if they haven't already, will move to multicore (in their next-generation phases, perhaps). More importantly, even UP systems have concurrency concerns, as we shall soon explore.

Preemptible kernels, blocking I/O, and data races

Imagine you're running your kernel module or driver on a Linux kernel that's been configured to be preemptible (that is, `CONFIG_PREEMPT` is on; we covered this topic in *Chapter 10, The CPU Scheduler – Part 1*). Again, referring to *Figure 12.6*, consider that a process, P1, is running the driver's read method code in process context, working on the global array. Now, while it's within the critical section (between time `t2` and `t3`), what if the kernel *preempts* process P1 and context switches to another process, P2, which is just waiting to execute this very code path? It's dangerous; it can lead to a data race. This could well happen on even a single CPU (UP) system! (Note that though we use the term "process" here, it's interchangeable with "thread.")

Another scenario that's somewhat similar (and again, could occur on either a single-core (UP) or multicore system): process P1 is running through the critical section of the driver method (between time `t2` and `t3`; again, see *Figure 12.6*). This time, what if, within the critical section, it hits a blocking call?

A **blocking call** is a function that causes the calling process context to be put to sleep, waiting upon an event; when that event occurs, the kernel (or the underlying driver) will "wake up" the task, and it will resume execution from where it left off.

This is also known as blocking on I/O and is very common; many APIs (including several user-space library and system calls, as well as several kernel APIs) are blocking by nature). In such a case, process P1 is effectively context-switched off the CPU and goes to sleep, which means that the code of `schedule()` runs and enqueues it onto a wait queue. In the interim, before P1 gets switched back, what if another process, P2, is scheduled to run? What if that process is also running this particular code path? Think about it – firstly, P2 now works on data that's in an in-between state perhaps, not fully updated. Furthermore, by the time P1 is back, the shared data could have changed "underneath it," causing all kinds of errors; again, a data race, a bug!

Hardware interrupts and data races

Finally, envision this scenario: process P1 is innocently running the driver's read method code; thus, it enters the critical section (between time `t2` and `t3`; again, see *Figure 12.6*). It makes some progress but then, alas, a hardware interrupt triggers on the same CPU core! (You can learn about hardware interrupts and their handling in detail in the *Linux Kernel Programming – Part 2* companion volume). On the Linux OS, hardware interrupts have the highest priority; they preempt all code (including kernel code) by default. Thus, process (or thread) P1 will be at least temporarily shelved, losing the processor, as the interrupt-handling code paths will certainly preempt it and run.

Well, you might be wondering, so what? Indeed, this is a completely commonplace occurrence! Hardware interrupts fire very frequently on modern systems, effectively (and literally) interrupting all kinds of task contexts (do a quick `vmstat 3` on your shell; the column under `system` labeled `in` shows the number of hardware interrupts that fired on your system in the last 1 second!).

Here, the key question to ask is this: is the interrupt-handling code (either the hardirq, ISR, or "top half," or the so-called tasklet or softirq "bottom half," whichever occurred), *sharing and working upon the same shared writable data as the process context that it just interrupted?*

If this is true, then *Houston, we have a problem* – a data race! If not, then your interrupted code is not a critical section with respect to the interrupt code path, and that's fine. The fact is that most device drivers do handle interrupts; thus, it is the driver author's (your!) responsibility to ensure that no global or static data – in effect, no critical sections – are shared between the process context and interrupt context code paths. If they are (which does happen at times), you must somehow protect that data from data races and possible corruption (this is typically effected by employing a spinlock; worry not, we shall get to it).

These scenarios might leave you feeling that protecting against these concurrency concerns is a really tall order; how exactly can you accomplish data safety in the face of critical sections existing, along with the various possible concurrency concerns just discussed: multicore (or SMP), kernel preemption, blocking I/O, and hardware interrupts? The good news is that it's really quite easy; the actual (locking) APIs that protect your critical sections are not hard to learn to use. Again, we emphasize that **recognizing – and then being able to protect – critical sections** is the key thing.

Without further ado, let's now begin our dive into the primary synchronization technology that will serve to protect our critical sections – locking.

Locking guidelines and deadlock

Locking, by its very nature, is a complex beast; it tends to give rise to complex interlocking scenarios. Not understanding it well enough can lead to both performance headaches and bugs – deadlocks, circular dependencies, interrupt-unsafe locking, and more. The following locking guidelines are key to ensuring correctly written code when using locking:

- **Locking granularity**
    - The "distance" between the lock and the unlock – in effect, the length of the critical section – should not be coarse (too long a critical section); it should be "fine enough." The points below expand on this.
    - You need to be careful here. When you're working on large projects, keeping too few locks is a problem, as is keeping too many! Too few locks can lead to performance issues (as the same locks are repeatedly used and thus tend to be highly contended).
    - Having a lot of locks is actually good for performance, but not good for complexity control. This also leads to another key point for you, the developer, to understand: with many locks in the codebase, you should be very clear on precisely which lock protects which shared data object. It's completely meaningless if you use, say, `lockA` to protect `mystructX`, but in a code path far away (perhaps an interrupt handler) you forget this and use some other lock, `lockB`, for protection when working on the same structure! Right now, these things might sound obvious, but (as experienced developers know), under sufficient pressure and complexity, even the obvious isn't always so!
    - Try and balance things out. In large projects, using one lock to protect one global (shared) data structure is typical. Now, *naming* the lock variable well can become a big problem in itself! This is why we often place the lock that protects a data structure within it, as a member.
    - Long critical sections that are atomic can cause high latency and become performance bottlenecks, especially in time-critical real-time systems. Several tools exist (in various states of stability) to help you understand where (and for how long) these sections occur. One from the eBPF stable is the tool named **criticalstat**[-bpfcc]. It can detect and report long critical sections (it also helpfully displays the kernel stack trace thus showing their origin) for long durations, in terms of when preemption and/or IRQs are disabled for long-ish durations.

The *criticalstat* utility's man page can be found here: `https://manpages.ubuntu.com/manpages/focal/man8/criticalstat-bpfcc.8.html`; here's its "examples" page: `https://github.com/iovisor/bcc/blob/master/tools/criticalstat_example.txt`. (Ftrace too has tracers that can catch long bouts of preemption/IRQ off times.)

- Only the "owner," the current holder, of a lock can release (unlock) it; attempts to release a lock you don't hold or to re-acquire it while held are considered bugs (the latter can be side-stepped by using recursive locking; see the next point though).
- Avoid recursive locking as much as possible; the kernel community in general frowns upon it.
- **Lock ordering** is critical, greatly mitigating deadlock. You must ensure that **locks are taken in the same order throughout**; these "lock-ordering rules" should be documented and followed by all the developers working on the project (annotating locks is useful too; more on this in the section on *lockdep* in the next chapter). Incorrect lock ordering often leads to deadlock.
    - On the other hand, the order in which locks are released doesn't matter (of course, you must at some point release all held locks to not cause starvation).
- Take care to prevent *starvation*; verify that a lock, once taken, is indeed released "quickly enough."
- **Simplicity is key**: Try to avoid complexity or over-design, especially concerning complex scenarios involving locks.

On the topic of locking, the (dangerous) issue of deadlock arises. A **deadlock** is the inability to make any progress; in other words, the application processes/threads and/or kernel component(s) appear to hang indefinitely. While we don't intend to delve into the gory details of deadlocks here, I will quickly mention some of the more common types of deadlock scenarios that can occur:

- Simple case, single lock, process context:
    - Attempting to acquire the same lock twice is considered a defect and results in **self-deadlock**. Think about it; while holding a lock, you attempt to re-acquire it. Now, as the lock is locked, you have to wait until it's unlocked. But *you* are holding the lock and waiting (and only you can unlock it) and thus cannot unlock it; the result is (self) deadlock! Recursive locking could solve this, but it's generally disabled and using it is frowned upon.
- Simple case, multiple (two or more) locks, process context – let's study this with an example:
    - On CPU `0`, thread A acquires lock A and then wants to acquire lock B.
    - Concurrently, on CPU `1`, thread B acquires lock B and then wants to acquire lock A.
    - Thus, each waits on the other, forever... The result is a classic case of circular deadlock, often called the **AB-BA deadlock**. An illustration of this is as follows (the timeline is vertically downward):

```
CPU 0 : Thread A                 CPU 1 : Thread B
...                                 ...
acquire lock A                   acquire lock B
...                                 ...
try to acquire lock B            try to acquire lock A
waits for lock B … … ...         waits for lock A … … …
  … forever … … … ...              … forever … … … …
```

    - This can be indefinitely extended; for example, the AB-BC-CA **circular dependency** (A-B-C lock chain) results in a deadlock.

- Complex case, single lock, process and interrupt contexts – let's again study this via an example (this case is actually meant for the *spinlock* one; we'll get to the details soon):
    - Process P1, running, say, the read method of your driver (module), acquires lock A on, say, CPU core 0.
    - A millisecond later, your driver's hardware interrupt occurs on the same core and thus your driver's interrupt handler immediately preempts P1. Now, what if the interrupt context attempts to take this same lock, lock A? As lock A is currently locked (by process context P1), the interrupt context is forced to wait on the unlock; but P1 is holding the lock and waiting to get back the CPU, which won't happen as long as the interrupt's running... thus, it cannot perform the unlock and the interrupt context too ends up waiting forever... again, *the result is (self) deadlock!*

  The prescribed solution to this case, of course, is to disable (mask) all hardware interrupts on the local core when acquiring the lock; then the process context cannot be interrupted and it runs the critical section to completion. In other words, it's now truly *atomic*. When it performs the unlock, this operation reenables all interrupts on the local core and all is well! Thus, locks acquired in an interrupt (or more generally, an atomic) context must always be used with interrupts disabled. (How exactly do we do this? With the spinlock; we shall of course look at these aspects in more detail when we cover spinlocks.)

- More complex cases, multiple locks, and process and interrupt (hardirq and softirq) contexts. (We don't take this further here; you'll find details in the upcoming *Locking and interrupts* section.)

In simpler cases, always following the *lock-ordering guideline* is sufficient: always obtain locks in a well-documented order (we will provide an example of this in kernel code in the *Using the mutex lock* section). However, as you are perhaps starting to realize, things can get very complex, and complex deadlock scenarios can trip up even experienced developers. Luckily for us, **lockdep** – the Linux kernel's runtime lock dependency validator – can catch (almost) every single deadlock case! (Don't worry – we shall get there, we'll cover lockdep in detail in the next chapter; just proceed step by step). When we cover spinlocks (in the *Using the spinlock* section), we'll come across process and/or interrupt context scenarios like the ones mentioned previously; the precise spinlock APIs to use (to avoid deadlock, among other things) are made clear there.

The reality is that even **livelock** situations can be just as deadly as deadlocks! Livelock is essentially a situation conceptually similar to deadlock; it's just that the state of the participating tasks is running and not waiting. As an example, an interrupt "storm" (hundreds or even thousands of hardware interrupts – and their associated softirqs – occur in bursts and need to be processed very rapidly, stressing the system) can cause a livelock; modern network drivers mitigate this effect by switching off interrupts (under interrupt load) and resorting to a polling technique called **New API (NAPI)**, switching interrupts back on when appropriate. (Well, it's more complex than that, but we'll leave it at that here.)

For those of you who've not been living under a rock, you will know that the Linux kernel has two primary types of locks: the mutex lock and the spinlock. Actually, there are several more types, including other synchronization (and "lockless" programming) technology, all of which will be covered in the course of this chapter and the next.

The kernel documentation refers to three lock categories implemented within the kernel: *sleeping, CPU-local, and spinning* locks. Sleeping locks include the (RT) mutex, the semaphore, and variations. Spinning locks include the spinlock, the reader-writer spinlock, and variations. The "local lock" is usually meant for RT usage, though the non-RT usage includes using it for lock debugging.

So now, let's delve into what exactly the mutex lock and spinlock really mean and which you should use in what circumstance!

Mutex or spinlock? Which to use when

The exact semantics of learning to use the mutex lock and the spinlock are quite simple (with appropriate abstraction within the kernel API set making it even easier for the typical driver developer or module author). The critical question in this situation is a conceptual one: what really is the difference between these two lock types? More to the point, under which circumstances should you use which lock? You will learn the answers to these questions in this section.

Taking our previous driver read method's pseudocode (*Figure 12.6*) as a base example, let's say that three threads – **tA**, **tB**, and **tC** – are running in parallel (on an SMP system) through this code. We shall solve this concurrency issue, while avoiding any data races, by taking (acquiring) a lock prior to the start of the critical section (time **t2**), and releasing the lock (unlock) just after the end of the critical section code path (time **t3**). Let's take a look at the pseudocode once more via a diagram (*Figure 12.7*), this time with locking to ensure it's correct:

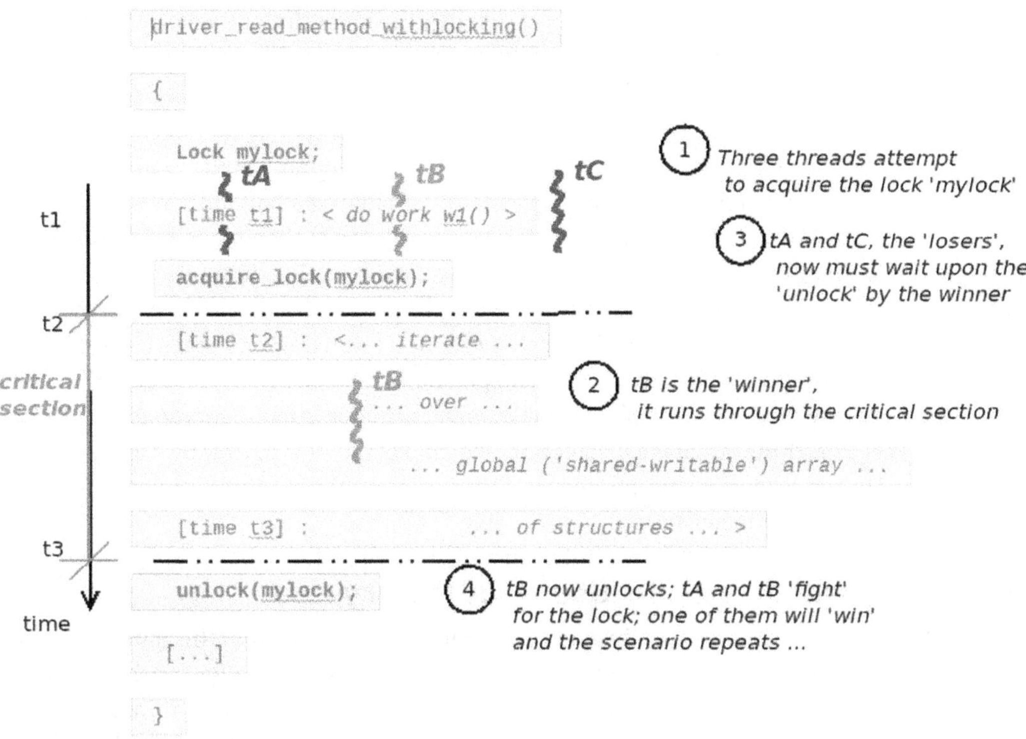

Figure 12.7: Pseudocode: a critical section within a (fictional) driver's read method; here it's done correctly, with locking

When the three threads attempt to simultaneously acquire the lock, the lock API semantics guarantee that exactly one of them will get it. Let's say that **tB** (thread B) gets the lock: it's now the "winner" or "owner" thread. This means that threads **tA** and **tC** are the "losers"; what do they do? *They wait upon the unlock!*

The moment the "winner" (**tB**) completes the critical section and unlocks the lock, the battle resumes between the previous losers; one of them will be the next winner and the process repeats. (Do understand that the "APIs" we show in *Figure 12.7* are merely pseudocode and not the actual (un)lock APIs; we cover those in the two main sections that follow this one.)

The key difference between the two lock types – the mutex and the spinlock – is based on *how the losers wait* upon the unlock event. With the **mutex** lock, the loser threads are put to sleep; that is, they wait by sleeping (in effect, when they attempt to lock the mutex and it's already locked, they see it as a blocking call; they are scheduled or context-switched off the CPU – they are now "sleeping"). The moment the winner performs the unlock, the kernel awakens the loser threads (all of them) and they run, again competing for the lock. (In fact, mutexes and semaphores are sometimes referred to as sleeplocks.)

With the **spinlock**, however, there is *no question of sleeping*; the loser threads wait by "spinning" upon the lock until it is unlocked. Conceptually, this looks as follows:

```
while (locked) ;
```

Note that this is *only conceptual*. Think about it a moment – this is actually polling. However, as a good programmer, you will understand that polling is usually considered a bad idea. Why, then, does the spinlock work this way? Well, it doesn't; it has only been presented in this manner for conceptual purposes. As you will soon understand, spinlocks only really have meaning on multicore (SMP) systems. On such systems, while the winner thread is away and running the critical section code, *the losers wait by spinning on other CPU cores!* In reality, at the implementation level, the code that's used to implement the modern spinlock is highly optimized (and arch-specific) and does not work by trivially "spinning" via a `while (locked);` type of semantic (for example, many spinlock implementations for ARM use the **wait for event** (**WFE**) machine language instruction, which has the CPU optimally wait in a low-power state. Do see the *Further reading* section for several resources on the internal implementation of the mutex and spinlock within the kernel).

Determining which lock to use – in theory

How the spinlock is implemented is really not our concern here; the fact that the spinlock has a lower overhead than the mutex lock is of interest to us. How so? It's simple, really: for the mutex lock to work, the loser thread has to go to sleep (and then be awoken on the unlock). To do so, internally, the `schedule()` function gets called, which means the loser sees the mutex lock API as a blocking call! A call to the scheduler will ultimately result in the thread being context-switched off the CPU. Conversely, when the owner thread eventually unlocks the lock, the loser thread(s) must be woken up; again, one of them, the next "winner," will be context-switched back onto the processor.

Thus, the minimal "cost" of the mutex lock/unlock operation is the time it takes to perform two context switches on the given machine. (See the *Information Box* designated as [1] in the next section.) By relooking at *Figure 12.7*, we can determine the time spent in the critical section (the "locked" code path); that is, `t_locked = t3 - t2`.

Let's say that `t_ctxsw` represents the time to context switch. As we've learned, the minimal cost of the mutex lock/unlock operation is two context switches (the first to "go to sleep" and the second when it's "awoken"): `2 * t_ctxsw`.

Now, let's say that the following expression is true:

```
t_locked < 2 * t_ctxsw
```

In other words, what if the time spent within the critical section is less than the time taken for two context switches? In this case, using the mutex lock is just wrong as this is far too much overhead; more time is being spent performing metawork than actual work – a phenomenon known as **thrashing**. It's this precise use case – the presence of very short critical sections – that's often the case on modern OSes such as Linux, that has us employ the spinlock in preference to the mutex. So, in conclusion, select the spinlock when you have short, non-blocking critical sections, and select the mutex when the critical section is long-ish and (possibly) blocking. Why the emphasis on "blocking"? The following section – the practical stuff – makes this clear; read on!

Determining which lock to use – in practice

So, while operating under the `t_locked < 2 * t_ctxsw` "rule" might be great in theory, hang on: are you really expected to precisely measure the context switch time and the time spent in the critical section of each and every case where one (a critical section) exists? No, of course not – that's unrealistic and pedantic.

Practically speaking, think about it this way: the mutex lock works by having the loser threads sleep (until the unlock occurs); the spinlock does not (the losers "spin"). Let's recall one of our "golden rules" of the Linux kernel: *the kernel cannot sleep* (*call* `schedule()`) *in any kind of atomic context*. Thus, we can never use the mutex lock in an interrupt context, or indeed in any context where it isn't safe to sleep; using the spinlock, however, would be fine. (Recall, a blocking API is one that puts the calling context to sleep by calling `schedule()`.) Let's summarize this:

- **Is the critical section running in an atomic (for example, interrupt) context, or in process context where it cannot sleep?** Use the spinlock.
- **Is the critical section running in process context and is sleep or blocking I/O in the critical section possible?** Use the mutex lock.

Of course, using the spinlock is considered lower overhead than using the mutex; thus, you can even use the spinlock in process context (such as our fictional driver's read method), *as long as the critical section does not block (sleep)*.

[1] The time taken for a context switch is varied; it largely depends on the hardware and the OS quality. Some previous (September 2018) measurements show that context-switching time is in the region of 1.2 to 1.5 us (microseconds) on a pinned-down CPU, and around 2.2 us without pinning (`https://eli.thegreenplace.net/2018/measuring-context-switching-and-memory-overheads-for-linux-threads/`).

Both hardware and the Linux OS have improved tremendously, and because of that, so has the average context-switching time. An old (December 1998) Linux Journal article determined that on an x86 class system, the average context switch time was 19 microseconds (us), and that the worst-case time was 30 us; those numbers are now outdated.

This brings up the next question: how do we know if the code is currently running in process or interrupt context? That's easy – use the `in_task()` macro to determine it (as our `PRINT_CTX()` macro within our `convenient.h` header does):

```
if (in_task())
    /* we're in process context (usually safe to sleep / block) */
else
    /* we're in an atomic or interrupt context (cannot sleep / block) */
```

Now that you understand which one – mutex or spinlock – to use and when, let's get into the actual usage. We'll begin with how to use the mutex lock!

Using the mutex lock

Mutexes are also called sleepable or blocking **mutual exclusion (mutex)** locks. As you have learned, they are used in process context if the critical section can sleep (block). They must not be used within any kind of atomic or interrupt context (top halves, bottom halves such as tasklets or softirqs, and so on), kernel timers, or in process context where blocking is not allowed.

Initializing the mutex lock

Prior to usage, every lock must be initialized to the "unlocked" state. A mutex lock "object" is represented in the kernel as a `struct mutex` data structure. Consider the following code:

```
#include <linux/mutex.h>
struct mutex mymtx;
```

To use this mutex lock, it must be explicitly initialized to the unlocked state. Initialization can be performed statically (declare and initialize the object) with the `DEFINE_MUTEX()` macro, or dynamically via the `mutex_init()` function (this is actually a macro wrapper over the `__mutex_init()` function).

For example, to declare and initialize the mutex object `mymtx`, we can use `DEFINE_MUTEX(mymtx);`.

We can also do this dynamically. Why dynamically? Often, the mutex lock is a member of the (global) data structure that it protects.

Keeping the lock variable as a member of the (parent) data structure it protects is a common (and clever) pattern that's used within Linux; this approach has the added benefit of avoiding namespace pollution and is unambiguous about which mutex protects which shared data item (a bigger problem than it might appear to be at first, especially in enormous projects such as the Linux kernel!).

For example, let's say we have the following global context data structure in our driver code (note that this code is fictional):

```
struct mydrv_priv {
    <member 1>;
    <member 2>;
    [...]
    struct mutex mymtx; /* protects access to mydrv_priv */
    [...]
} *drvctx;
/*  Then, in your driver module's init method, do the following: */
static int __init init_mydrv(struct mydrv_priv *drvctx)
{
    [...]
    mutex_init(drvctx->mymtx);
    [...]
}
```

Correctly using the mutex lock

One quite often finds very insightful comments within the kernel source tree. Here's a great one that neatly summarizes the rules you must follow to correctly use a mutex lock (`https://elixir.bootlin.com/linux/v6.1.25/source/include/linux/mutex.h#L35`); please read this carefully:

```
/*
 * Simple, straightforward mutexes with strict semantics:
 *
 * - only one task can hold the mutex at a time
 * - only the owner can unlock the mutex
 * - multiple unlocks are not permitted
 * - recursive locking is not permitted
 * - a mutex object must be initialized via the API
 * - a mutex object must not be initialized via memset or copying
 * - task may not exit with mutex held
 * - memory areas where held locks reside must not be freed
 * - held mutexes must not be reinitialized
 * - mutexes may not be used in hardware or software interrupt
 * contexts such as tasklets and timers
 *
 * These semantics are fully enforced when DEBUG_MUTEXES is
 * enabled. Furthermore, besides enforcing the above rules, the mutex
 * [ ... ]
```

As a kernel (or driver) developer, you must understand the following (we have covered most of these points in the *Locking guidelines and deadlock* section):

- A critical section causes the code path *to be serialized, defeating parallelism*. Due to this, it's imperative that you keep the critical section as short as possible. A corollary to this is **lock data, not code.**
- Attempting to reacquire an already acquired (locked) mutex lock – which is effectively recursive locking – is *not* supported and will lead to a self-deadlock.
- **Lock ordering:** This is a very important rule of thumb for preventing dangerous deadlock situations. In the presence of multiple threads and multiple locks, the order in which locks are taken must be *documented and strictly followed by all the developers working on the project*. The actual lock ordering itself isn't sacrosanct, but the fact that once it's been decided on it must be followed, is. While browsing through the kernel source tree, you will come across many places where the kernel developers ensure this is done, and they (usually) write a comment regarding this for other developers to see and follow. Here's a sample comment from the slab allocator code (`mm/slub.c`):

  ```
  /*
   * Lock order:
  ```

```
 *    1. slab_mutex (Global Mutex)
 *    2. node->list_lock (Spinlock)
 *    3. kmem_cache->cpu_slab->lock (Local lock)
 *    4. slab_lock(slab) (Only on some arches)
 *    5. object_map_lock (Only for debugging)
...
```

Now that we understand how mutexes work from a conceptual standpoint (and their initialization), let's learn how to make use of their business end, the actual lock/unlock APIs.

Mutex lock and unlock APIs and their usage

In the Linux kernel, the mutex lock and unlock APIs, respectively, are as follows:

```
void __sched mutex_lock(struct mutex *lock);
void __sched mutex_unlock(struct mutex *lock);
```

(Ignore the `__sched` here; it's just a compiler attribute that has this function disappear in the `WCHAN` output, which shows up in procfs and with certain option switches to `ps` (such as `-l`).)

Again, the comments within the source code in `kernel/locking/mutex.c` are very detailed and descriptive; I encourage you to take a look at this file in more detail. We've only shown some of its code here, which has been taken directly from the 6.1.25 Linux kernel source tree, `https://elixir.bootlin.com/linux/v6.1.25/source/kernel/locking/mutex.c#L260`:

```
[ ... ]
/**
 * mutex_lock - acquire the mutex
 * @lock: the mutex to be acquired
 *
 * Lock the mutex exclusively for this task. If the mutex is not
 * available right now, it will sleep until it can get it.
 *
 * The mutex must later on be released by the same task that
 * acquired it. Recursive locking is not allowed. The task
 * may not exit without first unlocking the mutex. Also, kernel
 * memory where the mutex resides must not be freed with
 * the mutex still locked. The mutex must first be initialized
 * (or statically defined) before it can be locked. memset()-ing
 * the mutex to 0 is not allowed.
 *
 * (The CONFIG_DEBUG_MUTEXES .config option turns on debugging
 * checks that will enforce the restrictions and will also do
 * deadlock debugging)
 *
```

```
 * This function is similar to (but not equivalent to) down().
 */
void __sched mutex_lock(struct mutex *lock)
{
    might_sleep();

    if (!__mutex_trylock_fast(lock))
        __mutex_lock_slowpath(lock);
}
EXPORT_SYMBOL(mutex_lock);
```

`might_sleep()` is a macro with an interesting debug property: its presence catches code that's supposed to execute in an atomic context but doesn't! (More explanation for `might_sleep()` can be found in the *Linux Kernel Programming – Part 2* companion volume.)

So, think about it: `might_sleep()`, which is the first line of code in `mutex_lock()`, is literally telling you that the code that follows "might sleep"! This further implies that this code path should not be executed by anything that's in an atomic context since it might sleep. This of course means that you should only use the mutex in the process context when it's safe to sleep!

A quick and important reminder: The Linux kernel can be configured with a large number of debug options; in this context, the `CONFIG_DEBUG_MUTEXES=y` config option will help you catch possible mutex-related bugs, including deadlocks. Similarly, when you configure via the usual `make menuconfig` UI, under the *Kernel Hacking* menu, you will find a large number of debug-related kernel config options; we discussed this in *Chapter 5, Writing Your First Kernel Module – Part 2*. There are several very useful kernel configs with regard to lock debugging; we shall cover these in the next chapter, in the *Lock debugging within the kernel* section.

Mutex lock – via [un]interruptible sleep?

As usual, there's more to the mutex than what we've seen so far. You already know that a Linux process (or thread) cycles through various states of a state machine. On Linux, sleeping has two discrete states – an interruptible sleep and an uninterruptible sleep. A process (or thread) in an *interruptible sleep* is sensitive – which means it will respond – to user-space signals, whereas a task in an *uninterruptible sleep* is not sensitive to user signals.

In a human-interactive application with an underlying driver, as a general rule of thumb, you should typically put a process into an interruptible sleep (while it's blocking upon the lock), thus leaving it up to the end user as to whether to abort the application by pressing *Ctrl* + *C* (or some such mechanism involving signals).

There is a design rule that's often followed on Unix-like systems: **provide mechanism, not policy.**

Having said this, on non-interactive code paths, it's often the case that you must wait on the lock indefinitely, with the semantic that a signal that's been delivered to the task should not abort the blocking wait (it should instead be kept pending). On Linux, the uninterruptible case turns out to be the most common one.

So, here's the thing: the `mutex_lock()` API, in the "loser" thread's code path, always puts this loser task into an uninterruptible sleep. If this is not what you want, use the `mutex_lock_interruptible()` API, to put the (loser) calling task into an *interruptible sleep*. There is one difference syntax-wise: the latter returns an integer value of `0` on success and `-EINTR` (remember the `0/-E` return convention) on signal interruption (by the way, `EINTR's` English error message is `Interrupted system call`).

In general, using `mutex_lock()` is faster than using `mutex_lock_interruptible()`; use it when the critical section is short (thus pretty much guaranteeing that the lock is held for a short while, which is a very desirable characteristic).

The 6.1.25 kernel codebase contains over 22,000 and just over 800 instances of calling the `mutex_lock()` and `mutex_lock_interruptible()` APIs, respectively; you can check things like this out by leveraging the powerful `cscope` utility on the kernel source tree.

In theory, the kernel provides a `mutex_destroy()` API as well. This is the opposite of `mutex_init()`; its job is to mark the mutex as being unusable. It must only be invoked once the mutex is in the unlocked state, and once invoked, that mutex cannot be used. This is a bit theoretical because, on regular systems, it just reduces to an empty function; only on a kernel with `CONFIG_DEBUG_MUTEXES` enabled does it become actual (simple) code.

Thus, we should use this pattern when working with the mutex, as shown in the following pseudocode:

```
DEFINE_MUTEX(...);          /* init: initialize the mutex object statically */
/* or dynamically, via */
mutex_init();              /* don't use both, use one of them... */
[ ... ]
 /* critical section: perform the (mutex) locking, unlocking */
 mutex_lock[_interruptible]();
 << ... critical section code ... >>
 mutex_unlock();
    [ ... ]
        mutex_destroy();       // cleanup: destroy the mutex object
```

Now that you have learned how to use the mutex lock APIs, let's put this knowledge to use. In the next section, we will build on top of a poorly written – no protection! – simple "misc" driver by employing the mutex object to lock critical sections as required.

Mutex locking – an example driver

We have created a simple device driver code example in the *Linux Kernel Programming – Part 2* companion volume in its first chapter, *Writing a Simple misc Character Device Driver*.

As stated, you can obtain a free download of the *Linux Kernel Programming - Part 2* companion volume. The PDF version is available here: `https://github.com/PacktPublishing/Linux-Kernel-Programming/tree/master/Linux-Kernel-Programming-(Part-2)`. (You can also download the Kindle edition from Amazon.)

There, we wrote a simple `misc` (miscellaneous) class character device driver named `miscdrv_rdwr` (the code can be found here, within its GitHub repository: `https://github.com/PacktPublishing/Linux-Kernel-Programming-Part-2/tree/main/ch1/miscdrv_rdwr`). We also wrote a small user-space utility program (`https://github.com/PacktPublishing/Linux-Kernel-Programming-Part-2/blob/main/ch1/miscdrv_rdwr/rdwr_test_secret.c`) to read and write a (so-called) "secret" from and to the device driver's memory. The *Linux Kernel Programming – Part 2* book's first chapter will show you in detail how to write and then try out this device driver.

What we glaringly (egregiously is the right word here!) failed to do in that project is protect shared (global) writable data from concurrent invocation! In other words, *we blithely ignored critical sections*. This will cost us dearly in the real world. I urge you to take some time to think about this: isn't it viable that two (or more) user-mode processes open the device file of this driver, and then concurrently issue various I/O: reads and writes? Here, the global shared writable data (in this particular case, two global integers and the members of the driver's context data structure) could easily get corrupted.

So, let's learn from and correct our mistakes by making a copy of this driver (we will now call it `ch12/1_miscdrv_rdwr_mutexlock/1_miscdrv_rdwr_mutexlock.c`) and rewrite some portions of it.

The key point is that we use mutex locks to protect all critical sections. Instead of displaying all the code here (it's in this book's GitHub repository of course), let's do something interesting: let's look at a "diff" (the differences – the delta generated by `diff`) between the older unprotected and the newer protected driver code versions (output truncated for readability):

```
$ cd <lkp2e-book-repo>/ch12/1_miscdrv_rdwr_mutexlock
$ diff -u <lkp-p2...>/ch12/miscdrv_rdwr/miscdrv_rdwr.c \
miscdrv_rdwr_mutexlock.c > miscdrv_rdwr.patch
$ cat miscdrv_rdwr.patch
[ ... ]
+#include <linux/mutex.h> // mutex lock, unlock, etc
 #include "../../convenient.h"
[ ... ]
```

```
+DEFINE_MUTEX(lock1);   /* this mutex lock is meant to protect the integers ga
and gb */
[ ... ]
+       struct mutex lock; // this mutex protects this data structure
 };
[ ... ]
```

In the output of `diff`, lines prefixed with a "+" symbol have been added in the new version whereas lines prefixed with a "-" symbol were deleted (in the new version).

Here, we can see that in the newer safe version of the driver, we have protected all shared writable data (aka shared state) from concurrent access, which is the whole point. To do so, we first declared and initialized a mutex lock variable called `lock1`; we shall use it to protect the (just for demonstration purposes) two global integers, `ga` and `gb`, within our driver. Next, importantly, we declared a mutex lock named `lock` within the global "driver context" data structure; that is, `drv_ctx`. This will be used to protect all access to members of that data structure. It is initialized within the module's `init` code path:

```
+       mutex_init(&ctx->lock);
+       /* Retrieve the device pointer for this device */
+       ctx->dev = llkd_miscdev.this_device;
+
+       /* Initialize the "secret" value :-) */
-         strlcpy(ctx->oursecret, "initmsg", 8);
-         dev_dbg(ctx->dev, "A sample print via the dev_dbg(): driver
initialized\n");
+         strscpy(ctx->oursecret, "initmsg", 8);
+       /* Why don't we protect the above strscpy() with the mutex lock?
+         * It's working on shared writable data, yes?
+         * Yes, BUT this is the init code; it's guaranteed to run in exactly
+         * one context (typically the insmod(8) process), thus there is
+         * no concurrency possible here. The same goes for the cleanup
+         * code path.
+         */
```

Next, the detailed comment clearly explains why we don't need to lock/unlock around the `strscpy()` in this particular case (as it's the module `init` code that can run only once in a single context). Similarly, and this should be obvious to you, local variables are implicitly private to each process context (as they reside in that process or thread's kernel mode stack) and therefore require no protection (each thread/process has a separate *instance* of the variable, so no one steps on anyone else's toes!). Before we forget, the *cleanup* code path (which is invoked via the `rmmod(8)` process context) must destroy the mutexes:

```
-static void __exit miscdrv_rdwr_exit(void)
```

```
+static void __exit miscdrv_exit_mutexlock(void)
 {
+        mutex_destroy(&lock1);
+        mutex_destroy(&ctx->lock);
         misc_deregister(&llkd_miscdev);
-        pr_info("LLKD misc (rdwr) driver deregistered, bye\n");
+        pr_info("LKP2E misc driver %s deregistered, bye\n", llkd_miscdev.name);
 }
```

Now, let's look at the relevant portion of the patch (diff) with regard to the driver open method. Here, in the old unsafe version, we simply operated upon our global integers ga and gb willy-nilly, with no protection; now, we do so correctly, under the aegis of the mutex:

```
+
+       mutex_lock(&lock1);
+       ga++; gb--;
+       mutex_unlock(&lock1);
+
+       dev_info(dev, " filename: \"%s\"\n"
        [ ... ]
```

But (there's always a but, isn't there?), all is not well! Take a look at the `printk` function (via the `dev_info()` wrapper) following the `mutex_unlock()` line of code here:

```
+ dev_info(dev, " filename: \"%s\"\n"
+           " wrt open file: f_flags = 0x%x\n"
+           " ga = %d, gb = %d\n",
+           filp->f_path.dentry->d_iname, filp->f_flags, ga, gb);
```

Does this look okay to you? No, look carefully: we are *reading* the value of the global integers, `ga` and `gb`. Recall the fundamentals: in the presence of concurrency (which is certainly a possibility here in this driver's *open* method), *even reading shared writable data without making it an exclusive access (via locking, typically) is potentially unsafe*. If this doesn't make sense to you, please think: what if, while one thread is reading the integers, another is simultaneously updating (writing) them; what then? This kind of situation can result in what's called a **dirty** or **torn read**; we might end up reading stale/incorrect/partially correct data that must be protected.

Well, to be honest, the fact is that this isn't really a great example of a dirty read as, on most modern processors, reading and writing single integer items does tend to be an atomic operation. However, we must not assume such things – we must simply do our job and protect shared writable data accesses – both reads and writes – from concurrent access.

In fact, there's another similar bug-in-waiting: in the code snippet just seen, did you notice that we read data from the open file structure (the `filp` pointer) without bothering to protect it? (Indeed, the open file structure has a lock; we're supposed to use it! We shall do so later.)

The precise semantics of how and when things such as dirty reads occur do tend to be very arch (machine)-dependent; nevertheless, our job as kernel or driver authors is clear: we must ensure that we protect all critical sections. This includes reads upon shared writable data.

For now, we shall just flag these as potential defects (bugs, on the to-do list). We will take care of them in the following chapter, in the *Using the atomic_t and refcount_t interfaces* section, in a more performance-friendly manner.

Moving along, look at the patch (the diff) of the driver's *read* method; it reveals something interesting, how we now use the driver context structure's mutex lock to protect the critical sections that turn up in the code path:

```
 static ssize_t read_miscdrv_rdwr(struct file *filp, char __user *ubuf,
                                  size_t count, loff_t *off)
 {
-        int ret = count, secret_len = strnlen(ctx->oursecret, MAXBYTES);
+        int ret = count, secret_len;
         struct device *dev = ctx->dev;
-        char tasknm[TASK_COMM_LEN];
+
+        mutex_lock(&ctx->lock);
+        secret_len = strlen(ctx->oursecret);
+        mutex_unlock(&ctx->lock);

         PRINT_CTX();
-        dev_info(dev, "%s wants to read (upto) %zu bytes\n", get_task_comm(tasknm, current), count);
+        dev_info(dev, "%s wants to read (upto) %zu bytes\n", current->comm, count);

         ret = -EINVAL;
         if (count < MAXBYTES) {
@@ -141,16 +144,19 @@
          * member to userspace.
          */
         ret = -EFAULT;
+        mutex_lock(&ctx->lock);
         if (copy_to_user(ubuf, ctx->oursecret, secret_len)) {
                 dev_warn(dev, "copy_to_user() failed\n");
-                goto out_notok;
+                goto out_ctu;
         }
         ret = secret_len;

         // Update stats
         ctx->tx += secret_len;  // our 'transmit' is wrt this driver
         dev_info(dev, " %d bytes read, returning... (stats: tx=%d, rx=%d)\n",
-                secret_len, ctx->tx, ctx->rx);
+                 secret_len, ctx->tx, ctx->rx);
+ out_ctu:
+        mutex_unlock(&ctx->lock);
  out_notok:
         return ret;
 }
```

Figure 12.8: The diff of the driver's read() method; notice the usage of the mutex lock in the newer version (lines prefixed with "+" have been added in the newer version)

Figure 12.8 is a (partial) screenshot of the diff, focused on our driver's read method. Study it; you will notice how we've now used the mutex lock to protect every access to shared writable data, in other words, to protect every critical section. (Some of it's a bit pedantic, no doubt. Also, why not use the `get_task_comm()` helper to get the thread name (as we learned in *Chapter 6, Kernel Internals Essentials – Processes and Threads*)? This will become clear in the coming chapter. Patience, please!) The same protection-via-mutex goes for the *open*, *write* and *close* (release) methods of the device driver (why not generate the patch for yourself and take a look?).

Note that the user-mode app remains unchanged, which means for us to test the new safer version, we must continue using the original user-mode app at `ch12/miscdrv_rdwr/rdwr_drv_secret.c`. (For your convenience, I've copied its code across to this book's GitHub repo here: `https://github.com/PacktPublishing/Linux-Kernel-Programming_2E/blob/main/ch12/rdwr_test_secret.c`.) Do try it out along with the misc driver!

Running and testing code such as this driver code on a debug kernel, which contains various locking errors and deadlock detection capabilities, is crucial (we'll return to these "debug" capabilities in the next chapter, in the *Lock debugging within the kernel* section).

In the preceding code diff (*Figure 12.8*), we took the mutex lock just before the `copy_to_user()` routine; the `copy_to_user()` helper macro is a blocking one (and might sleep!). This is fine; **using mutex locks for a possibly blocking critical section is one of its key use cases**, after all.

However, we only release the mutex after the `dev_info()` call. Why not release it before this routine, thus shortening the critical section? A closer look at `dev_info()` reveals why it's *within* the critical section. We are printing the values of three variables here: the number of bytes read - by `secret_len` - and the number of bytes that are "transmitted" and "received" by `ctx->tx` and `ctx->rx`, respectively. Now `secret_len` is a local variable and does not require protection, but the other two variables are within the global driver context structure and thus (pedantically) do require protection, even when "only reading" them, thus protecting against possibly dirty or torn reads!

The mutex lock – a few remaining points

In this section, we will cover a few additional points regarding mutexes.

Mutex lock API variants

First, let's take a look at a few variants of the mutex lock API; besides the interruptible variant (described in the *Mutex lock – via [un]interruptible sleep?* section), we have the *trylock*, *killable*, and *io* variants.

The mutex trylock variant

What if you would like to implement a **busy-wait** semantic; that is, test for the availability of the (mutex) lock and, if available (meaning it's currently unlocked), acquire it and continue with the critical section code path. However, if it's not available (meaning it's currently in the locked state), *don't wait* for the lock; instead, perform some other work and retry.

In effect, this is a non-blocking mutex lock variant and is thus called the *trylock*; the following flowchart approximates how it works:

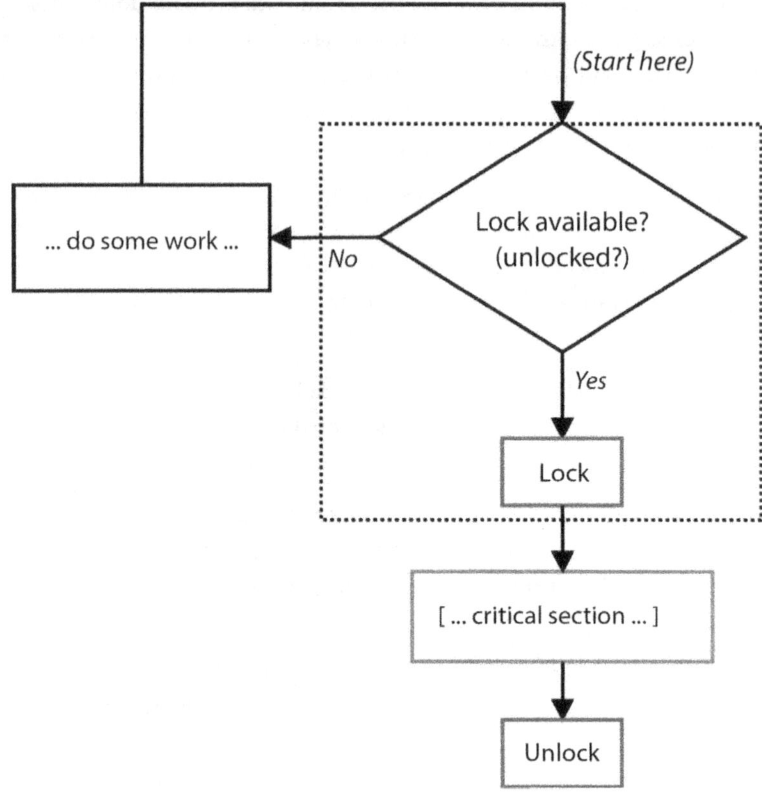

Figure 12.9: The "busy wait" semantic, a non-blocking trylock variant of the mutex lock

The dotted box in *Figure 12.9* signifies that the internal implementation – checking if the lock is unlocked and then locking it – is atomic. The API for this trylock variant of the mutex lock is as follows:

```
int __sched mutex_trylock(struct mutex *lock);
```

This API's return value signifies what transpired at runtime:

- A return value of `1` indicates that the lock has been successfully acquired.
- A return value of `0` indicates that the lock is currently contended (locked).

Though it might be tempting to, do not attempt to use the `mutex_trylock()` API to figure out if a mutex lock is in a locked or unlocked state; this is inherently "racy." Next, note that using this trylock variant in a highly contended lock path may well reduce your chances of acquiring the lock. The trylock variant has been traditionally used in deadlock prevention code that might need to back out of a certain lock order sequence and be retried via another sequence (ordering).

Also, with respect to the trylock variant, even though the literature uses the wording *try and acquire the mutex atomically*, it does not work in an atomic or interrupt context – it only works in process context (as with any type of mutex lock). As usual, the lock must be released by `mutex_unlock()` being invoked by the owner context.

I suggest that you now try working on the *trylock* mutex variant as an exercise. See the *Questions* section at the end of this chapter for an assignment!

The mutex interruptible and killable variants

As you have already learned, the `mutex_lock_interruptible()` API is used when the driver (or module) is willing to acknowledge any (user-space) signal interrupting it (and, if that happens, it returns `-ERESTARTSYS` to tell the kernel VFS layer to perform signal handling; the user-space system call will fail with `errno` set to `EINTR`). An example can be found in the module-handling code in the kernel, within the `delete_module(2)` system call (which `rmmod` invokes):

```
// kernel/module.c
[ ... ]
SYSCALL_DEFINE2(delete_module, const char __user *, name_user,
        unsigned int, flags)
{
    struct module *mod;
    [ ... ]
    if (!capable(CAP_SYS_MODULE) || modules_disabled)
        return -EPERM;
    [ ... ]
    if (mutex_lock_interruptible(&module_mutex) != 0)
        return -EINTR;
    mod = find_module(name);
    [ ... ]
out:
    mutex_unlock(&module_mutex);
    return ret;
}
```

Notice how the API returns `-EINTR` on failure. (By the way, the `SYSCALL_DEFINEn()` macro becomes a system call signature; `n` signifies the number of parameters this particular system call accepts. Also, notice the capability check – unless you are running as root or have the `CAP_SYS_MODULE` capability (or module loading is completely disabled), the system call just returns failure (`-EPERM`).)

If, however, your driver is only willing to be interrupted by fatal signals (those that *will kill* the user-space context), then use the `mutex_lock_killable()` API (the signature is identical to that of the interruptible variant).

The mutex I/O variant

The `mutex_lock_io()` API is identical in syntax to the `mutex_lock()` API; the only difference is that the kernel thinks that the wait time of the loser thread(s) is the same as waiting for I/O (the code comment in `kernel/locking/mutex.c:mutex_lock_io()` clearly documents this; take a look: `https://elixir.bootlin.com/linux/v6.1.25/source/kernel/locking/mutex.c#L1016`). This can matter accounting-wise.

You can find fairly exotic APIs such as `mutex_lock[_interruptible]_nested()` within the kernel, with the emphasis here being on the `nested` suffix. However, note that the Linux kernel does not prefer developers to use nested (or recursive) locking (as mentioned in the *Locking guidelines and deadlock* and *Correctly using the mutex lock* sections). Also, these APIs only get compiled in the presence of the `CONFIG_DEBUG_LOCK_ALLOC` config option; in effect, the nested APIs were added to support the kernel lock validator mechanism. They should only be used in special circumstances (where a nesting level must be incorporated between instances of the same lock type).

In the next section, we will answer a typical FAQ: what's the difference between the mutex and semaphore objects? Wait, does Linux even have a semaphore object? Read on to find out!

The semaphore and the mutex

The Linux kernel does provide a semaphore object, along with the usual operations you can perform on a (binary) semaphore:

- A semaphore lock acquire via the `down[_interruptible]()` (and variations) APIs
- A semaphore unlock via the `up()` API

In general, the semaphore is an older implementation, so it's advised that you use the mutex lock in place of it.

An FAQ worth exploring, though, is this: *what is the difference between a mutex and a semaphore?* They appear to be conceptually similar, but are actually quite different; one answer is summarized by the following points (of course, these are primarily wrt the Linux kernel):

- A semaphore is a more generalized form of a mutex; a mutex lock can be acquired (and subsequently released or unlocked) exactly once, while a semaphore can be acquired (and subsequently released) multiple times.
- A mutex is used to protect a critical section from simultaneous access, while a semaphore should be used as a mechanism to signal another waiting task that a certain milestone has been reached (typically, a producer task posts a signal via the semaphore object, which a consumer task is waiting to receive, in order to continue with further work).

- A mutex has the notion of ownership of the lock and only the owner context can perform the unlock; there is no ownership for a binary semaphore.

Priority inversion and the RT-mutex

A word of caution when using any kind of locking is that you should carefully design and code to prevent the dreaded *deadlock* scenarios that could arise (again, more on catching this is in the next chapter in the *The lock validator lockdep – catch locking issues early* section).

Aside from deadlocks, there is another risky scenario that arises when using the mutex: that of priority inversion (again, we will not delve into the details in this book). Suffice it to say that the unbounded **priority inversion** case can be a deadly one; the end result is that the product's high(est) priority thread is kept off the CPU for too long.

As I covered in some detail in my earlier book, *Hands-on System Programming with Linux*, it's precisely this priority inversion issue that struck NASA's Mars Pathfinder robot, on the Martian surface no less, back in July 1997! The usage of a hardware watchdog had the system reboot whenever the project's high-priority thread waited on the mutex for too long, past the deadline; the trouble is that this occurred often! Due to debug telemetry being enabled (yay!), the underlying issue was diagnosed from Earth and fixed (by using the **priority inheritance** (**PI**) mutex attribute to prevent lower-priority threads holding the mutex from being preempted before they released it). The firmware was then uploaded to the robot on Mars (!), and it all worked! See the *Further reading* section of this chapter for interesting resources about this, something that every software developer should be aware of!

The user-space Pthreads mutex implementation certainly has **priority inheritance** (PI) semantics available. But what about within the Linux kernel? For this, Ingo Molnar provided the PI-futex-based RT-mutex (a real-time mutex; in effect, a mutex extended to have PI capabilities. FYI, the `futex(2)` API is a sophisticated system call that provides a fast user-space mutex). It becomes available when the `CONFIG_RT_MUTEXES` config option is enabled. Quite similar to the "regular" mutex semantics, RT-mutex APIs are provided to initialize, (un)lock, and destroy the RT-mutex object. (This code has been merged into the mainline kernel from Ingo Molnar's `-rt` tree.) As far as actual usage is concerned, the RT-mutex is used for internally implementing the PI futex (did you know that on Linux, the `futex(2)` system call itself internally implements the user-space Pthreads mutex?). Besides this, the kernel-locking self-test code and the I2C subsystem use the RT-mutex directly.

Thus, for a typical module (or driver) author, these APIs are not going to be used very frequently. The kernel provides some documentation on the internal design of the RT-mutex at `https://docs.kernel.org/locking/rt-mutex-design.html` (covering priority inversion, priority inheritance, and more).

Internal design

A word on the reality of the internal implementation of the mutex lock deep within the kernel fabric: Linux tries to implement a *fast-path* approach whenever possible.

A **fast path** is the most optimized high-performance code path, typically, one with no locks and no blocking. The intent is to have code follow this fast path as far as possible. Only when it really isn't possible does the kernel fall back to a (possible) "mid path," and then a "slow path," approach; it still works but is slow(er).

This fast path is taken in the absence of contention for the lock (that is, the lock is in an unlocked state to begin with). So, the lock is locked with no fuss, pretty much immediately. If, however, the mutex is already locked, then the kernel typically uses a "mid path" optimistic spinning implementation, making it more of a hybrid (mutex/spinlock) lock type. If even this isn't possible, the "slow path" is followed – the process context attempting to get the lock may well enter the sleep state. If you're interested in the mutex's internal implementation, more details can be found within the official kernel documentation here: `https://docs.kernel.org/locking/mutex-design.html`.

The **LDV** (**Linux Driver Verification**) project: back in *Online Chapter, Kernel Workspace Setup*, in the section *The LDV – Linux Driver Verification – project*, we mentioned that this project has useful "rules" with respect to various programming aspects of Linux modules (drivers, mostly) as well as the core kernel.

With regard to our current topic, here's one of the rules: *Locking a mutex twice or unlocking without prior locking* (`http://linuxtesting.org/ldv/online?action=show_rule&rule_id=0032`). It mentions the kind of things you cannot do with the mutex lock (we have already covered this in the *Correctly using the mutex lock* section). The interesting thing here is that you can see an actual example of a bug – a mutex lock double-acquire attempt, leading to (self) deadlock – in a kernel driver (as well as the subsequent fix).

Another type of mutex that we will just mention is the so-called **w/w** – **wait/wound** (or simply the WW-mutex) – one (where *wound* rhymes with "doomed"). The term originates with RDBMS literature and is a way to handle deadlock (the documentation goes as far as to use the term "deadlock-proof"). With w/w in the Linux kernel, the usage is primarily within the graphics subsystem and, to some extent, DMA. Here, the task acquiring the mutex lock receives a unique reservation or ticket identifier. Now, the "age" of the task acquiring the WW-mutex is taken into account. In the case where deadlock is detected, the "oldest" task – the one that holds the reservation the longest – is given precedence. How? It's done by having the "younger" task(s) back off, by having (forcing) them to release the WW lock(s) they hold; hence, the younger task is "wounded" (poor fellow). Usage of the WW- mutex within the kernel is pretty low, especially when compared with the regular mutex. See the *Further reading* section for more on it.

Now that you've understood how to use the mutex lock, let's move on and look at the other very common lock within the kernel – the spinlock!

Using the spinlock

In the *Determining which lock to use – in practice* section, you learned – practically speaking – when to use the spinlock instead of the mutex lock and vice versa. For convenience, we have reproduced the key statements we provided previously here:

- **Is the critical section running in an atomic (for example, interrupt) context, or in process context where it cannot sleep?** Use the spinlock.
- **Is the critical section running in process context and is sleep or blocking I/O in the critical section possible?** Use the mutex lock.

In this section, we shall consider that you've now decided to use the spinlock.

Spinlock – simple usage

For all the spinlock APIs, you must include the relevant header file, that is, `#include <linux/spinlock.h>`.

Similar to the mutex lock, you *must* declare and initialize the spinlock to the unlocked state before use. The spinlock is an "object" that's declared via the `typedef` data type named `spinlock_t` (internally, it's a structure defined in `include/linux/spinlock_types.h`). It can be initialized dynamically via the `spin_lock_init()` macro:

```
spinlock_t lock;
spin_lock_init(&lock);
```

Alternatively, this can be performed statically (declared and initialized) via the `DEFINE_SPINLOCK(lock);` macro.

As with the mutex, declaring a spinlock within the (global/static) data structure it's meant to protect (against concurrent access) is typically a very good idea. As we mentioned earlier, this very idea is made use of within the Linux kernel often; as an example, the data structure representing an open file in the kernel is called the `struct file`:

```
// include/linux/fs.h
struct file {
    [...]
    struct path f_path;
    struct inode *f_inode; /* cached value */
    const struct file_operations *f_op;
    /*
     * Protects f_ep_links, f_flags.
     * Must not be taken from IRQ context.
     */
    spinlock_t f_lock;
    [...]
```

```
    struct mutex f_pos_lock;
    loff_t f_pos;
    [...]
```

Check it out: for the file structure, the spinlock variable named f_lock is the spinlock that protects (as the comment says) the f_ep_links and f_flags members of the file data structure; also, it has a mutex lock to protect another member, that is, the file's current seek position f_pos.

How do you actually perform the lock and unlock operations with the spinlock? There are quite a few variations on the API that are exposed by the kernel to us module/driver authors; the simplest form of the spin(un)lock APIs are as follows:

```
void spin_lock(spinlock_t *lock);
<< ... critical section ... >>
void spin_unlock(spinlock_t *lock);
```

Note that there is no spinlock equivalent of the mutex_destroy() API. Now, let's see the spinlock APIs in action!

Spinlock – an example driver

Similar to what we did with our mutex locking sample driver (refer to the *Mutex locking – an example driver* section if you need to), to illustrate the simple usage of a spinlock, we shall (again) make a copy of our – poorly written, deliberately-with-no-locking – simple "misc" character device driver from the *Linux Kernel Programming – Part 2* companion volume, from its first chapter, *Writing a Simple misc Character Device Driver* (ch1/1_miscdrv_rdwr) as a starting template and then place it in a new module driver location for this book; that is, here: ch12/2_miscdrv_rdwr_spinlock (of course, we expect you to clone the book's GitHub repo).

Again, here, for readability, we'll only show small, relevant portions of the *diff* (the differences, the delta generated by diff) between the original program and this one:

```
+#include <linux/spinlock.h>
[ ... ]
-/*
- * The driver 'context' (or private) data structure;
- * all relevant 'state info' regarding the driver is here.
+static int ga, gb = 1;
+DEFINE_SPINLOCK(lock1); /* this spinlock protects the global integers ga and
gb */
```

```
+
+/* The driver 'context' data structure;
+ * all relevant 'state info' reg the driver is here.
  */
 struct drv_ctx {
        struct device *dev;
        int tx, rx, err, myword;
        u32 config1, config2;
        u64 config3;
    [ … ]
+#define MAXBYTES    128
        char oursecret[MAXBYTES];
+       struct mutex mutex;  /* this mutex protects these members of this data
+                             * structure: oursecret
+                             * in the (possibly) blocking critical section
+                             */
+       spinlock_t spinlock; /* this spinlock protects these members of this
data
+                             * structure: oursecret, tx, rx
+                             * in the non-blocking / atomic critical sections
+                             */
 };
 static struct drv_ctx *ctx;
```

This time, to protect the members of our `drv_ctx` global data structure, we have both the mutex lock (from the previous version) and a new spinlock. It's quite common to have more than one lock within a data structure; typically, the mutex lock protects member usage in critical sections where blocking can occur, while the spinlock is used to protect members in critical sections where blocking (sleeping) *cannot* occur (you could use a spinlock to protect data members in process context as well, as long as the critical section's non-blocking).

Of course, we must ensure that we initialize all the locks so that they're in the unlocked state. We do this in the driver's `init` code path (so, continuing with the patch output):

```
+       mutex_init(&ctx->mutex);
+       spin_lock_init(&ctx->spinlock);
```

In this version, in the driver's open method, we replace the mutex lock with the spinlock to protect the increments and decrements of the global integers; a partial screenshot of the patch serves to show it with color highlighting:

```
 */
static int open_miscdrv_rdwr(struct inode *inode, struct file *filp)
{
        struct device *dev = ctx->dev;
-       char *buf = kzalloc(PATH_MAX, GFP_KERNEL);

-       if (unlikely(!buf))
-               return -ENOMEM;
+       PRINT_CTX();            // displays process (or intr) context info

-       PRINT_CTX();    // displays process (or atomic) context info
-       ga++;
-       gb--;
-       dev_info(dev, " opening \"%s\" now; wrt open file: f_flags = 0x%x\n",
-               file_path(filp, buf, PATH_MAX), filp->f_flags);
-       kfree(buf);
+       spin_lock(&lock1);
+       ga ++; gb--;
+       spin_unlock(&lock1);
+
+       dev_info(dev, " filename: \"%s\"\n"
+               " wrt open file: f_flags = 0x%x\n"
+               " ga = %d, gb = %d\n", filp->f_path.dentry->d_iname, filp->f_flags, ga, gb);

-       return nonseekable_open(inode, filp);
+       display_stats(1);
+       return 0;
}
```

Figure 12.10: Partial screenshot of the patch showing the code of the driver's open method (and the spinlock usage therein)

Again, it's worth repeating: wherever we use the spinlock(s), it implies that the critical section that it protects – the code path between the spinlock and the unlock – is definitely non-blocking code.

Next, in this version, within the driver's `read` method, we use the spinlock instead of the mutex to protect some critical sections:

```
 static ssize_t read_miscdrv_rdwr(struct file *filp, char __user *ubuf,
-                                 size_t count, loff_t *off)
+                                size_t count, loff_t *off)
 {
-       int ret = count, secret_len = strnlen(ctx->oursecret, MAXBYTES);
+       int ret = count, secret_len, err_path = 0;
        struct device *dev = ctx->dev;
-       char tasknm[TASK_COMM_LEN];
```

```
+
+        spin_lock(&ctx->spinlock);
+        secret_len = strlen(ctx->oursecret);
+        spin_unlock(&ctx->spinlock);
```

However, that's not all! Continuing with the driver's read method, carefully take a look at the following code snippet taking the mutex lock and the comment that follows:

```
         ret = -EFAULT;
+        mutex_lock(&ctx->mutex);
+        /* Why don't we just use the spinlock??
+         * Because - VERY IMP! - remember that the spinlock can only be used
when
+         * the critical section will not sleep or block in any manner; here,
+         * the critical section invokes the copy_to_user(); it very much can
+         * cause a 'sleep' (a schedule()) to occur.
+         */
         if (copy_to_user(ubuf, ctx->oursecret, secret_len)) { [ … ]
```

When protecting data where the critical section has possibly blocking APIs – such as `copy_to_user()` – we *must* only use a mutex lock! (Due to space constraints, we haven't displayed more of the patch here; we expect you to read through the spinlock sample driver code and try it out for yourself.)

Test – sleep in an atomic context

You have already learned that the one thing we should *not do is sleep (block) in any kind of atomic or interrupt context*. Let's put this to the test. As always, the empirical approach – where you test things for yourself rather than relying on other's experiences – is key!

How exactly can we test this? Easy: we shall use a simple integer module parameter, `buggy`, which, when set to `1` (the default value being `0`), executes a code path within our driver's spinlock-protected critical section that violates this rule. There, we shall invoke the `schedule_timeout()` API, which internally invokes `schedule()`; it's how we put the process context – *current* – to sleep in kernel space. (FYI, using this API, and a lot more, is covered in the *Linux Kernel Programming – Part 2* companion volume in *Chapter 5, Working with Kernel Timers, Threads, and Workqueues*.) Here's the relevant code:

```
// ch12/2_miscdrv_rdwr_spinlock/miscdrv_rdwr_spinlock.c
[ ... ]
static int buggy;
module_param(buggy, int, 0600);
MODULE_PARM_DESC(buggy,
"If 1, cause an error by issuing a blocking call within a spinlock critical
section");
[ ... ]
static ssize_t write_miscdrv_rdwr(struct file *filp, const char __user *ubuf,
                size_t count, loff_t *off)
```

```
{
    int ret, err_path = 0;
    [ ... ]
    spin_lock(&ctx->spinlock);
    strscpy(ctx->oursecret, kbuf, (count > MAXBYTES ? MAXBYTES : count));
    [ ... ]
    if (1 == buggy) {
        /* We're still holding the spinlock! */
        set_current_state(TASK_INTERRUPTIBLE);
        schedule_timeout(1*HZ); /* ... and this is a blocking call!
                 * Congratulations! you've just engineered a bug */
    }
    spin_unlock(&ctx->spinlock);
    [ ... ]
}
```

Now for the interesting part: let's test this (buggy) code path within our custom 6.1.25 "debug" kernel (the kernel where we have enabled several kernel debug configuration options (mostly from within the **Kernel Hacking** menu in the `make menuconfig` UI)).

Testing the buggy module on a 6.1 debug kernel

First of all, ensure you've built the custom 6.1 kernel (we're using version 6.1.25) and that all the required kernel debug config options are enabled (refer back to *Chapter 5, Writing Your First Kernel Module – Part 2*, the *Configuring a "debug" kernel* section, if you need to). Then, boot off your debug kernel (here, it's named `6.1.25-dbg`). Now, build the misc driver (in `ch12/2_miscdrv_rdwr_spinlock/`) against this debug kernel (the usual `make` within the driver's directory should do this; you might find that, on the debug kernel, the build is a bit slower and the binary module is bigger):

```
$ lsb_release -a 2>/dev/null | grep "^Description" ; uname -r
Description:    Ubuntu 23.10
6.1.25-dbg

# Before running make, ensure that the Makefile has the variable MYDEBUG set to
# y (I have kept it this way).
$ cd <lkp2e_book_src>/ch12/2_miscdrv_rdwr_spinlock
$ make
--- Building : KDIR=/lib/modules/6.1.25-dbg/build ARCH= CROSS_COMPILE= ccflags-
y="-DDEBUG -g -ggdb -gdwarf-4 -Wall -fno-omit-frame-pointer -fvar-tracking-
assignments -DDYNAMIC_DEBUG_MODULE" MYDEBUG=y DBG_STRIP=n ---
gcc (Ubuntu 13.2.0-4ubuntu3) 13.2.0
[ … ]
$ modinfo ./miscdrv_rdwr_spinlock.ko
```

```
filename:        /home/c2kp/lkp2e/ch12/2_miscdrv_rdwr_spinlock/./miscdrv_rdwr_
spinlock.ko
[ … ]
vermagic:        6.1.25-dbg SMP preempt mod_unload modversions
parm:            buggy:If 1, cause an error by issuing a blocking call within a
spinlock critical section (int)
$ sudo virt-what
virtualbox
kvm
$
```

As you can see, we're running our custom 6.1.25 "debug" kernel on an x86_64 Ubuntu 23.10 guest VM here.

How do you know whether you're running on a **virtual machine** (**VM**) or on the "bare metal" (native) system? `virt-what(1)` is a useful little script that shows this (you can install it on Ubuntu with `sudo apt install virt-what`).

To run our test case, insert the driver into the kernel and set the buggy module parameter to `1`. Invoking the driver's `read` method (via our user-space app; that is, `ch12/rdwr_test_secret`) isn't an issue, as the buggy code path isn't there, as shown here:

```
$ sudo dmesg -C
$ sudo insmod ./miscdrv_rdwr_spinlock.ko buggy=1
$ lsmod |grep miscdrv
miscdrv_rdwr_spinlock    20480  0
$ ../rdwr_test_secret
Usage: ../rdwr_test_secret opt=read/write device_file ["secret-msg"]
 opt = 'r' => we shall issue the read(2), retrieving the 'secret' form the
driver
 opt = 'w' => we shall issue the write(2), writing the secret message <secret-
msg>
  (max 128 bytes)
$ ../rdwr_test_secret r /dev/llkd_miscdrv_rdwr_spinlock
Device file /dev/llkd_miscdrv_rdwr_spinlock opened (in read-only mode): fd=3
../rdwr_test_secret: read 7 bytes from /dev/llkd_miscdrv_rdwr_spinlock
The 'secret' is:
 "initmsg"
$
```

Next, we issue a `write(2)` system call to the driver via the user-mode app; *this* time, the write method of the driver runs (in the context of our user-space `rdwr_test_secret` process of course), and thus the buggy code path gets executed!

Right, let's run our user-space app, writing the new "secret" to the device (triggering our kernel module bug):

```
$ ../rdwr_test_secret w /dev/llkd_miscdrv_rdwr_spinlock "When you have
exhausted all possibilities, remember this: you haven't"
Device file /dev/llkd_miscdrv_rdwr_spinlock opened (in write-only mode): fd=3
../rdwr_test_secret: wrote 70 bytes to /dev/llkd_miscdrv_rdwr_spinlock
$
```

As you saw in the code snippet shown earlier in this section, we issued a `schedule_timeout()` within a spinlock critical section (that is, between the lock and unlock). The debug kernel detects this as a bug and spawns (impressively large) debug diagnostics into the kernel log (note that bugs like this can quite possibly hang your system, so it's always recommended to test stuff like this on a test VM). We check it out via `sudo dmesg`:

```
[  152.312529] misc llkd_miscdrv_rdwr_spinlock: stats: tx=0, rx=0
[  152.312572] miscdrv_rdwr_spinlock:write_miscdrv_rdwr(): 005)  rdwr_test_secre :3066   |  ...0   /*
write_miscdrv_rdwr() */
[  152.312575] misc llkd_miscdrv_rdwr_spinlock: rdwr_test_secre wants to write 70 bytes
[  152.312577] misc llkd_miscdrv_rdwr_spinlock:  70 bytes written, returning... (stats: tx=0, rx=70)
[  152.312579] BUG: scheduling while atomic: rdwr_test_secre/3066/0x00000002
[  152.312582] Modules linked in: miscdrv_rdwr_spinlock(OE) isofs snd_seq_dummy snd_hrtimer binfmt_mis
c nls_iso8859_1 snd_intel8x0 snd_ac97_codec ac97_bus snd_pcm snd_seq intel_rapl_msr intel_rapl_common
crct10dif_pclmul crc32_pclmul polyval_clmulni snd_seq_device polyval_generic ghash_clmulni_intel aesni
_intel snd_timer crypto_simd cryptd snd vboxguest(OE) rapl i2c_piix4 soundcore video wmi joydev input_
leds mac_hid serio_raw vmwgfx drm_kms_helper syscopyarea sysfillrect sysimgblt fb_sys_fops drm_ttm_hel
per ttm drm msr parport_pc ppdev lp parport efi_pstore dmi_sysfs ip_tables x_tables autofs4 hid_generi
c usbhid hid psmouse e1000 ahci libahci pata_acpi
[  152.312678] Preemption disabled at:
[  152.312678] [<ffffffffc08fd930>] write_miscdrv_rdwr.cold+0xf5/0x1c8 [miscdrv_rdwr_spinlock]
[  152.312685] CPU: 5 PID: 3066 Comm: rdwr_test_secre Tainted: G           OE      6.1.25-dbg #2
[  152.312689] Hardware name: innotek GmbH VirtualBox/VirtualBox, BIOS VirtualBox 12/01/2006
[  152.312690] Call Trace:
[  152.312691]  <TASK>
[  152.312693]  dump_stack_lvl+0x5a/0x82
[  152.312696]  ? write_miscdrv_rdwr.cold+0xf5/0x1c8 [miscdrv_rdwr_spinlock]
[  152.312700]  dump_stack+0x10/0x18
[  152.312701]  __schedule_bug.cold+0x84/0xa4
[  152.312704]  __schedule+0xfaa/0x15b0
[  152.312706]  ? trace_hardirqs_on+0x36/0x100
[  152.312709]  ? _raw_spin_unlock_irqrestore+0x21/0x70
[  152.312711]  ? __mod_timer+0x276/0x440
[  152.312714]  schedule+0x66/0x110
[  152.312716]  schedule_timeout+0x95/0x170
[  152.312717]  ? __bpf_trace_tick_stop+0x20/0x20
[  152.312720]  write_miscdrv_rdwr.cold+0x1ae/0x1c8 [miscdrv_rdwr_spinlock]
[  152.312723]  vfs_write+0xee/0x460
[  152.312725]  ? debug_smp_processor_id+0x17/0x30
[  152.312727]  ksys_write+0x79/0x100
[  152.312747]  __x64_sys_write+0x19/0x30
[  152.312748]  do_syscall_64+0x5c/0x90
[  152.312750]  ? trace_hardirqs_on_prepare+0x2e/0xb0
[  152.312752]  ? irqentry_exit_to_user_mode+0xe/0x20
[  152.312753]  ? irqentry_exit+0x48/0x70
[  152.312755]  ? exc_page_fault+0xa9/0x1d0
[  152.312757]  entry_SYSCALL_64_after_hwframe+0x63/0xcd
[  152.312759] RIP: 0033:0x7f858591b214
[  152.312761] Code: c7 00 16 00 00 00 b8 ff ff ff ff c3 66 2e 0f 1f 84 00 00 00 00 00 f3 0f 1e fa 80
3d 35 b3 0e 00 00 74 13 b8 01 00 00 00 0f 05 <48> 3d 00 f0 ff ff 77 54 c3 0f 1f 00 48 83 ec 28 48 89 5
4 24 18 48
[  152.312762] RSP: 002b:00007ffcdb39f3b8 EFLAGS: 00000202 ORIG_RAX: 0000000000000001
```

Figure 12.11: Kernel diagnostics being triggered by the "scheduling while atomic" bug we've deliberately hit here

The preceding (partial) screenshot (*Figure 12.11*) shows what transpired; let's follow along while viewing the driver code in `ch12/2_miscdrv_rdwr_spinlock/miscdrv_rdwr_spinlock.c`:

1. Firstly, see the second line, our useful `PRINT_CTX()` macro's output (we've reproduced this line here):

   ```
   miscdrv_rdwr_spinlock:write_miscdrv_rdwr(): 005)  rdwr_test_secre :3066 |
   ...0   /* write_miscdrv_rdwr() */
   ```

   Clearly, the driver's write method code, `write_miscdrv_rdwr()`, is running in the context of our user-space process (`rdwr_test_secre`; notice how the name is truncated to the first 16 characters, including the `NULL` byte).

2. It copies in the new "secret" from the user-space writer process and writes it, for 70 bytes.
3. It then "takes" the spinlock, thus entering the critical section, and copies this data to the `oursecret` member of our driver's context structure.
4. After this, the `if (1 == buggy) {` condition evaluates to true.
5. So, it calls `schedule_timeout()`, which is a blocking API (as it internally calls `schedule()`), triggering the bug, which is helpfully highlighted in red:

   ```
   BUG: scheduling while atomic: rdwr_test_secre/3066/0x00000002
   ```

6. The kernel now dumps a good deal of the diagnostic output. Among the first things to be dumped is the **call stack.**

The call trace or stack backtrace of the kernel-mode stack of the process – here, it's our user-space app, `rdwr_drv_secret`, which is running our (buggy) driver's code in process context – can be clearly seen in *Figure 12.11*. Each line after the `Call Trace:` header is essentially a call frame on the kernel stack.

A couple of tips:

- Always read the call stack (or call trace) bottom-up (so, we can see we entered via a system call, quite clearly, `write(2)`, as it becomes `ksys_write()`, then `vfs_write()`, and so on).
- Ignore all stack frames that begin with the ? symbol; they are questionable call frames, in all likelihood "leftovers" from previous stack usage in the same memory region. It's worth taking a small memory-related diversion here: this is how stack allocation really works; stack memory isn't allocated and freed on a per-call frame basis as that would be frightfully expensive. Only when a stack memory page is exhausted is a new one automatically *faulted in!* (Recall our discussions in *Chapter 9, Kernel Memory Allocation for Module Authors – Part 2*, in the *A brief note on memory allocations and demand paging* section). So, the reality is that as code calls and returns from functions, the same stack memory page(s) tend to keep getting reused.

  Not only that, but for performance reasons, the call frame's memory isn't wiped each time, leading to leftovers from previous frames often appearing. (They can literally "spoil" the picture. However, fortunately, the modern stack call frame tracing algorithms are usually able to do a superb job at figuring out the correct stack trace.)

The `vfs_write()` calls `__vfs_write()`, which ends up invoking our driver's write method; that is, `write_miscdrv_rdwr()`! (That, of course, is how the character driver framework is designed to work. Also, notice the [module name] in square brackets on the right of any module function call frame – another useful clue.) This code, as we well know, invokes the buggy code path where we call `schedule_timeout()`, which, in turn, invokes `schedule()` (which calls the actual worker routine `__schedule()`), causing the whole `BUG: scheduling while atomic` kernel bug to trigger.

The format of the `scheduling while atomic` code path is retrieved from the following line of code, which can be found in `kernel/sched/core.c`:

```
printk(KERN_ERR "BUG: scheduling while atomic: %s/%d/0x%08x\n", prev->comm,
prev->pid, preempt_count());
```

Interesting! Here, you can see that it printed the following string:

```
BUG: scheduling while atomic: rdwr_test_secre/3066/0x00000002
```

After `atomic:`, it prints the (possibly truncated) process name, the PID, and then invokes the `preempt_count()` inline function, which prints the *preempt depth*; essentially, the preempt depth is a counter that's incremented every time a spinlock is taken and decremented on every unlock. So, if it's positive, this implies that the code is within a critical or atomic section; here, it shows as the value `2`, proving we're in an atomic, non-preemptible code path!

We don't interpret the rest of the useful kernel diagnostics here; do check out the *Linux Kernel Debugging* book for a deep dive into all these aspects.

It's key to note that this bug gets neatly served up by the kernel during this test run as the `CONFIG_DEBUG_ATOMIC_SLEEP` debug kernel config option is turned on (it's on because we're running a custom "debug kernel"!) This config option's details (you can interactively find and set this option in `make menuconfig`, under the **Kernel Hacking** menu) are as follows:

```
// lib/Kconfig.debug
[ ... ]
config DEBUG_ATOMIC_SLEEP
    bool "Sleep inside atomic section checking"
    select PREEMPT_COUNT
    depends on DEBUG_KERNEL
    depends on !ARCH_NO_PREEMPT
    help
      If you say Y here, various routines which may sleep will become very
      noisy if they are called inside atomic sections: when a spinlock is
      held, inside an rcu read side critical section, inside preempt disabled
      sections, inside an interrupt, etc...
```

What happens if we also perform the very same test on a standard "distro" system and kernel (like our Ubuntu 22.04 LTS VM), which, importantly, is typically *not configured as a "debug" kernel* (hence the `CONFIG_DEBUG_ATOMIC_SLEEP` kernel config option hasn't been set)? We'd perhaps expect that the kernel then doesn't catch this particular bug; well, being empirical, this is how it worked out:

- In this book's first edition, running the test on a standard Ubuntu 20.04 VM with the standard distro (5.4) kernel, it indeed didn't catch this bug.
- However, I find that on more recent distros, and hence the kernels they run, even though the `CONFIG_DEBUG_ATOMIC_SLEEP` kernel config option hasn't been explicitly set, it still does catch the *scheduling while atomic* bug! Kernel enhancements over the years seem to have ensured this forward progress... (I tested on both Ubuntu 22.04 with the 5.19.0-45-generic kernel as well as on a Fedora 38 distro with the more recent 6.5.6-200.fc38.x86_64 kernel.)

It typically helps to keep the `CONFIG_DEBUG_ATOMIC_SLEEP` kernel config option enabled even in production.

Again, the LDV project has this as one of its rules: *"It's not allowed to acquire spin_lock twice. It's not allowed to release not acquired spin_lock. At the end all spin_lock should be released. It's not allowed to re-release a lock by spin_unlock/spin_unlock_irqrestore functions"* (`http://linuxtesting.org/ldv/online?action=show_rule&rule_id=0039`). (We have of course covered these points in the *Locking guidelines and deadlock* section.) It mentions key points with regard to the correct usage of spinlocks; interestingly, here, it shows an actual bug instance in a driver where a spinlock was attempted to be released twice – a clear violation of the locking rules, leading to an unstable system.

Great, done with the spinlock? No, not by a long shot! The next section has us delve into a deeper understanding of the spinlock and its usage in the kernel and modules, where we work with the spinlock in interrupt contexts as well.

Locking and interrupts

So far, we have learned how to use the mutex lock and, for the spinlock, the basic `spin_[un]lock()` APIs. A few other API variations on the spinlock exist, and we shall examine the more common ones here.

In this section, there's a bit more advanced coverage, and it will definitely help if you understand at least the basics of writing a (char) device driver and hardware interrupt handling on Linux (typically in the device driver context). These topics are covered in depth in this book's companion volume *Linux Kernel Programming - Part 2*, in *Chapter 1, Writing a Simple misc Character Device Driver* and *Chapter 4, Handling Hardware Interrupts* (the LKP-2 e-book is freely downloadable). Also, as a quick guide/refresher, we have provided a sub section titled: *Interrupt handling on Linux - a summary of key points*, that comes later in this section. (If you're unfamiliar with interrupt handling concepts on Linux, we suggest you first take a look at the *LKP - Part 2* book and/or this section and then continue from here.)

To understand exactly why you may need other APIs for spinlocks, let's go over a scenario: as a driver author, you find that the device you're working on asserts a hardware interrupt; accordingly, you write the interrupt handler for it.

Now, while implementing, say, a `read` method for your driver, you find that you have a non-blocking critical section within it. This is easy to deal with: as you've learned, you should use a spinlock to protect it. Great! But, think, what if, while in the `read` method's critical section, the device's hardware interrupt fires? As you're aware, *hardware interrupts preempt anything and everything*; thus, control will go to the interrupt handler code, preempting the driver's `read` method.

The key question here: is this an issue? That answer depends on what both your driver's interrupt handler and your `read` method were doing and how they were implemented. Let's visualize a few scenarios:

- The driver's interrupt handler (ideally) uses only local variables, so even if the `read` method were in a critical section, it really wouldn't matter; the interrupt handling will complete very quickly (typically and broadly speaking, within a range of 10 to 100 microseconds) and control will be handed back to whatever was interrupted (again, there's more to it than this; any existing bottom-half mechanism, such as a softirq or tasklet, may also need to execute before your driver's `read` method is handed back the processor). So, as such, there is really no race in this case.
- The driver's interrupt handler is working on (global) shared writable data but *not* on the shared data items that your driver's `read` method is using. Thus, again, there is no conflict and no race with the `read` method's code. (What you should also realize, of course, is that as your interrupt handler's working on shared state, it *does have a critical section that must be protected* (typically via another spinlock).)
- The driver's interrupt handler is working on the *same* (or a portion of the same) shared writable data that your driver's `read` method is working upon. In this case, we can see that the potential for a data race definitely exists, so we need locking! (Recall our definition of a data race in the *Data races – a more formal definition* section.)

Let's of course focus on the third case, the one that has the potential for data races. Obviously, we *must* use a spinlock to protect the critical section within the interrupt-handling code (recall that using a mutex is disallowed when we're in any kind of atomic or interrupt context). Also, *unless we use the very same spinlock* in both the driver's `read` method and the interrupt handler code path, they won't really be protected at all!

As this discussion shows, be extra careful when working with locks; take the time to think through your design and code in detail.

Let's try and make this discussion a bit more hands-on (with pseudocode): let's say we have a (global) shared data structure named `gctx`; we're operating on it in both the `read` method as well as the interrupt handler (a top half and/or tasklet) within our driver. With regard to our `read` method, since this code path can run concurrently and it contains shared state, it is a critical section and therefore requires protection. Now, since we are running in process context, it might seem like using a mutex here would be fine.

No! Think... to be effective, we obviously need to protect the same shared state by using the very same lock in both our driver's read method and the interrupt handler. Now, the interrupt handler's critical section also requires protection, and as we've learned, we really don't have much of a choice here: we must use the spinlock. So, we find that both the driver's read method and the driver's interrupt handler must thus use the same spinlock! (Here, we call the spinlock variable slock; read it as "s lock").

The following pseudocode shows some timestamps (t1, t2, ...) for this situation; but first, let's do it the wrong way:

```
/* Driver read method ; WRONG ! */
driver_read(...)                      << time t0 >>
{
    [ ... ]
    spin_lock(&slock);
    <<--- time t1 : start of read method critical section >>
... << critical section: operating on global data object gctx >> ...
    spin_unlock(&slock);
    <<--- time t2 : end of read method critical section >>
    [ ... ]
}                                     << time t3 >>
```

(Why do we mark this driver read method as being "wrong"? You'll soon find out!)

Now, to make this discussion interesting, we consider *three scenarios* (going from less to more complex):

- Scenario 1 – wherein the driver's read method and interrupt run sequentially, one after the other (in other words, in a serialized fashion, no race)
- Scenario 2 – (more interesting) where they run in an interleaved fashion (potential for data race exists)
- Scenario 3 – where they run interleaved and some interrupts are masked (and some not; potential for data race exists)

We begin with the first scenario.

Scenario 1 – driver method and hardware interrupt handler run serialized, sequentially

The following pseudocode is for the device driver's interrupt handler. As we "know" they run serialized, sequentially here, we have the timeline continue from earlier (from the code snippet of our fictional driver's read method that was seen just before this section); hence, the first timestamp is shown as *time t4* following the earlier *time t3* one at the end of the read method):

```
handle_interrupt(...)           << time t4; hardware interrupt fires! >>
{
    [ ... ]
```

```
    spin_lock(&slock);
    <<--- time t5: start of interrupt critical section >>
    ... << critical section: operating on global data object gctx >> ...
    spin_unlock(&slock);
    <<--- time t6 : end of interrupt critical section >>
    [ ... ]
}                               << time t7 >>
```

This can be summed up with the following diagram; study it carefully:

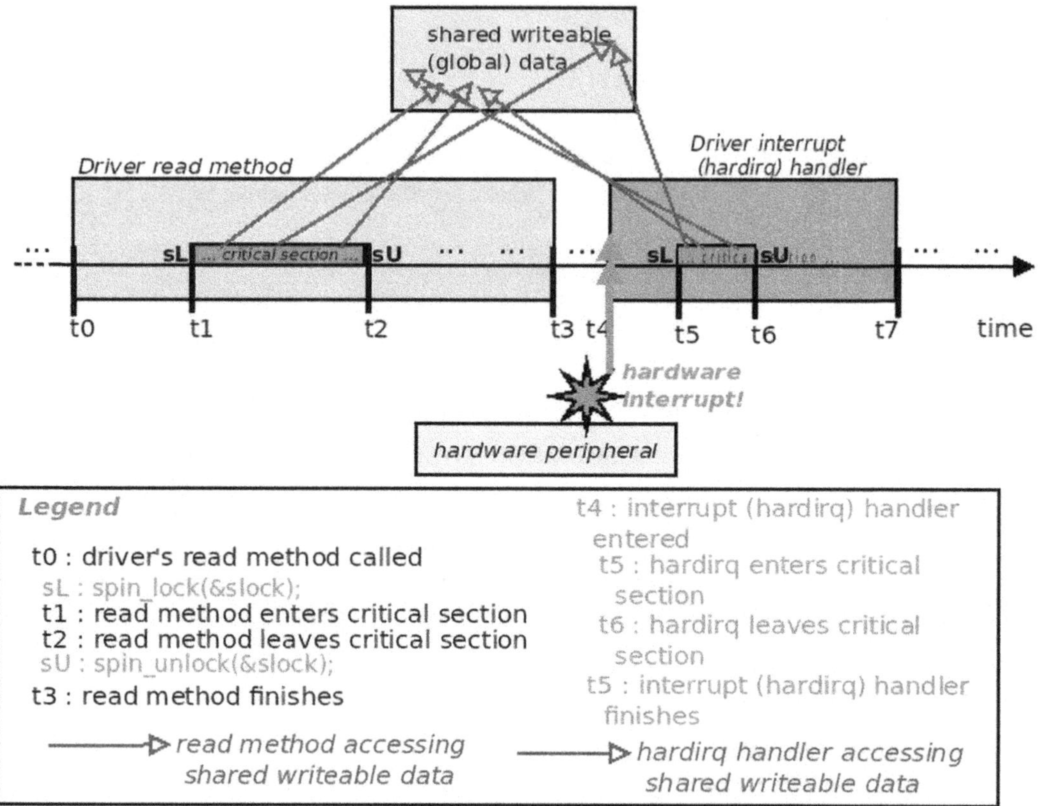

Figure 12.12: Timeline, scenario 1: the driver's read method and hardware interrupt handler run sequentially when working on global data; there are no issues here (diagram not to scale)

In *Figure 12.12*, sL and sU imply the `spin_lock()` and `spin_unlock()` APIs being used. Luckily, everything has gone well – "luckily" because, here, the hardware interrupt fired *after* (it could have fired before as well) the `read` function's critical section completed and not in between (as we shall consider in the following scenario!). Surely we can't count on luck as the exclusive safety stamp of our product!?

Scenario 2 – driver method and hardware interrupt handler run interleaved

Whether your code's running on a UP (**uniprocessor**, with only one CPU core) or SMP (multicore) system matters as well. Let's first consider this scenario on a UP system.

Scenario 2 on a single-core (UP) system

Hardware interrupts are, of course, asynchronous (they can arrive at any time). What if your peripheral chip's (that your driver is purpose-built to "drive") interrupt fired, on the same CPU core that the `read` method is running upon, at a less opportune time (for us) – say, while the `read` method's critical section is running, that is, between time `t1` and `t2` (see *Figure 12.12*)? Well, shouldn't the spinlock (taken at `t1`) simply do its job and protect our data?

No, think carefully: at this point, the interrupt handler *will preempt* your `read` method's critical section and it will soon enter its critical section; doing so, it will obviously attempt to acquire the same spinlock (`&slock`). But wait a minute – it cannot acquire (lock) it as it's currently locked (by the process context running the `read` method)! Thus, it "spins," in effect waiting on the unlock. But how can it ever be unlocked? The driver's `read` process context will have to unlock it (as it owns it) but can't as it's preempted by the hardware interrupt! Thus, it cannot be unlocked, and so the interrupt-side spinlock "spins" forever; so there we have it, a **(self) deadlock.**

So what's the solution? We'll reach it in a moment, so do read on...

Scenario 2 on a multicore (SMP) system

Interestingly, the spinlock is more intuitive and makes complete sense on an SMP (or multicore) system. So, here, let's now consider a slightly different scenario from the previous one – running on multicore. Assume that the `read` method is running on CPU core 1; the interrupt can be delivered on another CPU core, say core 2. Given that the `read` method's in its critical section, it implies the spinlock is locked (of course); thus, the interrupt code path, attempting to acquire the same spinlock, will "spin" on the lock on CPU core 2. Once the `read` method, on core 1, completes its critical section, it will unlock the spinlock, thus unblocking the interrupt handler, which can now "take" the spinlock and proceed forward; great!

But what about on **UP**? How will it work then? One can't, after all, be running logic in one (say, process) context and "spinning" simultaneously in another (say, interrupt) context on one CPU core. And what if, on a multicore (SMP) box, the driver's `read` method and the hardware interrupt handler happen to execute on the same core?

Worry not, there is indeed a solution!

Solving the issue on UP and SMP with the spin_[un]lock_irq() API variant

Ah, so finally, here's the solution to this conundrum: when "racing" with interrupts, regardless of it being on uniprocessor or SMP systems, simply use the `_irq` variant of the spinlock API:

```
#include <linux/spinlock.h>
void spin_lock_irq(spinlock_t *lock);
```

The `spin_lock_irq()` API internally **masks hardware interrupts** (except for non-maskable ones, like the (**Non Maskable Interrupt** (NMI)) on the processor core that it's running upon, that is, on the local core.

So, by using this API in our driver's `read` method, interrupts will be effectively disabled on the local core during the length of the critical section, thus making any possible "race" impossible via hardware interrupts. (If the interrupt does fire on another CPU core, the spinlock technology will simply work as advertised, as discussed previously!)

The `spin_lock_irq()` implementation is pretty nested (as with most of the spinlock functionality), yet fast; down the line, it ends up invoking the `local_irq_disable()` and `preempt_disable()` macros, thus **disabling both interrupts and kernel preemption on the local processor core** that it's running on. In other words, with this spinlock API variant, disabling hardware interrupts has the (desirable) side effect of disabling kernel preemption as well!

The `spin_lock_irq()` API pairs off with the corresponding `spin_unlock_irq()` API. So, the correct usage of the spinlock for this scenario (as opposed to the naive approach we saw previously, which could in fact result in self-deadlock) is as follows:

```
/* Driver read method ; CORRECT ! */
driver_read(...)                    << time t0 >>
{
    [ ... ]
    spin_lock_irq(&slock);          // Note: we're using the _irq version of the 
spinlock!
    <<--- time t1 : start of critical section >>
[now all interrupts + kernel preemption on local CPU core are masked 
(disabled)]
... << critical section: operating on global data object gctx >> ...
    spin_unlock_irq(&slock);
    <<--- time t2 : end of critical section >>
[now all interrupts + kernel preemption on local CPU core are unmasked 
(enabled)]
    [ ... ]
}                                   << time t3 >>
```

Now, it should work as, of course, with the `_irq` variant of the spinlock API, it's impossible for hardware interrupts to preempt the driver `read` method's critical section!

A sneaky problem still lurks, though (eye roll); the following scenario describes it and provides a solution.

Scenario 3 – some interrupts masked, driver method and hardware interrupt handler run interleaved

Before patting ourselves solidly on the back and taking the rest of the day off, let's consider yet another scenario. In this scenario, on a more complex product (or project), it's quite possible that, among the several developers working on the codebase, one has deliberately set the *hardware interrupt (bit)mask* to a certain value, thus blocking some hardware interrupts while allowing others (the `disable_irq()` / `enable_irq()` APIs allow you to selectively disable/enable individual IRQ lines).

For the sake of our example, let's say that this had occurred earlier, at some point in time `t0`. Now, as we described previously, another developer (you!) comes along, and in order to protect a critical section within the driver's `read` method uses the `spin_lock_irq()` API (as we just learned in the *Solving the issue on UP and SMP with the spin_[un]lock_irq() API variant* section). Sounds correct, yes? Well, yes, but this API has the power *to turn off (mask)* ***all*** *hardware interrupts* (*and* kernel preemption, which we'll ignore for now) on the local CPU core. It does so by manipulating, at a low level, the (very arch-specific) hardware CPU state register(s).

FYI, on the `x86[_64]`, the `spin_lock_irq()` API saves the EFLAGS (32-bit)/RFLAGS (64-bit) register content (more correctly, it saves the LSB 16 bits, the content that's called 'FLAGS'). On the AArch32, it saves the EFLAGS register content, and on the AArch64 it saves the **DAIF (Debug, Asynchronous (serror), Interrupts, and FIQ exceptions)** register content. The `spin_unlock_irqrestore()` API, of course, restores it.

Well, so what, you ask? Consider this scenario unfolding:

- time `t0`: The interrupt mask is set to some value, say, `0x8e (10001110b)`, enabling some and disabling some interrupts (via, say, careful usage of the `disable_irq()` API). This setting is important to the project (here, for simplicity, we're assuming it's an 8-bit mask register):

  ```
  [... time elapses ... ].
  ```

- time `t1`: Just before entering the driver `read` method's critical section, call `spin_lock_irq(&slock);`. As we now understand, this API will internally save some CPU state register *and* have the internal effect of clearing all bits in the interrupt mask register to `0`, thus effectively masking all interrupts.
- time `t2`: Now, hardware interrupts cannot be handled on this CPU core, so we go ahead and complete the critical section. That's fine. Once we're done with the critical section, we call `spin_unlock_irq(&slock);`. This API will have the internal effect of restoring certain (arch-specific) CPU state registers, which will also have the internal side effect of *setting* all bits in the interrupt mask register to `1`, thus **reenabling all interrupts**.

However, the interrupt mask register has now been wrongly "restored" to a value of `0xff (11111111b)`, *not the value* `0x8e` as the original developer(s) wants, requires, and assumes! This can (and probably will) break something in the project. Sounds difficult.

The solution, though, is quite straightforward: when saving the CPU state, don't assume anything; simply save and restore the existing CPU state, and thus, the interrupt mask state. This way, if the interrupt mask initially was the value `0x8e (10001110b)`, this is what gets saved, and later, restored; perfect. This save-and-restore-cpu-state (with the spinlock) can be achieved with the following spinlock API pair:

```
#include <linux/spinlock.h>
unsigned long spin_lock_irqsave(spinlock_t *lock, unsigned long flags);
void spin_unlock_irqrestore(spinlock_t *lock, unsigned long flags);
```

The first parameter to both of these lock and unlock functions is the pointer to spinlock variable to use. The second parameter, `flags`, *must be a local variable* of the `unsigned long` data type. This will be used to save and restore the arch-specific CPU register (and thus the interrupt mask state as well). So, finally, for this scenario, the correct (pseudo)code is this:

```
spinlock_t slock;
spin_lock_init(&slock);
[ ... ]
driver_read(...)
{
    [ ... ]
    spin_lock_irqsave(&slock, flags);
        <<--- time t1 : start of critical section >>
[now the CPU state is saved; as a side effect, all interrupts + kernel
preemption on the local CPU core are masked (disabled) ]
    << ... critical section ... >>
    spin_unlock_irqrestore(&slock, flags);
<<--- time t2 : end of critical section >>
[now the CPU state is restored; as a side effect, only the previously masked
interrupts + kernel preemption on the local CPU core are unmasked (enabled) ]
    [ ... ]
}
```

To be pedantic, `spin_lock_irqsave()` is not an API, but a macro; we've shown it as an API for readability. Also, although the return value of this macro is not void, it's an internal detail (the `flags` parameter variable is updated here).

Great, now let's round off this discussion by using spinlocks with regard to interrupt bottom halves.

Interrupt handling, bottom halves, and locking

Before embarking on the points regarding locking and bottom halves, a quick overview (summary, really) of some key points with regard to interrupt handling on Linux will help.

Interrupt handling on Linux – a summary of key points

Interrupt handlers must be non-blocking and complete their work quickly. How quickly? A rule of thumb is, within 100 microseconds. However, what if there is quite a bit of work to perform while in the handler? It might take longer... which can introduce latencies. A solution (which most modern OSes go with) is to "split" the interrupt handling into two halves - top and bottom. The "top half" (or hardirq) handler is the one invoked almost instantly when the hardware interrupt fires; you - the driver author - are to do the bare minimum required work here, keeping it as short as possible. If more work is required to be carried out, you (earlier) register and invoke the "bottom half," a deferred function that runs "later."

In reality, bottom halves are implemented as "tasklets," which in turn are built on the kernel's low-level softirq technology. The tasklet or softirq runs pretty much immediately after the top half completes.

But then where's the benefit in this "splitting" approach? Ah, here's the key thing, with important side effects with regard to locking: the top half (hardirq) always runs with all interrupts disabled (masked) on the current CPU and the IRQ it's handling disabled (masked) across all CPUs; the bottom-half handler runs with all interrupts enabled.

It's important to realize that even the bottom-half mechanisms - the softirq and tasklet - run in interrupt (and not process) context. Thus, as with the top-half or hardirq handler, you musn't do anything that could possibly block within them.

A hardirq handler will never run in parallel with itself (thus, it's not reentrant). A tasklet (which is built on the softirq mechanism under the hood) will also never run in parallel with itself; this property makes using the tasklet easier. A softirq, though, can run in parallel with itself (on other cores).

Next, using *threaded handlers* for interrupt processing is now quite popular with many types of drivers (especially for devices that are a bit slower, that don't have extremely high performance demands unlike many block, network, graphics devices, and is in fact the default when registering an interrupt handler). In this case, as the interrupt handler entity is really nothing but a kernel thread (at `SCHED_FIFO` scheduling policy and real-time priority 50), issuing blocking calls within it is allowed. Moreover, this removes the 'top-bottom half' dichotomy as well.

Bottom halves and locking

Now, having understood these points, let's move on to locking.

What if your driver's softirq or tasklet has a critical section that "races" with your top-half (hardirq) interrupt handler (by executing on a core different from the one where the top half executes)? Easy: use a regular spinlock in both to protect the critical section(s).

Next, what if your driver's softirq or tasklet has a critical section that "races" with your process-context code paths?

In such situations, using the `spin_lock_bh()` routine in your driver's methods is likely what's required as it first disables bottom halves on the local processor and then takes the spinlock, thus safeguarding the critical section (similar to the way that `spin_lock_irq[save]()` protects the critical section in process context by disabling hardware interrupts on the local core):

```
void spin_lock_bh(spinlock_t *lock);
```

The corresponding unlock API is the `spin_unlock_bh()`.

Of course, *overhead* does matter in highly performance-sensitive code paths (the kernel code implementing the network stack being a great example). Thus, *using the simplest form of spinlocks will help in performance terms, than with the more complex variants.*

Having said that, though, there are certainly going to be occasions that demand the use of the stronger forms of the spinlock API.

For example, with regard to the 6.1.25 LTS Linux kernel codebase, here's an approximation of the number of usage instances of different forms of the spinlock APIs we have seen:

- `spin_lock()`: Over 10,200 usage instances
- `spin_lock_irq()`: Over 3,800 usage instances
- `spin_lock_irqsave()`: Over 15,500 usage instances
- `spin_lock_bh()`: Over 4,400 usage instances

(Just for contrast, the numbers for the book's first edition with the 5.4.0 LTS kernel were 9,400, over 3,600, over 15,000, and over 3,700 usage instances, respectively.) We aren't drawing any major inference from these numbers and the different forms of spinlock API usage; we just wish to point out that using the stronger form of spinlock APIs is quite widespread in the Linux kernel.

Finally, a very brief note on the internal implementation of the spinlock: in terms of under-the-hood internals, the implementation tends to be very arch-specific code, often comprised of atomic machine language instructions that execute very fast on the microprocessor. On the popular `x86[_64]` architecture, for example, the spinlock ultimately boils down to an *atomic test-and-set* machine instruction on a member of the spinlock structure (typically implemented via the `cmpxchg` machine language instruction). On many ARM machines, as we mentioned earlier, it's often the `wfe` (Wait For Event, as well as the **SetEvent** (**SEV**)) machine instruction at the heart of the implementation. (You will find resources regarding its internal implementation in the *Further reading* section.) Regardless, as a kernel or driver author, you should only use the exposed APIs (and macros) when using spinlocks.

Using spinlocks – a quick summary

Let's quickly summarize spinlocks, API-wise:

- **Simplest, lowest overhead**: Use the non-irq spinlock primitives, `spin_lock()/spin_unlock()`, when protecting critical sections in process context (there are either no hardware interrupts to deal with or there are interrupts but we do not race with them at all; in effect, use this when interrupts don't come into play or don't matter).

- Also, this form can be used when protecting critical sections between a top-half and bottom-half handler.

- **Medium overhead:** Use the irq-disabling (as well as kernel preemption disabling) versions, `spin_lock_irq()/spin_unlock_irq()`, when hardware interrupts are in play and do matter (here, process and interrupt contexts can "race"; that is, they share global writable data).
    - Also, you can use the `spin_[un]lock_bh()` API pair when protecting critical sections between the process context and bottom half (it internally disables/enables bottom halves on the local core).
- **Strongest form, (relatively) high overhead:** This is the safest way to use a spinlock. It does the same as the one with medium overhead, except it performs a save-and-restore on the CPU state via the `spin_lock_irqsave()/spin_unlock_irqrestore()` API pair, so as to guarantee that, in effect, the previous interrupt mask settings aren't inadvertently overwritten, which could happen with the previous case.

As we saw earlier, the spinlock – in the sense of "spinning" on the processor it's running on when awaiting the lock – is impossible on UP (how can you spin on the one CPU that's available while another thread runs simultaneously on the very same CPU?). Indeed, on UP systems, spinlock APIs become a no-op (unless spinlock debug configs are on); the only real effect of the spinlock APIs here is that they disable (mask) hardware interrupts and kernel preemption on the processor! On SMP (multicore) systems, however, the spinning logic actually comes into play, and thus the locking semantics work as expected.

Hang on – these details should not stress you, budding kernel/driver developer; in fact, the whole point is that you should simply use the spinlock APIs as described and you will never have to worry about UP versus SMP, about kernel preemption and what-not; the details of what is internally done and what isn't are all hidden by the implementation.

A new feature was added to the 5.8 kernel from the **Real-Time Linux** (**RTL**, previously called PREEMPT_RT) project, which deserves a quick mention here: "**local locks.**" While the main use case for local locks is for (hard) real-time kernels, they help with non-real-time kernels too, mainly for lock debugging via static analysis, as well as runtime debugging via lockdep (we cover lockdep in the next chapter). Here's the LWN article on the subject: `https://lwn.net/Articles/828477/`.

Next, when a thread's holding an irq-off/preempt-off spinlock, it cannot, by definition, be preempted. This is a thorn in the side for the Linux **real-time kernel** (**RTL**); obviously, it needs to be able to guarantee that a lower priority thread is preempted when a higher priority real-time thread becomes runnable, no matter what. With the traditional spinlock, this design is defeated; hence, when RTL is enabled, the spinlock is actually re- implemented as a "**sleeping spinlock**"! This is achieved by substituting the spinlock with an `rt-mutex` lock, in effect making the critical section sleep-able and thus preemptable. Just something to be aware of.

With this, we complete the section on spinlocks, an extremely common and key lock used within the Linux kernel by virtually all its subsystems, including drivers.

Locking – common mistakes and guidelines

To wrap up, a quick reference or summary, if you will, covering the typical common mistakes made when locking, and (with some repetition), locking guidelines. (Note that some of the techniques mentioned here – like lock-free programming – are covered in the following chapter).

Common mistakes

- Not recognizing critical sections:
    - "Simple" increments/decrements (of the `i ++` or `i --` type): As we learned in the *A classic case – the global i ++* section, these too can be critical sections. In the following chapter, we show optimized and atomic ways to work with them.
    - *"Hey, I'm only reading the shared data"*: It's still a critical section if the two conditions for one are met; not protecting it can result in a dirty or torn read, inconsistent or corrupted data.
    - *Deadlock*: A situation where forward progress is impossible; carefully design your locking schema and follow the well-understood locking rules or guidelines that follow to avoid deadlock. It essentially comes down to these key points:
        - Document and always follow the *lock-ordering* rule.
        - Don't attempt to re-acquire a lock you already hold.
        - Only release a lock that you currently hold.
    - Prevent starvation.

Locking guidelines

An overall summary on locking guidelines follows:

- In the first place, try your best to avoid locks:
    - Now, this does not simply mean: "use no globals." Instead, you have to figure an overall architecture where, as far as is possible, no writer thread (to shared writable data) can run concurrently with any other read/write access to that data.
    - Next, if you are using shared writable data, say within a global structure, try to keep all, or as many (of the integer) members as possible as `refcount_t` or `atomic_t` (covered later).
    - Take into account memory barriers (when required).
    - Use lock-free technologies!
- If you must use locking, proceed in this order:
    - Try to use *lock-free* technologies:

        - Per-CPU variables
        - RCU
    - Where you cannot, use normal locking:
        - Mutex: In process context, when the critical section is long-ish, and/or blocking I/O (sleeping) in the critical section is required or can occur
        - Spinlock: When working in any atomic context (like interrupt handling), when the critical section is short; must be non-blocking (no sleep allowed) in the critical section. Or, can use it even in process context where there's no blocking in the critical section.
        - If either the mutex or spinlock is fine to use, prefer the spinlock (it not only gets you typically better performance, it also enforces stricter rules within the critical section that must be followed).
        - `refcount_t` for integer operations (the `atomic_t` atomic operators are the older interface for this).
        - Use the kernel's RMW bitwise operators when manipulating bits (covered in the following chapter).
        - reader-writer (spin)locks (going out in favor of RCU).
- Always keep in mind *lock ordering*: Always take locks in the same order; document the order and follow it strictly; this helps with deadlock prevention (the order of releasing the locks doesn't really matter).
- Lock data, not code:
    - Move toward fine(r) granularity locking, where possible.
    - There's (much) more to this point; at a deeper level, it means to *design* your locking schema by carefully looking at the data structure (or even the member(s) within) you're protecting, specifying exactly how it will be protected against concurrent access – in effect, using a data-centric approach rather than a code-centric one where you more or less randomly sprinkle around some mutex or spinlocks in the code until it seems to "work." These ideas, and more, are delved into in this article by Daniel Vetter: *Locking Engineering Principles*, July 2022: `https://blog.ffwll.ch/2022/07/locking-engineering.html`).
- Performance is affected by the length of the critical section (the code path between the lock and the unlock); keep it short! (Recall the *criticalstat* eBPF utility we mentioned earlier; you can use it to check for and report long atomic critical sections as they're encountered.)
- Prevent starvation.
- Keep cache effects and memory barriers in mind (false sharing and cache line bouncing; covered in the following chapter).
- Use a debug kernel to run all your test cases; with regard to locking, the debug kernel must have "lockdep" enabled; also, locking statistics being enabled can also help pinpoint hotspots (covered in the following chapter).

- Keep your locking schema as simple as possible.

Done.

Solutions

These small exercises were mentioned in the *What is a critical section?* section.

Solution to Exercise 2. Again, ask yourself, *what exactly constitutes a critical section?* It's when the code path in question both can possibly execute in parallel *and* works upon shared writable data. So, now, does the code in question (the lines between `t1` and `t2`) fulfill these two pre-conditions? Well, it *can* run in parallel (as explicitly stated), and it does work on shared writable data (the `mydrv` variable is shared, as it's a global variable, thus each and every thread within this code path will work upon that very same memory item in parallel).

So, the answer here is clearly yes, it is indeed a critical section. In other words, it should not be allowed to run without some kind of explicit protection.

Solution to Exercise 3. This one's interesting; the second condition – *does it work upon shared writable data?* – is true, but the first condition – *can it possibly execute in parallel?* – is false. This is because the init and cleanup "methods" (functions) of a kernel module will only ever execute once (thus there are no more possible instances to execute in parallel with). So, this does not constitute a critical section and thus requires no special protection.

Summary

Congratulations on completing this chapter!

Understanding concurrency and its related concerns is absolutely critical for any software professional. In this chapter, you learned key concepts regarding critical sections, the need for exclusive execution within them, and what atomicity and data races really mean. You then learned *why* we need to be concerned with concurrency while writing code for the Linux OS. After that, we delved into two very common locking technologies that are heavily used in the kernel – mutex locks and spinlocks – in detail. You also learned how to decide which lock to use when. Finally, and again, importantly, learning how to handle concurrency concerns when hardware interrupts (and their possible bottom halves) are in play was covered.

But we aren't done with kernel concurrency yet! There are many more concepts and technologies we need to learn about, which is just what we will do in the next, and final, chapter of this book. I suggest that you first digest the content of this chapter well by browsing through it, as well as the resources in the *Further reading* section, and work on the exercises provided, before diving into the last chapter!

Questions

As we conclude, here is a list of questions for you to test your knowledge regarding this chapter's material: `https://github.com/PacktPublishing/Linux-Kernel-Programming_2E/blob/main/questions/ch12_qs_assignments.txt`. You will find some of the questions answered in the book's GitHub repo: `https://github.com/PacktPublishing/Linux-Kernel-Programming_2E/tree/main/solutions_to_assgn`.

Further reading

To help you delve deeper into the subject with useful materials, we provide a rather detailed list of online references and links (and, at times, even books) in a *Further reading* document in this book's GitHub repository. The *Further reading* document is available here: `https://github.com/PacktPublishing/Linux-Kernel-Programming_2E/blob/main/Further_Reading.md`.

Leave a review!

Enjoying this book? Help readers like you by leaving an Amazon review. Scan the QR code below for a 20% discount code.

**Limited Offer*

13

Kernel Synchronization – Part 2

This chapter continues the discussion on the fairly complex topics of kernel synchronization and dealing with concurrency within the kernel in general, from the previous chapter. I suggest that, if you haven't already, you first read the previous chapter and then continue with this one.

Here, we shall continue our learning with respect to the vast topic of kernel synchronization and handling concurrency when in kernel space. As before, the material is targeted at kernel and/or device driver/module developers. In this chapter, we shall cover the following:

- Using the `atomic_t` and `refcount_t` interfaces
- Using the RMW atomic operators
- Using the reader-writer spinlock
- Understanding CPU caching basics, cache effects, and false sharing
- Lock-free programming with per-CPU and RCU
- Lock debugging within the kernel
- Introducing memory barriers

Technical requirements

The technical requirements remain identical to those for the previous chapter, *Chapter 12, Kernel Synchronization – Part 1.*

Using the atomic_t and refcount_t interfaces

You'll recall that in our original simple misc character device driver program's (available in the *Linux Kernel Programming – Part 2* companion volume, in the code here: `https://github.com/PacktPublishing/Linux-Kernel-Programming-Part-2/blob/main/ch1/miscdrv_rdwr/miscdrv_rdwr.c`) open method (and elsewhere), we defined and manipulated two static global integers, ga and gb:

```
static int ga, gb = 1;
[...]
ga++; gb--;
```

By now, it should be obvious to you that the place where we operate on these integers is a potential bug if left as it is: it's shared writable data (aka shared state) being accessed in a possibly concurrent code path and, therefore, qualifies as a critical section, thus requiring protection against concurrent access. (If this isn't clear to you, please first read the *Critical sections, exclusive execution, and atomicity* section in the previous chapter carefully.)

In the previous chapter, *Kernel Synchronization – Part 1*, to fix the issue, in our `ch12/1_miscdrv_rdwr_mutexlock/1_miscdrv_rdwr_mutexlock.c` program, we first used a *mutex lock* to protect the critical section. Later, you learned that using a *spinlock* to protect non-blocking critical sections such as this one would be (far) superior to using a mutex in terms of performance; so in our next driver, `ch12/2_miscdrv_rdwr_spinlock/2_miscdrv_rdwr_spinlock.c`, we used a spinlock to protect the critical section instead (here: `https://github.com/PacktPublishing/Linux-Kernel-Programming_2E/blob/05c6b2853ef053ad22621983f17d4f326c4de97f/ch12/2_miscdrv_rdwr_spinlock/miscdrv_rdwr_spinlock.c#L111`):

```
spin_lock(&lock1);
ga++; gb--;
spin_unlock(&lock1);
```

That's good, but we can do better still! Operating upon global integers turns out to be such a common occurrence within the kernel (think of reference or resource counters getting incremented and decremented, and so on) that it provides a class of operators, called the **refcount** and **atomic integer operators** or interfaces; these are very specifically designed to atomically – safely and indivisibly – operate on **only integers**.

The newer refcount_t versus older atomic_t interfaces

Before we start this topic, it's important to mention this: until version 4.10 (inclusive), a set of interfaces to operate atomically upon integers existed; they were called the `atomic_t` interfaces. Since the 4.11 kernel, there has been a newer and better set of interfaces christened the `refcount_t` APIs, designed for manipulating a kernel-space object's reference counters. This newer set greatly improves the security posture of the kernel – via much-improved **Integer Overflow** (**IoF**) and **Use After Free** (**UAF**) protection as well as memory ordering guarantees, which the older `atomic_t` APIs lack. The `refcount_t` interfaces, like several other security technologies used on Linux, have their origins in work done by the PaX Team – `https://pax.grsecurity.net/` (also called `PAX_REFCOUNT`).

Having said that, the reality is that (at the time of writing) the older `atomic_t` interfaces are still very much in use within the kernel core and drivers (they are slowly being converted, with the older `atomic_t` interfaces being moved to the newer `refcount_t` model and the API set). Thus, in this topic, we cover both, pointing out differences and mentioning which `refcount_t` API supersedes an `atomic_t` API wherever applicable. Think of the `refcount_t` interfaces as a variant of the (older) `atomic_t` interfaces, which are specialized toward reference counting.

A key difference between the older `atomic_t` operators and the newer `refcount_t` ones is that the former works upon signed integers and can be used for any type of counting, whereas the latter is designed to work for object reference counting only and has a strict valid range.

More details on the newer `refcount_t` follow:

- **It works only within a strictly specified range:** `1` to `INT_MAX-1`, that is, [1 .. 2,147,483,646].

  From `include/linux/refcount.h`:

```
[ ... ]
 *              INT_MAX         REFCOUNT_SATURATED        UINT_MAX
 *   0          (0x7fff_ffff)    (0xc000_0000)          (0xffff_ffff)
 *   +--------------+----------------+-----------------+
 *                  <---------- bad value! ----------->
 *
 * (in a signed view of the world, the "bad value" range corresponds to
 * a negative counter value). [ ... ]
```

- To quote the official kernel documentation: *"The counter saturates at REFCOUNT_SATURATED and will not move once there. This avoids wrapping the counter and causing 'spurious' use-after-free bugs. [...]"*

Attempting to set a `refcount_t` variable to `0` or negative, or to `INT_MAX` or above, is impossible; this is just what the doctor ordered to help prevent and/or catch integer underflow/overflow issues and thus prevent the UAF class bug in many cases! (This is because an object must only be freed when its reference count hits zero; so the last holder will have it at reference count 1 and will free it. If the reference count is ever out of the valid range, it's a likely use-after-free case, and the noisy warning emitted to the kernel log alerts us to this possibility.) Well, it's not impossible to under/overflow a reference counter; it's just that it now results in a (noisy) warning being fired via a `REFCOUNT_WARN()` – which translates to the `WARN_ONCE()` macro – and the developer is expected to notice and thus fix (or at least report) the issue. (Of course, the naming `WARN_ONCE()` supplies a hint: the particular warning will be emitted exactly once during the system's lifetime).

Again, `refcount_t` variables are meant to be used only for kernel object reference counting, and nothing else. Thus, this is indeed the required behavior; the reference counter must start at a positive value (typically `1` when the object is newly instantiated), is incremented (or added to) whenever code "gets" or takes a reference to it, and is decremented (or subtracted from) whenever code "puts" or drops a reference on the object in question. You are expected to carefully manipulate the reference counter (matching your "gets" and "puts"), always keeping its value within the legal range.

But you're perhaps wondering, how do I use these exotic interfaces? The following sections cover just this.

The simpler atomic_t and refcount_t interfaces

Regarding the `atomic_t` interfaces, firstly, understand that all the following `atomic_t` constructs are for 32-bit integers only; of course, with 64-bit integers now being commonplace, 64-bit atomic integer operators are available as well (we cover them in the upcoming *64-bit atomic integer operators* section).

Typically, they are semantically identical to their 32-bit counterparts, with the difference being in the name (`atomic_foo()` becomes `atomic64_foo()`). So the primary data type for 64-bit atomic integers is called `atomic64_t` (AKA `atomic_long_t`). The newer `refcount_t` interfaces, on the other hand, cater to both 32- and 64-bit integers.

Next, it's key to realize that *all accesses* to the `atomic_t` or `refcount_t` variable in question *must* be made only via these accessor "methods" or helpers. For example, to even read the content of an `atomic_t` variable, you must use the `atomic_read()` helper method and not simply read it in the usual way. Similarly, to set an atomic integer v to the value `i`, you must use (something like) `atomic_set(&v, i);` and not simply `v = i` (likewise for all `refcount_t` operations).

The complete set of all the `atomic_t` and `refcount_t` interfaces available within the kernel is large; to help keep things simple and clear in this section, we list only some of the more commonly used atomic (32-bit) and `refcount_t` interfaces in the following table (they operate upon generic `atomic_t v` and `refcount_t v` variables, respectively):

Operation	**Older (<= 4.10 kernel) 32-bit atomic_t interface**	**Newer (>= 4.11; 32- and 64-bit) refcount_t interface** **Range: [1 .. INT_MAX-1]**
Header file to include	<linux/atomic.h>	<linux/refcount.h>
Declare and initialize a variable v to 1	static atomic_t v = ATOMIC_INIT(1);	static refcount_t v = REFCOUNT_INIT(1);
Atomically read the current value of v	int atomic_read(atomic_t *v)	unsigned int refcount_read(const refcount_t *v)
Atomically set v to the value i	void atomic_set(atomic_t *v, int i)	void refcount_set(refcount_t *v, int i)
Atomically increment the v value by 1	void atomic_inc(atomic_t *v)	void refcount_inc(refcount_t *v)
Atomically decrement the v value by 1	void atomic_dec(atomic_t *v)	void refcount_dec(refcount_t *v)
Atomically add the value of i to v	void atomic_add(int i, atomic_t *v)	void refcount_add(int i, refcount_t *v)
Atomically subtract the value of i from v	void atomic_sub(int i, atomic_t *v)	void refcount_sub(int i, refcount_t *v)
Atomically add the value of i to v and return the result	int atomic_add_return(int i, atomic_t *v)	bool refcount_add_not_zero(int i, refcount_t *v) (Not a precise match; adds i to v unless it's 0)

Atomically subtract the value of i from v and return the result	int atomic_sub_return(int i, atomic_t *v)	bool refcount_sub_and_test(int i, refcount_t *r) (Not a precise match; subtracts i from v and tests; returns true if resulting refcount is 0, else false)

Table 13.1: The older atomic_t (the second column) versus the newer refcount_t (the third column) interfaces for common reference counting operations

You've now seen several atomic_t and refcount_t functions; let's quickly check out a few examples of their usage in the kernel.

Examples of using refcount_t within the kernel codebase

Again, in the *Linux Kernel Programming – Part 2* companion volume, within one of our demo kernel modules regarding kernel threads (the code's here: https://github.com/PacktPublishing/Linux-Kernel-Programming-Part-2/blob/main/ch5/kthread_simple/kthread_simple.c), we created a kernel thread and then employed the get_task_struct() inline function to mark the kernel thread's task structure as in use. As you can now guess, the get_task_struct() routine works internally by incrementing the task structure's *reference counter* – a refcount_t variable named usage – via the refcount_inc() function:

```
// include/linux/sched/task.h
static inline struct task_struct *get_task_struct(struct task_struct *t)
{
	refcount_inc(&t->usage);
	return t;
}
```

The converse routine, put_task_struct(), performs the subsequent decrement on the task structure's reference counter. The actual routine employed by it internally, refcount_dec_and_test(), tests whether the new refcount value has dropped to 0; if so, it returns true, and if this is the case, it implies that the task structure isn't being referenced by anyone. Then, the call to [__]put_task_struct() frees it up:

```
static inline void put_task_struct(struct task_struct *t)
{
	if (refcount_dec_and_test(&t->usage))
		__put_task_struct(t);
}
```

Another example of the refcounting methods in use within the kernel is found in kernel/user.c (which helps track the number of processes, files, and so on that a user has claimed via a per-user structure).

The following screenshot (*Figure 13.1*) shows us grepping this source file for lines containing the (case-insensitive) string `refcount`; thus, here, we can see how the code performs the initialization, increment/decrement, and setting of some `refcount_t` variables:

```
linux-6.1.25 $ grep -i -Hn -A1 refcount kernel/user.c
kernel/user.c:58:       .ns.count = REFCOUNT_INIT(3),
kernel/user.c-59-       .owner = GLOBAL_ROOT_UID,
--
kernel/user.c:100:      .__count        = REFCOUNT_INIT(1),
kernel/user.c-101-      .uid            = GLOBAL_ROOT_UID,
--
kernel/user.c:124:                      refcount_inc(&user->__count);
kernel/user.c-125-                      return user;
--
kernel/user.c:185:      if (refcount_dec_and_lock_irqsave(&up->__count, &uidhash_lock, &flags))
kernel/user.c-186-              free_user(up, flags);
--
kernel/user.c:204:              refcount_set(&new->__count, 1);
kernel/user.c-205-              if (user_epoll_alloc(new)) {
linux-6.1.25 $ _
```

Figure 13.1: Screenshot showing the usage of the refcount_t interfaces in kernel/user.c

Look up the `refcount_t` API interface official kernel documentation here: `https://www.kernel.org/doc/html/latest/driver-api/basics.html#reference-counting`. FYI, the `refcount_dec_and_lock_irqsave()` function seen (in *Figure 13.1*), has a spinlock as it's second parameter. If able to decrement the reference counter to 0, it returns true while holding the spinlock (with interrupts disabled), or false otherwise.

Exercise

Modify our earlier `ch12/2_miscdrv_rdwr_spinlock/miscdrv_rdwr_spinlock.c` driver code; it has the integers ga and gb, which, when being read or written, are protected via a spinlock. Now, make them `refcount` variables (initializing them to some value – say, 42), and use the appropriate `refcount_t` methods when working on them. Include a test case where you deliberately make them go outside the valid allowed range (`[1..INT_MAX-1]`).

You'll find a reference solution in the GitHub repo within `solutions_to_assgn/ch13`

Careful! Don't allow their values to go out of the (default) allowed range, `[1..INT_MAX-1]`! If you do, that is, you underflow or overflow it, this will typically trigger a kernel-level warning (via a `REFCOUNT_WARN()`, which translates to the `WARN_ONCE()` macro).

Deliberately causing an underflow or overflow with refcount variables

It's useful at times to play devil's advocate; what if a module (or the kernel) does cause a `refcount_t` variable to go out of range? As mentioned, this will have the kernel generate a noisy one-time warning to the kernel log. But, of course, let's be empirical and try it out!

So here's a code snippet (from a solution to the aforementioned exercise):

```
…
static refcount_t ga = REFCOUNT_INIT(42);        /* ga will be init to 42 */
static refcount_t gb = REFCOUNT_INIT(42);        /* gb will be init to 42 */
…
static int open_miscdrv_rdwr(struct inode *inode, struct file *filp)
{
…
        /* Tweaked the code here to _deliberately_ cause overflow/underflow
         * refcount bugs (resp) if the #if 1 is changed to #if 0 */
#if 1
        refcount_inc(&ga);
#else
        //refcount_add((int i, atomic_t *v); // adds i to v
        // Here we deliberately overflow it, leading to a warning (once)!
        pr_debug("*** Bad case! About to overflow refcount var! ***\n");
        refcount_add(INT_MAX, &ga);
#endif
...
```

If we change the `#if 1` to `#if 0`, then the deliberate bug will surface; clearly, we overflow the reference counter by adding `INT_MAX` to whatever value's already within it! Imagine we do so, and then build (in debug mode), load, and test this module.

Here's a partial screenshot showing the one-time warning that the kernel's reference counting code emits:

```
$ echo abc > /dev/llkd_miscdrv_rdwr_refcount ; sudo dmesg
[  137.143144] miscdrv_rdwr_refcount:miscdrv_init_refcount(): LLKD misc driver (major # 10) registered, minor# = 120,
 dev node is llkd_miscdrv_rdwr_refcount
[  137.143149] misc llkd_miscdrv_rdwr_refcount: A sample print via the dev_dbg(): driver initialized
[  142.155554] miscdrv_rdwr_refcount:open_miscdrv_rdwr(): 002)  bash :1474   |  ...0   /* open_miscdrv_rdwr() */
[  142.155559] miscdrv_rdwr_refcount:open_miscdrv_rdwr(): *** Bad case! About to overflow refcount var! ***
[  142.155560] ------------[ cut here ]------------
[  142.155561] refcount_t: saturated; leaking memory.
[  142.155567] WARNING: CPU: 2 PID: 1474 at lib/refcount.c:22 refcount_warn_saturate+0x148/0x150
[  142.155572] Modules linked in: miscdrv_rdwr_refcount(OE) binfmt_misc nls_iso8859_1 snd_intel8x0 snd_ac97_codec ac9
7_bus snd_pcm snd_seq snd_seq_device intel_rapl_msr snd_timer intel_rapl_common crct10dif_pclmul crc32_pclmul polyval
_clmulni polyval_generic snd ghash_clmulni_intel aesni_intel crypto_simd cryptd rapl video wmi i2c_piix4 soundcore vb
oxguest(OE) joydev input_leds mac_hid serio_raw vmwgfx drm_kms_helper syscopyarea sysfillrect sysimgblt fb_sys_fops d
rm_ttm_helper ttm drm msr parport_pc ppdev lp parport efi_pstore dmi_sysfs ip_tables x_tables autofs4 hid_generic usb
hid hid psmouse ahci e1000 libahci pata_acpi
[  142.155605] CPU: 2 PID: 1474 Comm: bash Tainted: G           OE      6.1.25-dbg #2
[  142.155607] Hardware name: innotek GmbH VirtualBox/VirtualBox, BIOS VirtualBox 12/01/2006
[  142.155608] RIP: 0010:refcount_warn_saturate+0x148/0x150
[  142.155610] Code: b8 77 01 8d c6 05 5e 36 6c 01 01 e8 e2 f2 9b ff 0f 0b e9 38 ff ff ff 48 c7 c7 90 77 01 8d c6 05
45 36 6c 01 01 e8 c8 f2 9b ff <0f> 0b e9 1e ff ff ff 90 8b 07 3d 00 00 00 c0 74 12 83 f8 01 74 1d
[  142.155611] RSP: 0018:ffffb7a14291baf0 EFLAGS: 00010246
[  142.155613] RAX: 0000000000000000 RBX: 0000000000000000 RCX: 0000000000000000
[  142.155614] RDX: 0000000000000000 RSI: 0000000000000000 RDI: 0000000000000000
[  142.155615] RBP: ffffb7a14291baf8 R08: 0000000000000000 R09: 0000000000000000
[  142.155616] R10: 0000000000000000 R11: 0000000000000000 R12: 0000000000000000
[  142.155617] R13: ffff96c0c36c9000 R14: ffff96c0d95e5028 R15: ffffffffc0671700
[  142.155618] FS:  00007f1129306740(0000) GS:ffff96c13dc80000(0000) knlGS:0000000000000000
[  142.155619] CS:  0010 DS: 0000 ES: 0000 CR0: 0000000080050033
[  142.155620] CR2: 0000558aa6ccde24 CR3: 000000000e2ea001 CR4: 00000000000706e0
[  142.155623] Call Trace:
[  142.155624]  <TASK>
[  142.155627]  open_miscdrv_rdwr+0x153/0x1d0 [miscdrv_rdwr_refcount]
[  142.155631]  misc_open+0x127/0x150
```

Figure 13.2: Partial screenshot showing the one-time warning emitted by the kernel reference counting code when we overflow the counter

You can see the `WARNING:` … line highlighted in *Figure 13.2*; lower in the kernel log we then see the following (not seen in the figure):

```
…
[  142.155685] ---[ end trace 0000000000000000 ]---
[  142.155687] misc llkd_miscdrv_rdwr_refcount:  filename: "llkd_miscdrv_rdwr_
refcount"
                 wrt open file: f_flags = 0x8241
                 ga = -1073741824(0xc0000000), gb = 41(0x29)
[  142.155690] misc llkd_miscdrv_rdwr_refcount: stats: tx=0, rx=0
```

(The code for this demo seen in *Figure 13.2* isn't included in our GitHub repository.) Look at the value of the `refcount` variable ga now; it's `0xc0000000`, which is the value `REFCOUNT_SATURATED`, as expected.

The modern kernel has several useful test infrastructures; one is called the **Linux Kernel Dump Test Module (LKDTM)**; see `drivers/misc/lkdtm/refcount.c` for many test cases being run on the `refcount` interfaces, which you can learn from. FYI, you can also use LKDTM via the kernel's fault injection framework to test and evaluate the kernel's reaction to faulty scenarios (see the documentation here: *Provoking crashes with Linux Kernel Dump Test Module (LKDTM)* – `https://www.kernel.org/doc/html/latest/fault-injection/provoke-crashes.html#provoking-crashes-with-linux-kernel-dump-test-module-lkdtm`.

Right, let's move on. The atomic interfaces we've mentioned so far (*Table 13.1*) all operate on 32-bit integers; what about on 64-bit? That's what follows.

64-bit atomic integer operators

As mentioned at the start of this topic, the set of `atomic_t` integer operators we have seen so far all operate on traditional 32-bit integers (note: this discussion doesn't apply to the newer `refcount_t` interfaces; they transparently operate upon both 32- and 64-bit quantities). Obviously, with 64-bit systems becoming the norm rather than the exception nowadays, the kernel community provides an identical set of atomic integer operators for 64-bit integers. The difference is as follows:

- Declare the 64-bit atomic integer as a variable of type `atomic64_t` (that is, `atomic_long_t`).
- For all operators, in place of the `atomic_` prefix, use the `atomic64_` prefix.

So, take the following examples:

- In place of `ATOMIC_INIT()`, use `ATOMIC64_INIT()`.
- In place of `atomic_read()`, use `atomic64_read()`.
- In place of `atomic_dec_if_positive()`, use `atomic64_dec_if_positive()`.

Recent C and C++ language standards – C11 and C++11 – provide an atomic operations library that helps developers implement atomicity more easily due to the implicit language support; we won't delve into this aspect here. A reference can be found here (C11 also has pretty much the same equivalents): `https://en.cppreference.com/w/c/atomic`.

Note that all these routines – both the 32- and 64-bit `atomic[64]_*` operators – are **arch-independent**. A key point worth repeating is that any operations performed upon an atomic integer must be done by declaring the variable as `atomic[64]_t` and *only* via the methods provided (including initialization and even an (integer) read operation).

A note on internal implementation

In terms of internal implementation, a `foo()` atomic integer operator is typically a macro that becomes an inline function, which in turn invokes the arch-specific `arch_foo()` function. (For example, the x86 employs the well-known `LOCK` assembly prefix as part of the internal implementation, while ARM families tend to use a more generic approach with memory ordering semantics.)

The x86 `LOCK` prefix isn't an instruction; it's a prefix to an instruction (for example, `lock orb %b1,%0`). It guarantees that the core the instruction following LOCK is running on, has exclusive ownership of the cache line being operated upon; also, the control unit won't allow hardware interrupts – as it's a single machine-level instruction executing – plus there are memory ordering guarantees.

As usual, glancing through the official kernel documentation on atomic operators is always a good idea (within the kernel source tree, which can be found here: `Documentation/atomic_t.txt`, or here: `https://www.kernel.org/doc/Documentation/atomic_t.txt`). It neatly categorizes the numerous atomic integer APIs into distinct sets. FYI, arch-specific *memory ordering issues* do affect the internal implementation (a quick note on this follows). If interested, refer to this page on the official kernel documentation site at `https://www.kernel.org/doc/html/latest/core-api/refcount-vs-atomic.html#refcount-t-api-compared-to-atomic-t`.

Quite non-intuitively, at least for the generic arch-independent refcount implementation, the `refcount_t` APIs are internally implemented over the (older) `atomic_t` API set. For example, the `refcount_set()` API – which atomically sets a `refcount` variable's value to the parameter passed – is implemented like this within the kernel:

```
// include/linux/refcount.h
/**
 * refcount_set - set a refcount's value
 * @r: the refcount
 * @n: value to which the refcount will be set
 */
static inline void refcount_set(refcount_t *r, unsigned int n)
{
    atomic_set(&r->refs, n);
}
```

We can see that it's simply a thin wrapper over `atomic_set()`. The obvious FAQ here is: then why use the `refcount` API at all? There are a few reasons:

- Unlike the atomic operators, the counter saturates at the `REFCOUNT_SATURATED` value (which is set to `INT_MIN/2` by default, which is `0xc0000000`) and will not budge once there. This is critical: it avoids wrapping the counter, which could cause weird and spurious UAF bugs; this is even considered as a key security fix (`https://kernsec.org/wiki/index.php/Kernel_Protections/refcount_t`).
- Several of the newer APIs do provide **memory ordering** guarantees; in particular, the `refcount_t` APIs – as compared to their older `atomic_t` cousins – and the memory ordering guarantees they provide are documented at `https://www.kernel.org/doc/html/latest/core-api/refcount-vs-atomic.html#refcount-t-api-compared-to-atomic-t` (do have a look if you're interested in the low-level details).

- Also, realize that arch-dependent `refcount` implementations (when they exist – for example, x86 does have them, while ARM doesn't) can differ from the previously mentioned generic one.

What exactly is *memory ordering* and how does it affect us? The fact is, it's a complex topic, and unfortunately, the intricate details of this are beyond the scope of this book. It's worth knowing the basics: I suggest you read up on the **Linux-Kernel Memory Model (LKMM)**, which includes coverage on processor memory ordering and more. We refer you to good documentation on this here: *Explanation of the Linux-Kernel Memory Model* (`https://github.com/torvalds/linux/blob/master/tools/memory-model/Documentation/explanation.txt`).

We haven't attempted to show all the atomic and `refcount` APIs here (it's not necessary). The following are references from the official kernel documentation on `atomic_t` and `refcount_t`; browse through if interested:

- `atomic_t` interfaces:
  - *Semantics and Behavior of Atomic and Bitmask Operations* (`https://www.kernel.org/doc/html/v6.1/core-api/local_ops.html`)
  - API ref: *Atomics* (`https://www.kernel.org/doc/html/latest/driver-api/basics.html#atomics`)
- (Newer 4.11+) `refcount_t` interfaces for kernel object reference counting:
  - *refcount_t API compared to atomic_t* (`https://www.kernel.org/doc/html/latest/core-api/refcount-vs-atomic.html#refcount-t-api-compared-to-atomic-t`)
  - API reference: *Reference counting* (`https://www.kernel.org/doc/html/latest/driver-api/basics.html#reference-counting`)

Let's move on to the usage of a typical construct when working on device drivers – **Read-Modify-Write (RMW)**. Read on!

Using the RMW atomic operators

A more advanced set of atomic operators called the RMW APIs is available as well. (Why exactly it's called RMW and more is explained in the following section.) Among its many uses (we show a list in the upcoming section) is that of performing atomic RMW *bitwise* operations (safely and indivisibly). As a device driver author operating upon device or peripheral *registers*, this is indeed something you will very likely find yourself using.

The material in this section assumes you have at least a basic understanding of accessing peripheral device (chip) memory and registers; we have covered this topic in detail in the *Linux Kernel Programming – Part 2* companion volume in *Chapter 3, Working with Hardware I/O Memory*. It's recommended you first understand this before moving further.

When working with drivers, you'll typically need to perform bit operations (with the bitwise `AND &` and bitwise `OR |` being the most commonplace operators) on registers; this is done to modify its value, setting and/or clearing some bits within it. The thing is, merely performing some C manipulation to query or set device registers isn't quite enough. No, sir: don't forget about concurrency issues! Read on for the full story.

RMW atomic operations – operating on device registers

Let's quickly go over some basics first (please control your eye rolling): a byte consists of 8 bits, numbered from bit `0`, the **Least Significant Bit (LSB)**, to bit `7`, the **Most Significant Bit (MSB)**. (This is formally defined as the `BITS_PER_BYTE` macro in `include/linux/bits.h`, along with a few other interesting definitions.)

Here, by the term **register**, we refer to a small piece of memory within a peripheral device (or chip); typically, its size, the register bit width, is one of 8, 16, or 32 bits. (Here, as is the case for driver authors, we refer to peripheral chip - not CPU - registers. These discussions, though, equally apply to CPU registers with even a 64-bit width, as we'll point out.) These device registers provide control, status, and other information and are often programmable. This, in fact, is largely what you as a driver author will do - program the device registers appropriately to make the device do something, and query it.

To flesh out this discussion, let's consider a hypothetical device that has two registers: a status register and a control register, each 8 bits wide. (In the real world, every device or chip has (at least it should have) a *datasheet* that will provide a detailed specification of the chip and register-level hardware; this becomes an essential document for the driver author.) Hardware folks usually design devices in such a way that several registers are sequentially clubbed together in a larger piece of memory; this is sometimes called *register banking*. By having the base address of the first register and the offset to each following one, it becomes easy to address any given register (here, we won't delve into how exactly registers are "mapped" into the virtual address space on an OS such as Linux; this is part of the coverage in the *Linux kernel Programming - Part 2* book). For example, the (purely hypothetical) registers for our purely hypothetical device may be described like this within a header file:

```
#define REG_BASE        0x5a00
#define STATUS_REG      (REG_BASE+0x0)
#define CTRL_REG        (REG_BASE+0x1)
```

Now, say that in order to turn on some feature in our fictional device, the datasheet informs us we can do so by setting bit 7 (the MSB) of the control register to `1`. As every driver author quickly learns, there is a hallowed sequence for modifying registers, as shown below:

1. **Read** the register's current value into a temporary variable.
2. **Modify** the temporary variable to the desired value.
3. **Write** back the temporary variable to the register.

This is often called the **RMW – Read-Modify-Write – sequence.** So great, we write the (pseudo) code like this:

```
turn_on_feature_x_dev()
```

```
{
    u8 tmp;

    tmp = ioread8(CTRL_REG);    /* read: the current register value into tmp */
    tmp |= 0x80;                              /* modify: set bit 7 (MSB) of tmp */
    iowrite8(tmp, CTRL_REG);     /* write: the new tmp value into the register
*/
}
```

(FYI, the actual routines used on Linux **MMIO – memory-mapped I/O** – to perform I/O are `ioread[8|16|32]()` and `iowrite[8|16|32]().`)

A key point here: *this isn't good enough*; the reason is **concurrency and data races!** Think about it: a register (both CPU and device registers) is in fact a *global shared writable memory location*; if the code path where it's accessed can run concurrently – and this is usually the case – then it *constitutes a critical section*, which you have to take care to protect from concurrent access! The *how* is easy; we could just use a spinlock (for now at least). It's trivial to modify the preceding pseudocode to insert the `spin_[un]lock()` APIs in the critical section – the RMW sequence.

However, there is an even better way to achieve data safety when dealing with small quantities such as integers; we have already covered it: *atomic operators*! Linux, however, goes further, providing a set of atomic APIs for both of the following:

- **Atomic non-RMW operations** (the ones we saw earlier, in the *Using the atomic_t and refcount_t interfaces* section).
- **Atomic RMW operations**: These include several types of operators that can be categorized into a few distinct classes: arithmetic, bitwise, swap (exchange), reference counting, miscellaneous, and barriers.

Let's not reinvent the wheel; the kernel documentation (`https://www.kernel.org/doc/html/v6.1/core-api/wrappers/atomic_t.html`) has all the information required. We'll show just a relevant portion of this document as follows, quoting directly from there:

```
[ ... ]
Non-RMW ops:
  atomic_read(), atomic_set()
  atomic_read_acquire(), atomic_set_release()

RMW atomic operations:

Arithmetic:
  atomic_{add,sub,inc,dec}()
  atomic_{add,sub,inc,dec}_return{,_relaxed,_acquire,_release}()
  atomic_fetch_{add,sub,inc,dec}{,_relaxed,_acquire,_release}()
```

```
Bitwise:
  atomic_{and,or,xor,andnot}()
  atomic_fetch_{and,or,xor,andnot}{,_relaxed,_acquire,_release}()

Swap:
  atomic_xchg{,_relaxed,_acquire,_release}()
  atomic_cmpxchg{,_relaxed,_acquire,_release}()
  atomic_try_cmpxchg{,_relaxed,_acquire,_release}()

Reference count (but please see refcount_t):
  atomic_add_unless(), atomic_inc_not_zero()
  atomic_sub_and_test(), atomic_dec_and_test()

      Misc:
  atomic_inc_and_test(), atomic_add_negative()
  atomic_dec_unless_positive(), atomic_inc_unless_negative()
[ ... ]
```

Good; now that you're aware of these RMW (and non-RMW) operators, let's get practical and check out how to use RMW operators to safely perform bit operations.

Using the RMW bitwise operators

Here, we'll focus on employing the RMW bitwise operators; we'll leave it to you to explore the others (refer to the kernel docs mentioned). So let's think again about how to code our pseudocode example more efficiently. We can set (to `1`) any given bit in any register or memory item using the `set_bit()` API:

```
void set_bit(unsigned int nr, volatile unsigned long *p);
```

This API, atomically, meaning safely and indivisibly, sets the `nr`th bit (**nr** is simply short for **number**) of `p` to `1`. (The reality is that the device registers (and possibly device memory) are mapped into kernel virtual address space and, thus, appear to be visible as though they are RAM locations – such as the address `p` here. This is called MMIO and is the common way in which driver authors map in and work with device memory. Again, we cover this (and more) in the *Linux Kernel Programming – Part 2* book.)

Thus, with the RMW atomic operators, we can safely and correctly achieve what we've (incorrectly) attempted previously – turning on our (fictional) device feature – with a single line of code:

```
set_bit(7, CTRL_REG);
```

The following table summarizes common RMW bitwise atomic APIs:

RMW bitwise atomic API	Comment
`void set_bit(unsigned int nr, volatile unsigned long *p);`	Atomically set (set to `1`) the `nr`th bit of `p`

`void clear_bit(unsigned int nr, volatile unsigned long *p)`	Atomically clear (set to `0`) the `nr`th bit of `p`
`void change_bit(unsigned int nr, volatile unsigned long *p)`	Atomically toggle the `nr`th bit of `p`
The following APIs return the previous value of the bit being operated upon (`nr`)	
`int test_and_set_bit(unsigned int nr, volatile unsigned long *p)`	Atomically set the `nr`th bit of `p`, returning the previous value (the kernel API docs can be found at `https://docs.kernel.org/core-api/kernel-api.html?highlight=test_and_set_bit#c.test_and_set_bit`)
`int test_and_clear_bit(unsigned int nr, volatile unsigned long *p)`	Atomically clear the `nr`th bit of `p`, returning the previous value
`int test_and_change_bit(unsigned int nr, volatile unsigned long *p)`	Atomically toggle the `nr`th bit of `p`, returning the previous value

Table 13.2: Common RMW bitwise atomic APIs

Careful: These atomic APIs are atomic with respect to the CPU core they're running upon, but *not* with respect to other cores. In practice, this implies that if you're performing atomic operations in parallel on multiple CPUs – that is, if they (can) race – then it's a critical section, and you must protect it with a lock (typically a spinlock)!

Trying out a few of these RMW atomic APIs will help build your confidence in using them; we do so in the section that follows.

Using bitwise atomic operators – an example

Let's check out a quick kernel module that demonstrates the usage of the Linux kernel's RMW atomic bit operators (the code's here: `ch13/1_rmw_atomic_bitops`). You should realize that these operators can work on *any memory*, both upon a (CPU or device) register or on RAM. Here, to keep it simple, we operate on the latter, a simple static global variable (named `mem`) within the example LKM. It's very simple; let's check it out:

```
// ch13/1_rmw_atomic_bitops/rmw_atomic_bitops.c
[ ... ]
#include <linux/spinlock.h>
#include <linux/atomic.h>
#include <linux/bitops.h>
#include "../../convenient.h"
[ ... ]
static unsigned long mem;        // the memory variable we operate upon
static u64 t1, t2;
static int MSB = BITS_PER_BYTE - 1;
DEFINE_SPINLOCK(slock);
```

We include the required headers and declare and initialize a few global variables (notice how our `MSB` variable uses the `BITS_PER_BYTE` macro). We employ a simple macro, `SHOW_MEM()`, to display the current value of the RAM variable we operate upon, `mem`, formatted via the `printk`. The module's `init` code path is where the actual work is done:

```
[ ... ]
#define SHOW_MEM(index, msg) do { \
    pr_info("%2d:%27s: mem : %3ld = 0x%02lx\n", index, msg, mem, mem); \
} while (0)
[ ... ]
static int __init atomic_rmw_bitops_init(void)
{
        int i = 1, ret;

        pr_info("%s: inserted\n", OURMODNAME);
        SHOW_MEM(i++, "at init");

        setmsb_optimal(i++);
        setmsb_suboptimal(i++);

        clear_bit(MSB, &mem);
        SHOW_MEM(i++, "clear_bit(7,&mem)");

        change_bit(MSB, &mem);
        SHOW_MEM(i++, "change_bit(7,&mem)");

        ret = test_and_set_bit(0, &mem);
        SHOW_MEM(i++, "test_and_set_bit(0,&mem)");
        pr_info("       ret = %d\n", ret);

        ret = test_and_clear_bit(0, &mem);
        SHOW_MEM(i++, "test_and_clear_bit(0,&mem)");
        pr_info("       ret (prev value of bit 0) = %d\n", ret);

        ret = test_and_change_bit(1, &mem);
        SHOW_MEM(i++, "test_and_change_bit(1,&mem)");
        pr_info("       ret (prev value of bit 1) = %d\n", ret);

        pr_info("%2d: test_bit(%d-0,&mem):\n", i, MSB);
        for (i = MSB; i >= 0; i--)
```

```
                pr_info("  bit %d (0x%02lx) : %s\n", i, BIT(i), test_bit(i,
&mem)?"set":"cleared");

        return 0;                /* success */
    }
```

Besides the `SHOW_MEM()` macro, the RMW atomic operators we use here are highlighted in a bold font. A key part of this demo is to show that using the RMW bitwise atomic operators is not only much easier but also much faster than using the traditional approach (where we manually perform the RMW operation within the confines of a spinlock). Here are the two functions for both of these approaches:

```
/* Set the MSB: optimally, via the set_bit() RMW atomic API */
static inline void setmsb_optimal(int i)
{
    t1 = ktime_get_real_ns();
    set_bit(MSB, &mem);
    t2 = ktime_get_real_ns();
    SHOW_MEM(i, mem, "optimal: via set_bit(7,&mem)");
    SHOW_DELTA(t2, t1);
}
/* Set the MSB: the traditional way, using a spinlock to protect
 *  the RMW critical section  */
static inline void setmsb_suboptimal(int i)
{
    u8 tmp;

    t1 = ktime_get_real_ns();
    spin_lock(&slock);
    /* critical section begins: RMW : read, modify, write */
    tmp = mem;
    tmp |= 0x80; // 0x80 = 1000 0000 binary
    mem = tmp;
    /* critical section ends */
    spin_unlock(&slock);
    t2 = ktime_get_real_ns();

    SHOW_MEM(i, mem, "set msb suboptimal: 7,&mem");
    SHOW_DELTA(t2, t1);
}
```

We call these functions early in our module's `init` method; notice that we take timestamps (via the `ktime_get_real_ns()` routine) and calculate and display the time taken via our `SHOW_DELTA()` macro (defined in our `convenient.h` header; do check it out: `https://github.com/PacktPublishing/Linux-Kernel-Programming_2E/blob/main/convenient.h`).

Right, here's the output (on a run I did; the time taken can vary, of course):

```
1_rmw_atomic_bitops: inserted
 1:                    at init: mem :   0 = 0x00
 2:optimal: via set_bit(7,&mem): mem : 128 = 0x80
delta: 29 ns
 3: set msb suboptimal: 7,&mem: mem : 128 = 0x80
delta: 125 ns
 4:             clear_bit(7,&mem): mem :   0 = 0x00
 5:            change_bit(7,&mem): mem : 128 = 0x80
 6:      test_and_set_bit(0,&mem): mem : 129 = 0x81
         ret = 0
 7:    test_and_clear_bit(0,&mem): mem : 128 = 0x80
         ret (prev value of bit 0) = 1
 8:test_and_change_bit(1,&mem): mem : 130 = 0x82
         ret (prev value of bit 1) = 0
 9: test_bit(7-0,&mem):
  bit 7 (0x80) : set
  bit 6 (0x40) : cleared
  bit 5 (0x20) : cleared
  bit 4 (0x10) : cleared
  bit 3 (0x08) : cleared
  bit 2 (0x04) : cleared
  bit 1 (0x02) : set
  bit 0 (0x01) : cleared
```

Figure 13.3: Screenshot of output from our ch13/1_rmw_atomic_bitops LKM on an x86_64 VM, showing off some of the atomic RMW operators at work (the dmesg timestamp is filtered out)

I ran this demo LKM on:

- An x86_64 Ubuntu 23.04 VM running our custom 6.1.25 kernel; these are the results:
    - Via a modern `set_bit()` API: 29 nanoseconds
    - Via the traditional approach: 125 ns (over 4 times slower!)
- A Raspberry Pi 4 AArch64 board running our custom 6.1.25 kernel; these are the results:
    - Via a modern `set_bit()` API: 240 nanoseconds
    - Via the traditional approach: 908 ns (over 3.7x slower!)

It's clear: the modern approach – via the `set_bit()` RMW atomic bitwise API – as opposed to the older, "manual" RMW approach using a spinlock for protection, is a lot faster and easier.

The coding effort, via a single call to `set_bit()`, as opposed to taking a spinlock, performing each step (the RMW sequence), and then releasing the spinlock, is so much simpler as well. Steps 4 to 9 (refer to *Figure 13.3*) show various typical bit operations being carried out on the memory variable (and their result), via the powerful and really easy-to-use RMW bitwise atomic operators.

Efficiently searching a bitmask

On a somewhat related note to the atomic bitwise operators, this section is a very brief look at the highly efficient APIs available within the kernel for searching a bitmask – a fairly common operation in the kernel, as it turns out.

Several algorithms depend on performing a really fast search of a bitmask; several scheduling algorithms (such as `SCHED_FIFO` and `SCHED_RR`, which you learned about in *Chapter 10, The CPU Scheduler – Part 1*, and *Chapter 11, The CPU Scheduler – Part 2*) often internally require this. Implementing this efficiently becomes important (especially for OS-level, very performance-sensitive code paths). Hence, as the following list shows, the kernel provides a few APIs to scan a given bitmask (these prototypes are found in the `include/linux/find.h` header):

- `unsigned long find_first_bit(const unsigned long *addr, unsigned long size)`: Finds the first set bit in a memory region and returns the bit number of the first set bit; otherwise (no bits are set), it returns `@size`.
- `unsigned long find_first_zero_bit(const unsigned long *addr, unsigned long size)`: Finds the first cleared bit in a memory region and returns the bit number of the first cleared bit; otherwise (no bits are cleared), it returns `@size`.
- Other routines include `find_next_bit()`, `find_next_and_bit()`, `find_last_bit()`, and several others. Also, looking through this header reveals other quite interesting macros as well, such as `for_each_{clear,set}_bit{_from}()`.

With that, we complete our coverage on RMW atomic operators and move on to another interesting synchronization technology (with pros and cons): the reader-writer spinlock.

Using the reader-writer spinlock

Visualize a piece of kernel (or driver) code wherein a large, global data structure – say, a doubly linked circular list with a few thousand nodes or more – is being searched. Now, since this data structure is global (shared and writable), accessing it concurrently constitutes a critical section, and that requires protection.

Assuming a scenario where searching the list is a non-blocking operation, you'd typically use a spinlock to protect the critical section.

A naive approach might propose not using a lock at all, since we're *only reading data* within the list, not updating it. But, of course, even a read on shared writable data has to be protected in order to protect against an inadvertent write occurring simultaneously (as you have learned – refer back to the previous chapter if required), thus resulting in a dirty or torn read.

So we conclude that we require the spinlock; we imagine the pseudocode might look something like this:

```
spin_lock(mylist_lock);
for (p = &listhead; (p = next_node(p)) != &listhead; ) {
```

```
    << ... search for something ...
        found? break out ... >>
}
spin_unlock(mylist_lock);
```

So what's the problem? Performance, of course! Imagine several threads on a multicore system ending up at this code fragment more or less at the same time; each will attempt to take the spinlock, but only one winner thread will get it, iterate over some amount of the list, and then perform the unlock, allowing the next thread to proceed. In other words, as expected, execution is now *serialized*, very likely dramatically slowing things down. But, you might say, it can't be helped – or can it?

Enter the **reader-writer spinlock**. With this locking construct, it's required that all threads performing reads on the protected data object will ask for a (shared) **read lock**, whereas any thread requiring write access to the data object will ask for an **exclusive write lock**. A read lock will be granted immediately to any thread that asks for it *as long as no write lock is currently in play*. In effect, this construct *allows all readers concurrent access to the data, meaning, in effect, no real locking at all*. This is fine, as long as there are only readers. The moment a writer thread comes along, it will request a write lock. Now, normal locking semantics apply: the writer **will have to wait** for all readers to unlock. Once that happens, the writer gets an exclusive write lock and proceeds. So now, if any readers *or* writers attempt access, they will be forced to wait, to spin upon, the writer's unlock.

Thus, for those situations where the access pattern to data is such that reads are performed very often and writes are rare, that is, *read-mostly*, and the non-blocking critical section is a fairly long one, the reader-writer spinlock would appear to be a good fit. (There are some downsides; what exactly they are, and the alternative, are examined in a later section.)

Reader-writer spinlock interfaces

Having used spinlocks, using the reader-writer variant is straightforward; the lock data type is abstracted as the `rwlock_t` structure (in place of `spinlock_t`), and in terms of API names, simply substitute `read` or `write` in place of `spin`:

```
#include <linux/rwlock.h>
rwlock_t mylist_lock;
```

The most basic APIs of the reader-writer spinlock are as follows:

```
void read_lock(rwlock_t *lock);
void write_lock(rwlock_t *lock);
```

As one example, the kernel's `tty` layer has code to handle a **Secure Attention Key (SAK)**. The SAK is a security feature, a means to prevent a Trojan horse-type credentials hack; it works by killing all processes associated with the TTY device when the user presses the SAK (`https://www.kernel.org/doc/html/latest/security/sak.html`).

When this happens (that is, when the user presses the SAK, mapped via the kernel's so-called *magic SysRq* facility to the `Alt-SysRq-k` key sequence by default), within its code path, this SAK code will iterate over all tasks, killing the entire session and any threads that have the TTY device open. To do so, it must take, in read mode, a reader-writer spinlock meant for iterating over the task list, named `tasklist_lock`. The (truncated) relevant code is seen as follows, with `read_[un]lock()` on `tasklist_lock` highlighted:

```
// drivers/tty/tty_io.c
void __do_SAK(struct tty_struct *tty)
{
    [...]
    read_lock(&tasklist_lock);
    /* Kill the entire session */
    do_each_pid_task(session, PIDTYPE_SID, p) {
        tty_notice(tty, "SAK: killed process %d (%s): by session\n", task_pid_
nr(p), p->comm);
        group_send_sig_info(SIGKILL, SEND_SIG_PRIV, p, PIDTYPE_SID);
    } while_each_pid_task(session, PIDTYPE_SID, p);
    [...]
    /* Now kill any processes that happen to have the tty open */
    do_each_thread(g, p) {
        [...]
    } while_each_thread(g, p);
    read_unlock(&tasklist_lock);
```

Trying out the reader-writer spinlock

Back in *Chapter 6, Kernel Internals Essentials – Processes and Threads,* in the *Iterating over the kernel's task lists* section, we mentioned a simple module in our codebase – `ch6/list_demo` – to perform some very simple operations on a linked list (leveraging the kernel's `list.h` header's routines). However, in that demo module, to keep it simple, we didn't use any kernel synchronization constructs whatsoever. This, of course, is just plain wrong; we certainly need to protect critical sections against concurrent access (to prevent data races)! So now, armed with the (basic) reader-writer spinlock, let's protect it.

The code's here: `ch13/2_list_demo_rdwrlock`. Code-wise, besides the actual usage of the reader-writer spinlock, we do make another fairly significant change: we use our *Linux Kernel Programming – Part 2* the companion volume's `misc` device driver code as a wrapper of sorts over the list demo (thus, here, we use the approach of building more than one source file: the `misc` driver code's in `ch13/2_list_demo_rdwrlock/miscdrv_wrapper.c`, and the actual list work is carried out in `ch13/2_list_demo_rdwrlock/list_demo_rdwrlock.c`).

You may wonder, why encapsulate this `rwlock` demo module within a `misc` driver? It's pretty useful; this way, we can perform I/O – reads and writes – on the device node (and, thus, on our list!) whenever we want. Otherwise, we'd be limited to manipulating the list only in the module's `init/cleanup` methods via the usual `insmod/rmmod` utilities. This way, we can perform reads, where the code iterates over the list, and writes, where it will insert a few nodes into the list tail, at our whim (we use a simple wrapper Bash script named `run` to do precisely this).

Code-wise, here's the relevant portion of the list manipulation (`ch13/2_list_demo_rdwrlock/list_demo_rdwrlock.c`):

```
[ … ]
int add2tail(int v1, int v2, s8 achar, rwlock_t *rwlock)
{
    struct node *mynode = NULL;
    u64 t1, t2;

    mynode = kzalloc(sizeof(struct node), GFP_KERNEL);
    if (!mynode)
        return -ENOMEM;
    mynode->ival1 = v1;
    mynode->ival2 = v2;
    mynode->letter = achar;

    INIT_LIST_HEAD(&mynode->list);

    /*--- Update (write) to the data structure in qs; we need to protect
     * against concurrency, so we use the write (spin)lock passed to us
     */
    pr_info("list update: using [reader-]writer spinlock\n");
    t1 = ktime_get_real_ns();
    write_lock(rwlock);
    t2 = ktime_get_real_ns();
    // void list_add_tail(struct list_head *new, struct list_head *head)
    list_add_tail(&mynode->list, &head_node);
    write_unlock(rwlock);
    pr_info("Added a node (with letter '%c') to the list...\n", achar);
[ … ]
```

Clearly, the addition of a node to the tail of the list (via the `list_add_tail()` routine) is now protected against concurrent access via the write-side routine of our reader-writer spinlock, via the `write_{un}lock()` pair!

You can safely ignore the `ktime_get_real_ns()` calls for now; we use them to figure out the time delta between asking for the write lock and actually getting it. Why? As you'll soon see, the reader-writer spinlock often suffers from a problem – that of the writer starving as many readers keep coming up. Here, we attempt to measure the possible time lag.

What about when reading the list? Here's the relevant code (`ch13/2_list_demo_rdwrlock/list_demo_rdwrlock.c`):

```
void showlist(rwlock_t *rwlock)
{
    struct node *curr;
    u64 t1, t2;

    if (list_empty(&head_node))
        return;
    [ … ]
    t1 = ktime_get_real_ns();
    read_lock(rwlock);
    list_for_each_entry(curr, &head_node, list) {
        pr_info("%9d %9d   %c\n", curr->ival1, curr->ival2, curr->letter);
    }
#if 1
    mdelay(750); // deliberately make the reads slower to demo write
starvation..
#endif
    read_unlock(rwlock);
    t2 = ktime_get_real_ns();
[ … ]
```

Again, it should be clear: we now employ the read-side routine of our reader-writer spinlock, via the `read_{un}lock()` pair! Of course, we understand that, in the absence of any writers, it will allow all reader threads to run concurrently! (Thus, we also use the read spinlock when carrying out the search operation.) Do build this module and try it out.

As an aside, you may recall that, again, back in the *Iterating over the kernel's task lists* section in *Chapter 6, Kernel Internals Essentials – Processes and Threads,* we did something kind of similar: we wrote a kernel module (`ch6/foreach/thrd_show_all`) that iterated over all threads in the task list, spewing out a few details about each thread. So now that we understand the deal regarding concurrency, shouldn't we have taken the `tasklist_lock`, the reader-writer spinlock meant for protecting the task list against concurrent access? Yes, we did try to, but it didn't work (the `insmod` failed with the message `thrd_showall: Unknown symbol tasklist_lock ...`). The reason, of course, is that this `tasklist_lock` variable is *not* exported and, thus, is unavailable to our kernel module. So then, how exactly should we protect the critical section there?

In this particular case, the "right" answer is: by using the very powerful and performant "lock-free" **Read-Copy-Update** (**RCU**) mechanism! (We cover RCU in an upcoming section.)

As another example of reader-writer spinlock usage within the kernel codebase, the `ext4` filesystem uses one when working with its extent status tree. We don't intend to delve into the details here; we will simply mention the fact that a reader-writer spinlock (within its `inode` structure, `inode->i_es_lock`) is quite heavily used to protect the extent status tree against data races (`fs/ext4/extents_status.c`).

Just as with regular spinlocks, we have a few typical variations on the reader-writer spinlock APIs: `{read,write}_lock_irq{save}()` paired with the corresponding `{read,write}_unlock_irq{restore}()`, as well as the `{read,write}_{un}lock_bh()` interfaces; they follow similar and expected semantics, as do the corresponding spinlock routines. Note that – as with the regular `spin_lock_irq*()` APIs – even the read IRQ lock disables hardware interrupts and kernel preemption on the local core.

Performance issues with reader-writer spinlocks

Serious performance issues do exist with reader-writer spinlocks. One typical issue with it is that, unfortunately, **writers can starve** when blocking on several readers. Think about it: let's take a hypothetical case where, say, three reader threads currently hold (i.e., have acquired) a reader-writer spinlock. Now, a writer comes along wanting the lock. It must wait, of course, until *all* three readers perform the unlock. But what if, in the interim, more readers come along (which is entirely possible)? This becomes a disaster for the writer, who must now wait even longer – in effect, it can quite easily starve. (Carefully instrumenting or profiling the code paths involved or using the kernel's locking statistics (we cover this in an upcoming section) might be necessary to figure out whether this is indeed the case.)

Not only that, *cache effects* – known also as false sharing or cache ping-pong – can and do occur quite often when several reader threads on different CPU cores are reading the same shared state in parallel (while holding the reader-writer spinlock); we, in fact, discuss this in the upcoming *Understanding CPU caching basics, cache effects, and false sharing* section). The official kernel documentation on spinlocks (`https://www.kernel.org/doc/Documentation/locking/spinlocks.txt`) says pretty much the same thing. Here's a quote directly from it: "*NOTE! reader-writer locks require more atomic memory operations than simple spinlocks. Unless the reader critical section is long, you are better off just using spinlocks.*" In fact, the kernel community is working toward removing reader-writer spinlocks as much as possible, moving to superior lock-free techniques (such as RCU, an advanced lock-free technology). Thus, gratuitous use of reader-writer spinlocks is ill advised.

The neat and spare kernel documentation on the usage of spinlocks (written by Linus Torvalds himself), which is well worth reading, is available here: `https://www.kernel.org/doc/Documentation/locking/spinlocks.txt`.

The reader-writer semaphore

We earlier mentioned the semaphore object (*Chapter 12, Kernel Synchronization – Part 1*, in the *The semaphore and the mutex* section), contrasting it with the mutex. There, you understood that it's preferable to simply use a mutex.

Here, we point out that within the kernel, just as there exist reader-writer spinlocks, so do there exist *reader-writer semaphores*. The use cases and semantics are similar to that of the reader-writer spinlock. The relevant macros/APIs are (within `<linux/rwsem.h>`) `{down,up}_{read,write}_{trylock,killable}()`. A common example within the `struct mm_struct` structure (which is itself pointed to from within the task structure) is that one of the members is a reader-writer semaphore: `struct rw_semaphore mmap_lock` (`https://elixir.bootlin.com/linux/v6.1.25/source/include/linux/mm_types.h#L580`).

Rounding off this discussion, we'll merely mention a couple of other related synchronization mechanisms within the kernel. A synchronization mechanism that is heavily used in user-space application development (we're thinking particularly of the Pthreads framework in Linux user space) is the **Condition Variable** (CV). In a nutshell, it provides the ability for two or more threads to synchronize with each other based on the value of a data item or some specific state. Its equivalent within the Linux kernel is called the ***completion mechanism***. Please find details on its usage within the kernel documentation here: `https://www.kernel.org/doc/html/latest/scheduler/completion.html#completions-wait-for-completion-barrier-apis`.

Then there is the ***sequence lock***; it's used in write-mostly situations (as opposed to the reader-write spinlock/semaphore locks, which are suitable in read-mostly scenarios), places where the writes far exceed the reads on the protected variable(s). As you can imagine, this isn't a very common occurrence; a good example of using a sequence lock is the update of the `jiffies_64` global.

The so-called `jiifies` value translates to the `get_jiffies_64()` function, which, on 32-bit systems, reads in the 64-bit `jiffies_64` global data quantum under the protection of a sequence lock (so as not to perform a dirty read). (A bit simplistically) the `jiffies_64` global is updated – written to – on every timer interrupt. For the curious, the `jiffies_64` global's update code begins here: `kernel/time/tick-sched.c:tick_do_update_jiffies64()`. This function figures out whether an update to jiffies is required and, if so, calls `do_timer(++ticks);` to actually update it. All the while, the `write_seq[un]lock(&jiffies_lock);` APIs provide protection over this write-mostly critical section.

Great; now that we're done with this section on reader-writer spinlocks, let's move to another interesting topic, that of CPU caching and cache effects that can severely impact your code performance.

Understanding CPU caching basics, cache effects, and false sharing

Modern processors on multicore symmetric multi-processing (SMP) systems make use of several levels of parallel cache memory within them, in order to provide a very significant speedup when working on memory (we briefly touched upon this in *Chapter 8, Kernel Memory Allocation for Module Authors – Part 1*, in the *Allocating slab memory* section). FYI, this kind of computer architecture is often classified as a **Multiple Instruction, Single Data (MISD)** stream (as instructions can run concurrently in several cores while working upon a single shared data item).

Here's a purely conceptual diagram (*Figure 13.4*) showing two CPU cores, each core having two internal caches (Level 1 and Level 2, abbreviated as L1 and L2, respectively), plus a shared or unified L3 cache and the main memory (RAM):

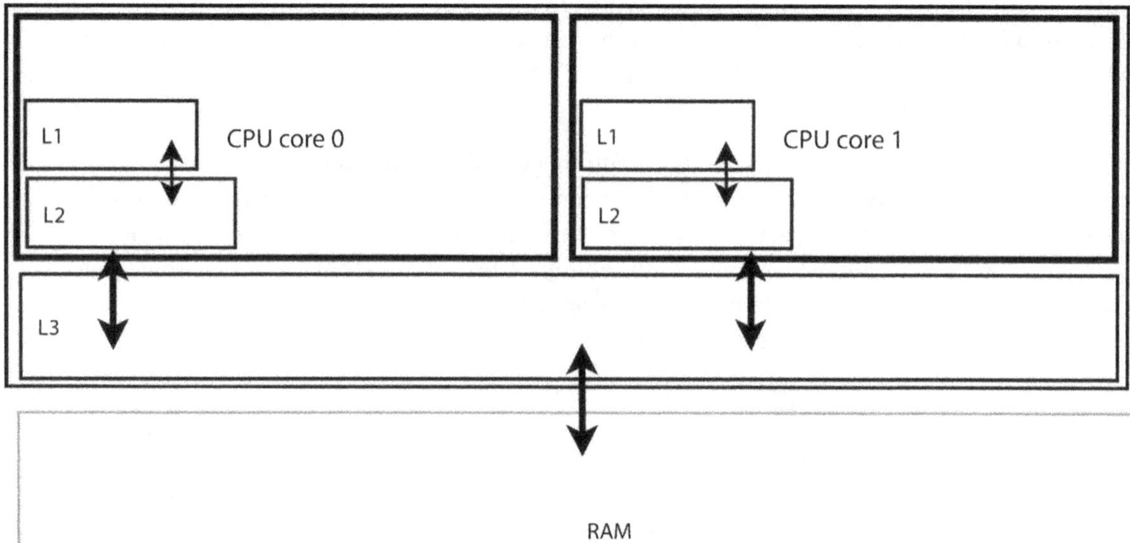

Figure 13.4: Conceptual diagram – 2 CPU cores with internal L1, L2 caches, a shared (unified) L3 cache, and RAM

Note how each CPU core has its own internal caches. We expand on its usage in the material that follows.

An introduction to CPU caches

We must realize that modern CPUs do *not* read and write (load and store) RAM content directly; no, when the software indicates that a byte of RAM is to be read starting at some address, the CPU atomically, at one shot, reads several bytes – a whole **cacheline** of bytes from the specified starting address into all the CPU caches (say, L1, L2, and L3). A typical processor cacheline size is 64 bytes (note carefully that it's *bytes*, not bits).

FAQ: What exactly is the size of my CPU's cacheline?

Answer: Several utilities can show this; a simple way on Linux is to use `getconf`; try `getconf -a |grep "CACHE_LINESIZE"` on your system (and grep for `CACHE_SIZE` to see the CPU cache sizes in bytes. Note that it may not work on a guest system though). Also recall, in *Chapter 7, Memory Management Internals – Essentials*, we mentioned that you can use the GUI *lstopo* utility to see your CPU and memory topography.

So think on this: say we have a code scenario like this:

```
char myarr[264], x; int i;
[ ... ]
```

```
for (i=0; i < 264; i++)
    x = myarr[i];
```

The naive thinking is that we're iterating over every single byte of the array in turn; the reality is more sophisticated! The moment we attempt to access the very first byte `myarr[0]`, the CPU firmware (for lack of a better term; perhaps the term microcode works as well) searches the CPU caches (L1, L2, and L3, in turn) for it; if found, we have a *cache hit* and the value is worked upon.

Let's say this array is being accessed by the process (or thread) for the very first time, thus it's not present in any of the CPU caches. Hence, we get a *cache miss;* the processor now issues a bus transaction: go across the bus to the memory controller and fetch the byte from RAM. Wait a minute! The processor actually has *an entire cacheline worth of bytes read (fetched) and they're then written into all the CPU caches!* So, assuming a CPU cacheline size of 64 bytes, when we access `myarr[0]` for the first time, bytes 0 to 63 of `myarr[]` are now fetched from RAM and then stored within (all) the CPU caches (see *Figure 13.5*).

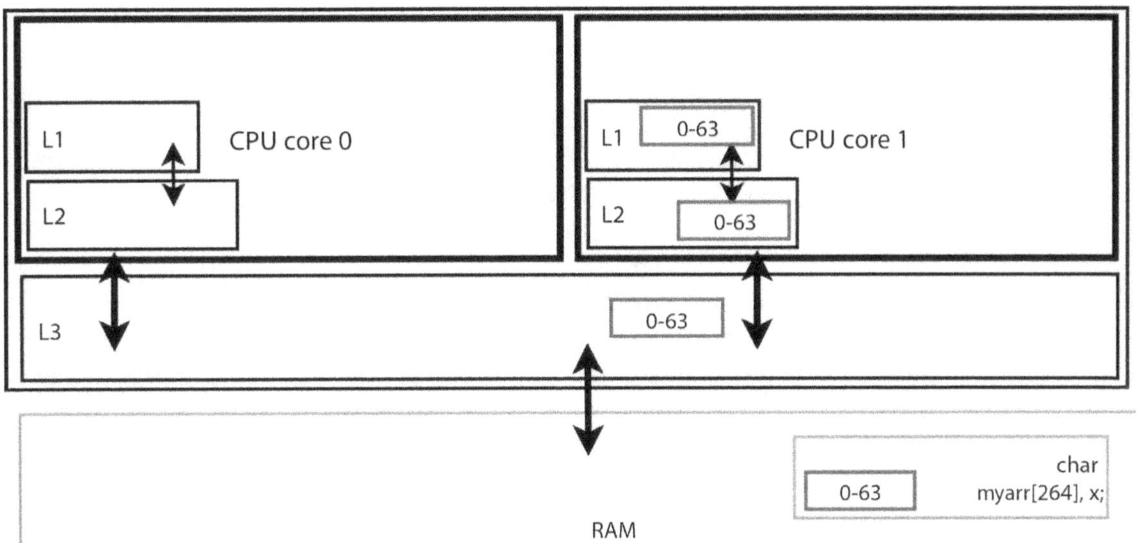

Figure 13.5: Conceptual diagram – a thread on CPU core 1 reads myarr[0] for the first time; the processor fetches all the data in the surrounding cacheline (typically 64 bytes, myarr[0] to myarr[63], simply shown as 0-63) into the shared L3 cache and the L2, L1 caches of core 1

What's the advantage? It should be clear: accessing the next few elements of sequential memory (right up to byte 63) results in a tremendous speedup, as it's first checked for in the caches (first in L1, then L2, and then L3, and a *cache hit* becomes likely).

The reason it's (much) faster is simple: accessing CPU cache memory typically takes one to a few (single-digit) nanoseconds (with L1 being the fastest and smallest, and the **Lowest Level Cache** (**LLC**), L3 here, being the slowest and largest (relatively speaking)). Accessing RAM can take anywhere between 50 and 100 nanoseconds (of course, the precise numbers depend on the hardware system in question and the amount of money you're willing to shell out!). This is the whole point of a cache of course: that the original memory being cached is much slower than the cache itself (think of a web browser caching web pages).

Furthermore, caching leverages the principles of *locality of reference*:

- Spatial locality of reference (in space): It's likely that software will access memory in sequence (sequentially).
- Temporal locality of reference (in time): It's likely that software will access memory that was recently accessed.

Practically speaking, we software folks take advantage of caching phenomena (and locality of reference) by doing things such as the following:

- Keeping important members of a data structure together (hopefully, within a single cacheline) and at the top of the structure
- Padding a structure member such that we don't "fall off" a cacheline (again, these points have been covered in *Chapter 8, Kernel Memory Allocation for Module Authors – Part 1*, in the *Data structures – a few design tips* section).

The risks – cache coherency, performance issues, and false sharing

However, risks are involved and things can and do go wrong. One risk is to do with what's termed cache coherency, and the other is often termed false sharing; the following sections delve into them.

What is the cache coherency problem?

Let's first understand the key notion of *cache coherency*. Multicore SMP systems have two or more CPU cores, each with its own private internal caches (and **Translation Lookaside Buffers (TLBs)**). The caches typically follow an optimized **read-ahead, write-behind policy** (with respect to main memory, RAM). This means, of course, that data can be read into the CPU caches in a read-ahead manner (as we just learned, reads and writes actually work in an atomic unit of a cacheline, not a single byte. So the read-ahead performed is to optimize likely subsequent fetches, taking advantage of the principles of both spatial and temporal locality of reference, which in turn simply means that "memory worked upon recently is likely to be worked upon again in the near future," both in terms of space (spatial locality) and time (temporal locality)).

What about writes to memory? They're governed via a *cache writeback policy*. When a data item (a variable in your program) is modified (written to), it's worked upon within the internal caches of the local core and *not* flushed to (the possible unified cache and) the main memory (RAM) until a later point in time, again optimizing performance; this is the so-called *write-behind* policy in action (as opposed to a *write-through* policy where it's immediately flushed).

We'll take a simple example. Take a multicore SMP system with 2 cores (again, like the one conceptually shown in *Figure 13.4*). Say there's a global integer *N* in RAM with the value 55. Now, at time t0.1 (see *Table 13.3*), let's say a thread running on core #0 wants to modify it to a value – say, 41; we now understand how it will work internally. Assuming there are CPU caches as shown in *Figure 13.4* (L1, L2, and shared L3), the value of N (55) is first read from RAM (across the buses) into the L3 cache (time t0.2) and into core 0's local internal (L1, L2) caches (time t0.3).

Then (at time t1.1), it's modified there – *not* in L3 or RAM – to the value 41.

Okay, but now, what if (at time t2.1) a thread on core 1 also wishes to work on the same data (remember, it's a shared memory system, following SMP and MISD semantics); it wants to increment its value by 1 (we understand this is a critical section. Lets assume we use some synchronization primitive to ensure its done safely; that's not the what the discussion here's about).

The problem now becomes obvious: to do so, it must read in the current value of N and add 1 to it. But if it reads in the value from RAM (55), it's stale; the "latest" value isn't 55, it's 41, the value in core 0's internal caches! So we now see that we can have a *data inconsistency issue* when the same data item is shared between different levels in the system's memory hierarchy (here, between multiple cores, local L1 and L2 caches, a unified L3 cache and RAM) and is modified in one cache. This is the heart of the *cache coherency problem.* Keeping all copies of this data object in sync (consistent) with main memory is a requirement for correct computing; this ability to keep the processor caches and RAM in sync is called *cache coherence.*

Maintaining CPU cache coherency is one of the most power-hungry processes (i.e., it has a significant impact on battery-backed systems and even otherwise).

A vertical timeline showing how the data values of our example variable *N* propagates across the memory hierarchy is seen in *Table 13.3*:

State	**Time/ value of N in:**	**CPU core 0 L1, L2**	**CPU core 1 L1, L2**	**Unified L3 cache**	**RAM**
Initial state (N); core 0 reads N (value 55)	t0.1	-	-	-	55
	t0.2	-	-	**55**	55
	t0.3	**55, 55**	-	55	55
Core 0: set N to 41	t1.1	**41, 41**	-	55	55
	t1.2	41, 41	-	**41**	55
	t1.3	41, 41	-	41	**41**
Core 1: N ++;	t2.1	41, 41	**41, 41**	41	41
	t2.2	41, 41	**42, 42**	41	41
	t2.3	41, 41	42, 42	**42**	41
	t2.4	41, 41	42, 42	42	**42**

Update core 0's L1, L2	t3	**42, 42**	42, 42	42	42

Table 13.3: Vertical timeline showing how data values propagate across the memory hierarchy for even simple operations, due to the need for cache coherence

A step-by-step account of the various transactions that occur in *Table 13.3* is shown below:

1. Time t0.1: In sync (N=55):
    - t0.2: The thread on core 0 reads in N from RAM – an L3 cache entry setup.
    - t0.3: An L1 and L2 cache entries setup (value 55 is now in all that core's internal caches and in L3).
2. Time t1.1: The thread on core 0 modifies N from 55 to 41; initially, only core 0's internal caches reflect the change.
3. Time t2: A thread on core 1 (safely) performs N ++.
4. Now, we need the updated value of N.
5. The processor (via its hardware cache coherence protocol) **detects** the caches/RAM are out of sync; so to sync, it does this:
    - t1.2: It invalidates and flushes core 0's caches to L3.
    - t1.3: It invalidates and flushes core 0's caches to RAM (A cache invalidation implies marking the cache memory/line(s) as invalid so that subsequent reads occur from the original memory, the one being cached (typically, RAM).)
6. Now that the **RAM and caches are in sync**, we can go ahead with the work on core 1; so the work and cache coherency process repeats there:
    - t2.1: Reads in N from RAM into core 1's internal caches.
    - t2.2: Increment N; core 1's internal caches are updated from 41 to 42.
    - t2.3: Flush the updated value (42) to the shared L3 cache.
    - t2.4: Flush (*N*=42) to RAM.
7. On core 1's caches, the stale value's now updated (from the new updated value in RAM), worked upon, and flushed again to reflect the new reality.
8. t3: Detects that core 0's internal caches are out of sync and performs the update.
9. Now that the new value's set, all CPU caches are coherent with each other and with RAM!

You, alert reader, will now start to see that this cache coherence handling is great but performance-wise (and power-wise) can be very expensive! You can literally see how, for even simple memory transactions (reads/writes, loads/stores) on SMP, it often necessitates multiple write-invalidate and/or flush sequences, just to keep everything in sync.

This cache behavior, where a write from cache-to-RAM then requires a corresponding RAM-to-cache update on other cores, is often called "**cache bouncing**" or **cache "ping-pong"**! (*Figure 13.7*, which we will see soon, shows how the issue is exacerbated on multicore SMP systems with large numbers of cores.) This is really not good for performance/power, and this really is the point being conveyed. (Worry not, the next section on false sharing goes into these aspects in a bit more detail.)

As an interesting aside, how exactly is the cache coherence problem handled by the system? A typical approach is to use a hardware-based *cache coherence protocol*; there are several (snooping or bus-watching protocols, named MESI, MOESI, Illinois, and so on, and then there are the so-called directory-based protocols). Suffice it to say that it's very arch-dependant, with the modern AMD64/x86_64 and most ARM processors typically using the bus snooping-based MOESI hardware protocol. (The letters are abbreviations for the different cache states that a cacheline can be in, so *MOESI* is *Modified Owned Exclusive Shared Invalid*. The protocol works by essentially detecting a shared copy on other caches over the shared internal bus. FYI, the ARM Cortex-A9 uses the MESI protocol.) Some processors use a software coherence scheme (usually implemented deep within the OS) to keep caches coherent (like MIPS); some use a mix of hardware and software. We won't delve further into these aspects; do see the links in the *Further reading* section for more on this.

The false sharing issue

So the cache coherence mechanism seems, to some extent at least, to be all well and good, keeping memory in sync across processor cores, caches, and RAM. Bad cases – performance- and power-wise – can arise though. As a way to understand the issue, consider two variables declared like so:

```
u16 ax = 1, bx = 2;
```

(where `u16` denotes an unsigned 16-bit integer value).

Now, as they have been declared adjacent to each other, *the compiler will, in all likelihood, have them occupy the same CPU cacheline at runtime* (their combined size is just 32 bits or 4 bytes; recall that a typical CPU cacheline size is 64 bytes).

So what? To understand what the issue is, let's take a common runtime scenario: consider, again, a multicore system with two CPU cores, with each core having two CPU internal caches, *L1* and *L2*, as well as a unified *L3* cache, and RAM (again, like *Figure 13.4* shows).

Now, visualize that, on core `0`, a thread, *T1*, is working on variable `ax`, and in parallel, on core `1`, another thread, *T2*, is working on variable `bx`. Note carefully: there's no data race here, as each thread's working on a different variable, a different memory item; hence, there's no critical section and, thus, no need for synchronization.

Now, when thread *T1*, running on core `0`, accesses (reads or loads) `ax` from main memory (RAM), it's *L1* and *L2* CPU-local caches (and *L3* as well) that get populated with the current values of `ax` *and* `bx`, as both variables fall within the same cacheline!

Similarly, and concurrently, when thread *T2*, on core 1, accesses (reads or loads) bx from RAM, it's CPU-local caches that will get populated with the current values of *both* variables as well. *Figure 13.6* conceptually depicts the situation:

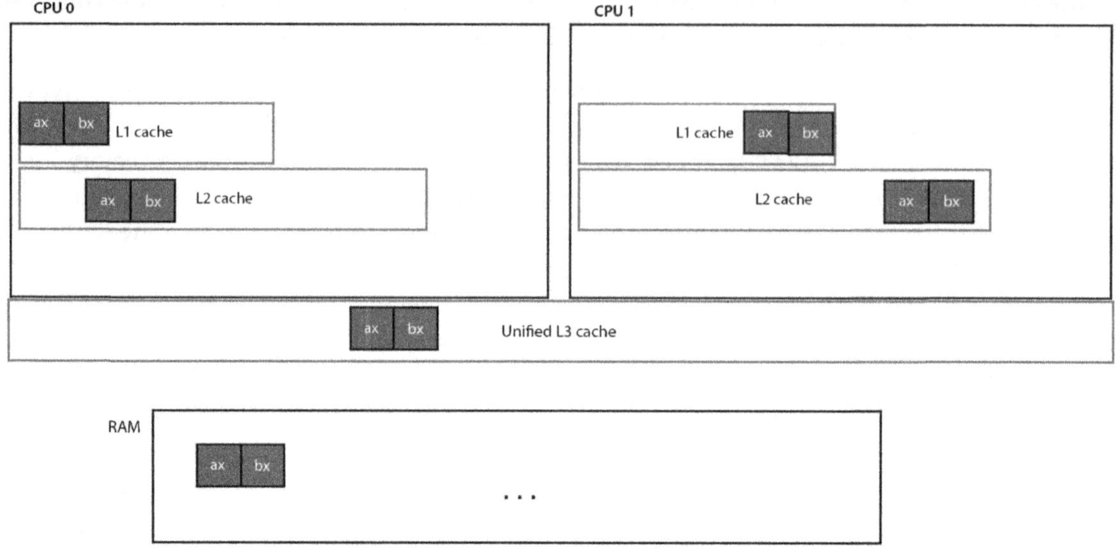

Figure 13.6: Conceptual depiction of the CPU cache memory when threads T1 and T2 work in parallel on two adjacent variables, each on a distinct one

Seems fine so far? Yes. But what if *T1* performs an operation, say, ax ++, while concurrently, *T2* performs bx ++? Well, so what? (As already mentioned, an interesting thing here is, synchronization (locking) isn't required, and is thus quite irrelevant to this discussion; there's no data race, as each thread accesses a different variable. The issue lies with the fact that the data items in question live in the same CPU cacheline.)

The issue lies in keeping the caches coherent. The processor and/or the OS in conjunction with the processor (this is all very arch-dependent stuff) *will have to keep the caches and RAM synchronized or coherent with each other*. Thus, the moment that thread *T1* modifies ax, that particular cacheline of CPU 0 will have to be invalidated, and, a core 0-caches-to-RAM flush of the CPU cacheline will occur to update RAM to the new value, and then immediately, a RAM-to-core-1-caches update must also occur to keep everything coherent!

Think: in a "normal" situation, we wouldn't invalidate and flush the caches the moment an update is done; this defeats the primary purpose of caching (which is performance). This is, of course, the whole point of the typical *cache write-behind* policy that most modern processors use. But here, there's a problem: a *shared cacheline* is being manipulated concurrently, necessitating that the invalidation+-flushes happen almost instantly, thus causing performance headaches.

It gets worse: the cacheline contains bx as well, and as we said, bx has also been modified on core 1 by thread *T2*. Thus, at about the same time, the CPU 1 cacheline will be invalidated and flushed to RAM with the new value of bx and subsequently updated to CPU 0's caches (all the while, the unified L3 cache too will be read from/updated as well).

As you can imagine, any updates on these variables will result in a whole lot of traffic over the caches and RAM; in effect, they will "bounce." In fact, this is often referred to as **cache ping-pong**! This effect is very detrimental to performance, significantly slowing down processing; furthermore, increased cache coherency work results in more power consumption. This phenomenon is (also) known as **false sharing** (perhaps a better term would be *oversharing*). Not just that, it will occur *every time* an update occurs on any variable within the cacheline concerned and can, thus, become terribly expensive.

Fixing it

Recognizing false sharing is the hard part; we must look for *variables living on a shared cacheline that are updated by different contexts (threads or whatever else) simultaneously.*

Interestingly, an earlier implementation of a key data structure in the memory management layer, `include/linux/mmzone.h:struct zone`, suffered from this very false sharing issue: as two spinlock variables were declared adjacent to each other and, thus, usually ended up sharing the same cacheline! This has long been fixed (we briefly discussed memory zones in *Chapter 7, Memory Management Internals – Essentials,* in the *Physical memory* section).

Measuring performance carefully, comparing against a "normal" baseline, can help reveal outliers and perhaps help you recognize that a slowdown is due to false sharing.

My highly able technical reviewer, Chi Thanh Hoang, adds that using an often overlooked but very useful processor feature – its performance measurement counters (PMCs) – to fine-tune cache-miss/flush events can definitely help in narrowing down performance issues; use tools like perf (and eBPF) to do this.

How do you fix this kind of false sharing issue? Easy: just ensure that the variables are spaced far enough apart to guarantee that they *do not share the same cacheline* (dummy padding bytes are often inserted between variables for this purpose). So if you did this, it would likely be fine (on a system with 64-byte cachelines):

```
u16 ax = 1;
char padding[64];
u16 bx = 2;
```

Can't the compiler help? Well, it depends; some recommend using pragma directives (GCC refers to them as compiler attributes, of the form `__attribute__(foo)`), which can help, but there seems to be no solid consensus on this. Do check out the references to false sharing in the *Further reading* section as well.

Next, to show how it can get even worse, consider a large (SMP) NUMA system with several sockets and physical (perhaps threaded) CPU cores within it (we covered NUMA in *Chapter 7, Memory Management Internals – Essentials*, in the *Physical RAM organization* section; see *Figure 7.22* as an example).

Now, visualize a multithreaded application running on the system; let's say we spawn one thread per CPU core (a fairly common technique). Even having each thread simply increment a global integer atomically on such a NUMA system can incur huge performance losses (and again, it's a very power-hungry thing). Why? Think about this: if every thread, and thus every core, operates on this same shared data item, even though the accesses may be carefully synchronized to be atomic and thus safe (using `refcount_t` / `atomic_t` operators perhaps), the variable still *has* to be cached within each core's local caches when working upon it, and, of course, cache coherence across every core that works on the same data item must be maintained! This can cause a huge amount of traffic in both directions: cache invalidations and flushes to RAM and updates from RAM to caches of other cores that are working on this data! So interestingly, here the issue is not about data safety (the thing we've been primarily concerned about) but about performance (and power consumption). Ironically, the more sockets and the CPU cores within them there are, the worse it scales.

Paul E. McKenney, in a presentation (*What is RCU, Prof Paul E. McKenney, 2013; YouTube video (at IISc, Bangalore):* `https://www.youtube.com/watch?v=obDzjElRj9c&list=PLIlI4QbmdBqH9-L1QOq6O5Yxt-YVjRsFZ`), shows this situation as follows:

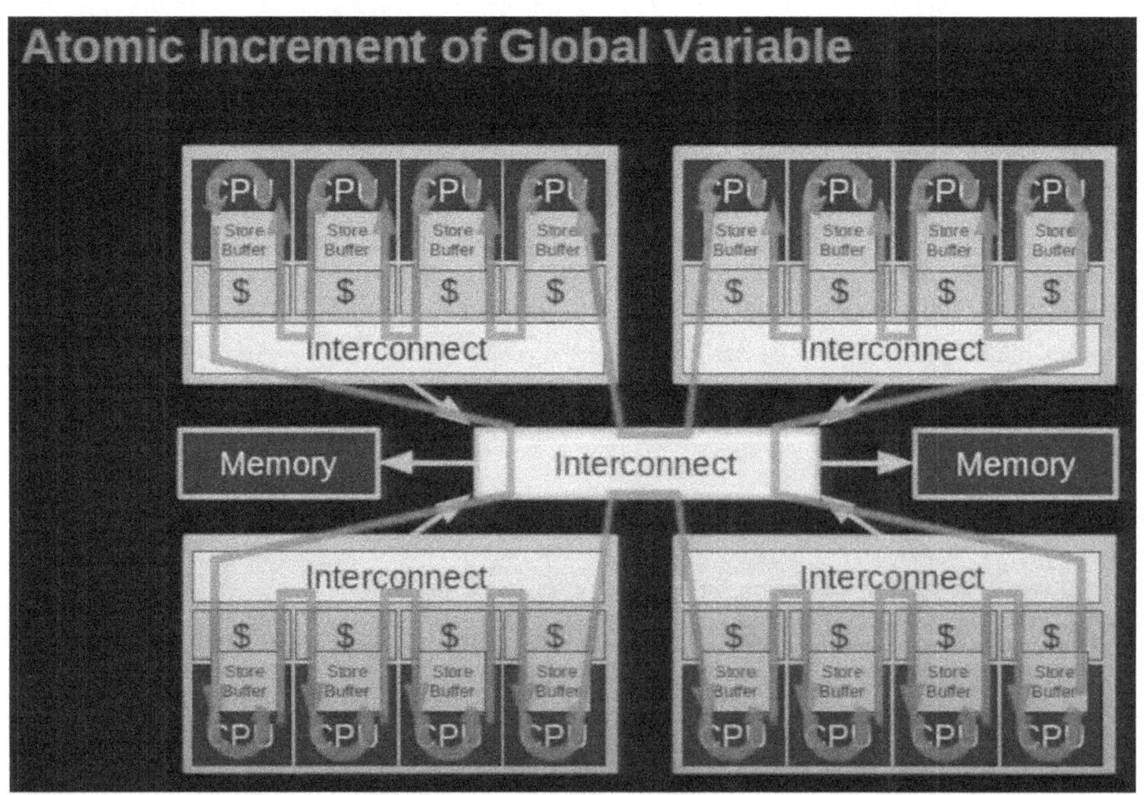

Figure 13.7: Working on shared data on a large NUMA system with many cores, although data safety can be assured, can incur large performance losses (and increased power usage) due to cache coherence/false sharing issues cropping up. Credit (c): What is RCU, Prof Paul E. McKenney

Now, as mentioned, the data accesses themselves may be atomic and thus safe, but the million-dollar question is: how on earth can we actually avoid these performance (and power) issues? *Lock-free* programming techniques can go a long way in helping, and the next section delves into just this.

Lock-free programming with per-CPU and RCU

As you have learned, when operating upon shared writable data, the critical section must be protected in some manner. Locking is perhaps the most common technology used to effect this protection. It's not all rosy, though, as performance can suffer.

To quite intuitively see why, consider a few physical-world analogies to a lock:

- One is a funnel, with the stem of the funnel – the "“critical section” – just wide enough to allow one thread at a time to flow through, and no more.
- Another is a single toll booth on a wide and busy highway, or a set of traffic lights at a busy intersection.

These analogies help us visualize and understand why locking can cause *bottlenecks*, slowing performance down to a crawl in some drastic cases. Worse, these adverse effects can be multiplied on high-end (SMP/NUMA) multicore systems (with a few hundred cores); in effect, locking doesn't scale well. Caching effects – as the previous section clearly revealed – add to the troubles.

Another issue is that of *lock contention*; how often is a particular lock acquired? Increasing the number of locks within a system has the benefit of lowering the contention for a particular lock between two or more processes (or threads). This is called **lock proficiency**. However, again, this is not scalable to an enormous extent: after a while, having thousands of locks on a system (the case with the Linux kernel, in fact) is not good news – the inherent complexity and, thus, the chances of subtle deadlock conditions arising significantly multiplies.

So, many challenges exist – performance issues, possible deadlock and/or priority inversion risks, convoying (due to lock ordering, fast code paths might need to wait for the first slower one that's taken a lock that the faster ones also require), cache effects, and so on. Evolving the kernel in a scalable manner while remaining performant has mandated the use of *lock-free algorithms* and their implementation within the kernel. These have led to several innovative techniques, among them being per-CPU data, lock-free data structures (by design), and **Read-Copy-Update** (**RCU**).

In this book, we will cover two very key, powerful, and popular lock-free technologies: per-CPU data and RCU. Others do exist of course (refer to the *Further reading* section of this chapter).

Per-CPU variables

As the name suggests, **per-CPU variables** work by keeping a copy of the variable, the data item in question, with one copy per (live) CPU core on the system. By doing this, in effect, we get rid of the problem area for concurrency, the critical section, *by avoiding the sharing of data between threads*. With the per-CPU data technique, since every CPU refers to its very own copy of the data, a thread running on that processor core can manipulate it without any worry of racing.

(This is very roughly analogous to local variables; as locals are on the private stack of each thread, they aren't shared between threads, there's no critical section, and thus no need for locking.) Here, too, the need for locking is thus eliminated – making it a *lock-free* technology!

So think about this: if you are running on a system with four live CPU cores, then a per-CPU variable on that system is essentially an array of four elements: element `0` represents the data value on the first CPU, element `1` the data value on the second CPU core, and so on (take a peek at *Figure 13.9*). Understanding this, you'll realize that per-CPU variables can be considered as also roughly analogous to the user-space Pthreads' **Thread Local Storage (TLS)** implementation, where each thread, at birth, automatically receives a copy of any and all (TLS) variables marked with the `__thread` keyword (there's more to it; do refer to the *Hands-On System Programming with Linux* book for details). There, and here with per-CPU variables, it should be obvious that typically, you should use per-CPU variables for small-sized data items only. This is because the data item is reproduced (copied) with one instance per CPU core (on a high-end system with a few hundred or even thousands of cores, the overheads do climb). So again, there is a trade-off: we get the ability to be lock-free at the cost of some more RAM usage.

In a nutshell, using per-CPU variables is good for performance enhancement on time-sensitive code paths because of the following:

- We avoid using costly, performance-busting locks.
- The access and manipulation of a given per-CPU variable is guaranteed to always occur on one particular CPU core; this, coupled with the fact that it's never shared, eliminates expensive cache effects such as cache ping-pong and false sharing (covered in the *Understanding CPU caching basics, cache effects, and false sharing* section). This also results in lower power consumption.

This latter point is worth delving into in more detail. As we saw in the previous section (refer to *Figure 13.7*), even simply incrementing a shared integer atomically via multiple threads on a NUMA system can incur huge performance penalties due to cache effects.

As already mentioned, Paul E. McKenney, in this presentation (*What is RCU, Prof Paul E. McKenney, 2013; YouTube video (at IISc, Bangalore)*: `https://www.youtube.com/watch?v=obDzjElRj9c&list=PLIlI4QbmdBqH9-L1QOq6O5Yxt-YVjRsFZ`), also shows that using the per-CPU lock-free programming technique can result in far superior performance, as this diagram of his indicates:

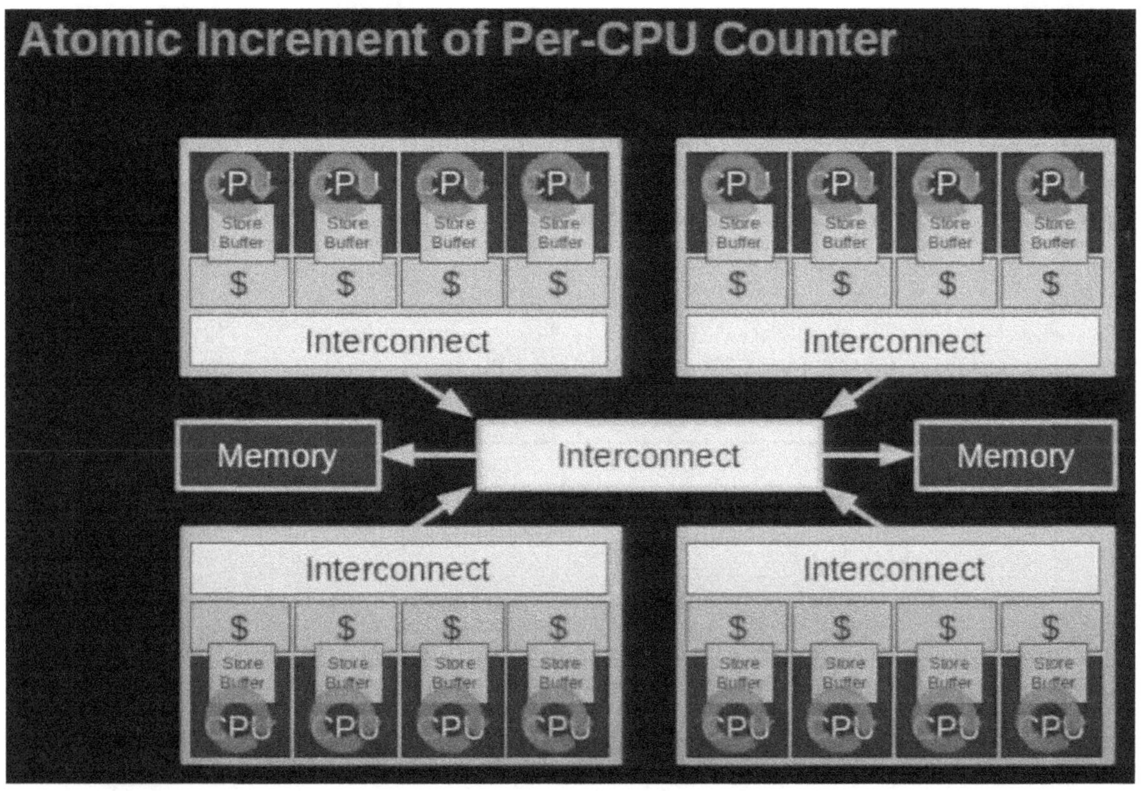

Figure 13.8: Working instead on per-CPU data on a large NUMA system with many cores assures both data safety and good performance, as caching effects cease to be a concern. Credit (c): What is RCU, Prof Paul E. McKenney

We shall mention some examples of per-CPU usage in the kernel codebase (in the *Per-CPU usage within the kernel* section).

Working with per-CPU variables

When working with per-CPU variables, you must use the helper methods (macros and APIs) provided by the kernel and not attempt to directly access them (much like we saw with the `refcount` and atomic operators).

Let's approach the helper APIs and macros (methods) for per-CPU data by dividing the discussion into two portions. First, you will learn how to allocate, initialize, and subsequently free a per-CPU data item. Then, you will learn how to work with it (reading/writing from/to it).

Allocating, initialization, and freeing per-CPU variables

There are broadly two types of per-CPU variables: statically and dynamically allocated ones. Statically allocated per-CPU variables have their memory allocated at compile time itself, typically via one of these macros: `DEFINE_PER_CPU()` or `DECLARE_PER_CPU()`. Using the `DEFINE` one allows you to allocate and initialize the variable in one shot. Here's an example of allocating a single integer named pcpa as a per-CPU variable; it will be auto-initialized to 0:

```
#include <linux/percpu.h>
DEFINE_PER_CPU(int, pcpa);        // signature: DEFINE_PER_CPU(type, name)
```

Now, on a system with, say, four CPU cores, this per-CPU variable would conceptually appear like this after initialization:

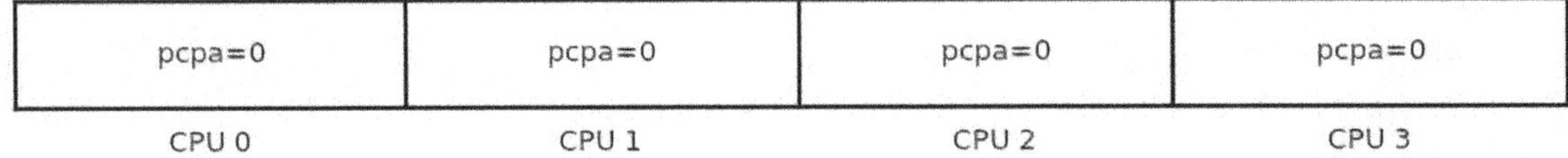

Figure 13.9: Conceptual representation of a per-CPU data item on a system with four live CPUs

The actual implementation is quite a bit more complex than this, of course; please refer to the *Further reading* section of this chapter to see more on the internal implementation.

Dynamically allocating per-CPU data can be achieved via the `alloc_percpu()` or `alloc_percpu_gfp()` wrapper macros, simply passing the data type of the object to allocate it as per-CPU and, for the latter, passing along the (now familiar from the dynamic kernel allocation APIs) GFP mask as well:

```
alloc_percpu[_gfp](type [,gfp]);
```

The underlying `__alloc_per_cpu[_gfp]()` routines (that these wrapper macros invoke) are exported via `EXPORT_SYMBOL_GPL()` (and, thus, can be employed only when the module is released under a GPL-compatible license).

The memory allocated via the preceding routine(s) must subsequently be freed using the `void free_percpu(void __percpu *__pdata)` API.

As you've learned, the resource-managed `devm_*()` API variants allow you (typically when writing drivers) to conveniently use these routines to allocate memory; the kernel will take care of freeing it, helping prevent leakage scenarios. The `devm_alloc_percpu(dev, type)` macro allows you to use this as a resource-managed version of `__alloc_percpu()`.

Performing I/O (reads and writes) on per-CPU variables

A key question, of course, is how exactly can you access/load (read) and update/store (write) to per-CPU variables? The kernel provides several helper routines to do so; let's employ a simple example to understand how.

We statically define a single integer per-CPU variable, and at a later point in time, we'll want to access and print its current value. You should realize that, being per-CPU, the value retrieved will be auto-calculated *based on (or indexed on) the CPU core the code currently runs on*; in other words, if the following code runs on core 1, then in effect, the pcpa[1] value is fetched (it's not done exactly like this; this is just conceptual, but you get the idea). So let's statically define an integer per-CPU variable named pcpa (it will be auto-initialized to 0):

```
DEFINE_PER_CPU(int, pcpa);
int val;
[ ... ]
val = get_cpu_var(pcpa);
pr_info("cpu0: pcpa = %+d\n", val);     // critical section, must be atomic!
put_cpu_var(pcpa);
```

The pair of macros, get_cpu_var() and put_cpu_var(), allows us to safely retrieve or modify the per-CPU value of the given per-CPU variable (its parameter). It's important to understand that the code between get_cpu_var() and put_cpu_var() (or equivalent) is, in effect, a critical section – an atomic context – where kernel preemption is disabled and any kind of blocking (or sleeping) is disallowed. If you do anything here that blocks (sleeps) in any manner, it's a kernel bug. Also, in the per-CPU critical section, unlike with the spinlock IRQ variants (and as with the mutex), the state of hardware interrupts (on the local core) remain unaffected.

As you've learned, the vmalloc() API is possibly blocking, so it might sleep (block). So as an example, let's see what can happen if you try to allocate memory via vmalloc() within the get_cpu_var()/put_cpu_var() pair of macros:

```
void *p;
val = get_cpu_var(pcpa);              // disables kernel preemption!
void *vp=vmalloc(1024*1024);          // vmalloc(), vfree() are possibly blocking!
mdelay(1);                            // mdelay(), printk() are non-blocking
vfree(vp);
pr_info("cpu1: pcpa = %+d\n", val);
put_cpu_var(pcpa);                          // enables kernel preemption
[ ... ]
```

```
$ sudo insmod <whatever>.ko
$ sudo dmesg
[ ... ]
BUG: sleeping function called from invalid context at include/linux/sched/
mm.h:274
in_atomic(): 1, irqs_disabled(): 0, non_block: 0, pid: 14135, name: thrd_1/1
preempt_count: 1, expected: 0
```

```
RCU nest depth: 0, expected: 0
Preemption disabled at:
[<ffffffffc06b64f2>] thrd_work+0xe2/0x310 [percpu_var]
[ ... ]
$
```

Ah, we can literally see from the diagnostic that since the code between the `get_cpu_var()`/`put_cpu_var()` pair runs in an atomic context, kernel preemption was disabled (this disabling of kernel preemption is done within the expanded code of the `get_cpu_var()` macro). Then, we tried to do something that's possibly blocking – the memory alloc via the `vmalloc()` API – causing a kernel bug (if you try it, the call trace will show that the bug originates by calling the `vmalloc()`).

By the way, calling the `printk()` (or `pr_<foo>()`) wrappers (as well as the `mdelay()`) as we do within the critical section here is fine, as they're non-blocking.

Internally, the `get_cpu_var()` macro invokes `preempt_disable()`, disabling kernel preemption, and `put_cpu_var()` undoes this by invoking `preempt_enable()`. As seen earlier (in the chapters on *CPU scheduling*), this can be nested, and the kernel maintains a `preempt_count` variable to figure out whether kernel preemption is actually enabled or disabled. (Here, you can see that the value of `preempt_count` is `1`, meaning the kernel is in a non-preemptible state, whereas the `vmalloc()` code path expected it to be `0`, that is, preemptible.)

The upshot of all this is that you must carefully match the `{get,put}_cpu_var()` macros when using them (for example, if we call the `get` macro twice, we must also call the corresponding `put` macro twice).

The `get_cpu_var()` is an `lvalue` and, thus, can be operated upon; for example, to increment the per-CPU `pcpa` variable, just do the following:

```
get_cpu_var(pcpa) ++;
put_cpu_var(pcpa);
```

You can also safely retrieve the current per-CPU value via the `per_cpu()` macro:

```
per_cpu(var, cpu);
```

So to retrieve the per-CPU `pcpa` variable for every CPU core on the system, use the following:

```
for_each_online_cpu(i) {
    val = per_cpu(pcpa, i);
    pr_info(" cpu %2d: pcpa = %+d\n", i, val);
}
```

FYI, you can always use the `[raw_]smp_processor_id()` macro to figure out which CPU core you're currently running upon; in fact, this is precisely how our `convenient.h:PRINT_CTX()` macro does it. (The *raw* variant avoids issues cropping up even when running it with kernel preemption turned off, so in fact, we use it.)

In a similar manner, the kernel provides routines to work with pointers to variables that need to be per-CPU, the `{get,put}_cpu_ptr()` and `per_cpu_ptr()` macros. These macros are heavily employed when working with a per-CPU data structure (as opposed to just a simple integer). We often use them to safely retrieve the pointer to the structure on the CPU core we're currently running upon (as the following example also demonstrates).

Per-CPU – an example kernel module

A hands-on session with our sample per-CPU demo kernel module will definitely help us learn how to use this powerful lock-free technology (the code's here: `ch13/3_lockfree/percpu`). Here, we define and use two per-CPU variables:

- A statically allocated and initialized per-CPU integer
- A dynamically allocated per-CPU data structure

As an interesting way to help demo per-CPU variables, let's arrange for our demo kernel module to spawn off a couple of kernel threads (aka *kthreads*). Let's call them `thrd_0` and `thrd_1`. Furthermore, once created, we shall make use of the task structure's CPU affinity mask feature by leveraging the `sched_setaffinity()` API to *affine* our `thrd_0` kernel thread on CPU `0` and our `thrd_1` kernel thread on CPU `1` (hence, they will be scheduled to run on only these cores; of course, we must test this code on a system (VM or native) with at least two CPU cores).

The following code snippets illustrate how we define and use the per-CPU variables (we have left out the code that creates the kernel threads and sets up their CPU affinity masks, as it's not relevant to this chapter; nevertheless, it's key for you to browse through the full code and try it out!):

```
// ch13/3_lockfree/percpu/percpu_var.c
[ ... ]
/*--- The percpu variables, an integer 'pcpa' and a data structure --- */
/* This percpu integer 'pcpa' is statically allocated and initialized to 0;
 * one integer instance will be allocated per CPU core! */
DEFINE_PER_CPU(int, pcpa);

/* This "driver context" per-cpu structure will be dynamically allocated
 * via alloc_percpu() */
static struct drv_ctx {
    int tx, rx; /* here, as a demo, we just use these two members,
                        ignoring the rest */
    [ ... ]
} *pcp_ctx;
[ ... ]

static int __init init_percpu_var(void)
{
    [ ... ]
```

```
    /* Dynamically allocate the per-cpu structures; one structure instance
     * will be allocated per CPU core! (If you want to specify the GFP flags,
then use the alloc_percpu_gfp(type, gfp) macro instead) */
    ret = -ENOMEM;
    pcp_ctx = (struct drv_ctx __percpu *) alloc_percpu(struct drv_ctx);
    if (!pcp_ctx) {
        [ ... ]
}
```

Why not use the resource-managed `devm_alloc_percpu()` instead? Yes, you should when appropriate; here, though, as we're not writing a proper driver, we don't have a `struct device *dev` pointer handy, which is the required first parameter for `devm_alloc_percpu()`.

By the way, I faced an issue when coding this kernel module; to set the CPU mask (to change the CPU affinity for each of our kernel threads), the kernel API is the `sched_setaffinity()` function, which, unfortunately for us, is not exported, thus preventing us from using it. So we look to perform what is definitely considered a hack: obtain the address of the uncooperative function (`sched_setaffinity()`) via the `kallsyms_lookup_name()` API (which works when `CONFIG_KALLSYMS` is defined) and then invoke it as a function pointer.

This technique worked well enough for this book's first edition using the 5.4 LTS kernel, but now, guess what? From 5.7 onward, the kernel devs unexported the `kallsyms_lookup_name()` API as well (commit ID 0bd476e6c671. Rationale: `https://lwn.net/Articles/813350/`)! With it gone, or not, we can now simply use this approach: a helper Bash script (named `run`) greps the `sched_setaffinity()` function's address (from `/proc/kallsyms`) and passes it to this module as a parameter! Here, we equate it to the expected function signature – that of `sched_setaffinity()` – and use it. Pedantically, it's not the right way to do things, but hey – it works. Don't do this in production.

Our design idea here is to create two kernel threads and have each of them differently, and in parallel, manipulate the per-CPU data variables. If these were ordinary global variables, this would certainly constitute a critical section and we would, of course, require a lock of some sort; but here, precisely because they are *per-CPU* and because we guarantee that our threads run on separate cores, we can concurrently update them with differing data without any locking whatsoever! Lock-free, as promised.

Our kernel thread worker routine is as follows; the argument to it is the thread number (`0` or `1`). We accordingly branch off and manipulate the per-CPU data (as a simple case, we have our first kernel thread increment the per-CPU integer three times, while our second kernel thread decrements it three times (the `THRD0_ITERS` and `THRD1_ITERS` macros are set to the value 3):

```
/* Our kernel thread worker routine [...] */
static int thrd_work(void *arg)
{
    int i, val;
    long thrd = (long)arg;
```

```
    struct drv_ctx *ctx;
    [ ... ]

    /* Set CPU affinity mask to 'thrd', which is either 0 or 1 */
    if (set_cpuaffinity(thrd) < 0) {
        [ ... ]
    } [ … ]
    SHOW_CPU_CTX();

    if (thrd == 0) {                /* our kthread #0 runs on CPU 0 */
        for (i=0; i<THRD0_ITERS; i++) {
            /* Operate on our perpcu integer */
            val = ++ get_cpu_var(pcpa);
            pr_info(" thrd_0/cpu0: pcpa = %+d\n", val);
            put_cpu_var(pcpa);

            /* Operate on our perpcu structure */
            ctx = get_cpu_ptr(pcp_ctx);
            ctx->tx += 100;
            pr_info(" thrd_0/cpu0: pcp ctx: tx = %5d, rx = %5d\n",
                ctx->tx, ctx->rx);
            put_cpu_ptr(pcp_ctx);
        }
    } else if (thrd == 1) {         /* our kthread #1 runs on CPU 1 */
        for (i=0; i<THRD1_ITERS; i++) {
            /* Operate on our perpcu integer */
            val = -- get_cpu_var(pcpa);
            pr_info(" thrd_1/cpu1: pcpa = %+d\n", val);
            put_cpu_var(pcpa);

            /* Operate on our perpcu structure */
            ctx = get_cpu_ptr(pcp_ctx);
#if 0
         /* If we try and run a blocking API within the per-CPU
             * 'critical section', we land up with a kernel bug; hence,
             * it's commented out by default. */
            void *vp=vmalloc(1024*1024); // vmalloc(), vfree() are possibly
blocking!
            mdelay(1);                                       // mdelay(), printk()
are non-blocking
            vfree(vp);
#endif
```

```
            ctx->rx += 200;
            pr_info(" thrd_1/cpu1: pcp ctx: tx = %5d, rx = %5d\n",
                ctx->tx, ctx->rx);
            put_cpu_ptr(pcp_ctx);
        }
    }
    disp_our_percpu_vars();
    pr_info("Our kernel thread #%ld exiting now...\n", thrd);
    return 0;
}
```

To have it run, simply `cd` to it's source folder and type `./run`. The effect at runtime is interesting; see the following kernel log, the output of `sudo dmesg` (on an Ubuntu 23.04 VM running our custom 6.1.25 kernel):

```
[65427.790479] percpu_var:init_percpu_var(): inserted
[65427.790637] percpu_var:thrd_work(): *** kthread PID 16994 on cpu 0 now ***
[65427.790662] percpu_var:thrd_work():   thrd_0/cpu0: pcpa = +1
[65427.790664] percpu_var:thrd_work():   thrd_0/cpu0: pcp ctx: tx =   100, rx =     0
[65427.790667] percpu_var:thrd_work():   thrd_0/cpu0: pcpa = +2
[65427.790683] percpu_var:thrd_work():   thrd_0/cpu0: pcp ctx: tx =   200, rx =     0
[65427.790685] percpu_var:thrd_work():   thrd_0/cpu0: pcpa = +3
[65427.790686] percpu_var:thrd_work():   thrd_0/cpu0: pcp ctx: tx =   300, rx =     0
[65427.790689] percpu_var:disp_our_percpu_vars():  cpu  0: pcpa = +3, rx =     0, tx =   300
[65427.790691] percpu_var:disp_our_percpu_vars():  cpu  1: pcpa = +0, rx =     0, tx =     0
[65427.790694] percpu_var:disp_our_percpu_vars():  cpu  2: pcpa = +0, rx =     0, tx =     0
[65427.790716] percpu_var:disp_our_percpu_vars():  cpu  3: pcpa = +0, rx =     0, tx =     0
[65427.790718] percpu_var:disp_our_percpu_vars():  cpu  4: pcpa = +0, rx =     0, tx =     0
[65427.790720] percpu_var:disp_our_percpu_vars():  cpu  5: pcpa = +0, rx =     0, tx =     0
[65427.790723] percpu_var:thrd_work(): Our kernel thread #0 exiting now...
[65427.790855] percpu_var:thrd_work(): *** kthread PID 16995 on cpu 1 now ***
[65427.790860] percpu_var:thrd_work():   thrd_1/cpu1: pcpa = -1
[65427.790862] percpu_var:thrd_work():   thrd_1/cpu1: pcp ctx: tx =     0, rx =   200
[65427.790865] percpu_var:thrd_work():   thrd_1/cpu1: pcpa = -2
[65427.790866] percpu_var:thrd_work():   thrd_1/cpu1: pcp ctx: tx =     0, rx =   400
[65427.790869] percpu_var:thrd_work():   thrd_1/cpu1: pcpa = -3
[65427.790870] percpu_var:thrd_work():   thrd_1/cpu1: pcp ctx: tx =     0, rx =   600
[65427.790873] percpu_var:disp_our_percpu_vars():  cpu  0: pcpa = +3, rx =     0, tx =   300
[65427.790875] percpu_var:disp_our_percpu_vars():  cpu  1: pcpa = -3, rx =   600, tx =     0
[65427.790878] percpu_var:disp_our_percpu_vars():  cpu  2: pcpa = +0, rx =     0, tx =     0
[65427.790881] percpu_var:disp_our_percpu_vars():  cpu  3: pcpa = +0, rx =     0, tx =     0
[65427.790883] percpu_var:disp_our_percpu_vars():  cpu  4: pcpa = +0, rx =     0, tx =     0
[65427.790885] percpu_var:disp_our_percpu_vars():  cpu  5: pcpa = +0, rx =     0, tx =     0
[65427.790888] percpu_var:thrd_work(): Our kernel thread #1 exiting now...
```

Figure 13.10: Screenshot showing the kernel log when our ch13/3_lockfree/percpu/percpu_var LKM runs on a Linux system with six CPU cores

Notice how our module has got the job done, here on a Linux system with six CPU cores. You can clearly see the cores labeled `cpu 0` through `cpu 5` and the value of the per-CPU data item in each of the cores at runtime (*Figure 13.10*)!

In the last few lines of output in *Figure 13.10*, you can see a summary of the values of our per-CPU data variables on CPU 0 and CPU 1 (as well as on the remaining cores; we show it via our disp_our_percpu_vars() function). Clearly, for the per-CPU pcpa integer (as well as for the rx and tx members of our pcp_ctx data structure), the values are *different* as expected, *without explicit locking*.

The kernel module that we just demonstrated uses the for_each_online_cpu(i) macro to display the value of our per-CPU variables on each online CPU. However, what if you have, say, six CPUs on your VM but want only two of them to be "live" at runtime? There are several ways to arrange this: one is to pass the maxcpus=n parameter to the system's kernel at boot; another is to manipulate the (root-writable) pseudofile /sys/devices/system/cpu/cpuN/online, where N is the CPU core number (this value is typically 1 by default, implying the core is online; writing 0 to it as root takes the core offline, although you can't do so for CPU core 0).

Per-CPU usage within the kernel

Per-CPU variables are quite heavily used within the Linux kernel; let's check out two quite interesting use cases.

Implementing the current macro on the x86 via per-CPU data

The current macro on the x86 architecture (recall that current is the canonical way to refer to the task structure of the thread that's currently executing; we covered its usage in some detail in *Chapter 6*, *Kernel Internals Essentials – Processes and Threads*, in the *Accessing the task structure with current* section) is looked up (and set) every so often; *implementing it as a per-CPU data item ensures that we keep its access lock-free* and, thus, high in performance! Here's the code that implements it:

```
// arch/x86/include/asm/current.h
[ ... ]
DECLARE_PER_CPU(struct task_struct *, current_task);
static __always_inline struct task_struct *get_current(void)
{
    return this_cpu_read_stable(current_task);
}
#define current get_current()
```

The DECLARE_PER_CPU() macro declares the variable named current_task as a per-CPU variable of type struct task_struct *. The get_current() inline function (which is the implementation of current on the x86) invokes the this_cpu_read_stable() helper on this per-CPU variable, thus reading the value of current on the CPU core that it's currently running on (read the comment at https://elixir.bootlin.com/linux/v6.1.25/source/arch/x86/include/asm/percpu.h#L214 to see in detail what this routine's about).

As an aside, now that we understand how the (per-CPU) value of `current` is fetched (read), an FAQ comes up: where does this `current_task` per-CPU variable get updated? Think about it: the kernel must change (update) `current` whenever it context-switches to another task. That's exactly the case – it is indeed updated within the context-switching code; on the x86, it's here: `arch/x86/kernel/process_64.c:__switch_to()` `https://elixir.bootlin.com/linux/v6.1.25/source/arch/x86/kernel/process_64.c#L558`:

```
__no_kmsan_checks
__visible __notrace_funcgraph struct task_struct *
__switch_to(struct task_struct *prev_p, struct task_struct *next_p)
{
    [ ... ]
        this_cpu_write(current_task, next_p);
        [ ... ]
}
```

Using per-CPU data on the network receive path

Another interesting use case of per-CPU data occurs deep in the kernel network stack. Very simplistically, when a network packet is received via a hardware interrupt, it's typically DMA'ed into kernel (i.e., network device driver) memory. The so-called bottom-half mechanism, the network receive `softirq`, then takes over processing, pushing the packet up the protocol stack (to higher layers). Many moons ago (read, decades), it enqueued the packets on the receive path into a global data structure. This created a huge performance bottleneck of course; access to this structure had to be protected via a lock (a spinlock); contention could become high and performance steadily dropped (especially on higher-end systems with many cores and network adapters driving network packets in parallel).

The networking developers, realizing this, switched to using per-CPU data for the global receive queue data structure (it's named `softnet_data`). So now, this is how it looks:

```
// net/core/dev.c
[ … ]
/*
 * Device drivers call our routines to queue packets here. We empty the
 * queue in the local softnet handler.
 */
DEFINE_PER_CPU_ALIGNED(struct softnet_data, softnet_data);
EXPORT_PER_CPU_SYMBOL(softnet_data);
[ … ]
```

The network receive softirq, `NET_RX_SOFTIRQ`, function can be seen here:

```
// net/core/dev.c
[ … ]
```

```
static __latent_entropy void net_rx_action(struct softirq_action *h)
{
    struct softnet_data *sd = this_cpu_ptr(&softnet_data);
    [ … ]
```

Exercise

Check out more examples of per-CPU usage within the kernel codebase via the `__alloc_percpu()` macro (as well as others you've learned about) by running `cscope -d` in the root of your 6.1 kernel source tree (this assumes you've already built the `cscope` index there via `make cscope`).

All right, good job! Let's now move on to learning about a critical lock-free technology, RCU!

Understanding and using the RCU (Read-Copy-Update) lock-free technology – a primer

RCU is a lock-free synchronization technology that allows readers to run concurrently with writers. It was merged into the Linux kernel back in October 2002.

Traditionally, or before RCU, when working upon, say, a large global linked list or tree data structure (a common occurrence in the kernel), kernel (and/or driver) developers used spinlocks or reader-writer (spin)locks to handle concurrency concerns. However, as we've learned, these locks can become very expensive very quickly and can cause large performance bottlenecks as well as high power consumption, especially on high-end (NUMA) systems (those with literally thousands of cores!). Thus, although they work, performance becomes worse and worse; in other words, it's not scalable (RCU has become an ideal replacement for reader-writer locks.)

With RCU, we can have readers running in parallel with each other *along with writers (or updaters)*. For reader threads, *no locks are required.* Yes, no lock acquire/release operations, no atomic instructions, no shared memory updates, and no memory barriers (except on the DEC Alpha processor, which requires memory barriers). As these very operations are typically the ones that hurt performance, relinquishing the need for them is what makes RCU very fast and valuable.

Hang on though – this "no locks" approach won't work with writers (updaters); they will still require protection from each other (typically effected via a spinlock). But the fact that readers will never require locking significantly simplifies the design and thereby can greatly reduce the potential for deadlock. So we now realize that RCU is brilliant at, and is designed to, provide synchronization in ***read-mostly*** situations, which are aplenty in the kernel! (As a thumb-rule, use RCU when reads account for 90% or more of the workload as opposed to writes).

Also, more technically, RCU works best in read-mostly situations *and* when inconsistent (or stale) data reads don't matter so much to the readers. The more "consistent" the data needs to be, and the more time spent doing updates, the less use there is for RCU. (To see this visually, take a peek ahead at *Figure 13.19* if you wish).

How does RCU work?

Especially for someone new to it, RCU begs the question: *how can I have readers and writers run concurrently and safely while working on shared data, along with absolutely no locking for readers?* We attempt to satisfy your thirst for this knowledge right here!

Paul E. McKenney is one of the key persons credited with implementing RCU synchronization technology within the Linux kernel and is one of the maintainers for RCU in the Linux kernel; he's also the lead maintainer of the Linux kernel memory model. (Thus, it makes sense to furnish this explanation with terms and phrases that he, as well as the larger community, uses.)

We approach this topic at three "levels", from a simple conceptual explanation (Level 1) to a Level 2 (which is a Level 1 plus more) explanation, which is a bit deeper and more detailed, and then onto a Level 3 explanation, which conveys Level 2 detail plus some RCU code primitives. So ladies and gentlemen, fasten your seatbelts, and let's go!

RCU Level 1 explanation (no code, only conceptual, and very simple)

A step-by-step account of our RCU Level 1 explanation follows; right now, don't worry regarding the details – we show diagrams illustrating these concepts in the next section:

1. RCU reader (or subscriber) threads can read the shared writable data object, in parallel, with no locking imposed; this makes RCU very lightweight and very fast (and suitable in *read-mostly* scenarios)!
2. When a writer (or publisher or updater) thread comes along, there's no RCU locking for it to do. What it must do is create a copy of the data object - the shared one being concurrently operated upon - and work on that copy. Note though that writers must protect themselves from concurrent access, from stepping on each other's toes (usually done by using a spinlock to serialize them).
3. The pre-existing readers – the ones that were running *before* a writer came along - continue to read the "old" data object; that's their view of reality, and that's fine. The writer, and any subsequent "new" readers (the ones that run after a writer came along), will work only upon the "new" data object, their view of reality! (So who's got the "correct" view of reality? Both do. Quite quantum physics slash metaverse stuff, no?)
4. The writer now updates (writes to and, thus, modifies) the new, copied, data object in question, modifying it. The pre-existing readers continue to work upon the old object in parallel.
5. The writer completes its work and atomically updates the (old) data object pointer to point to the new object, ensuring that it and any subsequent readers will see the new, updated version. Note that the pre-existing readers continue to work (read) from the old data object in parallel.
6. Once all the pre-existing RCU readers complete their reads, the writer destroys (frees) the old data object.

And the job's done, (mostly) lock-free. The only portion that isn't lock-free is when multiple writers exist; they must serialize themselves via a spinlock. But think – by definition, we use RCU in mostly-read situations; thus, writers account for a tiny percentage of the overall time, making it mostly lock-free.

RCU Level 2 explanation (no code, only concepts, and more details)

A step-by-step account of our RCU Level 2 explanation follows:

1. RCU readers (or subscribers) proceed unhindered, in parallel with each other; *there's no locking of any sort*, no atomic operations, and no memory barrier even (there is one exception: the DEC Alpha processor requires memory barriers). This is the whole point of RCU: to use it in mostly read scenarios, keeping reads parallelized and quick. Let's consider a sample data structure, `struct X`, with a pointer to it (`ptr_x`); *Figure 13.11* shows it being accessed – for reading – concurrently by multiple RCU readers running on different cores:

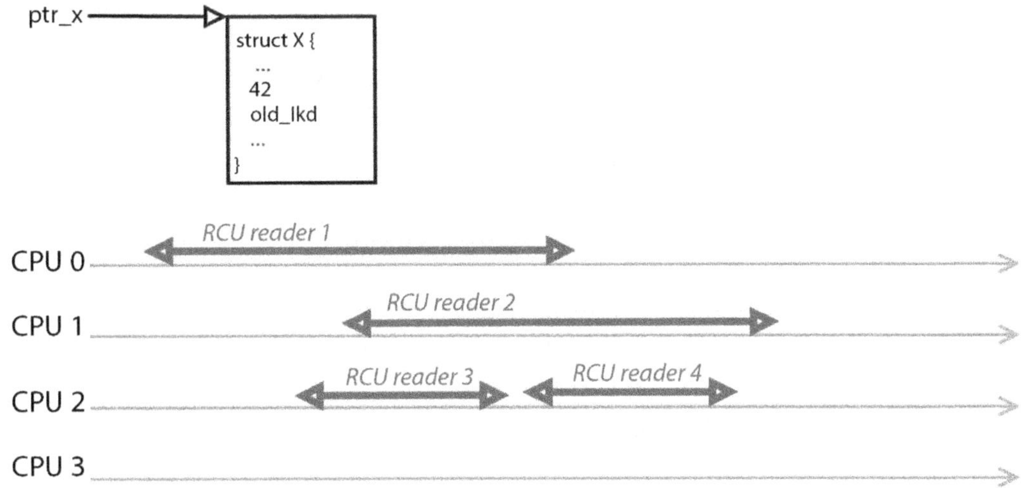

Figure 13.11: A few RCU readers run their read-side critical section in parallel; no locking is needed! The code in the read-side critical section (the double-arrow lines) needs to be non-blocking

Do note though: by default, these RCU **read-side critical sections need to be atomic and non-blocking; they run with kernel preemption disabled** (just as is the case with per-CPU critical sections). Also, as the diagram shows, several RCU readers can sequentially run on the same core (the case with readers 3 and 4 here on core 2).

Hard-copy book readers, you'll find all the color diagrams for the book here: `https://packt.link/gbp/9781803232225`.

2. A writer (or publisher or updater) thread comes along. It too proceeds forward, but the first thing it does is make a *copy* of the data object in question! (You could peek ahead at *Figure 13.14*, at the (red) circle with 1 in it.)

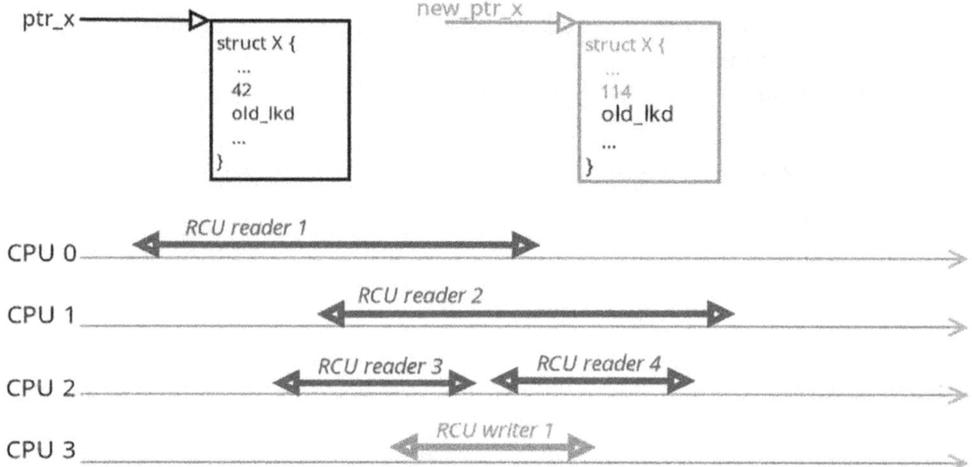

Figure 13.12: A writer thread comes along and creates a copy of the data object (notice that it's begun to modify its content); the vertical line – once a writer enters – separates the old RCU readers (1, 2, and 3) from the new one (4)

Note though that multiple concurrent writers must protect themselves from concurrent access, from stepping on each other's toes (usually done by using a spinlock to serialize them).

3. The pre-existing (or "old") RCU readers continue to work upon the old data object (RCU readers 1, 2, and 3); the writer, and any subsequent "new" reader (RCU readers 4), will work only upon the "new" data object. This way, they don't collide, there's no data race, and each sees its version of reality!
4. The writer now updates (writes to and, thus, modifies) the data object in question, modifying it. The pre-existing readers continue to work upon the old object in parallel. (See *Figure 13.14* and the (red) circle with **2** in it.)
5. The writer completes its work and atomically updates the pointer to point to the new object, ensuring that it and any subsequent readers (like *RCU readers 4* and *5* in *Figure 13.13*) will see the new, updated version. Note that the pre-existing readers continue to work (read) from the old data object in parallel. (See *Figure 13.14* and the (red) circle with **3** in it.)

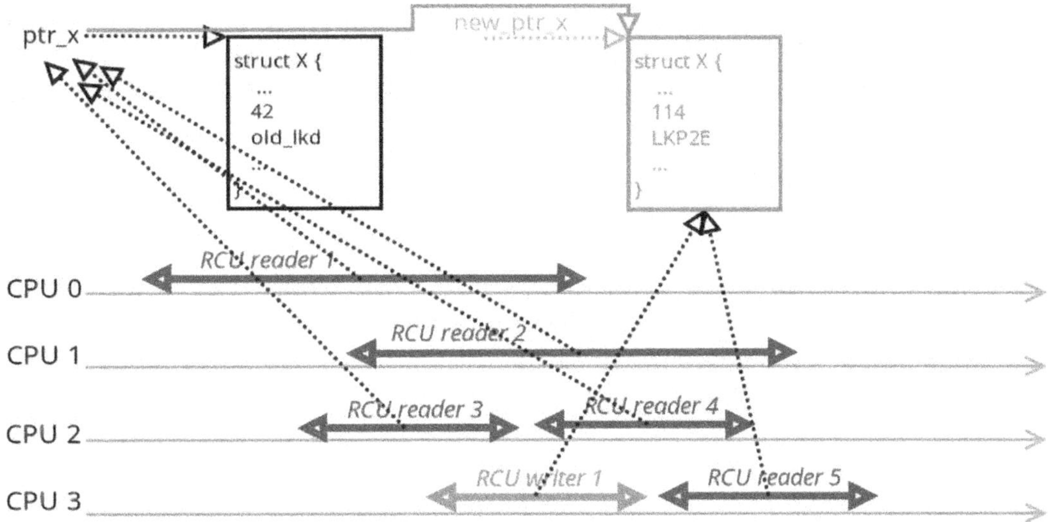

Figure 13.13: The pre-existing RCU readers (1 to 3) refer to the old reality; the writer and new readers (4 and 5) see the new reality. Who's right? Both are (a "quantum" feel, no?)

Before going further, there are a couple of important rules to follow, and a definition is required:

- *Rule 1:* Within an RCU read-side critical section, you cannot block, switch to user mode, or enter the idle state. (This is just as is the case with a spinlock's or per-CPU variable's critical section.)
- *Rule 2*: Just as with any other synchronization or locking primitive, the data object(s) that are protected by the lock must only be accessed while the lock is held. So with RCU reads, you must only read the data object(s) when within the RCU read-side critical section and update (write) them when within the RCU write-side critical section. The moment a reader leaves the read-side critical section (worry not, we'll come to how exactly this is defined), it's said to have entered a **quiescent state**. Any read performed when outside the RCU read-side critical section, in other words when in a queiscent state, is disallowed and considered illegal.
- RCU **Grace Period (GP)**: The time period when every reader thread has passed through at least one queiscent state. In effect, this is the time period after which it's guaranteed there are no pre-existing readers (as when they're in a quescient state, they're by definition outside the read-side critical section). This is important for the writer to know (we'll now see why).

6. The writer must, of course, free the old object (effectively destroying it and preventing memory leakage), *but it cannot do so until it's guaranteed that all pre-existing readers have finished their RCU read-side critical section.* Doing so would cause a subsequent **UAF** (**Use After Free**) bug! In fact, with RCU, it's now christened a **UAFBR** (**UAF by RCU**) bug; bugs like these can lead to exploitable vulnerabilities, so be careful! (By the way, a recent security issue named *StackRot* (June 2023) is based on precisely this vulnerability: `https://github.com/lrh2000/StackRot#stackrot-cve-2023-3269-linux-kernel-privilege-escalation-vulnerability.`)

   Back to the point: the writer waits for one RCU **grace period** (or **GP**) and then frees the old object! (See *Figure 13.14*, with the (red) circles with **4** and **5** in them, respectively. You might wonder how exactly the writer will know that a GP has elapsed. We cover this in the following section; for now, just believe it.)

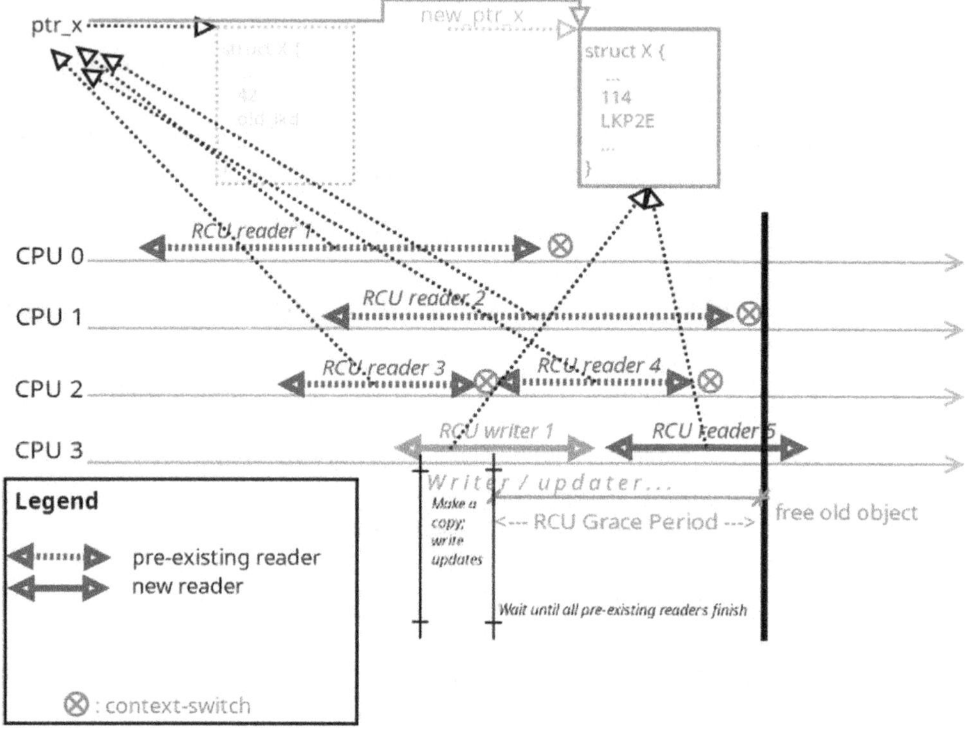

Figure 13.14: After updating the data object, the writer waits for one RCU GP (up to the thick vertical line), and it then frees the old object; subsequent readers work upon the new data object. The (red) circles with numbers in them show the writer's steps

Now, from our studies on CPU scheduling, you'll realize that every thread will eventually, for whatever reason, context-switch off the processor. Here, the pre-existing RCU reader threads will perform their reads in a non-blocking manner and will, of course, at some point, finish reading. At this point, as they're done, or perhaps a bit later, they will perform a context switch. In *Figure 13.14*, the (green) circle with the cross (the X within it) denotes the context switch that all pre-existing (and even new, perhaps) RCU readers will eventually perform.

The point by which all the older (pre-existing) RCU readers (here, 1 to 3) finish and context-switch off their core is shown via the thick vertical line. (In this particular example, even the new RCU reader 4 has finished and context-switched; it can happen. Note though that *it's* reading the new object.) At this point, *the elapse of an RCU GP* – the point where the last pre-existing RCU reader context-switches off the core – the writer frees the old data object. From now on, any subsequent RCU readers – and writers – will work upon the "new" data object (which will soon enough become the "old" one as the entire process repeats). The job's done!

So, now we can see: readers (R) run truly in parallel with absolutely no interference, not even from writers (making RCU lightweight and quick in the extreme). Writers can – in parallel with readers – modify data by making a copy (C) of the data object, updating (U) the copy, and then (when safe) freeing the original object and replacing it with the copy. Hence the term RCU.

RCU Level 3 explanation (Level 2 + more details)

While going through this section, you really must take a gander at (relevant portions of) the kernel's RCU code; what I really mean is at the *comments* in the codebase! They're excellent, often enough to document the salient features (to help with the RCU's basic, core APIs and their usage, I've constructed two tables as a kind of quick reference for a few core RCU APIs, in the upcoming *Quick summary of the RCU core APIs* section; you can refer to them as you read this section):

1. RCU readers (or subscribers) proceed unhindered, in parallel with each other; *there's no locking of any sort*, no atomic operations, and no memory barrier even (there is one exception: the DEC Alpha processor requires memory barriers). This is the whole point of RCU: to use in mostly read scenarios, keeping reads parallelized and quick:

    ```
    rcu_read_lock();
    < The RCU read-side critical section; non-blocking, non-preemptible
    Rule: don't block in any manner
    Read the data object(s)
    >
    rcu_read_unlock();
    ```

    The code between the `rcu_read_lock()` and the `rcu_read_unlock()` is, of course, the RCU read-side critical section. Now, it's key to understand that, besides kernel debugging stuff (like `lockdep` annotations, and so on), the code of `rcu_read_{un}lock()` is literally nothing! (It's a newline in the simplest case; however, as mentioned, due to the need for debug features, there can be some code, but it will evaluate to nothing (null) in typical production environments.)

    So this begs the question: how can you synchronize without changing the machine state in any manner whatsoever? That should be pretty much impossible. Paul E. McKenney brilliantly answers this (`https://youtu.be/obDzjElRj9c?list=PLIlI4QbmdBqH9-L1QOq6O5Yxt-YVjRsFZ&t=3401`). To paraphrase his answer: these RCU APIs do not need to change machine state; instead, *they're designed to act on the developer*. They must understand that it's a critical section, and thus, no blocking (and, thus, no context-switching) is allowed. If this rule is followed, it simply works (as explained earlier). In effect, *"RCU is synchronization by social engineering."* (Ha! Awesome!)

When you think about it, most (if not all) locking/synchronization tech relies on the developer *following the rules*. Merely taking a lock does not prevent another thread from working on the data object in question; it always can. The *rule*, though, is that it must only work on the data object when it holds the lock. Breaking the rules breaks the social contract and, thus, breaks the program. (As Paul E. McKenney says, typical locking technology is implemented via both machine state and social engineering components; RCU is unusual in that the read-side critical section is protected by only social engineering.)

At the risk of repetition, do note carefully: these RCU read-side critical sections need to be atomic and non-blocking; they run with kernel preemption disabled (just as is the case with spinlocks and per-CPU). Also, as *Figure 13.14* shows, several RCU readers can sequentially run on the same core (the case with readers 3 and 4 here on core #2).

There is a way to perform blocking ops within the RCU read-side critical section: by using the **Sleeping RCU** (**SRCU**) implementation. (These days, this feature is always built in to the kernel, but it isn't the default RCU implementation used, of course.) See the *Further reading* section for more details.

Next, that the reader (or subscriber) must dereference the (data) pointer via the `rcu_dereference(p)` routine. (Internally, assuming `CONFIG_PROVE_RCU=y`, it verifies that you're indeed reading from within an RCU read-side critical section.)

Continuing with the whole "no locking" semantics that RCU brings, see this comment in the kernel source (version 6.1.25):

```
// include/linux/rcupdate.h
[ … ]
 /* So where is rcu_write_lock()?  It does not exist, as there is no
  * way for writers to lock out RCU readers.  This is a feature, not
  * a bug -- this property is what provides RCU's performance benefits.
  * Of course, writers must coordinate with each other.  The normal
  * spinlock primitives work well for this, but any other technique may
be
  * used as well.  RCU does not care how the writers keep out of each
  * others' way, as long as they do so.
  */
```

2. A writer (or publisher or updater) comes along. It too proceeds forward, but the first thing it does is make a *copy* of the global data object in question (see *Figure 13.14* and the (red) circle with **1** in it)!

   If there's a chance that multiple writers will arrive (it can certainly happen), protect them from trampling on each other by (typically) taking a spinlock to protect the write-side critical section.

Moreover, as RCU has evolved, the kernel provides specialized APIs for working with (mostly, and almost always, read) lists, arrays, queues, trees, and a few other data structures that commonly use RCU. These specialized APIs will internally make the copy of the data object as required (typically allocating kernel memory via the "usual" slab APIs like `kmalloc()`). For example, with (linked) lists, here are a few typically employed RCU-list APIs (they're in `include/linux/rculist.h`. We don't show the full prototype here; it's more to gain a conceptual understanding of what's available):

- `list_add_rcu(new, head)`: Adds a new entry to an RCU-protected list after the head (good for stacks)
- `list_add_tail_rcu(new, head)`: Adds a new entry to an RCU-protected list before the head (good for queues)
- `list_del_rcu(element)`: Deletes an entry from the list without re-init
- `list_replace_rcu(old, new)`: Replaces the old entry with the new one

Similarly, the kernel provides numerous helpers for RCU-protected queues and trees. (Recall us asking you to read the code-level comments; they're very good.)

As steps 3 and 4 are identical to what we showed in the previous section (*RCU Level 2 explanation (no code, only concepts, and more details)*), we don't gratuitously repeat them here, we move directly to the remaining steps 5 and 6.

5. The writer completes its updates (see *Figure 13.14* and the (red) circle with **2** in it); it then atomically updates the data object pointer to the new data object's address (see *Figure 13.14* and the (red) circle with **3** in it), ensuring that it and any subsequent readers (like *RCU readers 4* and *5* in *Figure 13.14*) will see the new updated version. Note that any pre-existing readers continue to read from the old data object in parallel.

   To atomically update the data object pointer to its new value, the writer typically employs the `rcu_assign_pointer(p, v)` routine. Internally, it guarantees correctly updating the pointer. You might wonder, why all the complexity? Why not simply assign the new value to the pointer variable (`(p) = (v)`)? Don't forget, both hardware and compiler complexity (with arch-specific memory-ordering issues to add to the fun) can cause all kinds of trouble; hence, this routine inserts a write memory barrier where required and then performs the assignment (we have some coverage on memory barriers at the end of this chapter).

   At the beginning of this point, we stated: "Note that any pre-existing readers continue to read from the old data object in parallel." It does make you wonder, how do we precisely define a pre-existing or "old" RCU reader from a "new" one? This is important, as the pre-existing readers continue to work on the old object and the new ones on the new copy. Well, here's the answer: the defining moment is when the writer completes execution of the `rcu_assign_pointer(p, v)` routine. All readers running prior to this point see the old pointer, and any readers that begin after this point see the new pointer and, thus, the new object.

6. The writer must, of course, free the old object (effectively destroying it and preventing memory leakage), but it cannot do so until it's guaranteed that all pre-existing readers have finished their RCU read-side critical section. Thus, the writer waits for one RCU GP and then frees the old object (see *Figure 13.14* and the (red) circles with **4** and **5** in them, respectively)!

   How do we know when an RCU GP – in effect, that all the (older) RCU readers have completed their reads – has elapsed? Ah, there's a neat trick that the implementation employs, which we hinted at earlier: as we know that all RCU read-side critical sections are guaranteed to be non-preemptible, the moment a reader thread context-switches off its CPU core, switches to user-land, or enters the idle state, it has definitely finished the RCU read. So the kernel keeps track of just this, and when all RCU readers have context-switched (at least once), it *knows* that they are done. This is implemented in code via the `synchronize_rcu()` API. So the writer must call this routine, and as it's synchronous, once it returns, it can confidently free the data object!

   Figure 13.15 attempts to sum up all (well, most) of this discussion, along with the core RCU APIs employed (although, to avoid clutter, I've shown abbreviated API names (the *Legend* shows the full names); plus the `rcu_dereference()` API is seen just once, in *RCU reader 1*, as an example. You should visualize all RCU readers calling it soon after the `rcu_read_lock()`. Furthermore, possible writer `spin_{un}lock()` APIs aren't shown at all here):

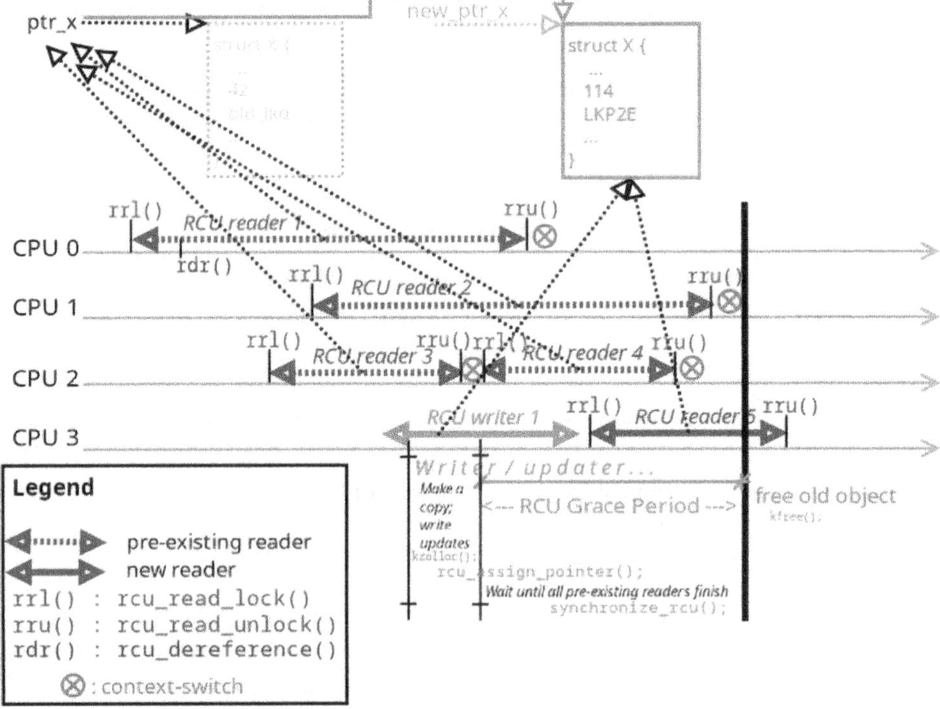

Figure 13.15: The same as Figure 13.14; plus the core RCU APIs are seen. After updating the data object, the writer waits for one RCU GP (up to the thick vertical line), and it then frees the old object; subsequent readers work upon the new data object

A "toy" (pseudocode) implementation of the synchronize_rcu() function (as shown by Paul E. McKenney) can be as simple as this:

```
for_each_online_cpu(cpu)
    run_on(cpu);
```

Merely running on each CPU core guarantees that a complete system-wide RCU GP has passed and that there are no more pre-existing RCU readers (as the thick vertical line in *Figure 13.15* shows). This works as, in order for the current context to run on a core, any prior thread on that core will (obviously) have to context-switch off it; so, if we run on all online cores, it by definition implies a system-wide RCU GP has indeed elapsed!

Okay, playing devil's advocate, what if one of the RCU readers takes longer (than is shown in *Figure 13.15*)? Will it now fail? No, a correctly implemented RCU will **extend the RCU grace period** to accommodate that long read! This does imply that:

- The writer might need to wait for a longer while before it can free up the old data object.
- More importantly, readers can keep going (for long periods), thus implicitly causing the GP to extend; this works as long as all readers are guaranteed to finish in a finite amount of time.

It does bring up the question: for just how long can readers extend the GP? It can't be too long, right? Well, yes... the kernel triggers a warning when an ***RCU CPU stall*** occurs, when it holds up the writer for what's considered to be a long time (somewhat akin to kernel warnings that occur as a result of soft or hard lockups, although RCU allows a really long time period). How long is long here? The kernel config RCU_CPU_STALL_TIMEOUT (defined here: https://elixir.bootlin.com/linux/v6.1.25/source/kernel/rcu/Kconfig.debug#L85) specifies the *"RCU CPU stall timeout in seconds"*; the range is from 3 to 300 seconds, and it defaults to a value of 21 seconds. (On this note of extending or batching the GP, having a so-called "lazy RCU" does tend to result in some trade-offs; McKenney explains the same here: https://youtu.be/obDzjElRj9c?list=PLIlI4QbmdBqH9-L1QOq6O5Yxt-YVjRsFZ&t=3722). The official kernel doc's here: *Using RCU's CPU Stall Detector*: https://www.kernel.org/doc/html/v6.1/RCU/stallwarn.html#using-rcu-s-cpu-stall-detector.

FYI, you can see an example of the "splat" caused by an RCU CPU stall (or several) in my *Linux Kernel Debugging* book: *Chapter 10, Kernel Panic, Lockups, and Hangs*, in the *RCU and RCU CPU stalls* section.

So there we are. Obviously, there's more to RCU than what's been explained here; these are the basics, conceptually (and to some extent code-wise) giving you a feel for how RCU works, which is the purpose.

Trying out RCU

As we've learned, critical sections need protection; using locks is a common, though expensive, way to do so. So let's learn the very basics of practically using RCU synchronization within the kernel. To do so, we take a very simple example where readers and writers run concurrently and work upon a shared global data structure (or shared state). We build three scenarios when working with it and show different ways in which it can be protected (or not!), ultimately via RCU (their code can be found in this folder: <book_src>/ch13/rdwr_concurrent/). Then, the *Quick summary of the RCU core APIs* section summarizes the five core RCU APIs.

Once this is done, we show two more RCU usage code examples: one is the list manipulation module that we built, earlier protected with the reader-writer spinlock; we now protect it via RCU. As another example, we simply provide the link and leave it for you to study and try out.

A simple example – concurrent readers and writers – and how we protect them

This is the example global data structure, the shared state, we shall work with:

```
static struct global_data {
        int gps_lock;
        long lat, longit, alt;
        int issue_in_16;
} *gdata;
```

Lets begin!

Case 1: Critical sections with no locking, thus no protection

Working like this, with no locking when in a critical section(s), is of course simply wrong. Still, it's informative to our buildup.

So, as a very simple example, see the following code snippets. We haven't shown all the code, just the most relevant portions of the `reader` and `writer` functions, which can run concurrently and work upon shared state (thus making them critical sections; again, don't forget the basics! We covered this in *Chapter 12, Kernel Synchronization – Part 1,* in the *What is a critical section?* section; go over it again if you're at all unclear).

We've also provided the full code for this example scenario here: `ch13/rdwr_concurrent/1_demo_rdwr_nolocks`. Do ensure you read it and run the demo module.

Reader	Writer

`static int reader(struct global_data *gd)`
`{`
`    long x, y, z;`
`    int stat = gd->issue_in_16;`
 	
`    if (gd->gps_lock) {`
`        x = gd->lat;`
`        y = gd->longit;`
`        z = gd->alt;`
`    }`
`    return stat;`
`}`
	`static void writer(struct global_data *gd)`
	`{`
	`    long x = 129780, y = 775952, z = 920;`
	
	`    gd->lat = x;`
	`    gd->longit = y;`
	`    gd->alt = z;`
	`    gd->issue_in_16 = 1;`
	`}`

Table 13.4: Our deliberately buggy Case 1: reader/writer code with no protection

Clearly, as the structure's global (shared state) and writable, and is accessed in a concurrent fashion, accessing it implies a critical section, and it thus requires protection. Here, we haven't provided any protection; thus, it's just wrong.

Let's further assume that the reads occur (much) more often than writes (even if this isn't the case, we'd still require locking, of course). Recall what we've learned; there are several locking technologies we could use to protect these read/write critical sections. Here, given that reads occur more often than writes, one way to provide protection is via the **reader-writer spinlock**.

Case 2: Critical sections protected with a reader-writer (spin)lock

So let's use the well-known *reader-writer spinlock* to provide protection (we don't show all the code here, just the key portions; find the full code here: `ch13/rdwr_concurrent/2_demo_rdwr_rwlock`).

Reader	Writer

<pre>static int reader(struct global_
data *gd) {
    long x, y, z;
    int stat;
 	
    read_lock(&gd->rwlock);
    stat = gd->issue_in_16;
    if (gd->gps_lock) {
        x = gd->lat;
        y = gd->longit;
        z = gd->alt;
    }
    read_unlock(&gd->rwlock);
    return stat;
}</pre>
	<pre>static void writer_work(struct
	global_data *gd) {
	    long x = 129780, y = 775952,
	z = 920;
	
	    write_lock(&gd->rwlock);
	    gd->lat = x;
	    gd->longit = y;
	    gd->alt = z;
	    gd->issue_in_16 = 1;
	    write_unlock(&gd->rwlock);
	}</pre>

Table 13.5: Our deliberately sub-optimal Case 2: reader/writer code protected via the reader-writer spinlock

(We've added a member to the global data structure – `rwlock_t rwlock`; the reader-writer spinlock.) It works; however, and again, as we have learned, the reader-writer (spin)lock does suffer from issues, particularly that of writer starvation and adverse cache effects, due to holding shared state in parallel on different cores. The kernel community has, thus, been aggressively moving away from the reader-writer spinlock to... guess what? RCU of course (as the use case is similar; both are used in *mostly read* situations)!

Case 3: Critical sections protected via RCU synchronization

All right, *now* let's finally get it right, protecting the critical sections within these code snippets **with RCU!**

The code might look like this (we've annotated it with numeric placeholders to help relate this code-level view with the RCU core APIs reference tables – *Table 13.6* and *Table 13.7* – that we show after this section. Be careful – the numbers in *Figure 13.16* do *not* denote the order in which code paths execute). Okay, let's look at the (relevant) code (as usual, the full code can be found here: `ch13/rdwr_concurrent/3_demo_rdwr_rcu`):

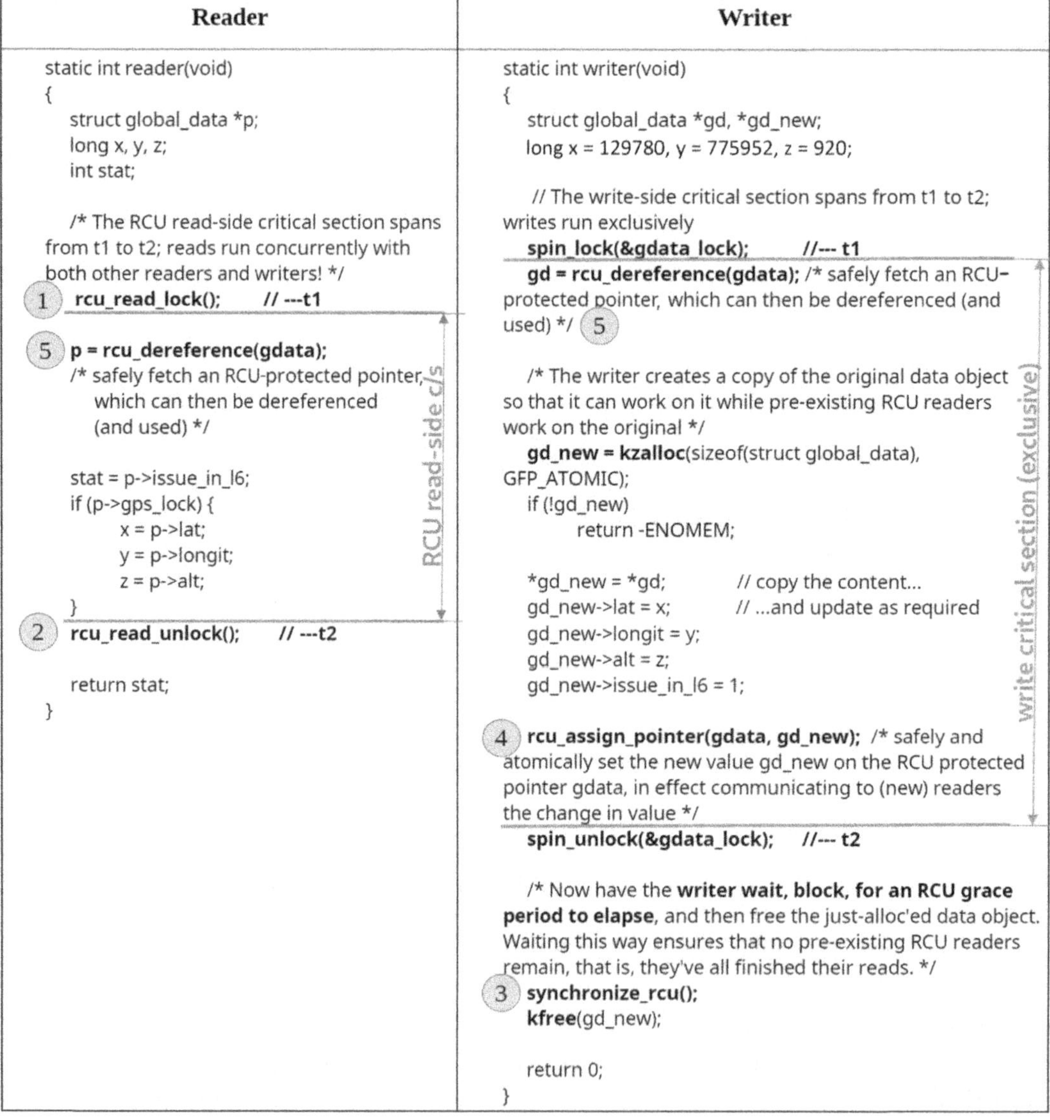

Figure 13.16: Annotated code-level example usage of the core RCU APIs (the numeric callouts refer to the core RCU API number shown in the following section; they're _not_ the order in which it runs)

This time, as RCU provides no protection at all for writers, we use a simple spinlock so that writers run in an exclusive (serialized) fashion (keeping them from stepping on each other's toes):

```
static spinlock_t gdata_lock; /* spinlock to protect writers.
```

```
     * Why not keep it inside the structure?
     * It's a bit subtle: we need the spinlock during the write critical
section;
     * however, it's in here that we create a copy of the data object and
     * modify/update it.
     * If the copy includes the spinlock (which it will) it won't work, as
     * we then violate the contract, using different locks...
     * Trying this, in fact, creates an interesting bug!
     * ...
     * pvqspinlock: lock 0xffff9e... has corrupted value 0x0!
     * ...
     * (The pvqspinlock is a paravirt one (as I ran it on a guest)).
     */
/* There is no 'read' RCU lock object; readers are meant to run
 * concurrently with each other *and* with writers. It's a 'socially
 * engineered' contract with the developer(s), is all it is!
 */
```

Do read the comments above very carefully! They're instructive... Run this demo via its `run` script and study the output.

The following section summarizes the five core RCU APIs.

Quick summary of the RCU core APIs

In this section, we show two quick summary tables of the core RCU APIs, essentially from the official kernel documentation here: `https://elixir.bootlin.com/linux/v6.1.25/source/Documentation/RCU/whatisRCU.rst#L137`.

There are just five core RCU APIs (and 18 other ones, shown later in the documentation); as the kernel docs itself says, do also read the source comments.

The first of these quick summary tables names the core RCU API, shows the prototype, and provides a very brief explanation of its purpose; the second table names the core API and provides links to its official kernel documentation and the useful comments regarding it in the kernel source tree.

Here's the first quick summary table of the five core RCU APIs with the function prototype and a very short explanation:

No.	RCU core API	Prototype and purpose
1	`rcu_read_lock()`	`void rcu_read_lock(void);` Reader: Marks the beginning of an RCU read-side critical section.
2	`rcu_read_unlock()`	`void rcu_read_unlock(void);` Reader: Marks the end of an RCU read-side critical section.

3	`synchronize_rcu()`
	
	`call_rcu()`
		`void synchronize_rcu(void);`
		
		Writer/updater calls this to initiate a GP. A GP is the time period after which it's guaranteed that all pre-existing RCU read-side critical sections that began before this call was invoked are done. This call blocks the writer until that occurs; the writer can then free the old data object.
		
		`void call_rcu(struct rcu_head *head, rcu_callback_t func);`
		
		`call_rcu()` is similar to `synchronize_rcu()` in what it does but is essentially an async version (a callback function's invoked asynchronously, when all readers complete); this is appropriate for when it's illegal to block (e.g., within interrupt contexts or preempt/IRQ-disabled code paths) or when write performance is critical. (Prefer the former in the normal case for overall system performance though.)

4	`rcu_assign_pointer()`	`rcu_assign_pointer(p, v);` Writer: Uses this macro to safely and atomically set the new value `v` on the RCU-protected pointer `p`, in effect communicating to readers the change in value (effectively, it's a safe way of doing `(p) = (v)`, taking into account CPU/compiler memory-ordering quirks).

5	`rcu_dereference()`	`typeof(p) rcu_dereference(p);`
		
		Reader: Safely fetch an RCU-protected pointer that can then be dereferenced (and used). Be careful – you must only use it within the enclosing RCU read-side critical section. Here's typical usage:
		
		`p = rcu_dereference(foo); x = p->whatever;`
		
		Please do see the detailed kernel documentation as well (shown in the next table; it's usually implemented as a macro).

Table 13.6: Quick summary of RCU core APIs: names and very brief explanations

Here's the second quick summary table of the five core RCU APIs, with links to their official kernel documentation and kernel source comments (the links are with respect to kernel version 6.1.25):

No.	RCU core API	Kernel documentation	Useful comments in the kernel source
1	`rcu_read_lock()`	`https://elixir.bootlin.com/linux/v6.1.25/source/Documentation/RCU/whatisRCU.rst#L156`	`https://elixir.bootlin.com/linux/v6.1.25/source/include/linux/rcupdate.h#L696`
2	`rcu_read_unlock()`	`https://elixir.bootlin.com/linux/v6.1.25/source/Documentation/RCU/whatisRCU.rst#L170`	`https://elixir.bootlin.com/linux/v6.1.25/source/include/linux/rcupdate.h#L762`

3	`synchronize_rcu()`
	
	`call_rcu()`
		`https://elixir.bootlin.com/linux/v6.1.25/source/Documentation/RCU/whatisRCU.rst#L178`
		
		<same>
			`https://elixir.bootlin.com/linux/v6.1.25/source/kernel/rcu/tree.c#L3461`
			
			`https://elixir.bootlin.com/linux/v6.1.25/source/kernel/rcu/tree.c#L2733`

4	`rcu_assign_pointer()`	`https://elixir.bootlin.com/linux/v6.1.25/source/Documentation/RCU/whatisRCU.rst#L231`	`https://elixir.bootlin.com/linux/v6.1.25/source/include/linux/rcupdate.h#L461`
5	`rcu_dereference()`	`https://elixir.bootlin.com/linux/v6.1.25/source/Documentation/RCU/whatisRCU.rst#L251`	`https://elixir.bootlin.com/linux/v6.1.25/source/include/linux/rcupdate.h#L650`

Table 13.7: Quick summary of RCU core APIs: names and links to the official kernel documentation and kernel source comments

A quick note: you'll at times come across different internal "versions" or implementations of RCU APIs within the kernel – typically, classic RCU, tree RCU, tiny RCU, real-time RCU, and **Sleepable RCU (SRCU)**. Practically speaking, the modern one, the one that's typically used, is the **hierarchical or "tree RCU" implementation** where scalability is vastly improved (allowing RCU to function well even on systems with thousands of cores). This LWN article by (who else) Paul E. McKenney explains it superbly: *Hierarchical RCU, Paul E. McKenney, Nov 2008*: `https://lwn.net/Articles/305782/`.

Trying out RCU: a simple list RCU demo module

Again, as mentioned earlier back in *Chapter 6*, *Kernel Internals Essentials – Processes and Threads*, in the *Iterating over the kernel's task lists* section, we showed a simple module – `ch6/list_demo` – to perform simple operations on a linked list (leveraging the kernel's *list.h* header's routines). There, we didn't use any kernel synchronization constructs whatsoever, which, of course, is wrong; we certainly need to protect critical sections against concurrent access (to prevent data races)! So, in the *Trying out the reader-writer spinlock* section in this chapter, we rewrote that module and employed the reader-writer spinlock to protect the critical sections (here's the code: `ch13/2_list_demo_rdwrlock/`).

After understanding the downsides of the reader-writer lock, and the fact that it's being aggressively replaced by RCU, we now (finally!) refactor even that earlier code, which employs the reader-writer spinlock, to use RCU lock-free synchronization instead; the code's available here: `ch13/3_lockfree/list_demo_rcu`. A brief look at a few relevant portions of the code follows:

```
// ch13/3_lockfree/list_demo_rcu/list_demo_rcu.c
[ … ]
int add2tail(u64 v1, u64 v2, s8 achar, spinlock_t *lock)
{
    struct node *mynode = NULL;
    [ … ]
    mynode = kzalloc(sizeof(struct node), GFP_ATOMIC);
    [ … ]
    INIT_LIST_HEAD(&mynode->list);

    /*--- Update (write) to the data structure in qs; we need to protect
     * against concurrency, we employ a spinlock
     */
    pr_info("list update: using spinlock\n");
    spin_lock(lock);
    // void list_add_tail(struct list_head *new, struct list_head *head)
    list_add_tail_rcu(&mynode->list, &head_node);
    spin_unlock(lock);
    pr_info("Added a node (with letter '%c') to the list...\n", achar);
[ … ]
```

First off, notice how the write-side critical section is protected not by RCU but via a regular spinlock (we talked about this; RCU has no way to protect concurrent writers, and they must ensure some protection themselves). Next, we use the RCU-safe version of `list_add_tail()`, called `list_add_tail_rcu()`, to append a node to the tail of the list. Its comment is instructive as well:

```
// include/linux/rculist.h
[ … ]
 * The caller must take whatever precautions are necessary
```

```
 * (such as holding appropriate locks) to avoid racing
 * with another list-mutation primitive, such as list_add_tail_rcu()
 * or list_del_rcu(), running on this same list.
 * However, it is perfectly legal to run concurrently with
 * the _rcu list-traversal primitives, such as
* list_for_each_entry_rcu().
[ … ]
```

Which is why we use a spinlock.

Now, let's look at the RCU read-side critical sections in the module code:

```
[ … ]
void showlist(void)
{
    struct node *curr;

    if (list_empty(&head_node))
        return;
    pr_info("          val1      |      val2     | letter\n");

    /* list_for_each_entry_rcu() is simpler than list_for_each_rcu();
         * it internally invokes __container_of() to get the ptr to curr
struct.
         * Also, this is the RCU-safe ver: from it's comments:
         * '... This list-traversal primitive may safely run concurrently with
         * the _rcu list-mutation primitives such as list_add_rcu()
         * as long as the traversal is guarded by rcu_read_lock(). ...'
         */
        rcu_read_lock();
        list_for_each_entry_rcu(curr, &head_node, list) {
                pr_info("%16llu %16llu     %c\n", curr->ival1, curr->ival2,
curr->letter);
        }
        rcu_read_unlock();
}
[ … ]
```

Aha, there it is: the read-side critical section "protected" via RCU. Well, as we've learned, there's really no locking; the reader threads run concurrently, in parallel, even with writers in play.

Let's try it out! Once built and inserted into kernel memory, you can use the `run` Bash wrapper script within the same folder to test the I/O – the read/write to the list.

We have the script write to the device node, which in turn has our driver's write method invoke the `list_demo_rcu.c:add2tail()` routine, writing (besides some integers) the characters `R`, `C`, and `U` to the list.

```
$ cd <book_src>/ch13/3_lockfree/list_demo_rcu
$ make
[ … ]
```

To make the demo a bit more interesting, I run the an script thrice: `./run ; ./run; ./run`. Here's a partial screenshot of the kernel log after doing so:

```
[ 2898.388316] list_demo_rcu_lkm:open_miscdrv_rdwr(): 004)  run :8588   |  ...0   /* open_miscdrv_rdwr() */
[ 2898.388344] list_demo_rcu_lkm:write_miscdrv_rdwr(): 004)  run :8588   |  ...0   /* write_miscdrv_rdwr() */
[ 2898.388354] list_demo_rcu_lkm:add2tail(): list update: using spinlock
[ 2898.388359] list_demo_rcu_lkm:add2tail(): Added a node (with letter 'R') to the list...
[ 2898.388390] list_demo_rcu_lkm:add2tail(): list update: using spinlock
[ 2898.388395] list_demo_rcu_lkm:add2tail(): Added a node (with letter 'C') to the list...
[ 2898.388399] list_demo_rcu_lkm:add2tail(): list update: using spinlock
[ 2898.388404] list_demo_rcu_lkm:add2tail(): Added a node (with letter 'U') to the list...
[ 2898.388412] list_demo_rcu_lkm:close_miscdrv_rdwr(): 004)  run :8588   |  ...0   /* close_miscdrv_rdwr() */
[ 2898.397096] list_demo_rcu_lkm:open_miscdrv_rdwr(): 000)  dd :8591   |  ...0   /* open_miscdrv_rdwr() */
[ 2898.397930] list_demo_rcu_lkm:read_miscdrv_rdwr(): 000)  dd :8591   |  ...0   /* read_miscdrv_rdwr() */
[ 2898.397971] list_demo_rcu_lkm:read_miscdrv_rdwr(): dd wants to read (upto) 1024 bytes
[ 2898.398048] list_demo_rcu_lkm:showlist():         val1     |     val2     | letter
[ 2898.398077] list_demo_rcu_lkm:showlist():            1               2        R
[ 2898.398151] list_demo_rcu_lkm:showlist():            3            1415        C
[ 2898.398175] list_demo_rcu_lkm:showlist():   4295616376      4295616451        U
[ 2898.398181] list_demo_rcu_lkm:showlist():            1               2        R
[ 2898.398189] list_demo_rcu_lkm:showlist():            3            1415        C
[ 2898.398195] list_demo_rcu_lkm:showlist():   4295616652      4295616727        U
[ 2898.398264] list_demo_rcu_lkm:showlist():            1               2        R
[ 2898.398270] list_demo_rcu_lkm:showlist():            3            1415        C
[ 2898.398354] list_demo_rcu_lkm:showlist():   4295616870      4295616945        U
[ 2898.398421] list_demo_rcu_lkm:close_miscdrv_rdwr(): 000)  dd :8591   |  ...0   /* close_miscdrv_rdwr() */
```

Figure 13.17: Screenshot showing the kernel log after performing some I/O on the RCU-protected list thrice (you can see the characters R, C, and U have been written to the list thrice each!)

Removing the module has its "exit" function invoke our function:

```
freelist(&writelock);
```

...which safely deletes and frees each node on the list like so:

```
void freelist(spinlock_t *lock)
{
        struct node *curr;
```

```
        if (list_empty(&head_node))
                return;

        pr_info("freeing list items...\n");

    // Wait for any pre-existing RCU readers to cycle off the CPU(s)...
        synchronize_rcu();

        // ... and now delete and free up the list nodes
        spin_lock(lock);
        list_for_each_entry_rcu(curr, &head_node, list) {
                list_del_rcu(&curr->list);
                kfree(curr);
        }
        spin_unlock(lock);
}
```

Here's the output from the sudo rmmod list_demo_rcu_lkm command:

```
[ … ]
[ 2902.546835] list_demo_rcu_lkm:freelist(): freeing list items...
[ 2902.574092] list_demo_rcu_lkm:list_demo_rcu_exit(): LKP misc driver lkp_
miscdrv_list_rcu deregistered, bye
```

Exercise

Make a copy of the ch13/3_lockfree/list_demo_rcu/list_demo_rcu.c module's code. Refactor it to allow passing a module parameter buggy; when passed and set to 1, have the module issue a blocking call in the RCU read-side critical section. See what happens; study the kernel log.

More RCU usage examples

Another, and a better, example of using basic RCU primitives is something we cover soon, in the upcoming *Fixing it – take 2: use RCU* section. There, we solve concurrency issues when iterating over all threads in the kernel task list, using RCU for lock-free synchronization. Ensure you check out the code and try it out!

An even better example of using RCU primitives for a list-based module can be found here: https://github.com/jinb-park/rcu_example/tree/master. Do check it out.

The next section shows you where you can access (much) more RCU documentation and, within it, learn about in-kernel RCU usage.

RCU: detailed documentation

Much deeper detail on RCU and its in-kernel implementation and usage can be obtained via the official kernel documentation, of course:

- The "root" document (the `index.rst` one) for RCU documentation on the 6.1 kernel series can be found here: *RCU Concepts*: `https://elixir.bootlin.com/linux/v6.1.25/source/Documentation/RCU/index.rst`. Within it, you'll see that it lists a *Design* directory with some documents, plus several other RCU documents (in the usual `.rst` format). To get an idea of what it looks like, see the following screenshot:

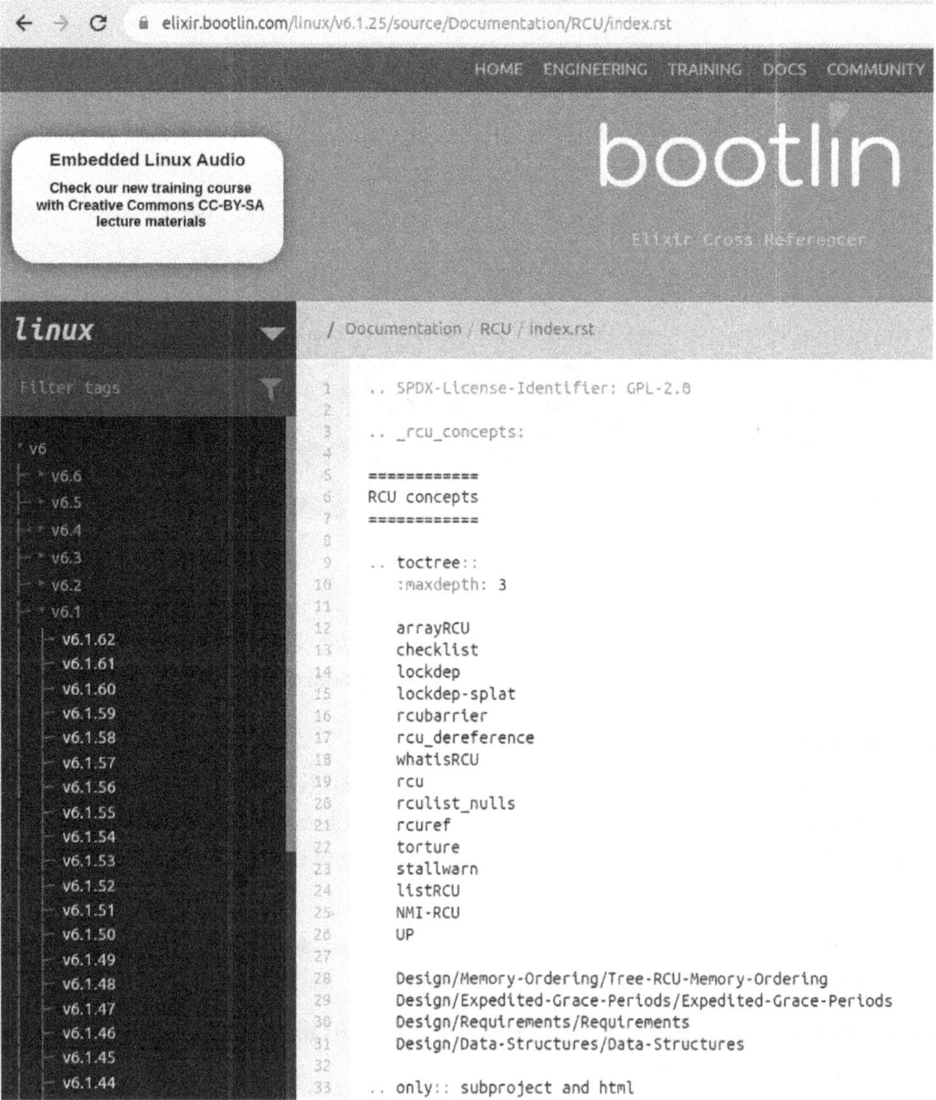

Figure 13.18: Screenshot showing the "root" (index doc) of the official (v6.1) kernel documentation on RCU

- Starting with the classic *What is RCU? – "Read, Copy, Update"* is a great idea (it's the well-known six-article LWN series collection on RCU): `https://elixir.bootlin.com/linux/v6.1.25/source/Documentation/RCU/whatisRCU.rst` (it's the one labeled *whatisRCU* in *Figure 13.18*).
- Next, to delve deeper into RCU design constructs, check out *(RCU) Design*: `https://elixir.bootlin.com/linux/v6.1.25/source/Documentation/RCU/Design`; within this, the *Requirements* section is especially useful (at least initially): *A Tour Through RCU's Requirements, Paul E. McKenney, IBM, 2015*: `https://elixir.bootlin.com/linux/v6.1.25/source/Documentation/RCU/Design/Requirements/Requirements.rst`.
- Here are details for RCU (specialized APIs) usage, focusing on particular data structures:
    - *Using RCU to Protect Read-Mostly Linked Lists*: `https://elixir.bootlin.com/linux/v6.1.25/source/Documentation/RCU/listRCU.rst`. Here, you'll find excellent step-by-step examples on how exactly RCU is leveraged within the kernel (ordered with the "best fits" first). The very first example in fact – traversing over the process list – is precisely what we learned about in *Chapter 6, Kernel Internals Essentials – Processes and Threads*, in the *Iterating over the kernel's task lists* section. One of the modules we wrote to demo this – `ch6/prcs_showall/prcs_showall.c` – in fact uses the `rcu_read_{un}lock()` primitives when iterating over the list, just as the documentation here shows.

A similar approach, using RCU to solve concurrency issues when iterating over all threads in the kernel task list, is what we cover in the upcoming *Fixing it – take 2: use RCU* section.

    - Of course, there's (a lot!) more to it; RCU APIs are regularly maintained, and they evolve over time (one major update series happened in 2010, one in 2014, and the most recent in 2019). Here's the table-based summarization of *RCU List APIs* (within the broader *The RCU API tables, 2019 edition)*, Paul E. McKenney, Jake Edge: `https://lwn.net/Articles/777165/#RCU%20List%20APIs`.
    - *Using RCU to Protect Read-Mostly Arrays*: `https://elixir.bootlin.com/linux/v6.1.25/source/Documentation/RCU/arrayRCU.rst`.
    - *A Tour through TREE_RCU's Data Structures [LWN.net]*: `https://elixir.bootlin.com/linux/v6.1.25/source/Documentation/RCU/Design/Data-Structures/Data-Structures.rst`.
- The *Further reading* section for this chapter has this and more on RCU, for you to peruse and digest at your leisure! (`https://github.com/PacktPublishing/Linux-Kernel-Programming_2E/blob/main/Further_Reading.md#chapter-13-kernel-synchronization-part-2---further-reading`).

With all these (some of which are quite advanced) materials at hand, many of us will be left wondering where to start, yes? Well, there's an easy answer to that – you've started really well already, with this book!

Okay, that aside, once the basic RCU concepts are understood (which we're hoping the previous few sections have achieved for you), looking into the **core RCU APIs** (or primitives) is a great idea, and for that, this is the place to visit: `https://elixir.bootlin.com/linux/v6.1.25/source/Documentation/RCU/whatisRCU.rst#L137` (again, you can see a quick summary in *Table 13.6*). A brief word on RCU-related kernel debug options follows.

A word on debugging RCU usage

There is help for situations when RCU usage within the kernel (or a module) needs to be debugged. To see the relevant menu items within the kernel configuration step, do the usual `make menuconfig`, navigate to the **Kernel hacking** > **RCU Debugging** sub-menu, and check out the various RCU debug config options therein. The relevant *Kconfig* file is `kernel/rcu/Kconfig.debug` (browsing through it lets you see all options along with their help snippets straight away! Here's the link: `https://elixir.bootlin.com/linux/v6.1.25/source/kernel/rcu/Kconfig.debug`). Enabling them in a debug/development kernel is very useful and helps to catch RCU-related bugs.

Also, setting `CONFIG_RCU_EXPERT=y` enables more RCU debug options (like `CONFIG_PROVE_RCU_LIST`). It can be found in the **General setup** > **RCU Subsystem** sub-menu.

RCU usage within the kernel

Finally, how has it been going for RCU usage in the Linux kernel? *Well*, is the short answer, all things considered. Paul E. McKenney provides this *RCU Linux Usage* (`http://www2.rdrop.com/~paulmck/RCU/linuxusage.html`) page; do check out the various graphs therein.

To give you a feel of it, here's a small table with just a few sample entries (shamelessly gleaned from Paul's painstaking statistics here: `http://www2.rdrop.com/~paulmck/RCU/linuxusage/rculocktab.html`:

Date	# of RCU Invocations	# of (Regular) Lock Invocations
31-Oct-2002	5	28,248
29-Nov-2006	1,147	53,463
08-Dec-2014	10,040	122,705
24-Dec-2018	15,167	143,428
30-Oct-2023	20,355	165,860
08-Jan-2024	20,761	165,602

Table 13.8: RCU vs regular locking API usage in the Linux kernel – a few sample entries

It's interesting, and quite correct, that regular locking primitives (spinlocks, mutexes, and semaphores, for exclusive, reader-writer, and rt-mutex variants) are used a lot more. It makes sense: RCU is a specialized synchronization primitive, to be used in the right set of conditions.

So what are the "right set of conditions" to use RCU? Though mentioned before, it's clearly meant to be employed in *read-mostly* scenarios, when updates (writes) are few, and when readers working with slightly stale or inconsistent data is fine.

So, here's a quick summarization of the best and not-so-good situations in which to use, or not use, RCU; again, it's best to show this via Paul's "RCU Area of Applicability" slide seen here: `https://youtu.be/obDzjElRj9c?list=PLIlI4QbmdBqH9-L1QOq6O5Yxt-YVjRsFZ&t=4036`:

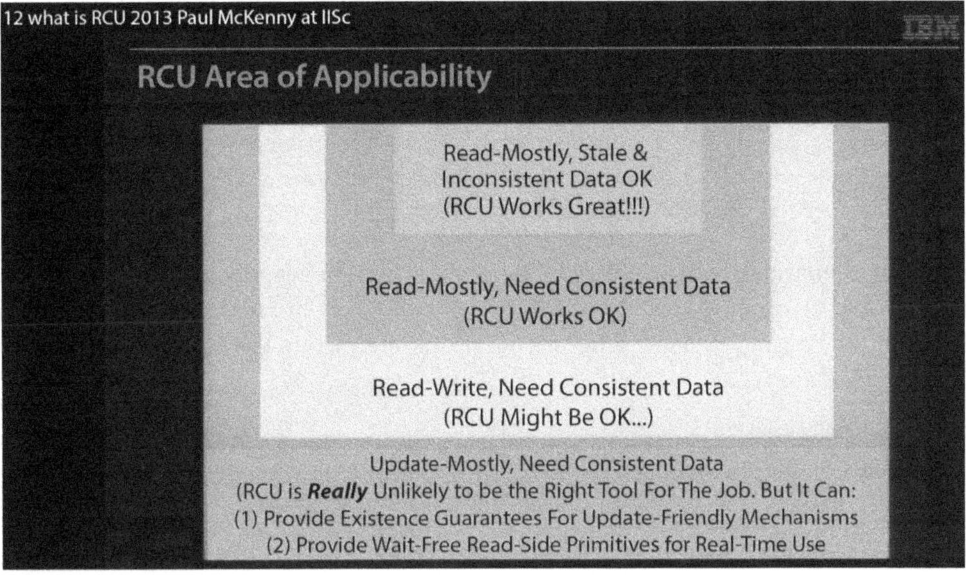

Figure 13.19: Screenshot from Paul E. McKenney's presentation, showing RCU's area of applicability. Credit: Paul E. McKenney

As with anything, it's key to use RCU where appropriate. (There's a saying along the lines of *If a hammer's the only tool you know to use, then everything looks like a nail*; so learn to use all tools!)

FYI: **Real-Time** (**RT**) and lock-free techniques

RT usually conflicts with lock-free programming; why? Most lock-free techniques involve non-preemptible code paths in kernel space. This is precisely what we don't want in a real-time environment to the maximum extent possible! This is because non-preemptible code paths can dramatically increase the (scheduling) latency of a real-time thread (in effect starving it of the CPU, perhaps just when it requires it the most!). With RCU, as we know, the read-side critical section *must* be non-preemptible; this clearly conflicts with RT. Hence, for the **RTL** (**Real-Time Linux**) kernel, a feature called *RCU priority boosting* is applied by default. This LWN article *Priority-Boosting RCU Read-Side Critical Sections, Paul E. McKenney, Feb 2007* (`https://lwn.net/Articles/220677/`) covers the deep details. (Also, enabling `CONFIG_PREEMPTION` takes care of using RCU in real-time Linux. Further, options like *General setup/RCU Subsystem/Offload RCU callback processing from boot-selected CPUs (CONFIG_RCU_NOCB_CPU=y)* can also help in RT scenarios.)

With this, having (almost) completed our detailed discussions on kernel synchronization techniques and APIs, let's move on to another key area: tools and tips when debugging locking issues within kernel code!

Lock debugging within the kernel

As we've learned, locking and synchronization design and implementation can tend to become complex, thus increasing the chances of lurking bugs. The kernel has several means to help debug these difficult situations regarding kernel-level locking issues, *deadlock* being a primary one.

Just in case you haven't already, do ensure you've first read the basics on synchronization, locking, and deadlock guidelines from the previous chapter (*Chapter 12, Kernel Synchronization - Part 1*, especially the *Critical sections, exclusive execution, and atomicity* and *Concurrency concerns within the Linux kernel* sections).

With any debug scenario, there are different points at which debugging occurs, and thus, perhaps differing tools and techniques that could/should be used. Very broadly speaking, a bug might be noticed, and thus debugged, at a few different points in time (within the **Software Development Life Cycle (SDLC)**, really):

- During development
- After development but before release (testing, **Quality Assurance** (**QA**), and so on)
- After internal release
- After release, in the field

A well-known and unfortunately true homily is the "further" a bug is exposed from development, the costlier it is to fix! So you really do want to try to find and fix them as early as possible!

As this book is focused squarely on kernel development, we shall focus here on a few tools and techniques for debugging locking issues at (kernel/module) development time.

Important: We expect that, by now, you're running on a debug kernel, that is, a kernel deliberately configured for development/debug purposes. Performance will take a hit, but that's okay - we're out bug hunting now! We covered the configuration of a typical debug kernel in *Chapter 5, Writing Your First Kernel Module - Part 2,* in the *Configuring a "debug" kernel* section. Specifics on configuring the debug kernel for lock debugging are in fact covered next.

Configuring a debug kernel for lock debugging

Due to its relevance and importance to lock debugging, we will take a quick look at a few key points from the *Linux Kernel patch submission checklist* document (`https://www.kernel.org/doc/html/v6.1/process/submit-checklist.html`) that are most relevant to our discussions here on enabling a debug kernel (especially for lock debugging):

```
[ … ]
12. Has been tested with CONFIG_PREEMPT, CONFIG_DEBUG_PREEMPT, CONFIG_DEBUG_
SLAB, CONFIG_DEBUG_PAGEALLOC, CONFIG_DEBUG_MUTEXES, CONFIG_DEBUG_SPINLOCK,
CONFIG_DEBUG_ATOMIC_SLEEP, CONFIG_PROVE_RCU and CONFIG_DEBUG_OBJECTS_RCU_HEAD
all simultaneously enabled.
13. Has been build- and runtime tested with and without CONFIG_SMP and CONFIG_
PREEMPT.
14. All codepaths have been exercised with all lockdep features enabled.
[ ... ]
```

I cannot fail to mention a very powerful dynamic memory error detector called **Kernel Address SANitizer (KASAN)**. In a nutshell, it uses compile-time instrumentation-based dynamic analysis to catch common memory-related bugs (it works with both GCC and Clang). **ASan** (**Address Sanitizer**), contributed by Google engineers, is used to monitor and detect memory issues in user-space apps (covered in some detail and compared with Valgrind in the *Hands-On System Programming for Linux* book). The kernel equivalent, KASAN, has been available since the 4.0 kernel for the x86_64, and from 4.4 Linux for the AArch64 (ARM64).

In the 5.8 kernel, another really useful tool, **KCSAN** – the **Kernel Concurrency Sanitizer** – was mainlined. It's a **data race detector** for the Linux kernel that works via compile-time instrumentation. The *Linux Kernel Debugging* book covers how to use it. You can also find details in these LWN articles: *Finding race conditions with KCSAN, LWN, October 2019* (`https://lwn.net/Articles/802128/`) and *Concurrency bugs should fear the big bad data-race detector (part 1), LWN, April 2020* (`https://lwn.net/Articles/816850/`).

Note though that KASAN conflicts with KCSAN; thus, you'll have to use them exclusive of each other. Detailed instructions and examples on enabling and using them can be found in the *Linux Kernel Debugging* book. I highly recommend you leverage them as well in your debug kernel.

Also, FYI, several tools do exist to catch locking bugs and deadlocks in user-space apps. Among them are the well-known `helgrind` (from the Valgrind suite), **TSan** (**Thread Sanitizer**), which provides compile-time instrumentation to check for data races in multithreaded applications, and lockdep itself; lockdep can be made to work in user space as well (as a library)! Moreover, the modern [e]BPF framework provides the `deadlock-bpfcc(8)` frontend. It's designed specifically to find potential deadlocks (lock order inversions) in a given running process (or thread).

As we saw back in *Chapter 2, Building the 6.x Linux kernel from Source – Part 1*, we can configure the Linux kernel specifically for our requirements. Here (within the root of the 6.1.25 kernel source tree), we perform `make menuconfig` and (following the advice of the kernel patch submission checklist) navigate to the **Kernel hacking > Lock Debugging (spinlocks, mutexes, etc...)** menu (see the screenshot in *Figure 13.20*, taken on our x86_64 Ubuntu 23.04 LTS guest VM):

```
.config - Linux/x86 6.1.25 Kernel Configuration
> Kernel hacking > Lock Debugging (spinlocks, mutexes, etc...)
                 Lock Debugging (spinlocks, mutexes, etc...)
 Arrow keys navigate the menu.  <Enter> selects submenus ---> (or empty submenus ----).
 Highlighted letters are hotkeys.  Pressing <Y> includes, <N> excludes, <M> modularizes
 features.  Press <Esc><Esc> to exit, <?> for Help, </> for Search.  Legend: [*] built-in
 [ ] excluded  <M> module  < > module capable

        [*] Lock debugging: prove locking correctness
        [ ]   Enable raw_spinlock - spinlock nesting checks (NEW)
        [*] Lock usage statistics
        -*- RT Mutex debugging, deadlock detection
        -*- Spinlock and rw-lock debugging: basic checks
        -*- Mutex debugging: basic checks
        -*- Wait/wound mutex debugging: Slowpath testing
        -*- RW Semaphore debugging: basic checks
        -*- Lock debugging: detect incorrect freeing of live locks
        (15) Bitsize for MAX_LOCKDEP_ENTRIES (NEW)
        (16) Bitsize for MAX_LOCKDEP_CHAINS (NEW)
        (19) Bitsize for MAX_STACK_TRACE_ENTRIES (NEW)
        (14) Bitsize for STACK_TRACE_HASH_SIZE (NEW)
        (12) Bitsize for elements in circular_queue struct (NEW)
        [ ] Lock dependency engine debugging (NEW)
        [*] Sleep inside atomic section checking
        [ ] Locking API boot-time self-tests
        < > torture tests for locking
        < > Wait/wound mutex selftests
        < > torture tests for smp_call_function*()
        [ ] Debugging for csd_lock_wait(), called from smp_call_function*()

          <Select>    < Exit >    < Help >    < Save >    < Load >
```

Figure 13.20: (Truncated) screenshot of the 'Kernel hacking > Lock Debugging (spinlocks, mutexes, etc...) menu with required items enabled for our debug kernel

Figure 13.20 is a (truncated) screenshot of the `< Kernel hacking > Lock Debugging (spinlocks, mutexes, etc...)` menu with required items enabled for our debug kernel.

Instead of having to interactively go through each menu item and selecting the `<Help>` button to see what it's about, a much simpler way to gain the same help information is to peek inside the relevant `Kconfig*` file (that describes the menu). Here, it's `lib/Kconfig.debug`, as all (well, most) debug-related menus are there. For our particular case, search for the menu `Lock Debugging (spinlocks, mutexes, etc...)` string, where the `Lock Debugging` section begins (see the following table). (FYI, the link is `https://elixir.bootlin.com/linux/v6.1.25/source/lib/Kconfig.debug#L1229`.)

The following table summarizes how each kernel lock debugging configuration option helps in debugging (we haven't shown all of them and, for some of them, have directly quoted from the lib/Kconfig.debug file. Also, the last two are with regard to RCU debugging - defined in kernel/rcu/Kconfig.debug - and typically require PROVE_RCU=y and, for some options, RCU_EXPERT=y):

Lock debugging menu title	What it does
Lock debugging: prove locking correctness (CONFIG_PROVE_LOCKING), aka **lockdep**!	This is the lockdep kernel option - turn it on to get "a rolling proof of lock correctness" at all times. Any possibility of locking-related deadlock *is reported even before it actually occurs*; it's very useful! (it's explained shortly in more detail).
Lock usage statistics (CONFIG_LOCK_STAT)	Tracks lock contention points (explained later in more detail).
RT mutex debugging, deadlock detection (CONFIG_DEBUG_RT_MUTEXES)	"This allows rt mutex semantics violations and rt mutex related deadlocks (lockups) to be detected and reported automatically."
Spinlock and rw-lock debugging: basic checks (CONFIG_DEBUG_SPINLOCK)	Turning this on (along with CONFIG_SMP) helps catch missing spinlock initialization and other common spinlock errors (and helps to turn on the non-maskable interrupts (NMI) watchdog as well).
Mutex debugging: basic checks (CONFIG_DEBUG_MUTEXES)	"This feature allows mutex semantics violations to be detected and reported."
RW semaphore debugging: basic checks (CONFIG_DEBUG_RWSEMS)	Allows mismatched RW semaphore locks and unlocks to be detected and reported.
Lock debugging: detect incorrect freeing of live locks (CONFIG_DEBUG_LOCK_ALLOC)	"This feature will check whether any held lock (spinlock, rwlock, mutex, or rwsem) is incorrectly freed by the kernel, via any of the memory-freeing routines (kfree(), kmem_cache_free(), free_pages(), vfree(), etc.), whether a live lock is incorrectly reinitialized via spin_lock_init()/mutex_init()/etc., or whether there is any lock held during task exit."
Sleep inside atomic section checking (CONFIG_DEBUG_ATOMIC_SLEEP)	"If you say Y here, various routines which may sleep will become very noisy if they are called inside atomic sections: when a spinlock is held, inside an rcu read side critical section, inside preempt disabled sections, inside an interrupt, etc..."
Torture tests for locking (CONFIG_LOCK_TORTURE_TEST)	"This option provides a kernel module that runs torture tests on kernel locking primitives. The kernel module may be built after the fact on the running kernel to be tested, if desired." (Can be built either inline with 'Y' or externally as a module with 'M')."

Kernel hacking | RCU Debugging RCU list lockdep debugging (CONFIG_PROVE_RCU_LIST)	"Enable RCU lockdep checking for list usages. By default it is turned off since there are several list RCU users that still need to be converted to pass a lockdep expression. To prevent false-positive splats, we keep it default disabled but once all users are converted, we can remove this config option."
Debug RCU callbacks objects (CONFIG_DEBUG_OBJECTS_RCU_HEAD)	"Enable this to turn on debugging of RCU list heads (call_rcu() usage)."

Table 13.9: Typical kernel lock (and RCU) kernel debug configuration options and their meaning

The last two entries in this table are with regard to RCU debug configs; we briefly covered setting up RCU-related debug options in the *A word on debugging RCU usage* section.

As suggested previously, turning on all or most of these lock debug options within a debug kernel used during development and testing is an excellent idea.

We've provided a small helper script (`ch13/4_lockdep/debugk_locking_configs_check`), which will show you the state of a few (mostly) locking related kernel configs. Configure your debug kernel and run it to verify the config.

Of course, as expected, enabling kernel debug configs might considerably slow down execution (and use more memory); as in life, this is a trade-off you must decide on: you gain detection of common locking issues, errors, and even the possibility of deadlock, at the cost of speed. It's a trade-off you should be more than willing to make, especially when developing (or refactoring) and testing the code.

The lock validator lockdep – catching locking issues early

The Linux kernel has a tremendously useful feature begging to be taken advantage of by kernel developers: a runtime locking correctness or locking dependency validator – in short, **lockdep**. The basic idea is this: the `lockdep` runtime comes into play whenever any locking activity occurs within the kernel – the taking or the release of *any* kernel-level lock, or any locking sequence involving multiple locks.

This is continually tracked or mapped (see a few paragraphs down for more on the performance impact and how it's mitigated). By applying well-known rules for correct locking, `lockdep` makes a conclusion regarding the validity of the correctness of what was done (recall that we learned concepts regarding both the self-deadlock and circular deadlock in *Chapter 12, Kernel Synchronization – Part 1*, in the *Locking guidelines and deadlock* section).

The beauty of it is that `lockdep` achieves 100% mathematical proof (or closure) that a lock sequence is correct or not. The following is a direct quote from the kernel documentation on the topic, `https://www.kernel.org/doc/html/v6.1/locking/lockdep-design.html#proof-of-100-correctness`:

The validator achieves perfect, mathematical 'closure' (proof of locking correctness) in the sense that for every simple, standalone single-task locking sequence that occurred at least once during the lifetime of the kernel, the validator proves it with a 100% certainty that no combination and timing of these locking sequences can cause any class of lock related deadlock.

Note there is a "conditions apply" disclaimer here: `https://www.kernel.org/doc/html/v6.1/locking/lockdep-design.html#id2`. Nothing's 100 percent perfect in every situation, as is the case with lockdep; do see the *A note on lockdep – known issues* section as well.

Furthermore, `lockdep` warns you (via the `WARN*()` macros) of any violation of the following classes of locking bugs: deadlocks/lock inversion scenarios, circular lock dependencies, and hard IRQ/soft IRQ safe/unsafe locking bugs. This information is precious; validating your code with `lockdep` can save hundreds of wasted hours of productivity by catching locking issues early.

We'll come across the term *lock sequence* or *lock chain* quite often; what does it mean? Say, on a certain code path, that lock A is taken, after which lock B is taken – this is a *lock sequence* or *lock chain*. FYI, `lockdep` tracks all locks and their lock chains (they can be viewed through `/proc/lockdep_chains`).

A word on *performance mitigation*: you might well imagine that, with literally thousands of lock instances floating around, it would be absurdly slow to validate every single lock sequence (yes, in fact, it turns out to be a task of order `O(N^2)` algorithmic time complexity!). This would just not work; so instead, `lockdep` works by verifying any locking scenario (lock sequence) **only once**, the very first time it occurs. (It knows this by maintaining a 64-bit hash for every lock chain it encounters.)

There are primitive **user-space** approaches to detecting a possible deadlock scenario: a very primitive – and certainly not guaranteed – way to try and detect possible deadlock is via user space by simply using GNU `ps(1)`; doing `ps -LA -o state,pid,cmd | grep "^D"` prints any threads in the `D` – uninterruptible sleep (`TASK_UNINTERRUPTIBLE`) – state. This could – but may not – be due to a deadlock; if it persists for a long while, chances are higher that it's a deadlock issue. Give it a try! Of course, `lockdep` is a far superior solution. (Note that this only works with GNU `ps`, not the lightweight ones such as `busybox ps`.)

Other useful user-space tools are `strace(1)` and `ltrace(1)` – they provide a detailed trace of every system and library call, respectively, issued by a process (or thread); you might be able to catch a hung process/thread and see where it got stuck (using `strace -p <PID>` might be especially useful on a hung process).

The other point that you need to be clear about is this: `lockdep` will issue warnings regarding (mathematically) incorrect locking *even if no deadlock actually occurs at runtime*! `lockdep` offers proof that there is indeed an issue that could conceivably cause a bug (deadlock, unsafe locking, and so on) at some point in the future if no corrective action is taken. It's usually dead right; take it seriously and fix the issue. (Then again, typically, nothing in the software universe is 100% correct 100% of the time: what if a bug creeps into the `lockdep` code itself? There's even a `CONFIG_DEBUG_LOCKDEP` config option. The bottom line is that we, the human developers, must carefully assess the situation, checking for false positives.)

Next, lockdep works upon a *lock class*; this is simply a "logical" lock as opposed to "physical" instances of that lock. For example, the kernel's open file data structure, struct file, has two locks – a mutex and a spinlock – and each of them is considered a lock class by lockdep. Even if a few thousand instances of struct file exist in memory at runtime, lockdep will track it as a class only. For more detail on lockdep's internal design, we refer you to the official kernel documentation on it here: *Runtime locking correctness validator* (https://www.kernel.org/doc/html/v6.1/locking/lockdep-design.html#runtime-locking-correctness-validator).

Catching deadlock bugs with lockdep – a few examples

Here, we shall assume that you've by now built and are running upon a debug kernel with lockdep enabled (as described in detail in the *Configuring a debug kernel for lock debugging* section). On my VM running a 'debug' kernel, let's minimally verify that lockdep is indeed enabled:

```
$ uname -r
6.1.25-lock-dbg
$ grep PROVE_LOCKING /boot/config-6.1.25-lock-dbg
CONFIG_PROVE_LOCKING=y
```

Okay, good! Now, let's get hands-on with some deadlocks, seeing how lockdep will help you catch them. Read on!

Example 1 – catching a self deadlock bug with lockdep

As a first example, let's travel back to one of the kernel modules from *Chapter 6, Kernel Internals Essentials – Processes and Threads*, in the *Iterating over the kernel's task lists* section, as detailed here: ch6/foreach/thrd_showall/thrd_showall.c. In this code, we looped over the kernel task list, over each thread alive, printing some details from within its task structure. With regard to this, here's one way to obtain the *name* of the thread (recall that it's a member of the task structure named comm):

```
// ch6/foreach/thrd_showall/thrd_showall.c
static int showthrds(void)
{
    struct task_struct *g = NULL, *t = NULL; /* 'g' : process ptr; 't': thread
ptr */
    [ ... ]
    do_each_thread(g, t) { /* 'g' : process ptr; 't': thread ptr */
        get_task_struct(t);     /* take a reference to the task struct */
        task_lock(t);
        [ ... ]
        if (!g->mm) {    // kernel thread
            snprintf(tmp, TMPMAX-1, " [%16s]", t->comm);
        } else {
            snprintf(tmp, TMPMAX-1, " %16s ", t->comm);
        }
        [ ... ]
```

From the 6.6 (very recent as of this writing) kernel, the `do_each_thread()/while_each_thread()` style macros have been removed in favor of the simpler and more readable `for_each_process_thread()` macro. Though our code takes this into account, to keep things simple, we don't show it in the above code snippet.

If the task's a kernel thread, we place its name within square brackets. Now, while retrieving the thread name this way works, instead of directly looking it up with `t->comm` (as we do here), there's a better way to do it: the kernel provides the `{get,set}_task_comm()` helper routines to both get and set the name of the task. (Recall that we covered similar stuff in *Chapter 6, Kernel Internals Essentials – Processes and Threads*, in the *Built-in kernel helper methods and optimizations* section.)

So now, let's do it this way! We rewrite the module (as `ch13/4_lockdep/buggy_thrdshow_eg/thrd_showall_buggy.c`) to use the `get_task_comm(buf, tsk)` helper macro; the first parameter is the buffer pointer to place the thread's name into (it's expected that you've allocated memory to it), and the second parameter is the pointer to the task structure of the thread whose name you are querying (see the following code snippet):

```
// ch13/4_lockdep/buggy_thrdshow_eg/thrd_showall_buggy.c
[ … ]
static int showthrds_buggy(void)
{
    struct task_struct *g, *t;                /* 'g' : process ptr; 't': thread 
ptr */
     char buf[BUFMAX], tmp[TMPMAX], tasknm[TASK_COMM_LEN];
     [ ... ]
#if LINUX_VERSION_CODE < KERNEL_VERSION(6, 6, 0)
    do_each_thread(g, t) {     /* 'g' : process ptr; 't': thread ptr */
#else
    for_each_process_thread(g, t) {   /* 'g' : process ptr; 't': thread ptr */
#endif
        get_task_struct(t);     /* take a reference to the task struct */
            task_lock(t);
            [ ... ]
            get_task_comm(tasknm, t);
            if (!g->mm)      // kernel thread
                    snprintf_lkp(tmp, sizeof(tasknm)+3, " [%16s]", tasknm);
            else
                    snprintf_lkp(tmp, sizeof(tasknm)+3, " %16s ", tasknm);
            [ ... ]
```

However, when compiled and inserted into the kernel on our test system (a VM, thank goodness), it can get weird, or even simply hang! (When I did this, running it on an x86_64 Ubuntu VM, I was able to retrieve the kernel log by running `sudo journalctl -k -f` in another terminal window.)

What if your system just hangs upon insertion of this LKM? Well, that's a taste of the difficulty of kernel debugging! One thing you can try (which worked for me when trying this very example on a x86_64 Fedora 29 VM) is to reboot the hung VM and look up the kernel log by leveraging systemd's powerful `journalctl(1)` utility with the `journalctl --since="1 hour ago"` command; you should be able to see the `printks` from `lockdep` now. Again, unfortunately, it's not guaranteed that the key portion of the kernel log is saved to disk (at the time it hung) for `journalctl` to be able to retrieve it. This is why using the kernel's `kdump` feature – and then performing postmortem analysis of the kernel dump image file with `crash(8)` – can be a lifesaver (see the resources on using `kdump` and *crash* in the *Further reading* section for this chapter).

Glancing at the kernel log, it becomes clear: `lockdep` *has caught a (self) deadlock issue!* We show relevant parts of the output in the screenshot (with the timestamp column eliminated):

```
============================================
WARNING: possible recursive locking detected
6.1.25-lock-dbg #2 Tainted: G           OE
--------------------------------------------
insmod/3395 is trying to acquire lock:
ffff9307c12b5ae8 (&p->alloc_lock){+.+.}-{2:2}, at: __get_task_comm+0x28/0x60

but task is already holding lock:
ffff9307c12b5ae8 (&p->alloc_lock){+.+.}-{2:2}, at: showthrds_buggy+0x11a/0x5a6 [thrd_showall_buggy]

other info that might help us debug this:
 Possible unsafe locking scenario:
       CPU0
       ----
  lock(&p->alloc_lock);
  lock(&p->alloc_lock);

 *** DEADLOCK ***
 May be due to missing lock nesting notation
1 lock held by insmod/3395:
 #0: ffff9307c12b5ae8 (&p->alloc_lock){+.+.}-{2:2}, at: showthrds_buggy+0x11a/0x5a6 [thrd_showall_buggy]

stack backtrace:
CPU: 0 PID: 3395 Comm: insmod Tainted: G           OE      6.1.25-lock-dbg #2
Hardware name: innotek GmbH VirtualBox/VirtualBox, BIOS VirtualBox 12/01/2006
Call Trace:
 <TASK>
 dump_stack_lvl+0x5a/0x82
 dump_stack+0x10/0x18
 __lock_acquire.cold+0xad/0x2f8
 lock_acquire+0xd0/0x2b0
 ? __get_task_comm+0x28/0x60
 ? vsnprintf+0x136/0x960
 _raw_spin_lock+0x37/0x90
 ? __get_task_comm+0x28/0x60
 __get_task_comm+0x28/0x60
 showthrds_buggy+0x2a3/0x5a6 [thrd_showall_buggy]
 ? 0xffffffffc04ec000
```

Figure 13.21: (Partial) screenshot showing the kernel log after our buggy module is loaded; lockdep catches the self deadlock!

It's pretty clear... furthermore, the stack backtrace, the kernel stack of `insmod` (as it was the process context), appears, confirming it was our module that called `get_task_comm()` (which became `__get_task_comm()`) that led to the bug (the bottom portion of *Figure 13.21*).

Although a lot more detail follows (in this case, register values, and even a soft lockup detected by the watchdog, RCU CPU stalls, and so on), what we see in the preceding figure is sufficient to deduce what happened. Clearly, `lockdep` tells us `insmod/3395 is trying to acquire lock:`, followed by `but task is already holding lock:`. Next (look carefully at *Figure 13.21*), the lock that `insmod` holds is `(p->alloc_lock)` (for now, ignore the `lockdep` notations that follow it; we will explain them shortly) and the routine that actually attempts to acquire it (shown after `at:`) is `__get_task_comm+0x28/0x60`.

So it seems like our calling `get_task_comm()` triggered the whole issue; now we're getting somewhere! Right, let's figure out what exactly occurred when we called `get_task_comm()`; we find that it's a macro (defined in *include/linux/sched.h*), a wrapper around the actual worker routine, `__get_task_comm()`. Its code is as follows:

```
// fs/exec.c
char *__get_task_comm(char *buf, size_t buf_size, struct task_struct *tsk)
{
        task_lock(tsk);
        /* Always NUL terminated and zero-padded */
        strscpy_pad(buf, tsk->comm, buf_size);
        task_unlock(tsk);
        return buf;
}
EXPORT_SYMBOL_GPL(__get_task_comm);
```

Ah, there's the problem, the root cause of the bug: the `__get_task_comm()` function straight away invokes the `task_lock()` API, in effect attempting to reacquire the very same lock that we're already holding, causing (self) deadlock!

Recall one of the "rules" we covered in the previous chapter: "don't attempt to re-acquire a lock you already hold."

Where did we acquire it? Recall that one of the very first lines of code in our (buggy) kernel module after entering the loop is where we call `task_lock(t)`; then, just a few lines of code later, we invoke `get_task_comm()`, which internally attempts to reacquire the very same lock: the result is *self deadlock*:

```
do_each_thread(g, t) {   /* 'g' : process ptr; 't': thread ptr */
    get_task_struct(t);      /* take a reference to the task struct */
    task_lock(t);
    [ ... ]
```

```
    get_task_comm(tasknm, t);
    [ … ]
```

Furthermore, finding which particular lock was double-acquired is easy; look up the code of the `task_lock()` routine:

```
// include/linux/sched/task.h */
static inline void task_lock(struct task_struct *p)
{
    spin_lock(&p->alloc_lock);
}
```

So, it all makes perfect sense now; it's a spinlock within the task structure named `alloc_lock`, just as `lockdep` informed us (see *Figure 13.21*).

But `lockdep`'s report has several puzzling notations. Take the following lines:

```
insmod/3395 is trying to acquire lock:
ffff9307c12b5ae8 (&p->alloc_lock){+.+.}-{2:2}, at: __get_task_comm+0x28/0x60

but task is already holding lock:
ffff9307c12b5ae8 (&p->alloc_lock){+.+.}-{2:2}, at: showthrds_buggy+0x11a/0x5a6
[thrd_showall_buggy]
```

Okay, let's break it down:

- The number in the leftmost column of the second line seen in the preceding output block (`0xffff9307c12b5ae8`) is the 64-bit lightweight hash value that lockdep uses to identify this particular lock sequence. Notice it's precisely the same as the hash in the following line, so we know it's the very same lock being acted upon!
- Clearly, the lock's named `alloc_lock`, and attempts have been made to take it twice in succession (the root cause of the bug): once by our module code calling `task_lock()` and once (again) by it calling `get_task_comm()`, which is a wrapper over the routine `__get_task_comm()`. Indeed, the name of our module shows up in square brackets, `[thrd_showall_buggy]`.
- Next, `{+.+.}` is lockdep's notation for what state this lock was acquired in (e.g., + implies a lock acquired with IRQs (interrupts) enabled, `.` implies a lock acquired with IRQs disabled and not in the IRQ context, and so on). These are explained in the kernel documentation (`https://www.kernel.org/doc/Documentation/locking/lockdep-design.txt`); we'll leave it at that.

A detailed presentation on interpreting `lockdep` output was given by Steve Rostedt at a Linux Plumber Conference (back in 2011); the relevant slides are informative, exploring both simple and complex deadlock scenarios and how `lockdep` can detect them:

Lockdep: How to read its cryptic output (`https://blog.linuxplumbersconf.org/2011/ocw/sessions/153`).

Fixing it

Now that we understand the issue here, how do we fix it? Let's look at two ways to do so.

Fixing it – take 1: refactor the code

By looking at `lockdep`'s report (*Figure 13.21*) and interpreting it, it's quite simple (this is one way to fix it; we discuss another shortly): since the `task_lock(t)` API in our module's code acquires the task structure spinlock named `alloc_lock` at the start of the `do-while` loop, ensure that before calling the `get_task_comm()` routine – which we now know internally attempts to acquire (and then release) this very same lock – you first unlock it, then perform `get_task_comm()`, and then lock it again.

The following screenshot (*Figure 13.22*) shows the code-level difference (via the `diff(1)` utility) between the older buggy version (`ch13/4_lockdep/buggy_thrdshow_eg/thrd_showall_buggy.c`) and the newer, fixed version of our code (`ch13/4_lockdep/fixed_thrdshow_eg/thrd_showall_fixed.c`):

```
        do_each_thread(g, t) {     /* 'g' : process ptr; 't': thread ptr */
                get_task_struct(t);     /* take a reference to the task struct */
-               task_lock(t);
+  (1)          task_lock(t);  /*** task lock taken here! ***/

                snprintf(buf, BUFMAX-1, "%6d %6d ", g->tgid, t->pid);
                /* task_struct addr and kernel-mode stack addr */
@@ -76,8 +75,17 @@
                snprintf(tmp, TMPMAX-1, "  0x%px", t->stack);
                strncat(buf, tmp, TMPMAX);

-               get_task_comm(tasknm, t);
-/*--- LOCKDEP catches a deadlock here !! ---*/
+           /* In the 'buggy' ver of this code, LOCKDEP did catch a deadlock here !!
+            * (at the point that get_task_comm() was invoked).
+            * The reason's clear: get_task_comm() attempts to take the very same lock
+            * that we just took above via task_lock(t);  !! This is obvious self-deadlock...
+            * So, we fix it here by first unlocking it, calling get_task_comm(), and
+            * then re-locking it.
+            */
+  (2)           task_unlock(t);
+  (3)           get_task_comm(tasknm, t);
+  (4)           task_lock(t);
+
                if (!g->mm)     // kernel thread
                        snprintf(tmp, sizeof(tasknm)+3, " [%16s]", tasknm);
                else
```

Figure 13.22: (Partial) screenshot showing the key differences between the buggy and fixed versions of our demo thrdshow LKM

(Do ignore the diff-generated line numbers; they can vary.) The four numbered circles in *Figure 13.22* tell the story of the fix:

1. The `task_lock(t)` acquires the task lock (which, as we know, is the `alloc_lock` spinlock within the task structure).
2. Understanding that the `get_task_comm()` also attempts to acquire this very same lock (by it calling `task_lock()`), we first unlock it, via the `task_unlock()`.
3. We call `get_task_comm()`, safely; it internally calls `task_lock()`, acquiring the `alloc_lock` spinlock, and then releases it (via `task_unlock()`).
4. Now, we call `task_lock()` to again acquire the required spinlock.

Great! It now seems to work. However, it's not a great solution; here's why:

- First, between the `task_unlock()` and `task_lock()` – steps 2 and 4 in *Figure 13.22* – we've introduced the possibility of a race!
- Next, the `task_{un}lock()` routines employ a (spin)lock that's effective for only some of the task structure members (it doesn't protect all of them).

So, a better solution would be to simply avoid using the `task_{un}lock()` routines completely and still provide protection.

One way to do so, and especially given the fact that this code is certainly of a *read-mostly* kind, we could use the `tasklist_lock` reader-writer spinlock (a reader-writer spinlock meant for the task list). This approach would work, but as we learned, the reader-writer lock can suffer from (pretty major) performance issues as well. In addition, it isn't exported and, hence, is unavailable to our kernel module.

Fixing it – take 2: use RCU

Again, as this kind of code – iterating over every task structure in a read-only manner – is clearly a *mostly read* candidate, let's (you guessed it!) use kernel RCU lock-free synchronization! Indeed, RCU proffers itself as the best, highly optimized solution here.

So we've refactored the code of this module to use an RCU read-side critical section; here's the new and better version: `ch13/3_lockfree/thrdshowall_rcu/`. Moreover, as here we don't ever modify any task structure's content – in other words, there are no in-place updates (writes) – we even eliminate the need for write protection via a spinlock.

So what does the code look like in the RCU version? As the majority of it is the same, instead of showing the entire code, we just show the difference between the iterate-over-task-struct version using the `task_{un}lock()` routines (our earlier "fix," the one explained in the previous section) and this new and improved RCU-based version:

```
+       rcu_read_lock(); /* This triggers off an RCU read-side critical section;
+                         * ensure you are non-blocking within it! */
[ ... ]
        do_each_thread(g, t) {      /* 'g' : process ptr; 't': thread ptr */
-               get_task_struct(t);     /* take a reference to the task struct */
-               task_lock(t);  /*** task lock taken here! ***/
+               g_rcu = rcu_dereference(g);
+               t_rcu = rcu_dereference(t);
+
+               get_task_struct(t_rcu); /* take a reference to the task struct */

[ ... ]
-               if (!g->mm) // kernel thread
+               if (!g_rcu->mm)   // kernel thread
                        snprintf(tmp, sizeof(tasknm)+3, " [%16s]", tasknm);
[ ... ]
-               put_task_struct(t);     /* release reference to the task struct */
+               put_task_struct(t_rcu); /* release reference to the task struct */
        } while_each_thread(g, t);
+       rcu_read_unlock();        /* This ends the RCU read-side critical section */
```

Figure 13.23: (Partial) screenshot showing, within the patch, the key differences between the original fixed version, using the task_{un}lock(), and the "better" version, using RCU, of our demo LKMs

The book's GitHub repo has the patch as well (`https://github.com/PacktPublishing/Linux-Kernel-Programming_2E/blob/main/ch13/3_lockfree/thrdshowall_rcu/using_rcu.patch`; do ignore the diff-generated line numbers; they can vary).

Study the code (and patch); notice how, as it's now RCU-based, we leverage the `rcu_dereference()` routine to "safely fetch an RCU-protected pointer that can then be dereferenced (and used)" (just as you learned in the *RCU Level 3 explanation (Level 2 + more details).*

All right, let's now move on to one more example of leveraging the power of lockdep, that of catching an AB-BA (circular) deadlock!

Example 2 – catching an AB-BA deadlock with lockdep

As one more example, let's check out a (demo) kernel module that quite deliberately creates a **circular dependency**, which will ultimately result in deadlock (we explained what exactly circular deadlock / AB-BA deadlock is in *Chapter 12, Kernel Synchronization – Part 1* in the *Locking guidelines and deadlock* section.) The code can be found here: `ch13/4_lockdep/deadlock_eg_AB-BA`. We've based this module on our earlier *percpu* module demo code (`ch13/3_lockfree/percpu/`). Just as you'll recall from that kernel module, we create two kernel threads (kthreads) and ensure (by using a hacked `sched_setaffinity()` function pointer) that each kernel thread runs on a unique CPU core (the first kernel thread on CPU core `0` and the second on core `1`).

This way, we have concurrency. Now, we have both these threads working with *two* spinlocks, `lockA` and `lockB`. Now, very importantly, recall the concepts we learned regarding both the self-deadlock and circular deadlock in *Chapter 12, Kernel Synchronization – Part 1*, in the *Locking guidelines and deadlock* section: *lock ordering* is critical.

So here, understanding that we're running in process context (kthreads) with two or more locks, we document and (must) follow a lock ordering rule – say, ***first take lockA, and then lockB***. The precise lock order doesn't matter (we can change it to *first take lockB, and then lockA* if we wish); the fact that it's always followed does.

Great! So one way it should *not* be done is like this (the timeline is vertically downward):

```
kthread 0 on CPU #0                         kthread 1 on CPU #1
 Take lockA                                  Take lockB
 <perform work>                             <perform work>
 (Try and) take lockA                         (Try and) take lockB
 < ... spins forever :                        < ... spins forever :
DEADLOCK ... >
 DEADLOCK ... >
```

Doing this, of course, causes the classic AB-BA deadlock! Because the module (it's *kernel thread (kthread) 1*, actually) **ignored the lock ordering rule** (which we will do in our upcoming demo module when the module parameter `lock_ooo` is set to `1; ooo = out-of-order`), it deadlocks.

Here's portions of the relevant code (we again don't show the whole program here, as we definitely expect you to clone this book's GitHub repository, carefully read the code, and try it out yourself):

```
// ch13/4_lockdep/deadlock_eg_AB-BA/deadlock_eg_AB-BA.c
[ ... ]
/* Our kernel thread worker routine */
static int thrd_work(void *arg)
{
    [ ... ]
   if (thrd == 0) {                        /* our kthread #0 runs on CPU 0 */
        pr_info(" Thread #%ld: locking: we do:"
            " lockA --> lockB\n", thrd);
        for (i = 0; i < THRD0_ITERS; i ++) {
                /* In this thread, perform the locking per the lock ordering
'rule';
                 * first take lockA, then lockB */
                pr_info(" iteration #%d on cpu #%ld\n", i, thrd);
                spin_lock(&lockA);
                DELAY_LOOP('A', 3);
                spin_lock(&lockB);
                DELAY_LOOP('B', 2);
                spin_unlock(&lockB);
                spin_unlock(&lockA);
        }
```

Our kernel thread 0 does it correctly, following our lock ordering rule (it first acquires the spinlock lockA and then the lockB spinlock). Now, the code relevant to our kernel thread 1 (continued from the previous code) is as follows:

```
    [ ... ]
  } else if (thrd == 1) {              /* our kthread #1 runs on CPU 1 */
  for (i = 0; i < THRD1_ITERS; i ++) {
    /* In this thread, if the parameter lock_ooo is 1, *violate* the
     * lock ordering 'rule'; first (attempt to) take lockB, then lockA */
    pr_info(" iteration #%d on cpu #%ld\n", i, thrd);
    if (lock_ooo == 1) {          // violate the rule, naughty boy!
            pr_info(" Thread #%ld: locking: we do: lockB --> lockA\n",thrd);
            spin_lock(&lockB);
                    DELAY_LOOP('B', 2);
                    spin_lock(&lockA);
                    DELAY_LOOP('A', 3);
                    spin_unlock(&lockA);
                    spin_unlock(&lockB);
            } else if (lock_ooo == 0) { // follow the rule, good boy!
                    pr_info(" Thread #%ld: locking: we do: lockA --> lockB\
n",thrd);
                    spin_lock(&lockA);
                    DELAY_LOOP('B', 2);
                    spin_lock(&lockB);
                    DELAY_LOOP('A', 3);
                    spin_unlock(&lockB);
                    spin_unlock(&lockA);
            }
     [ ... ]
```

Right, let's try it out.

Runtime – good case, no deadlock

Build and run it via our run Bash script wrapper (this script accepts a parameter; if you pass 1, it sets the module parameter lock_ooo to 1 at insmod!), first with the lock_ooo kernel module parameter set to 0 (the default).

We find that, obeying the lock ordering rule, all is well; lockdep detects no problems:

```
$ ./run
Usage: run 0|1
 0 : run normally, take locks in the right (documented) order
 1 : run abnormally, take locks in the WRONG order, causing an AB-BA deadlock
```

```
$ ./run 0
[30125.466260] deadlock_eg_AB_BA: inserted (param: lock_ooo=0)
[30125.466593] thrd_work():167: *** thread PID 17048 on cpu 0 now ***
[30125.466622]  Thread #0: locking: we do: lockA --> lockB
[30125.466652]  iteration #0 on cpu #0
[30125.466841] thrd_work():167: *** thread PID 17049 on cpu 1 now ***
[30125.466874]  iteration #0 on cpu #1
[30125.466878]  Thread #1: locking: we do: lockA --> lockB
[30125.466950] BBAAA
[30125.467137] deadlock_eg_AB_BA: Our kernel thread #1 exiting now...
[30125.467138] AAABB
[30125.467287] deadlock_eg_AB_BA: Our kernel thread #0 exiting now…
$
```

Runtime – buggy case, circular deadlock

Now, we (reboot and) run it with the `lock_ooo` kernel module parameter set to `1`; as expected, the system locks up! (Well, it may or may not hang.) We've disobeyed the lock ordering rule, and we pay the price as it deadlocks! This time, I got lockdep's report as follows in the kernel log (looked up via the usual `sudo dmesg` on my x86_64 VM):

```
[ … ]
[  134.153810] deadlock_eg_AB_BA: inserted (param: lock_ooo=1)
[  134.154110] thrd_work():167: *** thread PID 3578 on cpu 0 now ***
[  134.154181] thrd_work():167: *** thread PID 3579 on cpu 1 now ***
[  134.154321]  iteration #0 on cpu #1
[  134.154365]  Thread #1: locking: we do: lockB --> lockA
[  134.154503] BBAAA
[  134.154671] deadlock_eg_AB_BA: Our kernel thread #1 exiting now...
[  134.164618]  Thread #0: locking: we do: lockA --> lockB
[  134.164624]  iteration #0 on cpu #0
[  134.164631] AAA
[ … ]
```

Ah, do you see it? Thread 1 (on core 1) takes the locks in the wrong order! Shameful... just look at `lockdep`'s complaints in the screenshot (*Figure 13.24*):

```
[  134.164672] ======================================================
[  134.164678] WARNING: possible circular locking dependency detected
[  134.164702] 6.1.25-lock-dbg #2 Tainted: G           OE
[  134.164782] ------------------------------------------------------
[  134.164787] thrd_0/0/3578 is trying to acquire lock:
[  134.164855] ffffffffc06c80b8 (lockB){+.+.}-{2:2}, at: thrd_work.cold+0x248/0x270 [deadlock_eg_AB_BA]
[  134.164959]
               but task is already holding lock:
[  134.164964] ffffffffc06c8118 (lockA){+.+.}-{2:2}, at: thrd_work.cold+0x209/0x270 [deadlock_eg_AB_BA]
[  134.165120]
               which lock already depends on the new lock.

[  134.165125]
               the existing dependency chain (in reverse order) is:
[  134.165130]
               -> #1 (lockA){+.+.}-{2:2}:
[  134.165167]        _raw_spin_lock+0x37/0x90
[  134.165238]        thrd_work.cold+0xb8/0x270 [deadlock_eg_AB_BA]
[  134.165328]        kthread+0x194/0x1c0
[  134.165347]        ret_from_fork+0x22/0x30
[  134.165442]
               -> #0 (lockB){+.+.}-{2:2}:
[  134.165456]        __lock_acquire+0x1330/0x22b0
[  134.165465]        lock_acquire+0xd0/0x2b0
[  134.165547]        _raw_spin_lock+0x37/0x90
[  134.165556]        thrd_work.cold+0x248/0x270 [deadlock_eg_AB_BA]
[  134.165650]        kthread+0x194/0x1c0
[  134.165657]        ret_from_fork+0x22/0x30
[  134.165668]
               other info that might help us debug this:

[  134.165672]  Possible unsafe locking scenario:

[  134.165690]        CPU0                    CPU1
[  134.165694]        ----                    ----
[  134.165750]   lock(lockA);
[  134.165830]                                lock(lockB);
[  134.165839]                                lock(lockA);
[  134.165848]   lock(lockB);
[  134.165919]
                *** DEADLOCK ***
```

Figure 13.24: (Partial) screenshot showing the kernel log; lockdep detects and reports the AB-BA circular deadlock that our demo module causes

The .cold compiler attribute

By the way, why are some functions suffixed with `.cold` (like our `thrd_work` function here)? The short answer is that it's a compiler attribute specifying that a function is unlikely to be executed. These so-called *cold* functions are typically placed in a separate linker section to improve the code locality of the required-to-be-fast non-cold sections! It's all about optimization. (Note that, in some cases, the normal version of the function can co-exist with the cold version.)

The `lockdep` report is quite amazing. The initial portion makes us aware that a "possible circular locking dependency..." has been detected. It then goes on to show us what actually occurred at runtime: here, thread 0 (seen as `thrd_0/0/3578`, its name (`thrd_0/0`), followed by a forward slash and its PID) tries to acquire a lock (64-bit hash value `ffffffffc06c80b8`; it even shows its symbolic name, `lockB`!) from our module, "`but task is already holding lock:`". It goes on to show that it's being held in the same function (`thrd_work`) of the same module (`[deadlock_eg_AB_BA]`), clearly indicating a (circular/AB-BA) deadlock scenario. It then shows the (kernel) call stacks leading up to this (with a more detailed call stack that soon follows, not seen in *Figure 13.24*), which, of course, makes it crystal clear as to which code paths caused the deadlock.

Being a debug kernel, kernel symbols will certainly be available; also, even the frame pointers config is likely to be enabled. These help us get accurate and detailed call stacks, which, of course, can make debugging easier. What then in production? Well, while kernel symbols are likely enabled there, frame pointers are likely not; that, and many more complexities, result in it being harder... but don't lose heart, as plenty of useful kernel debug techniques and tools do exist (do check out the *Linux Kernel Debugging* book). Again, here, we're in a debug scenario and expect to use a debug kernel with various bug-catching configs enabled.

It then (from the line `other info that might help us debug this:`) goes on to show the generic root cause of this deadlock by showing a typical scenario (literally a small timeline diagram!). Check out the lines after this toward the bottom portion of *Figure 13.24*; it shows the kind of thing that actually occurred at runtime – the possible **out-of-order** (**ooo**) locking sequence. Note how lockdep reports the generic scenario here; the upper portion shows you what actually occurred at runtime. Here, clearly, it is precisely what actually occurred: our kthread (#0) on CPU core 0 took `lockA`; in parallel, concurrently, our other kthread (#1) on core 1 took `lockB` first - the bug! Then, *kthread1* attempted to take `lockA`, while in parallel, *kthread0* attempted to take `lockB`, and we have our lovely AB-BA circular deadlock.

By the way, this kind of thing might well lead to a soft lockup bug, which the kernel watchdog detects.

Also, recall that `lockdep` reports only once – the first time – that a lock rule on any kernel lock is violated.

Brief notes on lockdep – annotations and issues

Let's wrap up this coverage with a couple more points on the powerful `lockdep` infrastructure.

A note on lockdep annotations

In user space, you're likely familiar with using the very useful `assert()` macro. There, you assert a Boolean expression, a condition (for example, `assert(x == 5);`). If the assertion is true at runtime, nothing happens and execution continues; when the assertion is false, the process is aborted and a noisy `printf()` to `stderr` indicates which assertion failed and where it failed. This allows developers to check for runtime conditions that they expect. Thus, assertions can be very valuable – they help catch bugs! In effect, the `assert()` macro helps us *verify our (often faulty) assumptions!*

The kernel doesn't have an `assert()` macro; why not? Well, if it triggers, are we willing to abort the kernel itself? Interestingly enough, that's pretty much exactly what the `BUG[_ON]()` macros do – several end up emitting a noisy `printk` and then calling `panic()`. Don't use `BUG*()` unless there's no other recourse. (Also, some drivers do define their own version of an assert macro, which simply prints messages into the kernel log, alerting the developers.) Perhaps the closest equivalent to `assert()` in the kernel are the `WARN*()` macros; don't use them gratuitously either.

In a similar manner to `assert()`, `lockdep` allows the kernel developer to assert that a lock is held at a particular point, via the `lockdep_assert_held()` macro. This is called a **lockdep annotation**. The macro definition is displayed here:

```
// include/linux/lockdep.h
#define lockdep_assert(cond)                    \
        do { WARN_ON(debug_locks && !(cond)); } while (0)
[ … ]
#define lockdep_assert_held(l)                  \
        lockdep_assert(lockdep_is_held(l) != LOCK_STATE_NOT_HELD)
```

The assertion failing results in a warning (via `WARN_ON()`). This is very valuable, as it implies that though that lock `l` is supposed to be held now, it really isn't. Also, notice that these assertions only come into play when lock debugging is enabled (this is, of course, the default when lock debugging is enabled within the kernel; it only gets turned off when an error occurs within `lockdep` or other kernel locking infrastructure). The kernel codebase, in fact, uses `lockdep` annotations all over the place, both in the core as well as the driver code. (Try using `cscope` to search for occurrences of this macro being invoked. Also, there are a few variations on the `lockdep` assertion of the form `lockdep_assert_[not]_held*()` as well as the rarely used `lockdep_*pin_lock()` macros.)

So okay, by using `lockdep` annotations, we can annotate our code, checking for whether a certain lock's held or not. But what about RCU read-side locks? Is there a way to check whether, in effect, we're in an RCU read-side critical section? Yes, indeed: the `rcu_read_lock_held()` (and friends) API serves this very purpose. The official kernel documentation throws light on this aspect here: *RCU and lockdep checking*: `https://docs.kernel.org/RCU/lockdep.html#rcu-and-lockdep-checking`. Also, do read the detailed comment that goes along with the code here: `https://elixir.bootlin.com/linux/v6.1.25/source/kernel/rcu/update.c#L285`.

A note on lockdep – known issues

A couple of known issues can arise when working with `lockdep`:

- Repeated module loading and unloading can cause `lockdep`'s internal lock class limit to be exceeded (the reason, as explained within the kernel documentation, is that loading an `x.ko` kernel module creates a new set of lock classes for all its locks, while unloading `x.ko` does not remove them; it's actually reused). In effect, don't repeatedly load/unload modules; otherwise, you might need to reset the system.

- Especially in those cases where a data structure has an enormous number of locks (such as a large array of structures, each of which might have a few locks), failing to properly initialize every single lock can result in `lockdep` lock-class overflow.
- Lockdep works by developers carefully building lock(dep) annotations that tell lockdep what the "rules" are; when it detects they aren't being followed, it generates warnings (aka a lockdep "splat"). The trouble is, this approach can and sometimes does lead to false positives being generated by lockdep (still, it's better to generate a false positive than to not generate anything and have the system truly deadlock; then it's already too late). These issues are discussed in David Airlie's article *Lockdep False Positives, some stories about* (`https://blog.ffwll.ch/2020/08/lockdep-false-positives.html`). It talks about the reasons for the false positives (assumptions regarding the invariance of locking rules over time, and that all locks for the same object are following common rules) and how this should be tackled (essentially boiling down to, to quote, *"do not fix lockdep false positives, fix your locking."* Nice).
- The `debug_locks` integer is set to `0` whenever lock debugging is disabled (even on a debug kernel); this can result in this message showing up: `*WARNING* lock debugging disabled!! - possibly due to a lockdep warning`. This could even happen due to `lockdep` issuing warnings earlier. Reboot your system and retry.

Kernel lock statistics

A lock can be *contended*, that is, a context wants to acquire the lock, but it has already been taken, so it must wait upon the unlock. Heavy contention can create severe performance bottlenecks; the kernel provides detailed (kernel) lock statistics with a view to easily identifying heavily contended locks. This can be useful when performing performance analysis.

Enable lock statistics by turning on the `CONFIG_LOCK_STAT` kernel configuration option (it's in fact the highlighted menu item in *Figure 13.20*). Without it being selected – the typical case on most distribution kernels – the `/proc/lock_stat` pseudofile will not even be present.

The lock stats code takes advantage of the fact that `lockdep` inserts hooks into the locking code path (the `__contended`, `__acquired`, and `__released` hooks) to gather statistics at these crucial points. The neatly written kernel documentation on lock statistics (`https://www.kernel.org/doc/html/latest/locking/lockstat.html#lock-statistics`) conveys this information (and a lot more) with a state diagram.

Viewing and interpreting the kernel lock statistics

The `/proc/sys/kernel/lock_stat` pseudofile (with permission 0644) is the control interface for (kernel) locking statistics (the `sysctl` knob); *Table 13.10* shows you how to use it. As the mode of `/proc/lock_stat` is 0600, even reading the current lock statistics requires root access; we assume, of course, that `CONFIG_LOCK_STAT=y`:

Do what?	Command to run (as root)
Clear lock stats	echo 0 > /proc/lock_stat
Enable the collecting of lock stats	echo 1 > /proc/sys/kernel/lock_stat
Disable the collecting of lock stats	echo 0 > /proc/sys/kernel/lock_stat
See (read) the current lock stats	cat /proc/lock_stat

Table 13.10: Commands to view kernel lock stats (provided CONFIG_LOCK_STAT=y)

Next, here's a simple demo to see some kernel locking statistics: we write a simple Bash script, `ch13/4_lockdep/lock_stats_demo.sh`; a snippet from it follows, showing how we wrap the lock stats commands in tiny functions (of course, we perform validity checks, ensuring that `CONFIG_LOCK_STAT=y` and that the script runs as root):

```
clear_lock_stats() {
  echo 0 > /proc/lock_stat
}
enable_lock_stats() {
  echo 1 > /proc/sys/kernel/lock_stat
}
disable_lock_stats() {
  echo 0 > /proc/sys/kernel/lock_stat
}
view_lock_stats() {
  cat /proc/lock_stat
}
```

Our script first disables locking statistics and then runs a test case. What is it? We choose to run our previous section's `deadlock_eg_AB-BA` module (the code's here in the book codebase: `ch13/4_lockdep/deadlock_eg_AB-BA/`) but in the regular way – *not* causing circular deadlock. With this test case, we know that our two kernel threads will run concurrently, and both will attempt to first take the spinlock named `lockA`, and then the one named `lockB`, which is fine (relook at its code if you wish). So there will be some contention for the spinlocks, and the kernel lock statistics, being enabled, should pick it up. (We won't bother showing the details of the script, where it performs the `insmod` of this module, here.) Just prior to inserting the module, the script clears and enables lock stats; once done, it disables lock stats and writes them to a file (`lockstats.txt`).

Let's say we've run our `lock_stats_demo.sh` script; now, let's interpret the relevant portion of the resulting lock statistics (which are saved in the `lockstats.txt` file; to filter out unnecessary output, we grep for the lines containing our lock names, `lockA` and `lockB`, keeping one line of context).

Even so, the resulting output is very long horizontally and, thus, wraps around; so here's the same output, annotated in a color-coded fashion, making it easier to interpret the columns (hint: `con` is short for *contended*):

```
$ head -n3 lockstats.txt |tail -1

                                  class name    con-bounces    contentions   waittime-min    waittime-max waittime-total
waittime-avg      acq-bounces   acquisitions  holdtime-min  holdtime-max holdtime-total    holdtime-avg

$ sudo grep -E -C1 "lockA|lockB" lockstats.txt
                                      lockA:              2              2         100.41         1990.90        2091.31
1045.65                 4                 6          14.51        1988.76        4143.00          690.50
                                      -----
                                      lockA               2      [<000000011c458e5>] thrd_work.cold+0x141/0x270
                                                                                          [deadlock_eg_AB_BA]
                                      -----
                                      lockA               2      [<00000000603ef921>] thrd_work.cold+0x209/0x270
                                                                                          [deadlock_eg_AB_BA]

--
                           &mod->param_lock:              0              0           0.00            0.00           0.00
0.00                    0                 1           1.14           1.14           1.14            1.14
                                      lockB:              0              0           0.00            0.00           0.00
0.00                    4                 6           8.06         213.96         410.10           68.35
                               cgroup_mutex:              0              0           0.00            0.00           0.00
0.00                    2                 3           4.44           4.82          14.04            4.68

con-bounces                                                 acq-bounces
number of lock contention that involved x-cpu data          number of lock acquisitions that involved x-cpu data

contentions                                                 acquisitions
number of lock acquisitions that had to wait                number of times we took the lock
```

Figure 13.25: Partial annotated screenshot: kernel lock statistics of our "deadlock" demo module's lockA and lockB spinlocks only (note that the long lines wrap around)

The kernel docs (seen here: `https://docs.kernel.org/locking/lockstat.html#usage`) explains how to interpret the output in detail. Do refer to it to see the meaning of each of the columns.

Here's a few pointers to interpret the output:

- The majority of the lock stats columns apply only to the rows where the lock name is followed by a colon :' character. The ones where the lock name isn't followed by a colon typically implies a *lock contention point*.
- The time displayed is in microseconds. The first column `class name` is the lock class; the (symbolic) name of the lock is seen here.
- `contentions`: The number of lock contentions that had to wait.
- `con-bounces`: The number of lock contentions that involved x-CPU data; here, `con` is an abbreviation for contention; the `x` implies "cross" (so for cross-cpu data; this works analogously for the `acquisitions` and `acq-bounces` fields). In effect, this is overhead! The actual number of lock acquisitions might be less.

- The *lock contention points* are the ones that appear between the two short seperator lines (-----); their line content does *not* match the column headers. The format is lock name, *con-bounces* (which means the *number of lock contention that involved x-cpu data*), and the call site (shown as `<IP> func+x/y [module_name]`, where *x* is the start of offset and *y* is the length of the function). *Figure 13.25* makes it easier to interpret; you can clearly see the contention points (highlighted in a bold font as well) there.
- The various wait times (`waittime*`, fields 4 to 7) for the locks are seen next.
- The `acquisitions` field (#9) is the total number of times the lock was acquired (taken); it's positive.
- The last four fields, 10 to 13, are the cumulative lock hold time statistics (`holdtime-{min|max|total|avg}`).

The most contended locks on the system can be looked up via `sudo grep ":" /proc/lock_stat | head`. Of course, you should realize that this data is true from the point when the locking statistics were last reset (cleared).

Note that lock statistics can get disabled due to lock debugging being disabled; for example, you might come across this:

```
$ sudo cat /proc/lock_stat
lock_stat version 0.4
*WARNING* lock debugging disabled!! - possibly due to a lockdep warning
```

This warning might necessitate you rebooting the system.

All right, you're almost there! Let's finish this chapter with some brief coverage on memory barriers.

Introducing memory barriers

Finally, let's address another concern – that of the **memory barrier**. What does it mean? Sometimes, a program's flow becomes unknown to the human programmer, as the microprocessor, the memory controllers, and the compiler *can reorder* memory reads and writes. In most cases, these "tricks" remain benign and typically optimize performance. But there are cases where this kind of reordering of (memory I/O) instruction sequences *should not occur*; the original and programmer-intended memory load and store sequences must be honored. What cases? Typically, these:

- When working across hardware boundaries, such as across individual CPU cores on multicore systems
- When performing atomic operations
- When accessing peripheral devices (like performing I/O from a CPU to a peripheral device or vice versa, often via **Direct Memory Access (DMA)**)
- When working with hardware interrupts

The *memory barrier* (typically implemented as machine-level instructions embedded within the `*mb*()` macros) is a means to suppress such reordering; it's a way to force both the CPU/memory controllers and the compiler to order instruction/data in a desired sequence.

Memory barriers can be placed into the code path by using the following macros: you must include the `<asm/barrier.h>` header:

- `rmb()`: Inserts a read (or load) memory barrier into the instruction stream.
- `wmb()`: Inserts a write (or store) memory barrier into the instruction stream.
- `mb()`: A general memory barrier – here's a direct quote from the kernel documentation on memory barriers (`https://docs.kernel.org/core-api/wrappers/memory-barriers.html`), *"A general memory barrier gives a guarantee that all the LOAD and STORE operations specified before the barrier will appear to happen before all the LOAD and STORE operations specified after the barrier with respect to the other components of the system."*

That's the key point: **the memory barrier ensures that until the preceding instruction or data access executes, the following one will not, thus maintaining the ordering.** On some (fairly rare) occasions, DMA being the likely one, driver authors use memory barriers. When using DMA, it's important to read the kernel documentation (`https://www.kernel.org/doc/Documentation/DMA-API-HOWTO.txt`). It mentions where memory barriers are to be used and the perils of not using them; see the example that follows for more on this.

As the placement of memory barriers in the code path is typically a fairly perplexing thing to get right for many of us, we urge you to refer to the relevant technical reference manual for the processor or peripheral you're writing a driver for, in order to gain more detail. For example, on Raspberry Pi, the typical **system on-a chip (SoC)** is the Broadcom BCM2835 series; referring to its peripherals manual – the *BCM2835 ARM Peripherals* manual (`https://www.raspberrypi.org/app/uploads/2012/02/BCM2835-ARM-Peripherals.pdf`), specifically section 1.3, *Peripheral access precautions for correct memory ordering* – is helpful to sort out when and when not to use memory barriers.

An example of using memory barriers in a device driver

As one example, take the Realtek 8139 "fast Ethernet" network driver. In order to transmit a network packet via DMA, the driver must first set up a DMA (transmit) descriptor object. For this particular hardware (NIC chip), the DMA descriptor object is defined as follows `https://elixir.bootlin.com/linux/v6.1.25/source/drivers/net/ethernet/realtek/8139cp.c#L296`:

```
// drivers/net/ethernet/realtek/8139cp.c
struct cp_desc {
    __le32 opts1;
    __le32 opts2;
    __le64 addr;
};
```

As you can see, the DMA descriptor object, christened `struct cp_desc`, has three members. Each of them has to be initialized prior to performing a DMA transfer. Now, to ensure that the descriptor is correctly interpreted by the DMA controller, it's critical that the writes to the DMA descriptor are seen in the same order as the driver author intends. To guarantee this, memory barriers are used. In fact, the relevant kernel documentation – the *Dynamic DMA mapping Guide* (`https://docs.kernel.org/core-api/dma-api-howto.html`) – tells us to ensure that this is indeed the case. So, for example, when setting up the DMA descriptor, you must code it as follows to get correct behavior on all platforms:

```
desc->word0 = address;
wmb();
desc->word1 = DESC_VALID;
```

(I've highlighted the write memory barrier instruction, `wmb()`, in bold.) So here, the write memory barrier, `wmb()`, guarantees that the store to `desc->word0` will occur before the store to `desc->word1`, maintaining the order mandated in the source.

Thus, check out how the DMA transmit descriptor is set up in practice by the Realtek 8139 driver code, as follows starting at `https://elixir.bootlin.com/linux/v6.1.25/source/drivers/net/ethernet/realtek/8139cp.c#L731`:

```
// drivers/net/ethernet/realtek/8139cp.c
[ ... ]
static netdev_tx_t cp_start_xmit([...])
{
    [ ... ]
    len = skb->len;
    mapping = dma_map_single(&cp->pdev->dev, skb->data, len, PCI_DMA_TODEVICE);
    [ ... ]
    struct cp_desc *txd;
    [ ... ]
    txd->opts2 = opts2;
    txd->addr = cpu_to_le64(mapping);
    wmb();
    opts1 |= eor | len | FirstFrag | LastFrag;
    txd->opts1 = cpu_to_le32(opts1);
    wmb();
    [...]
```

The driver, acting upon what the chip's datasheet says, requires that the words `txd->opts2` and `txd->addr` are stored to memory first, followed by the storage of the word `txd->opts1`. As *the order in which these writes go through is important*, the driver makes use of the `wmb()` write memory barrier.

Also, as you've learned (in the *Understanding and using the RCU (Read-Copy-Update) lock-free technology – a primer* section), **RCU** is certainly a user of appropriate memory barriers to enforce memory ordering (in fact, for RCU read-side critical sections protected via the `rcu_read_{un}lock()` pair, barriers are only actually used when the processor is the DEC Alpha; otherwise, the "locking" evaluates to nothing at runtime (except when lock debugging is enabled)).

For more details on memory barriers, do refer to the official kernel documentation; it has a detailed section entitled *WHERE ARE MEMORY BARRIERS NEEDED?* (link: `https://elixir.bootlin.com/linux/v6.1.25/source/Documentation/memory-barriers.txt#L2340`). The four cases where memory barriers are required, which we mentioned at the beginning of this section, is covered in depth here.

The good news is that, in the majority of cases, as long as you correctly employ the correct methods (locking/lock - free APIs, the social contract required, and so on) to protect critical sections, all the memory barrier work is taken care of under the hood. For a driver author, it's typically only when performing operations such as setting up DMA descriptors or initiating and ending CPU-to-peripheral (and vice versa) communication that you might require explicitly a memory barrier.

A note on marked accesses

Using the `READ_ONCE()` and `WRITE_ONCE()` macros – the so-called ***marked accesses*** (as opposed to regular C language *plain accesses*) – on individual variables *absolutely guarantees that the compiler and the CPU will do what you mean.* It will preclude compiler optimizations and use memory barriers as required, guaranteeing cache coherency when multiple threads on different cores simultaneously access the variable in question.

Driver/module authors, please note! It's important to not get carried away by the `READ_ONCE()` and `WRITE_ONCE()` macros. It is *not* recommended for the typical driver author to "fix" a data race by using them.

But why not? The premise is that reads and writes to shared writable variables are not supposed to race. Now, if you mark every (or almost every) shared variable memory access with the `READ_ONCE()` or `WRITE_ONCE()` macros, this in effect prevents tooling like KCSAN from detecting buggy races that they may encounter! Thus, it's important that they not be protected via these macros and that the reads/writes they perform should be in plain C language. Instead, we expect that your memory accesses are correctly protected by design and via your code-level implementation (perhaps by using a mutex, spinlock, or `atomic_t/refcount_t` primitive, a lock-free technique such as RCU or per-CPU variables, and so on). Moreover, KCSAN reporting a data race is often a precursor, a hint, to the fact that a (severe) logic bug exists in code. Simply turning it off by using the `READ_ONCE()` or `WRITE_ONCE()` macros would do everybody a grave injustice.

One last thing – an (unfortunate) FAQ: will using the C language's `volatile` keyword magically make concurrency concerns disappear? Of course not. The `volatile` keyword merely instructs the compiler to disable common optimizations around that variable (things outside this code path could also modify the variable marked as `volatile`), that's all. This is often required and useful when working with **Memory Mapped I/O (MMIO)** in a driver. With regard to memory barriers, interestingly, the compiler won't reorder reads or writes on a variable marked as `volatile` with respect to other volatile variables. Still, atomicity is a separate construct, *not* guaranteed by using the `volatile` keyword.

Summary

Well, what do you know!? Congratulations, you have done it! You have completed this book!

In this chapter, we continued from the previous chapter in our quest to learn more about kernel synchronization. Here, you learned how to perform locking more efficiently and safely on integers, via both the `atomic_t` and the newer `refcount_t` interfaces. Within this, you learned how the typical RMW sequence can be atomically and safely employed in a common activity for driver authors – updating a device's registers. The reader-writer spinlock, interesting and conceptually useful, although with several caveats, was then covered. You then saw how easy it is to inadvertently create adverse performance issues caused by unfortunate caching side effects, including looking at the false sharing problem and how to avoid it.

A boon to performance – lock-free algorithms and programming techniques – was then covered in some detail, with a focus on understanding and learning how to use per-CPU variables and the powerful RCU kernel synchronization technology within the Linux kernel. It's important to learn how to use these carefully (especially the more advanced forms such as RCU). Finally, you learned how to debug common locking issues by leveraging the kernel's lock debugging constructs, by configuring, building, and working on a debug kernel (during development, refactoring and test, this is very important). lockdep, especially, is a very powerful way to detect locking issues; including deadly deadlocks. We finished with a word on what memory barriers are and where they are typically used.

Your long journey in working within the Linux kernel (and related areas, such as device drivers) has begun in earnest now. Do realize, though, that without constant hands-on practice and actually working on these materials, the fruits quickly fade away... I urge you to stay updated on these topics, and more. As you grow in knowledge and experience, contributing to the Linux kernel, or any open source project for that matter, is a noble endeavor, one you would do well to undertake.

Questions

As we conclude, the following link provides a list of questions for you to test your knowledge regarding this chapter's material: `https://github.com/PacktPublishing/Linux-Kernel-Programming_2E/blob/main/questions/ch13_qs_assignments.txt`. You will find some of the questions answered in the book's GitHub repo: `https://github.com/PacktPublishing/Linux-Kernel-Programming_2E/tree/main/solutions_to_assgn`.

Further reading

To help you delve deeper into the subject with useful materials, we provide a rather detailed list of online references and links (and, at times, even books) in a *Further reading* document in this book's GitHub repository. The *Further reading* document is available here: `https://github.com/PacktPublishing/Linux-Kernel-Programming_2E/blob/main/Further_Reading.md`.

Leave a review!

Enjoyed this book? Help readers like you by leaving an Amazon review. Scan the QR code below for a 20% discount code.

**Limited Offer*

`packt.com`

Subscribe to our online digital library for full access to over 7,000 books and videos, as well as industry leading tools to help you plan your personal development and advance your career. For more information, please visit our website.

Why subscribe?

- Spend less time learning and more time coding with practical eBooks and Videos from over 4,000 industry professionals
- Improve your learning with Skill Plans built especially for you
- Get a free eBook or video every month
- Fully searchable for easy access to vital information
- Copy and paste, print, and bookmark content

At `www.packt.com`, you can also read a collection of free technical articles, sign up for a range of free newsletters, and receive exclusive discounts and offers on Packt books and eBooks.

Other Books You May Enjoy

If you enjoyed this book, you may be interested in these other books by Packt:

The Software Developer's Guide to Linux

David Cohen

Christian Sturm

ISBN: 9781804616925

- Learn useful command-line tricks and tools that make software development, testing, and troubleshooting easy
- Understand how Linux and command line environments actually work
- Create powerful, customized tools and save thousands of lines of code with developer-centric Linux utilities
- Gain hands-on experience with Docker, SSH, and Shell scripting tasks that make you a more effective developer

- Get comfortable searching logs and troubleshooting problems on Linux servers
- Handle common command-line situations that stump other developers

Mastering Linux Security and Hardening - Third Edition

Donald A. Tevault

ISBN: 9781837630516

- Prevent malicious actors from compromising a production Linux system
- Leverage additional features and capabilities of Linux in this new version
- Use locked-down home directories and strong passwords to create user accounts
- Prevent unauthorized people from breaking into a Linux system
- Configure file and directory permissions to protect sensitive data
- Harden the Secure Shell service in order to prevent break-ins and data loss
- Apply security templates and set up auditing

Packt is searching for authors like you

If you're interested in becoming an author for Packt, please visit `authors.packtpub.com` and apply today. We have worked with thousands of developers and tech professionals, just like you, to help them share their insight with the global tech community. You can make a general application, apply for a specific hot topic that we are recruiting an author for, or submit your own idea.

Share your thoughts

Now you've finished *Linux Kernel Programming, Second Edition*, we'd love to hear your thoughts! Scan the QR code below to go straight to the Amazon review page for this book and share your feedback or leave a review on the site that you purchased it from.

`https://packt.link/r/1803232226`

Your review is important to us and the tech community and will help us make sure we're delivering excellent quality content.

Index

Symbols

A

B

C

D

M

S

T

U

V

W

X

Y

Z

Download a free PDF copy of this book

Thanks for purchasing this book!

Do you like to read on the go but are unable to carry your print books everywhere?

Is your eBook purchase not compatible with the device of your choice?

Don't worry, now with every Packt book you get a DRM-free PDF version of that book at no cost.

Read anywhere, any place, on any device. Search, copy, and paste code from your favorite technical books directly into your application.

The perks don't stop there, you can get exclusive access to discounts, newsletters, and great free content in your inbox daily

Follow these simple steps to get the benefits:

1. Scan the QR code or visit the link below

`https://packt.link/free-ebook/9781803232225`

2. Submit your proof of purchase
3. That's it! We'll send your free PDF and other benefits to your email directly

www.ingramcontent.com/pod-product-compliance
Lightning Source LLC
Chambersburg PA
CBHW081058241125
35880CB00028B/313

* 9 7 8 1 8 0 3 2 3 2 2 2 5 *